Springer Textbooks in Earth Sciences, Geography and Environment

The Springer Textbooks series publishes a broad portfolio of textbooks on Earth Sciences, Geography and Environmental Science. Springer textbooks provide comprehensive introductions as well as in-depth knowledge for advanced studies. A clear, reader-friendly layout and features such as end-of-chapter summaries, work examples, exercises, and glossaries help the reader to access the subject. Springer textbooks are essential for students, researchers and applied scientists.

More information about this series at https://link.springer.com/bookseries/15201

Juan A. Morales

Coastal Geology

Springer

Juan A. Morales
Department of Earth Science
University of Huelva
Huelva, Spain

ISSN 2510-1307 ISSN 2510-1315 (electronic)
Springer Textbooks in Earth Sciences, Geography and Environment
ISBN 978-3-030-96123-7 ISBN 978-3-030-96121-3 (eBook)
https://doi.org/10.1007/978-3-030-96121-3

This Springer imprint is published by the registered company Springer Nature Switzerland AG
The registered company address is: Gewerbestrasse 11, 6330 Cham, Switzerland

End of the Earth, unnamed place.

There starts the sea, no limit surface.

Where the foam is born by beating waves
crumbling rocks which the tides take away.

Whistle between clouds, music of the wind,
intense blue water that always is flowing,
red sunset that bathes a wide tidal flat,
sand on the beaches where children splat.

Coast you're called, poetry place
where land and sea fight every day.

"Eternal change"

E. T. Soro

This book is especially dedicated to my parents:
Alfonso and Mary.
They have lived for their sons.

Acknowledgements

I would like to acknowledge all the knowledge acquired during hours of fieldwork with my dear colleague, Dr. J. Borrego (Pepe). With him, I learned most of the things that I explain in this book. I also understood many concepts about coastal geology during my stays and surveys in research centers of different universities. Doctors Poppe L. De Boer, Richard A. Davis Jr., John Loder, José Ojeda, Pedro Proença e Cunha, Pedro Depetris, Paolo Ciavola, José Manuel Gutiérrez Mas, Mouncef Sedrati, David Menier, Erwan Garel and Cheikh Ibrahima Youm were my guides during these stays.

The writing of this book would have been impossible without the generosity during decades of my direct colleagues Berta Carro, Irene Delgado, Germán Flor Blanco, Inmaculada Jiménez, Nieves López, Claudio Lozano, Jesús Monterde and Antonio Rodriguez-Ramírez.

The true promotor of this book was Dr. Alexis Vizcaíno, who encouraged me to express my knowledge and helped me through the editorial process. Two people contributed significantly to this book. Dr. Edward Anthony did a thorough review of the thematic content, while Colette Folder did an intensive review of the English. Both of them notably helped to improve the final text and figures.

Finally, the most loving words are to my wife Mar Mateo. Mar *(Sea)* endured my absences for endless hours while writing this book during the quarantine confinement of the COVID-19 pandemic.

Introduction

Here is the first complete manual on Coastal Geology. A book that gathers the knowledge of more than a century of research on different geological aspects of the coast: dynamics of geological processes, geomorphology, sedimentology and stratigraphy. They are also reflected from the applications of these sciences to the social problems and challenges of the communities occupying coastal areas. This manual collects many of my research experiences on different shores of the world, but almost everything written in this book I learned in other books. That is why I thought this manual would not be the first on Coastal Geology. However, when I began collecting bibliography, before I began writing, I realized that all the books that had been written so far collected partial aspects of this science: coastal dynamics, coastal geomorphology or coastal sedimentary environments. None of them united all geological aspects into a single vision. They all reach greater thematic depth than this book. I wanted to write a student book that was complete. A book written in an easy-to-understand language and including very didactic figures (that worth more than a thousand words). A book that would cut to the chase and contain only the essentials. The book I would have liked to have had when I started realizing that my vocation was the coast. The final recipients are not advanced scientists, but students, well… also coastal researchers who start their careers. For this reason, the chapters will not be as deep as other books written for expert scientists and professionals. But then I thought maybe any of those advanced students will need a little more information on specifics. That is why I created the advanced boxes. A geology student can really understand the book without them, but advanced boxes are intended for those coastal geologists who begin their research, those who want to know more.

The structure of this book is a reflection of its objectives. Content is divided into thematic units (or parts) that include several related topics. Each of its parts is dedicated to completing information on different aspects of the coast. Some have a purely geological content, and others focus the applications of coastal geology toward other disciplines of knowledge.

Part I is a purely epistemological section that tells us about the geological vision of the coast. In this part, this science and its main concepts are defined, a history of science is outlined, the criteria of classification of the coasts are discussed, the visions that the different geological disciplines have about the coast are approximated and an inventory of the different methods and techniques used in research on Coastal Geology is made.

Part II focuses on the study of coastal processes. Some, like waves and tides, act on dynamics and evolution from a physical point of view. Other processes have a purely chemical or biological character. Sedimentary and hydrological contributions from the continent and the sea condition the rhythms and trends of evolution. All processes interact with each other. But in the end, it all translates into a movement of particles from, to or along the coast.

Part III is the most complex part of the book and the one that requires a broader vision. It focuses on the study of coastal environments. Some of these environments are erosive, but most are cumulative. Each of these topics speaks to us first of all of the geomorphological characteristics, to continue explaining the processes that give rise to that geomorphology and finally talk about its facies, facies sequences and facies models, in a more sedimentological vision.

Part IV analyzes the processes that act on a longer time scale and which in the end result in the preservation of a geological record. This thematic unit captures aspects such as climate, sea-level movements, paleoceanography and comparison with some coastal systems that were preserved in other periods of Earth's history.

Part V tells us about the problems that arise from the interaction between humans and the coast. On the one hand, human beings unbalance coastal processes, and on the other hand, the coast menaces the human being occupying the coastal areas. The end result is that man and coast have to live together, so we analyze the smartest coexistence strategy: Integral Coastal Zone Management.

Finally, Part VI consists of a single chapter that gives us a perspective of what Coastal Geology studies will be in the coming years.

In short, this is a book that aims to provide a firm foundation of knowledge to future coastal geologists. Then comes specialization and deepening, but for that there are already other books.

Contents

Part I
Geological Approaches to the Coast

Coast through it all
cause on the other-side i know
there's a better tomorrow.

"Coast"
Kaiden Patch

1 Coastal Geology as a Science Through Time

1.1 What is Coastal Geology?

Defining a scientific discipline is not an easy task, because each concept has had its own historical evolution and been submitted to multiple points of view which have led to different definitions or even to distinct concepts. So, the study of geology, and particularly coastal geology, has undergone an evolution which has been intimately entwined with the knowledge and beliefs of every epoch during its development.

Etymologically, geology is the science that studies the Earth. However, this definition corresponds today with the concept of geosciences: that is, all of the empirical natural sciences that study different aspects of the Earth, among which geology is only a part. In the past, geology was categorized within natural history, being the most ancient part of the natural world's history in its examination of the inanimate entities of the Earth. So, during the eighteenth and nineteenth centuries, geological studies centered on descriptive aspects and collectible samples to be included in museums. Since the second half of the nineteenth century, however, the geology linked with natural sciences acquired a more scientific focus and began dealing with the study of matter, its spatial distribution and the processes that operate on it, to try to explain these properties. Over time, accumulation of knowledge caused schisms in numerous branches, among them that which is responsible for geological aspects that have a close relationship with the coast: coastal geology.

The coast can be defined as the fringe of terrain where the continent meets the sea. So, **coastal geology** is the science that aims to study the characteristics, structure and origin of the geological materials that constitute this coastal fringe, from the emerged waterfront to the sublittoral areas, with special emphasis on the active geological processes that take place there. This focus on the active processes (weathering, erosion, transport and deposition) tends to explain the natural phenomena that contribute to the genesis of the geological materials deposited in the coastal environment and to the erosional features that appear on the already deposited geological formations (Figs. 1.1 and 1.2).

The fundamental idea of coastal geology is a simple one: there is no coast without geology. On the one hand, the interface between land and sea is strongly influenced by the geology of the land. On the other hand, how the sea is able to model (by sculpting or building) the coast is a product of combined geological processes through time. In consequence, the term coastal geology extends to the character and the history of the terrestrial fringe located on the strip of land bordering sea.

Its beginnings go back further, however, since despite having been born in the emerged littoral, the development of coastal geology arose from the domain of the marine realm, thanks to the development of indirect techniques that conducted scientific studies into very different aspects of natural science. In consequence, the greatest advances in coastal geology occurred at the hands of marine geologists, whose work resulted in a new vision of the coast derived directly from scientific investigations in shallow marine areas. From this, marine geophysics emerged with force as an essential instrument to understand the structure of the land covered by the sea, especially on the fringes adjacent to the continents. The techniques included gravity, heat flow and magnetism measurements, using seismic, sonic or electrical waves created artificially and passed through sediments and rocks to understand the basement of submerged coastal areas. Thus, geology and geophysics are closely linked, even though each has unique aspects that are sometimes very differentiated in studies of coastal geology.

From a geological perspective, the materials that constitute a coast can be highly resistant to erosional coastal processes or, on the contrary, easily erodible and transportable by fluid agents (wind, waves and currents). From this point of view, coastal geological formations may have an erosive character in the case of ancient consolidated sediments and rocks, but they can also be sediments. In the first case the littoral is sculpted by the coastal agents, while in the second materials are deposited under favoring conditions of wind, waves and tides. These sediments may vary

J. A. Morales, *Coastal Geology*, Springer Textbooks in Earth Sciences, Geography and Environment,
https://doi.org/10.1007/978-3-030-96121-3_1

Fig. 1.1 Erosional coast: cliffs sculpted on plutonic formations at Sines, Portugal

Fig. 1.2 Depositional coast: Guadiana River Delta at the southern border between Portugal and Spain

from fine grains of mud or sand (micrometric or millimetric) to bigger elements such as pebbles, cobbles or boulders (centimetric to metric) and are distributed in coastal sedimentary environments. The final geology in terms of geomorphology and sedimentology that results in a particular coastline is the product of combined physical, chemical and biological processes acting on this coastal track over decades, and hundreds or thousands of years.

Therefore, coastal geology may be considered as an extended form of active geology that is specifically developed on the shore and under the sea at the edge of continents. The broad spectrum of disciplines it covers can be summarized in the study of the more specific and fundamental aspects, as follows: structure, tectonics, stratigraphy, sedimentology, micropaleontology, petrology, geochemistry, mineral deposits, water dynamics and evolution of oceanic basin margins.

1.2 Observation of the Coast in Ancient Times

Through history, coastal populations have existed in close conjunction with their environment and its problems, developing an intuitive and pragmatic knowledge of the coastal functioning at the local level and adapting to the characteristics and peculiarities which the environment imposed upon them.

Observation of the coast, and by extension of the sea, was always related to its use as a resource. Although the use of the coast has its oldest records in the Upper Paleolithic shelly remains found along many coasts of the world, there were probably cultures and peoples that had previously exploited it in conditions of lower sea level. Consequently, the record of their activities would be currently underwater after the rise that followed the last Ice Age. A good example of that could be the ancient Briton communities that settled in the now-submerged Doggerland (or Dogger Bank). Throughout history, there have been people directly linked to the coast and that very proximity led them to dare to go into marine waters. On the other hand, other civilizations avoided the coast because they had a negative vision of the sea, seeing it as a permanent source of danger (from energetic processes but also from other people's invasions). The spectacular maritime undertakings of the Phoenicians, the Greeks, the Islanders of the Pacific and later the Vikings are well-known, but these adventures would probably have been a response to social stimuli such as population pressure or economic incentives rather than a direct consequence of the physical environment.

From this point of view, the knowledge of the sea was directly related to the expansion of territorial and economic domains throughout the history of humanity. In Western culture, this was dominated by Eastern Mediterranean peoples (until the Middle Ages) and Western Europeans during the time of the great geographical discoveries (late 15th and early sixteenth centuries), and then the countries of North America together with old Europe and Japan. This control has influenced the historical development of coastal geology as a relationship between cause and effect.

Almost certainly, the first observations on marine dynamics in the Western hemisphere were made by the Egyptians, who necessarily had to use knowledge about the oceanic currents and prevailing winds to travel to the coasts of Syria, Cyprus, Crete and the Middle East. Sometime later, the Cretans and the Achaeans also used, without a doubt, this kind of knowledge, and more broadly, the Phoenicians and Greeks had shown a thorough knowledge of the variation of these annual currents to circumnavigate the Mediterranean and so dominate the contemporary economy with their commercial transactions. At same time, in Western Europe were the Tartessos people from southwestern Iberia, who were able to control marine agents in reaching the coasts of Brittany and western Africa to get tin and thus manufacture bronze when melting it with copper from their own mines.

In the Asian realm, Persian populations operated a commercial trade across the Persian Gulf from the Sumerian period. Indian and Taiwanese sailors extended this trade in a network of navigation tracks known as the Maritime Jade Road. This network connected commercially the main points of the Indian Ocean, Persian Gulf and Arabian Sea coasts at least since 2000 BC. This same network was used later by Arab peoples, but was known then as the Maritime Silk Road. Further east, Chinese people dominated the routes of their surroundings, demonstrating an extensive knowledge of winds and currents since the Song dynasty. In the first century BC, they already used compasses and had drawn the first coastal charts, which even noted the position of sandy shoals to make coastal navigation easier (Fig. 1.3).

In Australia, the Aboriginal communities that had settled on the coasts also observed and interpreted the movements of the sea. The Aboriginals lived for millennia in equilibrium with the surrounding nature and this knowledge was transmitted orally as a part of their cultural traditions and heritage.

All of these examples demonstrate that ancient peoples had knowledge about coastal and marine functioning, but can be that considered coastal geology? Maybe not—but their knowledge was the base on which the true coastal geology was built.

Nevertheless, there are descriptions of the processes shaping the coast by Greek philosophers which one would consider to be real coastal science. These philosophers conducted the first observations and reflections about origins and effects on the shoreline. As early as the fifth century BC, Herodotus anticipated the idea of changes in the sea level, reflected in his observations on the Nile Delta, and described the existence of periodic tides in the Red Sea. It was also

Fig. 1.3 Illustration of an ancient nautical chart showing clouds blowing winds to generate sea waves

Herodotus who first defined the concept of a delta. At the same time, Aristotle made wide observations connecting physics, meteorology and geology. Among other matters, this philosopher described the origin of hurricanes and the interaction between wind and sea giving origin to waves. He observed a continuous flowing of the Mediterranean water, attributing its origin to changes of depth in the marine basin. He also deduced that coastal changes were caused by the sediment supply from rivers and described the existence of depositional and erosional coasts.

Two centuries later, Eratosthenes and Antigonus of Carystus related tides with reverse currents through the Strait of Messina. Just a century later, the Greek philosopher Posidonius attributed to the Moon the cause of tides, relating its range with lunar phases on the basis of observations made in Gadir (Cadiz, Spain). In the same century, the Babylonian astronomer Seleucus proved that tides were not uniform in all the coasts nor on all days, offering a meteorological interpretation. However, it was not until the second century AD when Pliny the Elder correctly attributed the inequality of the tides to the joint action of Sun and Moon.

Meanwhile, two centuries before had seen the first insights into the relationships between tectonics and dynamics of the coastline. Thus, Strabo deduced that collapses and elevations of land had taken place, proving the role of the related processes of fluvial erosion and transportation to the coast and the mechanism of formation of deltas. This Roman geographer also related earthquakes and volcanic eruptions to vertical movements of the continent and the subsequent invasions of the ocean.

During the centuries that followed the fall of the Roman Empire, there were no great scientific advances in Europe, except for some monks led by the Venerable Bede (seventh century) who were able to correctly interpret the astronomical control and weather influence of some coastal processes. There was also St. Albert the Great, who interpreted correctly that wide areas of Europe had in the past been under the sea and submitted to coastal action, and described transgressions/regressions and erosional–depositional processes and their influence on the coasts.

During this time, the Islamic world was the most direct heir of the classical cultures. During the medieval period they developed an accurate cartographic science that they used to make very precise coastal maps and course books by means of oriental or classical discoveries such as the compass or the astrolabe. Although their purpose was exclusively for navigation, these routes included information about waves, winds and currents that today can be used to reconstruct not only the geography but also the coastal dynamics of these past times. At the time, the wide expanse of medieval Islamic dominion extended knowledge to all coasts of the Mediterranean, the Black Sea and other coasts of Africa and Atlantic Europe. It was indeed an Arab geographer, Al Idrisi, who established in 1166 the existence of a relationship between the evolution of the coastline and the climatic zone where it developed.

At the end of this epoch, the rediscovery of ancient scientific texts like the volume "*Geographia*" by Ptolemy and its translation into Latin signified a recovery of the classical knowledge. In this realm, Leonardo da Vinci, following the texts by Pythagoras and Plato, began to formulate natural laws in mathematical terms, and thus was able to formulate other laws, such as the theory of the tides.

1.3 Age of Exploration: The Birth of Marine Geology

Observation of sediments from the seafloor linked to depth data, especially in the sublittoral fringe, began to become common in the sixteenth century in order to facilitate navigation in shallow areas. An instrument called a pole, consisting of a lead of 14 pounds (6.35 kg) attached to a rope of 200 fathoms (365.76 m), was used for this purposed (the method of lead and line). A hole was made in the lead to collect the sediment when touching the seabed. Unfortunately, this instrument only made accurate measures in depths of less than 50 fathoms. The information from the coastal bed began to be introduced in nautical charts, although this information never received a treatment or an interpretation from a geological perspective.

In the fourteenth century, the contribution of Galileo initiated a scientific revolution. Since then, the scientific method was the base of knowledge in all the sciences. Nevertheless, it was not until the end of the seventeenth and eighteenth centuries when society experienced a greater intellectual understanding of the principles associated with these examples. Then, the domination of a critical attitude regarding previous knowledge came to transform the philosophy in experimental sciences, with checked proofs required to demonstrate truths. This period was initiated in the eighteenth century and extended to the 19th, and it affected geological knowledge in general, and the coastal geological approach in particular. At that stage, there were important scientists travelling the world to describe coastal natural dynamics along exotic coasts, writing rich texts accompanied by accurate maps and excellent drawings. A compilation of the discoveries and contributions of the scientists and geographers involved in these expeditions was presented by Flor [3] in the introduction of his monograph *Marine Geology*.

Perhaps the best example was provided by James Cook, who explored Australia and New Zealand in the name of King George III of England, making geometrically reliable charts that included information about bathymetry, sediment nature, currents and geomorphological features (Fig. 1.4). Alejandro Malaspina contributed in the same way to the description of the coastal features of South America during a mission for the Spanish King Charles IV.

From a dynamic point of view, these studies correspond to the theories of Laplace from the same epoch (1775) about the mechanisms that generate the tides, their dynamics expressed in mathematic terms and the role of the Coriolis force on them.

Throughout these centuries, despite the multiple observations on dynamics of marine agents, coastal morphology and study of sediments, coastal geology was never defined as a science, and was not even referred to as a discipline of geology. The nineteenth century saw many expeditions that addressed scientific studies from different perspectives. During this period, many contributions focused on the drawing of detailed cartographies, of the emerged coastal strip as well as of the sublittoral fringe, sometimes extending these studies up to several kilometers into deep areas. In the same way, numerous advances were made in the characterization of coastal agents (waves, tides and currents). One view of these studies could be that they resulted in the birth of marine geology and of physical oceanography, both of which were defined as sciences in the middle of that century.

It was in this period when the German Alexander von Humboldt characterized surficial ocean currents of the South

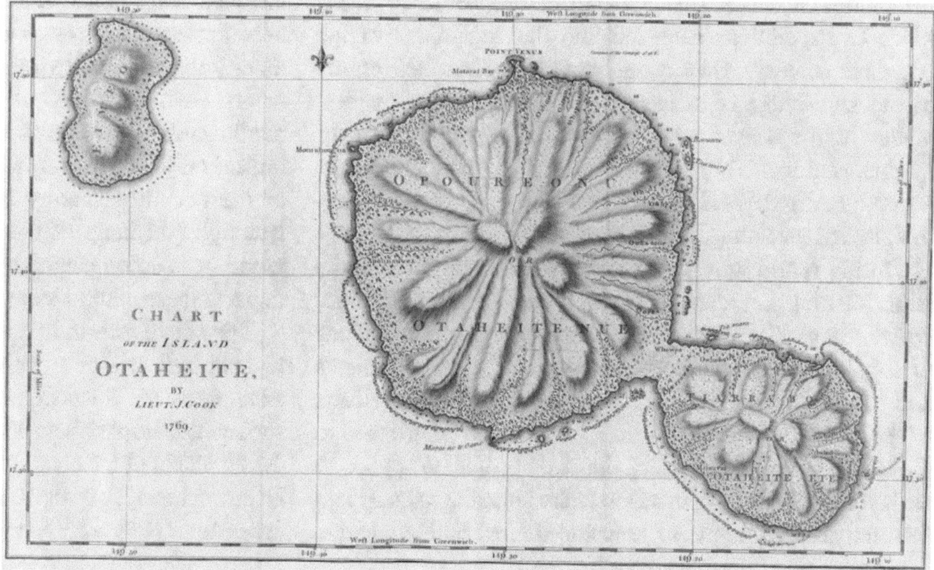

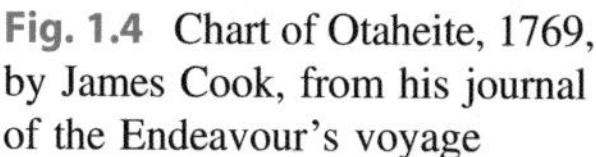
Fig. 1.4 Chart of Otaheite, 1769, by James Cook, from his journal of the Endeavour's voyage

Pacific and their influence on the oceanic islands. The Scottish explorer John Ross explored Baffin Bay, mapping the bathymetry and seabed sediments to a depth of almost 2 km. In the *Beagle* expedition, Charles Darwin, in addition to elaborating his evolution theory, carried out a characterization of the dynamics of coral reefs linked to sea level movements. The first studies of the beds located off the coast of the Antarctic continent, undertaken by Charles Wilkes and Clark Ross (nephew of John Ross), were also made during this period. On the side of the characterization of coastal processes, Gerstner (1802) and Stokes (1847) offered the first theories about wave dynamics expressed in mathematic terms.

The interest in the new science developing from the knowledge of the sea resulted in the emergence of some important books. The British geographer James Rennell published in 1832 his *Investigation of the Currents of the Atlantic Ocean*. In this work the author placed particular emphasis on the effect of these streams on some coasts in terms of sediment transport. Sometime later, in 1855, the American Matthew F. Maury brought a wealth of oceanographic information, publishing which is recognized as the first textbook of the history of oceanography, entitled *The Physical Geography of the Sea*. In the wake of this, Charles W. Thompson published his investigations of the North Atlantic and the Mediterranean seas in his book *Depths of the Sea*, published in 1873, which included for the first time the system of currents due to exchange of water masses in the Strait of Gibraltar.

From the emergence of marine geology as a science, knowledge of deeper environments was gaining importance. Thus, at the end of the nineteenth century, European governments financed expeditions such as the *Challenger* (British), *Gazelle* (German) and *Fram* (Swedish). The results of these expeditions are well-known and laid the foundations of the marine sciences, strengthening marine geology. These kinds of expeditions extended into the beginnings of the twentieth century, when many countries added their efforts to the knowledge of deep areas. During this time, a new vision of the seabed came to revolutionize the focus of marine geology. That was the theory of continental drift stated by Alfred Wegener in 1912, which signified the prelude to understanding plate tectonics.

During World War II, notable advances were made in the understanding of some specific coasts, especially those where troops planned or made landings or disembarkations of supplies. These studies included morphology, sediments and processes (wind, waves, tides and currents). The efforts of the main countries involved in the war also contributed to the development of many techniques that could be applied to seabed knowledge. So, acoustic (echo sounding, sonar) and seismic techniques were developed for anti-submarine warfare. After the war, these techniques were refocused to be used in science.

From the 1970s, the field of marine geology was shaken again by enormous knowledge. On the one hand, the development of plate tectonics theory in 1960 had offered a new vision, not only about the extension of the oceans but about a new framework for the coasts. That constituted the basis for the later tectonic classification of coasts stated by Inman and Nordstrom [6]. On the other hand, oil companies were contributing a large amount of data about marine basins through the development of seismic stratigraphy techniques which, in this case, were developed in greater profusion for use on the continental shelves.

However, these expeditions were not primarily focused on seismic stratigraphy, and studies on the geology of the coasts were dealt with in a collateral way or were not even considered. On the contrary, during this stage the new knowledge about coastal geology was linked to the application of new techniques of analysis of submarine areas on the submerged fringe of the coast. For this reason, this stage meant great development for oceanography, but no breakthroughs for coastal geology.

1.4 Explosion of Human Occupation on the Coast: The Birth of the Real Coastal Geology

Coasts were often historically avoided because of their potential danger for human beings and their infrastructures. For this reason, some civilizations never built in coastal areas. On the other hand, other peoples had occupied their coasts since the beginning of humanity, but typically coastal sites were used in natural conditions, even in some places where the occupation was high. However, in recent times, the increasing global population, industrialization and the need for larger harbors meant that human populations had to modify the coastal environment to adapt it to their requirements. Before the present century, the main parts of these modifications were carried out without causing a great impact on the coastal dynamics. However, at the beginning of the twentieth century that impact began to be greater, although still none of the coastal works, anywhere in the world, was accompanied by forecast studies of the possible impacts these could have on coastal dynamics.

The rate of growth in urbanized coastal areas has been too high in industrialized countries since the 1940s. One such example is the Spanish coast, where the urbanized coastal surface quadrupled between the 1960s and 1970s, with the development of residential complexes, tourist businesses and infrastructures, and the consequent destruction of coastal ecosystems (Fig. 1.5). Something similar had occurred in the

Fig. 1.5 Houses built on the coastline under an urban plan executed without taking into account coastal dynamics. Example shows the southwestern coast of Spain

United States' barrier island systems during the two previous decades and the same model would occur in the European Mediterranean countries (France, Italy and Greece). Soon imbalances caused by human modifications on the littoral were felt on many coastal fronts as erosional problems appeared in unexpected and unwanted ways. Surprisingly, the lack of studies about coastal dynamics as a form of urban planning continued up until the late 1970s.

Why did urban planners not predict or consider the impact of their structures? Perhaps because the knowledge about coastal dynamics was not well developed in these decades.

The story of the advances of the early studies of coastal dynamics is excellently described by Carter and Woodroffe [1]. The geographical studies of coastlines in the late nineteenth century were mainly focused on understanding how and why coastal changes occurred, while the best approaches to the coastal evolution principles took place in the first decades of the twentieth century. The first approaches to a real coastal geology from a geomorphological point of view were made in the first decades of the twentieth century in parallel with the study of Earth's other landforms (rivers, glaciers, deserts, mountains…). These first approaches were focused on the classification of the morphological features and their explanation in terms of processes. In 1919 the monograph which can be considered the first classic manual on coastal geomorphology was published: *Shore Processes and Shoreline Development*, written by Douglas [7]

It was also at this time when seminal works by the same Johnson, William M. Davis and Grove K. Gilbert were published. These can be considered as the first works of a real coastal geology. In the monograph *The New England-Acadian Shoreline* [8], Johnson starts from morphological features to interpret the evolution of a complex depositional coast. Some years later J. Alfred Steers developed his studies on British and Australian coasts that culminated in his book *The Coastline of England and Wales* (1946) [10].

It was not only American and British scientists who were developing research on the coastal fringe in these decades. In Denmark, Axel Schou published in 1945 [9] his monograph *Det Marine Forland*, a concise description and interpretation of the evolution of the Danish coast. A decade later, in 1954, the contribution of the French [5], *Coastal and Submarine Morphology*, approached coastal evolution from a wider oceanographic and geological point of view.

After the war, another notable advance for coastal geology occurred in parallel with the development of the study of sedimentology. Curiously, the first approaches to the coastal environments that were made by authors of the 1930s [4, 11] did not even use the word coast to define the system located in the border between land and sea, instead employing the term "transitional environments." During the 1950s, sedimentologists continued this trend. Nevertheless, the contributions to the knowledge of fluid mechanics, transport of particles and bedform dynamics were applied to understand the coast and explain the main sedimentary sequences that were also described in this period.

The decades of the 1950s and 1960s were under the influence of some very active coastal researchers: Richard J. Russell in Louisiana (USA), Cuchlaine A.M. King in Britain and Charles A. Cotton in New Zealand. They, among others, led the interpretations of coastal evolution as a response to the link between long-term factors (such as tectonic movements and climatic influences) and short-term variables (hydrodynamic agents). A more dynamic vision was that introduced by Davies [2] as a prelude to the advances of the next decade. So, during the 1970s, German and French geographers focused their studies of coastal geomorphology from a climatic point of view under the line of the newly created *climatic geomorphology*. In this sense, the contributions of Jean Tricart and André Cailleux demonstrated that the coasts evolve in different forms according to the climate where they are developed.

In parallel, the studies of coastal evolution beyond the "iron curtain" were well-represented by the works of the Russian Vsevolod P. Zenkovich. The importance of his contribution can be seen by the fact that his monograph, *Processes of Coastal Development* [12], was one of the few books translated into English. Curiously, this work referred to the studies of Gilbert, Davis and Johnson, but not always to agree with their interpretations of the coastal processes.

As we can see, when urban development exploded at a global level, the fundamentals of the coastal sciences were already well-known. Why, then, did the urban planners not apply this knowledge to predict the coastline evolution before constructing their buildings? The answer to this is not easy, but would relate to the enormous disconnect between science and society in these times. In any case, as was previously mentioned, this focus has changed since the late 1970s.

During those years, habitants of the East Coast of the United States and of the most urbanized areas of Australia created neighborhood associations with the aim of protecting against the coastal erosion which endangered their houses. These initiatives were funded by working groups on the protection of beaches. On the one hand, this created a discipline of coastal engineering that proposed the construction of seawalls, breakwaters and groins as beach replenishments that set precedents in the resolution of problems. In the future, these solutions will be extended to coasts all over the world. On the other hand, it created the next generation of coastal researchers, led by Richard A. Davis, Miles O. Hayes, Orrin H. Pilkey, Donald J.P. Swift and L. Don Wright, among others, who defended the studies of coastal geology as the real base to understand the problems and the development of realistic urbanizing plans which take into account coastal dynamics. The main contribution of this new generation of coastal geologists was to enhance the obvious critical links between the oceans and the shore, which at that moment were practically unstudied. Their works from the 1980s demonstrated the strong connections between the coastal systems (dunes, beaches, nearshores, estuaries and deltas) with deeper marine environments (shorefaces and shelves).

1.5 Recent Knowledge

In the last 30 years coastal geology studies have experienced a revolution. At present, legions of coastal researchers are integrated in research teams distributed in almost all the countries of the civilized world. The cause of this explosion of knowledge is due to several factors. Firstly, the rate of urban invasion of the coast that has continued in the last decades, increasing the problems of stability due to the coastal dynamics but also the problems created by the building of artificial structures. That created a social need that has to be solved by coastal researchers. In addition, the economic development of countries has made available funds for research, especially that which is focused on resolving social problems. In addition, international institutions have created various programs for funds that reinforce national politics.

In this way, UNESCO promoted the International Geological Correlation Program (IGCP), created in 1972. Under this program, the project IGCP274: Quaternary Coastal Evolution: Case Studies, Models and Regional Patterns (better known as: "Coastal Evolution in the Quaternary") connected the research developed by several teams by means of annual exchange meetings. At the beginning, in 1988, around 400 coastal researchers from more than 50 countries were involved. At the end, in 1993, the participants had increased to more than 600 from 70 countries.

The International Union for Quaternary Research (INQUA) was formed in 1928. The Commission for Coastal and Marine processes (formerly the Commission of the Quaternary Shoreline) whose aim is "to promote communication and international collaboration in basic and applied aspects of Coastal and Marine Quaternary research" was created in 1954. During the last 30 years this commission has been the center of the international focus of interest and also funds projects with concrete objectives.

The LOICZ project (Land–Ocean Interactions in the Coastal Zone) was established by the International Geosphere–Biosphere Programme (IGBP) in 1993. In its first stages LOICZ focused its investigation on biophysical approaches, but later increased attention to coast–human interactions, including themes such as the influence of coastal dynamics on coastal management.

In Europe, significant advances included the implementation in 1994 of the MAST (Marine Science and Technology) program and the ELOISE (European Land–Ocean Interaction Studies) thematic network by the Directorate-General XII of the European Union. The link between these two entities resulted in the creation of the ICZM (Integrated Coastal Zone Management) Demonstration Programme in 1996. The aim was "to provide technical information about sustainable coastal zone management, and to stimulate a broad debate among the various actors involved in the planning, management or use of European coastal zones." These programs have funded research projects about coastal research during the last 20 years, with special emphasis on coastal geology.

Within this framework, interdisciplinary teams of geologists and archaeologists have created a new discipline named *geoarchaeology*. The application of the knowledge supplied by this science in coastal areas has allowed the interpretation of the littoral in terms of coastal evolution. In a retrospective sense, many studies of coastal evolution from geological data allowed the reconstruction of the ancient geographies that had framed the archaeological discoveries.

As a direct consequence of these national and international developments, research centers have in the last decades found good sources of funding to develop research projects focused on the knowledge surrounding coastal geology. Hence, many research groups and laboratories were born during the 1990s and the beginning of the twenty-first century. Today, solid research teams study coastal geology in American nations, Western Europe, Asia and Australia. In parallel, the transmission of the recent knowledge to society has substantially improved during these years. Increased budgets to coastal researchers were accompanied by enormous developments in technology. Thanks to a remarkable development of the methodology and the appearance of new techniques and equipment, measurement processes became more and more accurate. With the introduction of new field equipment (accompanied by software for processing ever more up-to-date information), it could be said these advances have happened widely across all the sciences, but it is clear that this model of obtaining and processing digital data has had a deeper effect on coastal geology studies.

From a sedimentological point of view, the ways to interpret the geological record of coastal systems was increased by using different kinds of light corers (hand corers, piston corers and vibracorers). Equipment for acoustic echo sounding, multibeam sounds, side scan sonar (SSS), reflective seismic and ground penetrating radar (GPR) have been incorporated into coastal research and have successively very quickly improved. Coastal processes can be measured in an increasingly precise way by using current meters, acoustic Doppler current profilers (ADCP), water pressure tide gauges and wave buoys. Remote sensing signified a new frontier in coastal geomorphology, since it allowed a more global vision of the coastal areas and a comparison with accurate images from different dates. Drone and LiDAR flights have been recently incorporated into coastal research, contributing in the same way as remote sensing but supplying better image resolution. The positioning of geological, bathymetrical and geographical records could be acquired in a more accurate way using portable global positioning systems (GPS) to treat the information in geographic information systems (GIS). Finally, mathematical models have been employed as a very useful tool for the knowledge of the dynamic functioning of coastal systems.

Today, the foundations of coastal geology as science are firmly established, with clear links to other related sciences like marine geology, coastal engineering or coastal ecology. It is interesting to observe how coastal geology has become gradually more inclusive and all-embracing, encompassing previous visions like coastal morphodynamics, coastal sedimentology, even coastal environmental sciences. This is especially useful in the present climatic change framework. In the future, the world will turn towards an increasingly transversal science, for a more effective transmission of the latest advances to society. Let's go!

References

1. Carter RWG, Woodroffe CD (eds) (1994) Coastal evolution. Cambridge University Press, Cambridge, 517pp
2. Davies JL (1964) A morphogenic approach to world shorelines. Ann Geomorphol 8:27–42
3. Flor GS (2004) Marine geology. University of Oviedo, 576pp
4. Grabau AW (1913) Principles of stratigraphy. Seiler and Co., New York, 185pp
5. Guilcher A (1954) Coastal and submarine morphology. Methuen & Co. Ltd. London, 314pp
6. Inman DL, Nordstrom CE (1971) On the tectonic and morphological classification of coasts. J Geol 79:1–21
7. Johnson DW (1919) Shore processes and shoreline development. Wiley and Sons, New York, 584pp
8. Johnson DW (1925) The new England-acadian shoreline. Wiley and Sons, New York, 608pp
9. Schou A (1945) Det Marine Forland. Folia Geographica Danica, 4, 236pp
10. Steers JA (1946) The coastline of England and wales. Cambridge University Press, Cambridge, 644pp
11. Twenhofel WH (1939) Principles of sedimentation. McGraw-Hill, 610pp
12. Zenkovich VP (1967) Processes of coastal development. Oliver and Boyd, Edinburgh, 738pp

Defining Concepts of Coastal Geology

2

2.1 What is the Coast?

As seen in the first chapter of this book, the coast can be defined as the fringe of terrain where the continent meets the sea. However, the coast is actually not so easy to define, since there are multiple variables involved. On the one hand, there are land fringes that can be affected by marine processes under certain conditions; on the other hand, there are major submerged areas that are influenced by processes and sediments coming from land. So, the coast is a complex system which involves many factors that will be explained in the following chapters. This explains why it is not possible to supply concrete hypsometric or bathymetric coordinates, width values, lithological or environmental characteristics to rigorously delimit the coast. Classification of coasts is also complex since several criteria can be used, so there is not one classification system that is actually accepted from all the different perspectives from which coastal studies can be addressed. However, the purpose of this chapter is to clarify several concepts that we will encounter throughout this book, to give the coastal student a range of criteria enabling them to understand the global concept of the coast without the need for a precise definition.

As is the case with the proper term coast, there are other terms that are commonly used and sometimes incorrectly employed in popular usage. Some of these terms have been previously defined in scientific terms in works and books on coastal geomorphology (e.g. [7, 2, 14]). The following definitions tend to clarify all of these concepts (Fig. 2.1).

2.1.1 Coast

Although the coast is normally defined in two dimensions as the strip of emerged land meeting the sea, there are really three elements that come together in this coastal fringe, necessitating consideration of a third dimension—the air. In this sense, a more correct definition was provided by Carter [2]: "The coast is the terrestrial place where land, water and air interact." This definition is intentionally broad, since, as Carter says, water can be salty or sweet and to him it would include also lacustrine coasts where processes very similar to those on seashores occur.

According to this definition, though, the coast would be a really narrow strip of dry terrain, since it would correspond only to the area of contact between the three elements.

2.1.2 Coastal Zone

The concept of a coastal zone was defined in a more formal way, although its definition preserves the ambiguity of the concept of coast. Thus, [2] defines it as "the space in which the terrestrial environments influence marine ones and vice versa," while [7] defines it as an "area influenced by its proximity to the coast." These definitions deliberately avoided giving precise limits to the coast, because many of the environments are places of gradual transitions and can be defined by physical, biological or cultural criteria that rarely coincide.

The concept of a coastal zone is much broader than the concept of coast since, as well as the extension of the above-mentioned strip, it also includes the areas of influence of mixed land–ocean processes. The landward edge clearly depends on the adjacent environments, and can be very narrow in the case of erosive coasts or extend for several kilometers in areas of fluviomarine systems or plains of migration of coastal dunes. The seaward limit, meanwhile, is commonly regarded as the contact with the continental shelf or the last place reached by the base level of storm waves.

2.1.3 Littoral Zone

The littoral zone is defined more precisely. In this case, the definition can be made in biological or sedimentological terms. For the purposes of this book the most accepted

J. A. Morales, *Coastal Geology*, Springer Textbooks in Earth Sciences, Geography and Environment,
https://doi.org/10.1007/978-3-030-96121-3_2

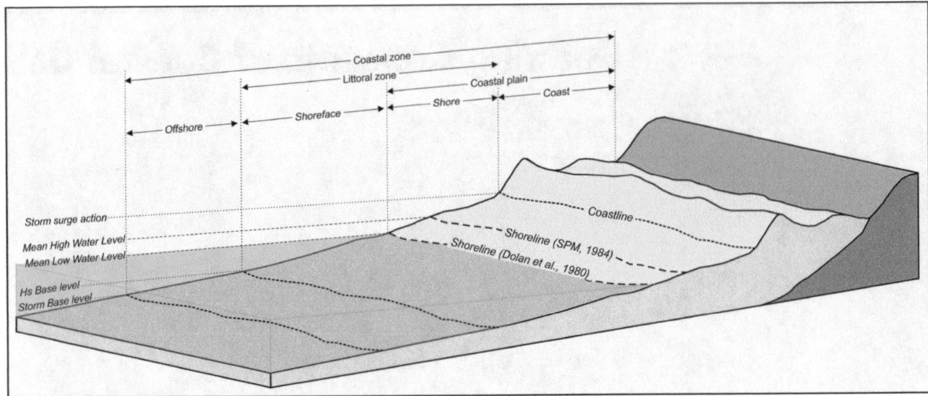

Fig. 2.1 Graphic definition of coastal terms, [3] adapted from the *Shore Protection Manual* CERC

definition in coastal geology will be used: "the littoral zone is the portion of the coastal zone whose sediment is susceptible to be transported due to the significant wave action" [7]. The landward limit would be then defined by the action of the significant waves during spring high tides, while the seaward limit would be defined by the base level of significant waves. The littoral zone is therefore an area which includes the intertidal areas, but is predominantly under water.

2.1.4 Coastal Plain

A coastal plain is defined as "a low and flat strip of land adjacent to the coast" [6]. This is a very broad geographical definition that would include the entire existing flat surface from the coastline to the first breaking of the relief located in the interior of the continent, without attending to the origin or the age of this plain or to the geological formations that constitute it.

A definition with a more genetic character is that provided by Cotton [4] which defines a coastal plain as "a flat of soft decline limited by a continental relief and a simple coast, whose origin is the rise of ancient marine deposits." This definition places the coastal plains in a particular environment: the coasts of emersion and their deposits are of marine origin but generated during a period prior to the position of the coastline that borders the plain.

From the landforms generated by coastal processes that act in a given period, the concept of coastal plain could be restricted to that "flat area adjacent to the coast generated by active coastal processes and occupied by current coastal environments" [10, 13]. This definition would restrict the coastal plain to a zone much more related to the interface between land–sea–air which defines coastal zones.

2.1.5 Shore

The shore is formally defined as a "zone that extends from the mean low water line to the effective limit of storm waves and storm surge" [12, 14]. From the point of view of a tourist it would be defined as "the zone where you can nail an umbrella," because there is a great possibility of arriving in an area in an exposed situation, even though it can be submerged under different conditions. Obviously, this area can be divided into two different strips: the lower fringe that is exposed and submerged every day by tides (the foreshore) and the upper one that is submerged only under storm conditions (the backshore).

2.1.6 Shoreline

The shoreline is defined as "the line along which water meets the land" [8] whereas for the Shore Protection Manual [3] it is the "line that separates the beach from the sea." One problem with these definitions is that a line represents only an instant in time and that can be a disadvantage, since the contact between sea and land is permanently in motion. So, defining an instantaneous position can lead to the error of considering this position as a stable situation in time or as an average position [1]. The first of the definitions would coincide effectively with the shoreline at mean low tide level, whereas the second would mean high water level. These references to average tide levels avoid regarding this line as an invariable position.

A more accepted concept is raised by authors such as [15], who considers shoreline to be "not only the line of contact between sea and land, but throughout the strip in which a short-term variation occurs." Others, like [11], considered it most appropriate to acknowledge the variability due to longer term processes. Obviously, this kind of definition is conceptually more appropriate, but it introduces a degree of uncertainty. What is clear is that today the concept of shoreline must be considered in a flexible and dynamic way that undergoes variability, and not as a static and unchanging concept.

2.1.7 Coastline

Waterfront is a term less used by coastal geologists, although it is widely used in stratigraphy. It is formally defined by Short [17]: "Coastline is the boundary between the coast and the shore." Having considered the coast as a permanently dry strip and the shore as an area submersible under certain conditions, the coastline is situated at an altitude higher than the shoreline. Thus, from a stratigraphic point of view, the position of this line locates all environments that act under the influence of waves and tides as marine-dominated environments.

2.1.8 Shoreface

The shoreface is the fringe located under the mean low tide and extends deeply to just reach the level of action on the bed of the fair weather waves (wave base level). It is a coastal fringe that is permanently submerged, except when the water is lower than the mean low tide, when its upper part can be exposed. This can also be understood as the lower part of the littoral zone [3].

2.1.9 Offshore

The offshore is the fringe located under the level of action on the bed of the fair weather waves (wave base level) and extends to the level of action on the bed of the storm waves (storm base level). It is a fringe permanently submerged and excluded from the action of normal waves, but during storms significant amounts of sediments can be transported on its surface. This fringe is normally understood as the transition between the coast and the inner shelf, but for some authors it is a part of the shelf and cannot be defined as a part of the coast.

2.1.10 Shelf

The continental shelf cannot be considered in any case as part of the coast, but "the expansion of the shallow seabed that runs along the coast to the deep" [14]. For coastal geologists, the bed of the platform extends from the boundary of the coastal zone (base level of storm waves) until the break of slope that is the edge of the continental slope. For marine geologists, however, this environment starts under the base level of the fair weather waves, since the waves only act on this extension of the bed during specific events.

2.2 A Datum Issue

An important aspect in addressing topographical or bathymetric data of the coastal zone, or representing graphically the coastal areas on a map, is the choice of a vertical datum. The vertical datum is the reference altitude used as level 0 in the topographical or bathymetric surveys. All the uprisings of the relief need a reference level. In addition, as explained in the previous section, the sea level is in constant motion responding to different period cycles, thus it is necessary to consider that any variation of water levels must also relate to a fixed reference level. Perhaps the clearest example of these cyclical variations is the tide, whose levels are subject to fluctuations with maximum and minimum heights varying in semidiurnal, biweekly and semiannual cycles, but also spatially between coasts with different tidal regimes from microtidal to megatidal. The averaged and extreme levels reached by the tide can be determined statistically, with some serving as vertical datum for the rest of the levels on certain occasions (Fig. 2.2). For example, international laws set the limits of national territories with respect to the level of the mean low water (MLW). However, nautical charts and some ports often use as vertical datum the lowest tide levels (spring low water or lowest low water), so the rest of the tidal levels will be considered as positive values. In this way, the depth values represented in nautical charts are located below this value and will be subtidal. Hence, these levels often receive the name of the hydrographical datum, nautical datum or datum of the nautical charts. A summary of the nautical reference datums used in different countries can be seen in Table 2.1 [3].

Apart from nautical charts, the most frequent types of representation of the relief are topographic maps. In this case, topographic maps often used as a reference datum the mean water level (MWL). This average level can be identified with the mean tide level, but usually occupies a different altitude, since there are level changes other than the tides.

Regarding the laws governing human activity in coastal areas, it is also necessary to establish cartographically those areas that are floodable by marine processes and thus be able to delimit accurately coastal areas. Many of these laws establish the boundary between the public domain and private property using as vertical datum the mean high water level (MHW) or the maximum flood level for extreme periods of storm high tides (extreme high water level, EHW).

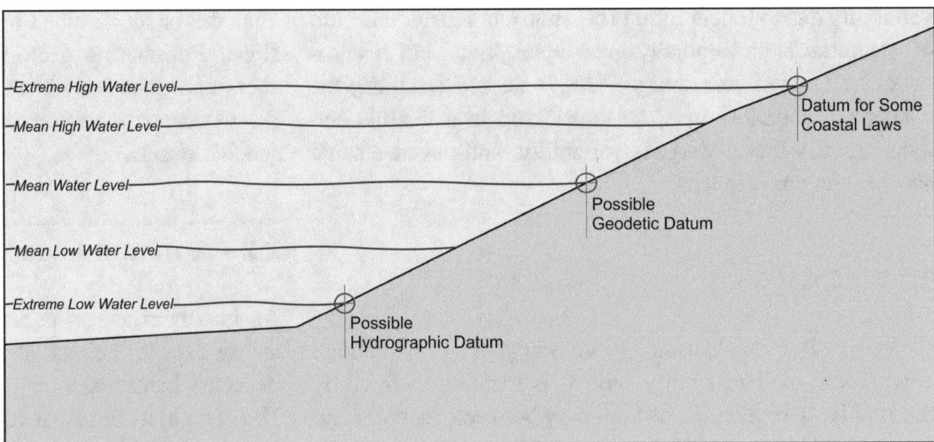

Fig. 2.2 Statistical levels achieved by tides that are frequently used by datums

Table 2.1 Hydrographical datums used by different countries

Hydrographical datum	Countries
Mean low water	US Atlantic Coast, Argentina, Sweden, Norway
Mean lower low water	US Pacific Coast
Mean spring low water	UK, Germany, Brazil, Italy, Chile
Historical low water	Great Lakes (US and Canada)
Lowest spring low water	Portugal
Indian spring low water	India, Japan
Lowest low water	France, Spain, Greece

What happens is that in any country that has a long enough coastline, the tide may experience a variation between each of its coastal sections, presenting each of them with different swings and different mean (and extreme) values. At a national level it will not be practical to have different reference levels and thus it is always one of these levels that will be chosen for use in all of the country's charts. This reference level is normally indicated in the margin of the maps.

One problem that comes from the use of datums that is different for terrestrial and nautical maps is that the coastal fringes ranging between these different datums are not represented in any mapping. The use of these different datums can also be a problem when trying to integrate data from land (topographic) with data from underwater (bathymetric). To solve this problem, the vertical relationship between the different datums should always be clear.

Advanced box 2.1

Relationships between different datums: case studies from Spain and the USA

The use of a datum which takes as its reference the vertical variations of the sea level along the coast of an entire country is not an easy task. First of all, to set a datum it is necessary to do numerous measures to establish statistical level variations on different segments of the national coast, determining statistical values. Secondly, choosing the level that suits best the needs for which the datum is intended needs to take into account that its position is as a statistical value, which will need to be reviewed periodically.

In this regard, it is better to try to establish a datum that presents a lesser possibility of change in time, if we want this value to be of lasting use. Either way, the establishment of a datum at the national level does not imply that it cannot coexist with other datums used at the local level. In this case, a clear vertical relationship between the different datums that can be used must be established, both with each other and with the general datum. On the other hand, it must be borne in mind that each datum was established for a specific purpose, but often used later for other purposes without understanding well how it can satisfy this secondary use.

A good example is the establishment of the datums of the Spanish coast, since this country has two coasts with distinctly different level variations: the Atlantic and the Mediterranean coasts. The Atlantic Coast is a coast of mesotidal and semidiurnal character, with very marked differences between the spring and neap tides, while the Mediterranean Coast is microtidal, with maximum variations below 0.25 m, which are always related to the conditions of wind, waves and atmospheric pressure. Taking into account these features, it is obvious that the Spanish Geographic

Fig. 2.3 **a** Location of tidal stations used as datum references. **b** Vertical relationships between topographic and hydrographic datums

Institute (IGN) established its geodetic datum on the Mediterranean Coast, using as altitude 0 the "mean water level" in Alicante (MWLA) (Fig. 2.3a).

On the other hand, for nautical charts it is necessary to establish a minimum level, and it is simple to understand that it is not practical for a datum from the coast with minor tidal oscillations, but in which the low tides can reach extreme minimum values, so that need was settled by the Atlantic Coast. To establish these statistical levels, the Navy Hydrographic Institute (IHM) used ahigh precision leveling procedure, with an aray of permanent tidal stations (NAP network). This procedure allows the linkage of the main milestones of the tidal stations with some of the landmarks of the NAP network (high-precision levelling) of the IGN, so that the relationships between hydrographic zeros in each local coast, with respect to the mean water level in Alicante (MWLA), can be obtained. An example of the result of these relationships can be seen in Fig. 2.3b.

In the USA, the National Oceanographic Surey (NOS) set up a permanent network of tidal stations. With this network it can establish the relative variation of the tidal levels along the coast of the entire country, taking into account the notable differences between the Pacific and the Atlantic coasts and also other differences between coastal tracks along each coast. Globally, they have established a national datum called NTDE (National Tidal Datum Epoch) using the average data of all the coasts for a period of 19 years. However, some coasts of the USA have established regional levels which are based on local measures from different periods. Good examples of local datums include the Mean Low Gulf Datum (MLGD), established for the Gulf of Mexico (microtidal) or the Pacific Coast Datum (PCD) established for all the Pacific Coast of the United States (mesotidal). For establishing charts, however, the Mean Low Water Level shoreline (MLWL), known as Datum EM1110-2-1003, is used as a datum on the Atlantic Coast, while the Pacific Coast is set by the Mean Lower Low Water Level (MLLWL), located 1.60 m below the EM1110-2-1003. At the same time, for the topographic maps the Geodetic Vertical Datum 1929 (NGVD1929) is used. The relationship between these datums can be seen in Fig. 2.4 [9].

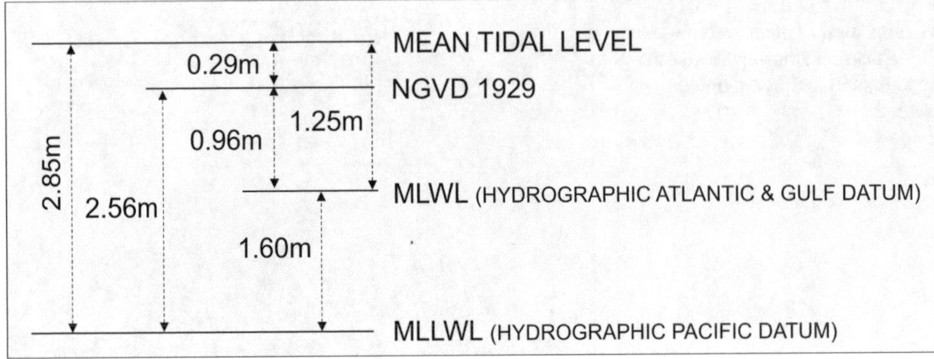

Fig. 2.4 Vertical relationships between datums used in the USA (data from Harris [9])

2.3 Coastal Framework

The morphological characteristics of a coast and the features that shape it should be known before tackling any more detailed study. Some concepts whose knowledge is essential to understanding the nature of the coast and its relation with the dynamic and geological environment are therefore addressed in this section.

2.3.1 Coastal Processes

Intuitively, when we talk about coastal processes, we are normally referring to processes that act in the short term, also called morphological processes. These processes may be defined as "a combination of environmental forces subject to the dynamics of fluids and sediments and which condition the morphological evolution of the coast" [5]. They could also be defined as "energy sources capable of moving sedimentary material, causing erosion or deposit and resulting in a reset of the pre-existing topography" [14]. This second definition makes clear the effect of the sediment transport processes and the consequences of sediment movement on the topography of the coastal environments. This relationship between topography and fluid and sediment dynamics acts in a circular manner and each change has a variable effect on the other. The link between processes and morphological evolution grounds the concept of morphodynamics. Morphodynamics is understood as "the mechanism by which morphology affects hydrodynamic processes influencing in a circular way the further evolution of the morphology itself" [19].

On the coast, there are also processes acting in the long term. The term "long-term process" is formally applied to a century timescale and distances in the order of 10 km, but informally can also be applied to longer periods of time and larger spatial scales (e.g., sea level movements, changes in the climate and tectonic movements). All short- and long-term processes will be explained in detail in Chap. 3.

2.3.2 Consolidated Versus Unconsolidated Coasts

Consolidated coasts are characterized by the presence of hard and consistent materials. Coastal areas consist of consolidated rocks typically found in hilly or mountainous terrain. In these, erosive processes are usually dominant. The degree of consolidation strongly influences the ability of the coastline to resist the processes of weathering and erosion. The resistance depends on the susceptibility of the rock to be altered by mechanical and chemical processes, as well as the hardness and solubility of the constituent minerals of their grains. The rock type, the disposition of its stratification and the orientation of the fracturing that affects the rocky formations significantly influence the erosional evolution of the coastline.

In contrast, **unconsolidated coasts** are dominated by a combination of sedimentary and erosive processes, usually with a depositional balance. In this type of coast, waves and currents are able to completely alter the features of relict reliefs. Along the unconsolidated coasts, there are usually large amounts of sediment available, and therefore the evolutionary changes occur rapidly. These are coasts that present normally low reliefs and whose coastline has been softened by the erosion of the preexisting headlands, by deposition of sand barriers or by the infilling of river mouths.

2.3.3 Erosional Versus Depositional Coasts

Coastal processes are the main suppliers of sediment to the coast but also of removing it to other places in the form of erosion. Both processes can be simultaneously present on a coastal stretch in different moments and places, but then there will be a balance between them which will tend towards one of two senses: erosion or sedimentation.

Erosional coasts are developed in places where there is an abundance of energy from coastal processes or there is a deficit in sediment available to be transported. As a result, erosion will be the dominant mechanism of coastal evolution. Erosional coasts are usually steep and are characterized by rocky coastlines that are exposed to high energy waves and that contribute a relatively small amount of sediment to adjacent coastal cells.

In contrast, **depositional coasts** are characterized by an abundant supply of materials, which results in a net sedimentation and in the creation of new depositional realms. They are very common along mature areas with broad drainage basins and rivers that are able to supply large amounts of sediments which will be redistributed by waves, tides and currents on the open coast. Along the depositional coasts a wide variety of landforms and sedimentary environments can be present, which will be described later.

If a coastal area does not have a long-term erosional or depositional balance it is a **coast in equilibrium**. To really speak of a coast in equilibrium should take into account the trends, because that equilibrium will be flexible, regarding a balance through short-term moments of erosion and recovery.

2.3.4 Open Versus Protected Coasts

When we talk about the coast, we always imagine a linear coastline dominated by wave action. Perhaps it is our relative ease in the holiday period, looking towards the sea as it gives a feeling of spaciousness that makes us nominate these coasts as **open coasts** in reference to their opening up to the direct influence of marine processes.

In contrast, there are coastal segments where landforms and environments are developed landwards from the coastal plain. In these instances, the dominant processes tend to be related to the tide, and wave action is absent due to existing coastal elements that limit or prevent their direct action. We refer to these as **protected coasts**, being those environments protected from the waves' action.

2.3.5 Mechanical Versus Biological Coasts

The main part of the coast (both erosional and depositional) is dominated by mechanical processes, and these can be named **mechanical coasts**. In addition, other coasts are built by the action of organisms. These are called **biological coasts**. Perhaps the best example of a biological coast is a reef. Coral and other organisms are able to build enormous volumes of calcareous solid structures. Note that the Great Barrier Reef in Australia is purported to be the largest coastal landform on Earth.

2.3.6 Emergent Versus Submergent Coasts

Global and regional sea level changes, slowly occurring over centuries, are also present on the coasts. Some of these changes include intervals of time over thousands or millions of years and have been caused by glacioeustatic, climatic or oceanographic factors, affecting the coasts at a global level. In addition to these, at the regional level changes may occur by slow vertical movements of the ground along the continental margins. Both types of changes, regional and global, are responsible for the long-term relative movements of the sea level and, therefore, for the displacements of the coastline.

Most of the movements of the land have a direct tectonic origin, but this is not their only source. Isostatic adjustments are vertical movements by which the Earth's crust seeks a gravitational balance to suit superimposed forces such as crustal thinning or thickening. Significant subsidence movements occur in the vicinity of deltas, where masses of fluvial origin sediment accumulate quickly. On the other hand, in the base of deltaic deposits occur phenomena of sediment compaction when grains with a low degree of packing by rapid sedimentation are adapted in a denser fabric, causing a decrease in volume.

A combination of global and regional effects results in the possibility that different coastal sections may be in emersion or immersion, and do so at different speeds. An **emergent coast** is a coastal area that has been exposed during a relative fall in sea level. On the contrary, the **submergent coasts** are those that have been inundated by ocean waters due to a relative rise in the sea level.

In the case of immersion and emersion, there may also be **stationary coasts**, where there is a balance of ascending and descending movements and the relative level remains stable in the long term.

Advanced Box 2.2

The Holocene Transgression

The Holocene transgression, also known as the Flandrian transgression, is a phenomenon of absolute rise in the sea level related to the global deglaciation that took place in the past thousands of years (Fig. 2.5). Glacial melting began about 19,000 years ago and accelerated 15,000 years ago. The best indication of this thaw was the quickly registered sea level rise from 11,700 years ago, which took place due to the rapid melting of the remaining ice sheets in Europe and North America. About 6000 years ago the rise began to slow down, reaching its maximum level at different times along the coasts of the world. In general terms it can be considered that the current level was reached about 4500 years ago.

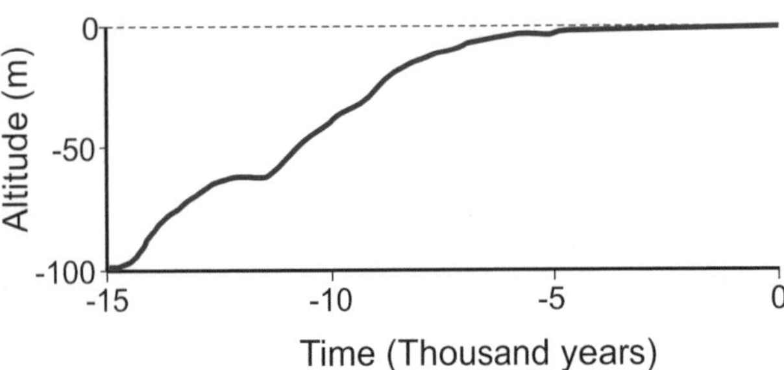

Fig. 2.5 Curve of sea level rise in the last 15,000 years (data from Waelbroeck et al. [18])

This rise led to a very rapid invasion of large continental areas that became continental shelves. Many fluvial valleys were flooded, turning the course of rivers in estuarine funnel-shaped areas. The arrival of the sea level to its current position can be considered the starting point of the present coastal dynamics and sedimentation of the regressive sequence that set the layering of the current depositional coastal environments.

2.4 Coastal Uses

2.4.1 Urban Use

The terrains of coastal areas are attractive and valuable for urban use, since an outlet to the sea makes them well-connected zones. Depositional coasts are, among all, the busiest, being normally a flat terrain where construction is easy. This attraction means that coastal areas were to become the most densely populated areas in the world. In developing countries, coastal areas are those that have the highest rates of population growth.

2.4.2 Industrial Use

Associated with urban use comes the need for the installation of industries supplying the products used by humans. Coastal areas are attractive for industries for the same reasons that drive urban use. In addition, there is the ease that is represented by the possibility of constructing harbor infrastructure for the transport of the freight produced in industry. All that means that the most important industrial complexes in the world are located in coastal areas.

2.4.3 Military Use

The potential provenance of hostile invasions from the sea meant that historically many military bases were located in coastal areas. That is the cause of the presence along the coasts of military naval ports, airfields and camps. Some of the military infrastructure, such as harbors, modified the natural equilibrium of the coast. On the other hand, in some cases, the presence of military in coastal areas has contributed to their conservation.

2.4.4 Recreational Use

The extensive and often insensitive marketing of the coastal zone for recreational uses in the last 40 years has not only led to the unpleasant disfigurement of many shores, which initially had a natural charm, but also to the reorganization of many local economies that rely on tourism.

Coasts such as California and Florida in the USA or the French and Spanish Mediterranean are very obvious examples. In these places, many small coastal towns were saturated by new touristic services, which displaced the traditional skills such as fishing, agriculture, livestock or crafts. Too often, the lack of foresight in the development of urban and recreational infrastructures has led to the destruction of fragile ecosystems, both in the emerged coast and along the submerged littoral. This is in fact an irony, since in many cases the appeal of wilderness that surrounded these localities or the charm of their traditional architecture were among the factors that initially attracted tourists.

2.4.5 Wetland Reclamation

A large area of low-energy coast has been claimed by humans in the last two centuries and this is a process that continues. The loss of wetlands associated with estuaries, delta plains and tidal flats (including shallow areas of the inter-, sub- and supratidal fringes) has had a marked effect on biological productivity, often for the simple fact that these zones act as nurseries for adjacent marine water species. In addition to the ecological impact, loss of tidal land inside the river mouths has affected the tidal prisms and resulted in a modification of the hydraulics and hydrochemistry in the balance between the tidal currents and river flows.

2.4.6 Waste Disposal on Coastal Environments

Since the Industrial Revolution of the nineteenth century, littoral and coastal areas have been seen as a perfect dump for fluids and solid waste materials. Physical and ecological disturbance caused by indiscriminate dumping results in chemical and physical changes in the environment. In extreme cases, such as for example radioactive waste, there is a direct threat to health, but most often there is only a deterioration of the environment that leads to a decrease of biological productivity and in extreme cases to its total destruction.

In the case of solid waste, these can also end up being stacked in piles on reclaimed wetlands, adding a further problem to the land rehabilitation. In most cases, uncontrolled drainage from these columns made of waste leaches straight into the natural coastal environment. The imposition of new environmental laws by international decrees since 2010 should lead to improvements in this matter, although the response to these international standards, as to national legislation, is always slow and ineffective.

2.4.7 Exploitation of Energy

Harnessing the power of waves and tides in coastal areas, such as by the installation of photovoltaic, wind or nuclear plants, is cause for concern since, as mentioned, these are areas with many other human uses.

The potential environmental impact of tidal energy turbines has been thoroughly investigated in some countries such as Canada, France and the United Kingdom, but there are still no major tidal power projects that have materialized. Similarly, devices for harnessing the energy of waves have been explored on a large scale. In both cases, the potential impacts are worthy of consideration, even if the projects are tailored to international laws.

The use of coastal areas for the settlement of photovoltaic or wind power plants presents parallel problems, as well as the influence that such use of these areas can pose to reclaimed wetlands.

The location of nuclear power plants in coastal areas has additional problems, since these plants may be affected by long-term and eventual coastal processes. The case of the Fukushima power plant, affected by the tsunami that struck Japan in March 2011, is perhaps the clearest example.

2.4.8 Coastal Conservation

Especially in the last decades, many coastal areas are preserved as natural reserves, natural parks or marine sanctuaries. These protection figures represent a guarantee, not only of conservation of the supported biological ecosystems, but also of the equilibrium of the dynamic processes.

2.5 Coastal Problems

The intensive use of coastal areas in all the forms that have been listed above has led to the emergence of certain problems. It must be remembered that the word "problem" involves the approach from a human point of view. As rightly stated by Orrin Pilkey, "there is no problem on the coast until the presence of a human infrastructure forces one to measure processes" [16]. This requires understanding that the presence of processes in themselves do not pose a threat to a natural system, since the system will adapt to this process in one form or another. It is the human being who understands these processes as a menace by standing in areas affected by the same. From this point of view, we should consider such issues as the following:

2.5.1 Sea Level Rise

Today it is accepted by the scientific community and also the common citizen that the global sea level is rising in general terms. Some models predict that the global rise will be closer to values of 1 m by 2050. The reasons for this increase in the sea level are complex, but one of the most influential factors would be the increase in atmospheric CO_2 and other trace gases. The presence of these gases induces an increase in the absorption of heat from the Earth's surface (the greenhouse effect). Planning for and managing this sea level rise is perhaps the biggest current environmental challenge for coastal scientists.

2.5.2 Shoreline Erosion

Due to multiple factors, some of them natural and others related to human activity, many coastal fringes present erosional trends. Traditionally, coastal engineers have used rigid structures which, in theory, protect the coast from the processes of erosion. These traditional engineering methods are being questioned today, particularly because they are nondefinitive solutions and due to the high economic cost in the long term. Now there are proposals for new, more flexible strategies for the control of erosion. These strategies will come about through precise knowledge of the processes, analysis and intervention on the causes of the erosion.

2.5.3 Storm and Tsunami Hazard

Destructive coastal storms are becoming more common. It is accepted that global change is causing an increase of storms and floods along many low and heavily populated coasts. Also, tsunamis represent potentially destructive events. In recent times, the increase of the population in coastal areas has brought about an increase in material and personal damage. Perhaps the implementation of better warning systems can reduce some risks, although adequate coastal planning could counteract the potential dangers of these high energy events.

2.5.4 Human Destabilization

Some of the problems described above are the product of the effects of natural processes on the development of human activity on the coast. However, there are numerous examples of imbalances in the coastal processes generated by any human action that directly affect other human activities or infrastructures, or even the natural environment. The most obvious disequilibrium may be that caused by port infrastructures on wave dynamics, but there are other actions that are detailed in further sections.

2.5.5 Saline Intrusion into Coastal Aquifers

The extraction of groundwater from freshwater coastal aquifers occurs extensively. This water is commonly used for human consumption, irrigation of the coastal plains or for use in industrial processes. As a consequence, it often results in a rise of the interface between fresh and salt waters that moves landwards, causing the salinization of aquifers and the loss of the ability to use these waters for the required purposes.

2.5.6 Subsidence

Subsidence is a slow sinking of coastal lands due to multiple causes. One of the most common is the compaction of old coastal sediments under the weight of more recent ones. Other causes may be related to extractive processes as described in the previous section. In natural environments, subsidence is normally compensated with an increase in the sedimentation rate, although in the environments under human use the collapse results in an increased susceptibility to coastal flooding (of marine, river or mixed origins).

2.5.7 Changes in Ecological Structure

The uncontrolled exploitation of many coastal ecosystems or imbalance in the process caused by human activities have led to changes in the physical environment that supports ecosystems and thus their ecological structure. These changes usually involve a decrease in biological diversity and productivity, as well as other more subtle and hard-to-predict changes. Many natural coastal systems are under a very sensitive equilibrium regarding the production of nutrients that support the food chain. Uncontrolled changes of the volume of nutrients can significantly alter any ecosystem, and these can be induced simply from a hydraulic or a geomorphological change in the physical environment. This idea underpins the concept of ecohydrology managed by UNESCO in the past decades.

2.6 Integrated Coastal Zone Management

The interaction of processes and coastal risks, natural environments, population growth and uses of the coast (often developed without planning), creates a series of problems involving the complex challenge of an adequate management of coastal areas (Fig. 2.6).

The concept of integrated coastal zone management (ICZM) is understood as a strategy for the sustainable development of the coast using as a tool the knowledge of dynamic and ecological functioning. The need to rely on an effective administrative structure for management is important, and for that, today, there are bodies at various levels and different structures that complement each other. National administrations are addressing the general problems of the coastal environments in coordination with inter- and supranational organizations. The overall objective of these agencies is to preserve sustainability against the growing use of resources.

On the other hand, the challenge to afford the consequences of global change has generated an intense debate

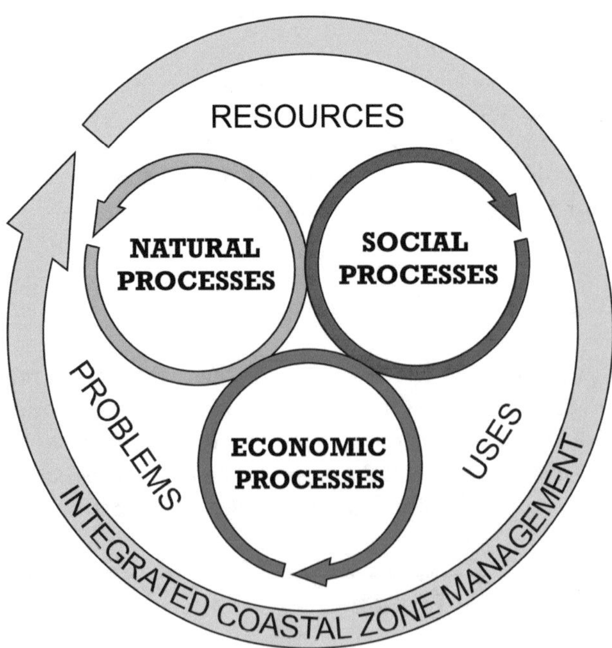

Fig. 2.6 Interactions between natural and anthropogenic processes (social and economic) that give rise to the need for an integrated management of the coastal zone for sustainable development

about the concept of "coastal protection." This is called the "debate of the coast," and it describes the challenge facing coastal engineers and geologists on their opposite sides as each one defends their strategy of protection [16]. The background of this debate was a dissatisfaction among geologists with the way engineers tackle coastal issues. Geologists believe that engineers often wrongfully modify the environmental balance of the coast with the construction of increasingly large and more expensive structures and defend the use of the knowledge of coastal processes and the ICZM to propose noninvasive solutions to coastal problems, enabling the coast to reach its own equilibrium.

The concept of ICZM and its development will be analyzed in more detail later in this book.

References

1. Boak EH, Turner IL (2005) Shoreline definition and detection: a review. J Coastal Res 21(4):688–703
2. Carter RWG (1988) Coastal environments: an introduction to the physical, ecological, and cultural systems of coastlines. Academic Press, London, 617pp
3. Coastal Engineering Research Center (CERC) (1984) Shore Protection Manual (2 volumes). US Army Corps of Engineers, Washington DC, 532pp
4. Cotton CA (1955) The theory of secular marine planation. Am J Sci 253:580–589
5. Cowell PJ, Thom FG (1994) Morphodynamics of coastal evolution. In: Carter RWG, Woodroffe CD (eds) Coastal evolution. Cambridge University Press, Cambridge, pp 33–86
6. Davis WM (1909) The outline of cape cod. Geographical essays. Ginn and Co., Boston, pp 690–724
7. Davidson-Arnott R (2010) Introduction to coastal processes and geomorphology. Cambridge University Press, Cambridge, 442pp
8. Dolan R, Hayden BP, May P, May SK (1980) The reliability of shoreline change measurements from aerial photographs. Shore and Beach 48(4):22–29
9. Harris DL (1981) Tides and tidal datums in the United States. Special Report No. 7. Coastal Engineering Research Center, US Army Engineer Waterways
10. Inman DL, Brush BM (1973) The coastal challenge. Science 181:20–32
11. Komar PD (1976) Beach processes and sedimentation. Prentice-Hall, New Jersey, 540pp
12. Mangor K (2004) Shoreline management guidelines. DHI Water and Environment. 294pp
13. Masselink G, Hughes M, Knight J (2011) Introduction to coastal processes and geomorphology. Routledge, London, 432pp
14. Morang A (2004) Coastal geology. University Press of the Pacific, US Army Corps of Engineers, 297pp
15. Morton RA (1991) Accurate shoreline mapping: past, present and future. Coastal Sediments 91(1):997–1010
16. Pilkey OH (1981) Geologists, engineers and a rising sea-level. Northeast Geol 3:150–158
17. Short AD (2012) Coastal processes and beaches. Nat Educ Knowl 3(10):15
18. Waelbroeck C, Alabeyriea L, Michela E, Duplessya JC, Mcmanusc JF, Lambeckd K, Balbona E, Labracherie M (2002) Sea-level and deep water temperature changes derived from benthic foraminifera isotopic records. Quatern Sci Rev 21:295–305
19. Wright LD (1995) Morphodynamics of inner continental shelves. CRC Press, Boca Raton, FL, 241pp

Factors Affecting Coastal Evolution: Spatiotemporal Scales

3

3.1 The Changing Coast

Coastal environments are probably the most variable systems on the Earth's surface. Many changes that occur on the coast are repetitive in space and in time, while others are slowly being produced by way of trends and others take place in the form of isolated events. It is also interesting to distinguish between the repetitive changes that occur in a periodic way and the cyclical changes. An example of periodic changes would be the effects of storms with a long return period, while examples of cyclical changes could include the effects of the succession of spring and neap tides on the bedforms of tidal environments or the well-studied seasonal changes on beaches (with erosion of the sand caused by winter storms and the spring and summer sedimentation because of the action of fairweather waves). The best example of isolated events would be the action of tsunamis, but this could also be considered as a periodic process with a geological-scale return, whereas a good example of evolutionary trends would be the long-term processes of transgression and regression.

Considering all this, it is clear that when a study of coastal geology is planned, the choice of temporal scale of research is important. Indeed, the scale of the observations in both space and time should be chosen carefully. A further aspect to take into account is that selected variables should be chosen according to those whose measurements are sensitive to the changes that happen within the chosen scales. The difficulty of this is high, because many of these variables are closely related to different scales. For geologists, habituated to working on very long timescales, understanding changes that occur in such short timescales as coastal change is not an easy task. In general terms, geologists are accustomed to interpreting the variables in facies sequences observed in the sedimentary record, but although the sedimentary record is a good information source to study and interpret the synthesis of long-term processes that have occurred in coastal systems, it is common that many of the changes at short time intervals are not reflected in that record, having been deleted by subsequent changes. In other words, many of the coastal facies which reflect changes in processes are destroyed if their preservation potential is low, and are reflected only in the presence of erosive unconformity surfaces. However, as described in the previous chapter, in recent decades coastal geology has experienced a revolution in its development, and short-term dynamic processes are among those that have received greater attention. Today, the factors affecting the changing coast are better known.

3.2 Factors of Coastal Dynamics

The previous paragraph makes it clear that coastal sedimentary environments are natural systems operating under very intense dynamics that induce morphological changes at different scales which occur under a situation of unstable equilibrium. It should be understood that all of these changes are caused by the movement of sedimentary material and, therefore, from a general point of view, the evolution of coastal systems is controlled by the same factors that control the sedimentary dynamics (Fig. 3.1). We must understand that the coastal fringe is affected by two sets of geological processes, which operate either separately or in conjunction [17]: exogenic processes that operate on the surface of the Earth and endogenic processes that operate in and below its crust, but can deeply affect the surficial behavior.

Knowledge of the phenomena of erosion, transport and deposition of sediment in the littoral environments needs, therefore, to include the analysis of the movement of sedimentary particles by wave action, tidal currents and other marine currents. This study should place special emphasis on the identification of hydrodynamic processes and sedimentary products (facies), as well as the longer-term influence of the sediment supply regime and the relative movements of the sea level.

J. A. Morales, *Coastal Geology*, Springer Textbooks in Earth Sciences, Geography and Environment,
https://doi.org/10.1007/978-3-030-96121-3_3

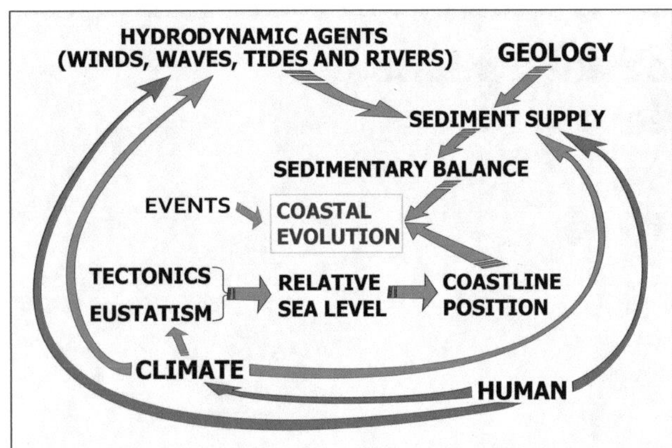

Fig. 3.1 Factors affecting coastal evolution

3.2.1 Hydrodynamic Agents

Hydrodynamic agents are most directly responsible for the functioning of coastal environments. First, waves are responsible for the erosion of rocky coasts, at the same time carrying out transport (longitudinal and transverse) between the coastal area and the sea and, finally, being the major cause of sedimentation of sand on the front of beach and barrier island environments. On the other hand, the tide is the main phenomenon that controls deposition in fluvio-marine interaction systems (estuaries and deltas), as well as in protected coastal systems such as bays and back-barrier areas (lagoons and tidal flats). Both of these have been individually considered as the main agent responsible for local coastal morphology [4, 5, 13].

The wind is responsible for movement of sedimentary material in the emerged coastal fringe, via the control it exerts over the genesis and migration of coastal dunes. At the same time, it is also the main agent causing waves and is fundamental in the genesis of the meteorological tides (surges). Finally, river flows are a fundamental component in the control of sedimentation in estuarine and deltaic channels, being responsible for the processes of mixing of waters and the phenomena of dispersion of suspended matter, both within channels and outside the mouth. The rivers are also the main agents of transport of material from the mainland to the coast.

The relationships of all of these agents in terms of energy create the main conditions that confer a determinate coastal morphology and also the distribution of coastal environments [6, 11, 12].

3.2.2 Sediment Supply

The contribution of sediments to the coastal system is a key factor in the balance of the coast. This supply can reach from land through the rivers and winds, as reflected in the previous paragraph. The sediment supply by rivers represents the most important connectivity between the continent and the coast; in fact, the main part of the coastal sediments enters the coast from rivers. The wind is a secondary source from the continent, but in some areas can constitute the main sediment supply. Additionally, the sediment can reach the coast from deeper marine areas. Sediment from the continental shelves can approach the coast via wave action. This sediment usually comes from ancient coastal sediments inherited from lower sea level stands. Erosion from submarine rocky outcrops can also be a frequent source of sediment from marine areas. Some coasts are also able to produce large amounts of sediments that then move along the coast due to littoral drift. Perhaps the best-known cases are the erosion and transport from cliffs, but in tropical areas the chemical and biochemical production of carbonates can be the main source of sediments.

The volume of sediment supply, in relation to the ability of coastal agents to rework material, is the determining factor in obtaining a sedimentary balance of the coast. Thus, if the sediment budget is higher than the transport capacity of marine agents, then the coast is depositional, while if the transportation capacity is greater than the external contribution, then the coast will be erosive; a balanced coastline is when both factors are in equilibrium.

It must be borne in mind that an important part of the contribution is seasonal, since the river flows, the force of the wind and the wave energy are seasonal, and all of them control the capacity of supply sediment to the coast and the ability to rework it in coastal systems. Therefore, sedimentary balance on the coast can vary in different seasons, and this is as important for understanding the functioning of a coastal stretch in establishing an annual balance as are the interannual trends.

3.2.3 Events on the Coast

Events are exceptional geological phenomena that display a great amount of energy. Events that occur on the coast tend to be considered geological events of a lower order or are not even regarded as events from a geological point of view. That is due to the frequency at which they occur, since they cannot be called exceptional, although they may be so on the human timescale [9].

On the coast, high energy events can reach from land or from sea. In coastal areas related to river mouths, floods represent a rapid supply of large amounts of sediment, which must then be distributed throughout the coastal systems by the processes that act on an ongoing basis.

Perhaps the most common high energy events in open coastal areas are storms. Storms represent a sudden shift in the distribution of coastal sediment, generating wide erosive zones in some places and rapid accumulation of sediments in others. Some extreme storms such as those associated with tropical cyclones, typhoons and hurricanes can produce large changes in coastal morphology, generating important temporary imbalances that lead to the dynamic processes achieving a new ordinary equilibrium.

However, the phenomenon that displays the greatest amount of energy on the coasts are tsunamis. In this sense, tsunamis, although with a tectonic origin, act on coastal systems in a way very similar to extreme storms.

All of these events are usually reflected in the geological record of the coastal environments in the form of a high energy facies sequence. So, the depositional record of a tsunami is nearly indistinguishable from the record of a storm. The record can also occur in the form of an erosional unconformity, overlapping the sequence of the ordinary coastal processes [10].

3.2.4 Global and Regional Sea Level Movements (Tectonics and Eustatism)

The relative movements of the sea level result from the combination of both tectonic and eustatic phenomena. Eustatic movements are those related to absolute rises and declines in the sea level. Most are due to variations in the volume of water in the oceans, and therefore take place globally. Eustatic variations can be associated with different geological phenomena, such as glacial periods (climate change), post-orogenic volcanism or changes in the morphological configuration of the seabed.

As a result of tectonic activity, coasts can also rise or sink regionally and locally. The elevation takes place normally as a result of orogenic compressive processes, while sinking is generated as a response to extensional movements, tilting or compaction of the oldest sediments by sedimentary load, constituting the phenomenon known as subsidence.

The term *relative sea level change* refers to variations in the depth of the sea at a point during a specific time interval, due to the combined action of eustatism and tectonics (global and regional factors). In this way, a relative fall in sea level is an observable effect caused by a eustatic lowering, a tectonic uplift, or the combination of both phenomena. On the contrary, a relative sea level rise may be due to eustatic rise, subsidence, or the combination of both. An episode of relative stability of the sea level will be an absolute stability or also a total counteraction of both phenomena.

A relative drop in the sea level will result in a loss of depth, overlapping shallower environments on deeper ones. If the lift is slow and progressive, overlapping facies will respect Walther's Law, which states that the facies will be found in the established order of the coastal sequences.

Subsidence can cause the collapse of the coast, resulting in a deepening. Thereby, in the coastal sequence the deeper facies are located on top of the shallower ones. If the subsidence occurs in a slow way, the facies remain in the order established by Walther's Law, but in reverse order. However, if a sudden pulse of subsidence occurs, then intermediate terms can be missing. Some coastal environments with high accumulation rates such as deltas are particularly susceptible to subsidence, especially where extractive human activities (exploitation of water or hydrocarbons) occur.

3.2.5 Changes in Position of the Coastline (Transgressions and Regressions)

The relative sea level movements combined with the sedimentary balance manifest in the coastal fringe as advances or setbacks of the coastline, which are known as transgressions and regressions.

The term transgression is applied to an invasion from the sea on an emerged zone, assuming an advance of the coast landwards. On the contrary, a regression is defined as a withdrawal of the sea, in such a way that a previously submerged area becomes part of the land, which implies a shift of the coastline to the sea. For coastal environments, a minimum variation of coastline means an amendment of the equilibrium and possibly a rough displacement.

So, the existence of transgressions and regressions depends on the balance between erosional and cumulative processes, rates of erosion or accumulation and the speed of relative sea level motion.

The combination of subsidence and sea level motion can create a space to accumulate significant amounts of sediments. That links with the concept of **coastal accommodation**, which is understood as the space able to accumulate a coastal sedimentary sequence which records the history of the environmental changes that generated this space. The registered sequence can be transgressive or regressive depending on the relationships between the accommodation and the accumulation rates.

3.2.6 Geological Influence

The geology of the coast and adjacent areas plays a secondary role in coastal evolution. On the one hand, the lithology of a coastline, as well as of the entire drainage basin of the rivers that drain into that coastal stretch, conditions the nature and volume of the sedimentary input. From an intuitive point of view, a coast composed mainly of soft (unconsolidated) lithologies can supply high amounts of sedimentary materials, whereas if the coast (or source area) is composed of hard lithologies, then the contributions will be minor. In both cases, the grain size of the source rocks may condition the nature of the sediment that reaches the coast and is finally deposited. Something similar happens with the coasts whose lithologies are easily alterable by chemical processes; in this case, the materials will be dissolved and enter coastal waters in ionic form to be subsequently deposited by chemical precipitation processes. These criteria would also apply to submerged areas that could potentially supply sediment to the coast.

On the other hand, the nature and orientation of geological structures in relation to the coastal relief have a remarkable influence on the geomorphologic processes that may occur. From this perspective, alternations of soft and hard strata, the presence of fractures and other tectonic structures and their geometric relationships (between them and with the coastline) condition the occurrence of gravitational processes such as rock falls, toppling, wedge failures or slides, all of which are usually associated with cliffs and rocky coasts.

3.2.7 Climatic Influence

Climate is the main control of coastal evolution because it acts on every variable influencing the coastal system. Climate is subject to cyclical changes that occur in different timescales. The shorter of these cycles are seasonal variations of the weather due to the tilt of the Earth's rotational axis and the relationships with the ecliptic plane during the Earth's orbit around the Sun. This fact, known as obliquity, makes the Sun's rays approach a determined latitude of the Earth from different angles at different times of the year.

Longer cycles are those related to the total amount of the heat energy emitted by the Sun's surface, which vary in periods of eight to 12 years. These cycles are responsible for the well-known phenomena of alternations between El Niño and La Niña.

Other even longer cycles are related to the radiation received by the Earth and are caused by the combination of the three main orbital cycles: precession, obliquity and eccentricity [1]. These phenomena cause climatic cycles of 11,000, 41,000 and 100,000 years, respectively, and were responsible for the four glaciations and their minor climatic oscillations that occurred during the Pleistocene.

So, it is clear that climate exerts a strong influence on factors controlling coastal dynamics in every level of influence. On the one hand, winds, waves, meteorological tides and river discharges are directly controlled by the shorter climatic variations. On the other hand, the sediment supply is also strongly influenced by climate because weathering processes and potential sediment transport from the source areas to the coastal sedimentary environments are controlled by climatic conditions.

At the higher level of influence, some eustatic mechanisms and relative sea level movements and coastline positioning are controlled by long-term climatic cycles.

3.2.8 Human Influence

The coastal environment has been greatly influenced by human action over the past decades. Anthropogenic action affecting coastal factors has occurred on three levels. In a direct way, humans have modified the action of coastal processes on the littoral such as altering the arrival of wave trains to the coastline by means of hard structures (jetties, groins and breakwaters), decreasing the velocity of the currents in the mouth of tidal systems acting on the tidal prism by means of dredging or damming tidal surfaces (saltpans and fisheries) and also decreasing the fluvial currents arriving at estuaries and deltas by means of water catchments in drainage basins. This last action has also had the effect of reducing the input of sediments supplied from rivers to the coastal systems, modifying directly the sedimentary balance of the coast.

In some of the world's coasts, the human influence has been present over centuries, even millennia. On the Mediterranean coasts, Egyptians controlled the fluvial supplies of the Nile River by modifying the delta dynamics. On the open coasts, Greeks and Carthaginians modified the coastal morphology to build harbors.

Human activities on beaches and in tidal channels have also directly modified the sediment budget in the coastal systems. Dredging activities, destruction of coastal dunes or artificial beach replenishments are perhaps the best examples of that.

But in an indirect way, humans are also responsible for this influence at a higher level, since it is human action that has been proved to be the cause of the current global warming. Thus, the Earth's climate is being modified by mankind with its entire influence on the coastal processes. Indeed, human action's influence on the natural environment has been so great since the second half of the twentieth century that some authors have considered that a new geological period has begun: the Anthropocene.

3.3 The Spatiotemporal Scale of Coastal Processes

As can be seen in the previous section, coastal environments present a wide variability in the scale of their phenomena and their time of response, since all the analyzed factors (or variables) act on different timescales.

Authors in the 1980s (e.g., [2]) and others in the early 1990s (e.g., [15, 16]) divided the action of these processes into two categories according to their timescale. Thus, on the short-term scale (minutes to years) are the actions of the hydrodynamic processes (waves, tides and coastal currents), all of them strongly influenced by the regime of sediment supply from the mainland, from the platform and along the coast. On a longer time-scale (centuries to millions of years), tectonics and eustatism condition the relative sea level movements and, consequently, have a significant influence on changes in position on the waterfront and in the secular evolution of coastal environments. Controlling these factors and exerting a direct influence on coastal dynamics at all timescales are the climate and anthropogenic action.

However, other authors of the early 1990s (e.g., [3, 18]) divided the main coastal processes into four spatiotemporal scales of action (Fig. 3.2).

At the lowest level are the **instantaneous** processes, linking fluid flow processes with forces that cause continuous sediment transport, generating micro- and mesoforms. At this level of action are also rapid fluctuations of energy (short-term cyclical oscillations). This scale of processes normally generates lithofacies and lithofacies sequences (depositional facies).

At the second level are the **events** that act in scales from dozens of meters to kilometers and occur from seasons to decades (may also be centuries). These phenomena can generate special facies sequences when they are depositional, but are normally erosional and are preserved in the form of erosive unconformities.

The third level was named **engineering** because engineers are especially interested in the phenomena of this scale. These are scales from decades to centuries and include cyclical processes and evolutionary trends. For that, this level would be named **historical**. The results of the action of these processes are depositional facies sequences.

The fourth level corresponds to the longer-term processes that operate from decades to millennia. These are controlled by global (climatic or tectonic) conditions and for that reason it is named **geological**. The record of the processes acting in this level is preserved as major sequences (parasequences and depositional sequences).

Instantaneous and event scales are considered as short-term factors by the authors of the 1980s, whereas engineering and geological scales are defined by them as long-term variables. Classical studies have been used as a method for the separate analysis of the variables involved in the evolution of media. This analysis facilitates the understanding of phenomena individually, especially of those which act in the short term, however it does not offer an understanding of the global dynamics of the coast that is presently required.

It is necessary to take into account that, conceptually, both processes that occur in the short- and long-term scales (instantaneous and geological) are regular and continuous processes, while those acting in intermediate-term scales (event and historical) are unique and unpredictable.

The representation in a spatiotemporal diagram of hydrodynamic and sedimentary coastal processes (Fig. 3a and b) shows a continuous transition between these processes rather than separate scales as suggested by Cowell and Thom [3], Stive et al. [18]. In this scheme, some instantaneous hydrodynamic processes like the action of wave trains, tidal cycles or wind regimes can be linked to short-term sedimentological processes such as migration of bedforms and sand bars (in beaches, deltas and estuaries). On the other side of the diagram, long-term processes like isostasy, eustatic and tectonic cycles or global oceanic currents would be connected to large-scale sedimentary processes such as barrier island formation, accretion of tidal-delta lobes, deltaic progradation, reef growth or cliff retreat. However, in the middle area of the diagram there is not a correspondence of concepts with those suggested by the original authors. In this sense, eventual hydrodynamic processes like fluvial floods, storms and tsunamis appear to be displaced with respect to the area suggested by Cowell and Thom [3] on the event scale. On the other hand, there are geomorphological processes such as inlet breaching that correspond with events, but others located in the spatiotemporal scales correspond with continuous sedimentary processes (dune migration, seasonal beach cycles and accretion of estuarine bodies). In any case, the processes of the engineering (historical) scale are not well distinguished from the processes of the geological scale.

A multifactor consideration regarding the end result of coastal evolution is really a matter of historical accidents [7, 8], because it is those events that control the geological

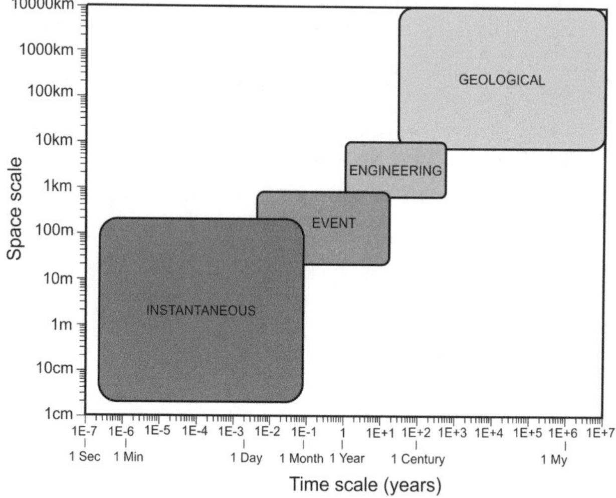

Fig. 3.2 Definition of spatial and temporal scales involved in coastal evolution (Adapted from Cowell and Thom [3])

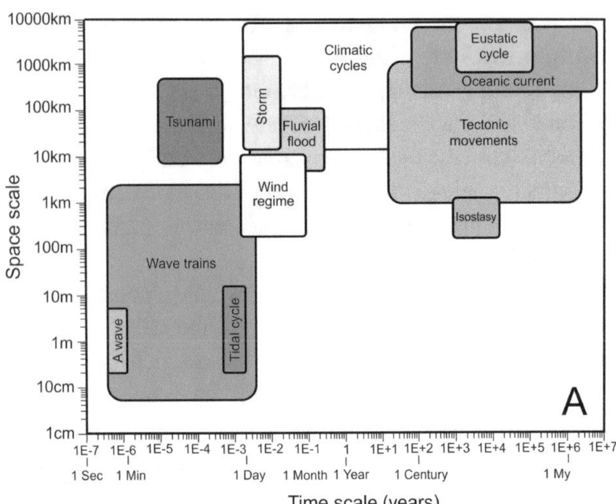

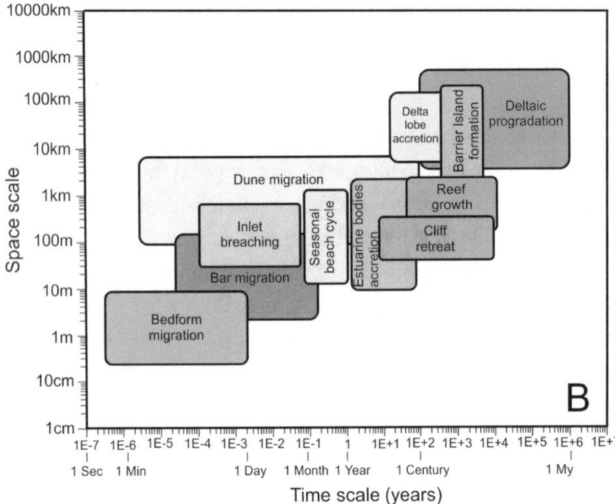

Fig. 3.3 Coastal processes, **a** Hydrodynamic processes, **b** Sedimentary processes (Adapted from Morang [14])

record that is finally preserved. Nevertheless, the possible combinations of factors acting on these different scales give rise to the existence of a variety of long-term dynamic models that subsequently generate several schemes of relationships among coastal environments and result in a finite number of architectural facies models that build the upper part of the marine system record. In this sense, the study of the coastal sequences preserved in the geological record is a significant help in understanding the global action of processes at all spatiotemporal scales.

References

1. Berger A, Loutre MF, Laskar J (1992) Stability of the astronomical frequencies over the Earth's history for paleoclimate studies. Sci 255:560–566
2. Bird ECF (1985) Coastline changes: a global review. John Wiley and Sons, Chichester, 219pp
3. Cowell PJ, Thom FG (1994) Morphodynamics of coastal evolution. In: Carter RWG, Woodroffe CD (eds) Coastal evolution. Cambridge University Press, Cambridge, pp 33–86
4. Davies JL (1964) A morphogenic approach to world shorelines. Ann Geomorphol 8:27–42
5. Davies JL (1980) Geographical variation in coastal development (2nd ed) Longman, New York, NY, 212pp
6. Davis RA, Hayes MO (1984) What is a wave-dominated coast? Mar Geol 60:313–329
7. De Vriend HJ (1991a) Mathematical modelling and large-scale coastal behavior: part 1: physical processes. J Hydraul Res Special issue Maritime Hydraulics, 727–740
8. De Vriend HJ (1991b) Mathematical modelling and large-scale coastal behavior: part 2: predictive models. J Hydraul Res Special issue Maritime Hydraulics, 741–753
9. Einsele G, Seilacher A (eds) (1982) Cyclic and event stratification. Springer Verlag, Heidelberg, 536pp
10. Einsele G, Ricken W, Seilacher A (1991) Cycles and events in stratigraphy. Springer Verlag, Heidelberg, 955pp
11. Hayes MO (1975) Morphology of sand accumulations in estuaries: an introduction to the symposium. In: Cronin LE (ed) Estuarine Research, vol 2. Academic Press, New York, pp 3–22
12. Hayes MO (1979) Barrier Island morphology as a function of tidal and wave regime. In: Leatherman SP (ed) Barrier Islands. Academic Press, New York, pp 1–27
13. Heward AP (1981) A review of wave-dominated clastic shoreline deposits. Earth-Sci Rev 17:223–276
14. Morang A (2004) Coastal geology. US Army corps of Engineers, University Press of the Pacific, 297pp
15. Pilkey OH, Morton RA, Kelley JT, Penland S (1989) Coastal land loss. Short course in geology. In: 28th international geological congress, vol 2. American Geophysical Union, Washington, DC, 73pp
16. Pilkey OH, Young RS, Riggs SR, Smith AWS, Wu H, Pilkey WD (1993) The concept of shoreface profile of equilibrium: a critical review. J Coastal Res 9(1):225–278
17. Rust IC (1991) Environmental geology of the coastal zone: a South African perspective. South African Journal of Marine Sciences 10:397–405
18. Stive MJF, Roelvink DJA, de Vriend HJ (1991) Large-scale coastal evolution concept. In: Proceedings of the 22nd international conference on coastal engineering, American Society of Civil Engineers, New York, pp 1975–1983

4 Coastal Elements: Types of Coasts and Criteria in Coastal Classifications

4.1 Introduction

Every coast presents a variety of landforms and environments which have been classically described and classified by geographers and whose dynamics have been studied by geologists and coastal engineers. In this way, some of the elements that make up our coasts can be characterized. Some of these elements can be defined as features or forms at scales of tens or hundreds of meters, while others are among the most common environments that can be found along a coast. It is necessary to take into account that many of the described elements are simply geomorphological features or landforms (mesoforms or megaforms), while others, in contrast, in addition to being defined as forms possess their own depositional realm and are capable of generating a characteristic sedimentary sequence. In the first case, they are considered as **coastal landforms**, while in the second they are **coastal environments**.

The landforms and environments that can be found on a particular coast, therefore, or the ensemble of environments and landforms that integrate on a coastal stretch, are controlled by the factors analyzed in Chap. 3. These factors act in different scales. So, the climatic, eustatic and tectonic factors exert a global large-scale first-order control, whereas the hydrodynamic and sediment supply factors exert a more local small-scale secondary control. The combination of these factors allows us to determine what type of coast it is. Nevertheless, throughout history there have been proposed multiple classifications of coasts using different criteria, with most of the classifications attempting to group the causes that influence the genesis of these landforms and environments.

In this chapter, landforms and coastal sedimentary environments will be defined and the various proposals for classification of the coasts will be analyzed from the perspective of the different criteria (local and global) used by the respective authors.

4.2 Coastal Landforms

Coastal forms that have developed along the coast (Fig. 4.1) will be addressed in this section. These are the result of a combination of processes that act on sediments and rocks forming the coastal zone. Some landforms are the result of an erosive combination of these processes and, therefore, do not have associated sediments. Other major landforms are depositional and develop facies and facies sequences, but can, however, group several environments. In this case, sedimentation is carried out under the typical dynamics of the respective set of environments and their facies sequences are not defining, and so they are regarded as landforms. This is the case for bays and other sandy landforms such as barriers, spits, barrier islands and tombolos.

4.2.1 Erosional Coastal Landforms

Cliffs

These are the most significant and recognizable erosional coastal landforms. Cliffs are formations of a certain height and high slope which are in contact with the sea and are subject to its dynamic processes. They are reliefs of erosional nature and their material constitutes one of the sources of supply to the adjacent depositional coastal segments.

The speed of erosion by marine agents and, therefore, the volume of material that they supply depends on its geological nature and the arrangement of the rock formations with respect to waves and currents. Actually, cliffs constitute a system (Fig. 4.2) which integrates other erosional forms of smaller dimensions (including abrasion platforms, caves, arches, the headlands) and some depositional environments (such as beaches and colluvial fans) formed at the foot of the cliff.

J. A. Morales, *Coastal Geology*, Springer Textbooks in Earth Sciences, Geography and Environment,
https://doi.org/10.1007/978-3-030-96121-3_4

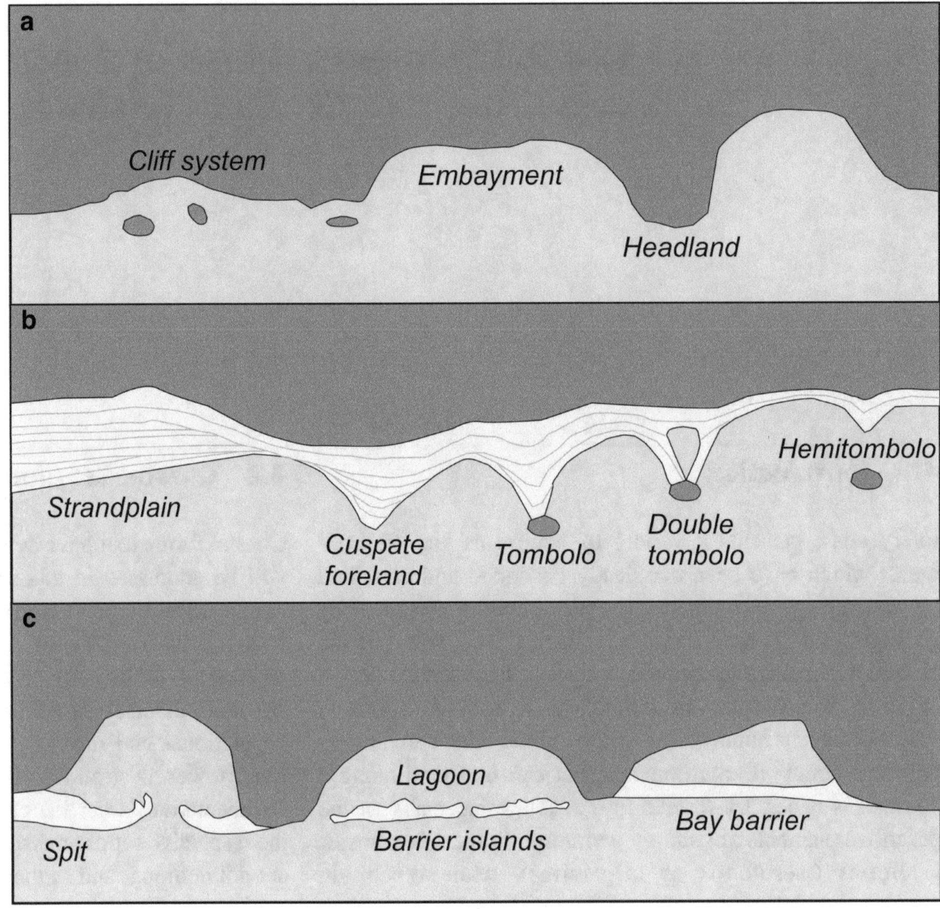

Fig. 4.1 Main coastal landforms. **a** Erosional landforms. **b** Depositional landforms. **c** Enclosed bays and different types of coastal barriers

Abrasion (or wave-cut) platforms

These are thus named because the dominant process that generates them is abrasion, but they are also called wave-cut platforms since it is the waves performing this abrasion. The platform is a flat surface located at the base of the cliffs below the mean tidal level. It is a surface of a transversal length to the coastline that can measure from a few meters to hundreds of meters from where the waves break up to the base of the adjacent cliff. The duration of the process that generates these platforms depends on the rheological nature of the rock. Thus, the existence of extensive abrasion platforms implies that the sea level has remained stable during their period of genesis. A rocky coast can present different levels of platforms, some of which can be located in the continental area, while others may be submerged. The presence of these platforms indicates previous stable positions of the sea level.

Caves

These are formed in the base of the cliffs (Fig. 4.2a, b) as a result of the existence of different rates of wave erosion on lithologies of variable resistance, taking also into account that the waves swash on the cliff base and do not reach its highest levels. This action determines the formation of cavities that can plunge tens of meters below the cliff front.

Sometimes the roof of these cavities can collapse and generate holes connecting the interior of the dome with the surface of the ground. These holes can relieve the pressure of the compressed air in the cavities by wave action and function as *blowholes*, especially in storm wave conditions.

Arches

This is a very striking erosional landform which is also a result of the differential erosion on lithologies of different resistance. These arches have an opening that extends below the water level and is covered by a solid vault that can have an arched or flat shape and whose height can be up to tens of meters above the sea level (Fig. 4.2a, b).

Stacks and stumps

As a result of the aforementioned cliff retreat, abrasion platforms can develop imperfectly, leaving not completely eroded remains protruding from the horizontal surface of the platform (Fig. 4.2a, c). These remains are called stacks and stumps, and present a distinctive type of coastal landform.

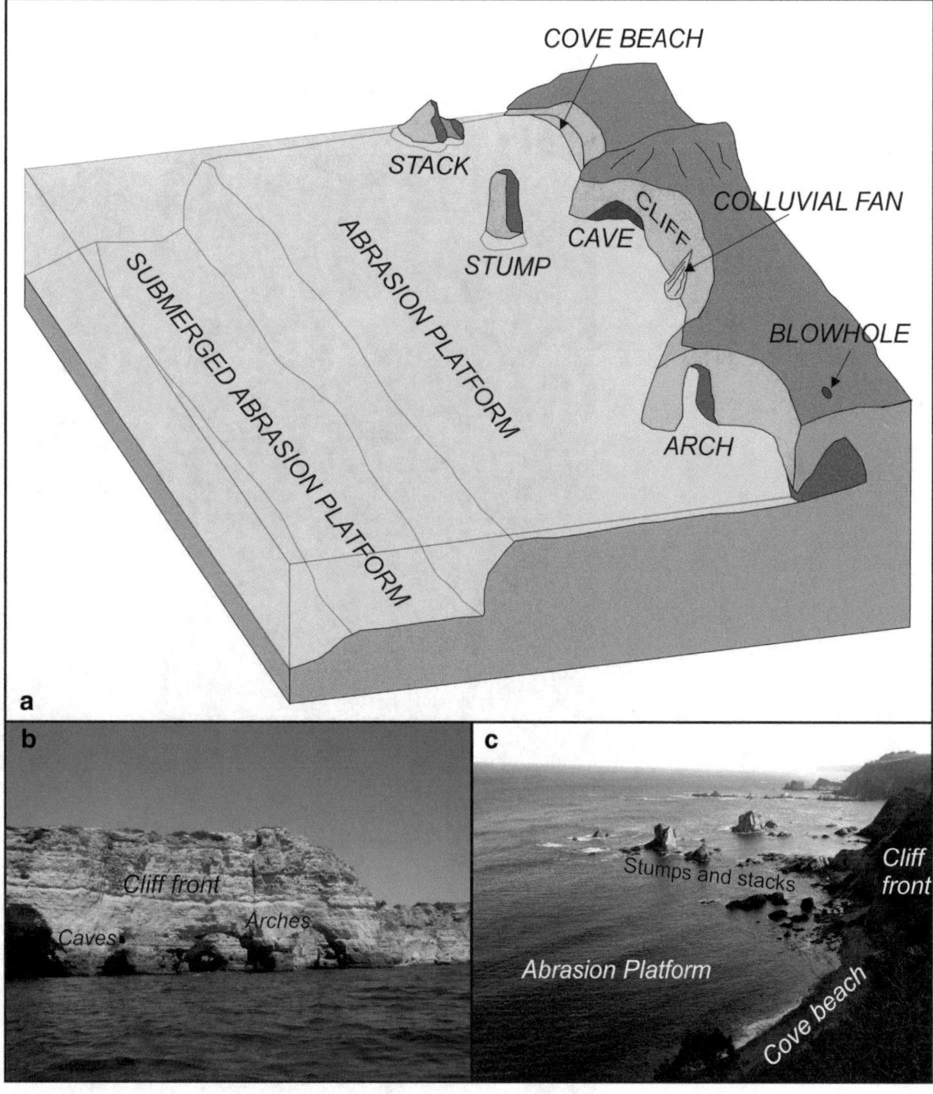

Fig. 4.2 **a** Main coastal forms linked in a cliff system. **b** Arches and caves in the cliff system on the Algarve Coast (Portugal). **c** Abrasion platform, stumps and stacks on the Asturias Coast (Spain)

Some of them may have several meters of altitude and appear as isolated pinnacles. These are forms that continue to be subject to erosion and, therefore, they are not permanent in time.

Capes

Capes are prominent on the coastline where a strip of land extends towards the sea (Fig. 4.1). A cape is simply a coastal morphology, which can have a consolidated and erosional nature (*headlands*, Fig. 4.1a) as well as an unconsolidated and depositional one (*cuspate forelands*, Fig. 4.1b) or it may have mixed features (*tombolos*). From the point of view of coastal dynamics, a cape is a hindrance to the longshore transportation of sediments, tending to limit littoral drift cells.

4.2.2 Depositional Coastal Landforms

Tombolos

These are sandy landforms which connect the mainland with an ancient emerged rocky island (Figs. 4.1b and 4.3a). The sand accumulation is generated by the phenomenon of deformation of the wave trains (refraction and diffraction) induced by the presence of the rocky element, initially away from the coast, which creates a shadow area of deposition. As a result of this phenomenon a process of sedimentation of sand begins, just linking the *tied island* with land.

Occasionally, the connecting element is submerged, while at other times the union between the tied island and the mainland is not complete (Figs. 4.1b and 4.3b). In most cases the tombolo is not symmetrical, though its axis leans

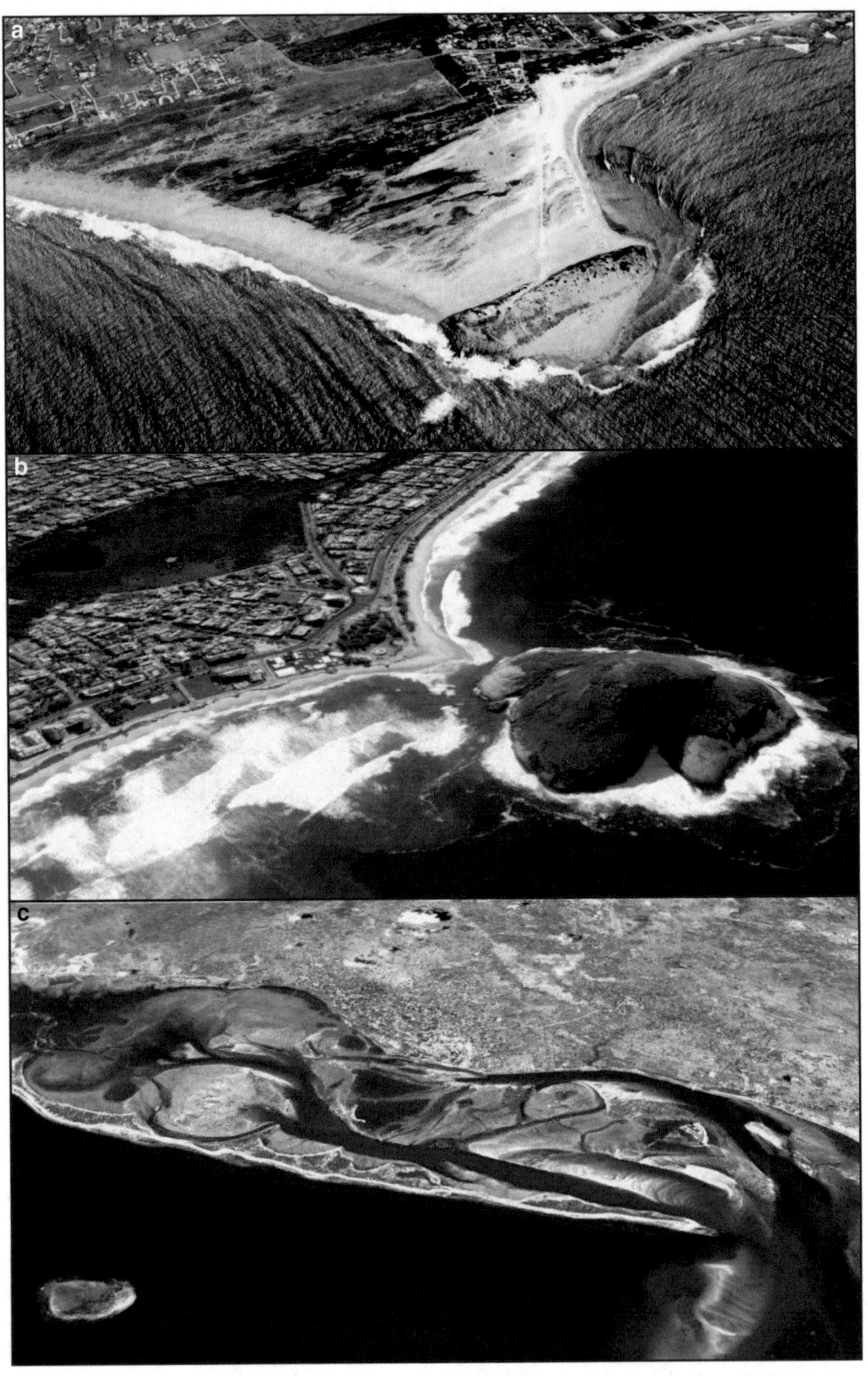

Fig. 4.3 Some depositional landforms. **a** Tombolo (example: Cape Trafalgar, Spain). **b** Incomplete tombolo (example: Pedra do Pontal, Brazil). **c** Spit (example: Zinga Coast, Tanzania). (Images Landsat/Copernicus from Google Earth)

towards the coastline. This occurs because the tail shaft tends to be parallel to the orientation of the crest of the wave trains that generate it. Sometimes, instead of generating a sandy landform, two sand barriers develop with a coastal lagoon between them. In this case we would talk of a *double tombolo*.

Embayments

This is defined as a coastal track where the sea curves into the shoreline towards the continent. They may have different sizes and shapes depending on the physical and geological characteristics of the coast that contains them. In many instances, the bay is open and their edges are rocky headlands (Fig. 4.1a), but in others the bays are partially closed by depositional sandy landforms (Fig. 4.1c). The bays within can be depositional or erosional and the shoreline can be formed by wave-cut platforms or beaches, depending on their sedimentary balance.

Barriers

Coastal barrier is a generic term to define those sedimentary landforms formed by sand or gravel that occur away from the continent and isolate from waves, totally or partially, a body of water that is subject to the exclusive action of the tide. This term is used regardless of the shape and the extension of the barrier (Fig. 4.1c) and it can be grounded at both ends (*bay barrier*) or only at one of them (*spit*) or be unconnected (*barrier island*).

Bay barriers

Coastal barriers that are connected with land at both ends, fully closing what was a bay, and subsequently developed in their back area a lagoon, pond, tidal flat, swamp, marsh or other wetland plain.

Spits

A spit is an elongated stretch of sandy material in a barrier shape that is attached to the mainland only at one end (Figs. 4.1c and 4.3c). A spit is normally originated by the action of the waves in an oblique way to the coastline, creating longshore transport by littoral drift in the direction of growth of the landform.

Barrier islands

These are coastal barriers that are completely separate from the mainland (Figs. 4.1c and 4.4a). Barrier islands represent a barrier with a wide beach, which commonly provides sufficient protection from the high tide to develop dune systems. The back region often supports vegetated and wetland areas.

Strandplains

These are coastal flats generated by accretion to transverse bars of sand or pebbles, transported by the littoral drift along a significant length of the coast (Figs. 4.1c and 4.4b). They consist only of beaches and dunes although may also have marsh areas. From the point of view of processes and facies, they are similar to the fronts of barrier islands but they are not far from the continent and lack the back-barrier elements (the lagoon and, therefore, tidal inlets). They correspond to the accretion of many mixed wind–wave systems.

Cuspate forelands

Cuspate forelands, also known as nesses in Britain, are pointed coastal landforms which are also generated by the accretion of sedimentary bars (Figs. 4.1c and 4.4c), transported by littoral drift in a well-supplied coast. The cumulative process results in progradation and extends the coast into the sea in a triangular-shaped form of a sedimentary nature. Their genesis occurs by a process very similar to the strandplains, but more localized.

4.3 Coastal Environments

The concept of a sedimentary environment was defined by Selley [13]. According to this author, "a sedimentary environment is a part of the Earth's surface that is physical, chemical and biologically different from the adjacent areas." The surrounding conditions of a sedimentary environment determine the processes occurring in it and, consequently, the nature of the sediments that are accumulated. Therefore, each sedimentary environment will be characterized by sequences of recognizable sediment that are indicative of the variability of processes that may have occurred in it. From this point of view, "a coastal sedimentary environment is a sector of the coastal zone that differs from the adjacent others due to its environmental characteristics and present distinctive facies sequences caused by coastal processes."

Some environments are grouped within other larger ones, and therefore these might be termed subenvironments, while others are presented grouped in systems. Some of these systems have particular physiographies which have been previously described as coastal landforms. The most characteristic coastal environments are defined below (Fig. 4.5).

Beaches

Beaches are widely distributed depositional coastal environments. They are primarily subject to wave dynamics although the influence of the tides can also be important. This active dynamic implies an environment where fine particles are absent, mainly consisting of sands but also of pebbles. Beaches may be present in different geographic locations, and can be attached to the mainland, in front of a barrier, barrier island or spit, and also form part of prograding landforms such as strandplains or cuspate forelands.

Coastal dunes

A dune is a hill of sand built up by the wind because of the presence of an obstacle. Coastal dune systems are located in the supratidal beach area and are aeolian environments where sedimentary materials are deposited on their way from

Fig. 4.4 Some depositional landforms. **a** Barrier island system (example: barrier islands closing Charlotte Bay, Florida, USA). **b** Strandplain (example: near King Sound, Northwest coast of Australia). **c** Cuspate foreland (example: Darss Cape, Germany). (Images Landsat/Copernicus from Google Earth)

sea to land. Dunes have a complex internal structure constituted by cross-stratification sets that reflect the stages in their evolution.

Washover fans

The washovers are sandy fan-shaped formations with convexity pointed towards the interior of the continent. They are developed during storms, when high energy waves erode the frontal sandy bodies (beaches and dunes), depositing this sand on the back area of coastal formations such as barriers.

Lagoons

A lagoon is a body of shallow water with a restricted connection to the sea. The term is generally applied to a subtidal

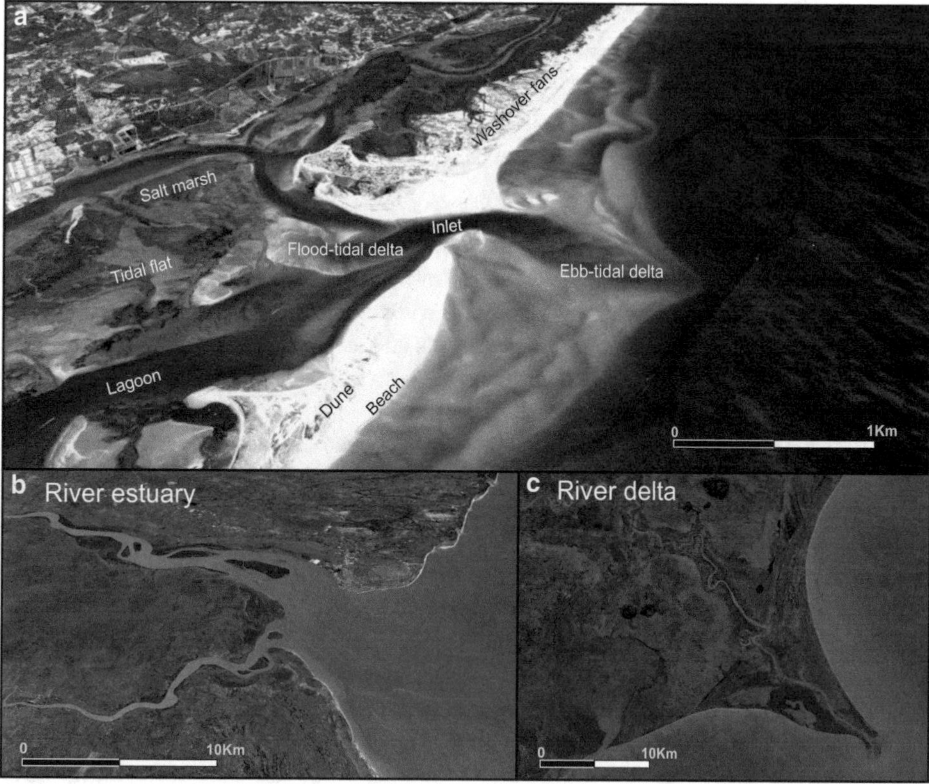

Fig. 4.5 Main clastic coastal environments. **a** Environments involved in a barrier island system (example: Algarve barrier islands, Portugal). **b** Fluvial estuary (example: Pungwe River estuary, Mozambique). **c** Fluvial delta (example: Coco River Delta, border between Nicaragua and Honduras). (Images Landsat/Copernicus from Google Earth)

zone semiconfined by a barrier. Associated with the lagoon are other environments such as intertidal marshes and tidal flats.

Tidal inlets

These are connecting channels through which the mechanism of tidal exchange between the open marine environment and the environments protected by a barrier, barrier island or spit takes place. They are environments that are subject to tidal currents of relatively high velocities that reverse in each tidal cycle.

Tidal deltas

Tidal deltas are one of the elements that constitute a barrier island system and are located in connection with the tidal inlets in both directions of the tidal current. Thus, ebb-tidal deltas are located on the front of the inlet, in contact with the nearshore area, while the flood-tidal deltas are located on the back of it and contact the coastal lagoon.

Estuaries

An estuary is a semi-enclosed coastal water body that has a free connection to the open sea and is also connected to a river stream, where salt water is diluted with fresh water from drainage of the land, with its innermost limit considered to be the last point of the tidal influence [6, 12]. This is one of the two most characteristic types of river mouths, in which the environment is developed in a coastal concavity that is commonly funnel-shaped (Fig. 4.5b). It is an environment that usually occurs after a sea level invasion by marine transgression and tends to total infilling due to the active sedimentation that occurs in its interior.

Fluvial deltas

The deltas are large, cape-shaped coastal sedimentary bodies linked to river mouths. The environment consists of a large accumulation of terrigenous clastic sediments which experience a progradation due to scattering phenomena taking place when flowing fresh water meets the marine water mass. Deltas are environments that are characteristic of coasts of emersion, but they can also develop as an evolution of an estuary, after infilling of the same during a long period of sea level highstand.

Tidal flats

Tidal flats are wide depositional areas formed mostly by fine sediments that are located in the space bounded by tidal fluctuations. These are present along open coasts with a low relief and are affected by low energy waves, as well as in coasts under higher wave energy, but are generally located in areas protected from wave action, behind barriers or reefs and linked to back area environments such as lagoons, estuaries and deltas.

Swamps and marshes

Marshes and swamps are vegetated wetland environments that are characterized by plants saturated in fresh, salt or brackish waters. In a marsh, herbaceous plants without any woody stems are the dominant vegetation. Grasses and reeds are thus the most common plants on marshes, while swamps are dominated by trees and woody plants.

Both types of wetlands are often found along river margins and lake shores (in these cases the water is fresh), but both environments can also appear on the coasts, associated with transitional environments such as deltas, estuaries, lagoons or tidal flats. In these cases, the waters can vary between salt, brackish and fresh, depending on the degree of mix between tidal and fluvial hydric volumes.

In many cases, the difference between the appearance of a marsh or a swamp in the supratidal fringe of sedimentary environments is marked by the climatic characteristics.

Reefs

Reefs are organic constructions built up by the stacking of skeletons of living organisms. These are rigid bio-constructed masses, forming a positive relief able to resist wave action and exert a sediment control over adjacent areas. They can be found in various places, but are often located on low latitude coasts with low terrigenous contribution. In their development, they influence coastal physical processes such as chemical and biological factors with an important climatic control.

4.4 Classifications of the Coast

The preceding sections have described many landforms and coastal sedimentary environments present on coasts all over the world. They also mentioned the importance of different factors controlling the morphology and the distribution of these forms, some with global and others with more local action, and all of them with controls over different time-scales. Since the beginning of research on coastal areas, there has been a need to classify the coasts with criteria that could explain the presence of the set of landforms and environments linked in any coastal stretch [10, 11]. Over time, different classifications using different criteria have been developed to respond to this need.

This section summarizes the classifications that have been more accepted over time, emphasizing the main descriptive criteria used for the categorization.

4.4.1 Genesis Under Relative Sea Level Movements

Early classification schemes were based on genetic aspects related to the relative movements of the sea level, including those caused by static variations as well as by tectonic movements (and also the combinations between both of them). It was Johnson [9] who synthesized the ideas of previous authors [2, 15, 4, 7] in proposing a classification using this context as the main criterion. Thus, he differentiated between:

- Coasts of submersion,
- Coasts of emersion,
- Neutral coasts and
- Compound coasts.

Within this, the shores of submersion were differentiated between those coasts that appeared after a marine inundation of an ancient river and those from a glacier valley, and these were named respectively ria coasts and fjord coasts. At the same time, Johnson defined the coasts of emersion as coastal plains.

In addition, this system marked a third type of coast known as neutral coasts, which corresponded to these coasts that had no apparent relative movements of the sea level. Within these shores could be distinguished deltaic coasts, alluvial coasts, flooded coastal plains, volcanic shores and coasts of faults.

A fourth type of coast includes compound coasts, which have experienced a transition between two or more types of those described above.

4.4.2 Origin of Processes

Almost half a century later, [14] delved into the roots of the previous classification, identifying coasts that did not correspond to any of the categories proposed by Johnson. To classify the coasts, this author also used genetic criteria, but attended to processes different from the sea level movements in differentiating between primary and secondary coasts (Table 4.1). Primary coasts were considered the result of the performance of non-marine processes, while secondary coasts were considered to occur through the direct action of marine processes or marine organisms. Within each of these types Shepard made an exhaustive partition according to all the possible origins of the coastal morphology.

Table 4.1 Classification proposed by Shepard [14]

Shepard's [14] Coastal Classification
1. Primary coasts: configuration due to non-marine processes
1.1. Land erosion coasts. Shaped by subaerial erosion and partly drowned by postglacial rise of sea level (with or without crustal sinking) or inundated by melting of an ice mass from a coastal valley
1.1.1. Ria coasts (drowned river valleys) *(A) Dendritic pattern; (B) Trellis pattern*
1.1.2. Drowned glacial erosion coasts *(A) Fjords; (B) Glacial troughs*
1.1.3. Drowned karst topography
1.2. Subaerial deposition coasts. Largely due to deposition, prograding the shoreline since the postglacial sea level highstand
1.2.1. River deposition coasts *(A) Digitate; (B) Lobate; (C) Cuspate; (D) Partially drowned deltas*
1.2.2. Glacial deposition coasts *(A) Partially submerged moraines; (B) Partially submerged drumlins; (C) Partially submerged drift features*
1.2.3. Wind deposition coasts *(A) Dune prograded coasts; (B) Dune coasts; (C) Fossil dune coasts*
1.2.4. Landslide coasts
1.3. Volcanic coasts
1.3.1. Lava-flow coasts
1.3.2. Tephra coasts
1.3.3. Volcanic collapse or explosion coasts
1.4. Shaped by diastrophic movements
1.4.1. Fault coasts *(A) Fault coasts; (B) Fault trough or rift coasts; (C) Overthrust coast*
1.4.2. Fold coasts
1.4.3. Sedimentary extrusions *(A) Salt domes; (B) Mud lumps*
1.5. Ice coasts. Glaciers form extensive coasts, especially in Antarctica
2. Secondary coasts: shaped primarily by marine agents or by marine organisms. May or may not have been primary coasts before being shaped by the sea
2.1. Wave erosion coasts
2.1.1. Wave-straightened cliffs *(A) Cut in homogeneous materials; (B) Hogback strike coasts; (C) Fault-line coasts; (D) Elevated wave-cut benches; (D) Depressed wave-cut benches*
2.1.2. Made irregular by wave erosion
2.2. Marine deposition coasts. Coasts prograded by waves, tides and currents
2.2.1. Barrier coasts *(A) Barrier beaches; (B) Barrier islands; (C) Barrier spits; (D) Bay barriers; (E) Overwash fans*
2.2.2. Beach plains. Sand plains, differing from barriers by absence of a lagoon
2.2.3. Mud flats or salt marshes
2.3. Coasts built by organisms. Are subclassified according to the dominant building organism
2.3.1. Coral reef coasts *(A) Fringing reefs; (B) Barrier reefs; (C) Atolls; (D) Elevated reef coasts*
2.3.2. Serpulid reef coasts
2.3.3. Oyster reef coasts
2.3.4. Mangrove coasts
2.3.5. Marsh grass coasts

Table 4.2 Classification proposed by Inman and Nordstrom [8]

Inman and Nordstrom [8] Coastal Classification
1. Collision coasts (convergent margins)
1.1. Continental collision coast: the margin of a thick continental plate colliding with a thin oceanic plate (e.g., west coasts of North and South America)
1.2. Island arc collision coasts: along island arcs where thin oceanic plates collide (e.g., the Aleutian island arc)
2. Trailing-edge coasts (divergent margins)
2.1. Neo trailing-edge coasts: new trailing-edge coasts formed near beginning spreading centers and rifts (e.g., the Red Sea and Gulf of California)
2.2. Amero trailing-edge coasts: the trailing edge of a continent having a collision coast on its opposite side (e.g., east coasts of the Americas)
2.3. Afro trailing-edge coasts: the coast on the opposite side of the continent is also trailing (e.g., the east and west coasts of Africa)
3. Marginal sea coasts: coasts fronting on marginal seas and protected from the open ocean by island arcs (e.g., Korea)

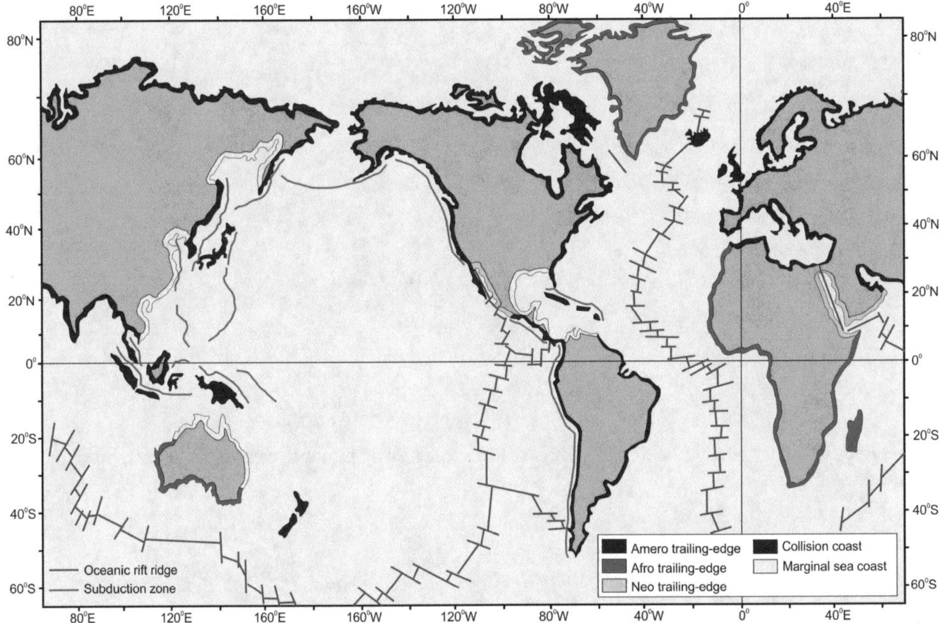

Fig. 4.6 Global distribution of coasts according to tectonic framework (adapted from and Nordstrom [8])

4.4.3 Tectonic Location

A proposal for the classification of coasts developed by Inman and Nordstrom [8] provided a novel conception to follow on from the previous classifications. It was a global classification based on the concepts of plate tectonics discovered in the 1960s. These authors highlighted that the general aspects of the physiography of a coast are in relation to their position at continental margins regarding the boundaries of plates and their relative vertical movements. From this perspective, the strong contrast between the east and west coasts of the United States could be explained, and so could the morphological differences between the northern and southern tracks within the US West Coast. Using the same criteria, they also explained the physiographic characteristics of those coasts all over the world, which were well studied in the last quarter of the twentieth century.

Inman and Nordstrom [8] suggested a classification of coasts into three categories according to their position on the boundaries of the plates to which they are associated (Table 4.2, Fig. 4.6).

The types of coasts included in this classification fit broadly with the physiography of the coast. In this way, the coasts of collision are presented as relatively narrow rocky and steep fringes that usually develop cliffs accompanied by different elevated flat levels of terraces, joining the front continental shelves of low amplitude and high slope, crossed by important submarine canyons.

The coasts of expansive plate margins (trailing edge coasts) show a greater variability. The American affinity type

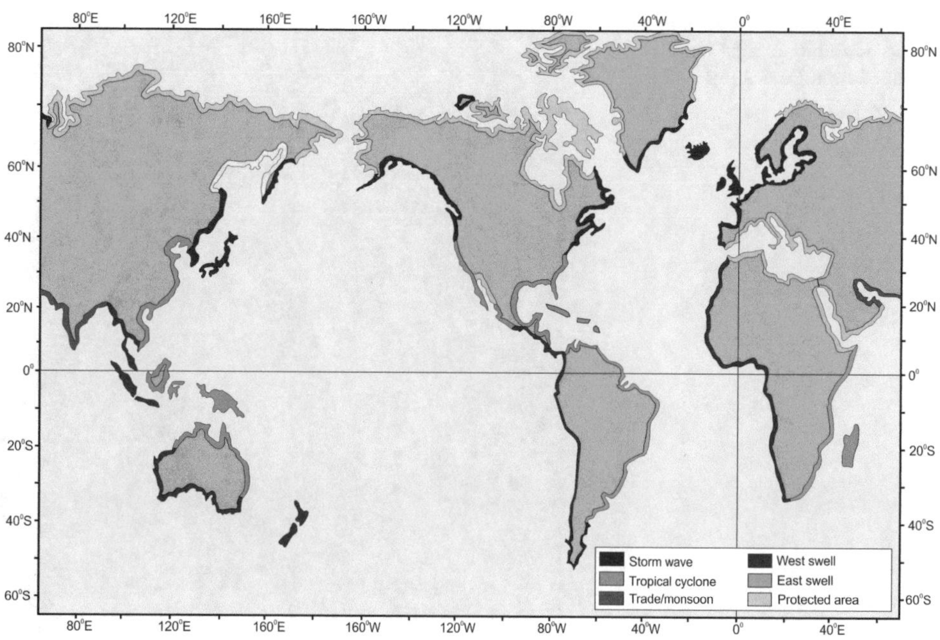

Fig. 4.7 Global distribution of coasts according to wave height (adapted from Davies [3])

(Amero) develop depositional coastal plains associated with extensive drainage networks which bring large amounts of sediment to sedimentary environments such as broad strands, barrier islands, tidal flats and deltas, associated with wide continental shelves.

Characteristics of the coasts of expansive margins of African affinity (Afro) are similar to the American affinity ones, although the drainage networks tend to be less extensive and, consequently, sedimentary input is minor, resulting in narrower coastal plains, deltas and barrier island systems in most common environments.

The coasts of expansive plate margins of recent formation (neo trailing-edge) are immature coasts undergoing important tectonic and gravitational processes. They are developed on the edge of shelves that belong to different plates for a short while. The coast is typically narrow and steep, with features similar to the coasts of collision. The cliffs are the more frequent systems, although they can develop marginal depositional environments.

The coasts of marginal seas present a great diversity of processes, forms and sedimentary environments, always in a tectonic margin of low activity. The continent may appear flat or hilly, although these coasts tend to be associated with large continental shelves that allow the development of well-supplied depositional environments such as estuaries and deltas.

4.4.4 Individual Hydrodynamic Processes

In the last decades of the twentieth century some coastal classifications that emphasized the importance of hydrodynamic processes in the development of coastal environments emerged. Most of these classifications categorized coasts using the dimensions of a single hydrodynamic process.

Thus, Davies [3] identified several types of coasts using only the wave height (Fig. 4.7), taking into account that the waves are generated by the wind and this is distributed latitudinally reflecting global climatic zones. This method differentiated **protected areas**, coasts **dominated by storms**, that would be well localized in areas of higher temperatures in the Arctic (and Antarctic) latitudes and **coasts dominated by swell waves**, located in middle latitudes. In the tropics alternate cycles of **fairweather trade-generated waves** dominated, with short periods in which the **tropical cyclone waves** would be important.

The same author subsequently developed another coastal classification using only the tidal range as the main criterion of classification (Fig. 4.8). In this regard, it should be taken into account that the tidal range increases from the center of the oceans to the coasts, amplifying according to the slope and width of the continental shelves and the coastal physiography. Global distribution of tidal ranges on the coasts is, consequently, controlled on a large scale by the geometry of the coastline and distribution of bathymetry. Davies distinguished between **macrotidal coasts** in those places where the tide range exceeds 4 m, **mesotidal coasts** which would have ranges between 2 and 4 m and **microtidal coasts** in those places where the range is less than 2 m. The first are mainly located in large funnel-shaped bays, where the tidal wave experiences a high magnification due to the effect of convergence, while the last are generally located in shallow seas and sheltered bays.

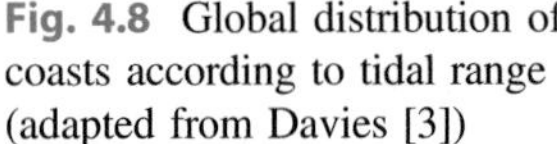

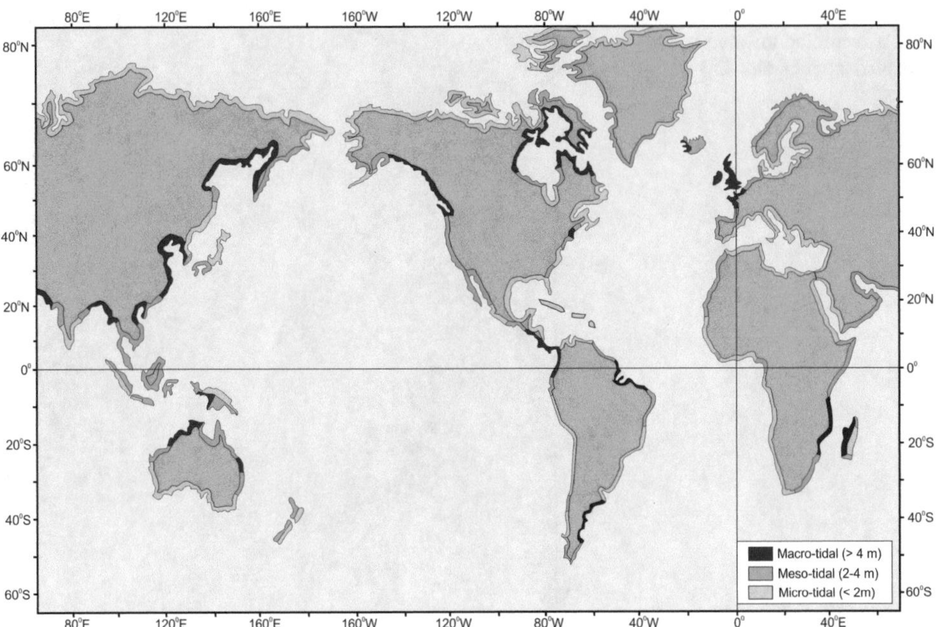

Fig. 4.8 Global distribution of coasts according to tidal range (adapted from Davies [3])

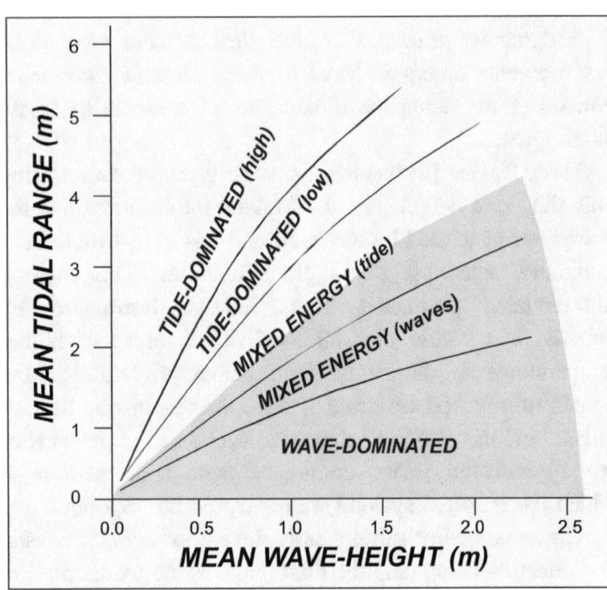

Fig. 4.9 Classification of coasts according to the relative dimensions of tide and waves (Davis and Hayes [5])

4.4.5 Relative Energy of Hydrodynamic Processes

Davis and Hayes [5] observed that the features (landforms and environments) of many coasts did not correspond to the expected physiography in relation to their characteristics of tidal range. So, for example, long barrier island systems expected on microtidal coasts also appeared in coasts with greater tidal ranges but also with higher waves. Conversely, open tidal flats expected on macrotidal coasts also appeared on coasts of smaller tidal range, but with scarce incidence of the waves. In this way, they proposed a classification that took into account the relative effect of waves and tides, more than the absolute values of both parameters separately. Their classification distinguishes between: **tide-dominated coasts**, **wave-dominated coasts** and **mixed energy coasts** (Fig. 4.9) and suggests that the distribution of landforms and sedimentary environments on the coasts would occur according to the relative importance of these parameters (Fig. 4.10).

4.4.6 Sediment Input and Evolving Time

The morphology of the depositional coasts, especially of coasts supplied by clastic sediments, responds to the volume of material contributed from a river source and the ability of the coast to rework this sediment by waves and tidal currents. Taking into account this fact, Dalrymple et al. [1] proposed a classification that included the relative importance of these three agents in their ability to mobilize a volume of sediments, developing a ternary diagram similar to the one proposed for the river deltas by Wright [16]. In this triangular diagram, coastal depositional environments would be located in accordance with the process that dominates their sedimentation. Thus, the deltas would be located at the apex dominated by the river, while the coasts prograding due to the performance of marine processes (strandplains and tidal flats) would be located on the opposite side, to match where sediments were being reworked by waves and tides, while the estuaries (wave- and tide-dominated) would occupy an intermediate place between the three processes.

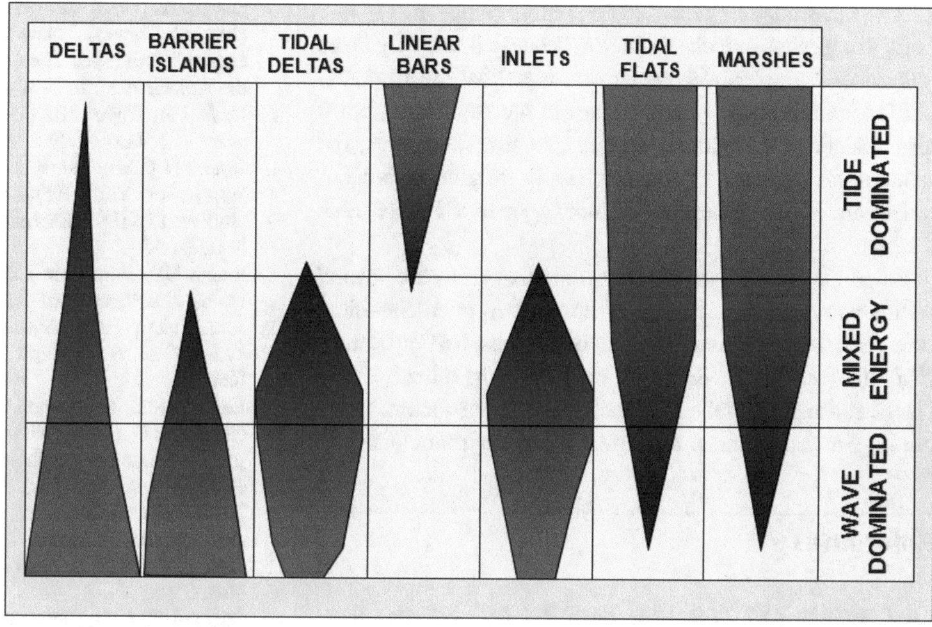

Fig. 4.10 Frequency of appearance of coastal environments in the three types of coasts distinguished by their relative dimensions of tide and waves (Davis and Hayes [5])

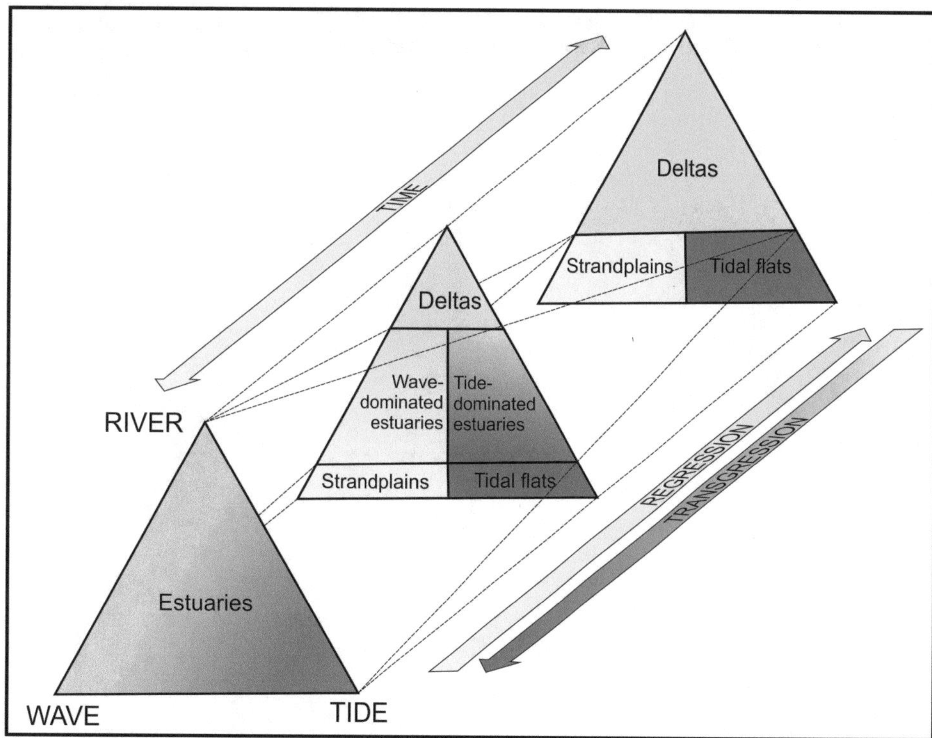

Fig. 4.11 Conceptual distribution of environments according to the dominant process and evolution over time (Dalrymple et al. [1])

The novelty of the classification proposed by these authors lies in the inclusion of the time evolution as a fourth dimension (Fig. 4.11). In this case, time is expressed in terms of trends: coastal regression or marine transgression.

Coastal regression occurs usually due to processes of sedimentary progradation under a standing sea level and/or a large volume of sediment contribution, while marine transgression takes place when the sea level floods coastal environments displacing the shoreline landwards. Progradation in river systems means the infilling of estuaries and their transformation into deltas, while in wave-dominated environments it implies the development of extensive strandplains and in tide-dominated realms an aggradation of open tidal flats occurs. On the other hand, a marine transgression is a movement in the opposite direction, transforming the deltas into estuaries by inundating the river valleys.

The classification proposed by Dalrymple et al. [1] is a fairly comprehensive classification, because it not only takes into account the hydrodynamic processes that are responsible for the morphology of the coast in the short term, but it also includes the sediment supply and time and, thus, the relative movements of the sea level, to give a broader reflection of the processes occurring over a longer time frame.

In addition, it is interesting to note that there are several sedimentary environments that have their own classifications. So, there are classifications of beaches, barrier islands, tidal flats, estuaries, deltas or reefs, among others. These classifications also take into account different factors (morphology and dynamics) that affect these environments.

References

1. Dalrymple RW, Zaitlin BA, Boyd R (1992) Estuarine facies models: conceptual basis and stratigraphic implications. J Sediment Petrol 62:1130–1146
2. Dana JD (1849) Report of the United States exploring expedition 1838–1842 (vol 10, Geology). Putnam American Agency, New York, 735pp
3. Davies JL (1980) Geographical variation in coastal development (2nd ed). Longman, New York, 212pp
4. Davies WM (1896) The outline of Cape Cod. In: Proceedings of the American academy of arts and sciences, vol 31, pp 303–332
5. Davis RA, Hayes MO (1984) What is a wave-dominated coast? Mar Geol 60:313–329
6. Dyer KR (1997) Estuaries. In: A physical introduction (2nd ed). Wiley, New York, 195pp
7. Gulliver FP (1899) Shoreline topography. Proc Am Acad Arts Sci 34:151–258
8. Inman DL, Nordstrom CE (1971) On the tectonic and morphological classification of coasts. J Geol 79:1–21
9. Johnson DW (1919) Shore processes and shoreline development. Wiley, New York, 584pp
10. Komar PD (1998) Beach processes and sedimentation. Prentice-Hall, Englewood Cliffs, New Jersey, 429pp
11. Masselink G, Hughes MG, Knight J (2003) Introduction to coastal processes and geomorphology. Routledge, London, 416pp
12. Pritchard DW (1967) What is an estuary? Physical viewpoint. In: Lauff GH (ed) Estuaries, vol 83. American Association for the Advancement of Science Publications, pp 3–5
13. Selley RC (1970) Ancient sedimentary environments. Chapman & Hall, 299pp
14. Shepard FP (1948) Submarine geology. Harper, New York, p 348
15. Suess E (1888) The faces of the Earth, vol II (English translation in 1906 by HB Sollas). Oxford University Press, London, 556pp
16. Wright LD (1977) Sediment transport and deposition at river mouths: a synthesis. Geol Soc Am Bull 88:857–868

5 Geological Approaches to the Coasts

5.1 Introduction

As described in the preceding chapters, coastal areas have been studied by a broad group of geological disciplines, each of which offers a different focus or approach to these areas and the environments developed therein. From this point of view, perhaps it is geomorphology that is the science from which there have been classically greater contributions to coastal geology, although in recent decades numerous studies have focused on hydrodynamic processes, bringing a perspective that is closer to physical oceanography. Similarly, and also recently, many studies have focused on the distribution of sediments, facies and sequences of facies in coastal sedimentary environments using the study of sedimentology. From a wider perspective, other studies have addressed the potential for preservation of these sequences in a broader time frame, especially taking into account the tectonic and eustatic framework, which means a contribution to the doctrinal body of stratigraphy. Finally, much knowledge about the coast supplied by its geology has contributed to an environmental approach, which entails the development of the environmental geology of the coasts.

This chapter details each one of the approaches on coastal research from these disciplines.

5.2 Geomorphological Approach

Geomorphology is a geological (and geographical) discipline that aims to study the landforms of the Earth's surface. Although this science was initially devoted to the quantitative description and classification of landforms from a more geographical approach (descriptive geomorphology and geomorphometry), it soon led to the study of the processes that generate these forms and their dynamic functioning with a predictive character (dynamic geomorphology), linking more directly with the geology [2]. These processes are understood as part of a genetic cycle that begins with the internal interactions that lead to the uprising of mountain ranges and continues with the dismantling of the same, and among these are especially addressed erosion, transportation and deposition.

In the coastal area, geomorphology focused on both the development of mesoforms (erosive and cumulative forms of meter scale; Fig. 5.1) and on the genetic mechanisms of the coastal landforms of kilometer scales (Fig. 5.2). In this case, studies of both erosive forms (such as cliffs) and cumulative forms (such as barrier islands, beaches and dunes) were also addressed.

In addressing the genetic phenomena of littoral landforms, coastal geomorphology had to deal with the characterization of the processes responsible for coastal dynamics, such as waves, tides and currents, whose combinations are what really shape the coast. It is clear that to approach this type of study, coastal geomorphology had to rely on other branches of science such as marine geology and physical oceanography.

Chapter 1 describes how the first steps of coastal geology were achieved through the approach of geomorphologies by almost legendary authors such as Douglas Johnson, Alfred Steers, Axel Schou and André Guilcher. The conceptual definitions set out in Chap. 2 of this book correspond to the field of coastal geomorphology, as do the descriptions of coastal landforms and coastal classifications described in Chap. 4.

5.3 Sedimentological Approach

Sedimentology is the science that studies both the sediments and their genesis [8]. In this way, sedimentology takes in the study of sediment formation and the processes of transformation into sedimentary rock, including the physical and chemical characteristics of sediments and sedimentary rocks, which are a result of the processes that originated them [1]. The objective of this geological discipline is to establish present facies sequences and models that allow us to recognize and interpret the sedimentary facies of the past by

J. A. Morales, *Coastal Geology*, Springer Textbooks in Earth Sciences, Geography and Environment,
https://doi.org/10.1007/978-3-030-96121-3_5

Fig. 5.1 Example of metric-scale landforms: coastal marmids (Sancti Petri, Spain)

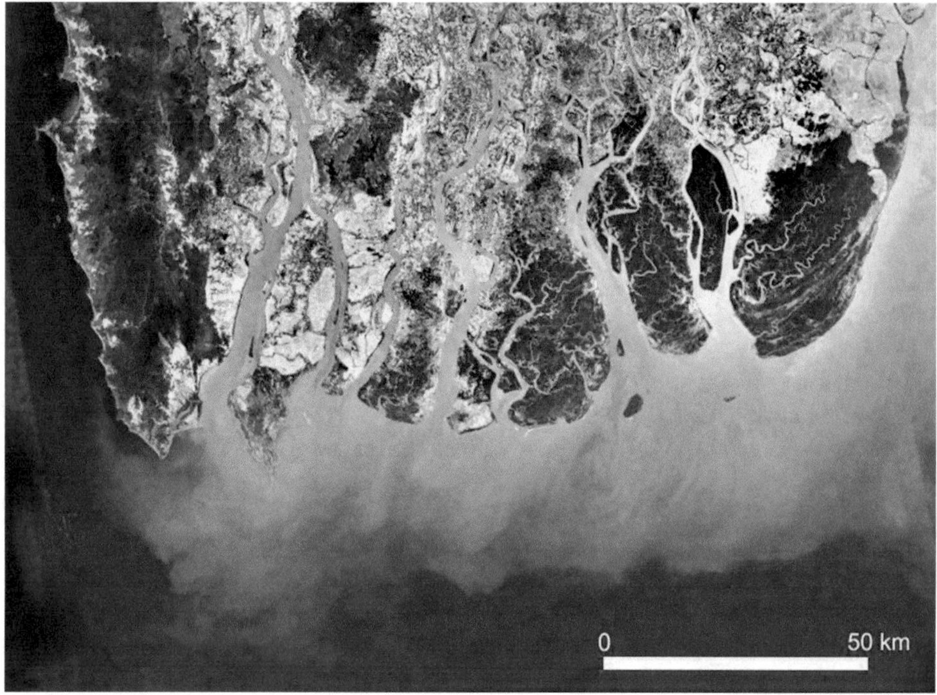

Fig. 5.2 Example of kilometer-scale landforms: Ayeyarwady River Delta (Myanmar). (Image Landsat/Copernicus from Google Earth.)

applying Charles Lyell's Principle of Uniformitarianism. Within the objectives of sedimentology, the coast is of particular interest, as it constitutes an unequivocal marker of the position of the sea level and the reference to this level is the basis for the establishment of transgressive and regressive sequences [7].

Since the application of Lyell's principles, sedimentologists have been especially concerned with establishing relationships between the facies of each of the coastal environments and the processes that generate them. Today's coasts have proved to provide an excellent opportunity to establish such relationships, as they can be observed, quantified and analyzed in detail, both in terms of facies and processes. Initially, thanks to their easy accessibility, the surficial sediments of many coastal environments (beaches, dunes, lagoons, tidal deltas, tidal plains, estuaries, deltas…)

Fig. 5.3 Example of a sequence of facies on a Holocene beach

were the object of study in numerous works regarding the process causality of the grain size distribution, bedforms and physical and organic sedimentary structures. These surface studies were approached by taking into account that, according to Walther's Law, the facies that were originally deposited adjacent to each other end up overlapping in a vertical succession in the form of facies sequences (Fig. 5.3).

Subsequently, the studies were expanded into deeper areas, incorporating trench data, manual cores and vibracores to establish more complex sequences and facies models. The results of these studies showed that it is in coastal environments where the preservation potential of sedimentary sequences can most easily be determined, thus establishing the models of facies to be preserved in the form of stratigraphic series.

5.4 Stratigraphical Approach

Stratigraphy is the geological discipline that deals with the study and interpretation of stratified sedimentary rocks, and the identification, description and interpretation of sequences as well as the correlation of stratified rocky units [9]. In the past, many sediment sequences were preserved in the geological record and transformed into sedimentary rocks. Ancient coastal records had a significant importance in establishing classical coastal sequences during the beginnings of the discipline. The results obtained from a sedimentological vision are easily applicable to the doctrinal body of stratigraphy, especially for the interpretation of stratified facies by means of application of the Principle of Uniformitarianism. This is among the fundamental principles of stratigraphy. The principle was stated by James Hutton and widely disseminated by Charles Lyell as a contrast against catastrophism, although the term was established by William Whewell. According to the principle, geological processes have acted in a uniform way throughout the Earth's history. Thus, if current processes are the same as those that acted in past geological periods, the interpretation of fossil facies can be made by comparison with current environments where both processes and resulting facies can be observed.

While sedimentology establishes sequences of lithofacies (depositional facies) for each of the subenvironments that constitute a sedimentary environment via models of depositional facies for each sedimentary environment, stratigraphy goes further, defining, through the analysis of facies, the relationships between different sedimentary environments in the face of phenomena occurring over a longer period of time (Fig. 5.4). These long-term processes include eustasy, tectonic and other changes in the sedimentary supply regime. These same coastal sequences also play an important role during the stages of development of sequence stratigraphy, which studies the relationships between processes that manifest in the record of sequences of a larger order (parasequences and depositional sequences) in which the facies deposited in the different sedimentary environments overlap.

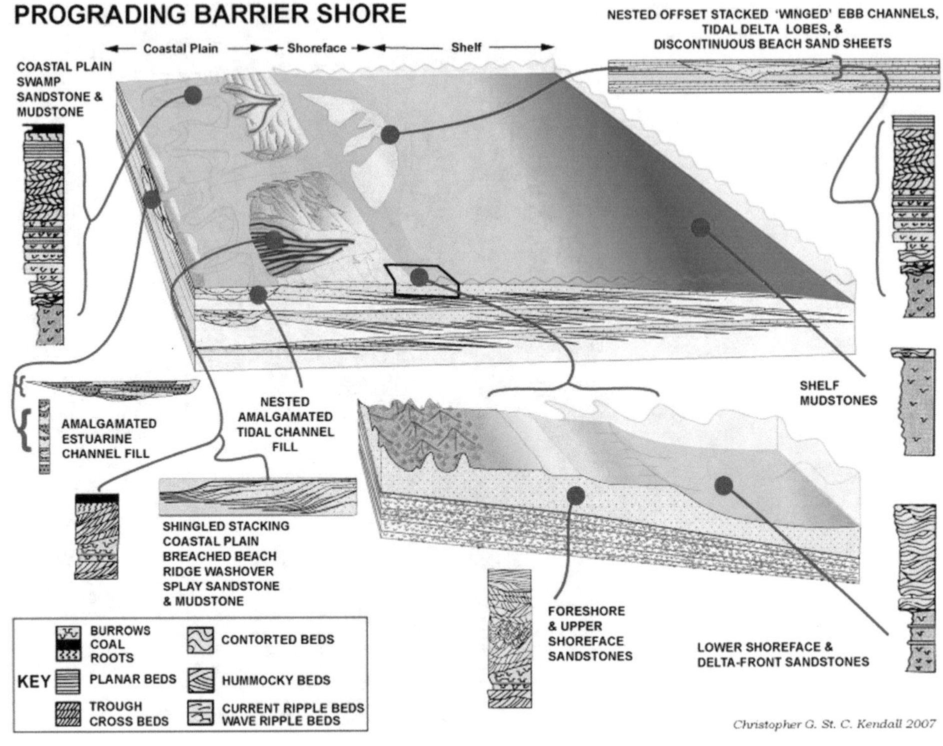

Fig. 5.4 Example of an architectural facies model of a prograding barrier island system [3]

Stratigraphy is also responsible for establishing relationships between coastal environments and their surroundings, both continental and marine, as a part of the same sequence tracking. These relationships are established not only from the point of view of the sedimentary successions that are preserved in the geological record, but also from the geometry of sedimentary bodies. In underwater environments, these geometric relationships are easily established thanks to the use of geophysical techniques such as reflection seismology, combined with the analysis of facies in sediment logs.

The establishment of geometric relationships other than the simple overlapping of parallel roof and base strata suggests a close relationship between this geometry and the original shape of some of the sedimentary bodies in which the facies are contained. This fact gives rise to a link between stratigraphy and geomorphology, so that numerous works use the word "Morphoratigraphy" to describe the studies in which the two disciplines are combined. These studies also tend to relate both the facies and their external form with the processes that generate both characteristics.

5.5 Oceanographical Approach

Oceanography is defined as the science that studies the seas and oceans and everything that happens in them [4, 5]. This science addresses research on the structure, composition and dynamics of ocean water bodies [1]. This broad approach includes the characterization of physical processes, such as currents, waves and tides, as well as small-scale geological processes, such as erosion and sedimentation, or large-scale ones, such as ocean floor expansion, as well as the chemical and biological phenomena occurring in these waters.

Oceanographic science considers that the coast, as the boundary of the ocean, is included within it and is affected, albeit with nuances, by the same processes as the rest of the sea [6]. Of all the approaches that oceanography addresses, it is clear that there are two of them that study aspects related to coastal geology: physical oceanography and marine geology.

Physical oceanography studies the physical processes that occur at sea, including oceanic currents, tides and waves. These studies are conducted from a dynamic approach by analyzing the movements of ocean water bodies and their causes, highlighting a discipline known as meteorological oceanography, which is responsible for studying the interactions between cells of atmospheric circulation and the movements of seawater.

Since its birth as a science, oceanography has brought important knowledge to the thematic development of coastal geomorphology, since it has applied a robust oceanographic methodology such that the global current system has come to be well understood, as has its genesis in relation to atmospheric circulation systems (Fig. 5.5). Similarly, oceanographic methods were used to explain the origin and dynamics of the waves, both in the open sea, in their area of approaching the coast and the dissipation of energy in their

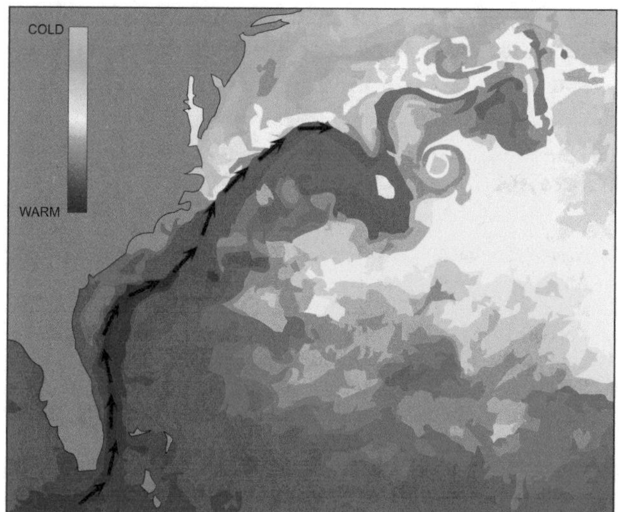

Fig. 5.5 Thermal Landsat image showing the Gulf Stream affecting the East Coast of North America

breaking. The tides were also described, so their origin and dynamic functioning in open marine systems and in coastal systems were also established in detail using oceanographic methods.

Once the fundamental dynamics of each one of the hydrodynamic processes were known, science was able to apply the characterization of these processes on each coast that was studied from a dynamical point of view, in order to explain the observed geomorphological features. In this way, oceanography and coastal geomorphology have gone hand in hand for decades in the geological characterization of the coasts. On the other hand, geological oceanography and marine geology focused on the characterization of the structure of the ocean floor and, within its objectives, the most coastal-related part is the study of the distribution of seabed sediments and the large sedimentary prisms generated on the margins of the ocean basins. With these aims, it is obvious that an immediate relationship was established with sedimentology and stratigraphy.

5.6 Environmental Geology Approach

Hydrodynamic processes (waves, tides and currents) control the physical and chemical variables of the sediment and condition the biological characteristics of the ecosystems that are established on sedimentary environments. These variables can introduce stressors for plants and animals that inhabit the coast and the shallow waters and, conversely, can also be important for the dispersion of larvae and adult organisms in coastal environments. The knowledge of the dynamics of erosional and depositional processes can be useful in the study of these living organisms. Many coastal environments have a sedimentary zonation that is related to the gradation of the physical processes that occur in them, and this zonation conditions the distribution of benthic organisms that live in surficial sediments of the environments near the coast.

Perhaps the clearest example is the interaction between plants and the action of waves and tidal currents. In this case, for studies on the meadows of marine phanerogams and coastal marshes it is important to know the effects, on the one hand, of the hydrodynamic processes and, on the other, of the control of the sedimentary substrate. Whereas waves and currents put stress on plants, the grain size of the sediment controls the degree of oxygenation of the rhizomes and roots. On the other hand, the presence of vegetation can alter the flow conditions near the bed and stabilize the substrate, and also modifies the chemical conditions of the water that surrounds the plants, causing reactions of chemical precipitation and flocculation. An additional effect is the important role of lagoons, estuaries, deltas and marshes as nurseries during the youth phases of fish development.

Another classic example of these interactions is that which occurs on coral reefs. The arrival of waves oxygenates the environment and provides the conditions for optimal coral development. In addition, the presence of corals contributes to protect areas so they are safe from the strongest wave breakers. Coastal areas protected by the action of the reef body provide an optimal habitat for the youthful development of many of the fish and mollusk species, so that many of the species that inhabit the open seawater usually reproduce in coastal waters.

This environmental vision of the hydrodynamic conditions of coastal sedimentary environments is closely related to sedimentology, as the action of organisms represents the genesis of many organic structures that are preserved in the sediment and that give rise to a whole science such as ichnology.

5.7 A Multidisciplinary Vision

Although the objectives of these geological disciplines may be somewhat different, they all coincide with a special interest in discerning in detail the performance of the physical processes that act on the coast, and so to understand their dynamics. Most of the time the interest is focused on these dynamics in relation to human activities, but also as generators of landforms, as well as facies and sequences that can be preserved in the geological record. On the one hand, it is clear that there is an overlap of objectives between geomorphology, physical oceanography and sedimentology in the study of coastal processes such as waves and tides, as well as erosion and sedimentation. But in addition, in recent years, many of the boundaries between these disciplines have been blurred, as there is now a tendency towards the transversality of science. When attending any of the international symposia on the coastal sciences, it can be observed that the groups of researchers and those behind the major projects are currently composed of collaborators from different geological disciplines (and some non-geological), which translates into a multidisciplinary focus of the works on coastal areas published in any of the journals in the field of geosciences.

Throughout the chapters of this book, all the points of view included in this one will be discussed. Some of the chapters offer a vision in which a single perspective dominates, as is the case of the chapters dedicated to the characterization of the hydrodynamic processes that act on the coast. However, in most chapters is offered a multidisciplinary treatment. In particular, in the chapters devoted to coastal environments, the aim will be to characterize the dynamic functioning through the performance of hydrodynamic processes within them, and then to carry out their geomorphological characterization and description, ending with the facies and facies models that are generated in them. In the last chapters, the most stratigraphic vision will be realized, analyzing the processes that act on the longest timescales, as well as the influence of environmental evolution on the ecosystems that settle in them, and on human activities.

References

1. Bates RL, Jackson JA (1987) Glossary of Geology (3rd ed). American Geological Institute, Virginia, p 788
2. Davis WM (1899) The geographical cycle. Geogr J 14:481–504
3. Kendall CGSC (2007) SEPM Strata: Barrier Island Morphology. http:// www. sepmstrata. org/ page. aspx?pageid=306
4. Krümmel O (1907) Handbuch der Ozeanographie. Band I. Die räumlich, chemise un physikalischen Verhältnisse des Meeres. Engelhorn, Stuttgart, p 526
5. Krümmel O (1911) Handbuch der Ozeanographie. Band I. Die Bewegungsformen Meeres (Wellen, Gezeiten, Strömungen). Engelhorn, Stuttgart, p 766
6. Shepard FP (1948) Submarine geology. Harper & Brothers, New York, p 348
7. Twenhofel WH (1926) Treatise on sedimentation. Dover, New York, p 926
8. Wadell HA (1932) Volume, shape and roundness of rock particles. J Geol 40:443–451
9. Weller JM (1960) Stratigraphic principles and practice. Harper, New York, p 725

6 Study Methods and Techniques

6.1 Introduction

Based on the different perspectives of geological disciplines on the objectives in the studies of coastal geology, a variety of research methods are used. Taking into account that the sedimentary dynamics of coastal environments are reflected, both in coastal processes and landforms of different scales, as well as in the distribution of sediments generated by these processes, it is necessary to study the present environments from all these visions, in order to characterize this dynamic in a global way by combining the evolutionary trends of the short and long terms (Fig. 6.1). The characterization of sedimentary dynamics in each of the present coastal sedimentary environments will allow us, by applying the Principle of Uniformitarianism, to interpret the generating processes from the recognition of sedimentary rocks deposited in ancient environments.

It is clear that not all studies about coastal geology have the same objective; on the contrary, some works have a more defined objective that is based on a single one of the perspectives that have been listed in the previous chapter.

6.2 Methods and Techniques to Study Physical Coastal Processes

In coastal areas, oceanographic methods for measuring dynamic agents are less complex and costly than those required for deeper marine areas, given the accessibility of coastal waters. In recent decades, oceanography has developed a variety of techniques dedicated to the measurement of these dynamic agents. Most of the measurements are carried out by direct methods from records made with the use of technological instruments. In this sense there are remarkable advances in the measurement of waves, tidal levels and currents (tidal and oceanic).

6.2.1 Instruments for Wave Measurements

In past centuries, wave dimensions were measured at sea from vessels by estimation of the height relative to the deck and from land by recording the dimensions in the breaking zone, including the directions of approach to the coast. For at least the last half century, a more precise characterization of the morphological parameters that define the waves has been performed through a complete analysis of each wave train approaching the coastal area. Currently, there are several methods for recording swell data: buoys, radars, pressure sensors and predictive models.

6.2.1.1 Wave Buoys

Buoys are measuring stations anchored to the seabed in areas far from the coast [1] that are used to measure the dimensions of the waves before they deform when they touch the bottom (Fig. 6.2 a, b). The simplest buoys are installed with an accelerometer that allows them to obtain the height, the period and the speed of each wave of the train. The most modern and advanced buoys also allow the measurement of the direction of origin of the wave trains through the use of an integral gyroscope.

There are other devices installed in the buoy which also allow us to measure atmospheric (such as barometers, thermometers and anemometers) and oceanographic (such as water temperature and conductivity sensors or current profiles by Doppler effect) variables (Fig. 6.2c).

6.2.1.2 Wave Radars

Swell radars measure the relief of the water surface by using electromagnetic waves that are modulated in amplitude or frequency when recorded in equipment once reflected in the moving water surface. There are several radar techniques used for measurements of wave dimensions. The different sensors available on the commercial market, as well as a

J. A. Morales, *Coastal Geology*, Springer Textbooks in Earth Sciences, Geography and Environment,
https://doi.org/10.1007/978-3-030-96121-3_6

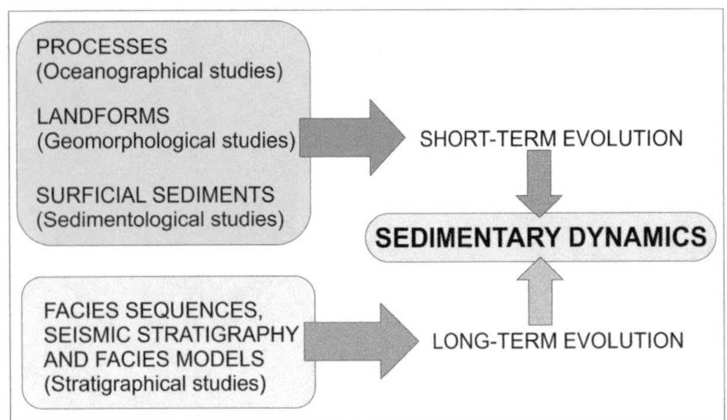

Fig. 6.1 Study objectives of coastal geology

critical review of them, were summarized, described and discussed by Grønlie [11]. According to the type of electromagnetic waves used, infrared, microwave and high frequency radars can be differentiated. They are distinguished depending on their location: platform radars, coastal radars, mobile navigating radars and remote sensing radars.

Platform radars are installed on fixed shelves and face vertically towards the water (Fig. 6.3a). Infrared light or microwaves may be used. Detection equipment (Fig. 6.3b) is usually cheap, although it may be expensive to install in a stable structure. These kinds of radars present the disadvantage of possible interference of the record by the structure on which the equipment is mounted. This interference can become important depending on the provenance of the waves if reliable directional measurements are required, although they are usually effective in determining scalar parameters. Due to this fact, the orientation of the equipment with respect to the tower must be carefully chosen. They present the advantage that having a fixed installation provides a continuous and prolonged recording over time.

Coastal radars are installed at fixed stations located on the coast and act in low grazing-angles using high frequency, so they can reach up to a distance of 70 km from the coast. Since the sensors are not in an upright position relative to the water surface, this method does not use the reflection of the waves, but instead dispersion in the roughness of the water surface (Bragg backscatter). In order to obtain a more reliable record, two radar installations are usually used to record the same sea surface.

Marine navigation radars are installed on board moving vessels. Since under large waves the boat will suffer oscillations, these usually act in conjunction with accelerometers and gyroscopes that correct the signal using complex algorithms. Either way, the obtained signal is unreliable in determining the dimensions of the waves, although it is useful to obtain data on the direction of propagation of the wave.

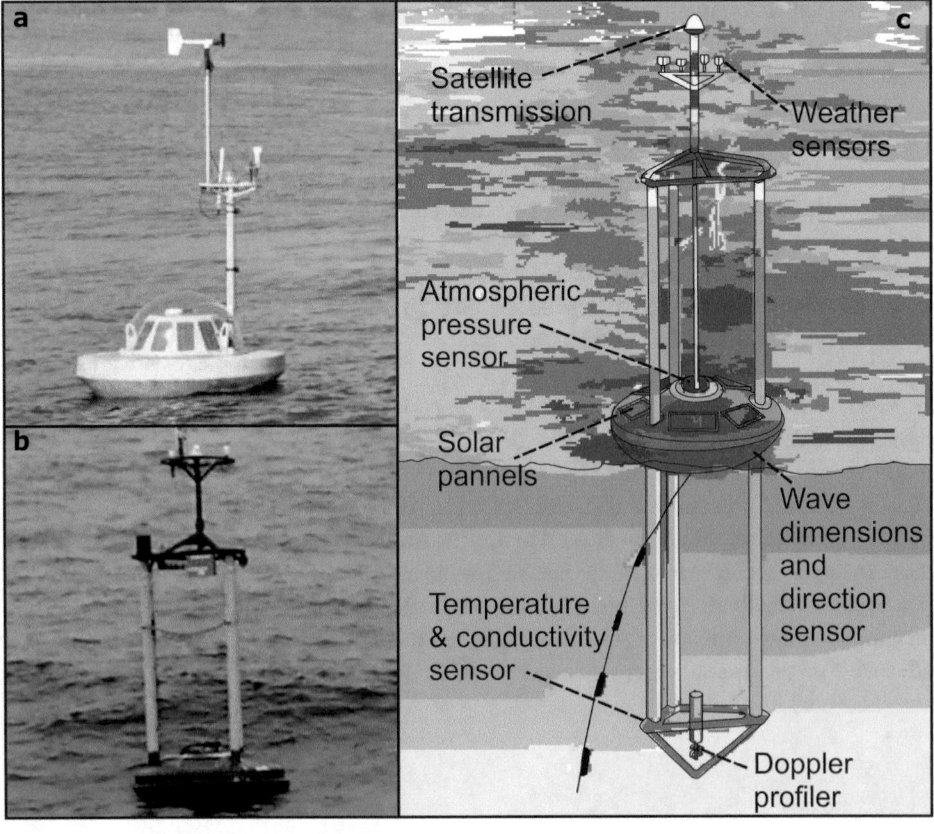

Fig. 6.2 **a** and **b** Different models of wave buoys. **c** Scheme of measuring instruments installed on and under water in a standard buoy

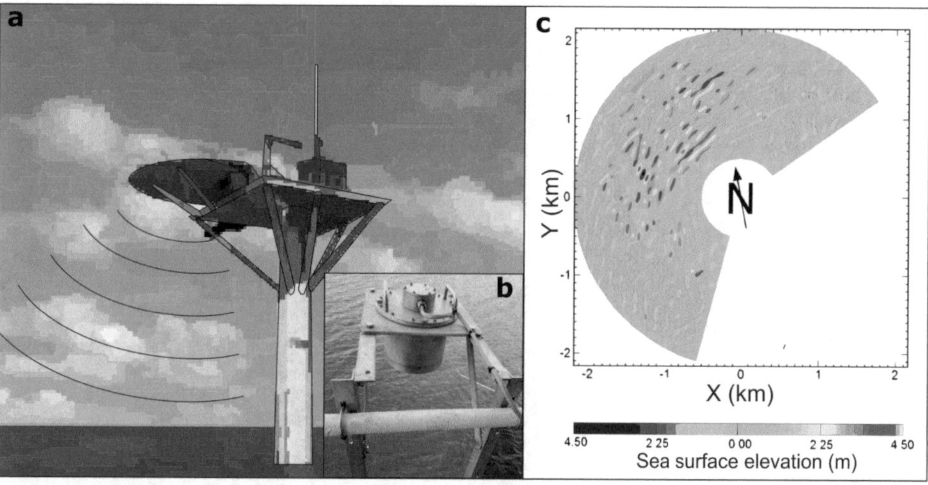

Fig. 6.3 **a** Example of a radar installation on a piloted fixed coastal structure. **b** Detail of a coastal wave radar. **c** Example of the registration obtained with a coastal wave radar (adapted from Reichert et al. [25]

There are also airborne radar sensors and radars installed on satellites (space-borne radars) responsible for characterizing the dynamics of the sea surface. These systems usually use very open radars that synthesize the signal (synthetic-aperture radars or SAR), so they can cover very wide surfaces, although this synthetic signal causes the record to lose resolution. In any case, it is appropriate to verify, calibrate and validate the measurements obtained by instruments installed on satellites using propagation models through real direct measurements obtained by coastal buoys or radars.

Whatever frequency and location of the equipment are used, a visual record of the results can be obtained immediately (Fig. 6.3c). The dimensional and dynamic characteristics of each wave train are obtained from the received signal through complex processing and statistical analysis of the data [17].

6.2.1.3 Wave Gauges

Wave gauges or wave sensors are compact and simple instruments that provide a reliable record of the instantaneous position of the water surface. This equipment can be installed on devices easily adaptable to the circumstances of the location which are able to measure water level variations caused by waves and also by tides. They are often combined with data acquisition and processing software that allow the calculation of the significant parameters of the wave trains and individual waves, although not its approaching directions. Their versatility makes them ideal for measurements in shallow waters such as those in coastal environments. Depending on the method used for the measurement, there are three types of sensors: electrical sensors (wave probes), pressure gauges and acoustic resistance sensors (Fig. 6.4).

The electric resistance meter (wave probe, Fig. 6.4a) works by measuring the resistance of water to the passage of the electric current between a pair of parallel rods, taking into account that the water resistance between the two rods is proportional to the depth of immersion [20]. The rods must be installed on the surface of the water in such a way that the sine and crest of the wave oscillate at the length of the rod. They are very economical devices that work well in laboratory experiments and channels where the waves are not of large dimensions.

Pressure gauges (Fig. 6.4b) are very simple, autonomous and programmable devices, which are commonly used in short- and medium-term measurement experiments in breaking areas [2]. Measurement instruments are mounted at the bottom in shallow water during a low tide. These devices use algorithms based on the linear theory of the waves to calculate the length of the water column above the sensor using the water pressure measurements as starting data [27]. Normally the data is stored in an internal memory that can be removed and read in a computer at the end of the experiment.

Acoustic sensors [18] are also mounted on the shallow water bed facing the surface (Fig. 6.4c). They are somewhat bulkier than pressure sensors, but equally versatile for their autonomy and data storage capacity.

6.2.1.4 Numerical Models of Wave Generation and Propagation

The numerical data modeling comes from mathematical simulations that calculate the wave dimensions from atmospheric conditions on the sea surface (sea and swell information). Databases (hindcast or reanalysis) are calibrated and developed from historical weather records. It should be noted that the data generated by numerical models are not instrumental data; on the contrary, these databases come from the use of mathematical models but they must be calibrated with instrumental wave measurements, and can be very useful to obtain data about the magnitude of the waves. There are models applied on a global scale and also models with higher resolutions for regional scales. These models provide information evenly at regularly selected points in space.

Fig. 6.4 Different types of wave gauge. **a** Wave probe (picture from the webpage of the College of Engineering of the University of Wisconsin). **b** Pressure gauge. **c** Acoustic wave sensor

The use of models began in the 1950s, but it was not until the 1970s that the first models were obtained on a global scale. The third generation of models is currently being implemented. The most widely used global model for open water today is the WAM [16], to which local data can be coupled to establish more accurate regional models. A good example can be the network of WANA points in Spanish waters, whose data, with a temporary periodicity of three hours, are obtained by applying the WAM model to the atmospheric data provided by the Spanish weather network.

In coastal waters, where the waves interact with the bed, it is necessary to apply other models. Another third-generation model such as SWAN is currently being used [4]. This global model can also be adapted at a regional level by applying detailed coastal bathymetric and weather data. In particular, in European waters, the analysis called HIPOCAS—Hindcast of Dynamic Processes of the Ocean and Coastal Areas of Europe—has been established [12]. The application of the HIPOCAS regional model to the Mediterranean and Atlantic waters of southern Europe has been used to create the database of the SIMAR point network (Fig. 6.5). All these databases are public and accessible via the internet.

6.2.2 Instruments for Tidal Level Measurement

Tidal characterization is essential work in coastal dynamics studies and is critical in establishing reliable water level datums. This characterization is carried out by statistical analysis of tidal height measurements. This is done using tide gauges which record water levels in a continuous way. The first wave gauges were deployed in fixed installations on the coast, usually in port facilities (Fig. 6.6). This type of gauge is still the main data source of the long-term measurement networks existing in most countries.

Regarding the measurement method, buoy tidal gauges have been classically used to measure the height of the water level. The mechanism consists of a small buoy that, introduced into a vertical pipe connected to seawater, floats on the surface, rising and falling with the tidal level oscillations. The buoy is hooked to a counterweight by a cable that passes through a system of pulleys and differentials. The pulleys transmit the movement to a needle that draws the curve on a recording paper that is in constant motion as it is driven by a reel mechanism. Today, the buoy system has been modernized and, although the system of buoy-and-pulley is maintained, a digital record is obtained (Fig. 6.7a). Digital data of water levels are stored in an internal memory or transmitted to control stations by radio frequency.

Currently, there are new techniques that can measure the tide through the use of ultrasound. These are the sonar sensors, which are characterized by a high-resolution record. The ultrasonic emitting antenna is installed at a coastal station at a certain water height, emitting pulses that are reflected in the water and returned to a receiving antenna. The distance to water is calculated through the ultrasonic pulse velocity. They are usually installed in fixed stations that emit the data by radio waves to the base where they are

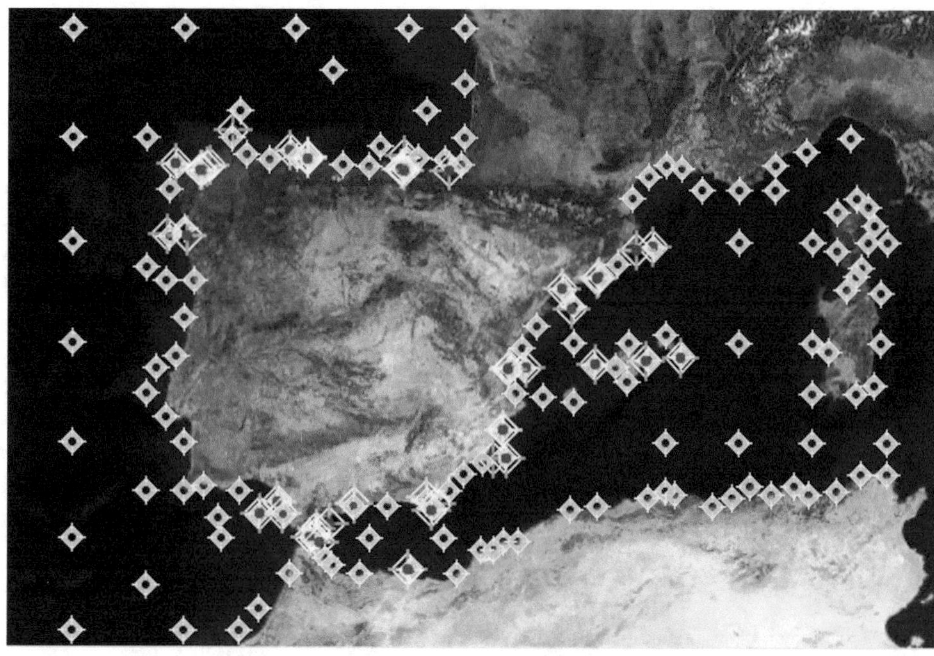

Fig. 6.5 SIMAR network database for coastal waves in SW Europe

Fig. 6.6 Installation of a tide gauge of the NOAA network in Virginia Key, Florida, USA

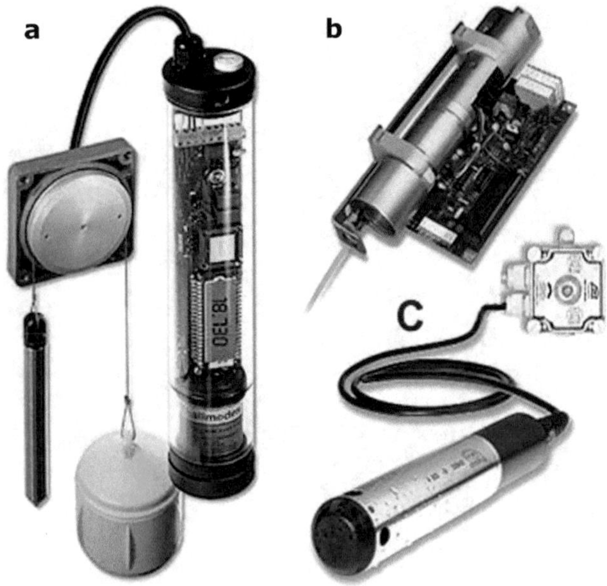

Fig. 6.7 Different types of tide gauge. **a** Buoy digital tide gauge. **b** Bubble pressure tide gauge. **c** Membrane pressure tide gauge

stored and analyzed. These are expensive systems that are part of long-term data networks in many countries.

In recent years, other sensors based on different principles of physics have been developed. Some of these instruments are based on the measurement of pressure variations driven by the rise and fall of water level (pressure sensors). The sensors of these instruments are usually installed in the background, anchored to a fixed structure, communicating the sensor with a memory or storage equipment that remains on the surface. Other sensors calculate the pressure through volume variations in an air bubble contained in a transparent tube. These are called bubble sensors (Fig. 6.7b). Other sensors measure pressure from pneumatic membranes or strain gauge cells. These are called membrane sensors (Fig. 6.7c). None of these instruments requires a fixed installation, so they are very versatile and are used in short- and medium-term characterization projects.

6.2.3 Instruments for Measuring Tidal and Ocean Currents

Current meters, flow meters and hydraulic reels have long been used to measure currents of different origins. The most classical of these instruments work by rotating a propeller (Fig. 6.8), which allows the calculation of the speed from the number of spins of the propeller per unit of time. The simplest model allows direct surface measurements to be performed by manually holding the instrument (Fig. 6.8a). More advanced models allow speeds to be measured at different depths in shallow channels, by holding the reel on a graduated rod to control the depth of the measurements (Fig. 6.8b).

There are also larger instruments that, operating on the same principles, can be installed at a fixed depth through anchoring. These instruments can record and store data continuously for months in an internal memory unit. Some of these instruments have a propeller arrangement with a horizontal axis, similar to the smaller instruments (Fig. 6.8c), while others are arranged in a vertical axis position (Fig. 6.8d).

All of these current meters have the disadvantage of needing a speed threshold to start the movement of the propeller, so currents below these thresholds are not recorded. However, they do have the advantage of their low cost and ease of use, when operating from the surface and not needing complicated installations.

In the 1990s, other current meters were developed which measure static electricity induced by water friction in an electrosensitive paddle. These are the so-called electromagnetic induction flow meters. These sensors can measure the current speed in the water sheet that is in contact with a flat sensor (Fig. 6.9a) or in the water volume around the sensor (Fig. 6.9b).

Current speed measurement systems using the acoustic wave Doppler effect are now being used. These acoustic profilers are called ADP (acoustic Doppler profiler) or ADCP (acoustic Doppler current profiler). These instruments involve the recording of a speed profile in one direction including its three components (Fig. 6.10). There is equipment on the market covering different measurement ranges and at different resolutions, allowing their application in shallow water environments such as coastal environments, but also in deeper systems. The most accurate profilers allow a continuous time determination of speed profiles in 20 × 20 cm cells along columns of water up to 100 m.

These systems can be installed on the surface (on a floating platform), focused to the bed or deployed on the bottom focused to the surface, and even on the sides of channels to measure cross-sectional speed profiles to the shores (Fig. 6.10a). There is also equipment with sensors oriented at different angles to measure in a two-dimensional way (Fig. 6.10b). On the bottom they can be installed using heavy structures that are usually braced and hoisted from a boat (Fig. 6.10c). In addition, to perform measurements from the surface the equipment can be installed on catamarans that remain anchored (Fig. 6.10d). There is also the possibility of towing these catamarans or installing them on board to make channel profiles where the two-dimensional distribution of currents—in depth and width—can be observed (Fig. 6.10e).

6.3 Study Methods of Coastal Landforms

The study of coastal landforms was the first topic addressed on the coast from a geological perspective. These early studies were conducted from a purely geomorphological

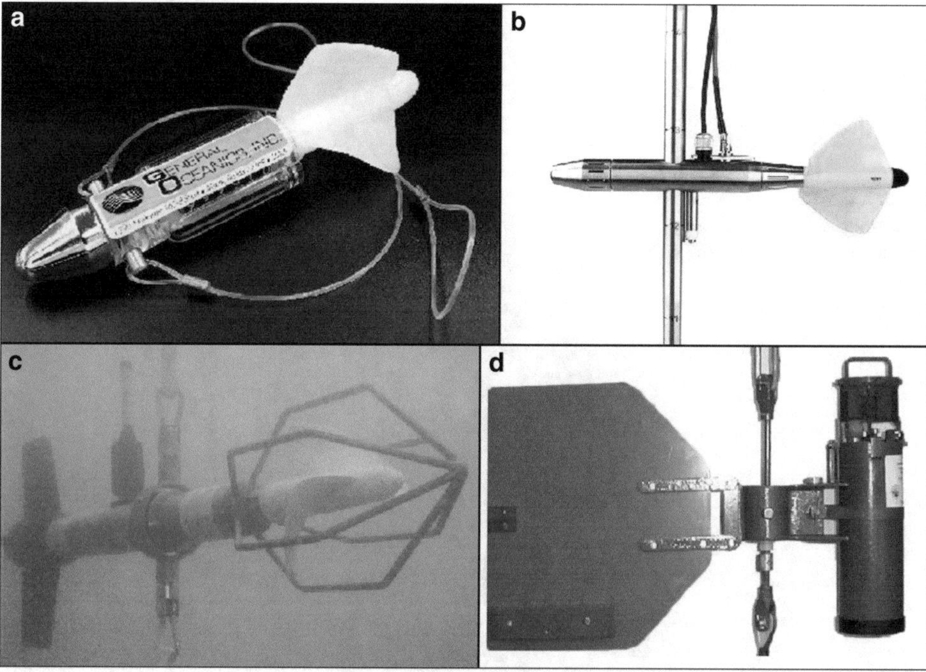

Fig. 6.8 Different types of propeller current meter. **a** Surface current meter. **b** Shallow water current meter. **c** Horizontal axis deep current meter. **d** Vertical shaft (rotor) current meter

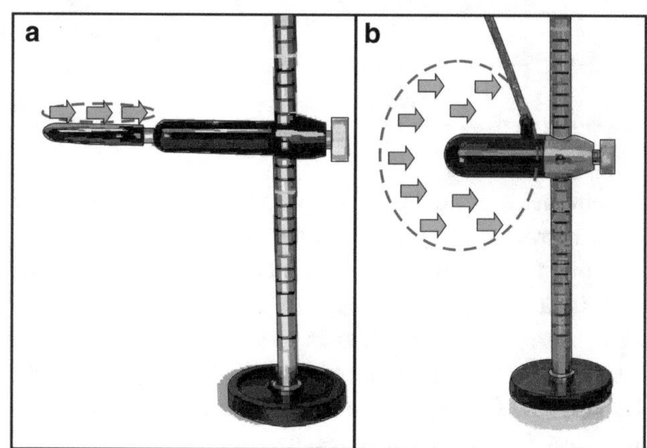

Fig. 6.9 Different models of electromagnetic induction flow meter. **a** Flat sensor. **b** Volumetric sensor

point of view and using the long-contested methods of this geological discipline. In more recent times, the emergence of increasingly sophisticated technologies has methodologically revolutionized coastal geomorphology studies. Nevertheless, the way in which the results are analyzed and interpreted has not undergone too much evolution. This section summarizes all the techniques that are used today, both the classic ones, which continue to be used, and the most modern, results of the advances of the last decades.

6.3.1 Topography: Classification and Measurement of Surficial Landform Dimensions

The geomorphological methodology is applied to the classification and measurement of landforms of different scales. On a small scale, direct measurements of quantitative morphological parameters are made. These are taken manually in the field according to the methodology described by Hammond [13]. The classification and measurement of the dimensions of current-generated bedforms (ripples, megarripples, sand waves and antidunes) is usually done in this way. It should be noted that there is a direct relationship between the presence and dimensions of these forms with their grain size and the speed of the current that generates them, so the genesis conditions can be deduced through the use of a simple diagram [14].

Another classical technique commonly used in coastal geomorphology is stereoscopic photointerpretation. The use of this methodology since the 1940s has allowed the measurement of landforms of meso- and macroscale. In recent decades, the resolution of the photographs has increased, allowing the analysis of even smaller forms.

In the late 1980s, the emergence of remote sensing and the use of satellite images allowed measurements of

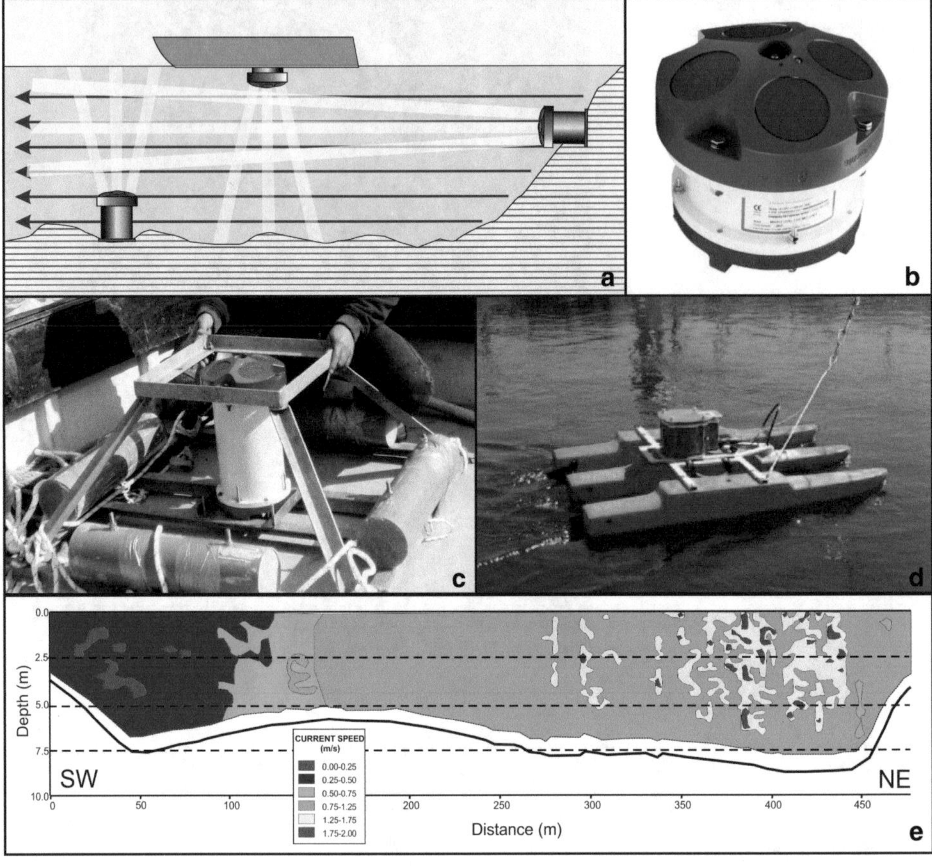

Fig. 6.10 Acoustic Doppler current profilers (ADCP). **a** Different positions in which the system can be installed. **b** Detail of the measuring instrument. **c** Deployment to install on the bottom and measure upwards through the water column. **d** Catamaran used to measure towards the bed through the water column. **e** Current speed cross-profile of an inlet

macroscale forms to be made [7]. Similarly, the increased resolution of the images and the use of multi-band sensors allowed for greater accuracy in the measurements, as well as enabling automation of both classification and measurements. Remote sensing continues to develop with airborne sensor logging. Also, photogrammetry techniques were developed in parallel. Using these techniques to link remote measurements with the position of previously established checkpoints allows high-resolution topographic surveying.

Since the end of the twentieth century, high-resolution surveying has been done using flights with LiDAR technology (laser imaging, detection and ranging). The LiDAR principle is to determine the distance between the laser emitter and the ground surface using laser pulses [5]. This is usually a device that launches a beam of pulses that open at a certain angle to both sides of the emitter. Being airborne sensors, the trajectory of the flight and its height condition the extent and resolution of the data, but in any case, a high-resolution topographic record is obtained, allowing the identification and measurement of mesoscale landforms (Fig. 6.11a).

The same principles apply in terrestrial LiDAR technology, with the difference being the data are taken from the ground from a point of known position. When the LiDAR sensor is found in a situation closer to the analyzed surface, measurements are obtained with a very high resolution due to the narrow spacing between the data [21].

Drone flights have also been used at intermediate scales. This is a cheap technology that, combined with photogrammetry techniques, allows one to perform the accurate measurement of landforms using the right software (Fig. 6.11b). It is a much cheaper technology than LiDAR and offers very satisfactory results.

Topographic data obtained through remote sensing techniques, whether satellite, airborne and land LiDAR or drone topography, are usually treated to obtain a digital elevation model (DEM). DEMs allow for very detailed, larger-scale studies with a larger number of variables. As numerical data, the results can be treated with algorithms that allow the automated identification of certain structures on the ground surface by statistical analysis of the main components [30] or of fuzzy logic [8].

All of these remote sensing techniques open up an enormous potential for work applied to the characterization of coastal zones at different scales and points of view. These applications continue increasing in all domains (e.g., the elaboration of digital terrain models, study of changes in the shoreline, shallow bathymetrical evolution, water mass movements and wave characterization).

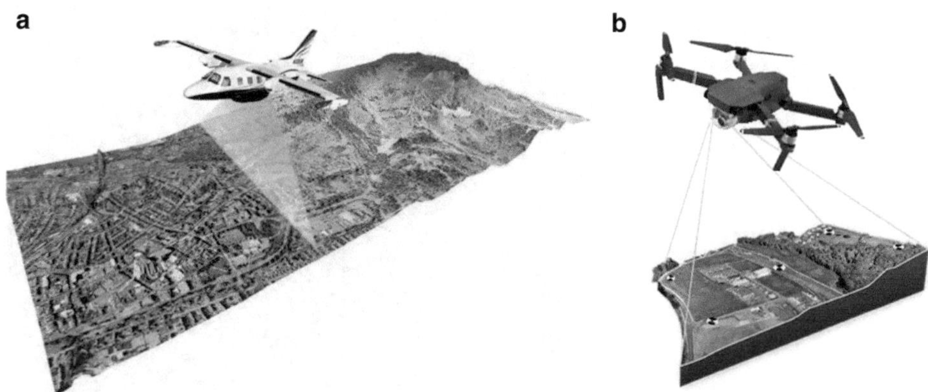

Fig. 6.11 Topographic surveys with airborne sensors. **a** Airborne LiDAR. **b** Drone flight

6.3.2 Monobeam Bathymetry

Bathymetry consists of the surveying of ocean floor depths by measuring and expressing them on maps. Initially, the depths were recorded using ropes knotted with stones attached to the lower end, although these evolved to be replaced by a graduated metal cable called lead-and-line.

In the 1920s, sonar techniques were developed and the first detailed profiles of coastal bottoms could be obtained. The bathymetric data are obtained from the sonic impulses emitted from an electrical source and, once reflected in the seabed, are collected by receiving hydrophones. The first probes recorded bathymetric profiles on continuous paper, but today you get a digital record that can be processed later.

The bathymetric echosounders currently work in a frequency ranging from 15 to 200 kHz, with a single transducer responsible for the emission and reception of acoustic pulses. Currently, any commercial, fishing and even recreational vessel is equipped with a single beam bathymetric echosounder, although most of these probes are not valid for bathymetry as they do not have a data storage system. For scientific purposes, bathymetric records must be able to be stored in XYZ files by combining the bathymetric data with the position obtained through a GPS system. The vessel from which registration is made must travel at low speeds, in order to allow the sound to go and return through the water column.

6.3.3 Multibeam Bathymetry

In the 1950s, probes that emitted a beam of acoustic pulses were developed, the so-called multibeam echosounders. These probes mapped the bed topography along bands extended on both sides of a navigating boat. Their data are very useful for the identification and measurement of submerged landforms.

The method consists of the emission of very high frequency ultrasonic pulses that are emitted in two beams that open several degrees on both sides of the transducer. The width of the bottom band being covered, as well as the data spacing, depends on the thickness of the water column.

The resulting data are numerical datasheets including the distances to the bed obtained from the pulse speed in the water after obtaining the return trip time of each signal. Normally, latitude and longitude coordinates are obtained when the equipment is connected to a differential GPS (DGPS) system. The XYZ coordinate files thus obtained are finally processed to obtain bathymetric maps of contours and three-dimensional diagram blocks showing the topography of the bed at very high resolution (Fig. 6.12).

6.3.4 Side-Scan Sonar

Side-scan sonar (SSS) is a geophysical method that uses acoustic frequency pulses ranging from 100 to 1000 kHz that are emitted by transducers [3]. The transducers are usually located in a submerged towfish that is connected by a cable to the data acquisition unit (DAU) installed on board a vessel (Fig. 6.13a).

Acoustic pulses are reflected in the submerged bed and returned to the receptors also located in the towfish (Fig. 6.13b). The towfish sends the received information to the DAU, which processes the information that will be transformed into a bed image (Fig. 6.13c). The intensity of the bed acoustic response depends on the nature (reflectivity and texture) of the bed material and the orientation of the surface of the material relative to the acoustic pulse.

The use of SSS is commonly applied to coastal geology for geomorphological mapping of submerged areas. These documents also allow sedimentological interpretations of the configuration of the submerged bed in coastal environments [23] and [31]. The use of this system is essential in coastal research because it allows the scanning of large areas quickly and efficiently, using images or records obtained for the interpretation of landforms.

Correct interpretation of SSS logs requires precise positioning of images. To do this, the equipment is normally

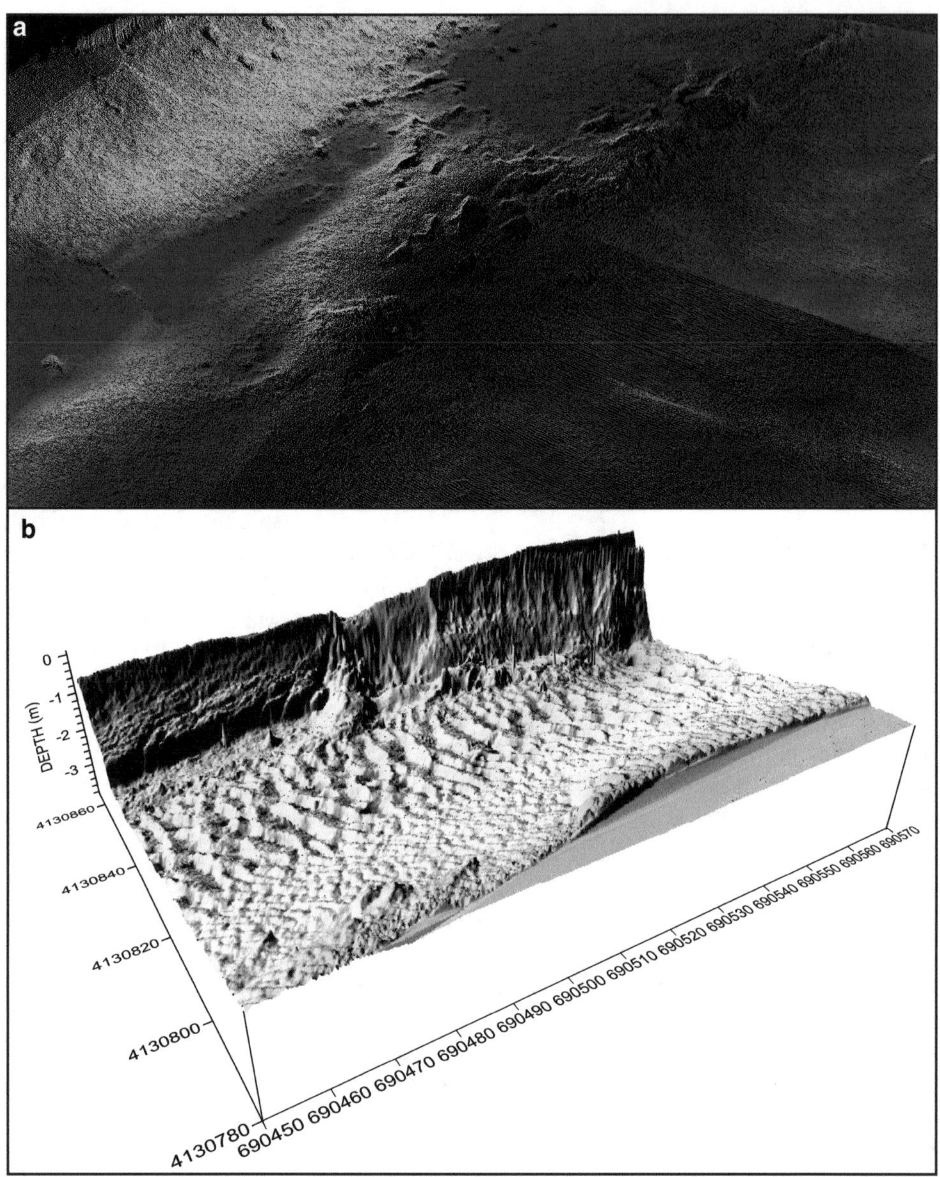

Fig. 6.12 Bathymetric surveys with multibeam echosounder. **a** Representation of information points by color according to their depth. **b** 3D shaded relief representation

connected to a DGPS system, obtaining an accurate georeferenced position of each point of the recorded image. Systematic navigation allows the scanning of wide underwater surfaces, and the precise positioning of successive records is the basis for building georeferenced photomosaics of the coastal bed. Images obtained by SSS can be as accurate as an aerial photograph and efficiently reveal sedimentary features and bedforms. The use of this technique allows the study of coastal beds from a bathymetric, morphological and lithological point of view. It also useful to determine the geometry, distribution, dimensions and orientation of the bedforms, and facilitates the analysis and interpretation of the flow regime.

6.3.5 Cartography of Coastal Environments

In coastal geomorphology, as a specific branch of geomorphology, the realization of cartography is vital to represent the distribution of the landforms on the land surface [24]. Maps are graphic documents in which the distribution of landforms of any kind is represented. The information on the relief that it represents will respond to the specific needs posed in the objective of its realization. Specifically, in coastal areas the distribution of morphogenetic systems is usually represented, which is for geomorphology a concept very similar to that of sedimentary environments in sedimentology. Thus, the distribution of beaches, dunes, subtidal

Fig. 6.13 Side-scan sonar. **a** Components of the equipment. **b** Scheme of the equipment elements during work. **c** Example of SSS registration on a bed with numerous bedforms

channels, intertidal flats or salt marshes would be represented. It is also common to represent the macro-, meso- and microforms that are present in each one of these coastal systems or environments, whether erosive or cumulative. From this point of view, dune ridges, coastal berms and bedform crests that acquire mapping dimensions are usually represented. For smaller shapes, whose dimensions do not allow mapping of individual elements, it is common to map shape fields.

In sedimentology, the characters of the geological materials can also be mapped. In these cases, it is usually represented by the type of facies of the sediment or rock. This type of map expresses distributions of grain sizes or content in certain lithological components. This representation can be easily interpreted in terms of polarity of the movement of materials across the coastal environments.

This method gave rise to a special type of map that is very useful in coastal areas: process mapping. This system of mapping not only represents fields with different shapes and bedform types, but includes symbols that represent specific processes. In particular, vector processes such as the trajectories followed by the transport of sediments are interesting.

It should be noted that any type of mapping in coastal areas can be made of both emerged and submerged areas, according to available sources of information. It is clear that, in most cases, coastal environment mapping overlaps with topographic and/or bathymetric information.

6.4 Study Methods of Short-Term Coastal Evolution

Since the coast is one of the most dynamic environments in existence, it produces changes in very short time spaces that can be observed on the human scale. As a result, many of these changes can be measured by comparing studied situations. Often these comparisons are made between two different dates, revealing evolutionary trends, but most of the time they compare situations on many successive dates, with which cyclical behaviors can be established and that are thus very characteristic of coastal dynamics. The intention of these measurements is to understand the causes of these changes by relating them to the processes that occurred in the period between the analyzed dates. Of these behaviors, reliable short- and medium-term predictions can often be established. This section describes the most frequently used methods to address the study of these morphological changes.

6.4.1 Topo-Bathymetric Comparisons

6.4.1.1 Beach Profiles

Perhaps one of the first methods addressed in coastal dynamics studies was the comparison of beach profiles carried out on different dates. In this case, in order to compare the profiles it is important that measurements begin from an easily identifiable point whose position in the three dimensions remains invariant. The topographic profile will thus be measured starting from the top of the backshore and ending at the lower beach. An important aspect is that the measurements are made during low tide to measure as much of the beach as possible. It is clear that it will also be necessary to take the course towards which the profile is measured, in order to ensure that all profiles are made along the same transect. Classically in the studies of the evolution of beaches, several profiles are usually made, separated at distances of about 50 m, in order to carry out three-dimensional surveys and perform volumetry.

The simplest and cheapest method is to use a fixed-length bar (often one meter or three feet) with a level to keep it horizontal. At the end of the bar, a graduated stake is placed to measure the topographical difference between the start and end points along the profile. A more precise method is that proposed by Kenneth O. Emery, which is to use an alidade or a sight whose horizontal position is assured by a bubble level (Fig. 6.14a). The optical horizontal level marks the increments between the starting point and a graduated stake that is progressively removed meter by meter from the reference point until reaching the lower beach (Fig. 6.14b). This method was improved with the entry into the market of topographical instruments such as the total station, which performs laser measurements of distance and angle between the station position and a mirror located at the end of a theodolite of the same height as the station optical instrument (Fig. 6.14c).

These last two methods can even be applied by continuing below sea level when moving stakes are placed on a wheeled buggy that can move under the water surface (Fig. 6.14d). The University of North Florida has recently developed a remote-controlled guided vehicle for measurements in the break zone, called Surf Rover.

The most modern method is the use of bars connected to DGPS systems (Fig. 6.14e). This method can be combined with the total station and is very useful for large area surveys, especially when the data obtained at reference points are also combined with a photogrammetric drone flight. The inconvenience is that its accuracy is lower than that obtained in the above methods.

6.4.1.2 Comparisons Between DEMs Obtained from Topographic and Bathymetric Data

DEMs obtained on different dates from topographic and bathymetric data can be compared if they cover exactly the same area. For the comparison to be possible, grids of the same characteristics must be set in the same size. That means they must have the same spacing and number of nodes, at times with the same start and end points, so that the entire X and Y points must completely match. Typically, the same software applications that can interpolate curves between node data can in turn subtract data from both grids. So, in places where the previous date surface is above the later one a negative value is obtained, while where the back date surface is above the previous date results in positive data.

This then allows the construction of a plot with curves of the same value, interpreting surfaces with negative values as areas where erosion has occurred and surfaces with positive values as depositional areas. Some functions of these software allow in turn the calculation of erosive and cumulative areas, as well as eroded and accumulated volumes.

6.4.2 Comparative Photointerpretation

For decades, comparative photointerpretation has been a very useful tool to visualize changes in the coastline, as well as changes in the distribution of coastal environments. From the 1950s to the mid-90 s, the main method was cartographic work on transparent paper over aerial stereo-photographs displayed through a stereoscope. In this way, all changes between two aerial photographs of different dates were qualitatively compared. However, this method has the disadvantage of not being able to quantify the observed changes, since the cartographic product obtained was not properly georeferenced.

Since the advent of geographic information systems (GIS) in 1962, these systems have become a very useful tool in comparative studies of coasts of different dates. Especially so, since the generalization of their use at the end of the 1980s meant that virtually all the comparative works are done with GIS software [10]. A GIS can be defined as a set of tools that can integrate, store, manipulate and organize layers with appropriately georeferenced geographic data [6]. These systems allow one to work with layers of information in meshes (raster), such as aerial photographs or digital

Fig. 6.14 Methods used for measuring beach profiles. **a** Emery method principle. **b** Alidade equipment with bubble level, graduated bars and rope. **c** Total station and theodolite. **d** Rover for shallow water measurements, positioned with a jet ski. **e** Measurements with DGPS

elevation models, and with vector information layers (thematic mappings), such as sedimentary environment maps or systems described in Sect. 6.3.5. In this way, the information corresponding to the coast on different dates can be easily quantitatively compared.

Today there are numerous applications to work with GIS on all operating systems and many of them come as free software. One can even find GIS applications that allow work online. A list of the most widely used programs can be found in the GIS hall of fame [28].

6.4.3 Numerical Models

The use of numerical models is a very useful procedure for forecasting the future behavior of coastal systems from the analysis of multiple variables, including comparing digital models of the successive date DEMs. Mathematical models are a very powerful and inexpensive tool that allow analysis and forecasting of the dynamic behavior of the coastal system in the short and medium terms. A first step in these models is the rapid and reliable calculation of the theoretical capacity of sediment transport by waves and currents.

A model actually consists of several mathematical analysis modules that can be coupled to get results more adjusted with reality. Each one of these modules applies the physical equations that regulate the analyzed process. In one of these models, for instance, would be the determination of the propagation and breaking wave patterns, the distribution of radiation vectors, the calculation of wave-induced currents and, finally, the transport of sediments induced by combined waves and currents. This is very useful when studying open coastal environments, such as beaches, that are completely dependent on the waves.

Another of these models is responsible for the modeling of tidal levels and tidal currents, in order to calculate the potential transport of sediments induced exclusively by this phenomenon. It is especially useful in the study of tidal systems, such as tidal flats, estuaries or deltas, where an additional variable such as the volume of water from the river system can be introduced. The most complex systems are those involving both waves and tides, such as tidal deltas and delta-front bars. In these cases, the models described above should be used together.

The numerical models start from an initial bathymetry and solve the sediment flow equations within the studied area, as well as changes in bathymetry associated with spatial variations in sediment transport. The model obtains a final bathymetry within the prescribed time frame. The correct functioning of the model requires the following as input data:

- Initial bathymetry,
- Wave data (dimensions and provenance),
- Tidal level data,
- Tidal current data (speed and direction), and
- Sediment characterization data (grain size parameters).

The potential transport of sediments can be expressed in terms of the total volume (or weight) of sediment transported. Bathymetric changes will be the result of such sediment transport, taking into account the areas with dominant erosion or deposition. However, these models must always be correctly validated and calibrated for reliable results. A common way of model calibration is to apply it to an old bathymetry to forecast at a time interval where another more recent bathymetry already exists. Then, check the degree of similarity between the actual bathymetry and the bathymetry predicted by the model. The model must be adjusted this way until you get a satisfactory match percentage. There are calibrated models that have reached more than 90% adjustment between forecast and reality in one-year periods.

Clearly, forecasts done by these models are of great uncertainty even if they are properly calibrated. It is necessary to take into account that they are subject to the arbitrariness of the wave regime which is completely unpredictable as it is linked to the maritime climate. Even so, they are a very effective tool for dynamic behavior analysis and evolutionary trends. Currently, the application to the transport of sediments on the coast by using numerical models is recent and these are in the period of improvement. The models for estimation of sedimentary dynamics are less well-developed.

6.5 Study Methods of Coastal Surficial Sediments

The study of coastal surface sediments is one of the first objectives addressed on the coast from a purely sedimentological approach. The first works were based on samples of the most superficial centimeters, but, little by little, techniques of sampling and testing to an increasing depth were appearing in studies of the top of the sedimentary sequence in current environments. This section presents the most common methods for sediment sampling and coring.

6.5.1 Sampling of Surficial Sediments

Sampling in the inter- and supratidal environments is usually done directly using a manual shovel, taking into account that, if the samples are to be analyzed from a geochemical point of view, plastic material should be used to avoid the metallic contamination of the sample.

Different types of bottom dredgers are used for sampling underwater sediments. There are numerous types of dredgers for this purpose, which are distinguished from each other by the mechanism used to trigger the closing of the buckets. The most frequently chosen for their ease of use are Van Veen (Fig. 6.15a), Petersen type (Fig. 6.15b) and Shipeck type (Fig. 6.15c) grabs. Less commonly used are Orange Peel (Fig. 6.15d) and Smith-McIntyre (Fig. 6.15e) grabs. Eckman type (Fig. 6.15f) has the particularity that a box is nailed to the sediment to allow the preservation of the internal structure of the sample taken. There are other types of dredgers, such as Lafond-Dietz, that in addition to taking sediments are able to sample the rocky substrates.

6.5.2 Surficial Coring

Unconsolidated sediments of tidal environments (supratidal, intertidal and shallow subtidal) can be manually testified using plastic pipes (mainly PVC). Any method of coring used should seek to not disturb the internal structure of the sediment since one of the sedimentological aims is the direct observation of the internal order.

6.5.2.1 Hand Corer

In those environments where the soil can be directly accessed, the pipe can be inserted directly by hitting with a hammer. To ensure the correct lifting of the sediment core, after inserting it, a rubber cap must be used to close the top end and prevent the entry of air, sucking the sediment into the PVC pipe.

6.5.2.2 Beeker Corer

In environments accessible by walking, where the consistency is soft and the sediment has high water content, Beeker type coring can also be used (Fig. 6.16). The Beeker corer is characterized by having a flexible rubber at one end that can be inflated from the surface once the pipe has been hammered into the sediment. When inflated, the rubber cuts the sediment at its base and prevents it from remaining in the ground when the pipe is removed. It can also be used

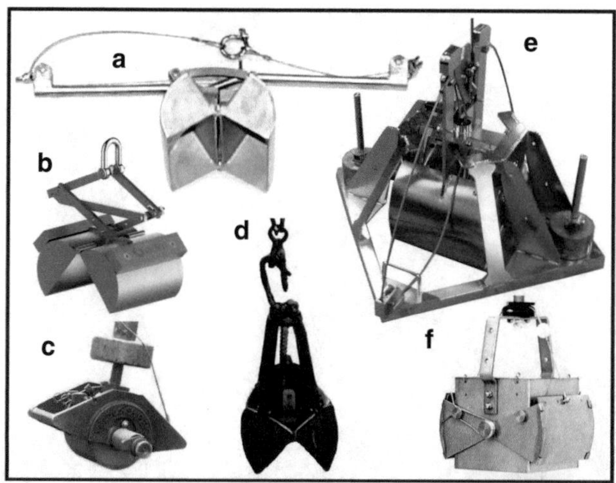

Fig. 6.15 Different types of dredger. **a** Van Veen type. **b** Petersen type. **c** Shipeck type. **d** Orange Peel type. **d** Smith-McIntyre type. **f** Eckman type

Fig. 6.16 Hand corer: Beeker sampler

through a column of water from a boat. In this case, the top end of the corer must be coupled with the number of steel rods needed to reach the bottom. In this sampler, the PVC jacket containing the sediment core needs to be replaced after each extracted core.

6.5.2.3 Suction Corer

The suction corer is a non-intrusive, economical and manual technique that allows quick and easy extraction of wet sediments without compacting. The first models were designed for the Netherlands Geological Survey by Jan Van der Staay's team in the early 1970s [29]. The method is based on the suction of a piston that ascends inside a tube thanks to the traction of a cable. This cable is connected to a winch that acts from the outside of the system through a pulley that reverses the direction of traction (Fig. 6.17).

The first models were made of metal, but it was very difficult to extract the sediment cores from the inside without disturbing the sediment, so in the 1980s design modifications were made to build it in plastic materials such as PVC [32].

The corer TESS-1 was designed in the early twenty-first century by researchers from the University of Vigo, Spain [22]. It is based on the Van der Staay suction machine, but allows the encapsulation of sediments in a PVC jacket that is inserted through the inside of a steel tube. This facilitates its subsequent opening and analysis without disturbing the ordering of the sediment.

6.5.2.4 Piston and Gravity Corers

When sampling in deeper subtidal environments, gravity corers are often used (Fig. 6.18). This method consists of dropping a metal pipe with a coupled weight that forces the tube to be inserted into the soft sediments of the bed. The metal tube contains an internal PVC coating that will house the extracted sediment core. The system used in the piston corers to keep the sediment core inside the pipe is usually an internal piston that makes the vacuum and prevents the return of sediments when the sediment column is extracted. In the case of gravity corers, the anti-return function is performed by a steel valve with an egg shape.

There are many models of piston corers and gravity corers, such as Baillie, Hydra, Lundqvist, Meteor, Phleger, Sjostedt or Strom models. All these types really only differ in their hydrodynamic design, pipe length and weight placement.

6.5.2.5 Box Corers

These kinds of samplers are used for the specific study of the sedimentary structures of the shallowest sediment layer (Fig. 6.19). This type of coring allows one to obtain samples of large size with minimally disturbed sediment. The removable characteristic of the system facilitates the study of sedimentary structures in two perpendicular planes that coincide with the sides of the box.

The box is fixed by several pins at the lower end of a cylindrical tube that slides through the inside of a guide, pushed by a weight of several hundred kilograms. This guide is welded to a leg-shaped structure that stabilizes the vertical movement of the box. When impacting with the bottom, the box goes down vertically until it is nailed to the seabed. This impact also acts on a safety display that releases the

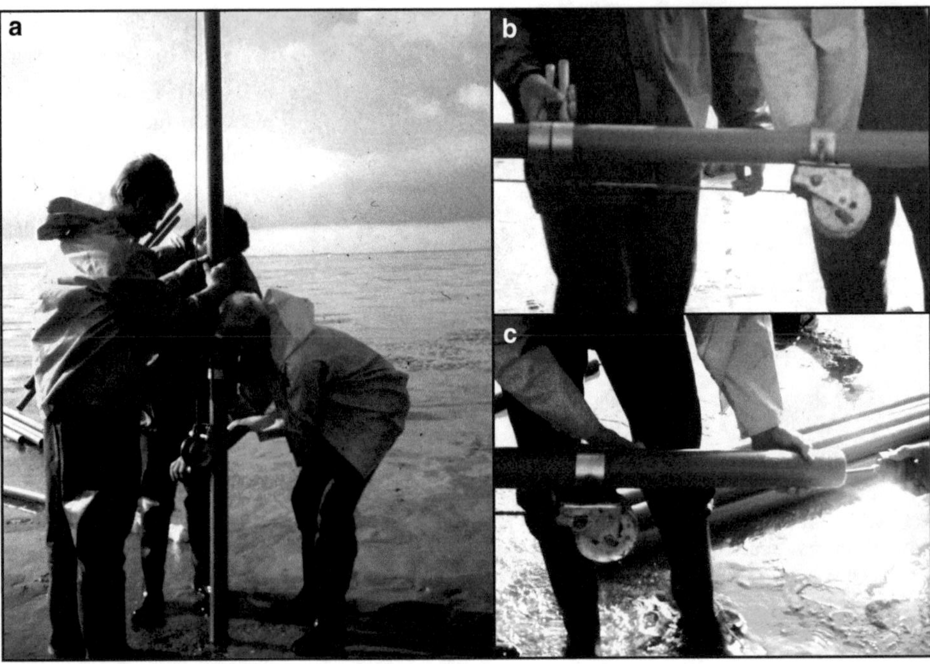
Fig. 6.17 Suction corer made in PVC. **a** During the thrust operation. **b** Detail of the push bars and the external winch. **c** Detail of the inner suction piston

Fig. 6.18 Gravity corer type Phleger, showing a detail of the egg-shaped valve

extraction instrument. When pulling the winch from the instrument to extract the sediment core, a flat shovel swings under the sample by plugging the mouth of the box to prevent sediment loss.

Once on board, the box can be removed from the rest of the machine by releasing the pins. Once the box is released, the water is extracted to minimize sediment disturbance during the opening process. The boxes are usually removable and the sediment inside is exposed when removing one or two of the sides.

6.5.3 Sediment Traps

Sediment traps are instruments used to capture particles that are being transported or are in the process of decanting. In both cases, the weight of the captured material allows calculations of the amount of material involved in sedimentary bypassing in a coastal environment and also estimations of deposition rates. According to the planned objectives, they can be classified into suspended matter traps and bed load traps.

Suspended matter traps (Fig. 6.20a) consist of opened-up section pipes or funnels. These are installed at a given distance from the surface and held in a capture position for a predetermined time. After that time the mechanism is retrieved and the amount of material decanted during that period is checked. The recovered sediment allows calculations of sedimentation rates and the amount of suspended material available in the water column, but, in addition, this sediment can be analyzed to characterize its nature from sedimentological and geochemical perspectives.

Surface traps are also used to measure accumulation rates (Fig. 6.20b). They consist of a square plastic surface equipped with a vertical cylinder that is used to fix it at the bottom and to visualize its position. They are installed in intertidal beds, matching the plastic surface with the surface

Fig. 6.19 Box corer. **a** Just before submersion. **b** Installing a new box before sampling

of the sediment, and are left in place for a certain time. After this time, the sediment thickness that has been deposited on the plastic surface is checked. This sediment can also be sampled in order to characterize the material accumulated during that period.

Traps for bed load-transported material are installed on the bottom and aim to capture the grains being transported by bearing, drag or saltation (bed load transportation). The objective is to calculate the real transport capacity of coastal currents, in relation to sediment availability. This type of trap consists of a metal inlet ramp allowing the bed load material to enter across a window where the sediment is held in an inner reservoir, while the water flow is diverted outward.

There are various trap models that are frequently used in coastal sedimentology. The first to be used was the Nesper, also called basket samplers, widely used before the 1940s. They have good efficiency for sand-type sediments and currents greater than 30 cm/s [15]. Recently, the most commonly used are box and basket models or Helley-Smith type (Fig. 6.20c). These traps have greater efficiency and are easier to install [9]. Somewhat less used are the tray traps, which present the appearance of a solid steel case. Polyakov traps belong to this type (Fig. 6.20d), presenting a slightly higher efficiency than those of basket type [26].

6.6 Methods to Study the Sedimentary Record of Coastal Environments

In the more continental area of the coasts, classic rotation cores are usually used. This type of sampling has the advantage of reaching to considerable depth, allowing one to analyze the lithological sequence in a fairly complete way. However, the cores obtained using this technique have the disadvantage that the internal order of sediment is disturbed, so sedimentary structures cannot be observed.

One method that allows obtaining cores of almost 7 m without disturbing the internal structure of the sediment is vibration coring. This is a fairly simple method that is based on the application of vibration at the top end of an aluminum pipe that is introduced into the sediment.

There are several methods for obtaining vibration cores. Lanesky *vibracores* are often used from land or in shallow water from a floating platform [19]. In this technique, the vibration of an eccentric needle connected to a conventional rotation motor is applied to the tube. This type of equipment is usually sold by construction material companies to be used in the compaction of concretes. The vibrating needle is connected to the aluminum tube through a piece of steel especially built for this purpose. The aluminum pipe with coupling piece and needle must be positioned upright before starting vibration. Once vertical, it is held in this position while it is introduced into the sediment using tensor straps (Fig. 6.21a). After being introduced, the core is removed with a differential pulley resting on a tripod, ensuring the sediment's recovery by using a rubber cap.

Vibration probes can also be performed from the boat, and in this case they are called *vibrocorers*. Here, the vibration mechanism is coupled directly at the end of the pipe, also exerting a weight on it. When operating underwater, the verticality of the tube should be adjusted through a leg structure supported at the bottom (Fig. 6.21b). The removal of the sediment core is performed from the vessel using a winch. It is necessary to ensure the preservation of the sediment inside the tube by using an egg-shaped sphincter valve similar to that used in gravity corers.

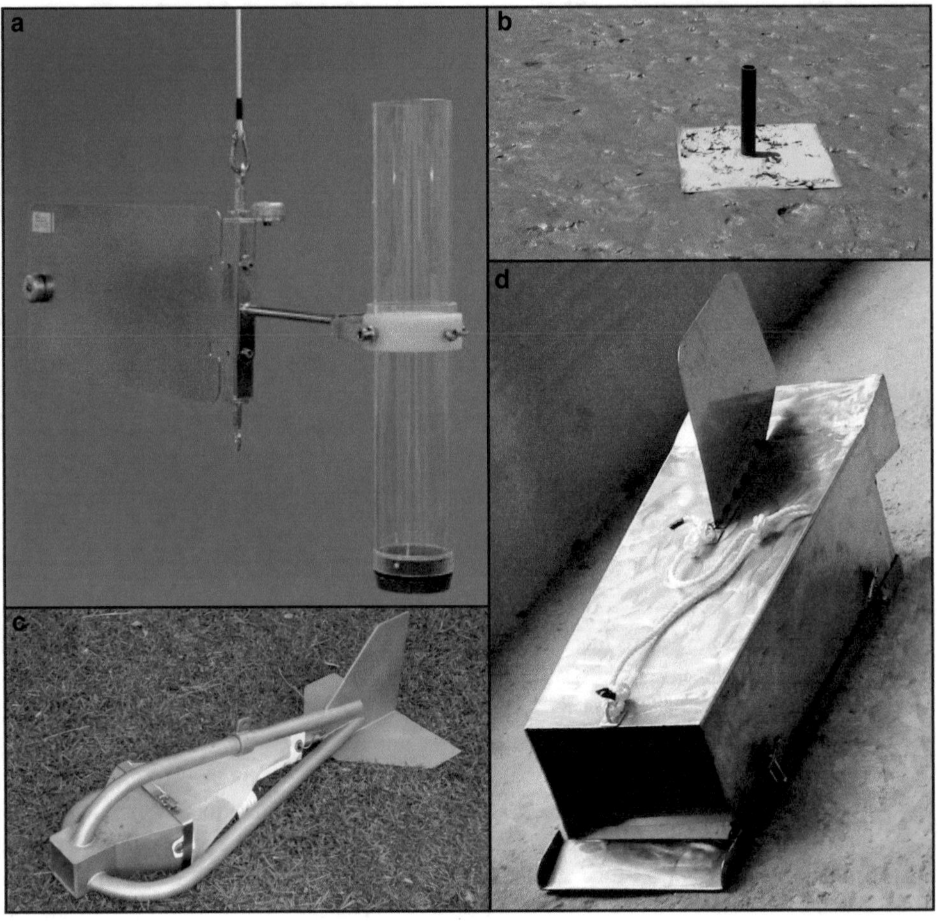

Fig. 6.20 Different types of sediment traps. **a** Suspended sediment traps. **b** Surface trap installed on an intertidal flat. **c** Box and basket trap (Helley-Smith type). **d** Tray trap (Polyakov type)

Advanced Box 6.1: Cutting the Cores

Once sediment cores have been obtained by any of the above-described methods, it is necessary to access the sediment inside the plastic or aluminum coating for the study of the facies and the corresponding sampling. It is very important that the sediment of the core is not altered or contaminated during the opening process, so the saw should cut only the coating without damaging the sediment.

When cutting on the vessel during the sampling survey, a manual grinder is usually used (Fig. 6.22a). At first, a longitudinal cut must be done to continue with a second one along the diametrically opposed side.

If the cutting is made in the laboratory, there is also the possibility of mounting the grinder on a stand with guides on which the pipe sits. A saw integrated into a cutting table can also be used in the laboratory. In this case, those used for cutting wood, aluminum or building material (Fig. 6.22b) may be appropriate.

Once the cylindrical coating has been cut into two valves, the sediment is cut by passing a guitar or nylon string through the two opposite slits. So, the cores are opened in two halves, already prepared for study (Fig. 6.22c).

6.7 Geophysical Methods for the Study of the Geometry of Coastal Sedimentary Bodies: Seismic Reflection

Geophysical methods are usually used to study the marine sub-bottom and, in particular, that located in coastal environments. These methods are based on measuring the physical properties of the materials that constitute the subsurface when they are traversed by a sound wave. Methods such as high-resolution seismics are often used for the study of coastal sedimentary records.

The objective of seismic methods is to determine the structure and geometry of sedimentary bodies through the physical properties of subsurface lithologies. Seismic reflection acquisition systems consist of three elements: an emitting (seismic) acoustic source, a chain of receivers (streamer hydrophones) and data processing and filtering software. Both the emitter and receiver are dragged by the ship along pre-set tracks to obtain the seismic profiles.

In the case of acoustic pulse emitters, marine seismic studies may use different energy sources that are classified according to the nature and frequency of the emitted pulses

Fig. 6.21 Vibration cores. **a** Lanesky vibracore. **b** Vibrocore VGK from Geomares (Colombia) operated from a vessel

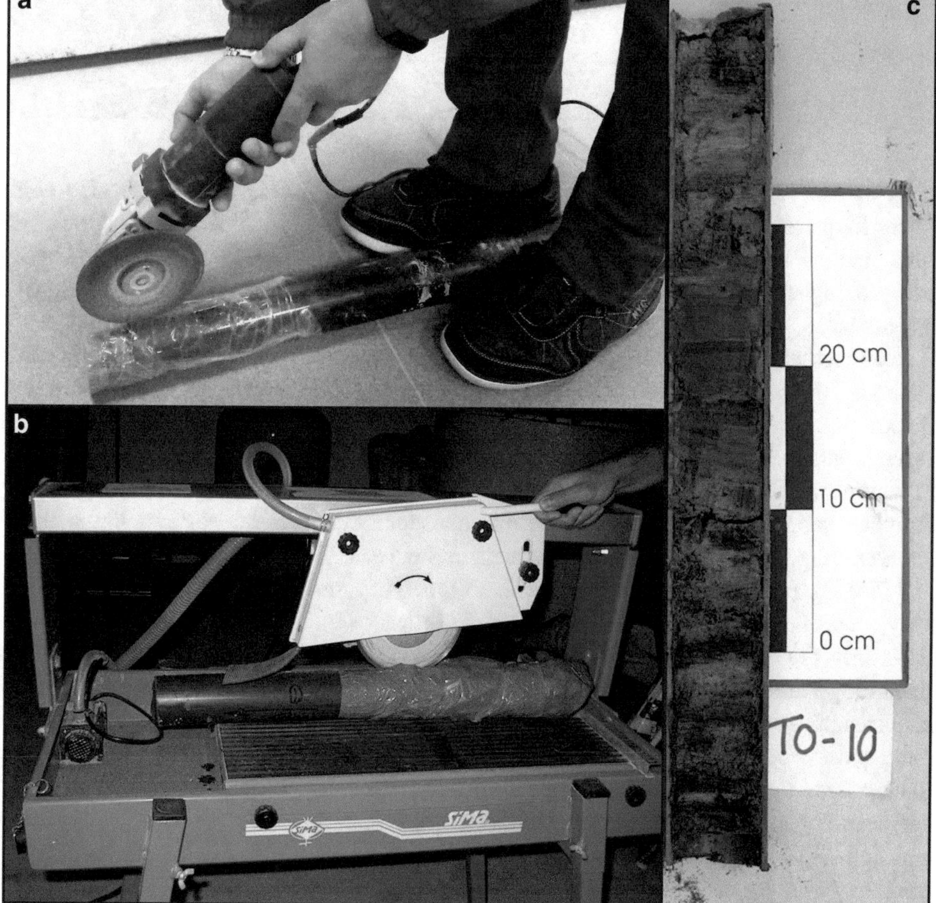

Fig. 6.22 Cutting the sediment cores. **a** With a manual grinder. **b** By using a cutting table. **c** Example of a sedimentary sequence observed in a core

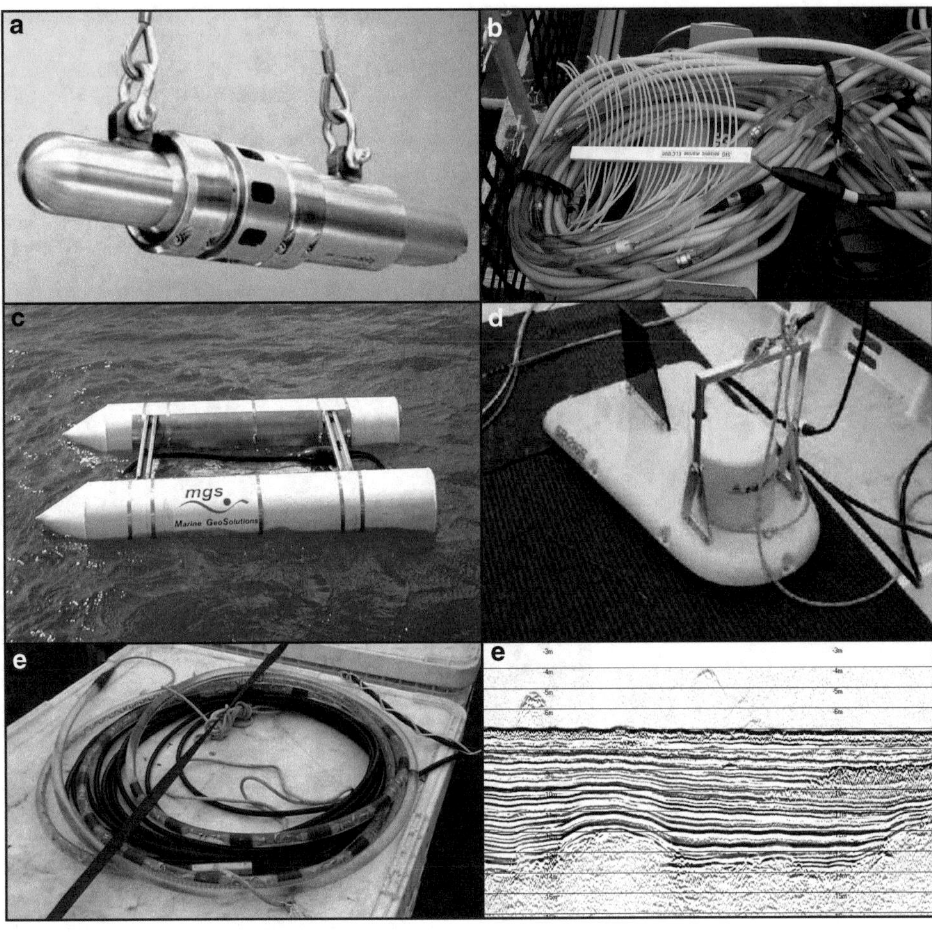

Fig. 6.23 Systems for the emission and reception of acoustic pulses during the acquisition of seismic profiles. **a** Small compressed air cannon. **b** Sparker type plasma transducer. **c** Boomer type piezoelectric system mounted on a floating catamaran. **d** Integrated emitter-receiver system in a sub-bottom profiler. **e** Small hydrophone streamer. **f** Example of a seismic record obtained with a parametric sub-bottom profiler

(Fig. 6.23). From this regard, it should be borne in mind that lower frequencies manage to record thicker subsurface profiles, but at lower resolution, while sharper frequencies achieve higher resolution but lower penetration into the subsurface.

Compressed *air cannons* (Fig. 6.23a) and also water cannons are often used to achieve a record of greater thicknesses. These emit frequencies so low that they can reach several kilometers thick, so they are often widely used in continental shelf studies, but not in coastal environments. Another source of emission are plasma transducers or *sparkers* (Fig. 6.23b). These work at frequencies between 20 and 200 Hz and can reach depths of up to 150 m. They are very useful in large-scale coastal stratigraphy studies.

In fact, the devices most commonly used in coastal sedimentology are piezoelectric transducers or (*boomers*) and acoustic chirps (*sub-bottom profilers*). Boomers (Fig. 6.23c) emit pulses between 700 and 2000 Hz and can reach thicknesses of about 50 m. They are frequently used in sedimentological studies of the complete record of estuaries and deltas. The most commonly used sub-bottom profilers are those working at 3.5 kHz, although parametric emitters with adjustable frequency between 2.4 and 5.5 kHz (Fig. 6.23d) are also common. They allow one to obtain profiles at a very high resolution, but do not reach across more than a dozen meters. So, their usefulness is limited to the geometry of the most superficial units of the record of estuaries, deltas, tidal flats, tidal deltas and beaches.

Regarding the reception of the acoustic signal, in the low-frequency systems the hydrophones are arranged in a long streamer (Fig. 6.23e). This streamer is transported behind the vessel to collect pulses that return with much delay after crossing a large sediment thickness. In higher frequency systems, such as boomers or sub-bottom profilers, there is not so much delay in the arrival of the signal reflected in the subsurface, so the emitters and receivers are integrated into the same transducer.

The wave field recorded by hydrophones consists primarily of the reflections generated in the geological discontinuities characterized by abrupt contrasts of the elastic parameters between lithologies (Fig. 6.23f). Seismic sequence analysis is the methodology consisting of the subdivision of the seismic section into sets of deposits limited by discontinuity surfaces and comprising groups of more or less concordant reflections with similar characteristics. On the other hand, the analysis of seismic facies consists of the description and interpretation of lithologies based on the seismic characteristics of the reflections that constitute the seismic sequence.

One problem of the method is that the reflected waves do not contain direct information of their propagation speed in the rocky environment. In this way, the vertical axis of the seismic sections is usually represented in time (there and back) and not in depth. There are techniques, called depth migration, that allow the transforming of seismic sections from time to depth. These are complex techniques to apply and interpretation requires a considerable level of subjectivity.

References

1. Barstow SF, Krogstad HE (1984) General analysis of directional ocean wave data from heave/pitch/roll buoys. Model Identif Control 5:47–70
2. Bishop CT, Donelan MA (1987) Measuring waves with pressure transducers. Coast Eng 11:309–328
3. Blondel P (2009) The handbook of Sidescan Sonar. Springer, Heidelberg, 278pp
4. Booij N, Ris RC, Holthuijsen LH (1999) A third-generation wave model for coastal regions, part I, model description and validation. J Geophys Res 104:7649–7666
5. Bracco Gartner GL, Schlager W (1997) Quantification of topography using a laser-transit system. J Sediment Res 67:987–989
6. Clarke KC (1986) Advances in geographic information systems. Comput Environ Urban Syst 10:175–184
7. Dikau R (1989) The application of a digital relief model to landform analysis in geomorphology. In: Raper J (ed) Three dimensional applications in geographical information systems. Taylor & Francis, London, pp 51–77
8. Dragut L, Blaschke T (2006) Automated classification of landform elements using objects-based image analysis. Geomorphology 8:330–344
9. Druffel L, Emmett WW, Shneider VR, Skinner JV (1976) Laboratory calibration of the Helley-Smith bedload sediment sampler. US Geological Survey Open-File Report 76–752, 63pp
10. Goodchild MF (2010) Twenty years of progress: GIScience in 2010. J Spatial Inform Sci 1:3–20
11. Grønlie Ø (2004) Wave radars—a comparison of concepts and techniques. Hydro Int 8(5):24–27
12. Guedes Soares C, Rute Bento A, Gonçalves M, Silva D, Martinho P (2014) Numerical evaluation of the wave energy resource along the Atlantic European coast. Comput Geosci 71:37–49
13. Hammond EH (1954) Small scale continental landform maps. Ann Assoc Am Geogr 44:34–42
14. Harms JC, Southard JB, Spearing DR, Walker RG (1975) Depositional environments from primary sedimentary structures and stratification sequences. Dallas, SEPM Short Course No. 2, 161pp
15. Hubbell DW, 1964. Apparatus and techniques for measuring bedload. US Geological Survey Water-Supply paper 1748, 74pp
16. Janssen P (2004) The interaction of ocean waves and wind. Cambridge University Press. Cambridge, 300pp
17. Kanevsky MB (2009) Radar imaging of the ocean waves. Elsevier Science, Alexandria, 191pp
18. Karaev VY, Kanevsky MB, Meshkov EM (2011) Measuring the parameters of sea-surface roughness by underwater acoustic systems: discussion of the device concept. Radiophys Quantum Electron 53:569–579
19. Lanesky DE, Logan BW, Brown RG, Hine AC (1979) A new approach to portable vibracoring underwater and on land. J Sediment Petrol 39:655–657
20. Liu HT, Katsaros KB, Weissman MA (1982) Dynamic response of thin-wire wave gauges. J Geophys Res 87:5686–5698
21. Lichti DD, Gordon SJ, Stewart MP (2002) Ground-based laser scanners: operation, systems and applications. Geomatica 56:22–33
22. Méndez G, Pérez-Arlucea M, Stouthamer E, Berendsen H (2003) The TESS-1 suction corer: a new device to extract wet, uncompacted sediments. J Sediment Res 73:1078–1081
23. Morales JA, Delgado I (2016) Side-scan sonar imaging of sediment bedload. In: Kennish MJ (ed) Encyclopedia of Estuaries. Encyclopedia of Earth Sciences Series. Springer, Dordrecht, pp 12–14
24. Peña JL (ed) (1997) Cartografía geomorfológica básica y aplicada. Geoforma Ediciones. Logroño, 243pp
25. Reichert K, Hessner K, Dannenberg J, Tränkmann I, Lund B (2005) X-band radar as a tool to determine spectral and single wave properties. In: 25th International conference on offshore mechanics and arctic engineering, vol 3, pp 1–9
26. Shingal HSS, Joshi GC, Verma RS (1981) Sediment sampling in rivers and canals. In: Erosion and sediment transport measurement. IAHS Special Publication, vol 133, pp 169–175
27. Smith MJ, Stevens CL, Gorman RM, McGregor JA, Neilson CG (2001) Wind-wave development across a large shallow intertidal estuary: a case study of Manukau Harbour, New Zealand. NZ J Mar Freshw Res 35:985–1000
28. Tomlinson R (2007) GIS hall of fame. URISA
29. Van De Meene EA, Van Der Staay J, Hock TL (1979) The Van der Staay suction corer: a simple apparatus for drilling in sand below groundwater table. Rijks Geologische Dienst. Haarlem, 24pp
30. Vosselman G, Gorte BGH, Sithole G, Rabbani T (2004) Recognising structure in laser scanner point clouds. Int Arch Photogrammetry, Remote Sens Spat Inform Sci 46:33–38
31. Wright LD, Prior DB, Hobbs CH, Byrne RJ, Boon JD, Schaffner LC, Green MO (1987) Spatial variability of bottom types in the lower Chesapeake Bay and adjoining estuaries and inner shelf. Estuar Coast Shelf Sci 24:765–784
32. Wallinga J, Van Der Staay J (1999) Sampling in waterlogged sands with a simple hand-operated corer. Ancient TL 17:59–61

Part II
Coastal Processes

Folding waves, in a foamy tide
of salt and brine,
washing in beds of opal shells,
drift over pools, of salt and clay.

"Alcan Road"
Ween

7 Wave Processes

7.1 Introduction

The waves are undulations of the surface between two fluids, water and air, that have a remarkable influence on coastal dynamics. This importance is manifested in the dynamic functioning and transport of sediments from open coastal areas. Waves are also the fundamental cause of the genesis and evolution of environments such as barrier islands and beaches, as well as the erosion of rocky coasts. Thus, understanding wave behavior becomes fundamental in interpreting the dynamics of these environments, as well as in correct decision-making regarding the design and construction of coastal structures such as groins, jetties, breakwaters, seawalls or docks.

Waves as a phenomenon have claimed people's attention from antiquity. In classical Greece, Aristotle was the first to reflect on the waves in his treatise on physics (335–322 BC) and he also linked the wind to the genesis of the waves. These reflections were expressed in deeper detail by Benjamin Franklin around 1774, who described how air moving when it passes over the surface of the water exerts a friction that deforms this surface into ridges and troughs that move in the direction of the wind. It was not until the early nineteenth century, though, that Gerstner [6] articulated the first wave theory, describing the circular movement of the internal liquid particles.

Since the works of Airy [1] and Stokes [10], it has been known that the dimensions and propagation of sea waves respond to basic theoretical equations that function under wave motion theory. However, the coincidence of these equations and reality do not fully fit, because marine waves constitute a complex phenomenon. So, the waves do not have a simple mathematical expression as they do not exactly match with sinusoidal functions. Since then, efforts have been made to simplify the phenomenon in a series of nonlinear differential equations. Thanks to this, today the swell is a well-studied and well-known phenomenon, where all the principles are very well established. Nevertheless, on a physical level there are still some questions regarding the genetic mechanism.

Perhaps the most dynamically interesting wave classification refers to the genesis of marine waves according to their period [9]. This author established a scale that considered capillary waves, gravitational waves and tidal waves according to their dimensions. He also established a connection between the types of waves with the primary forces that generate them, distinguishing between waves generated by surface tension with the air, induced by the wind, caused by the action of gravity, generated by the Coriolis force, produced by earthquakes (seiches and tsunamis), meteorological surges and astronomical tides (Fig. 7.1). On this scale, the wavelength and period would increase progressively, as would the volume of water displaced by the wave.

In this chapter we will refer only to wind waves, which are understood as ordinary gravity waves, and only wind waves will be analyzed here. Other undulations of different origins are studied in other chapters of this book. So, tides are analyzed in Chap. 8, whereas extreme waves (storm waves, rogue waves and tsunamis) will be characterized in Chap. 13.

7.2 Genesis of Wind Waves

Although there are different genetic models, it is well known that most waves are generated by the transmission of kinetic energy from the moving air to the surface of a water mass thanks to the viscous friction between the two fluids. When the wind speed exceeds a minimum of energy, the first waves begin to appear [11], especially when in this wind there are fluctuations or vortices of energy. It is these fluctuations that generate the primary waves. These waves usually have an abrupt surface, as well as a very random and chaotic arrangement (Fig. 7.2, left). This morphology presents high friction to the action of the wind, without producing a movement of water rather than at the surface level.

J. A. Morales, *Coastal Geology*, Springer Textbooks in Earth Sciences, Geography and Environment,
https://doi.org/10.1007/978-3-030-96121-3_7

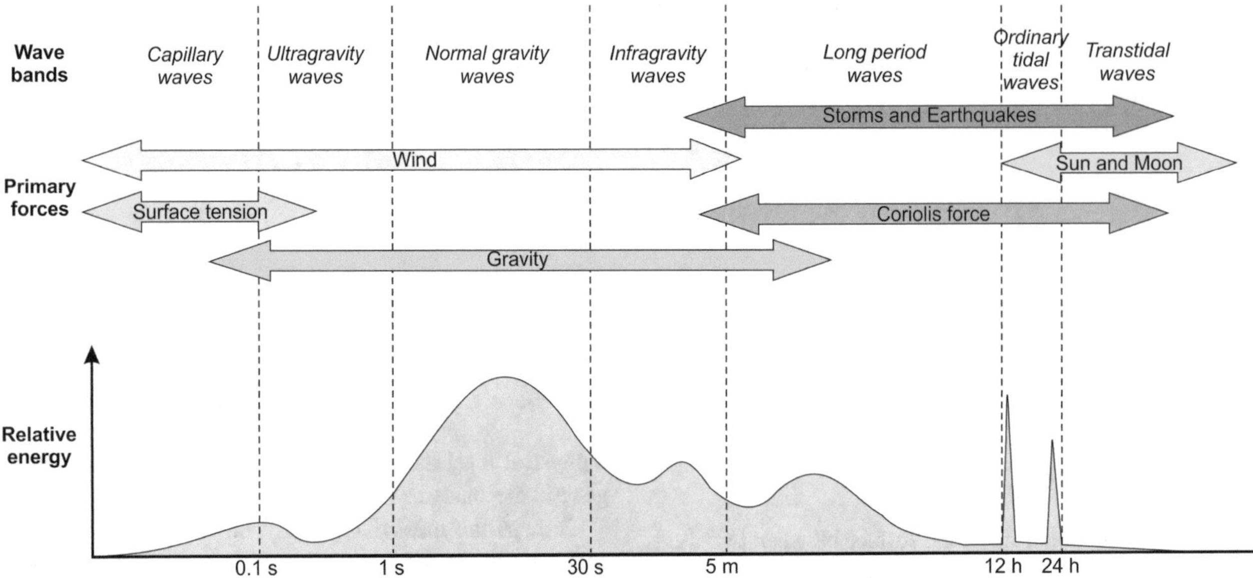

Fig. 7.1 Classification of sea waves according to their period. Adapted from Munk [9]

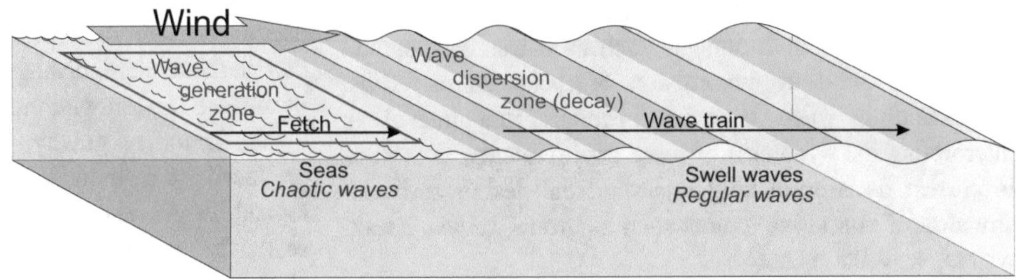

Fig. 7.2 Mechanisms of genesis and morphology of wind waves

Since these waves are directly related to the wind that is generating them, they are known as **waves in the generation zone** or by the name of **seas**.

Although it is the wind vortices that generate the irregularity in the sea surface, it is the action of the persistent wind speed in a single direction that pushes and moves the ridges to give them an orderly shape. In this way, the waves begin to move, ordering on wave trains and continuing to grow. As the waves move, they leave their generation zone and migrate to distant areas known as dispersal zones, propagation zones or decay zones. At this time, they grow in dimensions while at the same time becoming regular. In this process, the ridges are softened, adapting to a shape close to a sinusoidal function, thus decreasing the surface friction (Fig. 7.2, right). Propagation can be carried out through hundreds of kilometers. These types of waves that appear outside their generation zone are known as background waves or **swell waves**. The characteristics of the prevailing wind acting on them may be different from the generating wind, or there may even be no wind. The dimensions reached by these waves are based on the wind speed and the water surface on which it acts without changing its direction —a parameter known as the **fetch**.

Although the waves grow when regularized, this growth is limited by dimensions, as the heights and wavelengths must be in balance—the maximum ratio (steepness ratio) for the relationship between heights and wavelengths is 1/7. If the height exceeds the maximum elevation for its wavelength, a foam-shaped break occurs, which causes a loss of energy and maintains the height magnitude.

Linking these concepts, the surface of the sea on which a set of waves generated by the same wind spreads is known as the **wave field**, while the set of waves of similar dimensions being generated by the same wind is known as the **wave train**. The dimensions of the waves of the same train —i.e., genetically related—can be statistically analyzed through spectral analysis.

7.3 Morphology and Dimensions of Wind Waves

From a morphological point of view, it has already been said that the ridges and troughs of seas and swell waves have distinct morphologies. Usually seas have a trochoidal shape (with pointed ridges and rounded troughs), while swell

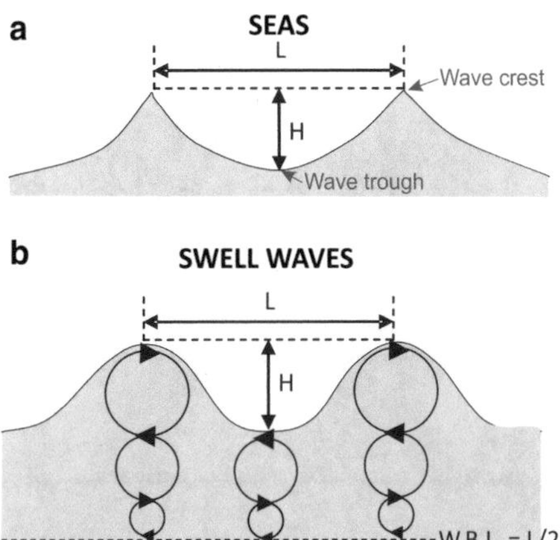

Fig. 7.3 Morphology of sea and swell waves and the main parameters characteristic of the waves

waves show sinusoidal morphologies (Fig. 7.3). The parameters that characterize both types of waves are the same as those involved in any undulatory phenomenon. In a characteristic wave, the elements characterized below can be distinguished.

- *Wave crest:* Higher level reached by the surface of the water at the passage of the wave.
- *Wave trough:* Lower level reached by the surface of the water at the passage of the wave.
- *Wave amplitude or wave height* (H): Vertically measured distance between the crest and trough of the same wave.
- *Wavelength* (L): Horizontally measured distance between two successive crests (or two troughs).
- *Wave period* (T): The time elapsed between the passage of two successive crests (or two troughs) by the same point.
- *Wave speed* (C): Distance traveled by the wave in a unit of time. It can be calculated through the ratio between the wavelength and the wave period.
- *Wave base level* (WBL): This is the depth at which a volume wave is able to move orbitally the water mass. It is set as half of the wavelength.

The crest and trough of the wave are morphological parameters, while the height and wavelength are known as dimensional parameters. Period, velocity and wave base level are dynamic parameters. In this regard, it must be borne in mind that swell waves are part of regular wave trains whose propagation responds to a periodic movement, so the dimensions of each wave do not depart too far from the average dimensions of the train. On the contrary, seas are irregular waves and their dimensions must be measured individually, without there being a relationship between the dimensions of the whole set.

In view of the movement of the water mass, the main difference between the two types of waves is that the sea wave is a surface wave that does not transfer movement to the mass of water in depth. However, the swell wave generates a movement of water particles in circular orbits until it reaches the WBL. Any coastal area can be affected at different times by the two types of waves, with one of them being able to dominate over the other during the time of action on the coast. It is interesting, then, to know the relationship between the two types of swell in a given coastal area.

Whatever the type of waves, as mentioned above, dimensional and dynamic parameters are usually set statistically for the spectrum wave set (Fig. 7.4). For this type of analysis, data obtained from measurement stations of different types, which have been described in Sect. 6.2.1, are used. The following parameters are often used to characterize the dimensions of the train:

- *Significant height* (Hs or $H_{1/3}$): This is the average height of the highest third of the waves in the spectrum. It is a measure that usually coincides with visual estimations and has been classically used by geologists to characterize the average energy of coastal environments and by coastal engineers to calculate the construction of structures.
- *Significant wavelength* (Ls or $L_{1/3}$): This is the wavelength corresponding to the greatest third of the major waves of the analyzed train. It is mainly used to calculate the depth of the wave base level, and is not a widely used parameter in coastal engineering.
- *Significant wave period* (Ts or $T_{1/3}$): This is the average time of passing between different ridges in the greatest third of the major waves of the train. It is a widely used parameter when determining the frequency of the waves acting on a given coast, i.e., the number of waves acting on a coast per unit of time.

In addition to the parameters of the significant waves, the mean wave parameters of the entire train and the parameters of the maximum wave of the train can be used, as well as other dimensions of interest:

- *Mean height* (H_m): Arithmetic average height of all waves in the spectrum.
- *Mean wavelength* (L_m): Average wavelength of all waves in the spectrum.
- *Mean period* (T_m or T_z): Average period of all waves in the spectrum. This parameter is often also referred to as a zero-crossing wave period.
- *Mean quadratic height* (H_{rms}): Obtained by calculating the square root at the mean of the heights. It is a

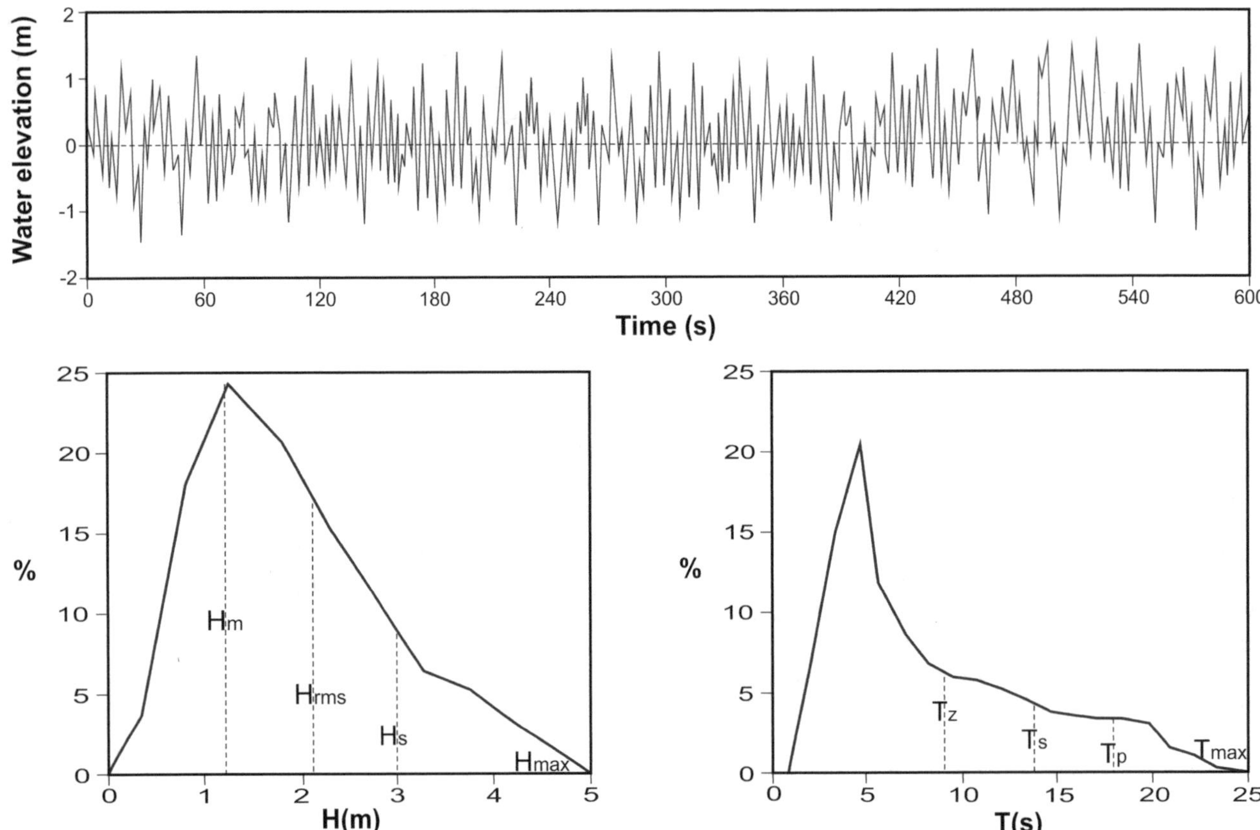

Fig. 7.4 Representation of the wave spectrum obtained in a season over a short period of 10 min and, below, curves of percentage of heights (H) and periods (T). The graphs indicate the statistical indices that are usually characterized on these dimensional parameters

parameter that is directly associated with the significant height through the ratio $H_s/H_{rms} = 1.41$.

- *Maximum wave height of the spectrum* (H_{max}): Corresponds to the height of the largest recorded wave.
- *Spectral peak period* (T_p): The peak period corresponds to the wave period that reaches the maximum energy level of the full wave spectrum. It is a widely used parameter in coastal engineering.

When performing a statistical analysis, it should also be taken into account that a wave spectrum can show a distribution of energy in different frequency bands, especially when it comes to the spectrum of a field of irregular waves, because this type of wave field often includes waves from several overlapping trains. This fact makes a wave spectrum more than one energy spike. Using a medium parameter as the only element to characterize the dimensions of a wave spectrum should be done with appropriate precautions.

In general terms, to avoid this problem so-called spectral analysis is performed. Keeping in mind that a wave record or irregular spectrum is composed of several harmonic wave trains of different frequency, a separation of these trains can be performed through harmonic or Fourier analysis. This analysis can be more easily performed by taking into account the propagation directions of wave trains (directional analysis) but it can also be performed in cases where the directions have not been recorded (scalar analysis). Actually, the application of spectral analysis equations is widely used in engineering, but it escapes the objectives of coastal geology, so they will not be detailed in this chapter. These equations, as well as all the theoretical development of the wave theory, can be found in the *Shore Protection Manual* [3].

Advanced Box 7.1
Studying a Wave Record

The characterization of a wave train whose registration has been obtained through any of the methods described in Sect.6.2 can be carried out dimensionally and also directionally. For dimensional study the record is usually divided into regular periods (of one or several hours). The full spectrum of that period can be represented in a time–height curve. For each time interval, the heights and periods will be represented in the form of frequency curves or histograms. These graphs represent the dimensional parameter on the *x*-axis, while the *y*-axis represents the percentage at which those values have been reached, taking as the reference value for the calculation the total number of waves, similar to what is shown in Fig. 7.4.

For each time interval, the dimensional parameters are statistically determined, commonly using the significant height and period as the most representative of the train, although the maximum height reached during that period is also used.

The significant height and period values of each time interval can be represented on time in monthly or annual spectra (Fig. 7.5). These are very representative of the temporary distribution of wave dimensions in the medium term. The data thus obtained are also used for statistical frequency analyses, in order to characterize the dimensions of the annual average swell. Analysis of the dimensional ratios between heights and periods can also be performed from this data.

Once the average and maximum annual data have been obtained, year-on-year comparative analyses are also carried out. This results in drop probability graphs and calculations of return periods of waves of certain dimensions.

For directional characterization, wind rose diagrams are used. These diagrams represent the dimensions of the waves in the form of columns of different width, oriented according to their direction of origin (Fig. 7.6a). As measurement stations are normally located in sectors far from the coast, waves from all directions are usually recorded. From this you can eliminate the ground-facing directions, leaving only the directions of the waves approaching the coast. In this type of study, a dimensional characterization can also be performed according to the provenance. This characterization will follow the steps marked in the previous paragraph and can end with a distance probability diagram for each of the directions (Fig. 7.6b).

7.4 Wave Energy and Power

In previous sections it has been mentioned that movement transmitted by the wind to the waves in its generation zone is not transmitted in depth, and therefore the movement of the water mass is limited to the surface. However, regular waves displace a larger volume of water in an orbital motion of particles that extends deep to half their wavelength. Either way, this orbital movement, as well as the spread of the wave in space, causes energy transfer in the shallowest layers of the seawater mass. This energy has two main origins. On the one hand, the waves have a potential energy associated with the elevation and descent of the water surface from an initial flat level (i.e., sine/wave crest). On the other hand, the orbital movement of water particles confers a velocity to the fluid particles that gives the phenomenon a kinetic energy. The sum of both energies is proportional to the wave height and water density according to Eq. (7.1).

$$E = 1/8\left(\rho g H^2\right) \tag{7.1}$$

In this equation, ρ is the density of water and g is the acceleration of gravity.

One of the consequences of this equation is that wave energy depends on the square of height. This means that an increase in height to double will result in a four-fold increase in energy. On the other hand, it is necessary to note that this concept actually corresponds to the amount of energy per unit area as it is measured in J/m^2, so it is also known as energy density. Thus, the total energy associated with a long-period wave is greater than that of a short-period wave because the long-period wave has a longer wavelength and is distributed over a larger surface area.

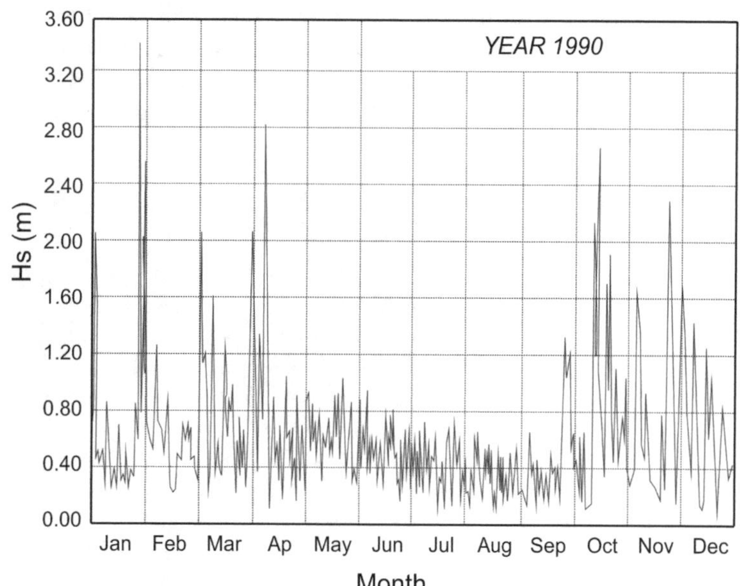

Fig. 7.5 Example of a wave spectrum during the year 1990 for a SW wave measurement buoy from SW Europe with data averaged in three-hour intervals

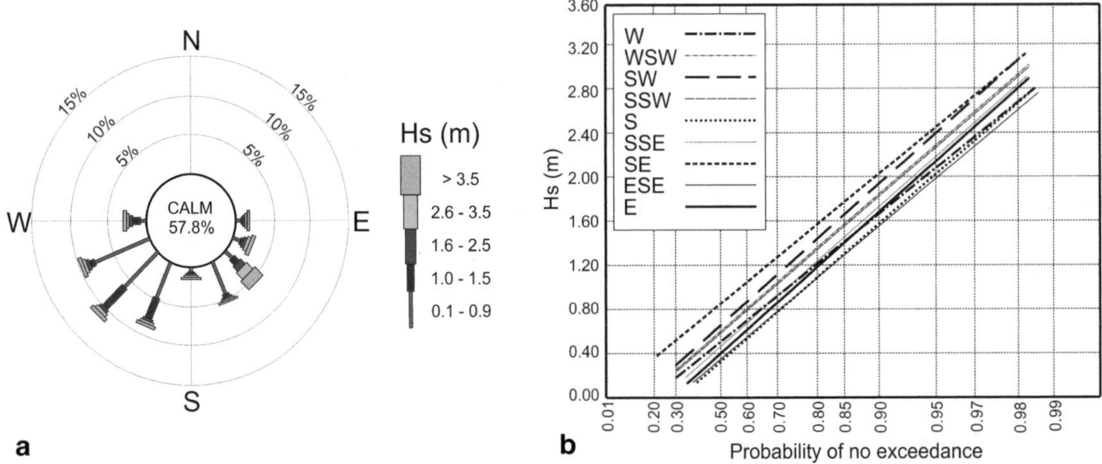

Fig. 7.6 Example of directional wave characterization. **a** Wind rose diagram. **b** Graph of exceedance probability

Another energy expression that is commonly used is energy flow (P) or wave power. This is expressed according to Eq. (7.2).

$$P = C E \tag{7.2}$$

In this equation C is the propagation velocity and E the total energy.

In this case it should be noted that C is the propagation velocity of the entire wave train, which may be different from the propagation speed of the individual waves.

7.5 Sea State and Dimensional Scale of Waves

The general conditions of the sea surface at a given place and time according to the absence or presence of waves is known as the state of the sea. This state is characterized according to the magnitude of the dimensional parameters (height, wavelength and period), as well as the state of energy and power. Being characterized by measurable parameters, the state of the sea can be accurately determined from direct wave measurements or also visually.

In 1920 British Navy Captain Henry Percy Douglas created a simple scale for the naming of sea states according to the rugosity of the sea, as a visual reflection of the height of the waves. This scale was quickly accepted by the international community, today being called the Douglas sea scale, also known as the "international sea and swell scale." The scale is simple and consists of ten values, listed from 0 to 9, accompanied by a description that can be applied in a very intuitive way (Table 7.1).

7.6 Wave Propagation and Shoaling

Previous paragraphs have described the propagation of waves outside their generation zone as a movement of wave trains, which can travel hundreds of kilometers on open ocean surfaces. This propagation involves a wave ordering, as well as an orbital movement of water beneath the surface to a depth that equals half the wavelength. According to

Table 7.1 International sea and swell scale suggested by Douglas in 1920

Degree	Average height (m)	Description
0	0.00	Calm
1	0.00–0.10	Rippled
2	0.10–0.50	Smooth
3	0.50–1.25	Slight
4	1.25–2.50	Moderate
5	2.50–4.00	Rough
6	4.00–6.00	Very rough
7	6.00–9.00	High
8	9.00–14.00	Very high
9	+14.00	Phenomenal

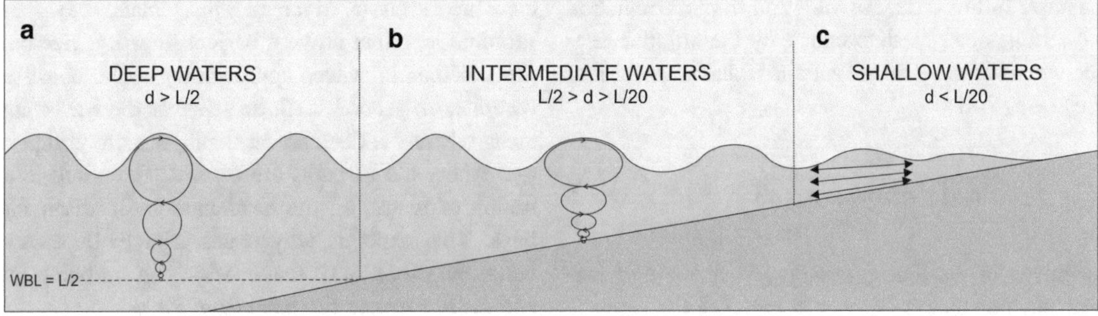

Fig. 7.7 Movement of water particles at the passage of the swell (d = depth; L = wavelength). **a** In deep water. **b** In intermediate waters. **c** In shallow waters

wave theory [1, 11], in deep waters, where the base level of the waves is above the bathymetry (d > L/2), the swell does not interact with the bottom and the orbital movements are circular, with orbits that become minor in depth (Fig. 7.7a). In this case, wave trains move at constant speed and the wave maintains its dimensions as it spreads.

As wave trains enter coastal areas with depths below the base level (d < L/2) an interaction with the bed begins. This process of propagation to shallower waters is known as shoaling. Interaction with the bottom results in a deformation of orbital movements, which transform into ellipses, where height movement tends to decrease faster than horizontal swing movements (Fig. 7.7b). The loss of depth makes the ellipses more and more eccentric (flatter and elongated). The waters where this depth decrease occurs are known as intermediate waters. In the depths where the orbital movement speed exceeds the threshold of particle movement, a portion of the wave energy begins to be transferred to the bed.

The process of elongation of the ellipses culminates in the shallow waters, where the ellipses finally disappear. Then, the vertical movements are removed and only the horizontal movements of swing remain (Fig. 7.7c). In general, it is estimated that shallow waters where this phenomenon occurs are below one 25th of the wavelength (d < L/25).

Interaction with the bed also results in a decrease in the speed of the waves from a depth less than the WBL. This decrease in depth follows the rules set out by Knauss [8], so that in each type of water the equation governing the speed of the wave responds to different functions and variables.

In deep water, where the depth is greater than half the wavelength, the velocity is regulated by Eq. (7.3). So, the speed of the waves depends simply on the acceleration of gravity and the wave period.

$$\mathrm{C} = \frac{gT}{2\pi} \tag{7.3}$$

where C is the speed of the wave, g is the acceleration of gravity and T is the wave period.

In intermediate waters, when the interaction with the bottom begins, the velocity becomes dependent on the relationship between gravity and wavelength with depth, through a more complex hyperbolic trigonometric function, expressed by Eq. (7.4).

$$\mathrm{C} = \sqrt{\frac{\mathrm{gL}}{2\pi}\tanh(\frac{2\pi d}{\mathrm{L}})} \tag{7.4}$$

where L is the wavelength and d is the depth.

The fact that in intermediate waters the velocity becomes directly dependent on the depth, even though a complex function, implies that a decrease in depth necessarily leads to a slowdown of the waves.

This dependence is further increased on shallow waters where speed becomes a direct function of depth in a simpler way, according to Eq. (7.5).

$$\mathrm{C} = \sqrt{\mathrm{gd}} \tag{7.5}$$

Shoaling directly results in a deformation of wave trains that are subject to depth loss according to two common effects of wave propagation: refraction and diffraction.

7.6.1 Wave Refraction

Since the wave speed is a ratio between wavelength and period, the progressive decrease in the speed of approach to the shoreline results in a decrease in wavelength [1]. When the direction of the wave train approaching the coast takes place in an oblique direction to the shoreline, the ridges begin to experience a greater approach in the shallower area than in the deeper area. You could then differentiate segments of the same crest of waves traveling to different depths and therefore at different speeds. In this way, the ridges of the tails take on a curved shape. This phenomenon is known as **wave refraction** and as such is regulated by principles complying with Snell's Law, which characterizes the change in speed of a wave when it passes through environments of different nature. In this case, the variation in the orientation of the wave ridges can be determined by the angular relationship between the wave ridges and the isobaths, according to Eq. (7.6) (Fig. 7.8).

$$\frac{d1}{\sin \alpha 1} = \frac{d2}{\sin \alpha 2} = \frac{d3}{\sin \alpha 3} \tag{7.6}$$

In a segment of rectilinear coastline with parallel isobathic lines, such as the one shown in Fig. 7.8, the result is that the wave ends up reaching the coast in an almost parallel way and, of course, at a much lower angle than the initial approach angle. However, in nature there are coastal stretches that are far from straight and with parallel isobathic lines. In fact, it is common that the presence of sandy shoals, bars and rocky elements distort the linearity of bathymetric curves at the bottom. Similarly, coastlines usually contain elements such as river mouths, inlets, bays, capes and promontories that prevent perfect linearity. Because of this, the schemes of waves approaching to the coast tend to be complex. In general, in these schemes the waves approach in areas where the decrease in depth is more abrupt and separate where the isobaths are distant. The result is a concentration of waves on the headlands and a dissipation in the bays. This explains why capes tend to be erosive zones, while the products of that erosion end up being transported and sedimented in the bays (Fig. 7.9).

Either way, in a coastal stretch where a bathymetric map is available, a wave refractive scheme can be built from the initial approach direction and by applying Snell's Law to each point of the bathymetry. The refractive schemes thus obtained are very useful for understanding coastal dynamics in open areas and are widely used by coastal geologists and engineers in areas where coastal works are projected.

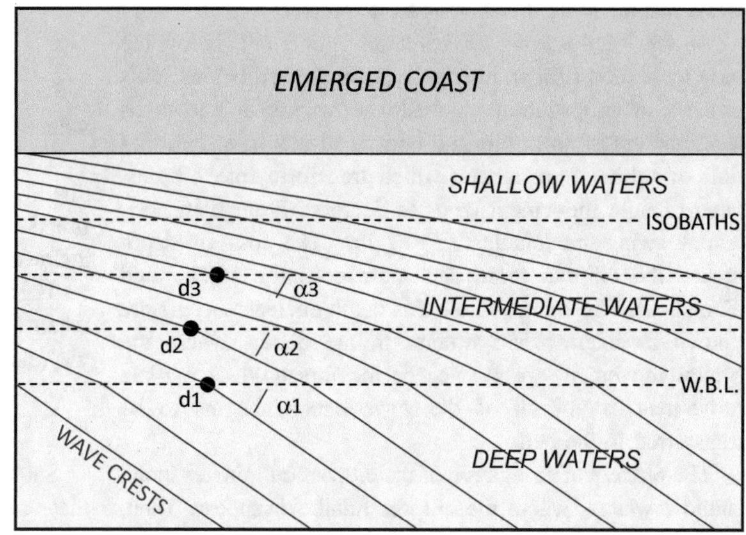

Fig. 7.8 Wave refraction by shoaling and the application of Snell's Law, where d1, d2 and d3 are the depths marked by isobaths and α1, α 2 and α3 are the values of the angles between the isobaths and the wave ridges

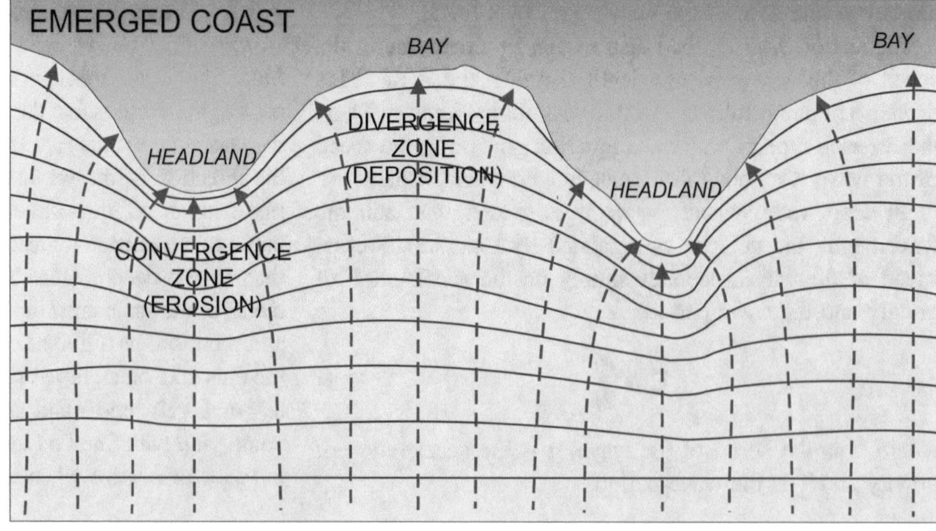

Fig. 7.9 Example of a wave refraction scheme in a zone of headlands and bays

7.6.2 Wave Diffraction

Diffraction is a phenomenon that occurs as a result of interference from a wave with an obstacle that stands in its way. When this occurs, after this obstacle there is a shadow zone to the swell. The obstacle causes a deformation of the wave that is incurved to access the shadow sectors behind it, so that a kind of envelope takes place (Fig. 7.10). The phenomenon is associated with a lateral transfer of energy at the end of the crest of the wave after the obstacle, extending to the shadow zone [7]. To understand this in an easy way, it is often said that by diffraction the waves "bend around corners."

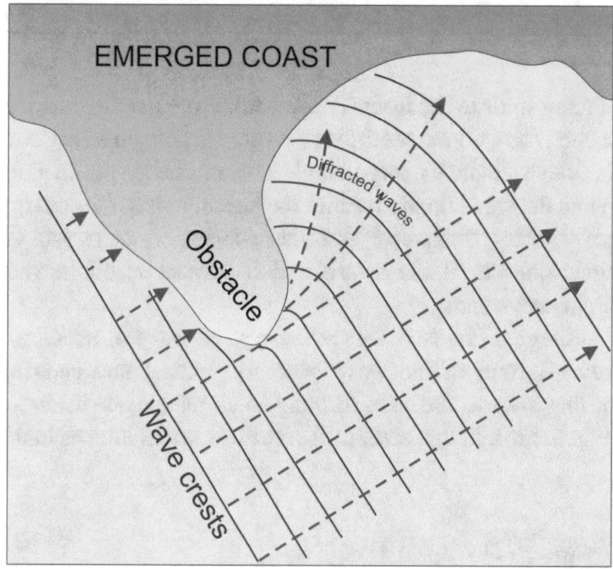

Fig. 7.10 Example of a wave diffraction scheme

This phenomenon allows the wave trains to enter restricted areas of the coast, either naturally, as restricted bays or lagoons, or artificially, as ports. For its effects on the coast, diffraction is a phenomenon inevitably linked to refraction, and wave propagation schemes in coastal areas are the joint result of both phenomena. An example can be seen in Fig. 7.11.

7.6.3 Littoral Drift

The oblique arrival of wave trains to the coast due to refraction and diffraction phenomena results in a zigzag movement in the sediment particles of the bed along the coastal fringe that is topographically above the WBL (intermediate and shallow waters). Each wave arrives with a direction whose transport vector acts perpendicular to the crest of the wave. During the swash the sediment particles are transported at a forward slope, while when the backwash occurs it does so perpendicular to the shoreline. The two components of the wave, swash and backwash, result in a movement of the particles parallel to the coast in the sense where the angle between the crest of the wave and the coastline opens. This results in the appearance of a transport component parallel to the coast known as **littoral drift** or **longshore transport** (Fig. 7.12). So, littoral drift (or longshore transport) is the term used to define the transport of sediment along the frontal area of the coast caused by the oblique action of breaking waves.

The magnitude of longshore sediment transport becomes of more or less relative significance depending on the magnitude of waves and the angle of incidence of wave trains. The variability over time of the incident waves causes

Fig. 7.11 Example of propagation scheme with refraction and diffraction acting together. **a** Original waves. **b** Diffracted waves. **c** Interference of two diffracted wave systems. **d** Diffracted and refracted waves. (Image Landsat/Copernicus from Google Earth.)

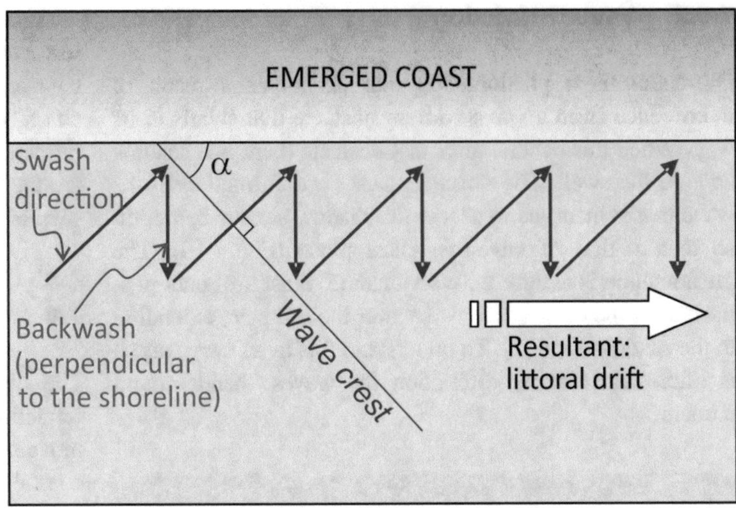

Fig. 7.12 Generation of littoral drift by the arrival of waves oblique to the coast

oscillations in the drift transport, both in its magnitude and in the sense of the same. The net balance of material transported will depend on the frequency in which the waves of different directions and dimensions act.

7.7 Dissipation of Wave Energy

7.7.1 Wave Breaking

It has been described how, in shallow waters, there is a dissipation of energy due to interaction with the bed. On the one hand, there is a deformation of the orbits, which become excessively eccentric causing the waves to lose height. On the other hand, the ridges approach each other so that the wavelengths are shortened. The relationship between the decrease in height and wavelength is of paramount importance, because if a factor of 1/7 is exceeded the wave breaks.

The breaker is manifested in the appearance of foam on the crest of the wave that forms when the orbital velocity of the wave crest exceeds the propagation velocity of the wave. It is clear that in coastal areas it occurs because the wave is slowed when interacting with the bottom, while the speed of the crest is maintained despite the deformation of the motion orbits. Once the breaker starts, it is maintained during the wave propagation until the swash reaches the shoreline. This breaker propagation area is known as the surf zone.

The length of the surf area is variable according to the dimensions of the waves in relation to the nature of the bottom and the slope of the coast. How this break occurs can also vary from one area to another. Most authors describe three different types of breakers: spilling, plunging and surging. In addition, other authors (e.g., Galvin [4]) include a fourth type: collapsing (Fig. 7.13).

Spilling breaker: Characterized by a progressive increase in wave height as its wavelength decreases until a soft crest of foam spills to the front (Fig. 7.13a). From the moment this occurs, the spill moves through a surf area of some length as the wave gradually loses height due to energy dissipation. When the wave finally reaches the beach, most of the energy has already dissipated and the wave has decreased its dimensions until the swash and backwash result in very slight movements.

Plunging breaker: This is the type of wave in which the forward front of the wave becomes vertical and ends up curling towards the base, forming an air tube inside the wave (Fig. 7.13b). In this case, the surf area is not as long as in the

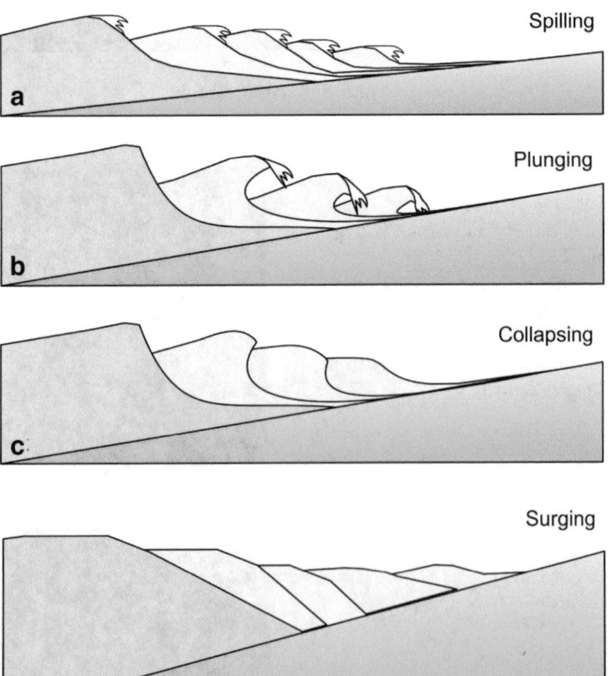

Fig. 7.13 Different types of wave breaker Galvin [4]

previous example and the wave energy dissipates into a much narrower fringe, in which the backwash between waves is the dominant movement. This is the most iconic type of breaker and, of course, is preferred by surfers.

Surging breaker: In this type of breaker, the crest of the wave remains with a relatively mild sinusoidal shape so that it reaches the coast without forming true breakers (Fig. 7.13c). The dissipation of energy in this type of break is minimal and usually a new wave that moves in the opposite direction is formed by reflection.

Collapsing breaker: In this case, the wave front also becomes vertical because the base of the wave is sharply slowed. Unlike the enveloping swell, the collapse of the top of the wave occurs in an abrupt and turbulent way, without forming a curl (Fig. 7.13d). Actually, in his definition Galvin characterizes it as a transitional type between plunging and surging.

The type of breaking wave has direct consequences on the erosional or depositional conditions of the beaches. The spilling breaker is dominated by swash, so produces landward sediment transport and is therefore a break that tends to deposit sediment. On the other hand, in plunging and collapsing type breakers, the backwash component and therefore the erosion predominate. Finally, reflection predominates in surging and this is a breaker that also generates very erosional conditions.

A factor of great importance when studying the interactive relationship between the waves and the coast is the slope of the coast, since the slope conditions the horizontal proximity of the first interaction of the wave on the bed. So, it is the main factor for the gradient of energy loss of the wave in its break, as well as the length of the surf area. The steeper or slighter slope of the beach directly conditions the type of breaker and has direct consequences on the direction in which the discharge of energy occurs on the sedimentary material of the beach. In this sense, Galvin [5] proposes the use of a breaker index (B), related to the dimensional parameters of the wave (height and period) and the slope of the beach according to Eq. (7.7).

$$B = \frac{\mathrm{H}}{\mathrm{gmT}^2} \tag{7.7}$$

where B is the breaker index, g is the acceleration of gravity, H is the wave height, T is the period and m is the beach slope.

Thus, for B values higher than 0.068 a spilling breaker will form; for values between 0.003 and 0.068 a spilling breaker occurs; finally, for B values less than 0.003 a surging or collapsing breaker will be formed. There are other equations that have tried to use these same parameters to narrow the genesis conditions of each of these breakers. A comparative synthesis of these equations was made by Camenen and Larson [2]. This work shows that, while some equations work better than others under certain conditions, they all use the ratio between slope and height as a defining variable. One of the consequences that can be derived from the use of these indices is that an increase in slope has the same effect as an increase in wave height. On the other hand, a wave of certain dimensions may present different breakers on beaches of different slopes. This explains the different erosive–cumulative behavior of beaches with different slopes in front of the same waves.

7.7.2 Wave Reflection

There are coasts whose bed morphology causes the waves to reach the coastline without dissipating its energy at the bottom, or with a minimal dissipation. In this case, the swell is reflected on the coast, forming trains that return to the sea. This phenomenon is associated with very high sloped coasts

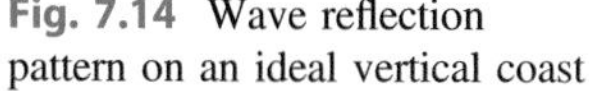

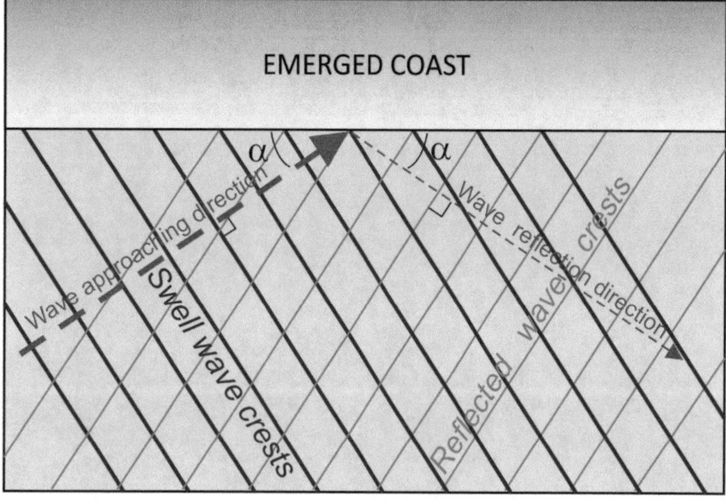

Fig. 7.14 Wave reflection pattern on an ideal vertical coast

or nearly very vertical areas such as cliff systems or artificial walls. In all cases, reflection is also associated with surge breakers. The direction of the reflected waves depends on the angle between the incident wave trains and the coast at which they break (Fig. 7.14). In general, wave reflection follows the laws of any kind of undulatory reflection.

Actually, the phenomenon of reflection is also present on lower sloped shores, although in these cases the reflected waves have much smaller dimensions than the incident waves, having dissipated most of their energy in the breakers. In these cases, although the phenomenon is not visible, it can be measured through the interaction between reflected waves and the incident swell.

References

1. Airy GB (1845) On tides and waves. Encyclopaedia Metro 5:241–396
2. Camenen B, Larson M (2007) Predictive formulas for breaking depth index and breaker type. J Coastal Res 23:1028–1041
3. CERC (1984) Shore protection manual. Coastal Engineering Research Center. US Army Corps of Engineers, Washington DC
4. Galvin CJ (1968) Breaker type classification on three laboratory beaches. J Geophys Res 73:3651–3659
5. Galvin CJ (1972) Wave breaking in shallow water. In: Meyer RE (ed) Waves on beaches. Academic Press, New York, pp 413–456
6. Gerstner FJ (1802. *Theorie der Wellen samt einer daraus abgeleiten Theorie der Deichprofile*. Abhand der König. Bömischen Gesel. d. Wiss. Prague
7. Iribarren R (1964) Obras marítimas. Oleaje y diques. Dossat, Madrid, 376 pp
8. Knauss JA (1997) Introduction to physical oceanography. Prentice Hall Inc.
9. Munk WH (1951) Origin and generation of waves. In: Proceedings of the conference of coastal engineering, vol 1. ASCE
10. Stokes GG (1847) On the theory of oscillatory waves. Trans Cambridge Philos Soc 8:441–455
11. Tricker RAR (1964) Bores, breakers, waves and wakes: an introduction to the study of waves on water. Mills and Boon, Cambridge, 250 pp

8 Tide Processes

8.1 Introduction

The sea surface, far from remaining static, experiences periodic oscillations related to the gravitational pull of the orbiting spheres. These oscillations constitute the phenomenon called the tide. The tide behaves like an oceanic wave; in fact, as mentioned in the introduction to Chap. 7, tidal waves are the waves with the longest wavelength. As a directly observable phenomenon, the tide consists of a cyclic ascent and descent of the sea surface. This movement is divided into two semi-cycles called rising tide and falling tide (Fig. 8.1). Each semi-cycle is limited by a moment of maximum level called high tide and a minimum level called low tide. Like other sea waves, tidal waves propagate by generating currents called flood and ebb.

Knowledge about and observation of the tides have been rooted in coastal towns since ancient times, although there are few records of the phenomenon in the classical texts, since the Greek populations settled on tidal-free coasts. However, there are several significant citations based on the contacts between Mediterranean civilizations and the peoples of the southwest of the Iberian Peninsula.

Pythagoras was possibly the first to describe the phenomenon and establish its cycles mathematically, although without reaching conclusions about its origin. The first early reflections on origin were made by Pliny the Elder in the first century of our age. Thus, this Roman sage correctly attributed the cycles to the joint action of the Moon and the Sun.

Despite these early observations, it was not until the seventh century that it was concluded that this was a predictable phenomenon and for the first tidal calendar to be drawn up. It is not surprising that it was a Benedictine monk who carried out these studies, after 19 years of thorough observations—the Venerable Bede knew Latin and Greek and was familiar with the texts of classical authors. His researches on the tides are collected in his work *De natura rerum*. The Anglo-Saxon friar's work had direct application in his own country six centuries later, when the first tide tables were recorded at London Bridge.

The strength of the tides has surely been known and exploited since ancient times, however, with the first to describe practical applications of their use being Islamic authors. At the end of the tenth century, Al Mohaddasi was the first to describe how to obtain a mechanical performance from the tidal force, by applying it to the movement of flour mills. Scientific studies about the tides culminate in the statement of dynamic tide theory, which we will discuss later.

The action of tides has a remarkable influence on the coasts and there are coastal environments clearly dominated by the action of the tides. However, only 27% of the time are tides able to transport sandy sediment [12], so tide-dominated coasts are classically associated with the deposit of lutitic materials. Among these environments that are heavily influenced by tides are marshes, tidal flats, inlets and tidal deltas, lagoons, estuaries and deltas. In order to understand the dynamic functioning of these environments, it is necessary to understand in depth the phenomenon, as well as its action on the coast, which differs from the dynamics in the open sea.

8.2 Genesis of the Tides

The genesis of tides is related to the gravitational action of the Moon and the Sun on the mass of oceanic water and is a well-known and studied phenomenon that is stated in the "theory of tides." The mathematical foundations of this theory were firmly established by Pierre-Simon de Laplace in 1778. Laplace described the equations that govern the dynamics of fluids in the mass of ocean water. The French physicist based his equations on the principles of universal gravitation established by Newton almost a century earlier, although he modified his formulation by introducing the strength of the Coriolis effect, and did so 60 years before the formal definition by Coriolis.

The phenomenon of tides can be understood intuitively if we look at the relative position with respect to the Earth of the

J. A. Morales, *Coastal Geology*, Springer Textbooks in Earth Sciences, Geography and Environment,
https://doi.org/10.1007/978-3-030-96121-3_8

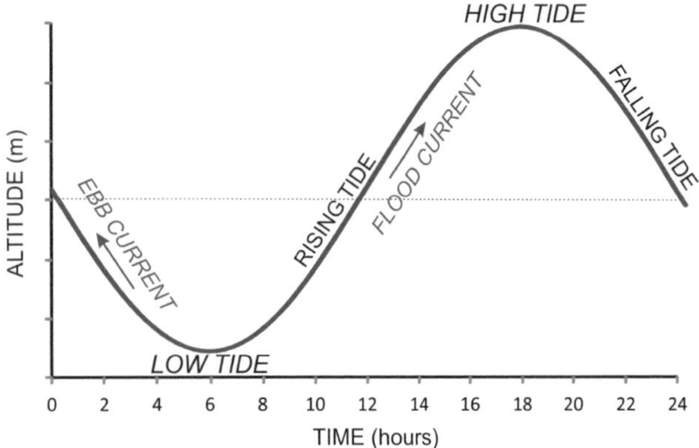

Fig. 8.1 The elements of a tidal wave

spheres that generate the gravitational forces that give rise to the variation of the surface of the sea: the Moon and the Sun.

8.2.1 Earth–Moon System

It is intuitive to understand that the attraction of the Moon draws water towards it, forming a bulge of the sea surface in that direction. However, the fact that water also accumulates on the opposite side requires a more detailed explanation. The Earth and Moon form a simple system of action and reaction with a common center of mass. This center of mass is known as the barycenter, and it approaches the center of the Earth but does not match it. The rotation around the barycenter causes the Earth to rotate eccentrically, generating a centrifugal force on the face farther from the Moon and acting counterclockwise to the sphere's force of gravity (Fig. 8.2). This is the force that generates the elevation of the sea on the opposite face of the lunar position.

These two forces (Moon gravity and centrifugal force) are responsible for two opposite bulges on the sea surface, one facing the Moon and one right on its opposite side, and two perpendicular depressions. The diurnal motion of the Earth causes each point of the planet to pass each ridge and depression once, generating two high tides and two low tides daily. These theoretical level variations are known as lunar tides. The different magnitude of the forces of lunar and centrifugal gravitational pull causes an inequality between the two daily lunar tides.

The Moon's rotation plane around the Earth does not match the normal plane of the axis of rotation of our planet, but both planes form an angle of 28º which is named **lunar declination**. This plane itself undergoes a variation in time, making the lunar declination vary in cycles of 18 years and 11 days. This phenomenon causes the maximum and minimum levels of lunar tide to vary over a long time on a given coast.

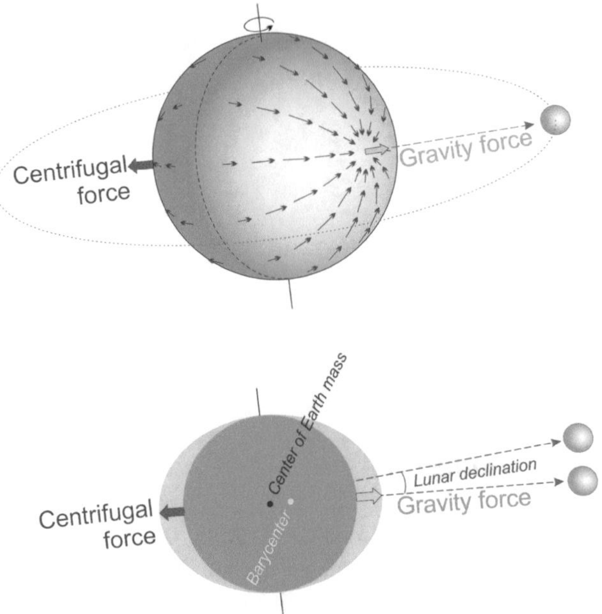

Fig. 8.2 System of forces that generates lunar tides

8.2.2 Earth–Sun System

Like the Earth–Moon system, the Earth–Sun system has a common center of mass that, in this case, is located near the center of the Sun. This also generates a centrifugal force on Earth in the opposite direction to the gravitational action of the Sun and a deformation of the surface of the water in the same direction. In this case, the action of this force is understood in a more intuitive way if we imagine the deformation of a balloon full of water turning tied to a rope. These forces generate two daily high tides and two low tides, which are called solar tides. Although the mass of the Sun is much greater than that of the Moon, its gravitational force on Earth is 2.73 times less due to the great distance that separates them, thus the solar tides are smaller than the lunar tides. There is also an angle of variation between the planes of Earth's rotation and the orbit around the Sun. This angle is known as solar declination.

8.2.3 Earth–Moon–Sun System

The rise and fall of the sea surface are actually controlled by the sum of lunar and solar tides, so that at each point of the ocean a force resulting from the gravitational and centrifugal action generated by both celestial bodies is obtained. When the three spheres act in conjunction, there are a number of relative positions between them that will cause a sum or subtraction of acting forces on the marine level to generate resultant tides. In this way, two extreme positions can be distinguished: in alignment (syzygy) and at right angles (quadrature).

When the three bodies are aligned they are in syzygy (Fig. 8.3a). This situation occurs during the days of the new moon and full moon. At these times, there is an addition of forces when the solar high and low tides coincide in space with the lunar high and low waters, giving rise to the **spring tides**. Conversely, when the three bodies form a right angle, they are in quadrature (Fig. 8.3b). This is the situation that occurs in the first quarter (waxing quarter) and in the third quarter (waning quarter). In both cases the forces are countered. The resulting high waters are less high because they are only generated by lunar high tide, while the low waters are less low because the lunar low tide coincides with the solar high tide. In this case, the **neap tides** occur.

The conjunction of the different relative positions between the three spheres results in the presence of biweekly cycles in the temporal distribution of tidal ranges, alternating two spring tides and two neap tides in a month, with the consequent transitions between them, which correspond to the transit between the different described positions. These relative positions actually result from an excessive simplification, since the planes of rotation of the Moon around the Earth, and this in turn around the Sun, differ between them and with the equatorial plane. The angles between these planes also undergo periodic variations.

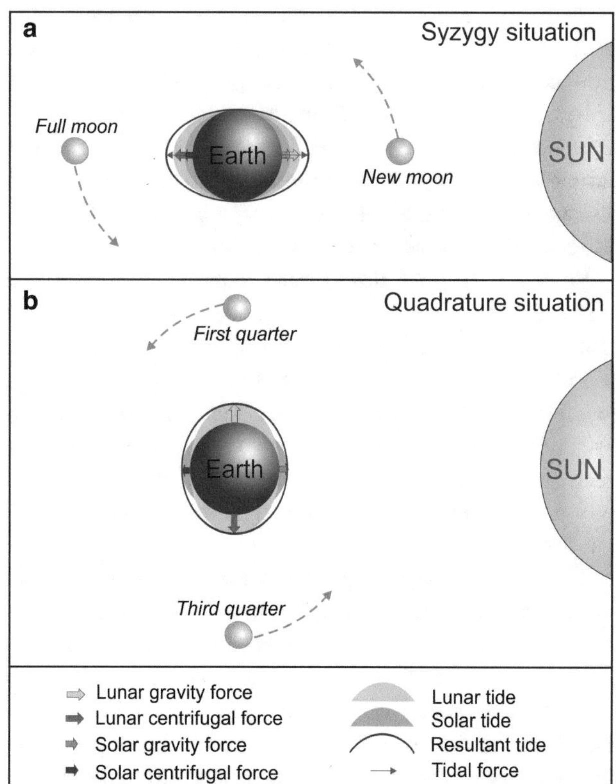

Fig. 8.3 Biweekly cycles induced by the relative Earth–Moon–Sun positions

There is also an influence that has not been mentioned above, and that is the influence of the distance between the spheres. The orbits of the Earth around the Sun and the Moon around our planet are elliptical. That causes the distances between each pair of spheres to vary over time.

In the case of the Moon around the Earth, the maximum distance is known as the **apogee**, and the minimum distance as the **perigee**. The difference between the apogee and perigee is 13% of the average distance between the spheres. Keep in mind that there is a turn of the Moon around the Earth passing through these two positions every 27.6 days (a lunar month).

In the same way, the orbit of our planet around the Sun also passes through two positions of maximum and minimum distance. In this case the maximum distance is known as **aphelion** and is reached in July, while the minimum distance is known as **perihelion** and is reached in January. As the mean distance is much greater, the differences between aphelion and perihelion are only about 4% of it. It takes 365.25 days to travel through the entire orbit (a year).

The concordance between the orbital cycles of the three spheres influences the magnitude of the solar and lunar tides and, therefore, the concordance between these tides in the cycles of spring and neap tides. This is reflected in the existence of six-monthly cycles where differences between spring and neap tide ranges vary between apogee, perigee and their intermediate positions. The result is the alternation over a year between two **solstice tides** and two **equinox tides**. Equinoxes occur during the apogee and perigee positions (end of March and September). In those months, there is a very marked difference not only between spring and neap tides but also between their two springs and their two neaps. Solstices, on the other hand, appear in intermediate orbital positions (end of December and June). So, the differences between their two spring tides and two neap ones are very small.

8.2.4 Dynamic Theory of Tides

In the seventeenth century Laplace established the equations that would theoretically follow the displacements of the ocean water mass due to the forces described above. Other hydrodynamic calculation equations concerning the deformations of fluid masses were drawn up almost at the same time by Bernoulli and Euler. The application of these equations together culminated almost a century later with the statement of William Thomson (Lord Kelvin), who modified the conceptions of attraction and centrifugal force expressed in the previous sections, taking into account the friction of

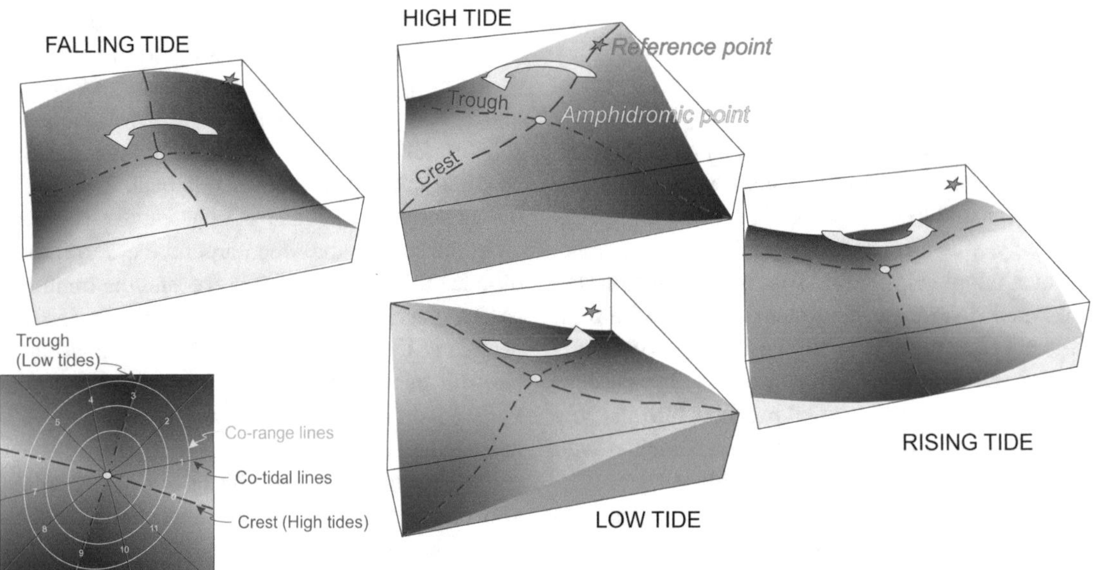

Fig. 8.4 Functioning of an amphidromic system. Adapted from von Arx [18]

water with the ocean floor and the distribution of continents on Earth.

In the simple models explained in the previous sections, the rotating Earth's mass seems to rotate freely inside a mass of water deformed by astronomical forces, and yet this assumption departs greatly from reality. On the one hand, the mass of water is in friction with the ocean floor in its relative displacement. On the other hand, the existence of the continents prevents the free passage of the ocean water mass in the rotation of the Earth. Actually, if we change the reference system and consider that it is the mass of water that moves over the oceans, its displacement would reproduce an undulatory phenomenon.

The dynamic theory of tides is based on the fact that the tidal wavelength is too large for the depth of the oceans. Considering a wavelength of several tens of kilometers, we should have an ocean more than 10 km deep so that there is no friction with the bottom; however, the average depth of the ocean is considerably less. Therefore, the tide has such a large wavelength that, when moving, it necessarily suffers friction with the bottom. If we use the equation of the propagation of a wave in shallow water (E. 7.5), it turns out that for the maximum depth of the ocean the highest possible speed for the displacement of the tidal wave is 230 m/s. This speed is much lower than the Earth's rotational speed, causing a delay relative to the position of the maximum lunar gravitational pull point. On the other hand, the presence of continents limits the rotation of the water prism and prevents the passage from one ocean to another.

In short, the speed of rotation of the Earth is too fast for the inertia of the water mass. That is the foundation for the existence of the Coriolis force, whose effect Laplace introduced into his dynamic theory. In the face of this force, the oceanic system responds by generating different rotation systems, known as **amphidromic systems**. In each of these systems the tides rotate around a tidal-free central point called an **amphidromic point** (Fig. 8.4).

Those rotating waves are often called **Kelvin waves**. Each system experiences a rotation that completes every 24 h and 50 min. Around this central point the tidal ranges grow concentrically defining **co-range lines** that join points of equal tidal amplitude. Normally, meter-to-meter lines are represented, where the value 0 corresponds to the amphidromic point, also called the **nodal point**. According to this scheme, the amplitude of the tide on a given coast depends on its distance to the corresponding nodal point.

Similarly, **co-tidal lines** define radially which tide is in phase. In this case, 12 numbered lines from 0 to 11 are usually represented, the order of which indicates the direction of rotation of the tides around the nodal point. The most significant co-tidal lines are those that join the minimum points (furrow) and the maximum points (ridge). The rotational passage of the ridge and furrow lines of the sea surface by a point of the coast coincides with the moments of high and low tides.

There are 15 major amphidromic systems in the world's oceans as well as other minor systems in inner seas such as the North, the Mediterranean and the South China Seas. Larger systems and their sense of rotation are represented in Fig. 8.5. This figure also depicts the distribution of tidal ranges on the coasts. You can see the relationship between the tidal range on the coast and the distance from the coast to the amphidromic point that controls the tides on that coast, although there are some discrepancies that are based on

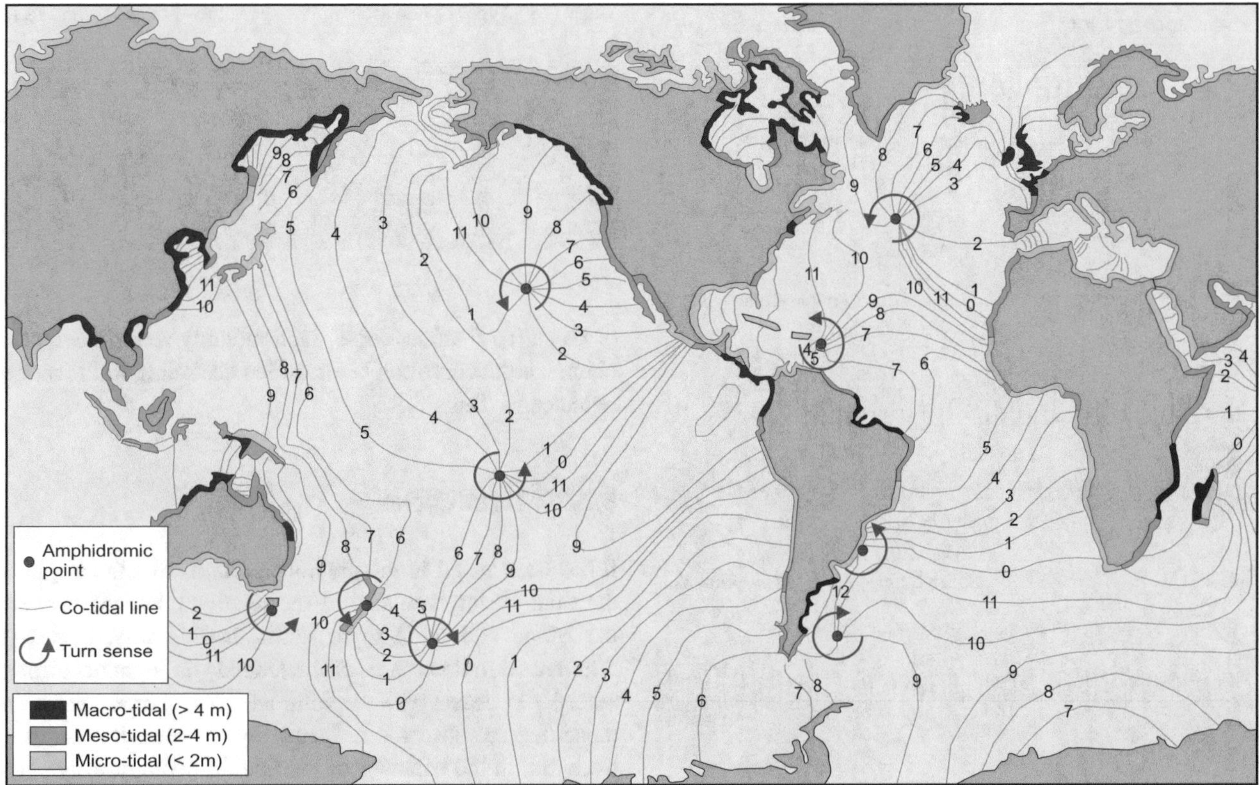

Fig. 8.5 Major amphidromic systems of the Pacific and Atlantic oceans, with indication of the location of the amphidromic point, the direction of rotation of the tide and the co-tidal lines. The tidal range of the coasts is also indicated

wave amplification due to the geometry of the coasts. Another observation that can be made in this scheme is that there are some coasts in the world that are influenced by two different amphidromic systems. These coasts are affected by tides of a mixed nature with characteristics induced by the phase between the two tidal waves.

8.3 Tidal Cycles and Tidal Levels

The tidal regime is one of the main factors of control of coastal sedimentary dynamics and in particular in the evolution of estuaries, deltas and tidal flats, because the number of hours of exposure and submersion of the intertidal fringe directly controls the bio-sedimentary zones existing in it [16].

8.3.1 Types of Tides Based on Their Periodicity

The theoretical approaches described in the preceding points mean semidiurnal behavior of the tide. In this way, two high and two low tides would occur throughout the day. However, this behavior does not adapt to the reality of many coasts of the world. On some coasts, the geometry of the bottoms and the physiography of the coastline modify the tidal wave, making it able to exhibit different behaviors. In general, there are three types of tides: semidiurnal, diurnal and mixed (Fig. 8.6).

Coasts with **semidiurnal tides** are the most common around the globe. They are the ones that adapt to the theory explained above, so there are two daily high waters and two daily low waters (Fig. 8.6a). The period of each tidal cycle is therefore 12 h and 25 min. Although there is always an inequality of amplitude between the two daily tides, in most cases the two tides result in a similar magnitude. However, there are coasts where the two tides are of a very different magnitude. An example of these tides can be found in San Francisco Bay, on the US Pacific Coast.

Coasts with **diurnal tides** are much sparser. In these, only one tidal cycle occurs each day, with a high and a low tide in periods of 24 h and 50 min (Fig. 8.6b). They are typical of restricted seas and large semi-closed bays, although they can also occur on open shores. The best examples can be found in the Gulf of Tonkin, between China and Vietnam, although some sections of the Gulf of Mexico and the Alaskan coast, north of the Bering Strait, also have such tides. Open coasts with diurnal tides can be found in southwestern Australia, northern New Zealand and the Antarctic continent.

Coasts with **mixed tides** have a complex record in which one part of the tidal cycle functions as a diurnal tide and

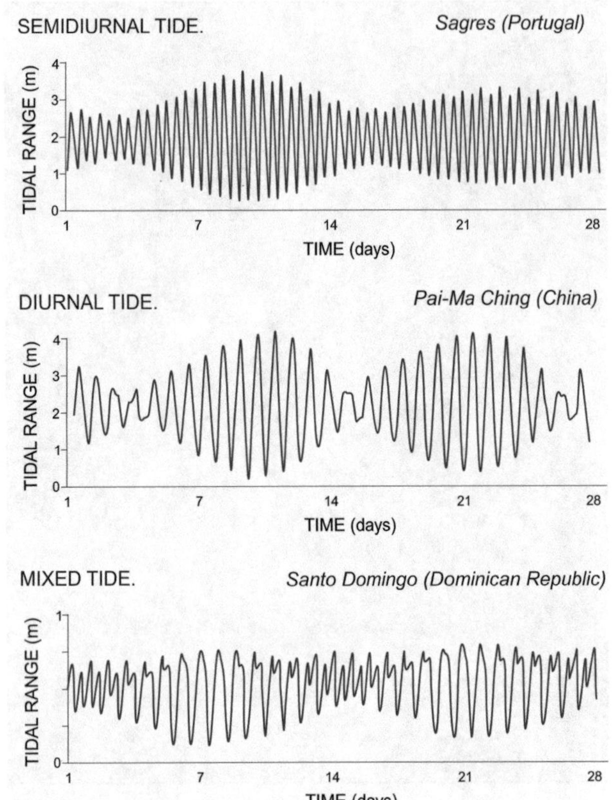

Fig. 8.6 Different types of tide according to their periodicity. Adapted from Defant [4]

another part is in semidiurnal mode (Fig. 8.6c). They usually occur in open marine areas influenced by more than one amphidromic point or in transit between open areas and semi-closed areas. The generic nature of its definition means that different models of tidal curve can actually exist depending on whether diurnal or semidiurnal behavior cycles dominate. There are very good examples of this type of coastline along the entire Pacific Ocean and also in the Caribbean Sea.

The analysis of tidal curves over a 28 day lunar cycle allows a quantitative way of classifying the tidal record among the types described above. The tidal form factor (F) was established by Defant [4] as a relationship between the sum of the amplitudes of the diurnal cycles and the sum of the amplitudes of the semidiurnal ones. This factor can be calculated according to Eq. 8.1.

$$F = \frac{a_{d1} + a_{d2}}{a_{s1} + a_{s2}} \tag{8.1}$$

Where F is the tidal form factor, a_{d1} and a_{d2} are the amplitudes of the diurnal tides and a_{s1} and a_{s2} are the amplitudes of the semidiurnal tides.

Table 8.1 Types of tide according to the value of the tidal form factor (F)

F factor	Type of tide
0–0.25	Semidiurnal
0.25–1.50	Mixed dominantly semidiurnal
1.50–3.00	Mixed dominantly diurnal
> 3.00	Diurnal

Once the F values for a given monthly record have been obtained, the curve can be classified according to the ranges reflected in Table 8.1.

8.3.2 Tidal Cycles

It has been noted in the previous sections that the origin of the tidal cycles is related to the relative positions between the Moon and the Sun to Earth. In this way, cycles of different durations are characterized throughout a tidal record. On coasts with semidiurnal tides, shorter cycles are those that are observed in inequality between the two daily tides, but in the behavior of the tide following astronomical patterns have become clear major cycles. Of a longer duration are the cycles of biweekly character [14], consisting of the alternation of spring and neap tides, so that each month there are two spring tides and two neap tides (Fig. 8.7).

In a higher temporal order, there are other variations of a six-monthly character whose consequence is the alternation over a year between two solstitial and two equinoctial tides. In solstices the differences between their spring and neap tides are small (Fig. 8.8a), while in the equinoxes there is a very marked difference between them (Fig. 8.8b). In successive years the tides differ because the angles of lunar and solar declinations vary over time, and these variations cause the tide values to be repeated in cycles of 17.6 years [2].

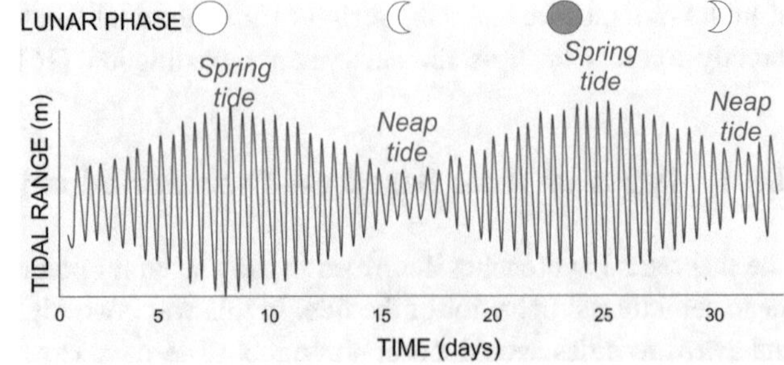

Fig. 8.7 Biweekly tidal cycles in a month

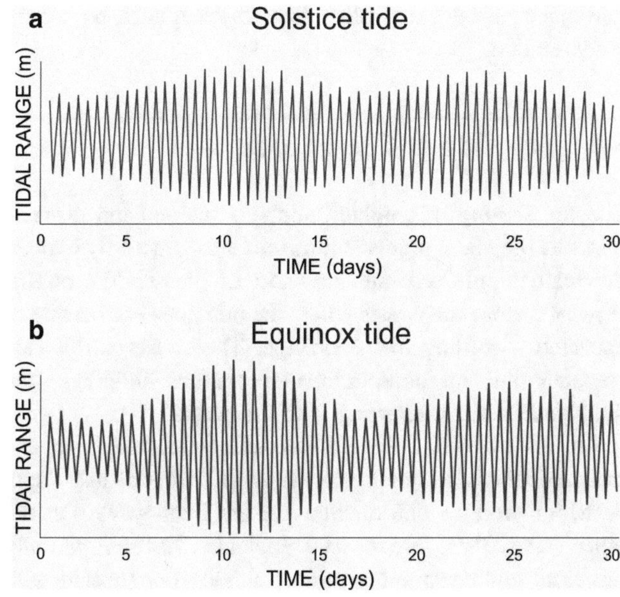

Fig. 8.8 Six-monthly tidal cycles. **a** Solstice tides. **b** Equinox tides

8.3.3 Critical Tide Levels

The cyclic variations experienced by the tide induce an area in the intertidal fringe whose zones are separated by what have been called **critical tidal levels** [6], and which are the topographic levels achieved by the high and low waters during one year (Fig. 8.9). In Doty's approach, which has been adopted by other authors [7, 16], they are called *critical* because they delimit areas that differ in both the number of monthly exposures and submersions, and in the time they remain exposed or submerged. This time of exposure and submersion is crucial for the tolerance of the species of organisms that inhabit the intertidal band, as well as for the transport of sediment in each of these zones.

The critical tidal levels statistically characterized by Doty are: Extreme Equinoctial High Water (EEHW), Mean Spring High Water (MSHW), Mean High Water (MHW), Mean Neap High Water (MNHW), Mean Water Level (MWL),

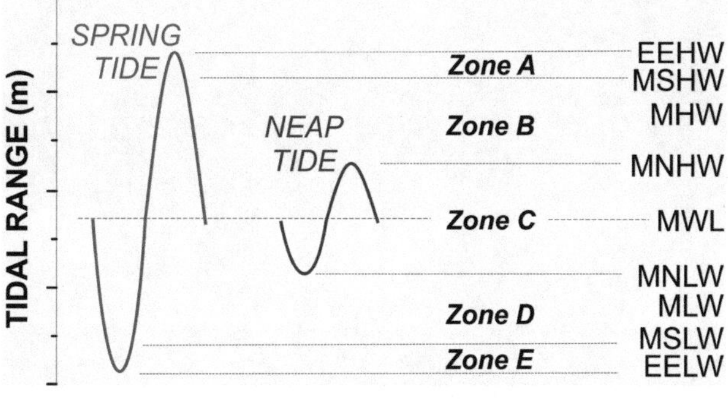

Fig. 8.9 Critical tidal levels and tidal zonation after Doty [6]

Mean Neap Low Water (MNLW), Mean Low Water (MLW), Mean Spring Low Water (MSLW) and Extreme Equinoctial Low Water (EELW).

The intertidal strip is thus divided into zones by these critical levels (Fig. 8.9):

- *Zone A:* Between the Extreme Equinoctial High Water (EEHW) and the Mean Spring High Water (MSHW). It is not submerged more than ten times per month and in total not more than 20 min per month of submersion (Fig. 4.19).
- *Zone B:* Between the Mean Spring High Water (MSHW) and the Mean Neap High Water (MNHW). Its lower level is exceeded by 95% of the tides, although only just under 17% are those that reach or exceed its upper level; this is an average of about 600 h per month of exposure, with the time that remains exposed being slightly less than triple that of the submerged time.
- *Zone C:* This is the real daily intertidal zone, since it has two daily immersions and exposures, when both levels exceed all tides except the below-average neap tides (less than 10% of the total tides).
- *Zone D:* This is located between the limits of the Mean Neap Low Water (MNLW) and the Mean Spring Low Water (MSLW). Its upper limit is exceeded by 95% of the low waters, while in contrast not even 2% of the number of low waters reach the lower level. Thus, it experiences about 550 h per month of average submersion versus about 190 h of exposure.
- *Zone E:* Under the Mean Spring Low Water (MSLW) and over the level of the Extreme Equinoctial Low Water (EELW). It is only affected by about five low tides per month that total no more than five minutes of monthly exposure.

Advanced Box 8.1. The Harmonic Constants

According to the periodicity of the orbital changes of the three spheres involved in the genesis of the tides, and the angular changes between the planes of these orbits, the tides over time may be considered to be composed of an overlap of different harmonic waves. A harmonic wave is a variation that is regulated by a trigonometric function of cosine type. Each one of these harmonics is regulated by one of the periodic astral variations (Moon orbital changes around Earth, Earth orbital changes around Sun, changes in lunar declination, changes in solar declination, changes in the distance between Earth and Moon, changes in the distance between Earth and Sun, and relationships between all these changes).

Therefore, any tidal wave can be calculated through the decomposition of its harmonic constants. Such calculations were already raised by Laplace in his general tide theory, but later development was due to George [3] and the current statement was formulated by Doodson and Warburg [5]. Depending on the period of the harmonic wave, the constants described in Table 8.2 [9] can be distinguished in a composite tidal wave.

The main constants are:

- The constant M2 corresponds with the given values of the tide if the effect of the Sun is neglected and considering the Moon orbit as a perfect circle exactly located around the plane of the Equator.
- The constant S2 corresponds to the solar tide considering the plane of the orbit of the Earth around the Sun is a perfect circle in a plane that coincides with the Earth's equatorial plane.
- The combined M2 + S2 form the effect of the ideal Sun and the ideal Moon on tides.
- The constant N2 considers the non-circularity (elliptical) of the Moon's orbit. So, the tides will be higher in the perigee and lower in the apogee.
- The diurnal constants (K1, O1, P1) consider other small variations of the ideal conditions given by the described constants. So, aspects of the lunar and solar declinations are considered.

Applying the periodicity of at least nine of these harmonics constituents can be used to obtain a fairly accurate forecast of the tide at a given time [15]. These calculations are based on the establishment of tidal coefficient tables. To obtain a height table from these coefficients in a given location, a more precise calibration must be performed that must take into account other local factors, such as coastal configuration.

8.4 The Tide and the Coast

The functioning of amphidromic systems and the distance between the nodal points and the coast satisfactorily explain the tidal regime recorded on most of the world's coasts. However, there are coasts where the tide does not behave as expected according to its location. The cause of this discrepancy must be found in the deformations suffered by the tidal wave when interacting with the coast.

On the one hand, there are coastlines located in special morphological configurations whose gradients of loss depth or width produce deformations in the tidal wave. On the other hand, there are coastal stretches that are partially restricted and form small basins, although connected with the open ocean. In the interior of these inner seas, and in their watersheds with the major ocean basins, there are also particular phenomena that force the tidal wave propagation. These causes result in three phenomena that need to be studied separately: dissipation by friction with the bottom, amplification by convergence and resonance.

8.4.1 Dissipation by Friction

As mentioned above, the tide behaves like a propagating wave and therefore responds to the same equations as the waves. In the open ocean, the wave propagates at a rate of

Table 8.2 Main harmonics constituting a complete tidal wave [9]

Harmonic constant	Symbol	Character	Period (solar hours)	Amplitude (%)
Principal lunar	M_2	Semidiurnal	12.42	100
Principal solar	S_2	Semidiurnal	12	46.6
Larger lunar elliptic	N_2	Semidiurnal	12.66	19.2
Lunisolar semidiurnal	K_2	Semidiurnal	11.97	12.7
Larger solar elliptic	T_2	Semidiurnal	12.01	2.7
Smaller lunar elliptic	L_2	Semidiurnal	12.19	2.8
Lunisolar diurnal	K_1	Diurnal	23.93	58.4
Principal lunar diurnal	O_1	Diurnal	25.82	41.5
Principal solar diurnal	P_1	Diurnal	24.07	19.4
Larger lunar elliptic	Q_1	Diurnal	26.87	7.9
Smaller lunar elliptic	M_1	Diurnal	24.84	3.3
Overtides of principal lunar	M_4	Quarterdiurnal	57.97	6.2
Shallow water quarter diurnal	MS_4	Quarterdiurnal	59.02	6.1
Lunisolar fortnightly	Mf	Bi-weekly	372.86	8.6
Lunar monthly	Mm	Monthly	661.3	4.6
Solar semi-annual	Ssa	Semi-annual	2191.43	4

more than 700 km/h, but entering shallow waters there is a sharp increase in friction with the bottom and the rate of propagation towards the coast decreases very quickly. Generally speaking, tidal wave speeds on continental platforms are reduced to values ranging from 10 to 20 km/h. In this process there is also a dissipation of the wave energy, which results in a reduction of its dimensions. In other words, the tide loses amplitude when it enters into shallow waters as it approaches the coast. The rate of decline of speed depends on the variations in the slope in the shallow water bottom.

8.4.2 Amplification by Convergence

The geometry of coastal areas has a second effect on the tidal wave, which is amplification by convergence. This effect occurs in continental areas whose coastline morphology is funnel-shaped. As it enters these areas, the tidal wave is forced to pass through increasingly narrow areas and the only way the water mass can do so is to amplify the tidal range. Good examples can be found in the Bay of Bengal and the English Channel (Fig. 8.10).

The case of the Bay of Bengal (Fig. 8.10a) is the most extreme case, in which the tidal range rises from an amphidromic point (null range) located at the entrance of the bay to 9 m in the Ganges–Brahmaputra delta.

In the case of the English Channel (Fig. 8.10b), amplification occurs from the entrance to the channel, starting from 4 m. In this case, the configuration of the French coast, with funnel-shaped bays, further amplifies the wave, reaching 11 m at Mont-Saint-Michel. On the other front, the English coast does not present this form, but there is a gradual decrease in depth. Consequently, on this coast the friction effect described in the previous section dominates and the tidal ranges decrease. The amplification effect is again noticed towards the interior of the channel, reaching 9 m in the area of Pas-de-Calais.

Normally, the effects of friction with the bottom and convergence work together, especially in coastal inlets such as bays and estuaries. The decrease or amplification of the tidal range will depend on what is the dominant effect. Thus, if the decrease in depth dominates over the narrowing of coastal margins, it will dominate frictional energy dissipation and the tidal range will decrease. On the contrary, in coastal inlets where pronounced confinement occurs, while the depth decreases more progressively, the convergence effect will dominate and the tide will undergo amplification. This is what explains the difference between the two coasts of the English Channel.

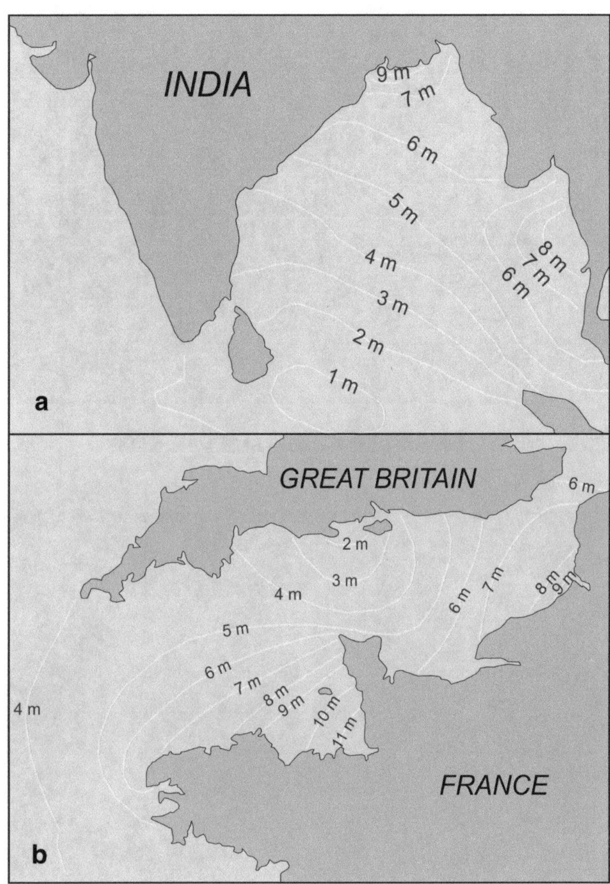

Fig. 8.10 Examples of coasts with tidal amplification due to entry into funnel-shaped coasts. **a** Bay of Bengal. **b** English Channel

8.4.3 Resonance

When a tidal wave enters a semi-confined basin, the reflection of the wave in the margins of the basin causes a wave interaction known as resonance. The final oscillations of the water mass are the result of the initial wave propagated into the basin and the wave reflections on its shores. When the phenomenon takes place on a small scale, the result is a stationary wave that is governed by the equations of the principle derived by Merian [13]. According to this principle, in a small-scale resonance system, there is a central point called a node on which the stationary wave oscillates (Fig. 8.11).

In wider systems there may be different nodal points, and even in wide basins the oscillation of these nodal points can be subject to the Coriolis force. In this way, resonance can generate amphidromic points inside a semi-closed basin. A clear example is the North Sea, in which the progressive tidal wave enters from the north and through the English

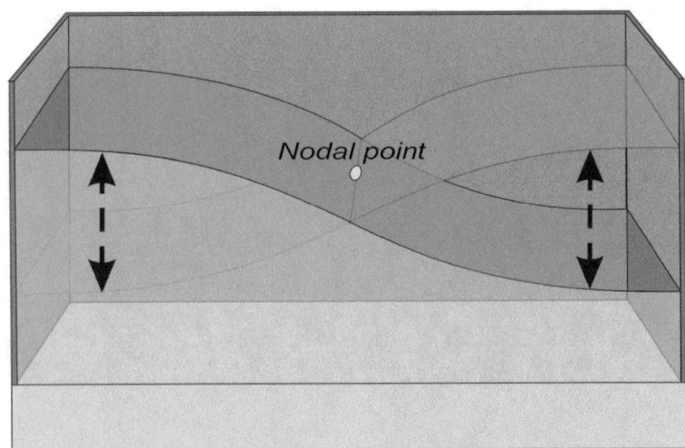

Fig. 8.11 Resonance effect in a closed ideal system [13]

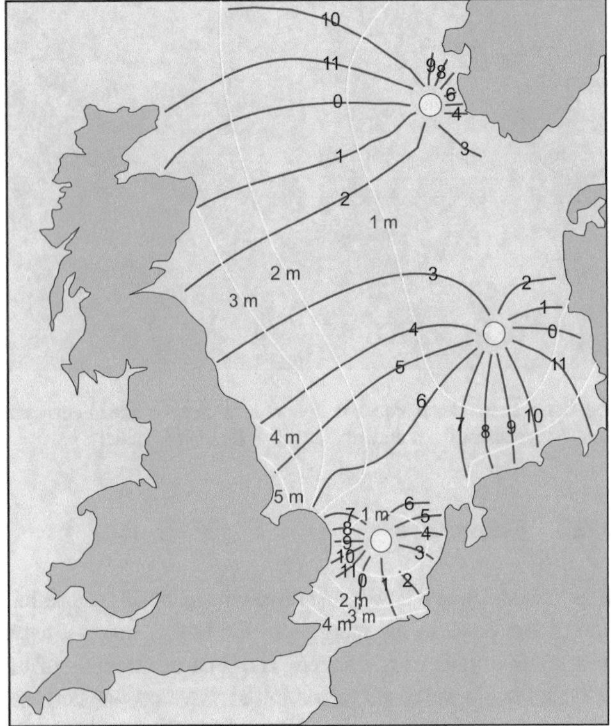

Fig. 8.12 Example of the North Sea resonance system with three nodal points affected by the Coriolis force, thus generating three amphidromic systems

Channel, and where resonance on the British and European coasts has generated a system with three amphidromic points coexisting in a very small space (Fig. 8.12).

8.5 Tidal Currents

The displacement of tidal waves generates movement in the mass of water, similar to the circular displacement that occurs in the phenomenon of wind waves. In this way, the water particles move in orbits, which would be circular if the tidal wave had no friction with the bottom. However, due to its large wavelength, for this friction to not take place, an ocean at least 22 km deep would be required [8]. When moving, this wave interferes with the seabed and the orbits followed by the particles become ellipses elongated on their horizontal axis. These ellipses will be more and more elongated in a direction parallel to the coast when the tide propagates towards shallower bottoms. Thus, in coastal areas, friction is such that vertical movements are negligible against horizontal movements. These horizontal movements of the water mass are known as tidal currents.

As stated, it is seen that the magnitude of tidal currents is directly related to the tidal range; however, this is not always the case, since other factors such as bottom morphology, horizontal viscous stress on the bed–water interface, water density and the presence of other non-tidal currents also intervene. There are known cases where strong tidal currents are recorded for small tidal ranges and vice versa. This fact makes it difficult to predict magnitudes of tidal currents, when some parameters that are difficult to evaluate come into play (e.g., Shields 17; White 19; Bagnold [1]). However, general guidelines appear to exist that are normally satisfied, with some exceptions (Fig. 8.13).

In the open ocean, tidal current velocities maintain a certain constancy, hovering around 0.28 m/s, and keep to rotating patterns in terms of their sense [11]. In this case, they approach a pure undulatory phenomenon, where currents should experience an inversion when the marine level is at the midpoint between the high and low water levels; that is, three hours after the previous high tide or the corresponding low tide (Fig. 8.13a).

In open coastal areas there is a turn in the direction of the currents that usually happens about one to two hours later than the respective high or low tide (Fig. 8.13b). This implies that the currents rotate earlier than would be expected in a pure undulatory phenomenon. This is due to the friction effect with the bottom. In fact, the reversal of currents is produced earlier near the bed than on the surface of the water. At peak times, values greater than those obtained in open oceanic areas can be achieved due to an amplification effect [8].

In coastal bays and channeled areas, due to increasing friction with the bottom and the induction of tides from open areas, the reversal of currents occurs more or less simultaneously with the high and low waters (Fig. 8.13c). In these cases, the speed of tidal currents is directly related to the volume of tidal water that exchanges the open coast with the restricted system [8]. This volume of water is known as the **tidal prism**.

It is very interesting to study the tidal current interaction models in the contact area between channelized and open coastal systems. Inlets, deltaic distributaries and estuarine mouths present in and out currents which are perpendicular

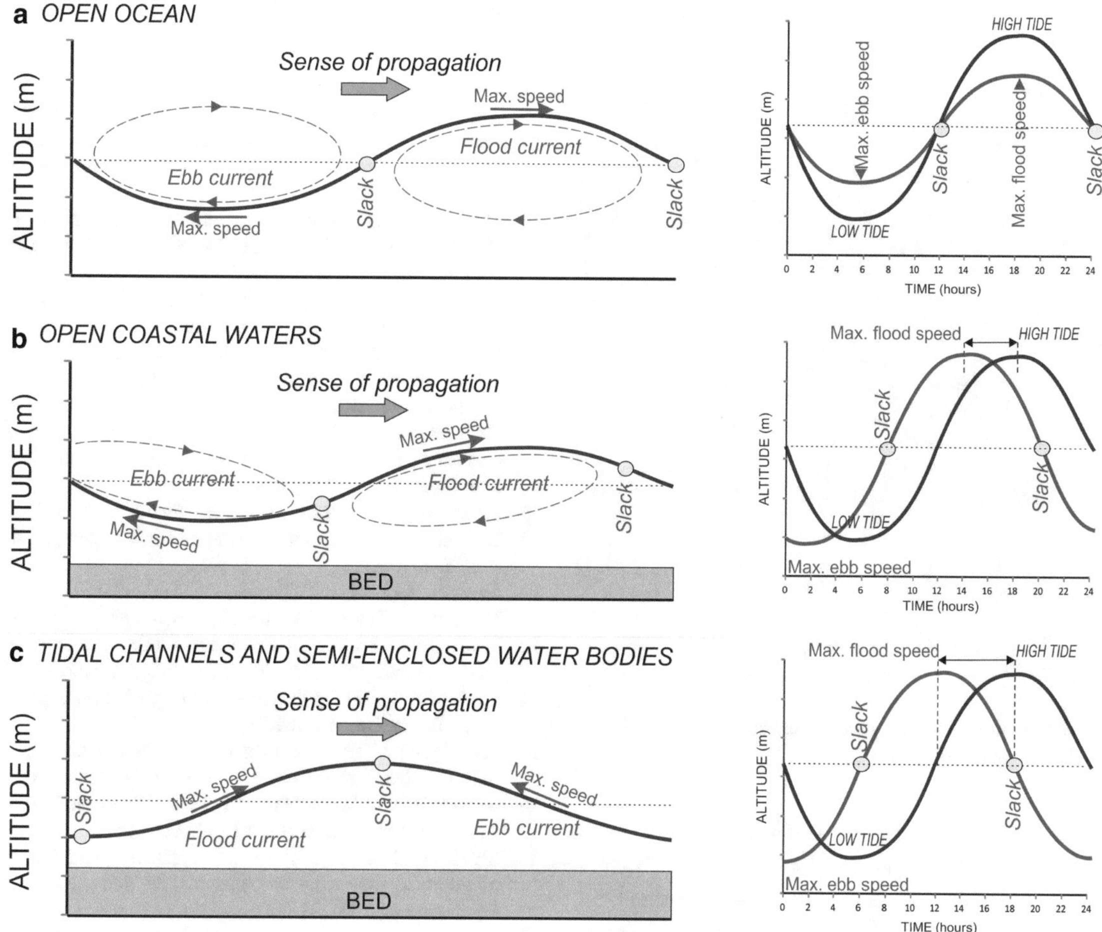

Fig. 8.13 Relationships between tidal currents and tidal levels in different environments

to the coastline, whereas the tidal currents of the adjacent open coast are parallel to the shoreline. These interaction models, linked with the wave patterns, condition the morphology of the coastal sandy bodies existing at the mouths of these channels.

Advanced Box 8.2. Genesis of Tidal Bores

Tidal waves of large amplitudes as they approach confined and shallow morphological environments are able to sharply increase their height and produce breakers. This phenomenon is known as a **tidal bore** in English. The term corresponds to the words *macareo* in Spanish, *mascaret* in French and *pororoca* in Portuguese.

The genetic mechanism of a tidal bore is simple. It requires a tidal wave with an amplitude of more than 5 m, propagating towards a coast of high slope, that generates in a small space with a very shallow bottom and that presents a narrowing funnel-shaped morphology. In these conditions, the narrow fringe in which the shoaling occurs prevents the dissipation of frictional energy on the bottom (Fig. 8.14). The original wave, when entering the shallow bottom, undergoes a refractive process, sharply shortening its wavelength while increasing its height. In these circumstances a breaking ridge is formed that penetrates the coastal inlet, further increasing in height due to the convergence phenomenon. Under normal conditions, this crest is usually around 0.5 m in height, but in many cases it can reach heights greater than 5 m.

This type of wave is typical of estuaries and macrotidal bays, although they can also occur during spring tides in systems whose average tides are usually less than 5 m. When the phenomenon occurs in an estuary, the waves can travel up to several kilometers into the river course.

Some case studies are very well documented. Well-known examples along the American continent are in Fundy Bay (Canada), Bristol Bay (Alaska), the head of the Gulf of California (USA) and Ría de La Plata (Argentina–Uruguay). In Europe there are examples in the estuary of the River Seine (France), as well as in the Rivers Severn and

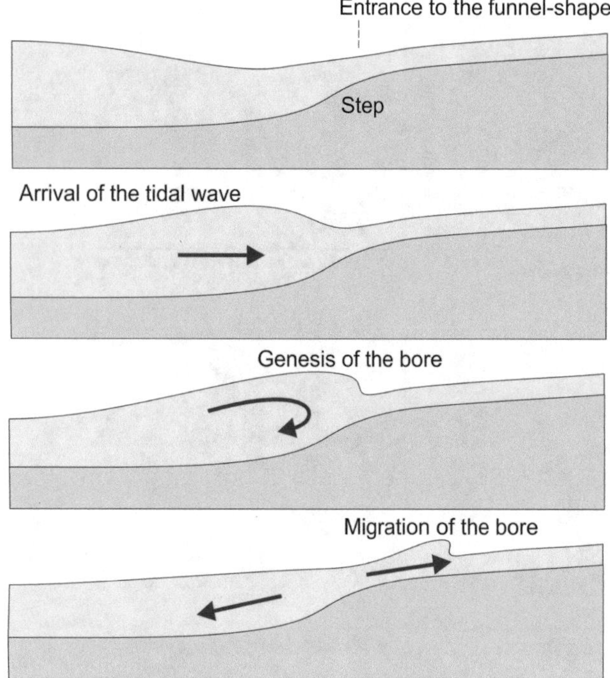

Fig. 8.14 Scheme illustrating the genetic mechanism of a tidal bore. Based onased on Lynch [10]

Trent (England). In Asia, the most well-known cases are the Gulf of Khambhat (India) and the Ganges Delta (Bangladesh). Finally, in Australia it occurs in the Cambridge Gulf.

A singular case is that of the mouth of the Amazon (Brazil), where the wave penetrates more than 500 km into the river course. The world's largest tidal bore is the one that occurs in the Qiantang River estuary (China), which reaches 5 m in height and exceeds 28 km/h in its penetration upriver.

References

1. Bagnold RA (1942) Beach and nearshore processes. In: Hill MN (ed) The sea. Wiley & Sons, New York, pp 507–582
2. Cherniawsky JY, Foreman MGG, Kang SK, Scharroo R, Eert AJ (2010) 18.6-year lunar nodal tides from altimeter data. Cont Shelf Res 20:575–587
3. Darwin GH (1914) Las mareas. El correo gallego, El Ferrol
4. Defant A (1958) Ebb and flow. University of Michigan Press, Ann Harbor
5. Doodson AT, Warburg HD (1941) Admiralty manual of tides. Her Majesty's Stationery Office, London
6. Doty MS (1946) Critical tide factors that are correlated with the vertical distribution of marine algae and other organisms along the Pacific Coast. Ecology 27:315–328
7. Featherstone RP, Risk MJ (1977) Effect of tube-building Polychaetes on intertidal sediments of the Minas Basin, Bay of Fundy. J Sediment Petrol 47:446–450
8. Grant M (1987) Oceanography. Prentice Hall, New Jersey, 406 pp
9. Knauss JA (1997) Introduction to physical oceanography. Prentice Hall Inc.
10. Lynch DK (1982) Tidal bores. Sci Am 247:46–57
11. Marmer HA (1926) The tide. Appleton, New York, 282 pp
12. Masselink G, Hughes MG, Knight J (2003) Introduction to coastal processes and geomorphology. Routledge, London, 416 pp
13. Merian JR (1928) On the motion of drippable liquids in containers. Ph.D. thesis. Basilea Switzerland
14. Pattullo JG (1966) Seasonal changes in sea level. In: Hill MN (ed) The Sea. Inter-Science, New York, pp 485–496
15. Russel RC and MacMillan DH (1952) Waves and tides. Hutchinson's Scientific and Technical Publications
16. Swinkbanks DD, Murray JW (1981) Biosedimentological zonation of boundary bay tidal flat, Fraser river delta, British Columbia. Sedimentology 28:201–237
17. Shields A (1936) Anwendung der Ähnlichkeits-Mechanil und der Turbulenz-forschung auf die Geschiebewegung. In: Preussische Versuchsanstalt für Wasserbau und Schiffbau, vol 26
18. Von Arx WS (1962) An introduction to physical oceanography. Addison-Wesley, Reading
19. White CM (1940) The equilibrium of grains on the bed of a stream. Proc Royal Soc (London) 174:322–338

9 Continental Processes and Sediments on the Coast

9.1 Introduction

The composition of the materials deposited on the coast depends on the clarity of the waters, the amount of suspended matter and the distribution of sediments by the currents that reach the coast, but, above all, on the sedimentary supply. The amount of sediment that coastal hydrodynamic agents (waves and tides) distribute throughout coastal systems is based not only on the carrying capacity of these agents, but also on the budget of available sediment. This sedimentary contribution can reach the coast from the continental shelf, but the material that arrives from the mainland is of much greater importance. These materials can reach the coast carried by rivers and also by the action of the wind. Of these, the sediment provided by the rivers is much greater. The amount of sediment supplied annually by the world's rivers was estimated to be between 10 and 20 billion tons [50], cited in Anthony [4], but more recently the same author calculated an amount of more than 20 billion tons [48].

The amount and nature of the terrigenous contribution from the mainland is influenced by the geology and climate of the adjacent continental area. The climate has a direct influence on all the factors, controlling both the hydrodynamic agents and the volume of input from different sources. In general terms, it can be said that there are greater contributions in areas where rainfall is higher and where the weathering is more active. Globally, there are four key river basins that make a contribution of sediment to the coastal areas (Fig. 9.1).

These are:

- Amazon basin: contributes 1438 tons per km^2 per year to the coast;
- China and Indochina basins: 3228 tons per km^2 per year;
- Indonesia–-New Guinea basins: 3000 tons per km^2 per year;
- East Asia basins: 1738 tons per km^2 per year.

To put the importance of these basins in perspective, the rest of the basins associated with coasts provide less than 300 tons per km^2 per year, and most of them even less than 100 tons.

When observing the distribution of these basins, we can see that all of them are located around the Tropics. The explanation for this is found in the combination of factors discussed above. Not only is there more sediment available in these areas, but the higher rainfall induces higher flow and gives rivers greater transport capacity to the coast.

The second of the sources of sediment contribution from the mainland to the coast is the wind. In this case, the wind can also have a secondary effect that acts in the opposite direction, removing sediment from the coast to transport it to continental areas. In relative terms, the contribution of wind to the coast has a much smaller influence than the rivers. However, its influence is not negligible and should also be analyzed.

Either way, the sediment that reaches the open coast from the mainland does not arrive directly [71]. There are transitional processes that mean some of the materials are trapped in the coastal systems of greater continental influence, while some can go ahead and eventually reach the open coast. This chapter will discuss the processes associated with the contribution of sediment from continental areas and all the transitional processes involved. These processes are not only linked to continental agents, but to the combination of these with the coastal processes that will finally distribute the sediment along coastal systems.

9.2 River-Induced Processes and Sediments

All of the sediment contribution that arrives from the rivers to the sea is transported through the river mouths (estuaries and deltas). The river mouths make up the transition zone between the mainland and the coastal area. At river mouths there is a conflict between river currents and the movements of the seawater mass which entails a series of processes and

J. A. Morales, *Coastal Geology*, Springer Textbooks in Earth Sciences, Geography and Environment,
https://doi.org/10.1007/978-3-030-96121-3_9

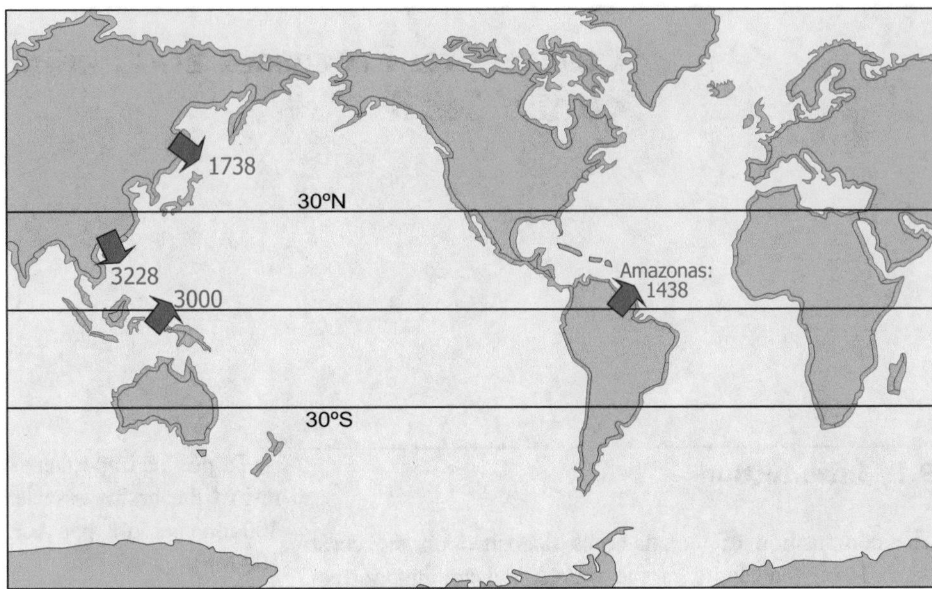

Fig. 9.1 Distribution of continental sediment supply from the main river basins of the world. Adapted from Milliman and Meade [49]

mechanisms that determines sedimentary transit from the mainland to the open coast [17]. This transit occurs in relation to numerous changes in the depositional regime. On the one hand, there is a change in the chemical conditions of water, from the fresh waters of rivers to normal sea waters. In addition, there is a change in confinement conditions, from a channeled environment to a fully open coast. Finally, there is a change in the origin of the energy that causes the movement of water and, with it, the transport of particles [4]. These processes must be understood as a series of intermediate steps that occur between purely continental and purely marine processes.

Moreover, on most occasions hydrodynamic processes that occur at river mouths do not only function as transit mechanisms between the continent and the ocean, but also often involve permanent depositional processes. This implies that some of the material from the continent will be embedded in the confined areas, generating a reservoir that is preserved as a sedimentary record of these transitional environments. In a sense, this sedimentation process inside fluviomarine systems limits the bypassing of sediment towards the open coast.

Classically, the processes that happen in the river mouth channels have been explained as a result of the mixing of two water bodies with different densities [65]. This mechanism emphasizes the sedimentation of fine particles transported in suspension, as well as other particles that are "born" within these systems as a product of processes such as flocculation [57], leading to the phenomenon known as the turbidity maximum [56]. However, the water mixing also means a stop in the currents, which causes the sedimentation of the material transported as bed load. These mechanisms occur together, so that each of them influences the rest, although in this chapter they will be studied independently in order to facilitate understanding.

9.2.1 Water Mixing Processes

Perhaps the most obvious process that occurs when a river meets the sea is the mixture of river and sea water. This mixture would implicitly mean a certain stratification of water induced by differing salinity. In this way, the less dense fresh water would move along the top of the water column, while the higher-density salt water would pass through the bottom. In the water column there would be a gradient of salinity from the surface to the bed. The relative importance of river and sea water bodies in river mouth channels determines whether this gradient occurs more or less sharply. The concentration of the gradient in a particular area or its distribution along the mouth gives rise to the existence of different water mixing models. Pritchard [65] distinguished three different models in terms of the shape of water mixture (Fig. 9.2), with the relationship between the volume of river water and the volume of tidal water being the variable that determines the transition from one model to another.

- *Well-stratified mixing model (salt wedge):* According to Simmons [69], this model occurs when the volume of river water is of the same magnitude or greater than the volume of tidal water, although for Postma [62] the volume of fresh water must be significantly higher than the tidal prism, so that the effect of the tide would be negligible. In any case, the good stratification of the waters is reflected in the "double layer" flow, with very

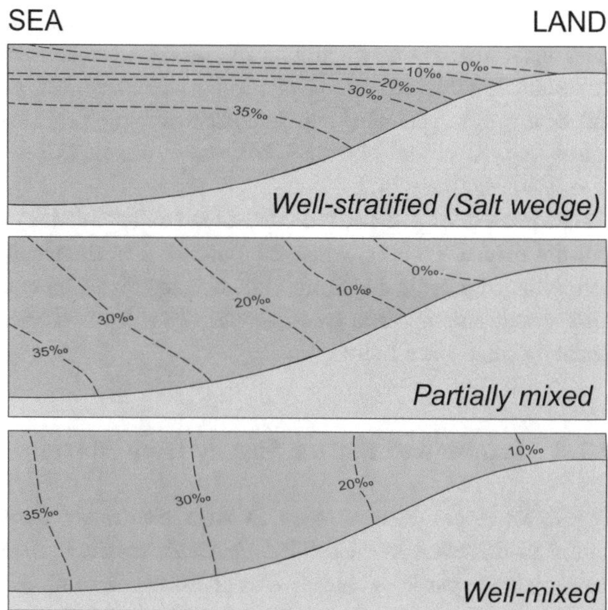

Fig. 9.2 Conceptual scheme representing the three possible types of water mixing at a river mouth. Adapted from Postma [62]

pronounced longitudinal and vertical gradients. These gradients are at their maximum at peak times of tidal flow of spring tides. In this situation the salt water mass circulates to land at the bottom of the water column, while fresh water can continue to circulate towards the sea across the surface. This implies that the contact of fresh and salted water occurs in a fairly horizontal and abrupt way and without any mixture of water in a narrow strip called the **halocline**. The inclination of this wedge is quite pronounced and, therefore, it has a good surface extension, being able to cover longitudinally tens of kilometers from its contact with the bottom of the channel to the surface [37].

- *Partially mixed model:* In this mixing model the volume of water moved by the tides is between 20 and 200 times the volume of fresh water [69], so it is the tide that dominates the mixing process [62]. The flow also occurs in a double layer, although there is an intermediate strip in which a greater mixture of water takes place (brackish water), making the gradients less pronounced. In this case it is the tide that is responsible for moving upstream and down the contact strip between both types of water, achieving a greater mixture and a smoother salinity gradient.
- *Well-mixed model:* At the mouths with this type of mixture, the volume of salt water is more than 200 times greater than the volume of river water [69]. In this case the tide produces so much movement in the water masses that it is able to eliminate stratification and produce a homogenization of salinity, which presents a smooth transition to seawater.

Intuitively, it can be thought that each mouth acts according to one of these models; however, in each river the circulation model can vary over time, taking into account seasonal variations in river flow and sea level variations that respond to long-period tidal cycles. In this way there are mouths that can vary between a saline wedge model and a partial stratification model, and others that vary between partial mixing and total mixing. There are even estuaries that can present all three models according to the conditions.

It is also clear that the area of water convergence is displaced along certain areas depending on the state of the tide and the river flow. This becomes important, as the water mixture carries certain sedimentation processes that are preserved in the sedimentary record of these areas.

9.2.2 Estuarine Genesis of Particles: Flocculation and Aggregation

River waters and also seawater are normally loaded with a special type of solution called **colloidal solutions**. Colloidal solutions are a particular state of dissolution involving two phases, a fluid phase and a molecular phase. Large organic molecules, such as water-soluble proteins, are able to form these types of solutions. Colloidal solutions undergo a transport that is neither ionic nor physical, with molecules forming a kind of agglomerate (looking like mucus) that is a mixture of water with long chains of electrically charged molecules.

Depending on the relative proportion of molecules and water, colloidal solutions will be able to be in two states: sol or gel. In the case of sol, water predominates and the solution is more fluid. In the gel state, the protein predominates and the solution is more viscous. The change from one state to another is reversible depending on physical and chemical factors that can cause a solution to change without the need to vary the fluid concentration.

The molecular structures of the sols and gels may be of different type depending on their morphology, but in all cases, each of these molecular structures has charged atoms located on its surface, with a hydrophilic and a hydrophobic part. The hydrophilic part is usually located on the outside and causes the particles to be electrostatically suspended in the water (Fig. 9.3a). Chemical changes in water mixing areas at river mouths can cause hydrophilic face surface loads to be null and void by joining negatively charged molecules with others that are positively charged (Fig. 9.3b). This process occurs when these molecules get close enough to collide, thus adhesion occurs between them. In these cases, there is a change in the density of molecules that will no longer be stable and will pass into the solid state. This process is called **flocculation** [60].

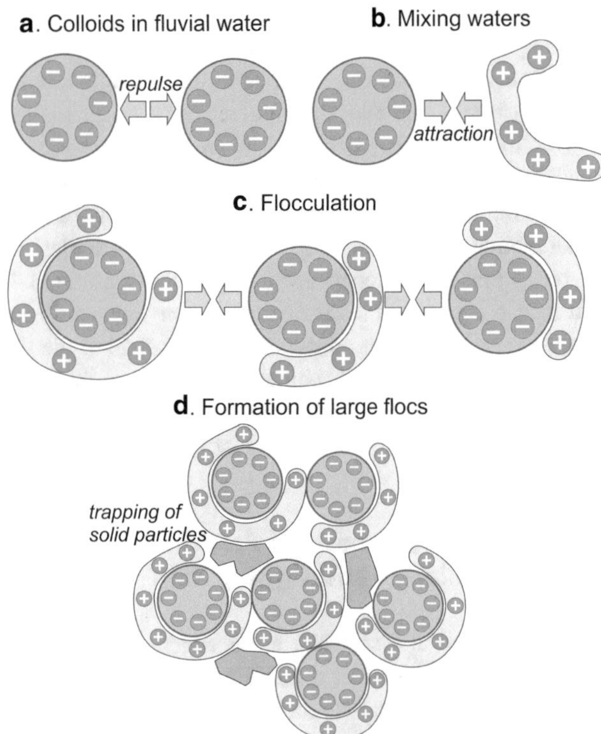

Fig. 9.3 Conceptual scheme that represents the flocculation process. **a** The colloids in the river water are in balance with water thanks to the hydrophilic charge of their surface. **b** River water colloids encounter the colloids of seawater in the mixing area, canceling their negative charge. **c** Transition to the solid state in the flocculation process. **d** Formation of large flocs by aggregation, wrapping other solid particles previously in suspension

Flocculation results in the "birth" of new particles associated with the water wedge in such a way that the higher the degree of mixing, the greater the volume of flocculated particles generated [22]. The direct flocculation process is able to generate particles of than 16 microns (Fig. 9.3c). The composition of flocculated particles can be diverse. On the one hand, organic molecules can form that give rise to the flocculation of solid organic matter, but on the other hand inorganic molecular structures can form with crystalline structure. This is the case of the formation of clay minerals (hydrated aluminum phyllosilicates), including illite and the groups of kaolinites and smectites.

These particles pass from their creation to be part of the suspended load, joining with particles that were already being transported in suspension by both the river and the tides. Flocculated particles also play an important role in the hydrodynamic behavior of sediment in transition zones between the river and sea, as they tend to trap larger suspended particles forming agglomerates called flocs (Fig. 9.3 d). Particles trapped between flocs can be mineral grains, small bioclasts or organic fecal pellets. Each floc can group dozens of individual solid particles reaching millimeter sizes (e.g., Graham and Manning [35]; Manning and Dyer [46]).

The size and density of flocs depend on numerous variables including: the concentration of suspended matter and its residence time, current speeds, water chemistry (Eh, pH and salinity), temperature and influence of biological processes (e.g., Dyer and Manning [24]; van Leussen [73]; Hill et al. [39]; Anthony [4]).

The process of **peptization** (also called deflocculation) is just the reverse process, when the particles are chemically destabilized by again acquiring ionic characteristics on their surface and entering into balance with water to be electrostatically lifted once more.

9.2.3 Suspended Matter Supply from Rivers

The finest grains from erosion in river basins are transported to the coast as suspended matter in the river water mass. These particles have sizes between 2 and 125 microns in the case of clays, silts and even very fine sands, so low current speeds are enough to keep them suspended in the water column [11]. High concentrations of suspended matter contribute to increased water turbidity. It is well known that suspended matter influences the primary production of organisms, controlling the development of phytoplankton and bacterioplankton species. Thus, indirectly it also controls the secondary production of the species that feed on these microorganisms (zooplankton and fish larvae).

Terrigenous suspended matter is generated in river drainage areas by erosion of solid particles during precipitation. Normally, the concentration of these particles in river waters suffers intermittent oscillations, increasing after heavy rains [14]. The pulses of higher concentration coincide with the high flow rates, especially during flooding events [42]. During these times, not only the materials from the tributaries arrive in the river, but also the fine sediments previously deposited in the intermediate sections of the river are remobilized.

There are large differences in concentration, size and nature of the suspended load grains between different rivers. It should be borne in mind that the variability of the grain size characteristics of the suspended matter and its nature depends not only on the characteristics of the geologic material in its source area, but also on the ability to be selected for erosion and transport processes [75]. Among the variables that condition the size and composition of the suspended matter are [79]:

- Latitude (which controls both the climatic factor of weathering processes and the primary production of aquatic organisms),
- River flow (fluvial regime),
- River longitude,

- Percentage of bare rocks in the drainage basin, relative to forest cover,
- The relief of the basin and
- The profile of the river in relation to the equilibrium profile.

As an example, a comparative study among the tributaries of the River Exe in the UK showed the enormous spatial variability that can occur in a relatively small area due to differences in the associated variables described above [75].

The suspended matter is of a diverse nature and consists of a mineral phase and an organic phase. Some of the components of the suspended matter are naturally present in river water. However, other components (organic and inorganic) come from human activity. It is clear that their quantity and composition, understood as the relative percentage between these components, change seasonally in parallel with the variation of the flows.

As for the mineral phase, grain size has an important influence on its mineralogy. Thus, for example, the fraction less than 16 microns is usually composed of phyllosilicates, while quartz usually dominates the fraction between 16 and 125 microns. This occurs because minerals from other compositions are easily alterable (chemically and physically) and are reduced during transport, so that it is quartz that predominantly gets to reach the coast [28].

The organic phase is composed of fragments of plants and some planktonic organisms, both phytoplankton and zooplankton. Diatoms dominate (genera *Navicula, Pinnularia, Asterionella* and *Tabellaria*), although dinoflagellates (genera *Peridinium* and *Ceratium*), flagellates (genera *Euglena, Colponema* and *Spiromona*), cyanophyceae (genera *Oscillatoria* and *Rivularia*) and chlorophytes (genera *Spirogya, Oedogonium* and *Zignema*) are also common. Zooplankton is represented by species of various phyla—protozoa, ctenophores, rotifers, bryozoans—and, above all, by some groups of crustaceans such as cladocerans, copepods and ostracods as well as insect larvae and eggs and fish larvae. In this case, not all organic components of the suspended matter manage to achieve confluence with seawater in their initial state. Far from this, during transport organic matter undergoes decomposition and fermentation processes, mostly passing to the colloidal state. Only the shells of organisms such as diatoms and ostracods get to reach the mixing area in their solid state.

To reach the open coast, the suspended matter transported by the river must pass through the mouth area where the water mixing processes occur. For suspended particles, the mixing area means a dissipation of the energy of the river currents and a loss of the capacity to transport the mechanical load when the water mass slows. This means an increment of the concentration of suspended particles around the contact between river and marine waters, not because of their different chemical nature, but because of their differential displacement.

Matching this accumulation of suspended load with the area where flocculation processes are occurring results in many of the particles being engaged in floc agglomerates. The sum of both processes—concentration of suspension and flocculation—results in the formation of a phenomenon known as **turbidity maximum**. The dynamics of this cloudy water mass will be discussed next.

9.2.4 The Turbidity Maximum

Turbidity is a concept for naming the optical perception of water clarity. The presence of suspended solids as well as the dissolved phase reduce the clarity of the water, creating an opaque, misty or muddy appearance. It has been previously said that when the river current meets the mass of seawater (usually displaced by the tide) turbidity levels increase, resulting in a phenomenon known as the turbidity maximum. There are three different mechanisms that contribute to the increase in turbidity in the area where river water converges with seawater: (1) the concentration of the solid suspended matter of the river due to the slowdown of the currents; (2) the flocculation processes in the mixing area; and (3) the resuspension of previously deposited bottom sediments.

The first of these processes is due to the concentration of solid material that occurs when the river water, which transports suspended matter, slowly decelerates when it meets the seawater mass. Since most of the suspended material is supplied by the river current, the highest concentrations will occur on the surface, where fresh water circulates over the halocline.

The second process involved in the formation of the turbidity maximum is flocculation. It is obvious that the birth of new particles by flocculation in the water mixing area contributes significantly to the formation of the turbid cloud [41]. However, it is in the lower part of the halocline where most of the flocculated particles are concentrated. The water mass under the halocline also receives the particles that come from the top and pass through the halocline due to decanting. The aggregation processes associated with flocculation trap the falling solid particles, increasing the size of the agglomerates and favoring settlement [40].

The third of the processes involved in the turbidity maximum is the resuspension of the fine sediments from the bed [57]. This phenomenon is associated with tidal dynamics and occurs independently of saline mixing processes. Either way, the resuspended particles become part of the lower area of the turbid cloud formed by the other two mechanisms.

In the formation of a turbidity maximum, both its magnitude and the relative importance of each of these three phenomena depend on river variables such as the flow and concentration of suspended matter of the river waters, but also on other variables that depend on marine processes such

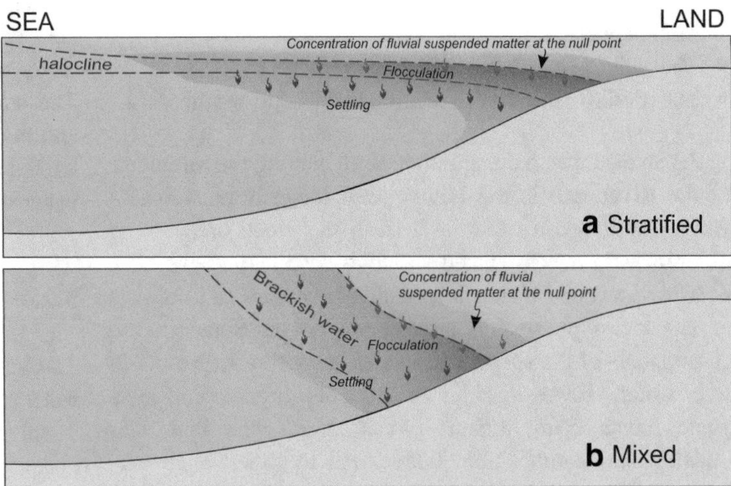

Fig. 9.4 Conceptual models of the genesis of a turbidity maximum. **a** With a halocline in a salt wedge (stratified). **b** With mixture of water masses in a volume of brackish water (mixed)

as the tidal regime, the amount of suspended matter introduced from the sea and the type of water mixing that occurs [61]. In general, two models that describe the importance of the water mixing phenomenon can be distinguished (Fig. 9.4).

In the first of the models (Fig. 9.4a), the presence of a halocline located in the same position as a null point of the water current speeds is of great importance. In this case, the contact between waters with different nature occurs practically on an abrupt surface. The existence of the null point results in the concentration of particles of river origin on the halocline as described above. On the other hand, the halocline concentrates the flocculation process, while below it occur the settlement processes favored by the aggregation associated with flocculation.

The second model (Fig. 9.4b) groups any type of saline intrusion where the waters are mixed in a volume of brackish water. In this case, there is no defined halocline, although the null point of the current speeds is still present. The concentration of river particles also occurs on the null point, with the difference that in this case the surface with speed 0 is arranged at a less horizontal inclination and covers less extent. On the contrary, as there is a greater mix of water, the flocculation process becomes more important and has a wider distribution.

Evidently, these models conceptually group together the different ways in which a cloud of turbidity can be generated. However, they do not represent the dynamics of the plumes in different tidal situations. Over a tidal cycle, the mechanisms of turbidity formation change while moving along transition zones (Fig. 9.5), and going upstream or downstream depending on the displacement of the water mass induced by the tidal circulation [1].

We can start the explanation at the moment of maximum tidal flood current (Fig. 9.5a). At this time, the suspended matter is concentrated in the terrestrial limit of the salt water intrusion, coinciding with the null point of convergence between the river and the tidal flood currents. In the mixing area, the flocculation processes contribute to the increase in turbidity and the aggregation results in the settlement in areas where the current speed is lower. The result is a dispersion of the suspended matter to the lower parts of the water column. The turbidity plume migrates landwards as the tide penetrates the river system, reaching its peak at the time of high water [70].

When the tide reaches its maximum height, the water mass stops its displacement for a few moments and the aggregation and decanting processes increase (Fig. 9.5b). At this time, in the innermost parts of the system, where the turbidity maximum is closest to the bed, the deposition of the particles of the plume occurs. Because of this, this area is usually a place of rapid shallowing. During these moments the plume reaches its greater thickness, although, due to the settlement of the particles closest to the bed, the concentration of matter decreases.

The beginning of the ebb current (Fig. 9.5c) implies a disappearance of the null point of the current convergence zone, since during the cycle the entire water mass moves seawards. During the ebb cycle, the water mixing area moves downstream, while increasing the volume of brackish mixed water. At this point, the mechanism that concentrates suspended matter from the river loses importance, while flocculation increases and the dispersion of matter in the water column gains importance. At moments of maximum ebb flow, the bed material is resuspended, causing the cloud to gain concentration in the part nearest the bottom. These are the moments of greatest extent of the turbidity maximum.

The arrival at the low tide slack (Fig. 9.5d) implies a new halt of currents at the bottom of the water column. This provides an opportunity for the aggregation and decanting flocculation processes to gain importance again, by

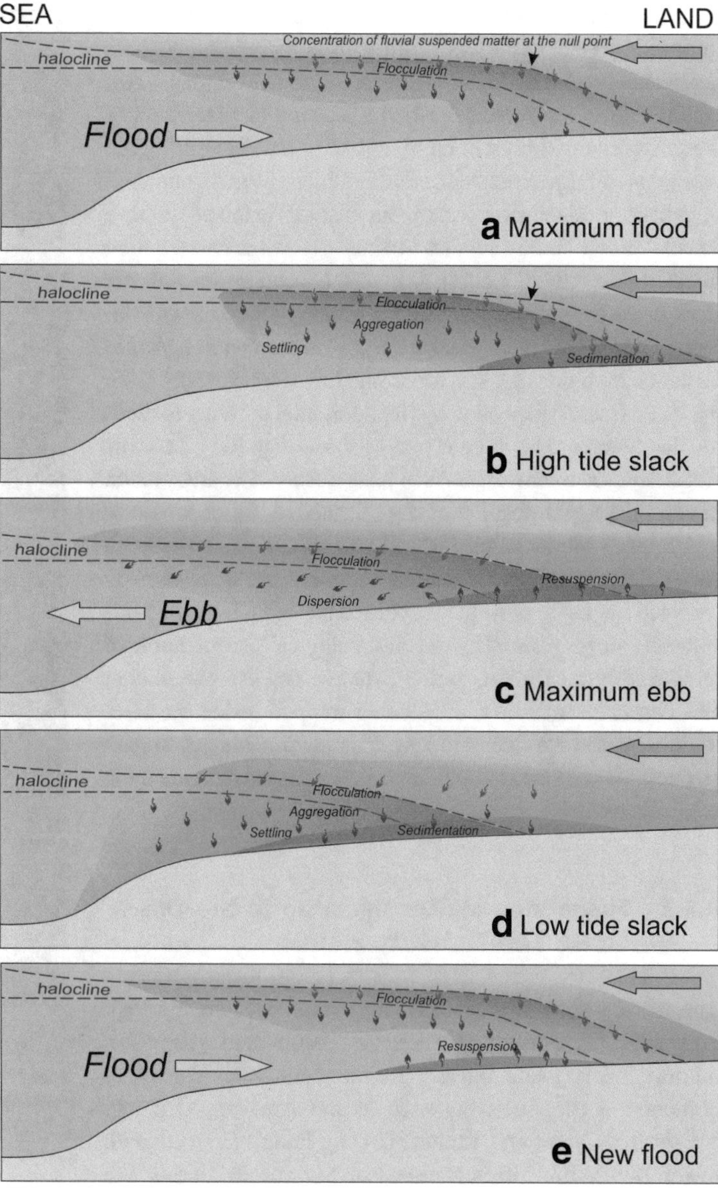

Fig. 9.5 Turbidity maximum dynamics over a full tidal cycle. Based on Allen et al. [1]

sedimentation of fine particles from the part of the turbidity maximum closest to the bed.

When a new tidal flow cycle begins, it returns to the initial situation with the emergence of a new current convergence point (Fig. 9.5e). In this case, the upstream circulation of the seawater mass is significant, producing a resuspension of some of the particles deposited during the low water slack.

This dynamic model includes only changes in processes associated with a turbidity maximum during short tidal cycles. However, the same authors also highlighted the dynamics of fortnightly cycles. Thus, due to the alternation between spring and neap tides, the position of the turbidity maximum and its concentration also evolve. In the transit between a spring and a neap tide, there is a decrease in the maximum speed of the currents, as well as a longer duration of the slacks. During the week that this semi-cycle lasts, net sedimentation exceeds the reworking processes. On the contrary, during the semi-cycles of increased tidal range that occur between a neap and a spring tide, just the opposite process occurs, with reworking dominating over accumulation. However, the succession of a complete fortnightly cycle results in net sedimentation. It should be noted that the fine sediments accumulated during the neap tide are cohesive. They need to be remobilized at a much higher lifting threshold and thus withstand spring tide currents without this occurring. This results in an infilling of the confluence zone by net accumulation of fine sediments.

Other dynamic changes of the turbidity maximum are imposed by river dynamics. Numerous authors (e.g., Nichols [56]; Allen et al. [1]; Dobereiner and McManus [20]; Gelfenbaum [31]) have described a seaward displacement of the turbidity maximum of up to tens of kilometers during the moments of large river floods. During these events, a process of general resuspension occurs, but in addition the position of the cloud can be so low that in times of ebb it can even be injected into the open coast. Conversely, during the summer river flows the tide can displace the mixing area many kilometers inside the river systems. To this phenomenon it should be added that, during these times, the low river currents favor the processes of dispersion by the flood tidal currents to areas located upstream of the position of the wedge [61]. In some rivers, marine microfossils have been observed in areas much beyond the usual position of the salt wedge [9].

These dynamics of the turbidity maximum mean that both suspended matter and decanted sediment are continuously recycled in each new tidal cycle. This results in an uniformization of grain size and mineralogical composition of the turbidity maximum. In this dynamic process, the thicker and denser sediments end up being trapped inside the river mouths, while the finer and less dense sediments get to pass through the mixing area and are finally injected into the open coastal system.

9.2.5 Suspended Matter Injection to the Open Coast: The Turbidity Plumes

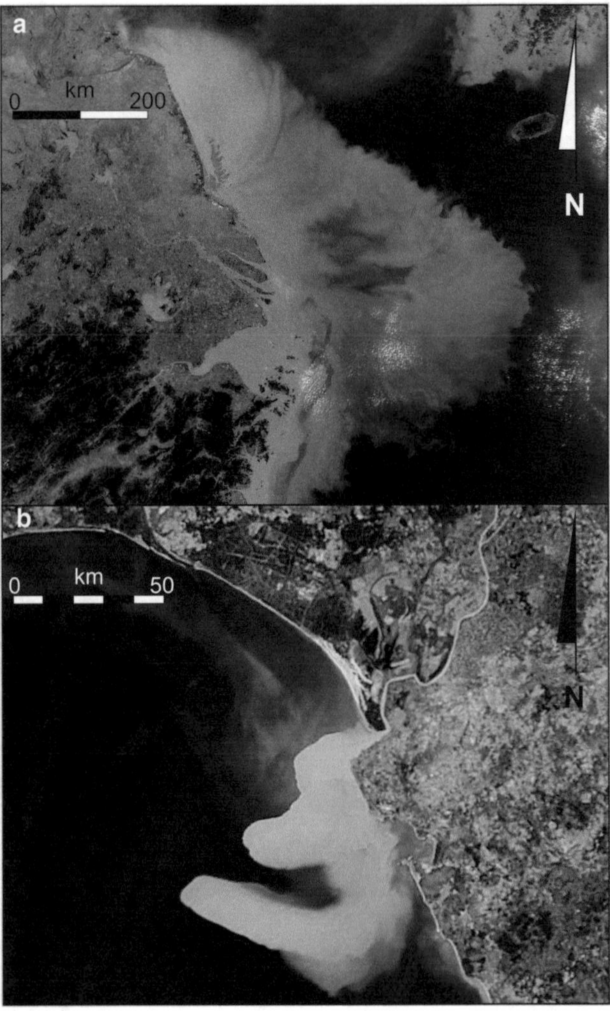

Fig. 9.6 Examples of turbidity plumes injected into coastal waters from river systems of different importance (NASA Earth Observatory images). **a** Plume from the mouth of the Yangtse River (China). **b** Plumes generated at the mouth of the Guadalquivir (SW Spain) in two successive floods (Images Landsat/Copernicus from Google Earth)

The world's most important river systems have such a high flow that fresh water reaches the mouth and the water mixture takes place directly in the open sea. This phenomenon occurs in rivers such as the Amazon, Orinoco, Mississippi, Niger and Yangtze (Chang Jiang). In rivers with lower flows, this phenomenon can also occur during the moments of large floods. Even in these smaller rivers, the finest material from the turbidity maximum generated in the interior of the channels can reach the coast during the moments of tidal ebb, as explained in the previous section. In any of the three cases, the result is the formation of a turbidity plume that is injected to a greater or lesser extent into the open coast (Fig. 9.6).

The dimensions and dynamics of these turbidity plumes differ depending on which of the three cases described is occurring, but they will also be influenced by the tide and wave dynamics of the coastal area where the plume is being injected. These dynamics control the place and the way in which this fine material will finally settle. Thus, the final destination of the lutitic material from the plumes may differ from one system to another. In many cases, the sediment can be redistributed by coastal systems and end up in nearby tidal systems. In other cases, the material may move to the lower water mass and settle in the deeper coastal areas at the front of the channels. This is the case of many fluvial prodeltas.

In the case of large rivers, the presence of the plume is associated with water mixing processes that take place in the front area of the mouth and not in the channeled areas. The case of the Amazon is very significant, where fresh water is able to form a tongue that extends about 150 km from the mouth to the sea [33]. Similar cases occur in the Yangtze River (Fig. 9.6a) [43, 76] and the Huanghe (Yellow River) [81]. In most cases, the mixing processes that occur in open areas correspond to a good mixing model [13, 32]. An example of this type of mixture is observed at the mouth of the Mississippi (USA), where in the absence of tides, the dynamics of the mixture are controlled by weak river flows and the displacement of the water mass with the wind [2]. In cases of large rivers with more extreme river currents, situations of good stratification can be identified in this marine plume. An example of this type of plume is that of Song Hong (Red River, Vietnam) described by van Maren [74].

In the cases of plumes associated with good mixing conditions, the movement of the water mass generated by the tide and/or wind, linked with the remobilization by waves, cause a process of diffusion of turbidity in the water column. This results in a greater distribution of suspended sediment that can be shifted to the prodelta or to the coast, or even to both sites.

Moreover, good stratification conditions imply greater importance of the suspended matter from the river. In these cases, the grain size of the suspended matter is usually larger and they then decant very quickly in areas near the mouth without even reaching the prodelta area. The latter case has been documented in Po (Italy) and Song Hong (Vietnam) [29, 74, 77].

Despite this, in most river mouths the mixing processes occur inside the channels and only on occasions of maximum discharge do situations such as those described above take place. In these situations, the finest fraction of the suspended matter generated in the turbidity maximum can go out continuously to the outside, while the thickest fraction is contained in the channels until a river flood can remove it to the open sea. At this type of river, the fine material from the turbidity maximum is distributed twice. On the one hand, the thickest fraction becomes part of the coastal systems, while the finer fraction associated with flocculation is dispersed on the coast and finally settles in the prodelta [17].

These smaller but more common rivers play an important role in the supply of fine material in coastal and continental shelf systems. The sediment reaches these systems in particular during periods that follow the large floods in which tides and waves have distributed the sediment of the plumes. Numerous examples of such rivers have been described in European estuaries, such as the Seine [19], the Guadiana [51], the Guadalquivir (Fig. 9.6b) [10], the Var [3] and the Po [77], and also in Australia as is the case of the Daly Estuary [78].

9.2.6 Bed Load Supply from Rivers

At river mouths, the dynamics of the material transported as bed load presents some parallels with the suspended matter. In the innermost areas of the estuary and delta channels, a current convergence zone usually forms. In this area, a low-energy point develops, as the river currents that circulate towards the coast are slowed when they encounter seawater.

In tidally influenced systems, this convergence is at its maximum during flow semi-cycles, so that the current is completely stopped allowing the immediate sedimentation of the bed load carried by the river (Fig. 9.7, Stage 1). From the beginning of the tidal ebb semi-cycle, the sediments begin to be transported seawards. Through many tidal cycles, the bed load of fluvial origin is blended with that of marine origin, so that the sediments of the central part of the confined mouths have a mixed composition. This sediment may have a tendency to residual seawards movement, or may become entrapped within the fluviomarine systems. In this case, open estuaries work differently from highly evolved estuaries, bedrock-confined estuaries and deltas.

In open estuaries, tidal circulation and current asymmetry often create a domain of flood currents. This implies a net transport that displaces the bed load from sea to land, introducing it to the central sector of the estuary where it converges with the bed load of river origin [23]. In these estuarine areas, the sands also coincide with the cohesive materials that settle from the turbidity maximum during the moments of slack. The cohesive material present at the bottom wraps the grains of sand, preventing them from returning to movement in the following tidal cycles. Both mechanisms (residual inland component and cohesive entrapping) contribute to the bed load sediment remaining

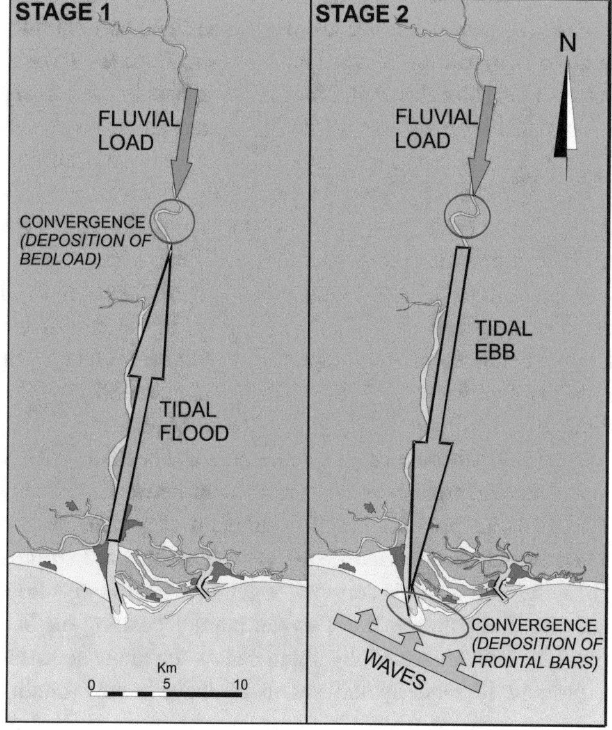

Fig. 9.7 Example of the discharge of bed load material in two phases in the Guadiana Estuary in Spain (bedrock-confined estuary). Stage 1: During the tidal flood there is a convergence with the river current which implies a deposit of the bed load. Stage 2: Successive tidal cycles take the material to the mouth. During the ebb semi-cycles, the output current collides with wave trains by depositing the load on a front delta bar (Morales and Borrego [52])

within the estuaries, giving these environments a very high sedimentation rate. This active sedimentation inside the estuaries causes them to evolve very quickly, contributing to greater shoaling and confinement, so that in a few thousand years they can evolve and transform into deltas. This effect is usually much greater in systems where the river is small and provides little material in bed load. In these cases, the central area of the estuary would be mostly filled by sediments of marine origin mixed with the autochthonous sediments of the turbidity maximum (e.g., Anthony and Dobroniak [5]). In very open estuaries, the action of the waves, at least in the marine sector of the estuary, can also be important. In these cases, the waves not only introduce sediment into the estuary channels but also contribute to increasing the magnitude of flood currents, especially during storms [4].

Highly evolved estuaries, as well as confined estuaries (bedrock-controlled estuaries) and deltas, have a different functioning. In these systems, the convergence of river and tidal currents is usually located more towards the sea (e.g., Dalrymple et al. [16]). In addition, in these systems it usually dominates the ebb tidal current, giving the sediment a net movement towards the sea [17]. In the channels of these systems, active sedimentation and bypassing mechanisms are usually distributed laterally in different areas. In this way, cohesive sediment traps a significant part of the bed load on the tidal bars, while the deepest part of the channels continues to allow a bypassing of bed load sediment from the river to the sea (e.g., Morales et al. [53, 54]). The bars of the estuary are covered by bedforms of different dimensions, so that in these we can observe the double sense of bed load transport (e.g., Lobo et al. [44]; Morales et al. [54, 55]). In these systems, the bed load sediment ends up reaching the exit to the sea after numerous tidal cycles (Fig. 9.7, Stage 2).

At the front of the mouth channels of these evolved systems described above and, especially, in the deltas, the sediment ends up forming important mouth bars. The case of deltas is significant, as these bars are almost entirely constituted by bed load of river origin (e.g., Maillet et al. [45]; Drexler and Nittrouer [21]). These bars have been classically named deltaic front bars and are located between the delta plain and the prodelta [7, 80], although they can also be called estuarine ebb-tidal deltas [18]. The bars are formed when the output river currents, together with the ebb tidal currents, dissipate into the seawater mass depositing the bed load. The effect is especially significant when the water mass is moving to land by the action of wave trains running towards the shore (Fig. 9.7, Stage 2). The result is the formation of an important sand body with linguoid morphology (Fig. 9.8). The dimensions, development and subsequent evolution of this body will be conditioned by the balance of forces between the output jet from the confined systems and the ability of waves to redistribute the sediment on the outside. Evidently, the balance of forces is influenced by the available grain size of the sediment (e.g., Garel et al. [30]; Edmonds and Slingerland [25]).

The deltaic front bars may be subjected to strong reworking by the waves that generate on the surface bars that move to the shoreline. These bars can end up joining the delta plain to form barrier islands. This is the case of mixed deltas such as the Nile, Senegal and Danube, among many others [72]. In any case, it is worth noting the role that these bars play as a source of material in the open coastal systems (beaches and barrier islands) adjacent to deltas and confined estuaries, because once the sand reaches land, its material can be redistributed by coastal drift currents along the entire coast (e.g., Garel et al. [30]; Sabatier et al. [67]).

9.3 Wind Supply and Erosional Effect

The action of the wind on the coast is not restricted to the generation of waves, but also acts as an effective transport agent that moves material from the mainland to the sea and vice versa, giving rise to wind supply processes and wind erosion. Conceptually both processes are involved in the overall sedimentary balance of the coast [63]. The influence on the coast of material movement by the wind is evident on some sandy European coasts such as Aquitaine and the Pertuis d'Antioche (France), Rosslare (Ireland), Doñana (Spain) and the entire coast of the Netherlands [27].

Interactions between wind systems and coastal aquatic systems have been studied in some restricted coastal cells where the material is continuously recycled from the dunes to the beach and vice versa (e.g., O'Connor et al. [59]). However, most of the knowledge about the exchange of material between wind systems and submerged coastal systems remains conceptual (Fig. 9.9). Although these processes can act together, in this chapter they are studied separately.

9.3.1 Sediment Supply from Continental Winds

In most manuals, when talking about the continental contributions to the coast it is noted that 95% of this is due to the rivers. However, the products of weathering in the continental rocks can also be transported by the wind. In fact, the remaining 5% of the material that reaches the coast comes from wind transport. However, the influence of these contributions to the coast is not frequently addressed and most monographs on coastal dynamics do not include a chapter dedicated to the influence of wind input on the coast.

Regarding the wind processes involved in this transport, there are not many details to comment on. It should be pointed out, however, that wind transport is the one that best classifies the grains by their size. On the one hand, the silty sediments removed from land can pass far above the coastline

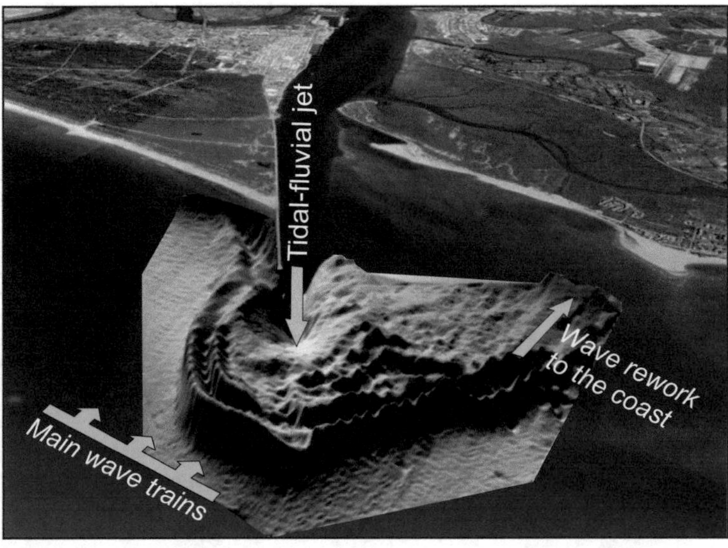

Fig. 9.8 Linguoid morphology of the deltaic front bar at the mouth of the Guadiana Estuary, indicating the dominant processes involved in its formation and its influence on the feeding of the coast. Based on Garel et al. [30]

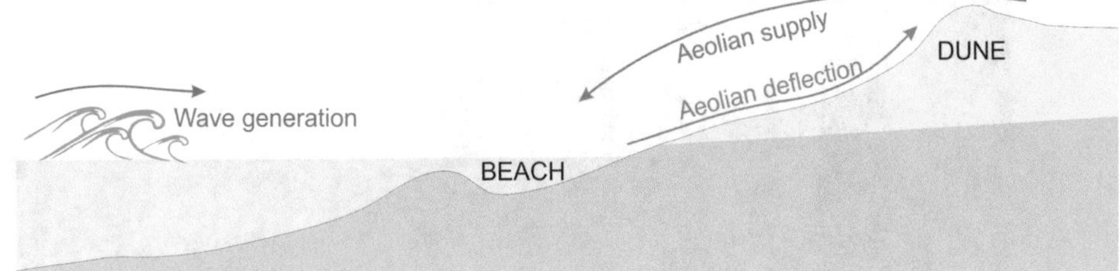

Fig. 9.9 Conceptual diagram showing the influence of the wind on the coast

when moving through the high layers of the atmosphere [15] to end up as sediments in the open sea (Fig. 9.10). On the other hand, the sand grains move closer to the ground and when they reach the coast they can get trapped in the water and contribute to the budget of the beaches.

In the case of sand, it must be borne in mind that wind transport exerts significant particle wear, so that along the journey from the mainland to the coast mineral grains acquire rounding, while polymineralic fragments are disaggregated and less resistant minerals are destroyed. Quartz is not only the most abundant mineral in sedimentary, igneous and metamorphic rocks that outcrop on the continent, but it is also one of the most chemically stable minerals and is able to withstand the abrasion processes suffered during transport. This explains why quartz makes up the vast majority of the grains provided by wind to the coasts.

There are some coastal areas in the world where persistent winds operate from the mainland to the sea. For example, on the west coast of Mauritania, the trade winds displace the dune systems of the Sahara Desert until they reach the coast [47]. The ergs of Azefal and Akchar continuously feed a strong coastal drift that redistributes these sediments south along the coast. Despite this wind contribution, this coast experiences a net erosion by not receiving any type of river input.

Other coasts of the world have a significant wind contribution because they are subjected to a persistent wind regime. Examples include some stretches of the coast of Libya and Tunisia in the Mediterranean. In the case of the Tunisian coast, the wind supply is sufficient to power some barrier island systems and coastal sabkhas [26].

9.3.2 Wind Deflation

The wind can intercept the sandy particles from the beaches and circulate them to the mainland, especially during low tide times. This process is known as wind deflation. This occurs on most of the world's shores when the wind blows persistently towards the coast. The amount of sediment provided by foreshore and backshore to wind domain areas depends on several factors, including: the amount of sand on the beach and its grain size, the range and tidal regime, the type of beach, the percentage of sand moisture and, of course, the direction and strength of the wind (e.g., Pye [64]; Hesp [38]). In the long term, the balance of sedimentary exchange between the coast and the continent is controlled by the distribution of erosive and cumulative zones over time. This distribution is determined by the transport mechanisms that take place in the sediment–air interface [8]. Any method that attempts to quantify the material displaced by these transport mechanisms must take into account the measurement of variables such as wind speed at the interface, air density,

Fig. 9.10 Satellite image showing the injection of dust from the Sahara Desert into the ocean through the high layers of the atmosphere. Image NASA Earth Observatory

grain size and grain density (e.g., Bagnold [6]). The sediment output from the beaches to the dunes represents a negative entry in the beach balance. In parallel, a transformation of the distribution of the grain size into its sediment occurs, since the wind selects only the sizes it can transport, leaving behind the largest grains in the deflation zone.

Although the wind acts with the same intensity throughout the beach, the amount of material transported varies along its sections. The particles begin to be incorporated progressively into the wind flow: there are very low values in the area near water and maximum values towards the backshore. In this way, the wind flow tends to incorporate grains along the beach profile until particle saturation is reached in the higher areas. This saturation will be achieved if the beach area is wide enough [58]. This incorporation of particles along the profile is called the fetch effect [12] or, more precisely, fetch erosion effect [34].

The moisture content of the sand tends to increase the fetch effect, since moisture retains the particles by conferring cohesion to the individual grains and increasing their lifting threshold of movement [34]. Either way, the first effect of wind when the tide falls is to contribute to increase the evaporation of interstitial water. On the other hand, the water table descends from centimeters to decimeters, releasing the particles from this humidity. Therefore, transport becomes more effective over time from the moment of immersion of the beach, although only the grains above the water table will start transport.

Another variable that influences the wind contribution in these areas is the slope of the beach. The slope, and also the morphological changes that occur throughout the year due to the beach dynamics, causes changes in wind speed. These changes are due to the greater or lesser deceleration that occurs by wind friction when attacking the beach at different angles. Higher sloped beaches tend to have a shorter fetch, although wind tends to create turbulence that contributes to exceeding the lifting threshold of particle movement. So, the sand grains will be incorporated into the flow more quickly. Conversely, on very shallow beaches, the fetch is much higher, although it will have less resistance to wind, which facilitates sedimentary bypassing to the higher parts [68].

Normally, the arrival of sediment in the upper area of the beach contributes to the formation of coastal dunes. The dune that immediately forms over the backshore is called the

foredune. On prograding beaches there may be trains of successive dunes towards the continent that show the generation of successive foredunes on the prograding beach. On other well-supplied coasts, where the transition to the mainland is very flat and under the right wind regime, coastal dunes several meters high can migrate several kilometers inland. Examples of these cases are in the Doñana National Park, in southwestern Spain [66]. In areas of transition to steeper continent, these dunes can be retained and become large by stacking. This is the case of the Dune of Pilat, on the coast of Aquitaine in western France [36].

References

1. Allen GP, Salmon JC, Bassoulet P, Du Penhoat Y, De Grandpre C (1980) Effects of tides on mixing and suspended sediment transport in macrotidal estuaries. Sed Geol 26:69–90
2. Allison MA, Kineke GC, Gordon ES, Goñii MA (2000) Development and reworking of a seasonal flood deposit on the inner continental shelf off the Atchafalaya River. Cont Shelf Res 20:2267–2294
3. Anthony EJ (2007) A review of inter-connected hazards on a steep, urbanised coastal river floodplain: the Var floodplain and delta, French Riviera. Méditerranée 108:91–97
4. Anthony EJ (2009) Shore processes and their Palaeoenvironmental applications. Elsevier, Amsterdam, 519 pp
5. Anthony EJ, Dobroniak C (2000) Erosion and recycling of estuary-mouth dunes in a rapidly infilling macrotidal estuary, the Authie, Picardy, northern France. In: Pye K, Allen JRL (eds) Coastal and estuarine environments: sedimentology, geomorphology and geoarchaeology, (Special Publication 175). Geological Society, London, pp 109–121
6. Bagnold RA (1941) The physics of blown sand and desert dunes. Methuen, London, 265 pp
7. Bates CC (1953) Rational theory of delta formation. Am Asso Petrol Geol Bull 37:2119–2162
8. Bauer GO, Davidson-Arnott RGD (2002) A general framework for modeling sediment supply to coastal dunes including wind angle, beach geometry, and fetch effects. Geomorphology 49:89–108
9. Brockmann C (1929) Das Brackwasser der Flussmundungen als Heimat und Vernichter des Lebens. Senckenberg am Meer, 29. Nat Mus 59:401–414
10. Caballero I, Morris EP, Ruiz J, Navarro G (2014) Assessment of suspended solids in the Guadalquivir estuary using new DEIMOS-1 medium spatial resolution imagery. Remote Sens Environ 146:148–158
11. Charlton R (2008) Fundamentals of fluvial geomorphology. Routledge, New York, 234 pp
12. Chepil WS, Milne RA (1939) Comparative study of soil drifting in the field and in a wind tunnel. Scientia Agricola 19:249–257
13. Cobb M, Keen TR, Walker ND (2008) Modeling the circulation of the Atchafalaya Bay system. Part 2: River plume dynamics during cold fronts. J Coastal Res 24:1048–1062
14. Crawford CG (1991) Estimation of suspended-sediment rating curves and mean suspended-sediment loads. J Hydrol 129:331–348
15. D'Almeida GA (1986) A model for Saharan dust transport. J Climate Appl Meteorol 25:903–916
16. Dalrymple RW, Baker EK, Harris PT, Hughes MG (2003) Sedimentology and stratigraphy of a tide-dominated, foreland–basin delta (Fly River, Papua New Guinea). In: Sidi FH et al (eds) Tropical Deltas of Southeast Asia: sedimentology, stratigraphy, and petroleum geology, vol 76. SEPM Special Publication, Tulsa, pp 147–173
17. Dalrymple RW, Choi K (2007) Morphologic and facies trends through the fluvial–marine transition in tide-dominated depositional systems: a schematic framework for environmental and sequence-stratigraphic interpretation. Earth Sci Rev 81:135–174
18. Dalrymple RW, Zaitlin BA, Boyd R (1992) Estuarine facies models: conceptual basis and stratigraphic implications. J Sediment Petrol 62:1130–1146
19. Deloffre J, Lafite R, Lesueur P, Verney R, Lesourd S, Cuvilliez A, Taylor J (2006) Controlling factors of rhythmic sedimentation processes on an intertidal estuarine mudflat: role of the turbidity maximum in the macrotidal Seine estuary, France. Mar Geol 235:151–164
20. Dobereiner C, McManus J (1983) Turbidity maximum migration and harbor siltation in the Tay Estuary. Am J Fish Aquat Sci 40:117–129
21. Drexler TM, Nittrouer CA (2008) Stratigraphic signatures due to flood deposition near the Rhône River: Gulf of Lions, northwest Mediterranean Sea. Cont Shelf Res 28:1877–1894
22. Dyer KR (1989) Sediment processes in estuaries: future research requirements. J Geophys Res 97:14327–14339
23. Dyer KR (1997) Estuaries: a physical introduction. Wiley, Chichester, 95 pp
24. Dyer KR, Manning AJ (1999) Observation of the size, settling velocity and effective density of flocs, and their fractal dimensions. J Sea Res 41:87–95
25. Edmonds DA, Slingerland RL (2007) Mechanics of river mouth bar formation: implications for the morphodynamics of delta distributary networks. J Geophys Res 112(F02034):1–14
26. El-Magsodi MO, Haddoud DA (2011) The salt marsh (sabkha) in the western part of Libya. In: Öztürk M et al (eds) Sabkha ecosystems. Volume III: Africa and Southern Europe. Springer, pp 79–84
27. EUROSION (2004) Living with Coastal Erosion in Europe: sediment and space for sustainability. A guide to coastal erosion management practices in Europe. Directorate General environment, 162 pp
28. Fleming G, Poodle T (1970) Particle size of river sediments. J Hydraul Div ASCE 96:431–439
29. Fox JM, Hill PS, Milligan TG, Ogston AS, Boldrin A (2004) Floc fraction in the waters of the Po River prodelta. Cont Shelf Res 24:1699–1715
30. Garel E, Sousa C, Ferreira Ó, Morales JA (2014) Decadal morphological response of an ebb-tidal delta and down-drift beach to artificial breaching and inlet stabilization. Geomorphology 216:13–25
31. Gelfenbaum G (1983) Suspended-sediment response to semidiurnal and fortnightly tidal variations in a mesotidal estuary: Columbia River, USA. Mar Geol 52:39–57
32. Geyer WR, Hill PS, Kineke GC (2004) The transport, transformation and dispersal of sediment by buoyant coastal flows. Cont Shelf Res 24:927–949
33. Geyer WR, Kineke GC (1995) Observations of currents and water properties in the Amazon frontal zone. J Geophys Res 100:2321–2339
34. Gillette DA, Herbert G, Stockton PH, Owen PR (1996) Causes of the fetch effect in wind erosion. Earth Surf Proc Land 21:641–659
35. Graham GW, Manning AJ (2007) Floc size and settling velocity within a *Spartina anglica* canopy. Cont Shelf Res 27:1060–1079
36. Grandjean G, Paillou P, Dubois-Fernandez P, August-Bernex T, Baghdadi NN, Achache J (2001) Subsurface structures detection by combining L-band polarimetric SAR and GPR data: example of the Pyla Dune (France). Trans Geosci Remote Sens 39:1245–1258

37. Hansen DV, Rattray M (1967) New dimensions in estuary classification. Limnol Oceanogr 11:319–326
38. Hesp PA (1999) The beach backshore and beyond. In: Short AD (ed) Handbook of Beach and Shoreface Morphodynamics. John Wiley and Son, Brisbane, pp 145–170
39. Hill PS, Milligan TG, Geyer WR (2000) Controls on effective settling velocity in the Eel River flood plume. Cont Shelf Res 20:2095–2111
40. Kranck K (1981) Particulate matter grain-size characteristics and flocculation in a partially mixed estuary. Sedimentology 28:107–114
41. Krone RB (1972) A field study of flocculation as a factor in estuarial shoaling processes. Technical bulletin, vol 19. Committee on Tidal Hydraulics, US Army Corps of Engineers, 62 pp
42. Lewis J (1996) Turbidity controlled suspended sediment sampling for runoff-event load estimation. Water Resour Res 32:2299–2310
43. Liu JP, Li AC, Xu KH, Velozzi DM, Yang ZS, Milliman JD, DeMaster DJ (2006) Sedimentary features of the Yangtze River-derived along shelf clinoform deposit in the East China Sea. Cont Shelf Res 26:2141–2156
44. Lobo J, Plaza F, Gonzáles R, Dias J, Kapsimalis V, Mendes I, Diaz del Rio V (2004) Estimations of bedload sediment transport in the Guadiana Estuary (SW Iberian Peninsula) during low river discharge periods. J Coastal Res 41:12–26
45. Maillet GM, Vella C, Berné S, Friend PL, Amos CL, Fleury TJ, Normand A (2006) Morphological changes and sedimentary processes induced by the December 2003 flood event at the present mouth of the Grand Rhône River (southern France). Mar Geol 234:159–177
46. Manning AJ, Dyer KR (2007) Mass settling flux of fine sediments in Northern European estuaries: measurements and prediction. Mar Geol 245:107–122
47. Michel J, Westphal H, Hanebuth TJ (2009) Sediment partitioning and winnowing in a mixed eolian-marine system (Mauritanian shelf). Geo-Mar Lett 29:221–232
48. Milliman JD, Farnsworth KL (2011) River discharge to the Coastal Ocean. Cambridge University Press, Cambridge, 384 pp
49. Milliman JD, Meade RH (1983) World-wide delivery of river sediment to the oceans. J Geol 91:1–21
50. Milliman JD, Syvitski JPM (1992) Geomorphic/tectonic control of sediment discharge to the ocean: the importance of small mountainous rivers. J Geol 100:525–544
51. Morales JA (1995) Sedimentologia del Estuario del Rio Guadiana. Ph. D. Thesis 1993. Huelva University. Serv. Publicaciones UH, 322 pp
52. Morales JA, Borrego J (2009) El litoral de Huelva: fisiografía y dinámica. In: Olías M et al (eds) Geología de Huelva. Lugares de interés geológico. Serv. Publicaciones UH, pp 28–34
53. Morales JA, Ruiz F, Jiménez I (1997) Papel de la sedimentación estuarina en el intercambio sedimentario entre el continente y el litoral: el estuario del Río Guadiana (S.O. España-Portugal). Revista de la Sociedad Geológica de España 10:118–135
54. Morales JA, Delgado I, Gutierrez-Mas JM (2006) Sedimentary characterization of bed types along the Guadiana Estuary (SW Europe) before the construction of the Alqueva dam. Estuar Coast Shelf Sci 70:117–131
55. Morales JA, Lozano C, Sedrati M (2019) Calculated potential bedload versus real transported sands along the Guadiana River Estuary (Spain-Portugal). J Marine Sci Eng 7:393–413
56. Nichols MM (1977) Response and recovery of an estuary following a river flood. J Sediment Petrol 47:1171–1186
57. Nichols MM, Biggs RB (1978) Estuaries. In: Davis RA (ed) Coastal sedimentary environments. Springer Verlag, pp 77–186
58. Nickling WG, Davidson-Arnott RGD (1991) Aeolian sediment transport on beaches and coastal sand dunes. In: Davidson-Arnott RGD (ed) Proceedings of the symposium on Coastal Sand Dunes. National Research Council of Canada, Ottawa, Canada, pp 1–35
59. O'Connor M, Cooper JAG, Jackson DWT (2007) Morphological behaviour of headland embayment and inlet-associated beaches, northwest Ireland. J Coastal Res 50:626–630
60. Parker DS, Kaufman WJ, Jenkins D (1972) Floc breakup in turbulent processes. J Sanitary Eng Div 98:79–99
61. Postma H (1967) Sediment transport and sedimentation in the marine environment. In: Lauft GH (ed) Estuaries, vol 83. American Association for the Advance of Science Publ, Washington DC, pp 158–179
62. Postma H (1980) Sediment transport and sedimentation. In: Olausson E, Cato I (eds) Chemistry and biogeochemistry of Estuaries. Wiley, Chichester, pp 153–186
63. Psuty NP (1988) Sediment budget and dune/beach interaction. J Coast Res 3(Special Issue):1–4
64. Pye K (1983) Coastal dunes. Prog Phys Geogr 7:531–557
65. Pritchard PW (1955) Estuarine circulation patterns. Proc Am Soc Civ Eng 81(717):1–11
66. Rodríguez-Vidal J, Rodríguez-Ramírez A, Cáceres LM, Clemente L (1993) Coastal dunes and post-Flandrian shoreline changes. Gulf of Cádiz (SW Spain). INQUA Mediterr Black Sea Shoreline Subcommission Newslett 15:12–15
67. Sabatier F, Maillet GM, Provansal M, Fleury TJ, Suanez S, Vella C (2006) Sediment budget of the Rhone delta shoreface since the middle of the 19th century. Mar Geol 234:143–157
68. Short AD, Hesp PA (1982) Wave, beach and dune interactions in southeastern Australia. Mar Geol 48:259–284
69. Simmons HB (1966) Estuary and coastline hydrodynamics. McGraw-Hill, Inc. 737 pp
70. Simmons HB (1972) Effects of man-made works on the hydraulic, salinity, and shoaling regimens of estuaries. In: Nelson BW (ed) Environmental framework of Coastal Plain Estuaries, vol 133. Geological Society of America Memoir, pp 555–570
71. Syvitski JPM, Milliman JD (2007) Geology, geography, and humans battle for dominance over the delivery of fluvial sediment to the coastal ocean. J Geol 115:1–19
72. Stutz ML, Pilkey OH (2002) Global distribution and morphology of deltaic barrier island systems. J Coastal Res 36:694–707
73. van Leussen W (1999) The variability of settling velocities of suspended fine-grained sediment in the Ems Estuary. J Sea Res 41:109–118
74. van Maren DS (2007) Water and sediment dynamics in the Red River mouth and adjacent coastal zone. J Asian Earth Sci 29:508–522
75. Walling DE, Moorehead PW (1989) The particle size characteristics of fluvial suspended sediment: an overview. Hydrobiologia 176(177):125–149
76. Wang Z, Li L, Chen D, Xu K, Wei T, Gao J, Zhao Y, Chen Z, Masabate W (2007) Plume front and suspended sediment dispersal off the Yangtze (Changjiang) River mouth, China during non-flood season. Estuar Coast Shelf Sci 71:60–67
77. Wheatcroft RA, Stevens AW, Hunt LM, Milligan TG (2006) The large-scale distribution and internal geometry of the fall 2000 Po River flood deposit: evidence from digital X radiography. Cont Shelf Res 26:499–516
78. Wolanski E, Williams D, Hanert E (2006) The sediment trapping efficiency of the macro-tidal Daly Estuary, tropical Australia. Estuar Coast Shelf Sci 69:291–298
79. Wood MS (2014) Estimating suspended sediment in rivers using acoustic Doppler meters. In: US geological survey fact sheet, pp 2014–3038
80. Wright DL (1977) Sediment transport and deposition at river mouths: a synthesis. Geol Soc Am Bull 88:857–868
81. Wright DL, Wiseman WJ Jr, Yang ZS, Bornhold BD, Keller GH, Prior DB, Suhayda JN (1990) Processes of marine dispersal and deposition of suspended silts off the modern mouth of the Huanghe (Yellow River). Cont Shelf Res 10:1–40

10 Marine Processes and Sediments on the Coast

10.1 Introduction

In the previous chapter we analyzed the processes that control the sedimentary contribution that reaches the coast from the mainland. However, significant amounts of sedimentary material can also come from the continental shelf through processes of a purely marine origin. These act in combination with the processes associated with tides and waves that have already been discussed in previous chapters. The arrival of this material in coastal formations is influenced by the dynamics of the *offshore* and *shoreface* environments (see definitions in Chap. 2). In this regard, the functioning of the subtidal zone of the coastal front is a fundamental factor in the long-term evolution (hundreds and thousands of years) of emerging coastal systems. This area plays a key role in sedimentary exchange between the continental shelf and the coast, especially on wave-dominated coasts.

The influence of these processes is not limited to the flow of sediments of marine origin towards the coast, but in parallel there is also a flow of sediments of coastal origin towards the sea. Both flows are part of the sedimentary balance of the coast and both are linked to the sedimentary exchange with the continent described in the previous chapter, as well as to sediment movements between different cells along the coast.

Several syntheses on the sedimentary exchange between the continental shelf and the coast have been published in recent decades [10, 20, 26, 35]. These have all made an effort to characterize the dynamics of the shoreface as an element of connection between the shelf and the coast. Recent papers have highlighted the role of the lower part of the shoreface as a store of sediments that is fundamental for the connectivity between the continental shelf and the coast [5]. However, all of them also demonstrate that the processes operating in this area are not sufficiently studied, and that therefore the precise mechanisms that govern the dynamics of the bed load material in this sector are not known in detail. A recent analysis of literature developed by Anthony [4] indicates that papers in this sector have focused on very specific places in the United States, Northern Europe and Australia, leaving the rest of the world's coasts uncovered. Similarly, not all processes have been addressed, focusing most of the efforts on characterizing the storm relaxation currents and the remobilization of fine sediments from the prodelta areas.

This chapter analyzes the marine sources of sediment that can reach the coast and be part of its sedimentary bodies, as well as the processes of marine origin that, together with the tide and swell, are responsible for the movement of these sediments.

10.2 Marine Sources of Sediment

Marine sediments are grains of unconsolidated material, both inorganic and organic, that are deposited on the seabed. The marine sediments that can reach the coast are those that were previously deposited on the continental shelves; thus, sediments from the deepest environments such as slope deposits, underwater canyons, deep underwater fans and abyssal plains will be excluded from this section.

The marine sediments deposited near the coast contribute about 25% of the seabed surface, although, from a volumetric point of view, they are of much greater importance, accounting for approximately 90% of the volume of all sediments deposited in marine environments [29].

Coastal sediments of marine origin are generally made up of a combination of several components. Most of the grains are silicates and come from the weathering and erosion of the continents (Fig. 10.1a) or from volcanic eruptions (Fig. 10.1b) and were previously deposited on the continental shelf. Another fraction of the grains of the shelf sediments comes from the chemical and biological processes that occur in seawater (Fig. 10.1c, d).

Normally the marine-derived siliciclastic material that reaches the coast comes from the reworking of relict sediments on the platform and is related to sea level movements.

J. A. Morales, *Coastal Geology*, Springer Textbooks in Earth Sciences, Geography and Environment,
https://doi.org/10.1007/978-3-030-96121-3_10

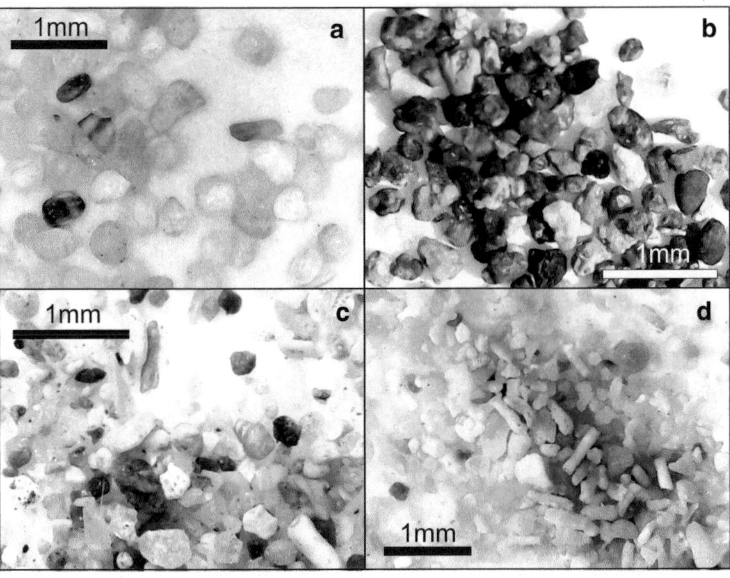

Fig. 10.1 Microscope images of coastal sediments with different compositions. **a** Sand composed of quartz grains (Cíes Islands, Spain). **b** Sand composed of silicates of volcanic origin (Batangas, Philippines). **c** Mixed siliciclastic–bioclastic sand (Île aux Nattes, Madagascar). **d** Bioclastic sand (La Cueva de Las Golondrinas, Dominican Republic)

Sediments inherited from previous sea level positions can be reworked to build coastal shapes. During lowstand periods there are erodible areas of old continental shelves whose materials can enter a new sedimentary cycle when transported by continental agents to the new coast. There can also be this type of contribution in times of highstand, when **coastal system tracts** are submerged. At these moments the old shelf sediments can be reworked by waves, tides and other currents, supplying material available to build the new coast.

Chemical sediments are generated by precipitation of minerals from seawater. Normally, for these precipitation processes to occur, large changes in the chemical parameters of the water need to occur. In marine environments near the coast these changes are usually associated with increasing salt concentration, as water evaporates when the saturation threshold is exceeded, especially in arid climates. The most common composition of precipitation sediments are calcite crystals, although dolomite, gypsum and halite also appear. This type of component is usually a minority in the sediments of siliciclastic platforms, although in carbonated platforms it may be the most common. Chemical grains are also abundant in mixed platforms, where siliceous and carbonated contributions are balanced. Precipitation processes, especially those associated with evaporation, can occur directly in coastal environments, especially in restricted subtidal and intertidal areas. Sometimes chemical precipitation processes occur in pores, resulting in cementation as part of the digenesis. In any case, chemical sediments deposited on continental platforms can be remobilized and shifted to the coast during high-energy moments in the form of **intraclasts**, while chemical sediments deposited in coastal environments can also be redistributed on the coast during these events. Low energy processes can concentrate layers of precipitated small crystals around a central detrital nucleus—these are called **oolites**. Elements and processes of chemical origin will be specifically addressed in Chap. 11.

Regarding sediments of biological origin, most of these come from the erosion of clastic marine deposits, which in turn are the result of the dismantling of **organic bioconstructions** such as coral reefs or sessile mollusk banks. These sediments are called bioclastic sediments and are composed of hard fragments of crystalline material secreted by organisms. The most abundant composition of these skeletal fragments is calcium carbonate (calcite or aragonite crystals), although there may also be **bioclasts** of other compositions (dolomite or silica). Coasts located in areas of high biological productivity usually have sediments composed almost entirely of bioclasts. This sediment is typical of subtropical areas, where grains correspond mostly to skeletal material from the dismantling of reefs or shell fragments. In these areas of high productivity, another biological component of sediment may come from the fecal secretions of organisms. This is the case of the carbonated **pellets**.

There are also elements of biological origin from the soft parts of organisms, such as the remains of dead plants or animals; these are called biogenic components. On the coasts of great continental contribution, the organic components are in the minority; however, in many cases, the components of biological origin can constitute a substantial fraction of the coastal sedimentary material, especially where the supply of terrigenous material is scarce. The processes governing the origin and distribution of these sediments will be discussed in detail in Chap. 12.

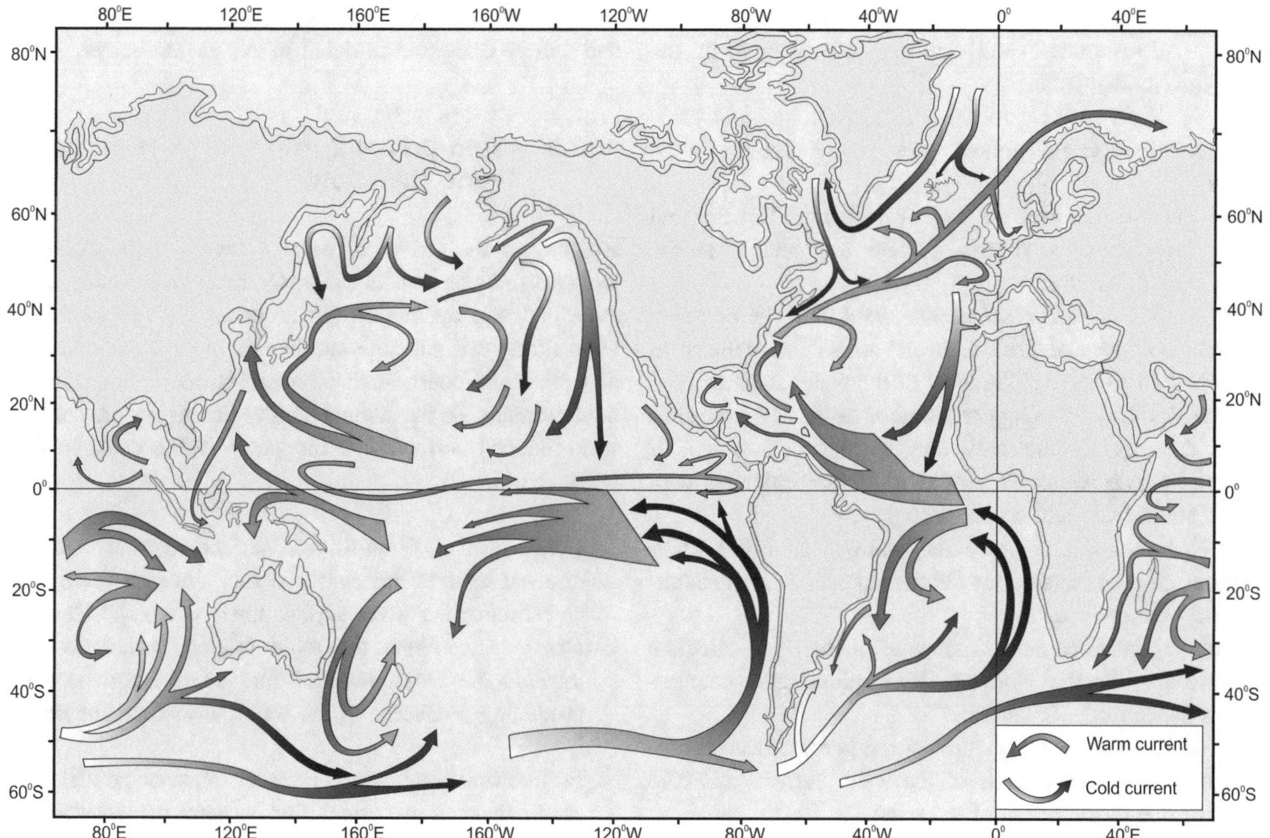

Fig. 10.2 Map of general oceanic circulation showing the different cells formed by the currents

Biological and chemical processes can act together. This is the case of the **stromatolites**, which are built in the intertidal areas by the trapping by cyanobacteria of grains of previously precipitated sediment.

10.3 Marine Processes on the Coastal Front

Tides and, above all, waves dominate the dynamics of the sediment on the coastal front. However, there are a number of processes that act in conjunction with waves and tides and hence have an important influence on the transfer of sediments between marine environments near the mainland and the coastal environments. The currents of the cells of general oceanic circulation approach some coasts, participating in the mobilization of sedimentary material and interacting with the tides. The tides themselves present a system of currents in the open coastal area with a particular dynamic that differs from that presented in the interior of the coastal environments. Finally, the winds, in addition to generating waves, can induce movements in coastal water bodies that result in permanent and ephemeral currents which are able to remobilize the sediment to and from the continental shelf, as well as along the coast.

10.3.1 General Oceanic Circulation: Currents in the Coastal Front

Ocean currents are large-scale waterway movements with different origins. There are two different circulation systems in which currents are generated by different causes. In the deepest water layer, there is a circulation system whose origin is based on the differences in density of the water bodies due to differing salinity and temperature. This deep system has no direct interaction with the coast and will not be addressed in this chapter.

On the surface, ocean currents are the result of viscous wind friction over the upper tens of meters of the ocean. These movements of the ocean water mass result in surficial currents that, although not as visible as waves and tides, act on a larger scale. Thus, the displacement of seawater bodies globally follows the same movement patterns as atmospheric circulation. Generally speaking, this current system is the ocean's response to the flow of energy in the atmosphere from the Tropics to the polar regions and is influenced by the Coriolis force and the presence of the continents. These forces generate a regular system of currents with a permanent character that affects large areas of the oceans (Fig. 10.2).

The system of general oceanic currents is now well known and responds to well-defined patterns based on the following fundamentals:

- Circulation systems follow very similar patterns in all oceans.
- Trade winds displace the water towards the Equator and the Coriolis force diverts it westward generating the equatorial currents.
- Equatorial currents circulate westward until the presence of the continents deflects them and turns them to the north and south to balance the level of the water surface.
- Currents flowing to higher latitudes result in a new shift of currents to the east due to the persistence of high-latitude winds. In parallel, the water interacts with the atmosphere and becomes cool.
- In each hemisphere the colder waters of the highest latitudes circulate again towards the Equator when encountering the continents.
- This circulation model generates cells that circulate clockwise in the Northern Hemisphere and counterclockwise in the Southern.
- In some areas, contrary currents are generated that compensate for rises or falls of the water surface that take place due to the action of the wind.

These persistent currents have a clear influence on some coasts, where they are among the main agents acting on sediment mobilization. Other currents affect a small area and act for a certain time, especially with seasonal frequencies. These are temporary phenomena and are the response of the sea to local conditions.

The interaction of these currents with the coasts has been studied throughout the 1970s and 80s. During these decades, some coastal currents induced by the general dynamics on the East Coast of the USA were characterized (e.g., Bennett and Magnell [7]; Schwing et al. [30]; Scott and Csanady [31]), although studies on other coasts were scarcer. A classic example is the research carried out in the Arctic by Wiseman and Rouse [34]. Also in this period, some authors were able to carry out synthesis work on the dynamics of these currents at the coastal boundary [3, 14, 15].

The three basic ways in which a current can interact with the coast are: (1) heading towards the coast; (2) moving away from the coast; and (3) circulating parallel to the coast. In all three forms, the loss of depth that occurs from the sea to the coast in relation to these currents influences a deviation of the surface of the water up or down, resulting in the appearance of secondary currents circulating at depth, in contact with the bottom, to compensate for this gradient of the sea surface. These effects are the same as those that arise when winds act directly on coastal waters, generating local currents. These currents are known as geostrophic currents and will be discussed in detail in the section below.

10.3.2 Wind-Driven Currents in the Coastal Front: Geostrophic Currents

Apart from the arrival of ocean currents at the coast, the wind also acts on coastal waters by transferring energy to the mass of water by friction on the surface of the coast. This wind thrust can generate currents or modify other currents that reach the coast, such as ocean currents themselves or tidal currents. To the same extent as the rest of the currents, wind-induced currents can interact with the coast in three ways:

1. *Convergence*: Wind-driven surface currents can be directed against the coast. In this situation there is an elevation of the water surface towards the shoreline.
2. *Divergence*: When the wind propels surface currents towards the sea it moves water away from the coast, producing a descent of the water surface on the coastal front.
3. *Parallelism*: If the wind drives the currents parallel to the coast, there is a current that sweeps the coast in the direction of the wind. This current is known as a coastal jet.

Each of these configurations in the vicinity of the coast results in a sea surface that ceases to be horizontal, inducing a pressure gradient in the water mass which in turn causes a movement of the deeper water layers to compensate for that gradient [15]. Secondary currents resulting from this gradient affect the deepest waters and can extend down the continental shelf to depths close to 100 m. The in-depth transmission of this sea surface deformation through the currents results in an offset of the stratification of the water column according to its density. The layers of this density stratification are regulated by their temperature and salinity, usually there is a contact surface between the warm and less dense waters of the surface and the deeper dense waters, called the **pycnocline**.

The three models of interaction of surface currents with the coast result in three forms of internal displacement of lines of equal water density (Fig. 10.3). In this way, the surface elevation of the convergent model results in descending bottom currents that depress the pycnocline. This phenomenon is known as **downwelling**. Conversely, the surficial depression associated with the model of divergence of surface currents results in ascending deep currents that raise the pycnocline, and this phenomenon is known as **upwelling**. In the third model, the **coastal jet** parallel to the shoreline is transmitted to

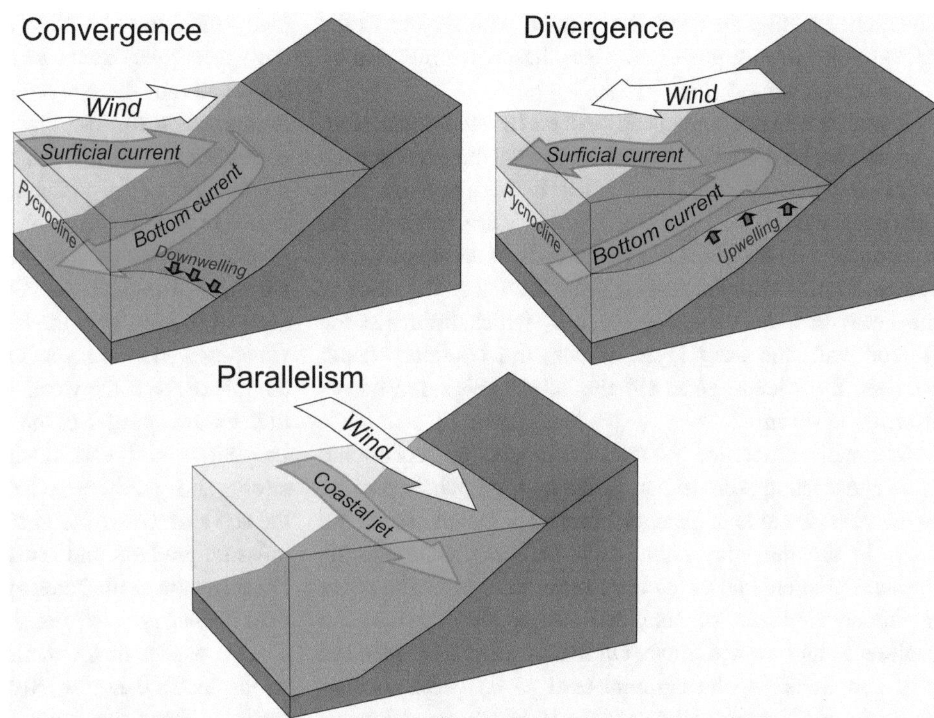

Fig. 10.3 Simplified conceptual scheme of the three different types of interaction between wind-driven currents and the shallow coast

the deep layer by a drag effect due to the internal viscosity of seawater through a laminar flow [21].

This laminar flow is not only linked to the coastal jet, but appears in all surficial currents. The depth transmission of the currents is affected by the Coriolis effect, so that at depth the speed of each layer is deviated at an angle. When looking at the velocity vectors in the water column together, a spiral turn is observed (Fig. 10.4). This turn is known as the **Ekman spiral** [16].

According to this theoretical model [8], for a constant wind, the angle between the direction of the wind and the surface current varies between 45º in deep water and 15º in coastal waters. This turn in the Northern Hemisphere would occur to the right of the wind, while in the Southern Hemisphere it would occur on the left. The combined displacement of the entire water mass of the flow sheets that make up the Ekman spiral results in a net movement perpendicular to the wind direction called **Ekman transport**. This net water movement occurs 90º to the right of the wind direction in the Northern Hemisphere and to the left in the Southern one.

If we take into account the data provided by Niedoroda et al. [26], in shallow areas the water column is not sufficient to develop the Ekman spiral, due to the influence of friction with the bottom. In this case, the currents of the upper and lower layers tend to align according to the convergence and divergence model in Fig. 10.3. Conversely, from a certain depth the water column offers enough thickness for the Ekman spiral to develop. In this coastal strip the currents of the surface layer may circulate differently from the deep layer, as there is a layer of shear between them. In this case, depending on the orientation of the coast with respect to the prevailing winds, the theoretical model of convergence and

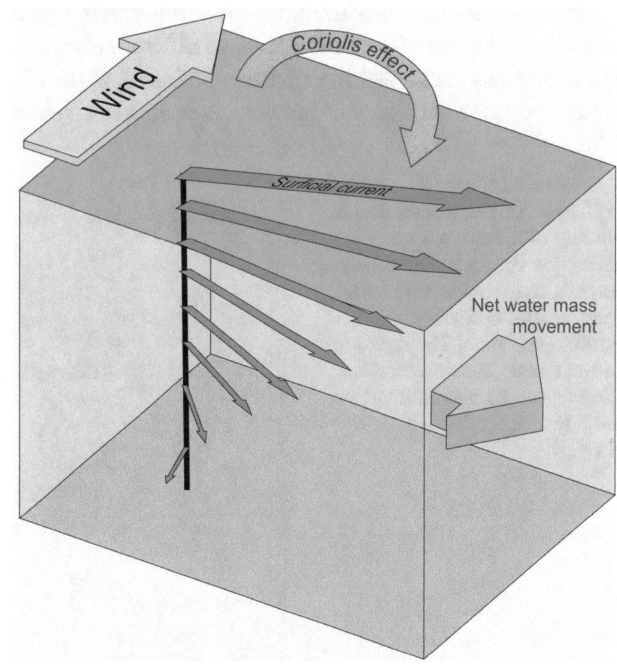

Fig. 10.4 Conceptual model showing the deviation of the wind-driven currents at depth by the Coriolis effect, indicating the direction of the net water mass movement (Ekman transport)

divergence is slightly more complicated with respect to the background currents generated, since these are influenced by the Ekman transport (Fig. 10.5).

These two zones have been defined by Niedoroda et al. [26] as the **friction-dominated zone** and the **geostrophic zone**, respectively, and there may be a transition zone between them (Fig. 10.6a). The limits and dimensions of this zone depend on the speed and persistence of the wind, which control friction with the surface layer, as well as the slope of the coast and the nature of the bed, which influence the friction with the deep layer. Hence, the boundary depth between the friction zone and the geostrophic zone ranges from 20 to 40 m.

The main difference between the models of geostrophic zone and friction zone is that, in the first, the winds parallel to the coast are able to generate currents perpendicular to the coast in the deep layer that arise to compensate for the pressure gradient caused by the Ekman transport. The result is the development of parallel flows to the coast in the shallowest and nearest fringe. On the contrary, in deeper and near-bottom areas, the currents tend to be perpendicular, either directed towards land or towards the sea depending on the direction of the wind.

Another consequence of the existence of this varying zonal behavior is that the same wind will have different effects on the geostrophic zone and the friction-dominated zone. The presence of the transition zone between them does not have its own behavior, but is simply an area in which the behaviors of the water layers adapt to the circulation of the surrounding areas and thus there is progressive behavior between them. It is interesting to note that, where the depth allows, vertical currents that cross the shear band appear to compensate for the movements of the water mass. These vertical currents may close some convection cells (Fig. 10.6b, c).

The position of these vertical currents depends on where the layer of friction with the atmosphere converges with the layer of friction with the bed. Therefore, the thickness of the water–bed friction layer is relative to the coastal slope, while the thickness of the water–air friction strip is directly related to the wind speed. So, the width of the coast affected by this convective cell varies with the weather conditions. With strong and persistent winds, the topographical anomaly of the sea surface will tend to be very pronounced, raising the pressure gradient and accelerating the bottom currents. This phenomenon contributes to wider turns and goes on to affect ever deeper areas of the shoreface.

One aspect that remains to be analyzed is the effect of winds that are not strictly perpendicular or parallel to the coast. In these situations, the shallow and deep currents are intertwined in a loop which circulates parallel to the coast (Fig. 10.7). These turns represent the 3D equivalent of the cells described in the previous paragraph, but they are affected by a longitudinal displacement to the coast parallel to the coastal jet.

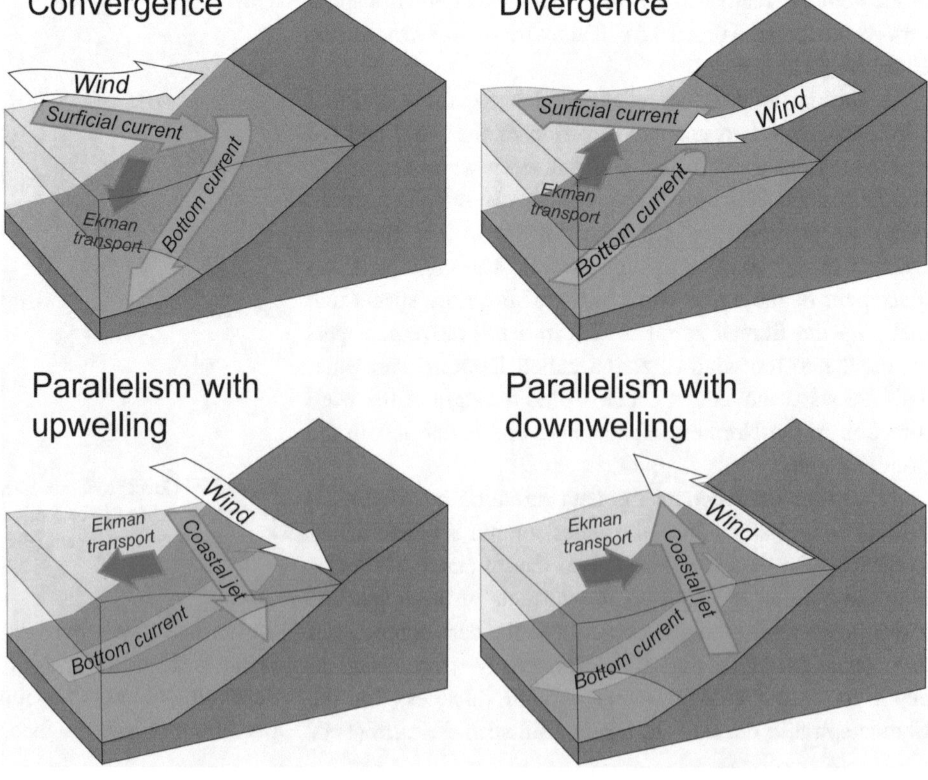

Fig. 10.5 Conceptual scheme of the four different types of interaction between wind-driven currents and the shoreface areas deep enough to develop an Ekman circulation. The angles between wind direction, surficial currents and Ekman transport are those typical for the Northern Hemisphere

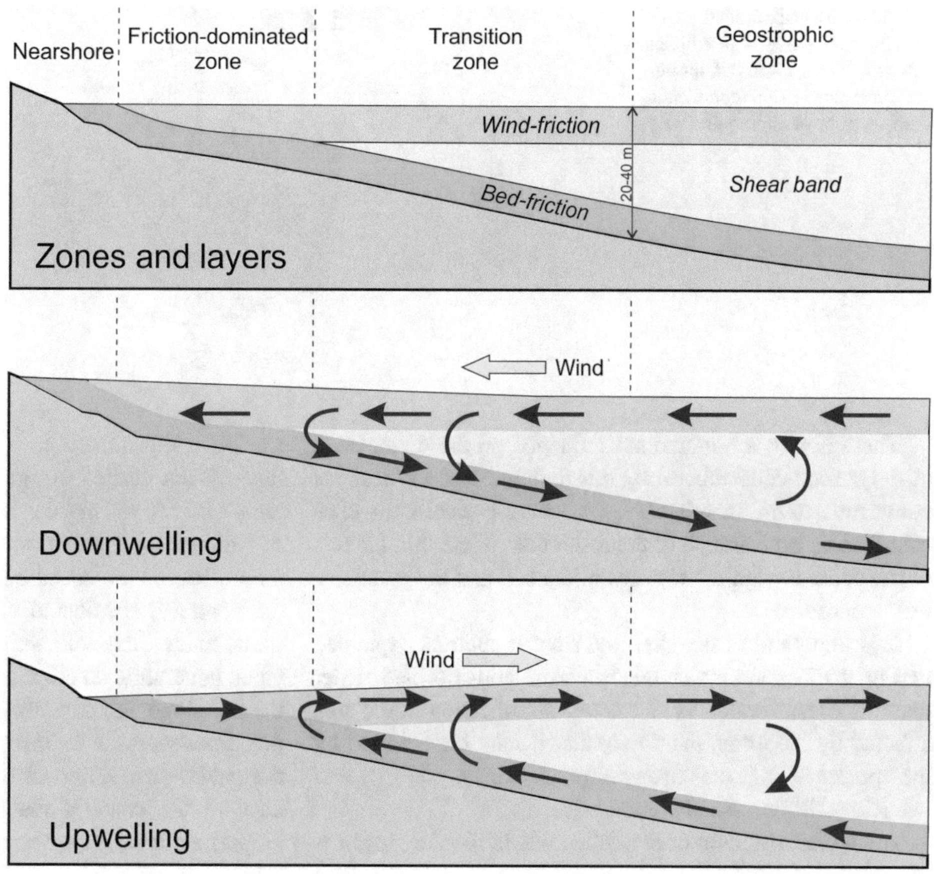

Fig. 10.6 Scheme of the main dynamic zones and layers of the coastal boundary layer and modeled geostrophic circulation schemes (adapted from Niedoroda et al. [26])

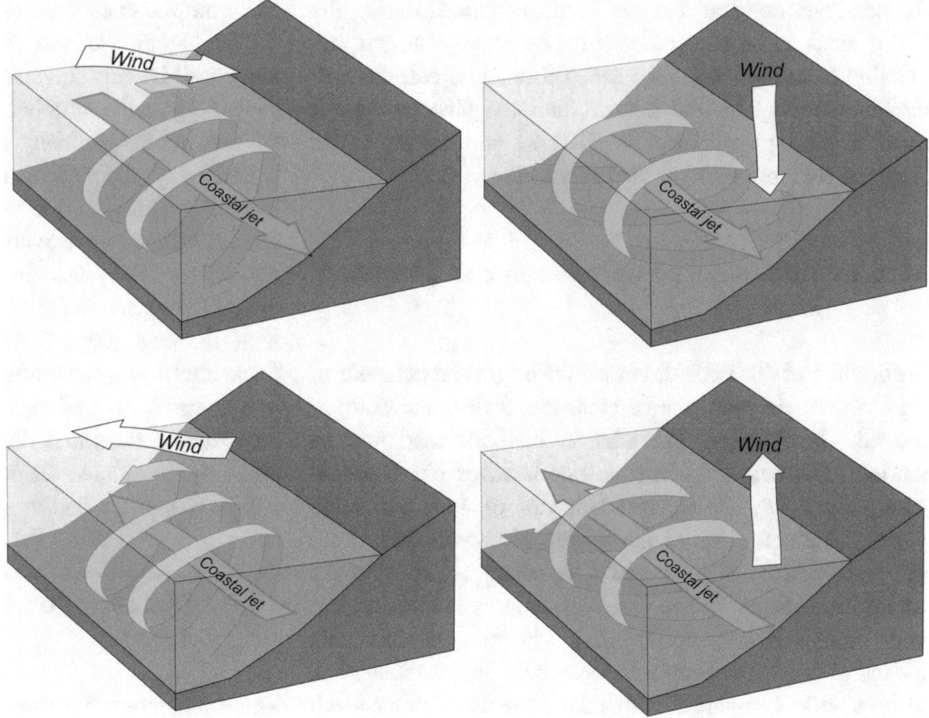

Fig. 10.7 Scheme of the gyres of currents constituting the coastal jet under winds oblique to the coast (based on Niedoroda et al. [26]). The senses of the coastal jet and the gyres are typical for the Northern Hemisphere

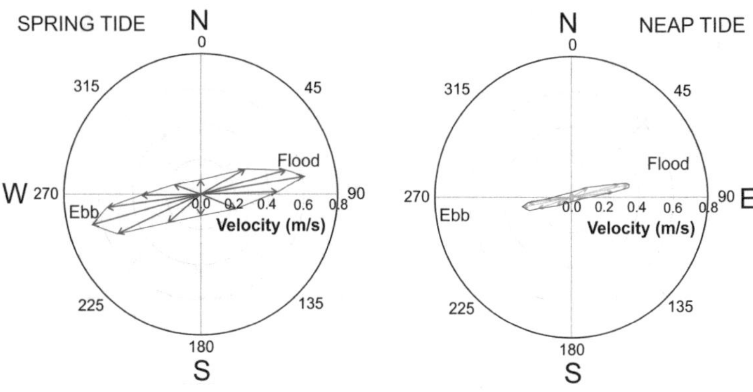

Fig. 10.8 Distribution of surficial tidal currents in elliptical patterns during cycles of spring and neap tides in the continental shelf in front of Plymouth (UK)

There is also a longitudinal influence on the dimensions of the coastal cells affected by this dynamic and the scale of convective turns. In this way, very long coastal cells contribute to an acceleration of the result due to less interference of the currents with the features of the bed and the geometry of the coastline.

It is important to note that the current patterns explained in Fig. 10.7 represent circulation in a non-stratified water mass. If coastal waters have marked stratification in the area affected by the turns, the flow patterns may be distorted by the presence of the pycnocline; however, the overall behavior will be essentially the same.

A singular situation occurs when winds blowing towards the coast stop, especially if the winds have been strong, such as in stormy conditions. When calmed, the abnormal rise of the sea level (surges) that has been maintained during the storm tends to be compensated for by strong currents circulating from the land to the sea. These dense currents will circulate through to the bottom, and are able to deposit material on the continental shelf. These currents are called **storm return currents** or also **relaxation currents**.

10.3.3 Tidal-Induced Currents in the Coastal Front

Section 8.5 of this book discussed the origin and behavior of tidal currents in open marine areas and their relationship to the tidal level curve. This chapter has described how in marine areas near the coast the tide behaves like a propagating wave and the internal behavior of the water mass closely resembles that of the waves. However, due to the proximity of the bed and the long wavelength of the tide, the orbital movements of the water mass are not circular, but very elongated ellipses [18, 28]. The analysis has only looked at this behavior two-dimensionally in a vertical plane to be able to compare it with tidal currents in channelized systems. In marine areas, however, the amplitude of the system allows tidal currents to move three-dimensionally in any direction. These currents may also be affected by Coriolis turns similar to those shown in wind-induced currents. Work carried out in the 1980s showed that tidal currents in open marine areas do not simply have a two-dimensional flood and ebb pattern that passes through a moment of zero speed before the currents are reversed. On the contrary, considered in a horizontal plane, the currents rotate their direction while accelerating and decelerating, again generating an elliptical pattern (Fig. 10.8). In this case, the bed slope, linked to the tidal range, make up the variables that determine the intensity of these currents [6]. Although the patterns are elliptical rather than bidirectional, the presence of the coast is manifested in the orientation of the ellipse, since its maximum axis coincides with the averaged orientation of the coast. In this way, the maximum tidal currents run parallel to the coastline, while the currents to and from the coast are minimal.

The variation of this elliptical pattern when approaching the coast is very diverse, in time as well as in space, depending on the morphology of the shoreface, the nature of the bed and the geometry of the coastline. When the process is underway, the currents can be amplified when approaching the coast if the tide wave enters narrow bays, while in open areas the opposite effect occurs if the energy dissipates by friction. In any case, in the lower layers of the water mass there is always a decrease in current speeds due to friction with the bed (e.g., Fjeldstad [19]; Sverdrup [32]). The thickness of this shear-affected bottom layer depends on the roughness of the bed and the mean speed in the water column, which in turn is determined by the tidal range. It is precisely in this lower layer, close to the water–bottom interface, that the bed load transport of sediment occurs.

10.4 Dynamics of Sediment in the Coastal Front

The sedimentary dynamics in the shoreface area have been synthesized in recent work [13]. This work highlights the importance of the papers of the end of the last century (e.g., Niedoroda et al. [25]; Wright [35]) and recent efforts to determine sediment transport patterns on the coastal front through the use of models (e.g., Aagaard and Hughes [1]). Despite this, the authors themselves declare that these patterns are simple conceptual models and the complex principles that govern sedimentary processes in this area remain definitively unestablished.

Part of the complexity of conducting studies in this area lies in the difficulty of directly measuring the currents that have been described in the previous sections. Any measure of the speed of the currents can never measure the currents individually according to their origin (wind-driven or tidal), but the final component of a combination thereof [10, 20]. Another part of the complexity can be attributed to the difficulty of measurement during processes that move a greater amount of sediment [35]. Intuitively, it will be understood that reliable measurements of turbulent processes that occur during a storm are an impossibility [22]. Finally, it should be noted that on the few occasions when there have been attempts to directly measure the processes on coastal fronts, these have always been carried out on a short timescale, and have not addressed the performance of the long-term processes [23].

One observation that aroused interest from the beginning of the studies of the coastal front was the concave profile, a characteristic observed by authors in the late nineteenth and early twentieth centuries [9, 17]. Since then, several attempts have been made to explain this phenomenon. It seems that the most commonly accepted explanation is based on the erosive nature of the contact fringe between the sea and the coast (e.g., Aagaard and Sorensen [2]; Ortiz and Ashton [27]). Thus, sediment deposited in this strip during times of a different sea level tends to be used by currents and, especially, by waves to feed coastal and also the deeper marine environments.

From the point of view of magnitude, wind-induced currents have been measured on coasts where the tidal influence is minimal. The currents in the water–bed interface range from 5 to 20 cm/s, reaching up to 70 cm/s on the surface [24]. These speeds are enough to move sediments with sizes smaller than medium sand. In this case, there are works that have measured specific transport rates (e.g., Wright [35]), although there are few papers that determine sediment balances at longer timescales (decades to millennia) from a general point of view. This is significant, because these net sediment movements are responsible for determining the actual contribution of this area to the building of coastal forms [13].

Regarding the tidal currents, it can generally be said that speeds are usually small and therefore have a very limited sandy transport capacity. Moreover, as can be seen in Fig. 10.8, these currents have a high symmetry, so that net residual transport in a given direction is usually almost negligible and, if any, would be longshore. However, tidal water mass displacement vectors may interact with the other currents already described in this area, contributing to increasing or diminishing their potential for the net transport of the bed material. Only on coasts with the right bathymetry and geometry can tidal currents in themselves be an important agent for the transport of sandy material on the open coast. However, these currents can be considered an effective mechanism for the movement of the suspended material in the water mass, mainly because they contribute to diverting and redistributing suspended sediment that has arrived from the mainland rivers via turbid plumes.

The difficulty that exists in understanding these processes through the direct measurement of the currents acting on the coastal front can be sidestepped by analyzing the processes from a morpho-sedimentary point of view. The analysis of changes in the topography of the bed and, above all in the distribution and dynamics of bedforms appearing on the coastal front, can be a useful tool in understanding the dynamics of sediment transport in this area. This is the type of approximation made by Cowell et al. [11, 12]. In this regard, although some authors claim that the current resolution of the equipment is not enough to characterize these changes [4], other authors such as [33] highlight the use of underwater acoustics (side-scan sonar and multibeam echosounder) in the two most recent decades for the study of sedimentary dynamics in this area.

The use of these techniques for the characterization of bedforms and the calculation of transport vectors in bed load deducted from these structures is well known. But, in addition, acoustic signals can be used to determine the dynamics of the turbulent flows at the bottom, as they form submerged plumes of highly concentrated suspended matter that disperses the acoustic signal as it passes through the water column. The analysis of the acoustic signal with the appropriate instruments allows vertical profiles of the concentration of the suspended load and particle size to be obtained, as well as visualization of three-dimensional blocks of the turbulent flow. The application of this technique is not intrusive and can be carried out with the spatiotemporal interval deemed most appropriate to finally understand the transport of sediments in this complex area.

References

1. Aagaard T, Hughes MG (2017) Equilibrium shoreface profiles: a sediment transport approach. Mar Geol 390:321–330
2. Aagaard T, Sorensen P (2012) Coastal profile response to sea level rise: a process-based approach. Earth Surf Proc Land 37:354–362
3. Allen JS (1980) Models of wind-driven currents on the continental shelf. Annu Rev Fluid Mech 12:389–433
4. Anthony EJ (2009) Shore processes and their palaeoenvironmental applications. Elsevier, Amsterdam, 519pp

5. Anthony EJ, Aagaard T (2020) The lower shoreface: morphodynamics and sediment connectivity with the upper shoreface and beach. Earth Sci Rev 210:10334
6. Battisti DS (1982) Estimation of nearshore tidal currents on nonsmooth continental shelves. J Geophys Res 87:7873–7878
7. Bennett JR, Magnell BA (1979) A dynamical analysis of currents near the New Jersey Coast. J Geophys Res 84:1165–1175
8. Broström G, Rodhe J (1996) Velocity shear and vertical mixing in the Ekman layer in the presence of a horizontal density gradient. Cont Shelf Res 10:1245–1257
9. Cornaglia P (1889) On beaches. In: Fisher JS, Dolan R (eds) Beach processes and coastal hydrodynamics. Benchmark papers in Geology, 39. Dowden, Hutchins and Ross, Stroudsberg, pp 11–26
10. Cowell PJ, Hanslow DJ, Meleo JF (1999) The shoreface. In: Short AD (ed) Handbook of beach and shoreface morphodynamics. Wiley, Chichester, pp 39–71
11. Cowell PJ, Stive MJF, Niedoroda AW, de Vriend HJ, Swift DJP, Kaminsky GM, Capobianco M (2003) The coastal-tract (part 1): a conceptual approach to aggregated modeling of low-order coastal change. J Coastal Res 19:812–827
12. Cowell PJ, Stive MJF, Niedoroda AW, Swift DJP, de Vriend HJ, Buijsman MC, Nicholls RJ, Roy PS, Kaminsky GM, Cleveringa J, Reed CW, de Boer PL (2003) The coastal-tract (part 2): applications of aggregated modeling of lower-order coastal change. J Coastal Res 19:828–848
13. Cowell PJ, Kinsela MA (20180 Shoreface controls on barrier evolution and shoreline change. In: Moore LJ, Murray AB (eds) Barrier dynamics and response to changing climate. Springer Nature, Cham, pp 243–275
14. Csanady GT (1977) The coastal jet conceptual model in the dynamics of shallow seas. In: Goldberg ED, McCave IN, O'Brien JJ, Stede JH (eds) The sea, vol 6. Wiley, New York, pp 107–114
15. Csanady GT (1982) Circulation in the coastal ocean. Reidel Publishers Co., Boston, 280pp
16. Ekman VW (1905) On the influence of the Earth's rotation on ocean currents. Arkiv för Matematik, Astronomi och Fysik 2 (11):1–53
17. Fenneman NM (1902) Development of the profile of equilibrium of the subaqueous shore terrace. J Geol 10:1–32
18. Fjeldstad JE (1929) Contribution to the dynamics of free progressive tidal waves, Norwegian North Polar Expedition with the Maud, 1918–1925. Sci Results 4(3):1–80
19. Fjeldstad JE (1936) Results of tidal observations, Norwegian North Polar Expedition with the Maud, 1918–1925. Sci Results 4 (4):1–88
20. Kleinhans MG (2002) Sediment dynamics on the shoreface and upper continental shelf, a review. Sandpit Report, EC MAST Project no. MAS3-CT97-0086, 67pp
21. Kämpf J, Chapman P (2016) Upwelling systems of the world. In: Chapter 2: The functioning of coastal upwelling systems. Springer Nature, Cham, pp 31–65
22. Kleinhans MG, Grasmeijer BT (2006) Bed load transport on the shoreface by currents and waves. Coast Eng 53:983–996
23. Nicholls RJ, Birkemeier WA, Lee GH (1998) Evaluation of depth of closure using data from Duck, NC, USA. Mar Geol 148:179–201
24. Niedoroda AW (1980) Shoreface surf-cone sediment exchange processes and shoreface dynamics. In: NOAA Tech. Memo. OMPA-l, 89pp
25. Niedoroda AW, Swift DJP, Hopkins TS, Ma CM (1984) Shoreface morphodynamics on wave-dominated coasts. Mar Geol 60:331–354
26. Niedoroda AW, Swift DJP, Hopkins TS (1985) The shoreface. In: Davis RA (ed) Coastal sedimentary environments (2nd ed). Springer Verlag, New York, pp 533–624
27. Ortiz AC, Ashton AD (2016) Exploring shoreface dynamics and a mechanistic explanation for a morphodynamic depth of closure. J Geophys Res Earth Surf 121(2):442–464
28. Rattray M Jr (1957) On the offshore distribution of tide and tidal current. Trans Am Geophys Union 38:675–680
29. Seibold E, Berger W (2009) The sea floor. In: Chapter 4: Sources and composition of marine sediments. Springer Nature, Cham, pp 45–61
30. Schwing FB, Kjerfve BJ, Sneed JE (1983) Nearshore coastal currents on the South Carolina continental shelf. J Geophys Res 88:4719–4728
31. Scott JT, Csanady GT (1976) Nearshore currents off Long Island. J Geophys Res 81:5401–5409
32. Sverdrup HU (1927) Dynamics of tides on the North Siberian shelf. Geofysiske Publikasjoner 4(5):1–75
33. Thorne PD, Bell PS (2009) Acoustic measurement of near-bed sediment transport processes. In: Steele JH (ed) Encyclopedia of ocean sciences (2nd ed). Elsevier, pp 38–51
34. Wiseman WJ Jr, Rouse LJ (1980) A coastal jet in the Chukchi Sea. Arctic 33:21–29
35. Wright LD (1995) Morphodynamics of inner continental shelves. CRC Press, Boca Raton, 241pp

11 Chemical Processes and Sediments on the Coast

11.1 Introduction

In the previous chapters, the origin of siliciclastic sediments has been discussed, including how they reach the coast from the continent through the mechanical transport of particles generated by weathering processes. In addition to these solid particles, inland waters transport a dissolved ionic charge that comes from the chemical disintegration of the rocks. Throughout the Earth's history, this ionic charge has enriched marine waters with salts, which contain much higher concentrations of almost all soluble elements than inland waters. However, this ionic charge does not remain permanently dissolved in seawater. Through some purely chemical processes, as well as others related to the activity of organisms, a part of this charge precipitates to form chemical and biochemical sediments. A typical example of a deposit produced by chemical processes is the salt deposit resulting from the evaporation of a volume of seawater.

The basic principles of chemistry state that, for chemical precipitation or dissolution processes to occur, the equilibrium conditions of the solution must be altered. Therefore, the occurrence of these processes in natural environments requires environmental changes. In the marine environment these changes rarely occur, because the mass of water containing the ionic charge is enormous. That is why most of the precipitation and dissolution reactions are related to the activity of organisms. However, coastal environments are especially changeable and theoretically present the ideal conditions for the chemical changes necessary for precipitation–dissolution processes to occur.

River mouths are environments that are well known for the processes of diffusion of suspended matter from the turbidity maximum, as described in previous chapters. However, they are also environments that are particularly conducive to the processes of precipitation and dissolution. For example, the increase in salinity of the water from mixing induces the precipitation of carbonates, as well as iron and manganese (Fe and Mn) oxides, which are also associated with other trace elements such as rare earths. These processes often go unnoticed, because their precipitation is volumetrically much lower than that produced by the diffusion of turbidity, but this does not mean that they are not important.

Lagoons, tidal flats and salt marshes are restricted environments where significant changes in the conditions regulating chemical and biochemical processes are common. As these are quiet environments in terms of energy, they are also the most suitable environments for high activity of the organisms that results in the possibility of accumulation of organic sediments. A good example of this is the formation of peat associated with supratidal environments. It is in these systems associated with arid climates where evaporation is the factor that regulates the relationship between solvent and solute that causes a strong precipitation of salts, giving rise to the characteristic deposits of chemical sediments called evaporites.

This chapter will study in detail all the processes that develop the sediments that will later become the chemical components of rocks, as well as certain elements that will be included in terrestrial sediments in a lesser way.

11.2 Solubility: Control Factors

The processes of precipitation and dissolution are linked to the concept of **solubility**. Solubility is defined as the capacity of a substance to contain another. The substance that is found in greater quantity is called the **solvent**, while the one that is mixed in smaller proportion is called the **solute**. In natural environments, the most common solvent is water, which has a great capacity to dissolve mineral substances. In this chapter, water is always considered as a solvent and the solid mineral phases as a solute. Thus, dissolution is understood as the process of breaking mineral molecules and incorporating their elements into the dissolved phase in an ionic form. In the opposite sense, precipitation is understood as the union of different ions dissolved in water to form mineral compounds that pass

J. A. Morales, *Coastal Geology*, Springer Textbooks in Earth Sciences, Geography and Environment,
https://doi.org/10.1007/978-3-030-96121-3_11

directly into the solid state. In this sense, precipitation is a form of crystallization that occurs at atmospheric temperature.

There are several concepts related to solubility that should be defined in order to understand the functioning of precipitation–dissolution reactions. These are the concepts of concentration, threshold of solubility, saturation and oversaturation.

The **concentration** is the proportion of solute contained in the solvent. It is usually expressed as amounts of solute in relation to the amount of solvent. The units used to express concentration vary depending on the units used to express the amounts of solute and solvent. In aqueous solutions, the ratio of weight of solute to volume of solvent is usually used, although percentages (%), parts per thousand (%o) and parts per million (ppm) are also used if concentrations are very small.

The **solubility threshold** is linked to the concept of solubility itself, since it is the maximum concentration of solute that can be dissolved in a given solvent. In other words, it is the maximum capacity that a solvent can contain of a given solute. Mathematically this threshold is expressed as a **solubility constant** (also solubility coefficient or solubility product). Solubility threshold is notably affected by the environmental conditions.

Saturation is the relationship between concentration and the solubility threshold. When a solvent contains all the solute it can, it will be right at the threshold. It is then said to be saturated and therefore **in equilibrium**. Under these conditions it will neither dissolve nor precipitate. If the concentration of a solute is below the threshold it is said to be **unsaturated**. In these conditions, the solvent can continue to dissolve solute until it reaches the threshold. If, on the other hand, the concentration is above the threshold it is said to be **oversaturated**. It should be noted that in theory a solvent can never contain a concentration of solute greater than its capacity. Only if there is a change in environmental conditions which decreases the threshold of solubility can oversaturation occur and then precipitation will be induced.

The solubility of two substances and therefore the saturation threshold depend directly on the balance of intermolecular forces between these two substances. The factors that control this equilibrium in coastal environments are the same as those that regulate reactions in the laboratory from a chemical point of view [7]. Some are linked to the nature of the substances to be dissolved (ionic potential), others depend on the conditions of the environment in which the reaction takes place (temperature, pH and Eh) and others are part of the environmental conditions of the natural environments (salinity, pressure of dissolved gases and organic activity). Although in natural environments it is normal for all these factors to act together, and changes in some affect others, the best way to understand the influence of these factors on chemical reactions is to analyze them individually.

11.2.1 Ionic Potential

Ionic potential can be defined as the ability of ions to surround themselves with water molecules. It is, therefore, an intrinsic property of ions that depends on their charge and their size (ionic radius). Thus, the ionic potential is directly proportional to the charge and inversely proportional to the ionic radius. The ionic potential is the property that gives the elements their soluble or insoluble character.

According to their ionic potential the ions can be classified into:

- *Soluble cations*: their ionic potential is less than 3.
- *Insoluble hydrolysable ions*: with ionic potential between 3 and 12.
- *Soluble anionic complexes*: their ionic potential is greater than 12.

11.2.2 Temperature

The temperature of the water controls its solubility constant and therefore its saturation capacity. An increase in temperature implies an increase in the solubility of the substances in water that will favor the dissolution processes. This is because the movement of water molecules decreases their cohesive forces and solute ionization is faster and more effective. A decrease in this will have the opposite effect, favoring oversaturation and, therefore, chemical precipitation processes (Fig. 11.1).

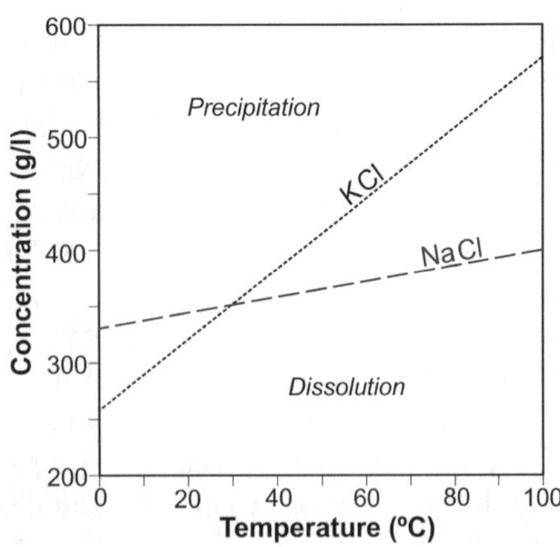

Fig. 11.1 Solubility thresholds of NaCl and KCl as a function of temperature

11.2.3 pH

The pH is the negative log 10 of the hydroxyl ion in a solution and represents the quantitative measure that determines whether a solution is alkaline or acidic. Therefore, in natural environments, as the pH depends on the concentration of H^+ ions dissolved in water, this concentration influences the dissolution of other substances due to the common ion effect. The pH also influences the solubility of salts containing an anion that can undergo hydrolysis, as these substances have negative ions that can react with the protons in the water in an acidic environment to produce other compounds. Thus, the pH can alter the balance in the dissolution of mineral substances depending on their composition, so that the solubility threshold is sometimes raised and sometimes lowered.

In coastal environments there are frequent changes in pH. It should be noted that the pH of river water is slightly alkaline to slightly acidic, while the pH at the seawater surface tends to be almost constant at 8.3 (slightly basic). It is evident that the processes of water mixing involve a change in pH for both seawater and river water, altering the chemical balance and meaning a move towards situations of oversaturation or unsaturation depending on the substance in question.

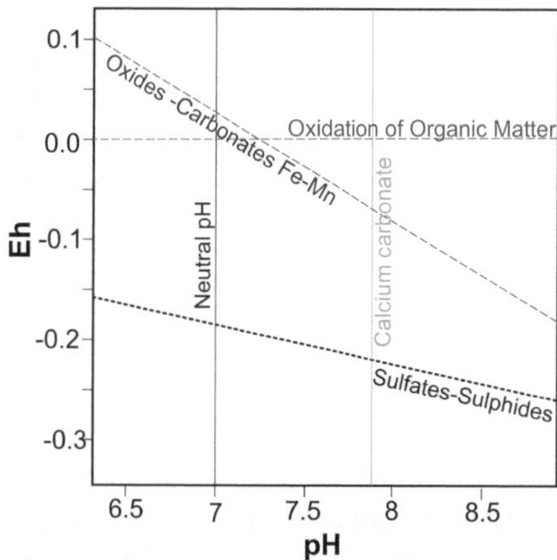

Fig. 11.2 Geochemical barriers in Eh–pH relationships

11.2.4 Redox Potential (Eh)

Oxidation–reduction potential (redox potential or Eh) is the energy required to lose or gain electrons in a given oxidation state. It is, therefore, a relative measure of the oxidation or reduction intensity in a solution—i.e., the concentration of electrons in a solution. Eh is directly dependent on the ionic potential and, in natural environments, also on the pH, so that changes in Eh greatly influence the conditions that set the equilibrium thresholds for solubility.

Both pH and Eh establish what are known as **geochemical barriers**. Each geochemical barrier corresponds to a value of Eh and pH, or a relationship between the two, and limits the formation of a certain mineral. Thus, barriers divide fields of mineralogical stability. In sedimentary environments, Eh and the Ph are interdependent, so, knowing the stability limits of water, it is possible to draw it in a diagram with Eh as an ordinate and Ph as an abscissa and show the fields for each of them (Fig. 11.2).

11.2.5 Salinity

In solutions where there is more than one dissolved substance, it is common for there to be some ion (anion or cation) that is part of more than one of the dissolved compounds. In this case, the solubility of one of the substances influences the other. This is called the **common ion effect** in chemistry. In the case of waters from natural environments, salinity is an important factor of influence, since it is a reflection of the concentration of dissolved salts, which may have ions in common with other mineral compounds that may dissolve or precipitate. Thus, variations in salinity can alter the balance in the saturation state of other compounds.

In coastal environments a constant supply of fresh water from the continent, through rivers and rains, necessarily reduces salinity, while evaporation associated with low water exposure increases it, especially in arid climates. These changes in salinity produce a shift in solubility thresholds that can lead to imbalances which cause precipitation and dissolution processes. The case of carbonate precipitation in fluviomarine systems will be analyzed in detail later on.

11.2.6 Pressure of Dissolved Gases

Marine waters contain a wide variety of dissolved gases. Equilibrium processes at the surface of seawater bodies normally lead to saturation in dissolved gases and this will not change unless non-conservative processes occur. In coastal systems, there are several processes that lead to variation in dissolved gas concentration, mostly associated with organic activity. Perhaps the most intuitive example is the variation in O_2 and CO_2 concentrations due to respiration of organisms and plant activity. In open coastal systems, water movements (waves, tides and other currents) renew gases at the water–atmosphere interface; however, in more restricted systems, these variations may not be compensated for and may induce chemical processes of precipitation. A clear example is the influence of CO_2 concentration on water pH. An increase of this gas in solution can acidify the water with an immediate consequence in the dissolution of calcium carbonate. Conversely, the decrease of CO_2 due to processes of photosynthesis can produce the precipitation of calcite. This phenomenon is often associated with coastal lagoons in tropical climates, such as the lagoon on Andros Island in the Bahamas [6].

11.2.7 Organic Activity

Physiological processes of organisms, such as photosynthesis, respiration or nutrition, directly affect the balance of solubility in coastal waters, as they affect many of the variables that have been characterized in the previous paragraphs. Examples that have already been cited are changes in the concentration of dissolved gases that cause the processes of respiration and photosynthesis. These changes in gas concentration directly influence the Eh and pH values of the sedimentary environment. Other activities carried out by organisms, such as some mechanisms used by plants to maintain their osmotic balance during tidal immersion processes, are capable of directly modifying the pH of the water surrounding plant tissues. In this way, the organisms cause imbalances in the environmental conditions that are capable of inducing processes of precipitation and dissolution. All these changes will be analyzed in detail at the end of this chapter.

11.3 Precipitation and Dissolution of Carbonates

Among the most common processes of dissolution and precipitation in nature are those involving calcium carbonate. Many books on geomorphology explain these processes in relation to karst phenomena occurring in continental environments, but these processes are equally important in marine waters and especially in coastal waters, where significant amounts of calcium carbonate (and also magnesium) can be formed in relation to changes in salinity, pH, Eh and dissolved gases. Many of these changes occur in relation to organic activity, but many others have a purely chemical origin.

11.3.1 Chemical Equilibrium of Calcium Carbonate

The main control of solubility of calcium carbonate is pH, which is controlled by the partial pressure of carbon dioxide dissolved in water according to the following reactions, all of which are reversible:

1. The dissolution of gaseous CO_2 in liquid water leads to the formation of carbonic acid:

$$CO_2 + H_2O \leftrightarrow H_2CO_3$$

2. Carbonic acid dissociates. This dissociation generates the calcium bicarbonate anion and a hydroxyl cation:

$$H_2CO_3 \leftrightarrow H^+ + HCO_3^-$$

This is why the dissolution of CO_2 leads to the acidification of the environment.

3. The bicarbonate dissociates to form the carbonate anion and another hydroxyl cation:

$$HCO_3^- \leftrightarrow H^+ + CO_3^{2-}$$

4. Finally, the carbonate anion can combine with the calcium cation to precipitate calcium carbonate:

$$CO_3^{2-} + Ca^{2+} \leftrightarrow CaCO_3$$

The combined result of all these reactions is summarized in the following reaction:

$$CO_2 + H_2O + Ca^{2+} \leftrightarrow CaCO_3 + 2H^+$$

This reaction shows how the dissolution of carbon dioxide is responsible for balancing the solubility of calcite in water, as it controls the pH of the environment in a secondary way. If the reaction occurs to the right, precipitation of calcite or aragonite will occur, while the reaction to the left will produce dissolution.

It is considered that there are five basic mechanisms through which calcium carbonate precipitation is achieved:

- *Temperature increase*: All gases are less soluble in warm water. A warming of the water will cause a loss of CO_2 pressure and the pH will increase: both effects combined will cause precipitation. Conversely, a cooling of the water will be responsible for dissolution. For this reason, carbonate sediments are formed in warm environments, with tropical and shallow subtropical areas being the preferred places for precipitation.
- *Increase in pH*: The loss of acidity causes an imbalance in the environment by modifying the pressure of dissolved CO_2. This environmental change can cause the solubility threshold to be exceeded and oversaturation will occur (Fig. 11.3).
- *Increased salinity*: Carbon dioxide is less soluble in saline water than in fresh water; therefore, as salinity increases through evaporation, the inhibition of calcium carbonate precipitation increases (Fig. 11.4).

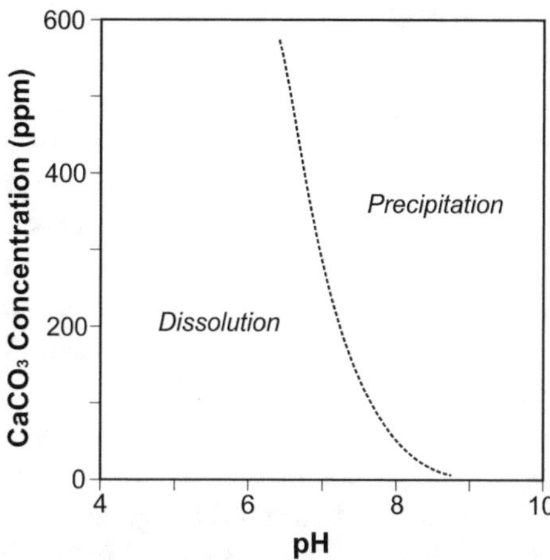

Fig. 11.3 Solubility of calcium carbonate as a function of pH in marine waters (adapted from Langmuir [15])

- *Water agitation*: When the water is agitated by the wind, saturation in dissolved gases is favored. The movement of the water favors the dissolution of CO_2 and has a secondary influence on the ionic imbalance of the calcium carbonate. On the other hand, agitation favors the oxygenation of the water and the development of organisms whose activity also influences this balance.
- *Organic activity*: Plants and animals have a metabolism that influences the concentrations of dissolved CO_2 in water. The activity of both types of organism has an influence that shifts the threshold of solubility in opposite directions. During photosynthesis, plants take up CO_2 and release O_2 through the reaction:

$$6CO_2 + 6H_2O \rightarrow C_6H_{12}O_6 + O_2$$

where $C_6H_{12}O_6$ is a glucose molecule.
Conversely, both animals and plants emit CO_2 and consume O_2 when they breathe. The organic reactions generate diurnal cycles of precipitation and dissolution. During the day, when photosynthesis takes place, the CO_2 produced causes the precipitation of $CaCO_3$. On the other hand, during the night, photosynthesis ceases but respiration continues, so that the CO_2 content increases, which favors dissolution.
- *Changes in pressure*: Under the seawater column, the partial pressure of carbon dioxide is much higher than the atmospheric pressure. In general terms, for every 10 m of water depth there is a pressure increase equivalent to 1 atm. As a result, deep seawater is enriched in CO_2 relative to surface water. If deep water rises to the surface as in upwelling processes, the pressure of the CO_2 will decrease and the gas will be released from the water, resulting in the precipitation of calcium carbonate crystals.

The dissolution process is also favored by the same factors, which act in the opposite direction, so a decrease in pH will raise the solubility threshold and favor the ability to dissolve. Similarly, an increase in pressure similar to that produced in a downwelling process will favor the dissolution of $CaCO_3$. The influence of pressure is notable at high depths in the marine environment where there is a carbonate compensation level of approximately 4500–5000 m in which the carbonate is completely dissolved; however, this process occurs in areas far from the coast.

11.3.2 Precipitation of Calcium Carbonate in Coastal Environments

According to the above, the chemical precipitation of carbonates occurs under very precise conditions. The temperature of the water cannot deviate much from the average temperature, as water that is too cold or too hot favors dissolution processes. The same applies to the pH, which cannot deviate more than one degree from neutrality, and to the Eh, to salinity, to depth and water clarity. This means that the environments in which carbonates can precipitate are very specific and localized. These environments are found in

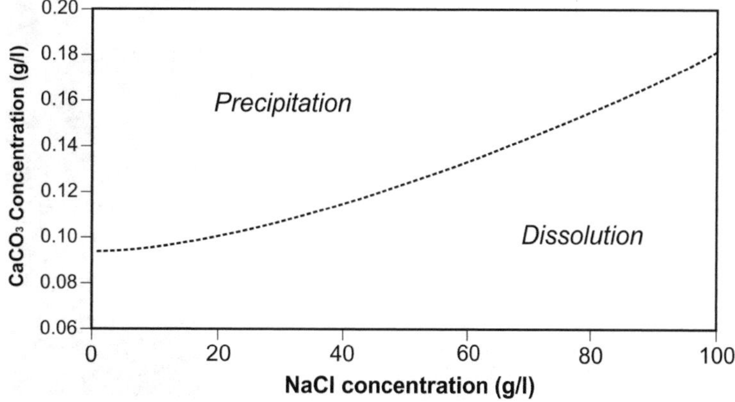

Fig. 11.4 Solubility of calcium carbonate as a function of NaCl concentration in an open system [1]

shallow subtidal waters with high clarity. This area has been called the **carbonate factory** by Schlager [18] and, although in stratigraphic terms it is located on the carbonate platforms, from a sedimentological point of view it can be considered that most of the conditions refer to coastal environments.

It is in these areas that the chemical changes necessary to achieve oversaturation with the consequent precipitation occur most frequently. This precipitation will normally result in aragonite crystals, which are the most stable mineral form at atmospheric temperatures. This type of elements formed in an abiotic way are called **orthochemicals** when they appear in sediments and sedimentary rocks. This process can take place in two different ways. Normally, these changes occur progressively. In that case, there is no high oversaturation, but rather the ions combine to form crystals slowly. Under these conditions, crystal growth predominates over nucleation, since the latter process requires a lot of chemical energy. The result will be the precipitation of large crystals that are known texturally as **sparite** (Fig. 11.5a). If, on the other hand, there is a sudden change in any of the conditions that control the soluble equilibrium, the state of oversaturation is very far from the threshold of solubility. Then, in a very short time, there may be a large number of insoluble ions available for precipitation. The most effective process for removing all these ions from solution is nucleation, which will dominate crystal growth. Under these conditions the result will be a large amount of crystals and all of them very small, giving rise to a type of orthochemical known as **micrite** (Fig. 11.5b).

When the oversaturation process is slow, aragonite crystals often try to fix themselves on other pre-existing carbonate elements. In this way, they can use carbonate particles suspended in water [11]. These particles will begin to grow in a radial way until they reach a size sufficient to enter into imbalance with the flow that transports them and then they decant. These spherical particles are known as **radial oolites**. Once decanted, the oolith can continue to be transported by traction. Its spherical shape will allow it to roll along the bottom. If the oolith decants on a carbonate mud base, the small crystals can remain adhered to the exterior of the oolith, continuing to grow in the form of concentric sheets similar to a snowball, giving rise to oolites with a radial nucleus and a concentric exterior. Other carbonate elements can also be transported in bearings on this micritic bottom. In that case, **concentric oolites** with a different core would be formed (Fig. 11.5c).

The concept of a carbonate factory is not limited to the production of chemical and abiotic precipitates, but Schlager [18] also considers two forms of precipitation that occur thanks to organic activity: bio-induced precipitation (where the activity of the organism acts as a catalyst for the process) and bio-controlled precipitation (where the organisms are the ones that cause the precipitation process as one of their organic functions). Examples of bio-controlled precipitation would be the formation of **bioclasts** (Fig. 11.5d) and fecal **pellets** (Fig. 11.5e). From this point of view, carbonate factories located in shallow subtidal areas would be well suited to support all three types of precipitation.

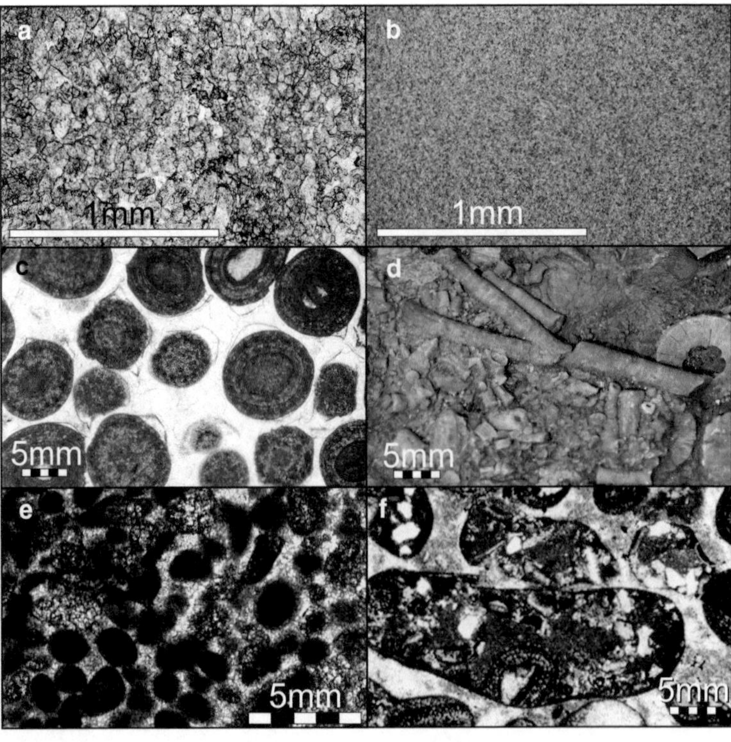

Fig. 11.5 Textural elements of carbonate rocks. **a** Orthochemical sparite. **b** Orthochemical micrite. **c** Oolites. **d** Bioclasts. **e** Pellets. **f** Intraclasts

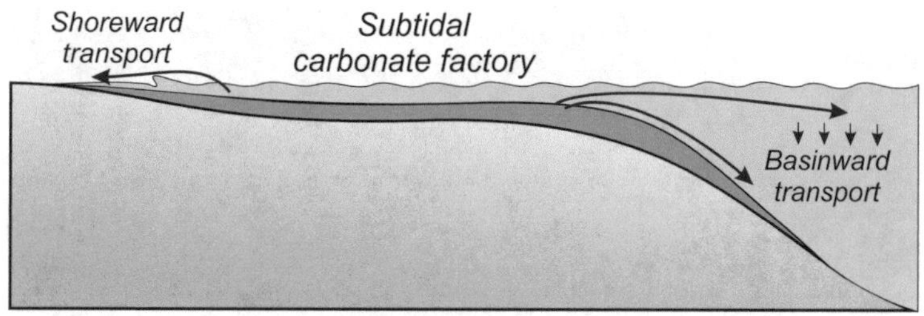

Fig. 11.6 Physical transport of the sediment produced in the carbonate factory to other environments of the sedimentary basin [13]

A study by the same author carried out on a global scale [19], proposes the existence of three types of environments that can be considered carbonate factories:

1. *Shallow waters of tropical areas*: These are dominated by abiotic and bio-controlled precipitation (mainly by photoautotrophic organisms).
2. *Cold water subtidal environments*: These are dominated by bio-controlled precipitates (mainly heterotrophic organisms).
3. *Subtidal mud piles*: These are dominated by abiotic and bio-induced precipitates (mainly microorganisms).

One of the changes related to abiotic precipitation that occurs in cold water coastal environments is related to pH variations. When seawater enters coastal environments, the pH decreases and calcium carbonate dissolves. This process is not very effective at higher temperatures. This is one of the reasons why the dissolution of aragonite and calcite crystals by inorganic processes hardly occurs in warm shallow seas.

However, not all of the processes of precipitation and dissolution of carbonates in the coastal fringe are linked to the carbonate factory. One of the most characteristic changes in the coastal zone is related to the mixing of river and marine waters. Actually, the processes of precipitation–dissolution of carbonates in the salt wedge are difficult to quantify. Recent laboratory experiments by Singurindy and Berkowitz [23] determined that salinity changes related to water mixing in river mouth channels induced oversaturation of carbonate and calcium ions, which then precipitate in the mixing zone according to a salinity-dependent geochemical barrier (Fig. 11.4). This precipitation depends on the type of water mixture and the volume of water to be mixed. In well-mixed systems, oversaturation is progressive throughout the brackish water zone. In this case, a large volume of precipitated crystals is obtained, which reach a good size when precipitation occurs slowly. Conversely, in systems with a model of mixture by saline wedge with a well-marked halocline, a smaller volume of crystals is produced, although the oversaturation is more abrupt and very small crystals are precipitated.

11.3.3 Physical Processes Moving Chemically Created Grains

Precipitated limestone elements of any of the forms described above in the carbonate factory can be transported by currents to other deeper or shallower areas of the same sedimentary basin (Fig. 11.6).

Waves are one of the main agents that move the precipitates in the form of grains towards the coastline, where they can be redistributed by tides. Pellets and ooliths are easily transported and will become part of the facies in many coastal environments. The carbonate skeletons and shells of organisms can also be transported and deposited as bioclasts. Micritic and sparitic carbonate muds have different destinies if their crystals remain disaggregated or if, on the contrary, they have undergone aggregation processes. If they are disaggregated, the micritic sludge can be incorporated into the suspended transport of some coastal currents and then be directed towards the coast or towards the basin. On the coast, they usually decant in the less energetic restricted environments such as lagoons or peritidal environments. In the basin, they will end up as part of the pelagic rain that feeds the deep oceanic basins.

The fragments of carbonate sediments that were deposited on the basin later become part of the carbonate crusts that can be fragmented and reworked by energy currents to give new sedimentary grains [9]. These grains are known as **intraclasts** (Fig. 11.5f). The morphology and composition of intraclasts can be very varied, since these aggregates can contain any of the carbonate elements described above. Intraclasts behave like any grain in water flows and can be displaced to the coast by tides and waves or to deeper areas by other currents.

11.3.4 Early Cementation and Beachrock Genesis

Cementing is one of the main processes that produce a reduction in porosity (and permeability) in carbonate sedimentary rocks. Immediately after the carbonate sediment has

Fig. 11.7 Beachrock developed in a warm climate foreshore

been deposited, seawater begins to circulate through the pores. The water that circulates through the pores has a chemical composition very similar to the water in the environment; however, different environmental conditions can occur in the pores. Therefore, changes can occur within the pores that lead to dissolution and precipitation processes, especially near the water–sediment interface. In the case that the conditions generate precipitation, carbonate cements are formed that give cohesion to the sediment. This early cementing process can build hard crusts or even contribute to the transformation of the sediment into a sedimentary rock.

The cement in carbonate rocks in coastal environments has very varied origins. Invariably, it is a carbonate cement whose composition can be of aragonite or calcite rich in magnesium (Mg). The crystals normally present fibrous or acicular characteristics, radiating from the walls of the pores on which they crystallize. In these cases, the pores can become totally filled due to the continuous precipitation from water circulating through them. Cementation may also occur under conditions where the pores are not completely filled with water, leading to processes similar to those that occur in the vadose zone of inland aquifers. Aggregates of sparitic calcite of varying grain size, although generally large, are then formed. Although the cement is usually sparitic, in some cases it can be micritic when situations of sudden chemical imbalance occur.

In coastal areas, these processes are often very pronounced in intertidal zones. When they occur in open beach areas the formation of cements can solidify the beach sediment. The result is the very fast genesis of a hardened rock. This hardened sediment is known by the generic name of **beachrock** (Fig. 11.7). The formation of a beachrock stiffens the beach and prevents the processes that normally remove sand from eroding the beach below its level. There are exceptions to this rule, as sometimes extreme events can disintegrate the beachrock, resulting in the formation of intraclasts that are redistributed on the coast by coastal agents.

Similar cementing processes can also occur in the topographically higher area of tidal flats. In these environments, the interstitial water that fills the pores during periods of tidal exposure circulates through them, giving rise to superficial capillary precipitation of aragonite and dolomite that act as cement and lead to the formation of calcareous crusts [22]. Accompanying this capillary cementation, evaporite crystallization of gypsum, anhydrite and also dolomite may appear [17]. These crusts come to form authentic intertidal *beachrocks* [5].

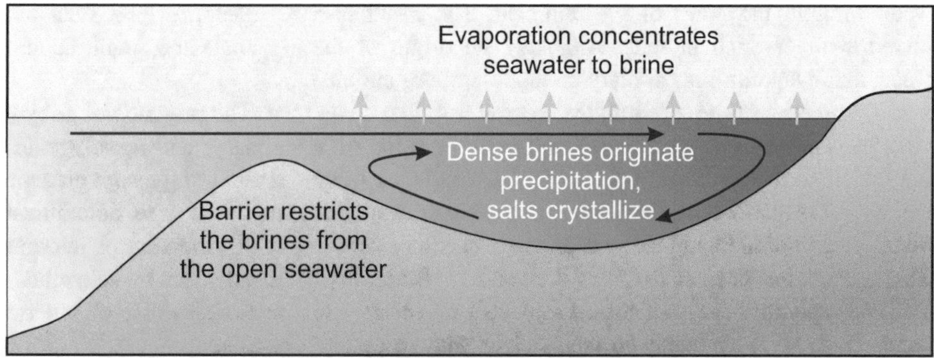

Fig. 11.8 Scheme of the classic "barred" model for evaporite genesis (adapted from Selley [20])

11.4 Evaporite Genesis

Coastal environments in arid climates present such extreme evaporation conditions that the loss of solvent increases the proportion of solute, so that the waters are always close to saturation, at least with respect to sodium chloride [4]. The term **evaporite** describes sedimentary deposits precipitated from brines supersaturated by evaporation. Evaporites can be generated in both marine and inland environments [10], but coastal environments are capable of generating large volumes of evaporite sediments.

The composition of evaporative precipitates is varied, with halite, gypsum and anhydrite (chlorides and sulphates) being the most common precipitates. Although the evaporite precipitation of aragonite, calcite and dolomite (carbonates) is also very important, the chemical processes that give rise to these minerals have been treated in detail in the previous section, as they are distinguished compositionally independently of the origin of the precipitation.

From the point of view of processes, it is necessary to differentiate between coastal evaporites formed in subaqueous environments and those formed by capillary evaporation in the pores of the sediment [17]. The former would preferably be formed in subtidal environments such as those related to lagoons. These environments are usually called **Salinas**. The second type is restricted to the higher areas of the tidal plains and are called **coastal sabkhas**. The processes, dynamics and evolution of both environments are closely related.

Not all intertidal evaporites correspond to sabkhas. The intertidal environments of tropical zones and some of the mid-latitudes can also generate evaporite precipitation [3, 21]. However, in this case they are not called sabkhas, since this type of sedimentation is not dominant throughout the environment.

11.4.1 Subtidal Evaporites: Salinas

The subtidal areas located on the coasts of these arid climates concentrate the salts to form brines, especially when the renewal of water is scarce, as is the case with environments restricted to the action of the waves. Because of this, the genesis of subtidal evaporites is linked to the formation of sandy barriers that favor the appearance of these restricted environments. Inside the barriers, the water will be transformed into a brine until it reaches a state of oversaturation. Then, the process of chemical precipitation of the salts will begin (Fig. 11.8).

Tidal currents are responsible for the distribution of these brines by other coastal environments such as subtidal channels and intertidal flats [12]. Brine can also be released to the outside world, being injected into open marine waters through inlets during tidal ebb. These outgoing currents move along the bottom towards the shoreface, due to the high density of these fluids. This process does not normally contribute to the crystallization of evaporites in open coastal waters, as the brines are permanently diluted in the large volume of seawater.

The aggradation of subtidal environments due to the sedimentary filling evolves towards the domination of the capillary processes; in this way, the evaporite lagoon would end up becoming a sabkha. In the transition period, it is possible to observe cycles where the domination of direct precipitation and capillary precipitation alternate.

11.4.2 Intertidal Evaporites: Coastal Sabkhas

The coastal evaporites have been named Gavish Sabkhas due to the important studies carried out by Eliezer [8] on the coast of the Sinai Peninsula. Sabkhas are characterized by salt precipitation associated with capillary circulation of water through the pores of the sediment. The result is a mixed sediment rich in salts. Although the origin of the water that evaporates is capillary circulation, there are also surface intakes of brine from the nearby subtidal zone. Under storm conditions, the wind may push the brine from the lake onto the supratidal flats. In this environment, brine sheets of less than 2 cm accumulate and can either evaporate directly or infiltrate and then evaporate by capillary action. This process has been called *flood recharge* by Butler [4].

The evaporative facies of coastal sabkhas are generated in bands (Fig. 11.9) according to their height with respect to tide levels [14]. Each of these bands is characterized by different levels of evaporation and recharge, and the result is a different mineralogy and texture of the evaporite precipitates, as well as of the sediments that accompany the evaporites. The bands from the bottom to the top are:

- *Upper intertidal*: Blue-green algae mats interspersed with aragonitic mud. It usually has gypsum cement and dolomitic cement may also appear.
- *Lower supratidal*: Soft masses (*mush*) of plaster and dolomitic and aragonite mud. Much microbial activity in the form of organic salts.
- *Mid-supratidal*: Corresponds to the zone of maximum chlorinity and maximum evaporation, occasionally of mixed waters of marine and continental origin. Halite precipitates with parallel lamination, altered by the diagenetic formation of gypsum crystals and anhydrite nodules. Dolomitization is also frequent.
- *Upper supratidal*: This area is controlled by inland waters and is only flooded once every few years, so it has a very low precipitation rate. It is dominated by diagenetic processes that totally modify the precipitation by evaporation. The gypsum is totally replaced by anhydrite which can have a nodular or *chicken-wire* structure. Deformations such as disharmonic folds of decimeter scale and small diapirs may also occur.

The presence of anhydrite in these bands is usually secondary, with a polygenic origin. One part is due to dehydration of gypsum precipitated by evaporation [16]. Another part is due to dolomitization and dissolution of aragonite with formation of intermediate gypsum, which later passes to anhydrite by dehydration. The anhydrite is usually formed as nodules that vary between the millimeter and decimeter scales.

All of these processes of capillary precipitation are accompanied by early diagenetic transformations that result in a hard salt crust of varied composition that fills the sabkha in a sedimentary sequence. When it rains, part of this saline sediment is dissolved and the waters of the lagoon are recharged with salt. The result of this cyclical process is a dynamic equilibrium in the ionic balance of the lagoon's brine. This salt recycling is a consequence of the early diagenesis of the evaporites. On the other hand, the evaporation of inland waters after rainfall generates carbonate and gypsum cements. These minerals can also be remobilized towards the lake, saturating its waters and ending up being deposited at the bed when evaporation resumes.

11.4.3 Reworked Evaporites

The minerals that result from evaporation precipitation can be reworked and transported as grains by physical processes in the same way as any of the components of siliciclastic or carbonate rocks. A typical example is the wind transport to form dunes (*eolianites*) that occurs in the intertidal zone of coastal sabkhas. Mass transport processes (landslides and slumps) have also been described, facilitated by the high

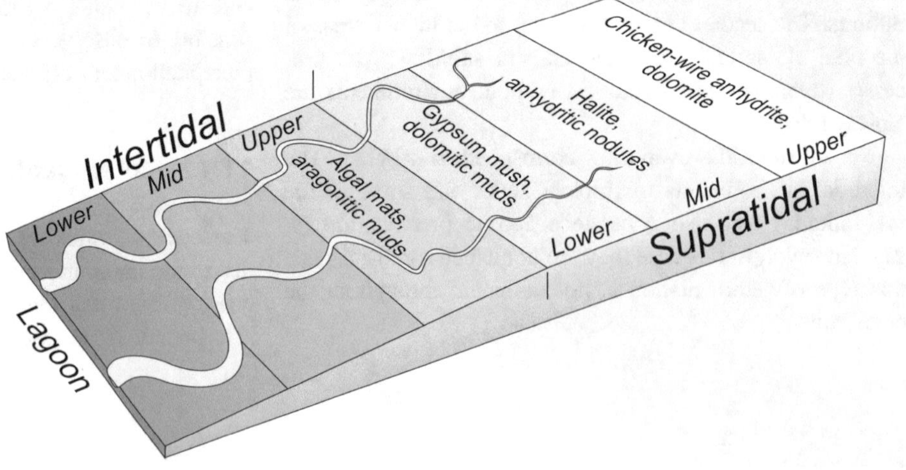

Fig. 11.9. 3D-diagram synthesizing the zones of capillary evaporate precipitation in a typical sabkha (based on Purser [17])

Fig. 11.10 A metric-scale cross-stratification in a Miocene gypsum arenite (Conil, SW Spain)

plasticity of these sediments. However, wind and gravity are not the only ways these grains are transported. Normally, aqueous transport is excluded when considering the transport of evaporite grains due to their high solubility, but they can be transported by fluid flow in saturated brines.

For example, in southern Spain, metric-scale sedimentary structures have been described in sandstones whose grains consist exclusively of gypsum crystals (Fig. 11.10). In this case, the origin of the brine is associated with the Messinian salinity crisis. Megaripples, ripples and other minor structures formed by the waves and tides may also be made up of evaporite mineral grains [2].

References

1. Berkowitz B, Singurindy O, Lowell RP (2003) Mixing-driven diagenesis and mineral deposition: $CaCO_3$ precipitation in salt water–fresh water mixing zones. Geophys Res Lett 30(5):1253
2. Boggs S (1995) Principles of sedimentology and stratigraphy (2nd ed). Prentice Hall, New Jersey, 744pp
3. Borrego J, Morales JA, Pendón JG (1995) Holocene estuarine surface facies along the mesotidal coast of Huelva, SW Spain. In: De Boer PA, Davis RA (eds) Tidal signatures in modern and ancient sediments, vol 24. IAS Special Publication, pp 151–170
4. Butler GP (1969) Modern evaporite deposition and geochemistry of coexisting brines, the Sabkha, Trucial Coast, Arabian Gulf. J Sediment Petrol 39:70–90
5. Davies GR (1970) Algal laminated sediments, Gladstone Embayment, Shark Bay, Western Australia. In: Davies GR (ed) Carbonate sedimentation and environments, Shark Bay, vol 13. Western Australia Memoirs of the AAPG, pp 169–205
6. De Mora SJ (2007) Chemistry of the oceans. In: Harrison R (ed) Principles of environmental chemistry. Royal Society of Chemistry Publishing, London, p 263
7. Drever JI (1997) The geochemistry of natural waters: surface and groundwater environments (3rd ed). Prentice Hall, New Jersey, 436pp
8. Gavish E (1974) Geochemistry and mineralogy of a recent sabkha along the coast of Sinai Gulf of Suez. Sedimentology 21(3):397–414
9. Hardie LA (1986) Stratigraphic models for carbonate tidal flat deposition. Q Colorado Sch Min 81:59–74
10. Hardie LA (1991) On the significance of evaporates. Annu Rev Earth Planet Sci 19:131–168
11. Heller PL, Komar PD, Pevear DR (1980) Transport processes in ooid genesis. J Sediment Petrol 50(3):943–952
12. Hsü KJ, Schneider J (1973) Progress report on dolomitization. Hydrology of Abu Dhabi sabkhas, Arabian Gulf. In: Purser EH (ed) The Persian Gulf: Holocene carbonate sedimentation and diagenesis in a shallow epicontinental sea. Springer Verlag, Heidelberg, pp 409–422
13. James NP (1984) Shallowing-upward sequences in carbonates. In: Walker RG (ed) Facies models. Geological Association of Canada Reprint Series vol 2, pp 213–229
14. Kinsman DJJ (1969) Modes of formation, sedimentary association and diagnostic features of shallow water supratidal evaporites. Bull AAPG 53:830–840
15. Langmuir D (1997) Aqueous environmental geochemistry. Prentice Hall, New Jersey, p 600
16. Moiola LM, Glover SI (1965) Recent anhydrite from Clayton Playa, Nevada. Am Mineral 50:2063–2069
17. Purser EH (1985) Coastal evaporite systems. In: Friedman GM, Krumbein WB (eds) Hypersaline ecosystems the Gavish Sabkha. Springer, Berlin, pp 72–102
18. Schlager W (2000) Sedimentation rates and growth potential of tropical, cool water and mud mound carbonate factories. In: Insalaco E, Skelton PW, Palmer TJ (eds) Carbonate platform systems: components and interactions, vol 178. Geological Society of London, Special Publication, pp 217–227
19. Schlager W (2005) Carbonate sedimentology and sequence stratigraphy. SEPM Concepts in Sedimentology and Paleontology Series no 8, 200pp
20. Selley R, 2000. *Applied Sedimentology* (2nd ed.). Academic Press, London. 523pp
21. Shinn EA (1964) Recent dolomite, Sugarloaf Key. Guidebook for GSA Field Trip No. 1, South Florida Carbonate Sediments. Geological Society of America, pp 62–67
22. Shinn EA (1983) Tidal flat environment. In: Scholle PA, Bebout DG, Moore CH (eds) Carbonate depositional environments, vol 33. Memoirs of the AAPG, pp 171–210
23. Singurindy O, Berkowitz B (2004) Carbonate dissolution and precipitation in coastal environments: laboratory analysis and theoretical consideration. Water Resour Res 40(W04401):1–12

12 Biological Processes and Sediments on the Coast

12.1 Introduction

Biological processes occur on all coasts to a greater or lesser extent. In many coastal environments these processes are overshadowed by the importance of the energy deployed by physical processes, and in others by the magnitude of chemical processes. However, in both cases organisms always play a role and have a major influence on the effect the dominant processes have on sediments. This influence has already been described when explaining some of these processes in earlier chapters. In addition, the presence of organisms in the sedimentary substrate can influence the sedimentation and physical reworking processes. A classic example is the western Wadden Sea, where the activity of microorganisms in the pores of sediment generates mucous membranes that give it a cohesive character, making it difficult for tidal currents to rework it [2].

On the other hand, there are biological processes capable of generating sediments by themselves: they are the bio-induced chemical processes and the bio-controlled processes [13]. Through bio-induced chemical processes, organisms are able to provoke changes in the surrounding environment which induce the precipitation or chemical dissolution of certain components of coastal waters. Organisms are also capable of producing chemical processes within their forms as part of their biological functions, these are the so-called bio-controlled processes. In this sense, many aquatic animals and plants are able to organically extract dissolved mineral matter to form their shells and skeletons. Once dead, these hard parts of the organisms accumulate in the coastal facies as sediment of biochemical origin. Organic carbon from the soft bodies of the organisms can also accumulate in the sediment and even constitute a type of sediment itself. Through this type of process, the organisms come to constitute their own environments and build significant sedimentary bodies. Both processes generate a type of sediment known as **biogenic sediments**.

The activity of organisms is not only expressed in the genesis of the sediment. Beyond that, there are animal and plant activities which alter the internal structure of the sediment and others, even more extreme, which can contribute to the disintegration and erosion of coastal formations. These are the processes of bioturbation and bioerosion. All these aspects will be discussed below.

12.2 Bio-Induced Precipitation and Dissolution

The influence of some organic processes, such as respiration and photosynthesis, on the concentrations of dissolved O_2 and CO_2 in the aqueous environment and, secondarily, on chemical precipitation processes has already been described in previous chapters. In restricted coastal environments, there are large numbers of microorganisms that make up the zooplankton and phytoplankton, as well as many others that are living on the bottom in benthic communities. These include macrofauna such as fishes, crustaceans, cephalopods, gastropods, bivalves and polychaetes; microfauna such as larvae of these animals, foraminifera and ostracods; and also plants such as upper phanerogams (Fig. 12.1a), green algae, diatoms and cyanobacteria, which populate the waters and beds of coastal environments [14]. All of these organisms modify the pressure of dissolved gases in the water with their physiological activity.

Another example of the influence of these organisms is the direct modification of the chemical characteristics of their immediate environment to maintain their osmotic balance. In this regard, it should be noted that the concentration of salts in water in coastal environments is much higher than the concentration of the same salts in intracellular fluids. In nature there must be a balance in the ionic concentrations of the fluids that are located on both sides of an osmotic membrane such as the cell wall. In this case, organisms cannot choose to increase the salts of their intracellular fluids, because they cannot pass through the cell wall. On the other hand, to increase the concentration, the cell must not expel the intracellular water, because this would lead to dehydration that would be lethal

J. A. Morales, *Coastal Geology*, Springer Textbooks in Earth Sciences, Geography and Environment,
https://doi.org/10.1007/978-3-030-96121-3_12

Fig. 12.1 Examples of bio-induced precipitation. **a** The photosynthesis of marine phanerogams decreases the concentration of dissolved CO_2. **b** Submerged roots of mangroves alter the pH of the surrounding water, inducing precipitation of salts

for the organism. Therefore, since they cannot increase the internal concentration, the strategy of most aquatic organisms is to modify some of the factors that control the balance of solubility, in order to reduce the external ionic concentration. Usually, the mechanism chosen is the modification of the pH or the combined Eh–pH equilibrium. In this way, flocculation processes and chemical precipitation are induced in the waters surrounding the organisms. This process is very common in higher halophyte plants (Fig. 12.1b) that are submerged in inter- or supratidal environments (marshes and mangroves) and generally occurs in both terrestrial [5] and carbonate environments [9].

The action of the underwater plants also plays a role in the biomechanical modification of the environment, since they act as a screen for the currents, reducing their speed by friction and contributing to the siltation of the suspended matter.

12.3 Bioconstructions

One of the most typical chemical reactions that acts as a physiological function inside organisms is the use of chemical elements dissolved in water to form organic skeletons. Any of the classes of mollusks as well as many other microorganisms are able to build external shells to cover their bodies. Sometimes the protection of these shells allows the animals to be mobile in the environment; however, in many cases the shelly organisms adopt a sessile way of life and fix themselves to the substrate. This fixation often has a colonial character. Thus, the skeletons of these organisms are left accumulated after their death and massive structures are generated that become true buildings due to the accumulation of successive generations of organisms at the same place. These buildings are known generically as **bioconstructions** or **bioherms** (Fig. 12.2).

Perhaps the most typical example of this type of organic construction are coral **reefs** (Fig. 12.2a). In this case, corals are small colonial polyps that secrete an external shell to fix themselves and live by capturing the organic particles circulating above them in the water flows with their tentacles. Actually, the formation of reefs also requires the participation of an algae that lives in a symbiotic way with the corals. The algae often provide food for the corals, while the corals provide the algae with nutrients from their organic waste, as well as protection.

There are other organisms capable of forming simpler bio-constructions. Sessile bivalves such as ostreids (Fig. 12.2b) are capable of forming **massive banks** in tidal areas protected from waves, even in estuarine environments [11]. This is less common in the case of gastropods, although a typical example includes vermetid banks (Fig. 12.2c) which are associated with mean tidal level in intertropical zones [12].

Another mechanism by which bioconstructions can be formed is by the action of organic activity as a sediment trap. The best-known case is the formation of **stromatolites** (Fig. 12.2d), which is a seasonal process. It begins in spring periods, when blue-green algae (cyanobacteria) cover a stable solid surface on the bottom of very well-lit coastal waters. The adherent character of these algae means that, during periods of maximum water agitation, the carbonate particles that are transported in suspension become stuck to their surfaces. When many particles adhere, the algae eventually die, solidifying a sheet of carbonates on the surface of the structure. A new spring will bring a new sheet of algae, starting the cycle again. These structures have been present in coastal environments since the Proterozoic. Nowadays, they are linked to intertidal environments, although in the past they have also developed in subtidal coastal environments.

A similar mechanism is that developed by Rhodophyceae algae, which also adhere carbonate particles to their structure of seasonal sheets. The difference is that in this case the algae are capable of resisting movement and shock, thus colonizing any solid body transported in the well-lit waters. The algae can thus adhere to carbonate nuclei in traction, making them grow concentrically as they roll along the bottom. These structures are called **rhodolites** and are characteristic of restricted carbonate zones such as subtidal lagoons.

Fig. 12.2 Different bioconstructions in coastal environments. **a** Coral reef (Dominican Republic). **b** Ostreid bank (Atlantic Spanish coast). **c** Vermetid bank (Mediterranean Spanish coast). **d** Stromatolites (Shark Bay, Australia)

The same phenomenon of particle adhesion can also be developed by green algae in intertidal environments. This is the case of the so-called ***eelgrass***. In this instance, the importance of chemical processes is much less, with the combined mechanism between biological and mechanical processes dominating.

12.4 Bioclastic Sediments

The rigid shells of the organisms generated through bio-controlled precipitation can be transported by currents when the organism dies. In the case of free-living organisms, incorporation into the detrital fraction is immediate. Thus, it is usual to find sediment composed of microorganism shells, which are part of the bioclastic sand fraction, but the skeletons of macroorganisms or their fragments can also constitute an important part of the sand and gravel sediments. Bivalve shells and gastropod or scaphopod shells are the most common examples.

In the case of the bioconstructions described in the previous section, the incorporation into the bioclastic fraction of the sediment is not so easy nor so immediate. However, the rapid currents and strong waves that develop during storm periods are able to pull fragments of organisms out from the banks and reefs and incorporate them into the coarser fraction of the sediment by reworking and distributing this bioclastic sediment throughout other coastal environments.

12.5 Organic Sedimentation

The accumulation of soft tissues of animal or plant organisms can give rise to a type of biogenic facies known as **organic sediments**. These organic components can be mixed with particles of other origins constituting one more phase of the sediment. When the organic carbon component of the facies constitutes a fraction greater than 3%, this forms a sediment type called *organic-rich sediments* that by lithification give rise to a group of sedimentary rocks of the same name.

The organic matter may appear dispersed in the sediment, giving it a dark gray or blackish color, although it may also appear as differentiated elements due to differing densities or textures. One interesting aspect is the origin of the organic matter, which can be animal or vegetable, and this will determine the first step of degradation in the transformation of the initial organic matter. In general terms, this step can generate **sapropelic acids** or **peat**. Sapropelic acids originate in anoxic environments from the organic remains of zooplankton, phytoplankton, spores and fragments of higher plant tissues. These sapropelic acids are subsequently transformed into **hydrocarbons**. If the origin of the organic matter is mainly composed of plant tissues, there is a greater variety of possibilities, since this matter can be deposited after being remobilized or it can also be found **in situ**. In this second case, the first step of the process is peat and its later transformations give rise to **carbonaceous rocks**.

12.5.1 Hydrocarbons

Hydrocarbons are molecular compounds consisting of carbon bound by covalent bonds to hydrogen and oxygen atoms. Most hydrocarbons have a natural origin and are derived from sapropelic acids. Petroleum is a viscous liquid generated from the decomposition of organic matter of sapropelic character accumulated in the sediments. Some of the hydrocarbons are in a gaseous state and are separated from the oil by migrating vertically through the pores. The interest of hydrocarbons lies in their use as the main source of energy in the world today.

The most common sediments which may include animal organic matter include black shales, oil-bearing limestones, oil shales and sapropelites [15]. If organic accumulation takes place in highly reducing environments of high biological productivity, the percentage of organic matter accumulated and preserved can reach significant values, much higher than 3%. Black shales may contain more than 10% of organic carbon, oil shales may exceed 25%, oil shales and oil-limestone may exceed 50% and sapropelites may even exceed 70%.

The transformation of organic matter into crude oil occurs thanks to bacterial activity in anoxic environments. There are a large number of bacteria that participate in this decomposition process. It is estimated that more than 160 genera are responsible for the biochemical combination of the C, H and O atoms in the long molecular chains of hydrocarbons.

Essentially, the organic matter in oil sediments is of four types: kerogen, asphalt, crude oil and natural gas. These are highly complex organic compounds with a poorly defined nomenclature, since these types cover a continuous spectrum of hydrocarbons. The most abundant compound is kerogen, and it is estimated that more than 80% of the organic matter in petroleum sediments is found in this form [3]. Kerogen is a dark and very dense substance, insoluble in water as well as in acids, bases and organic solvents. At depth, temperature and pressure transform kerogen into other hydrocarbons through a process called **catagenesis** [10]. Asphalt, also called bitumen, is very similar to kerogen, but is soluble in organic solvents, acids and bases. Asphalt can migrate, although with difficulty, and is found filling pores and fractures in sedimentary rocks. Petroleum is liquid and natural gas is obviously gaseous. Both have a lower density that allows them to migrate more easily due to discontinuities in the rocks.

Although hydrocarbons are generated in sedimentary rock rich in animal organic matter, their volatile nature causes them to migrate from the **mother rock** and accumulate in other porous sedimentary rocks that are called **storage rocks**. Finally, the presence of an impermeable rock at the top of the storage rock is necessary to seal the storage and prevent the migration of the oil to the surface, where it would volatilize into the atmosphere. This rock is known as the **seal rock**.

The genesis of hydrocarbons is linked in the collective mind to the continental shelves, where the conditions of accumulation of organic matter can occur, as well as anoxia and the high rates of burial necessary for the transformation of this organic matter into petroleum sediments. However, some coastal environments are very conducive not only to the formation of hydrocarbons, but also to their migration and storage in suitable geological traps. Perhaps the deltas are the best example, since environments rich in organic matter can occur in the prodelta areas, where they will quickly be buried by the progression of the deltaic front sand bars producing favorable burial conditions. The sandy sediment of the deltaic front is also an excellent storage rock, while the clays of the deltaic plain have the right characteristics to act as a seal rock.

Other ideal coastal environments for the accumulation of organic matter are those located on waterfronts where *upwelling* processes occur. Thus, *shoreface* areas associated with this phenomenon develop bodies of organic silty sands suitable for transformation into hydrocarbons.

The rise and fall of sea level mean that these coastal environments have been developed throughout geological history in different positions on the present continental shelves. Thus, they form part of the *system tracks* that constitute the genetic units that can be found there today. That is why numerous studies developed by the world's oil companies have been dedicated to the exploration of these areas.

12.5.2 Carbons

Organic accumulation of plant tissues can lead to the formation of peat, which will later be transformed into carbon through a process known as **maceration**. These plant accumulations can be of two types:

- *Autochthonous accumulations*: These are produced in the same place where the plant has lived, including the roots or rhizomes as well as the stems and leaves, although only the roots wrapped in their inorganic matrix can be preserved. These accumulations can be formed in continental and coastal environments. In coastal systems, there are several ideal environments for the formation of this type of accumulation. Marshes and mangroves are the most typical example, since they can accumulate significant levels of peat made up of the remains of higher halophyte plants (Fig. 12.3a). This is the most common case for the formation of coals in coastal environments,

Fig. 12.3 Vegetal accumulations. **a** Peat in the Everglades swamp (USA). **b** Seagrass flat (SW Spain). **c** Algal mat (SW Spain). **d** Algal allochthonous accumulations on a beach (S Portugal)

since the pores of the clays that surround the organic plant matter are isolated from environmental water and the conditions of anoxia necessary for maceration can occur there. Native accumulations of marine phanerogams (*seagrass*) and algae (*algal beds* and *algal mats*) may also develop (Fig. 12.3b, c). However, the decomposition of such accumulations usually results in humic and sapropelic acids, which normally oxidize and do not lead to the formation of carbonaceous rocks.

- *Allochthonous accumulations*: These occur after plant mass transport and accumulation in residual areas. In this way, algae accumulate on beaches (Fig. 12.3d) and some areas of tidal flats, but also exposed fragments of continental halophytes that can be uprooted from their place of origin by very energetic events and reach estuaries and deltas. Although these accumulations can generate carbon deposits, they rarely do so. Most often they decompose, generating volatile sapropelic acids. What happens is that these allochthonous accumulations are usually not very powerful and are dispersed, almost never giving rise to economically profitable deposits; however, they can be good paleogeographic markers.

As has been observed, not all accumulations of organic plant matter are transformed into carbon. For an organic accumulation to result in a carbonaceous sediment, other conditions must also be met:

1. That the accumulations are thick enough.
2. That no oxidation processes of organic matter occur after burial. In order for this condition to be met, the accumulation needs to occur in a sub-aqueous environment, since water, containing less oxygen than air, is usually less effective in the oxidation process. Oxidation occurs less when there is less dissolved oxygen in the water and it is usually null in anoxic environments. For example, in water with 10 mg/l of dissolved oxygen, the oxidation time is almost 11 times longer than in the atmosphere. With half that level of oxygen (5 mg/l), the oxidation time is 100 times longer. In supratidal environments, surface water usually has around 6 mg/l of oxygen, with less than 1 mg/l at the bottom.
3. The content of nutrients and dissolved salts will determine the volume of plants, the activity of the roots and the volume of inorganic impurities that the coal will contain.

Taking these factors into account, there are two types of environments for the formation of native plant accumulations that can give rise to the maceration process necessary to generate charcoal:

- *Oligotrophic environments*: Also called ombrogenic or ombrotrophic. These are environments with moderately well-oxygenated water, since it comes directly from rainwater or water from the tidal *circulation*. Large communities of vascular plants generally develop in this type of environment. The degree of waterlogging is usually so high that the roots do not need to penetrate deeply to obtain the water and they develop horizontally, giving rise to a very dense network with hardly any interstitial inorganic matter and with a practically null edaphic horizon. Therefore, the feeding of the plant is produced by capillarity, at the expense of the nutrients supplied by the plant accumulation itself. This fact contributes to a decrease in the plant biodiversity of the system, so much so that these populations can even become monospecific. The decomposition of plants produces a high quantity of humic acids, which lower the pH to such an extent that they inhibit bacterial activity and tend to preserve the internal structure of the plant.
  The plant accumulations produced in these environments have the following characteristics: (1) water content over 90%; (2) low quantity of mineral impurities; and (3) cellular structures with a high level of preservation. Mangroves are an excellent example of this type of environment.
- *Eutrophic environments*: Also called tropogenic or minerotrophic. In this case, the water has good circulation before it is dammed, and it can even involve environments where the water is not completely dammed and has a relative renewal. In coastal systems, these conditions can occur in environments such as deltas, estuaries and supratidal flats. This circulation makes the water carry abundant nutrients and dissolved minerals, which determine the high plant biodiversity of the system and the development of a greater humic horizon than in the previous case. After burial, the porewater is isolated from the overlying water. Then, the great amount of decomposing tissue quickly consumes the oxygen dissolved in the porewater, supplying favorable conditions for the formation of coal.
  The coals formed under these conditions present a large quantity of inorganic impurities and the plant structures are usually less well preserved due to the possibility of reworking the plants by the currents before they are definitively accumulated.

The transformation of plant organic matter into peat is an essentially biochemical process carried out by anaerobic bacteria. These bacteria are capable of decomposing plant tissues by draining impurities out of the peat. This peat is the basis for the maceration processes that begin after the increase in pressure and temperature that burial entails. The maceration processes give rise to the rocks of the coal series, which increase in carbon content at the same time as their purity and their calorific power increase as they lose H, O and S. In this process, they also lose the gases associated to sapropelic acids that accompany the first phases of the formation of coals. The components of the coal series are: lignite (brown coal), hard coal (bituminous coal) and anthracite.

12.6 Alteration of the Sediments by Organisms: Bioturbation

The activities of organisms in the sedimentary substrate of coastal environments involve an alteration of the primary structures of the sediment. This physical alteration is known as **bioturbation**. Both animals and plants can produce bioturbation. Annelids (Fig. 12.4), bivalves (Fig. 12.5) and crustaceans (Fig. 12.6) that live as infauna in the subtidal and intertidal flats drill the sediments with their galleries; gastropods, crustaceans and the organisms that are part of the infauna develop tracks on the sediment surface and generate a lower degree of bioturbation, while the roots of the higher plants intensely bioturbate the sediment of the supratidal environments such as marshes and coastal dunes.

Within coastal environments, it is the tidal systems that have the greatest potential for organisms to develop bioturbation. Lagoons, tidal flats, channels and channel margins of estuaries and deltas display a range of tracks and galleries. Also, *shoreface* environments, where wave energy and sediment remobilization are lower, have many organisms present at the bottom that alter their original structure. In more energetic and mobile environments such as *nearshore* or tidal deltas, bioturbation by bivalves is also important; however, the high mobility of sediment in these systems negatively influences the preservation potential of the galleries. In supratidal areas such as tidal flats, it is the roots of the upper halophytes that are responsible for most of the bioturbation (Fig. 12.7a).

Physical processes and sediment characteristics have a great influence on the distribution of benthic organisms. In intertidal systems, critical tidal levels, which control the degree of exposure and submergence of the tidal band and the strength of the tidal currents, mark the degree of tolerance of the organisms and their vertical distribution in the intertidal zone. In this case, the bioturbation degree has been described as one of the most important features in characterizing the tidal facies [14]. In general terms, bioturbation increases towards the highest part of tidal environments.

In fluviomarine systems such as estuaries and deltas, the chemical parameters of the water (temperature, salinity and pH, among others) control the inland distribution of both animals and plants. In these systems, there is a clear seasonality of bioturbation [4] due to biomass variations and changes in the volume of sedimentary input [7]. In general terms, bioturbation in the internal zone of estuaries and deltas increases towards land.

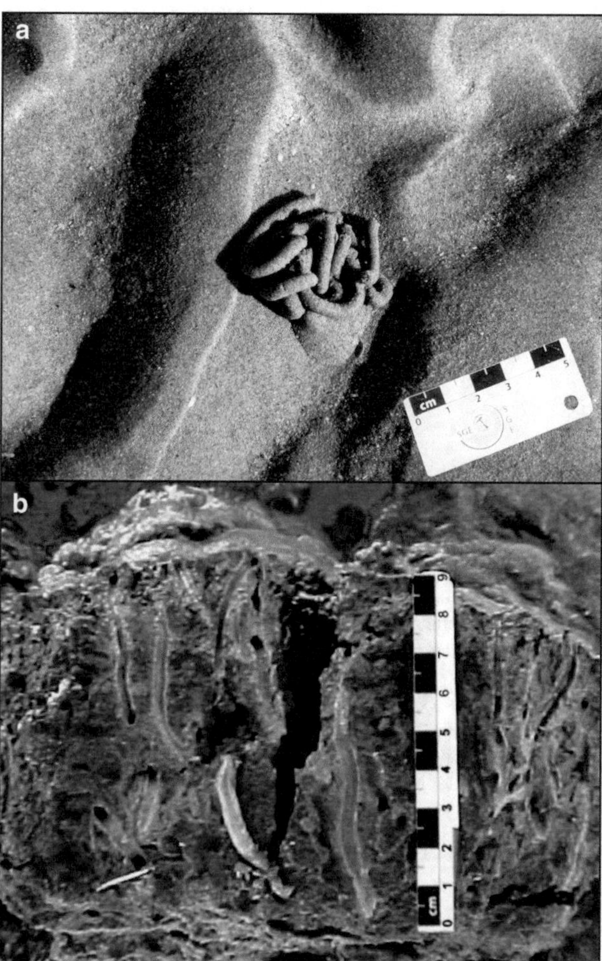

Fig. 12.4 Examples of bioturbation by annelids in coastal environments. **a** Surficial expression of *Arenicolides ecaudata* in a sandy tidal flat. **b** Profile of annelid bioturbation by *Arenicolides ecaudata* in a muddy tidal flat

Organisms capable of generating galleries also show their potential to mix the sediment of different lithologies. It is common to find in one sediment galleries filled by sediment of another type coming from the upper or lower stratum. There is also a clear influence of sedimentation rates on the filling of galleries by those organisms that live permanently near the surface to capture the suspended particles as a way of feeding. The gradual growth of the bed means that they have to re-grow the gallery upwards by filling the bottom. In this way, an interior lamination of the gallery develops, whose lines have changes in thickness that show the long-lasting tidal cycles. These structures have been called tubular tidalites [8].

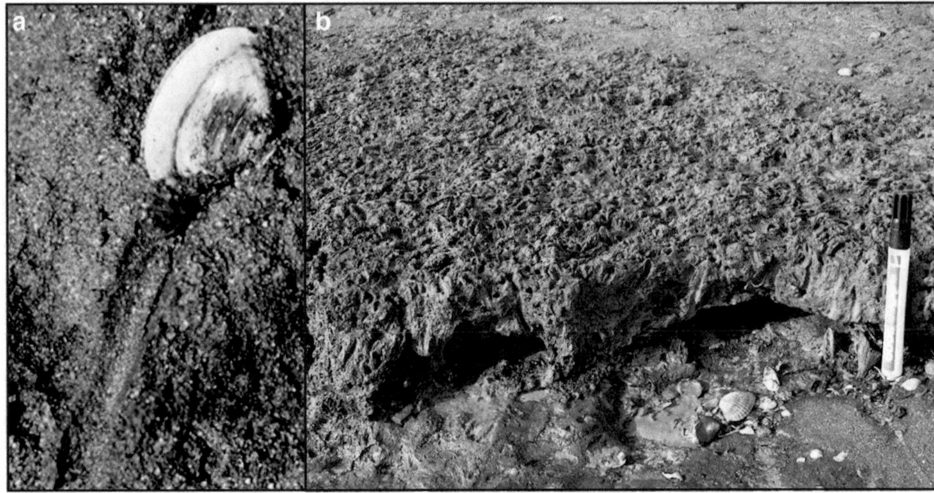

Fig. 12.5 Examples of bioturbation by bivalves in coastal environments. **a** Bivalve burrow in a mixed tidal flat. **b** Muddy sediments completely bioturbated by burrowing of *Cerastoderma glaucum*

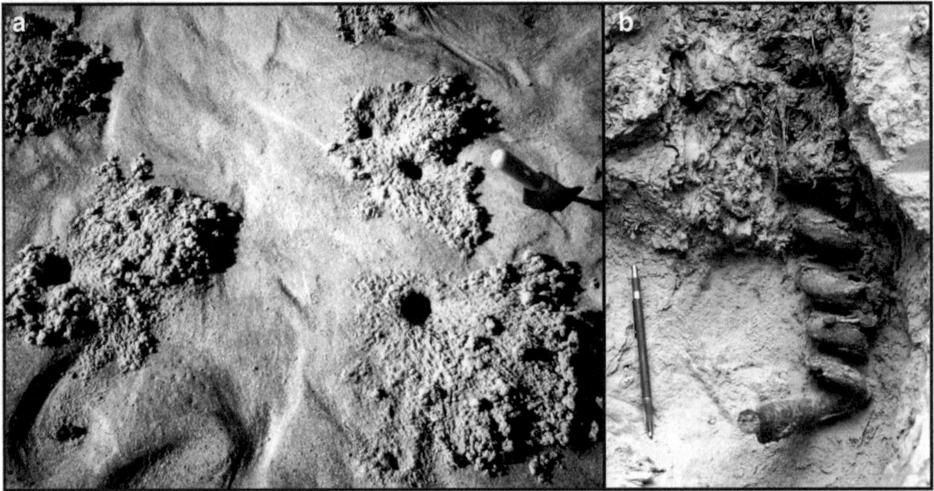

Fig. 12.6 Examples of bioturbation by crustaceans in coastal environments. **a** Surficial aspect of crustacean burrows in a sandy tidal flat. **b** Geometry of a crustacean burrow (*Gyrolites*) in a sandy tidal flat (photograph courtesy of E. Mayoral)

Bioturbation is also significant in coastal dune systems. The activity of macroorganisms is marked as footprints that deform the internal structure. In some dunes, the passage of goats and other animals has been fossilized as a deformation of the characteristic cross-layering of the dunes. The root activity is even more significant, as it is able to completely erase the evidence of the internal dune structure (Fig. 12.7b).

Bioturbation is shown to be a preserved remnant of the activity of organisms in coastal sedimentary beds. Therefore, the potential of ichnofacies as indicators of coastal environments is very remarkable [6].

12.7 Bioerosion

The physical and chemical activity of organisms in the coastal substrate can cause disintegration and erosion of the substrate. This process is known as **bioerosion**. In coastal systems, there are numerous organisms that cause bioerosion. Annelids, sponges, echinoderms, mollusks, crustaceans and fishes include numerous genera whose activity causes this process. On the coast, there are numerous environments where bioerosive organisms carry out their activity. However, it is the rocky coasts and reefs where this process is particularly important.

On rocky coasts, abrasion platforms and cliffs, there are good examples of chemical bioerosion. Higher plants, algae, lichens and some animals living on the rocks produce chemical secretions of an acidic nature that degrade the less resistant minerals and contribute to the disintegration of the more resistant grains, which are eroded by the waves into sand. Also characteristic of these coasts are lithophagous organisms, which wear down the rocks using a mixed process of chemical attack and mechanical wear. There are animals such as mollusks (bivalves and gastropods) and polychaetes that perform internal drilling activity to the

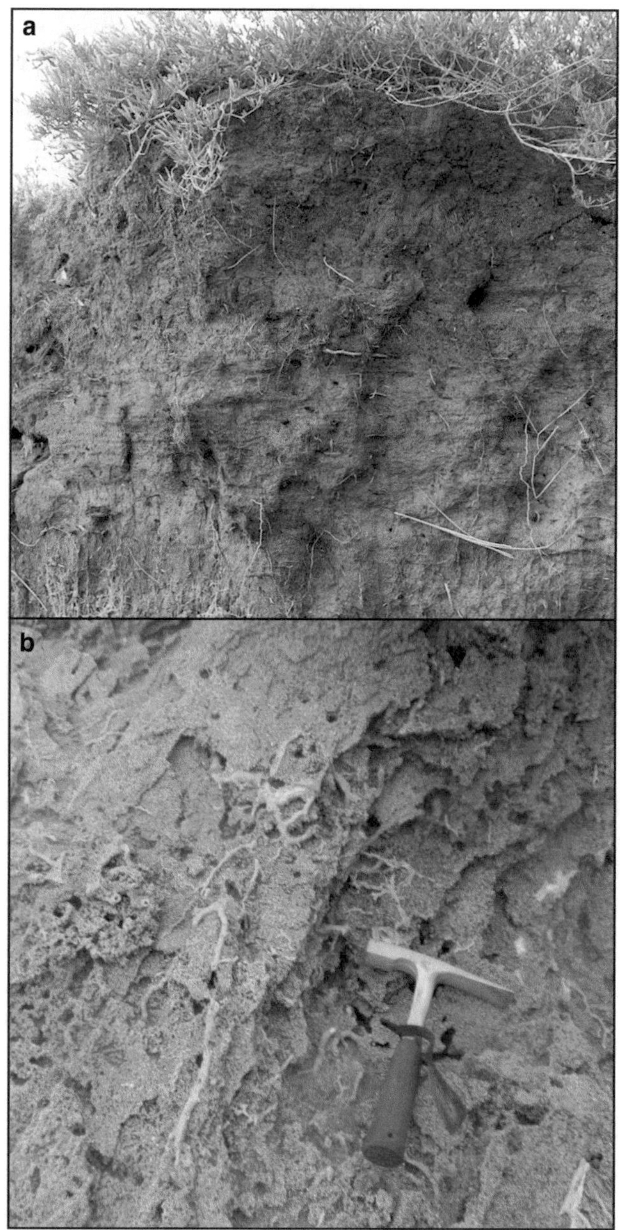

Fig. 12.7 Examples of bioturbation by roots in supratidal environments. **a** Root bioturbation in salt marsh facies. **b** Root bioturbation in dune facies

holes they develop, while others such as urchins or starfish perform this erosion externally to the rocks, without developing cavities.

The activity of these lithophagous organisms is also present in the reefs, although there it is linked to the behavior of others, such as sponges, algae, fungi and bacteria, that degrade coral formations through chemical attack. Also important is the activity of the fish that feed on the algae that live on the corals. Parrotfish and surgeonfish grind the carbonate material ingested by eating the algae to return it to the sea in the form of sand-sized bioclasts. Each of these fish is capable of generating almost half a cubic meter of carbonate sand per year [1].

References

1. Bellwood DR (1995) Direct estimate of bioerosion by two parrotfish species, *Chlorurus gibbus* and *C. sordidus*, on the Great Barrier Reef, Australia. Mar Biol 121(3):419–429
2. De Boer PL (1981) Mechanical effects of micro-organisms on intertidal bedform migration. Sedimentology 28:129–132
3. Boggs S (1995) Principles of sedimentology and stratigraphy (2nd ed). Prentice Hall, New Jersey, 744pp
4. Dalrymple RW, Makino Y, Zaitlin BA (1991) Temporal and spatial patterns of rhythmite deposition on mud flats in the macrotidal Cobequid Bay-Salmon River estuary, Bay of Fundy, Canada; clastic tidal sedimentology. Can Soc Pet Geol 16:137
5. Frey RW, Basan PB (1985) Coastal salt marshes. In: Davis RA (ed) Coastal sedimentary environments, 2nd edn. Springer Verlag, Heidelberg, pp 101–169
6. Gingras MK, MacEachern JA (2012) Tidal ichnology of shallow-water clastic settings. In: Davis RA, Dalrymple RW (eds) Principles of tidal sedimentology. Springer, Heidelberg, pp 57–78
7. Gingras MK, Pemberton SG, Saunders T, Clifton HE (1999) The ichnology of modern and Pleistocene brackish-water deposits at Willapa Bay, Washington; variability in estuarine settings. Palaios 14:352–374
8. Gingras MK, Bann KL, MacEachern JA, Waldron W, Pemberton SG (2007) A conceptual framework for the application of trace fossils. In: MacEachern JA, Bann KL, Gingras MK, Pemberton SG (eds) Applied ichnology, vol 52. SEPM Short Course Notes, pp 1–25
9. Hardie LA (1986) Stratigraphic models for carbonate tidal flat deposition. Q Colorado Sch Min 81:59–74
10. Hunt JM (1996) Petroleum geochemistry and geology (2nd ed). WH Freeman, New York, 340pp
11. Petuch EJ, Myers RF (2014) Molluscan communities of the Florida Keys and adjacent areas. Their ecology and biodiversity. CRC Press, Boca Raton, 300pp
12. Safriel UN (1974) Vermetid gastropods and intertidal reefs in Israel and Bermuda. Science 186:1113–1115
13. Schlager W (2005) Carbonate sedimentology and sequence stratigraphy. SEPM concepts in sedimentology and paleontology series no 8, 200pp
14. Swinbanks DD, Murray JW (1981) Biosedimentological zonation of boundary bay tidal flat, Fraser river delta, British Columbia. Sedimentology 28:201–237
15. Yen TF, Chilingarian GV (1976) Introduction to oil shales. In: Yen TF, Chilingarian GV (eds) Oil shale. Elsevier, Amsterdam, pp 1–12

13 Extreme Events

13.1 Introduction

The action of waves, tides and marine currents, as well as their interaction with continental agents, is a continuous phenomenon of erosion, transport and deposition of material in cycles that mark the ordinary sedimentary dynamics of coastal systems. However, these ordinary dynamics are interrupted at certain times by specific phenomena that display enormous energy on the coast. These are the so-called extreme events or high-energy events. High-energy events are processes that occur in the very short term (minutes, hours, days). These short time intervals could be considered instantaneous from a geological point of view.

The release of these enormous amounts of energy on the coast is manifested in the large volumes of ocean water that move landwards. At the same time, large areas of land are covered by seawater. The distribution of this energy on the coast involves very radical morphological and dynamic changes. These changes represent great threats of damage to property and ultimately a high risk to human lives.

Normally, the high energy of these phenomena is used in the transport of sediments. However, the increased transport capacity is not compensated for by a sedimentary input of the same magnitude. Therefore, the material displaced by the flows in these events comes from the reworking of sedimentary material from previous coastal formations.

The time of their arrival and their magnitude are unpredictable, but when a coast is affected by this type of event every certain period, the fact of their arrival should not be unpredictable. Coastal inhabitants must necessarily take measures to minimize the effects of these phenomena on infrastructure and their cost in human lives. There are different types of possible response to extreme events. Some measures are of an active nature, as is the case with protection structures. However, the most effective measures are those that come from adequate management and planning, which include public education measures and evacuation protocols. Even in extreme cases of reoccurrence, definitive evacuation and relocation of populations at risk cannot be ruled out. The social and political impacts of events depend on the development of countries and their response capacity. Typically, more developed countries have been more able to establish effective strategies to deal with such events. One only has to compare the difference in the number of deaths between two tsunamis with similar characteristics: that of Indonesia (2006) and Japan (2011). The countries affected by the 2006 tsunami were less prepared, with a cost in lives that almost reached 228,000 deaths. In contrast, in Japan, with a well-educated population, the figure was one order of magnitude lower, at below 16,000 deaths.

Knowledge of these events is important in order to understand their dynamics and behavior, but also for their impact on society. Perhaps because of this social impact, the processes associated with high-energy phenomena have been the object of scientific interest in recent decades, especially since media coverage of such events as the aforementioned tsunamis or Hurricane Katrina in 2005. In spite of this, the processes that occur during these phenomena and the mechanisms of transport are not very well studied due to the practical impossibility of carrying out measurements and observations during their development.

13.2 Extreme Storms

The term storm is defined in the Beaufort Scale as a situation that occurs above force 8, which is classified as a strong gale. Situations of force 9–12 would be defined by the intensity of the storm. The situations of strong gale, storm, violent storm and hurricane are defined respectively. This means winds of over 72 km/h and up to 110 km/ in category 1 cyclones and up to 250 km/h in category 5 cyclones. In colloquial terms, a storm situation is understood to be one in which the winds and waves reflect conditions that are much more intense than usual (called good weather conditions). In any case, storm situations always occur in conjunction with low pressure systems. At latitudes above 30°, these low pressure systems are called storms, while in inter-tropical

J. A. Morales, *Coastal Geology*, Springer Textbooks in Earth Sciences, Geography and Environment,
https://doi.org/10.1007/978-3-030-96121-3_13

areas they take on different names depending on the ocean they are acting on. Thus, in the central and eastern Pacific Ocean, they take the generic name of tropical cyclones, while in the Atlantic they are called hurricanes (from the word used by Carib Indians) and in Asia they are known as typhoons (a word taken from Chinese).

All coastlines are occasionally hit by storms or cyclones. Each storm represents an increase in energy above the average of the processes that normally act on it. However, since each coast suffers several seasonal storms per year, these can be considered as part of the ordinary dynamics of the coast. There are even coasts whose dynamics are dominated by storms [2]. However, on some occasions, an exceptional storm develops that involves an increase in energy that exceeds normal storms by an order of magnitude. The behavior of these exceptional storms and their sedimentary record are much more complex than those of normal storms [6].

If a storm is defined by an increase with respect to the energy that normally acts on the coast, it must be taken into account that the energy is growing in a quadratic manner with respect to the height of the waves (Chap. 7, Eq. 7.1). Thus, a double increase in wave height means a quadrupling of the energy, while a triple increase in height means that the energy is multiplied by nine. Data from *Project Stormfury* allowed the calculation that the energy reaching the coast in one of these extreme events can reach the equivalent of 600 terawatts [22]. This energy is 5000 times higher than the energy generated in an average nuclear power plant. Some authors [15] have estimated that the transport capacity of one of these events with a return period of tens of years is much higher than the energy dissipated on the coast by the waves during the period between two successive events. This gives an idea of the extent to which the arrival of high-energy events on a coast can transform its dynamics and condition its sedimentary record. More storm data can be found on the National Oceanic and Atmospheric Administration (NOAA)'s Atlantic Oceanographic and Meteorological Laboratory website.

The action on the surface of the sea of a wind situation above force 8, together with the low pressures that accompany the passage of cyclones, make two effects generated by a storm. The first, and most obvious, is the increase in the size of the waves generated by wind friction in the water mass. The second is a combined effect of low pressure and wind action, and is the topographic elevation of the sea surface which is called a meteorological tide or **surge** [8].

13.2.1 Surges

The influence of low pressures combined with the action of strong winds characteristic of cyclones is manifested by a rise in the sea surface (Fig. 13.1). The influence of low pressure is quantifiable [12]. In general terms, for every millibar that the barometer drops, the sea surface rises 14 mm [17]. This would mean that a 40 millibar drop, characteristic of a typical Atlantic storm, would be able to raise the sea level by about 54 cm, whereas a hurricane like Katrina, with a 109 millibar drop in pressure, would mean a sea level increase of 1.47 m due to pressure alone.

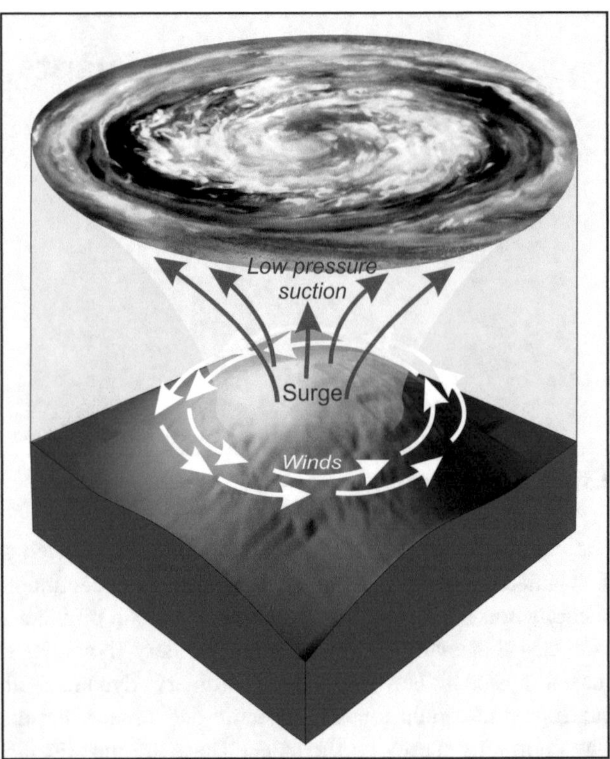

Fig. 13.1 Diagram showing the origin of a surge in the center of a low-pressure system

It is more difficult to quantify the effect of the wind. In the open sea there is practically no effect of vertical displacement of the water surface due to friction with the moving air. On the contrary, in coastal areas the presence of the bottom and the coastline cause interference with the free displacement of the water mass due to wind-directed currents (Fig. 13.2). This raises the sea level when the air blows perpendicular to the coast. The magnitude of these vertical movements is not directly quantifiable, as it is in the case of pressure influence, since the elevation depends on geometrical factors of the coast, such as the angle between the wind and the coastline or the relationships of slope and width of the *shoreface*.

In certain conditions, wind uplifts can even be greater than those caused by atmospheric pressure. A well-known case is the elevation caused by the southeast winds over the Argentine coast, especially in the estuary of the Río de la Plata, where the funnel shape increases the water level due to

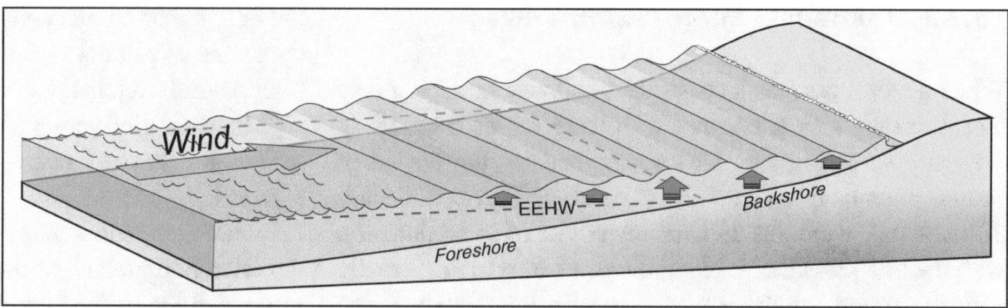

Fig. 13.2 Diagram showing the influence of the wind in generating a coastal surge. *EEHW* Extreme equinox high water

the convergence effect [11]. This phenomenon is known as the ***sudestada***. The largest wind flood in the history of Buenos Aires occurred on April 15, 1940, raising the water level by 4.4 m.

During a surge, the elevations due to the pressures are combined with those induced by the wind. The displacement of a low pressure system is usually slow and so this elevation can be sustained for several days [14]. The duration of this phenomenon ensures that at certain points the rise will coincide with high tides. During these times, flood peaks can be particularly significant.

13.2.2 Extreme Storm Waves

During storms, the action of the wind produces a notable increase in the size of the waves acting on the coast. It must be taken into account that the waves acting on coastal systems do not only correspond to the ones being generated by the wind in this same place, but they also include those propagated waves that the same cyclone generates offshore that now reach the coast. The result is that, during the duration of the storm, the waves discharge an enormous amount of energy on the coast. This energy is used to move sedimentary material from one place to another within the coastal system, but also to transfer material from the coast to other neighboring environments, such as the shelf or the continent.

The larger dimensions of the wave are manifested in the generation of more energetic breakers. A very frequent example occurs on a coast of moderate slope, where the waves normally break in a *spilling type* way; during storms these may become *plunging* type breakers or even more energetic types such as *collapsing* or *surging*. This implies that the dissipation of energy is lessened when compared with other phenomena such as the reflection of the waves. At the same time, the appearance of more energetic breakers favors erosional processes in the wave-breaking zone.

Under these conditions, it is common for significant erosive scars to appear in the *foreshore* and *backshore* areas. If the wave height does not reach the maximum height of the beach, eroded sediment from these zones is displaced to the *shoreface* by the backwash and reflected waves. If, on the other hand, the waves exceed the height of the dune system, an *overwash* phenomenon will occur and the sediment will be displaced landwards.

Chapter 7 explained how the dynamics of a wave-dominated coast can be described as a function of the probability of waves of a certain size acting on it. Normally, this probability is calculated as an annual exceedance probability (Fig. 13.3). The graph in Fig. 13.3 shows an example for the west coast of Europe: it can be seen that there is a probability of 1 that a wave will not exceed 9 m, which means that waves of these dimensions will not occur in a normal year. It is clear that ordinary years also produce storms, although these ordinary storms do not produce waves larger than 9 m on the coast analyzed in this graph.

However, as mentioned earlier, there are exceptional storm situations where this normal situation is overcome. These situations occur at intervals of many years. In order to calculate the probability of these storms occurring, and with them larger waves, it is necessary to establish the return period. The return period is the space of time that exceptional waves of certain dimensions take to return. Typically, the return period for exceptional wave sizes is measured in tens or hundreds of years.

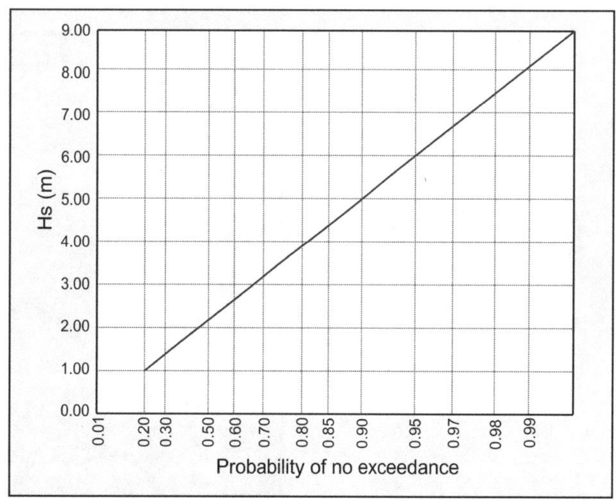

Fig. 13.3 Graph of annual probability of exceedance for significant wave heights (Hs). Example of a coast from NW Spain

13.2.3 Combined Surge–Storm Waves

When a storm reaches a coast, the wind acts in such a way that it usually surges together with large waves (Fig. 13.4). The case shown in Fig. 13.4 corresponds to the arrival of Storm Emma on the southwest coast of Europe in early 2018. In this event, the dominant winds acted in conjunction with the low pressures, causing the surge to coincide exactly with the timing of the larger waves. In this situation, the height reached by the surge places the height of the wave breakers in the areas of greater slope that are located above the *backshore*. The action of the big waves occurs at these moments at very high levels, but, in addition, the energy of the wave is applied to areas with a greater slope. Both phenomena—larger waves and steeper slopes—favor stronger erosion in these areas. The first line of dunes in natural systems is located there, while in anthropized systems it is the first line of human constructions. Therefore, knowing the dynamics of storms is fundamental for the preservation of both natural systems and human infrastructures.

Although wind surges and waves are usually simultaneous, sometimes they are not. The internal structure of cyclones locates low pressure zones in the center of the system, while the more energetic winds are located around the storm. The direction of approach of the cyclones to the continent can be made at different angles to the coastline. This implies a range of possibilities. On the one hand, when a cyclone reaches a coast, the center of the cyclone does not always pass over it. On the other hand, the cyclone can arrive from the sea and run along the coast or enter the continent. According to the route, and taking into account the direction of rotation of the cyclone (counterclockwise in the Northern Hemisphere and clockwise in the Southern Hemisphere), three situations can occur (Fig. 13.5; Carter [7] and Woodroffe [28]).

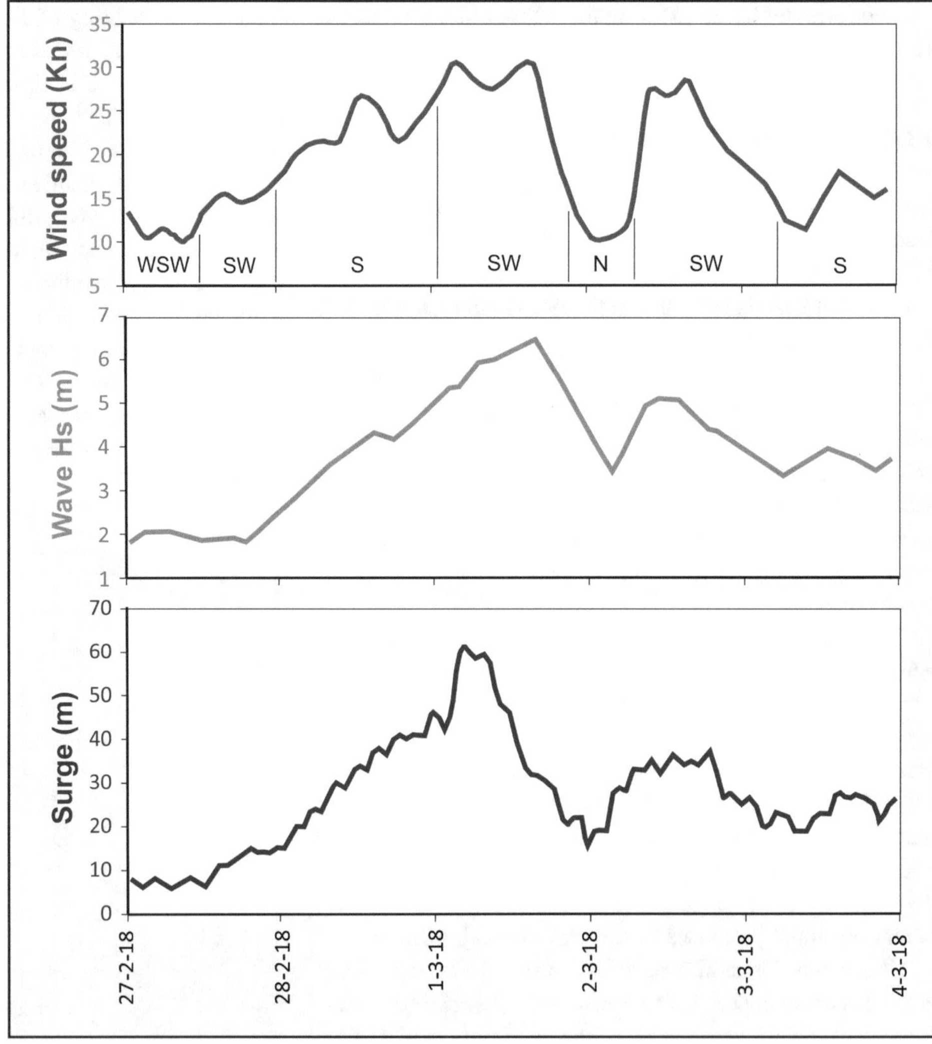

Fig. 13.4 Graphs of wind speed, significant wave height and surge that occurred during the arrival of Storm Emma (February 28 to March 1, 2018) and a subsequent storm (March 2–3). The depression in the curves early on March 2 corresponds to the interval between the storms

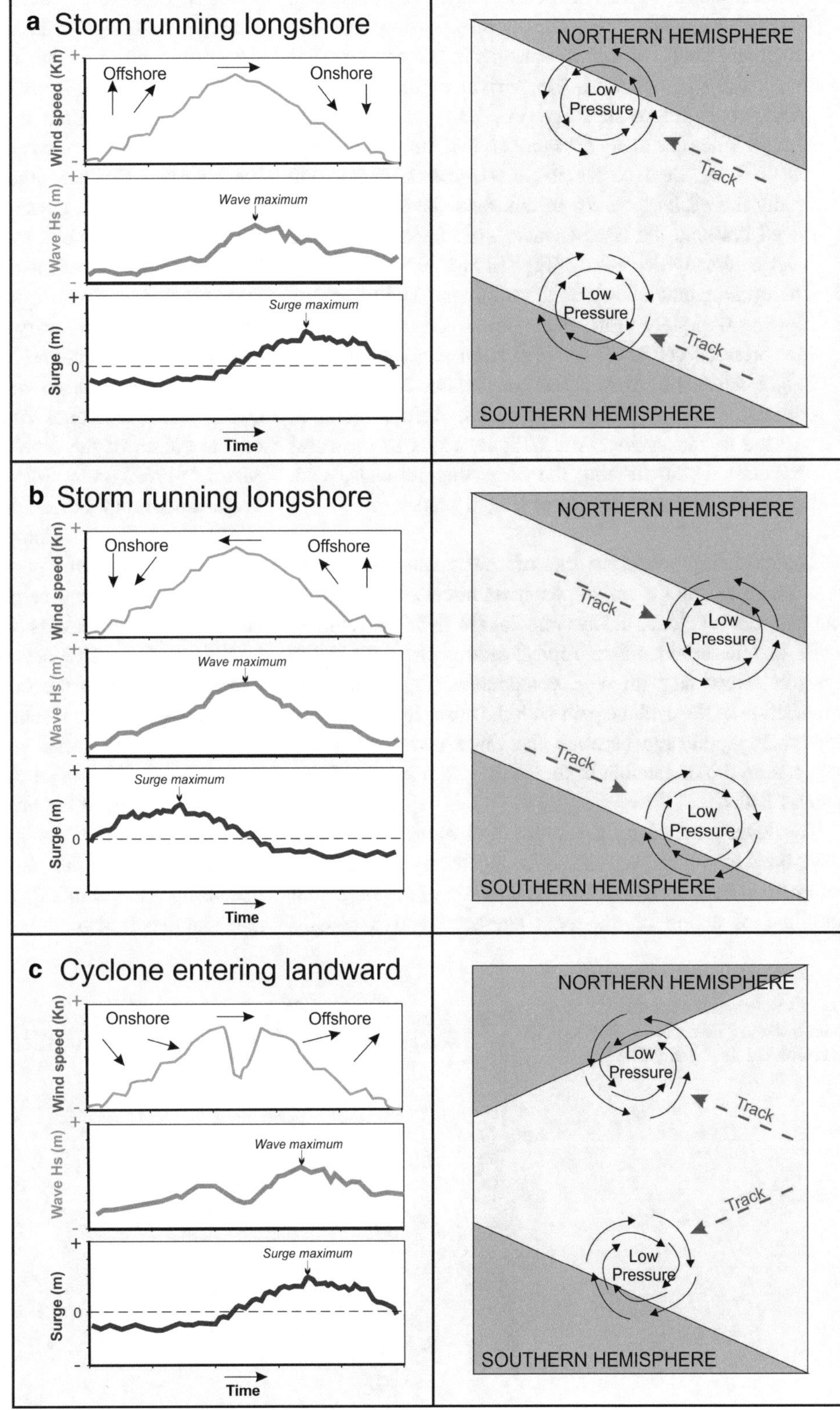

Fig. 13.5 Graphs of wind speed, significant wave height and surge for the three possibilities of storm pass (Based in Carter [7]; Woodroffe [28])

1. The cyclone runs along the coast in such a way that first the *offshore winds* blow, then comes the low pressure center and finally the *onshore winds*. In this case, there is first a drop in sea level. The arrival of the maximum swell occurs before the surge (Fig. 13.5a).
2. The cyclone runs along the coast so that first the *onshore winds* blow, then comes the low pressure center and finally the *offshore winds*. In this case, the surge is produced first and the largest waves arrive when the surge level is already decreasing (Fig. 13.5b).
3. The cyclone enters towards the continent. The winds are blowing obliquely to the coast before the arrival of the low pressure center. There is first an increase in wave height before the arrival of the low pressure center, which coincides with a decrease in winds. Before the final passage of the cyclone, the oblique winds to the coast accelerate again, forming the surge that coincides with the arrival of the major waves (Fig. 13.5c).

The most common trajectories of low pressure centers are determined by the general atmospheric circulation. At tropical latitudes, cyclones always circulate in E–W trajectories, while at mid-latitudes extratropical storms circulate in the opposite direction, with W–E components (Fig. 13.6). The intersection of the cyclone path with the coast and, therefore, the model applicable between the three described above depends on the orientation of the coast with respect to these general paths.

The level reached by the surge, and the effects of the wave breakers on these levels, also depend on the tide. We commented in Sect. 13.2.1 that the duration of the surge and the time of action of the most energetic waves usually exceeds 12 h—this guarantees that at some moment the extreme conditions will coincide with high tidal waters. But taking into account the long tidal cycles, flooding can be especially severe if it also coincides with spring tides. For example, the conditions of Storm Felix in March 2018 were more energetic than those of Storm Emma, which occurred a week earlier. However, the effects of Emma on the coast of Europe were much greater, because Emma coincided with spring tides while Felix took place during neap tides.

This coincidence depends on chance and does not happen very often. In this case, to analyze the frequency of these phenomena it is necessary to use the return periods as well. Extreme storm events are marked in the collective memory by the damage caused on the coasts. Examples of these events were hurricanes Andrew (1992) and Katrina (2005) on the coasts of the Gulf of Mexico, the "Great Storm of March" (1962) on the Meso-Atlantic coast of the USA, The Watersnoodramp (1953) in the Netherlands, storm Daria (1990) in Northern Europe and tropical cyclones Bhola (1970) in the Bay of Bengal and Tracy (1974) in Australia.

It is clear that the same event does not have the same effect on different coasts, so the return period must also be established for each location. The closer the studied coast is to the place where the cyclone touches the continent, the more energy will be discharged on the coast and the greater the effects of the storm.

Long-term factors such as global climate change are affecting the frequency of occurrence of extreme storms, as well as the timing of their arrival [3]. Thus, for example, in tropical latitudes the number of cyclones per year is increasing. In parallel, the number of storms in the mid-latitudes is also increasing. Another effect that has been

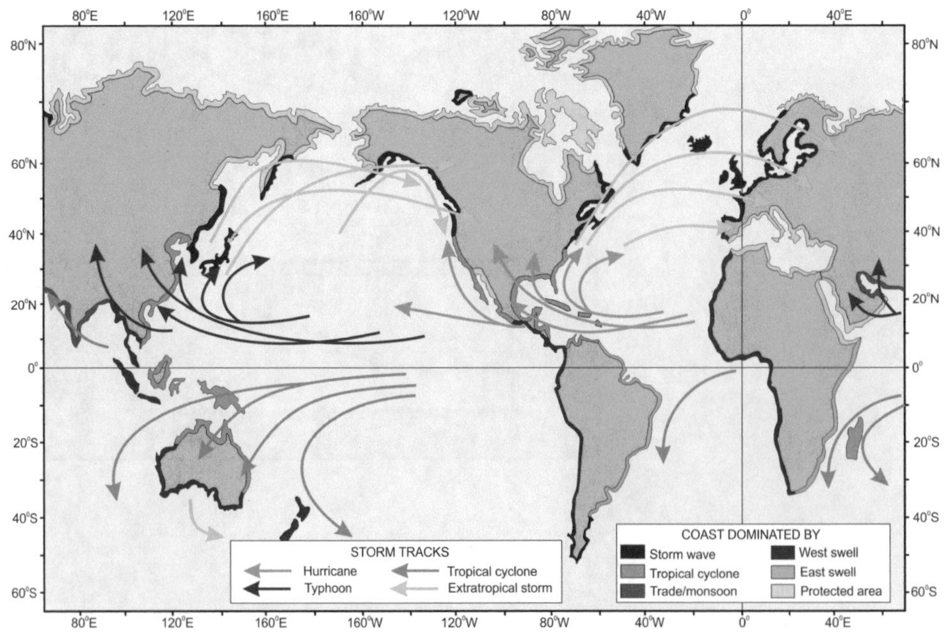

Fig. 13.6 Map showing the main tracks of the different kinds of storm and their dominance on the coasts

observed is that the season for the genesis of storms is changing, so that an increasing number of storms are occurring on abnormal dates. The displacement in time of these anomalies has different signs depending on the coast being studied.

13.2.3.1 Relaxation Currents

When the winds blowing towards the coast during a storm cease, the sea level rise that has been maintained during the storm tends to be compensated for by strong currents flowing from the coastline to the sea (Fig. 13.7). The body of water that moves with these currents is loaded with sediments eroded from the upper areas of the coast. The presence of the sediment gives the water a high density, so these currents will circulate on the bottom through the entire *foreshore* and even *shoreface*, discharging this material on the continental shelf. These are the **storm return currents**, also called **relaxation currents**.

13.2.4 Sedimentary Record of a Storm: Tempestites, Washovers and Cheniers

In the open coastal areas where the most energetic waves attack directly, the evidence of the storm is usually an erosive surface that clearly cuts the sedimentary formations. However, the sedimentary material produced by this erosion is transported and deposited in less energetic environments, where maximum wave energy dissipation occurs. The deposits left by the direct action of the storm and also those left by the relaxation currents are called **tempestites**. The name of these storm deposits comes from the word *tempest*, which is the Latin word used to define a storm. In formal sedimentological terms, a tempestite represents the preservation in facies form of a storm event within the sedimentary record.

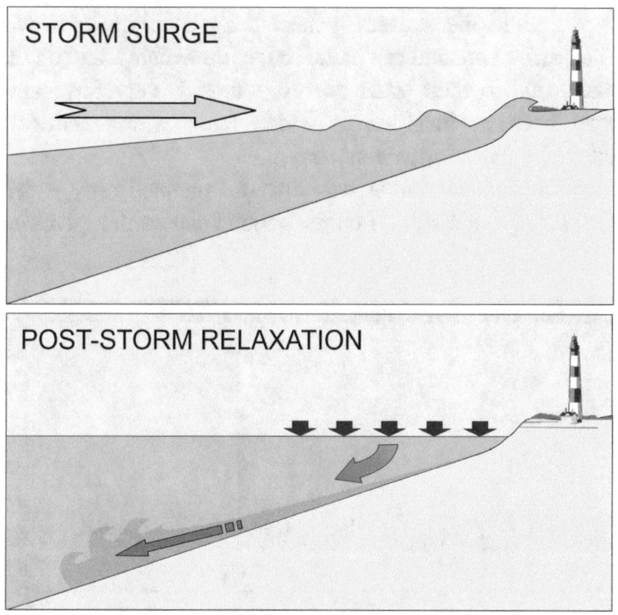

Fig. 13.7 Genesis of relaxation currents at the shoreface

The term was introduced into the scientific literature by Dott and Bourgeois [10]. The enormous energy developed during storms is manifested in a deposit that has a coarse grain size, much larger than the sediments that represent the average energy of the system and are located below and above the tempestite strata.

The tempestites can be preserved in the sequences of a variety of coastal sedimentary environments, including back-barrier environments, lagoons, estuaries and deltas. They can also be preserved in inland systems near coastal areas, such as lacustrine environments near the sea. However, the sequences where tempestites most frequently appear are those corresponding to marine waters located below the base level of fairweather waves and above the base level of the storm surge [20]. This places them offshore, although relaxation currents can cross this level and carry them to continental shelves, as an indication of the sedimentary exchange between the coast and these deeper systems.

When the surge breaks the *foredune* line and reaches the areas located at the rear of a sand barrier, a deposit with a fan geometry is created; this is known as a ***washover fan***. It is the most recognizable type of tempestite because it is exposed and accessible in an area that is much used by humans.

Another place where tempestites are often preserved is on the inter- and supratidal facies of tidal flats and deltas. In this case, the preservation of the coarse sediment characteristic of the tempestite is arranged with an elongated geometry parallel to the coast on the fine tidal facies. These formations are called **cheniers**, as they were first described in the delta plains near the Mississippi (*Chenier Plain*) and interpreted as the deposits left by successive hurricanes. *Washover* and *chenier* deposits will be described in more detail in the chapters on the barrier islands and the deltas, respectively.

However, the typical storm described by Dott and Bourgeois [10] is that developed in *shoreface* and *offshore* waters and is commonly characterized by the presence of *hummocky-type* cross-layering. These structures are formed in a regime of currents in what are known as combined flows. The origin of these flows is the interaction between strong unidirectional currents and wave oscillation. The following base–top order presents an idealized sequence (Fig. 13.8): (a) erosive base which may also develop molds of transported clast markings; (b) coarse residual deposit generally composed of disjointed bioclasts (L level) which may present positive gradation; (c) *hummocky* cross-layering level representative of combined flows (H level); (d) parallel rolling

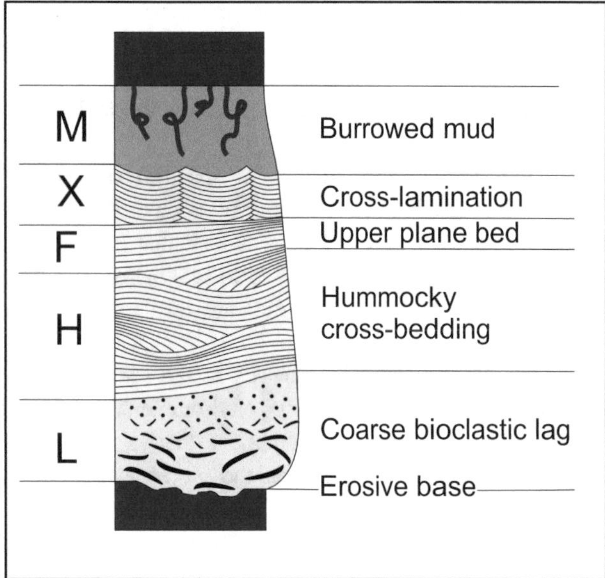

Fig. 13.8 Idealized sequence of a tempestite (based in Dott and Bourgeois [10])

level (F level) representative of upper plane bed; (e) spike crossbed level (X level) representative of wave action; and (f) muddy level (M) corresponding to the post-storm settling of the finest material carried in suspension.

The three typical deposits described above can be affected by subsequent bioturbation, when the organisms living in the sedimentary beds of these environments recover their activity after the return of the fairweather regime. However, if the bioturbation is too intense it can end up destroying the sedimentary structures, mixing the different levels and finally erasing the evidence of the storm.

13.3 Rogue Waves

Chapter 7 described how waves travel in trains that were formed by winds with certain characteristics. As their genesis is linked to the same process, most wave trains have very similar dimensions. However, a much larger wave can appear between them in an unpredictable way, displaying an enormous amount of energy (Fig. 13.9). These waves are known as **rogue waves,** although they can also be called episodic waves, extreme waves, monster waves and freak waves and, much more informally, killer waves. From a scientific point of view, they are defined as those waves whose height is greater than twice the significant height of the train in which they are propagated [24].

Rogue waves have been described by sailors since the eighteenth century, but there was no evidence of them beyond those fantasy tales. Nor was their arrival to the coast known until 1861, when one of them was documented at Eagle Island Lighthouse (Ireland). This wave broke the optical elements of the lighthouse, which were 40 m high. The first rogue wave recorded instrumentally was a 25.6-m wave that affected the Norwegian oil platform of Draupner, in the North Sea, in January 1995 [25]. More recently, in January 2009, a buoy located north of Santander (Spain) recorded a 26.13-m wave within a wave train generated during a storm, with a significant height of 14 m (Fig. 13.9).

Since these waves were documented, some descriptive advances have been made. Some authors have suggested that there are actually three types of rogue wave: one corresponds to what has already been described and is an individual wave that propagates inside the train in the form of a huge wall of water; a second type is formed by waves that travel in groups of three inside the train [19]; and finally, the third type is waves that are generated by wave interaction that collapse immediately and do not propagate.

On the high seas, rogue waves are very dangerous for navigation due to their large size and unpredictability [4]. However, the structures of today's large ships are designed to withstand the stresses generated by this type of wave. Their arrival on land can have a huge immediate effect on the coast, due to their great energy, although very few cases have been documented in which these waves seriously affected human infrastructures.

As this phenomenon was first documented only a few decades ago, the studies on the genetic causes that give rise

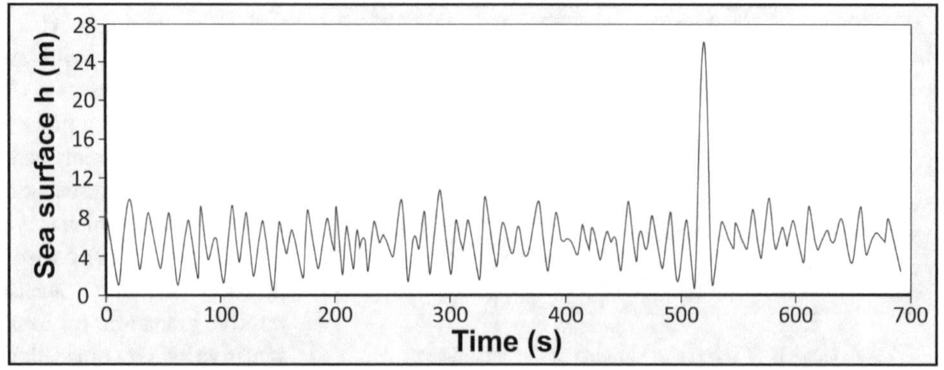

Fig. 13.9 Record of a wave train including a rogue wave of 26.13 m

to it are not entirely clear and are still under investigation. There is probably more than one cause that can generate such exceptional waves. There seems to be consensus, however, that more than one factor must come together for one of these waves to appear. Currently, the hypothesis that gains the most support is the one that attributes its origin to the propagation of the wave train over a strong current that acts in the opposite direction. In this way, some waves would slow down until they were captured by the next wave, adding their mass of water to it. Once formed, this larger wave would capture other waves, thus increasing its dimensions. Although it seems clear that the phased entry and capture of waves is associated with this phenomenon, the existence of a current does not explain all known cases. This phasing with capture of other waves by a single wave can be explained by other phenomena such as diffraction, non-linear instability effects, interaction between two wave trains and particular effects of interaction with the wind [1]. Some of these hypotheses have been successfully reproduced in the laboratory [16].

13.4 Tsunamis

Tsunamis are tectonically induced ocean waves characterized by long wavelengths and high speeds. They are usually associated with earthquakes whose epicenter is under the sea. In the past, they have been called seaquakes and sea surges, but these names have been discarded from a scientific point of view because they do not adequately describe the phenomenon. The name *seaquake* is not appropriate because not all earthquakes with an underwater epicenter are capable of generating a tsunami. The term *sea surge* is usually associated with meteorological phenomena and not with seismic phenomena, so it is not appropriate either. Finally, the term *tidal wave* that is so often used in the Anglo-Saxon world should be discarded, because a tsunami does not respond at all to tidal phenomena. The word **tsunami** comes from the Japanese language in which it means "*harbor wave*," referring to the fact that these are the only waves that are capable of breaking through port defenses.

From a dynamic point of view, a tsunami represents a high-energy event with a powerful capacity to erode and transport marine materials to the continent. It is a rare phenomenon on a human timescale, but it is highly spectacular and occasionally causes large numbers of casualties and material damage. We all remember the Indonesian tsunami of December 26, 2004, which affected the coasts of the entire Indian Ocean, causing an estimated 230,000 deaths (a huge number, if we compare it to the Twin Towers disaster, which did not reach 3000). The tsunami in Japan on March 11, 2011 also had a considerable impact. Although it did not cause as many deaths as the Indonesian tsunami, it did produce the greatest material damage in history.

In short, this is a phenomenon that should be considered as a risk for populations located in coastal areas and is deserving of attention by scientists and authorities.

13.4.1 Mechanism of Tsunami Genesis

The genesis of tsunamis is related to a sudden displacement of a large volume of water induced in turn by the rapid movement of a solid mass. This mass movement can have several possible causes. Small tsunamis are generated during the melting of glacial fronts when icebergs break off or in steep valleys caused by rock falls. On a much larger scale, mega-tsunamis with waves of tens or hundreds of meters in height can originate from meteorite impacts or large underwater landslides. However, the most frequent origin of tsunamis is associated with earthquakes, and more than 90% of these are concentrated at the plate boundaries. Of all types of plate boundaries, most are associated with subduction zones, although they can also be generated in other plate boundaries whenever there is a significant displacement of land mass that is transmitted to the sea.

In subduction zones, the movement of the subducting oceanic plate generates an accumulation of energy that results in elastic deformation of the upper plate (Fig. 13.10a, b). The release of this energy during an earthquake causes an elastic rebound of the upper plate, which moves over the subducting oceanic plate (Fig. 13.10c). This displacement of the ground causes a sudden movement of the water mass, generating a large wave that will propagate over the ocean surface to the nearby coasts (Fig. 13.10d).

Although this is the most frequent mechanism, and is the one that took place in the case of Indonesia (2004) and Japan (2011), it is not the only possible one. There are other known cases of large gravitational landslides on steep ocean floors. This is the case of the Storegga Slides, which occurred in Norway 5000 years BC and spread into the North Sea, reaching the coast of Scotland [5]. There are also known cases in which large explosions in, or large eruptions of, underwater volcanic structures have generated huge tsunamis (the case of Krakatoa in 1883). Similarly, tsunamis associated with earthquakes have occurred on transform faults when a significant underwater relief is displaced in one of their blocks.

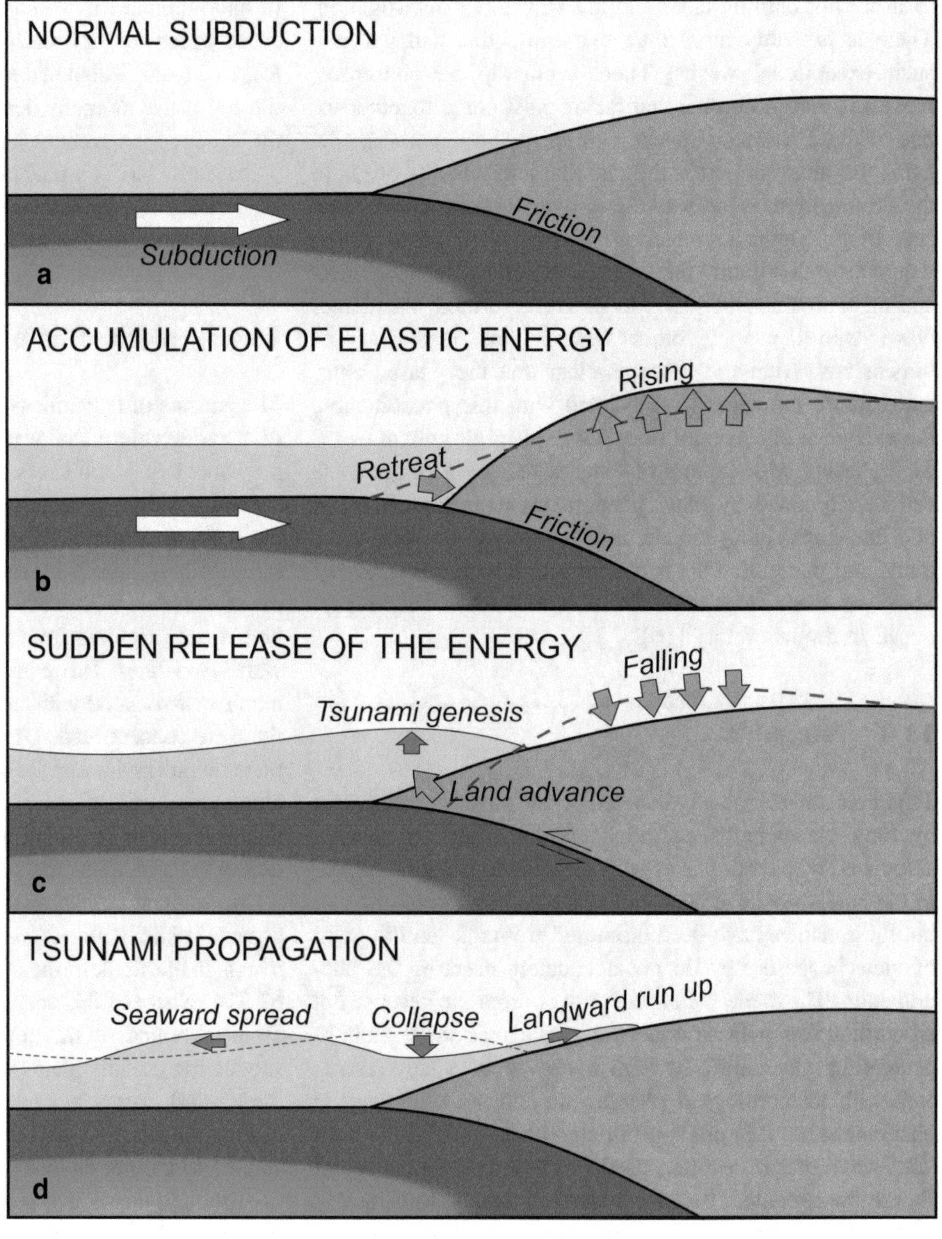

Fig. 13.10 Scheme showing the most common mechanism of tsunami genesis in a subduction zone

13.4.2 Propagation of Tsunami Waves Across the Open Sea

From the point of view of its displacement in the open ocean, a tsunami wave responds to the properties of any progressive wave movement. Theoretically, a wave propagated in a fluid would interact with the bottom at a depth that would be half the distance of the wavelength. As tsunami wavelengths are greater than 50 km, this implies that they touch the ocean floor at depths of less than 25 km. In this regard, it should be borne in mind that the maximum depth of the ocean is 11 km, thus causing a displacement of the entire mass of water, and generating constant friction with the bed.

In this way, the speed of the tsunami is predictable and calculable, as it depends exclusively on the depth. Remember that in Chap. 7, Eq. 7.5 stated that:

$$C = \sqrt{gd} \tag{7.5}$$

where C is the wave propagation speed, g is the acceleration of gravity and d is the depth.

From the application of this equation, the propagation velocity at different depths can be calculated (Table 13.1).

Table 13.1 Velocity of propagation of a tsunami wave in waters of different depths

Depth (m)	Speed (km/h)
1000	356.6
2000	504.3
3000	617.6
4000	713.2

It is clear that when the wave reaches the continental shelf and heads towards the coast, the friction with the bottom causes it to deform and be subject to the common effects of shallows: refraction, diffraction and reflection.

As a wave interacts with the bottom along its entire length, the orbits follow very eccentric elliptical paths, unlike smaller waves. Actually, the horizontal axis of the ellipses is tens of kilometers, while the vertical axis is of metric scale. This makes the inflow and outflow currents on the coast enormous in relation to the vertical displacement of water. As for the dimensional parameters of the wave, it should be noted that, in the case of a tsunami, the maximum height is known as the **set-up**, while the currents to and from the coast are known as the **run-up** and **run-down**.

13.4.3 Tsunami Breakers

In shallow water, the wavelengths are shortened and the wave rises, producing a suction of water from the front of the first wave. Therefore, the first effect on the coast is usually a drop in the water level. Although this phenomenon is commonly known, it does not always occur, since it depends on the situation of the coast with respect to the movement of the land mass that caused the phenomenon. Either after the water level drops or directly, the first break occurs when the first wave reaches the coast. However, due to its large dimensions, the break does not occur at the height of the crest as in common waves, but on its front. Thus, when the wave breaks, the backwash (return current) does not occur immediately, but the water continues to enter the coast and rises in a similar way to the bore of a spring tide (Fig. 13.11).

The entrance of the first wave towards the coast (Fig. 13.12a) generates very strong currents that introduce a huge mass of water towards the continent. This water becomes denser as it incorporates coarse sediment, which increases its destructive power. As mentioned, this movement towards the land is called the **run-up**. The currents towards land continue until the maximum height is reached at the passage of the crest of the wave, and then the current is inverted and begins to go out again towards the sea (the outgoing current called a **run-down**). The outgoing current is maintained until a second wave arrives. The encounter between this run-down and the second wave generates a second breaker (Fig. 13.12b). The breaker of the second wave carries the remaining water from the first flood that has not yet managed to break through, advancing even faster than the first wave. This phenomenon can be repeated up to three times. It is common for the second wave to cause more casualties than the second, as many people are left unprotected when they see the water from the first wave coming down, believing that the tsunami is already over.

The energy dissipated on the coast by a tsunami is several orders of magnitude greater than that of a similar wind–wave height.

13.4.4 Sedimentary Record of a Tsunami: Tsunamites

A tsunami normally generates a complex layer of sediment that is called a **tsunamite** by sedimentologists. When this layer is found on the coast it can easily be confused with storm layers, although in restricted environments and protected bays, such as lagoons, deltas, estuaries or tidal flats, it is easily recognizable as being very different from the muddy sediments that are normally deposited in these environments. In places close to the coastline, a tsunami can form deposits in the same places as extreme storms: washovers and cheniers and the run-down currents can also form very thick layers in the shoreface and offshore.

Various authors (e.g., Dawson et al. [9]; Fujiwara et al. [13]) have differentiated four types of layers generated by tsunamis:

1. Fining upwards sequences of shells and shell fragments with sandy-muddy matrix and ending with layers of bioturbated sand (Fig. 13.13a). The common characteristic of these sequences is a high number of mollusks, including open sea species mixed with others typical of environments protected from waves. They are typical of inland systems such as estuaries, deltas and lagoons.

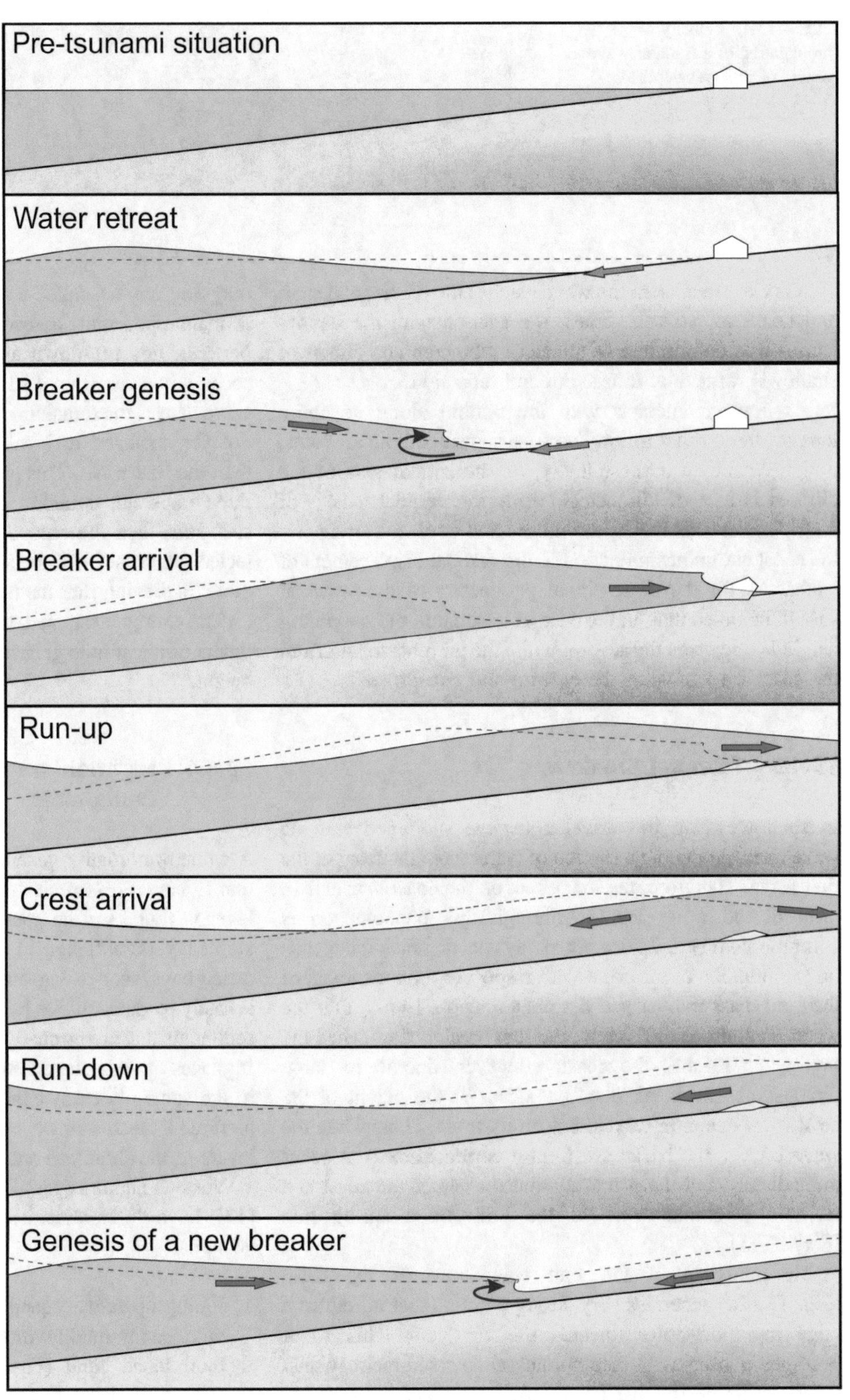

Fig. 13.11 Scheme showing the evolution of the arrival of a tsunami wave at a coast

2. Massive accumulations of shells and shell fragments These accumulations present similar faunistic characteristics to the base of the previous type (Fig. 13.13b). They are characteristic of open supratidal systems. They are also observed in the position and geometry of cheniers.

3. Layers of sand that also have an erosive base and normally contain plant fragments and soft pebbles (Fig. 13.13c). In these sands, the abundance of organisms is also present in microfaunal associations, especially of diatoms and ostracods. Sometimes, the sands of

Fig. 13.12 Photographs showing the breakers of the Indian Ocean tsunami, 2006. **a** First wave breaker entering on a shoreface drained by the water retreat. **b** Second wave breaker entering on waters moved by the run-down current. This current removed various objects from land which are now projected by the second breaker. The breaker height is representative of the tsunami set-up, because the real wave crest is coming several kilometers seaward (Images captured from TV videos)

these layers have cross-laminations that mark the direction of the outgoing and incoming currents of the tsunami. They are observed in barrier island areas, usually associated with *washover* geometries.

4. Layers of highly disorganized rock edges that often include fragments of marine organisms (Fig. 13.13d). They also have an erosive base. They appear in systems where bioclastic material is scarce.

In all cases, the tsunamigenic deposits present a characteristic enrichment in heavy metals (Pb, Cu, Ni, Fe and/or Cr). Scientists explain this as an accumulation of dense minerals from the erosion of formations adjacent to coastal areas [26].

When the large wave reaches an eroded coast where there is no loose sediment to form a tsunami layer, it usually carries large blocks of rock. This can result in dispersions of large blocks on the seafloor or wave-cut platforms, and also accumulations of overlapping blocks on coastal rock escarpments [27].

All these tsunamites bear a remarkable resemblance to those deposits developed by storms. There is an abundance

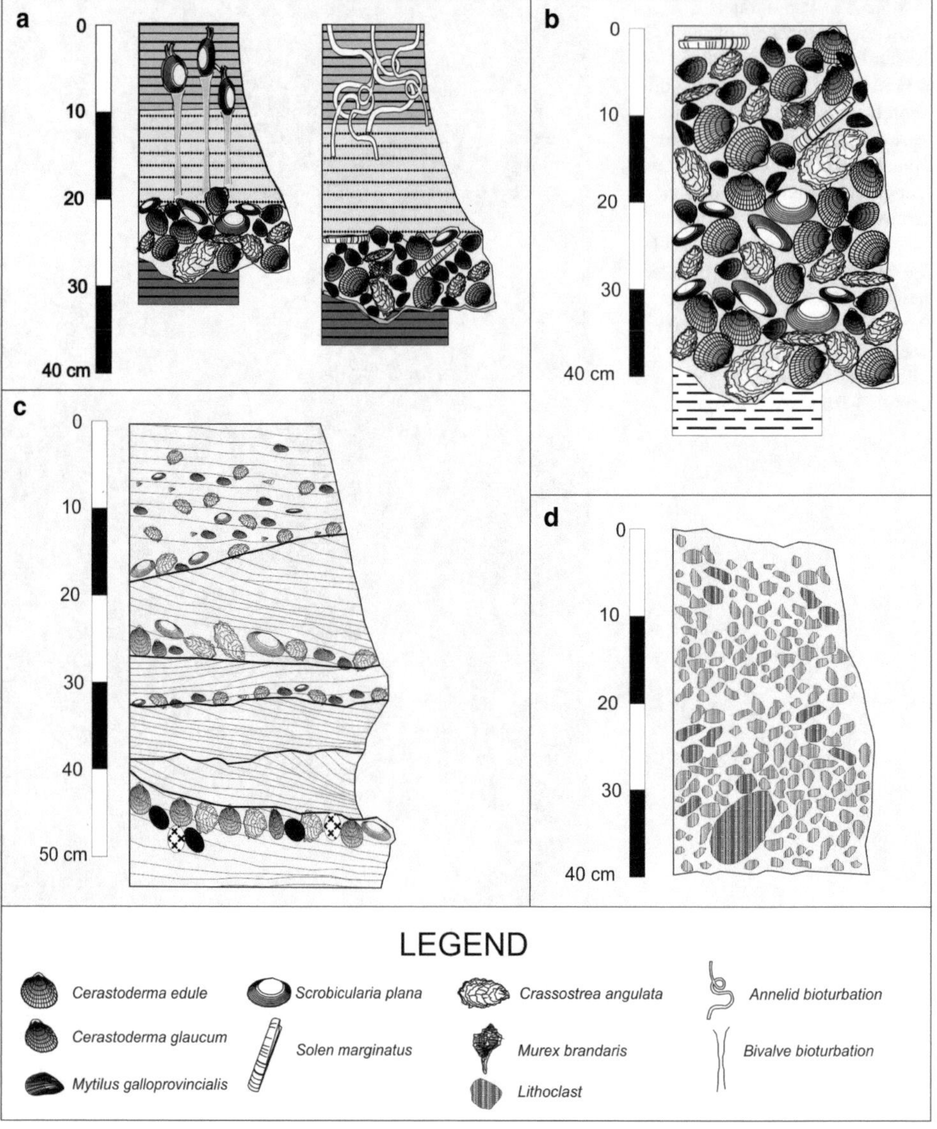

Fig. 13.13 Different types of tsunamites (adapted from Morales et al. [18])

of literature discussing possible criteria for differentiating storms and tsunamis (e.g., Nanayama et al. [21]; Sawai [23]) but there are really no clear rules that can be applied in all cases. Normally, to differentiate between these deposits, one must resort to locally established facies and geometric relationships with the environment.

References

1. Adcock TAA, Taylor PH (2014) The physics of anomalous (rogue) ocean waves. Rep Prog Phys 77(10):105901
2. Aagaard T (1990) Infragravity waves and nearshore bars in protected storm-dominated coastal environments. Mar Geol 94 (3):181–203
3. Bell RG, Goring DG, De Lange WP (2000) Sea-level change and storm surges in the context of climate change. IPENZ Trans 27 (1):1–10
4. Birkholz S, Brée C, Demircan A, Steinmeyer G (2015) Predictability of rogue events. Phys Rev Lett 114:213901
5. Bondevik S, Svendsen JI, Mangerud J (1997) Tsunami sedimentary facies deposited by the Storegga tsunami in shallow marine basins and coastal lakes, western Norway. Sedimentology 44:1115–1131
6. Bourrouilh-Le Jan FG, Beck C, Gorsline DS (2007) Catastrophic events (hurricanes, tsunamis and others) and their sedimentary records: introductory notes and new concepts for shallow water deposits. Sed Geol 199:1–11
7. Carter RWG (1988) Coastal environments: an introduction to the physical, ecological, and cultural systems of coastlines. Academic Press, London, 617pp
8. Chapman S, Lindzen R (1970) Atmospheric tides. Reidel Publishing Company, Amsterdam, 179pp
9. Dawson AG, Foster IDL, Shi S, Smith DE, Long D (1991) The identification of tsunami deposits in coastal sediment sequences. Sci Tsunami Haz 9:73–82
10. Dott RH Jr, Bourgeois J (1979) Hummocky stratification: significance of its variable bedding sequences. Bull Geol Soc Am 93:663–680

11. Escobar G, Vargas W, Bischoff S (2004) Wind tides in the Rio de la Plata estuary: meteorological conditions. Int J Climatol 24 (9):1159–1169
12. Flather RA (2001) Storm surges. In: Steele JH (ed) Encyclopedia of ocean sciences. Academic Press, London, pp 2882–2892
13. Fujiwara O, Masuda F, Sakai T, Irizuki T, Fuse K (2000) Tsunami deposits in Holocene bay mud in southern Kanto region, Pacific coast of central Japan. Sed Geol 135:219–223
14. Hubbert GD, McInnes KL (1999) A storm surge inundation model for coastal planning and impact studies. J Coastal Res 15:168–185
15. Hume JD, Shalk JD (1967) Shoreline processes near Barrow, Alaska: a comparison of the normal and the catastrophic. Arctic 20:83–103
16. McAllister ML, Draycott S, Adcock TAA, Taylor PH, van den Bremer TS (2019) Laboratory recreation of the Draupner wave and the role of breaking in crossing seas. J Fluid Mech 860:767–786
17. Macmillan DH (1966) Tides. CR Books Ltd., London, 240pp
18. Morales JA, Gutiérrez-Mas JM, Borrego J, Rodríguez-Ramírez A (2011) Sedimentary characteristics of the Holocene tsunamigenic deposits in the coastal systems of the Cadiz Gulf (Spain). In: Mörner NA (ed) The Tsunami threat: research and technology. Rijeka, pp 237–258
19. Moreau F (1963) The glorious three. Report to IFREMER Center of Brest, France. Reproduced in Olagnon M, Prevosto M (eds) Proceedings of the IFREMER meeting on Rogue Waves, 2004
20. Myrow PM, Southard JB (1996) Tempestite deposition. J Sediment Res 66:875–887
21. Nanayama F, Shigeno K, Satake K, Shimokawa K, Koitabashi S, Miyasaka S, Ishii M (2000) Sedimentary differences between the 1993 Hokkaido-Nansei-Oki tsunami and the 1959 Miyakojima typhoon at Taisei. Southwestern Hokkaido, northern Japan. Sed Geol 135:255–256
22. Neumann CJ, Jarvinen BR, McAdie CJ, Elms JD (1993) Tropical cyclones of the North Atlantic Ocean, 1871–1992. Prepared by the National Climatic Data Center, Asheville, NC, in cooperation with the NHC, Coral Gables, FL, 193pp
23. Sawai Y (2002) Evidence for 17th-century tsunamis generated on the Kuril-Kamchatka subduction zone, Lake Tokotan, Hokkaido, Japan. J Asian Earth Sci 20:903–911
24. Tayfun MA (1980) Narrow-band nonlinear sea waves. J Geophys Res 85:548–552
25. Taylor PH (2005) The shape of the Draupner wave of 1st January 1995. Unpublished memory, Department of Engineering Science, University of Oxford
26. van den Bergh GD, Boer W, de Haas H, van Weering TCE, van Wijhe R (2003) Shallow marine tsunami deposits in Teluk Banten (NW Java, Indonesia), generated by the 1883 Krakatau eruption. Mar Geol 197:13–34
27. Whelan F, Kelletat D (2005) Boulder deposits on the southern Spanish Atlantic coast: possible evidence for the 1755 AD Lisbon tsunami? Sci Tsunami Haz 23(3):25–38
28. Woodroffe CD (2002) Coasts: form. Cambridge University Press, Cambridge, Processes and Evolution, 623pp

14 Particle Transport

14.1 Introduction

Weathering residues and pyroclastic particles that reach the coast from the mainland, as well as any carbonate class generated in the marine environment, can be reworked, transported and deposited within the coastal system. Thus, these particles of sedimentary material end up being part of the sequence deposited in one of the coastal environments. In sub- and intertidal environments, this transport can be carried out by the flow of water set in motion by currents, tides or waves. Meanwhile, in supratidal environments it is the wind that plays the main role in the transport of particles.

The particular processes that occur in environments dominated by waves, tides or other marine currents all involve erosion, transport and deposition of particles by fluids. Understanding these processes requires a precise understanding of how the physical mechanisms of particle transport work. These mechanisms involve three different aspects: (1) the processes by which a fluid sets particles in motion by pulling them from the bottom; (2) the factors that cause particles to continue to be transported within the fluid; and (3) the processes that contribute to particles ceasing to be transported and being deposited on the bed. In this chapter we will explore the fundamental dynamics of the initiation and continuation of particle transport by fluids, as well as their sedimentation mechanisms.

14.2 Processes of Particle Transport by Fluid Flow

Between the late 1960s and early 1980s, numerous authors made important contributions to the understanding of the details involved in the processes of putting particles into motion by fluids (e.g., Bagnold [7]; Sternberg [44]; Moss [33]; Middleton [28]; Moss et al. [34]). This knowledge was disseminated didactically by the authors Middleton and Southard [30] and almost immediately included in sedimentology manuals (e.g., Middleton [29]; Allen [2]; Dyer [12]).

Some aspects of particle movement are already in our minds almost unconsciously. Everyone can easily understand that, as the speed of a fluid increases, a critical moment is reached when the movement of the grains begins. It is also intuitively understood that the smaller grains begin to move first and, conversely, the larger ones do not start moving until the speed of flow is very high. Similarly, it is easy to understand that the density of the grains, as well as the density of the fluid that transports them, are variables that significantly influence the movement of the grains. Perhaps one of the first things we learned as children about the movement of grains is that, when a stream slows down, grains are deposited in an orderly fashion, with the coarsest grains being the first to be deposited and the finest grains being the last. All this knowledge that we already have, almost without knowing it, will help us understand the variables involved in the movement of particles through fluids.

Actually, the action of a moving fluid on the grains of sediment at the bottom can be seen as a conflict between two types of forces acting on each of the grains. On the one hand, the kinetic forces (Fm) caused by the movement of the fluid applied to the grain try to make the grain move. On the other hand, the inertial forces (Fi) try to make the grain remain motionless. Thus, movement or rest is presented as a balance between both types of force. Any of the forces acting on a grain must be broken down into its two components: the vertical component, upwards or downwards, and the horizontal component, in the particle's plane of motion.

There are only two inertial forces that act in a general way: gravity and friction. Gravity acts downwards and is the result of the attraction of particles to our planet; therefore, it is a function of their mass and acts as a vertical force. Friction with the bottom is the result of the resistance to the movement of the grain that is in contact with the other grains; therefore, it has a horizontal component. To these

J. A. Morales, *Coastal Geology*, Springer Textbooks in Earth Sciences, Geography and Environment,
https://doi.org/10.1007/978-3-030-96121-3_14

two forces, a third one must be added in the case of the finest particle grains: the electrostatic attraction between particles.

The kinetic forces are those that the flow must generate to try to put the particles in motion. These are: the frictional thrust of the flow, buoyancy and the Bernoulli effect. The thrust of the flow on the particle is the result of the viscous friction between the fluid and the grain and acts horizontally. In contrast, buoyancy and the Bernoulli effect have a predominantly vertical component (Fig. 14.1).

Gravity acts on the grain as a function of its size and density, so that it exerts the same force on particles that have the same **hydraulic equivalence**. In order to calculate the hydraulic equivalence, the product density by volume of the grain is used, and those particles that have the same product are considered hydraulically equivalent ($\delta_1 V_1 = \delta_2 V_2$). In this way, grains of smaller size and greater density would be equivalent to others with larger size and lower density. As an example, in a deposit formed by hydraulically equivalent particles, the carbonate grains will be larger than the quartz grains, as they have a lower density. Conversely, grains of heavy minerals (e.g., pyroxenes and amphiboles) would be smaller.

Electrostatic forces are usually not important for particles larger than 63 μ; however, smaller particles, especially phyllosilicates, have a charged surface due to the nature of their electrochemical bonds. These particles, when deposited, are subjected to attraction forces that bestow on the sediment a property known as **cohesiveness**.

Among the kinetic forces, the one that is most easily understood is the thrust of the fluid on the grains. Thrust is the vector that appears due to the friction of the moving fluid on the particle and is related to the shear stress at the bottom. This component is greater in turbulent flows. Although in principle this vector is parallel to the flow and therefore horizontal, the pivot angle (α) between the particle and the bottom (or between different grains) must be taken into account. Due to this angle, the fluid thrust may also acquire a vertical component.

The buoyancy force appears to be due to Archimedes' principle. According to this principle, a body immersed in a liquid experiences an upward thrust related to the volume of fluid it displaces. Thus, this force is related to the contrast of densities between grain and fluid. It is a force that is directly opposed to the action of gravity and it is intuitively understood that bodies weigh less when they are immersed in water.

The Bernoulli effect generates a vertical force that is known as the **hydraulic lifting force**. This force is a function of the speed and shape of the grain. The force is the result of the convergence of the flow lines accelerating the speed over the particle (Fig. 14.2a). As a result of this increase in speed, there is an increase in pressure in the front area of the grain and a decrease in the upper area (Fig. 14.2b).

This drop in pressure on the grain tends to pull the grain upwards through a "sucking" effect. This effect depends on the shape of the particles. In general terms, it is much greater for particles with hydrodynamically efficient shapes. It should be noted that this is the same effect that is responsible for airborne support during flight.

The inertial forces act as vectors which combine into a single force called the fluid force (Fig. 14.3a). This force is normally broken down into its two components (horizontal and vertical) called the *drag component* and the *lift component*, respectively. These horizontal and vertical components are those which directly oppose the inertial forces of gravity and friction with the bottom (Fig. 14.3b).

14.2.1 Grain Entrainment Threshold

Let's consider a situation in which a grain is just about to move by the action of a flow. The force of the fluid must be great enough to overcome the static inertia of the grain. The particle will rise when the sum of the kinetic forces exceeds the inertial forces (Fm > Fi). For each size of particle, there

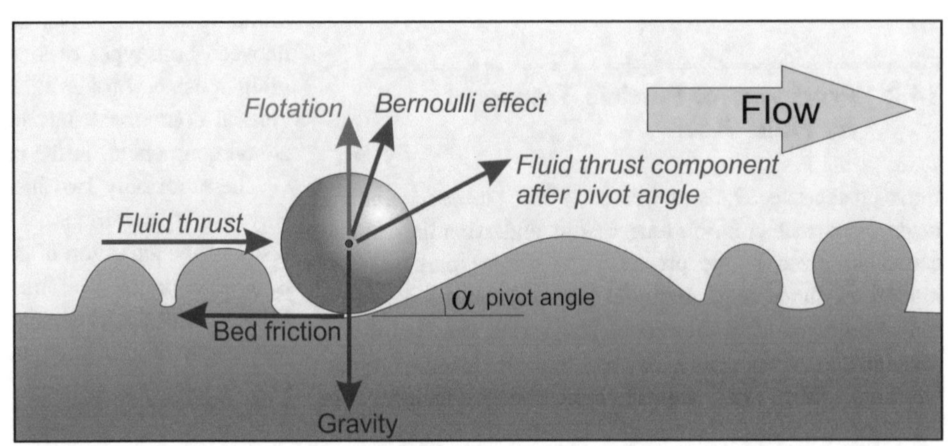

Fig. 14.1 Balance of forces acting on a particle under the action of an aqueous flow

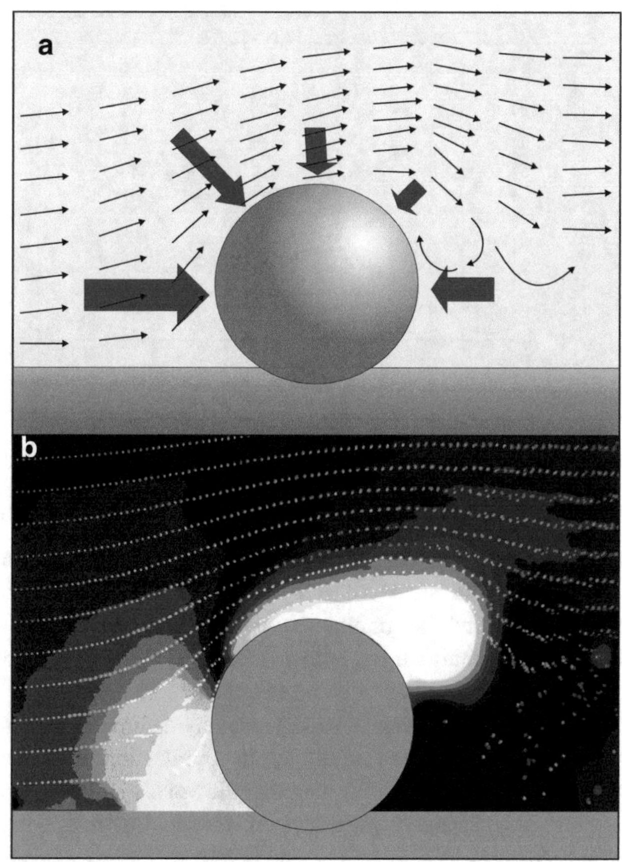

Fig. 14.2 Bernoulli effect on a spherical sediment particle **a** Flow lines (black) and pressure vectors (blue). **b** Pressure diagram in a flow model (wind tunnel). Red and yellow colors show high pressures while white and blue colors show low pressures

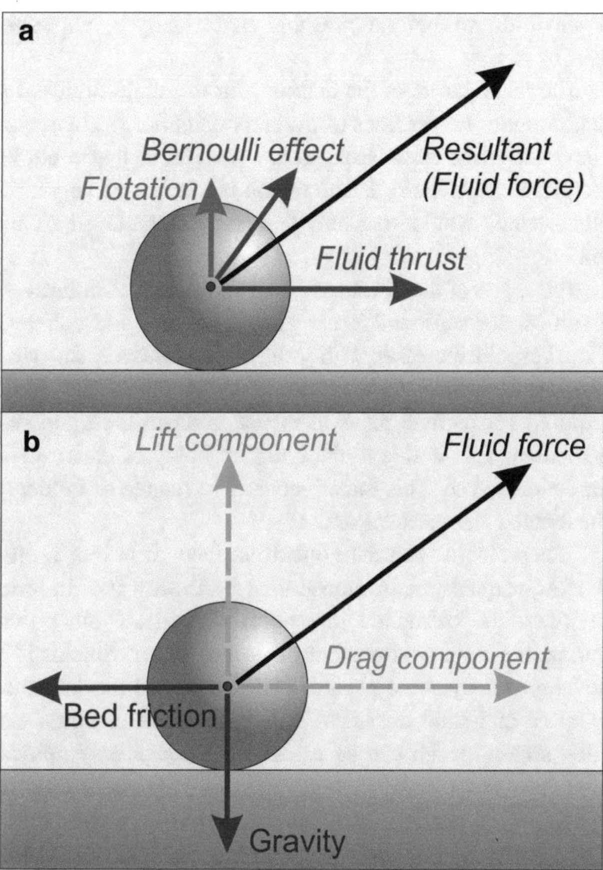

Fig. 14.3 Scheme of fluid force. **a** Combination of forces acting on a particle in the resultant fluid force. **b** Decomposition of the fluid force into its two components which oppose inertial forces

is a speed at which it is put into motion. This velocity is called the threshold velocity. The entrainment threshold is the curve joining all the points corresponding to the threshold velocity for each size of particle.

The above statement is supported by all the reasoning expressed in the theoretical approach of the previous paragraphs. These paragraphs are explained with the aim of being easy to understand, but they are enormously simplified. In fact, the movement of particles involves a large number of variables, among which are the size, shape and density of the particles, the roughness of the bed, the presence of cohesive particles, the vertical distribution of densities of the fluid, and the presence and degree of vortices in the flow. The degree of intervention of all of these variables does not make it easy to obtain a reliable equation to determine the thresholds of particle movement. Therefore, the determination of this threshold had to be done experimentally. The graphic expression of this threshold is the Hjulström diagram (1935).

The Hjulström diagram (Fig. 14.4) shows the empirical relationship between grain size and stream velocity, and defines the critical speed threshold at which particles are pulled from the bottom and start to move. The diagram was made from experiments using quartz grains in water at 25 °C and with a flow depth of 1 m. The results established a range of data that can slightly shift the threshold upwards or

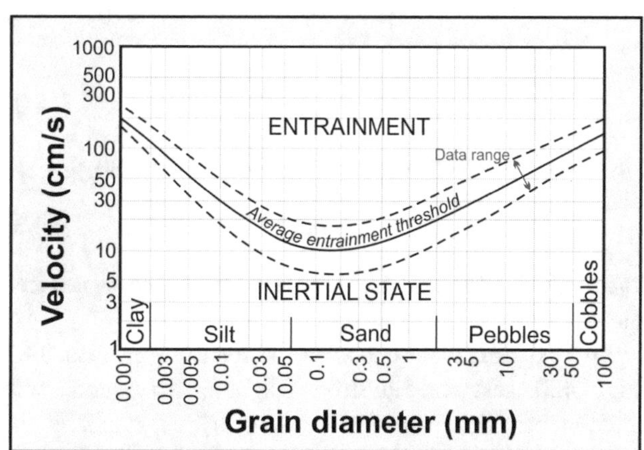

Fig. 14.4 Extraction threshold in the Hjulström diagram (adapted from Hjulström [21])

downwards, so that the threshold can actually be expressed as a band.

The mean curve of the entrainment threshold divides the diagram into two regions of different conditions. The region above the curve would express flow conditions that could lift the grains from the bed. The region below the curve, on the other hand, would maintain the inertial condition of the grains.

The shape of the threshold curve shows that, as intuitively assumed, for sand and larger grains (pebbles and cobbles), the threshold increases with grain size. However, for particles smaller than 63 μ, contrary to intuition, the speed required for particle removal increases when the grain size decreases. This is due to the entry into play of electrostatic attraction forces. This force increases its degree of influence the smaller the particles are.

The problem with the Hjulström diagram is that it only works for quartz grains transported by fresh water. In order to know the extraction threshold of particles other than quartz in a fluid environment other than water, Shields [40] developed a dynamic parameter that included the influence of grain and fluid densities. This is the dimensionless **viscous stress**, which can be calculated through an empirical equation (14.1).

$$\theta_t = \frac{\tau_t}{(\delta s - \delta f)gD} \tag{14.1}$$

where θt *is* the non-dimensional viscous stress, τ_t is the viscous stress of the fluid on the bed, δs is the density of the grains, δf is the density of the fluid, g is the acceleration of gravity and D is the average diameter of the grain. This equation also shows θ_t as a function of grain size, but now includes other variables that can influence the hydraulic equivalence. In this case, the viscous stress of the fluid on the bed (τ_t) is defined as the force exerted by the fluid per unit area of the bed [3]. As this is a tangential force per unit area, it is given in units of pressure. This variable is a function of fluid density, bottom slope, water depth (14.2) and also by flow velocity (14.3).

$$\tau_t = \delta_f ghS \tag{14.2}$$

where h is the water depth and S is the bottom slope.

$$\tau_t = \delta_f U_*^2 \tag{14.3}$$

where U_* is the velocity of the fluid at the water–bed interface.

An experimental curve is derived from Eq. 14.1 (Fig. 14.5). Like the Hjulström diagram, two regions with different behaviors can also be seen in this graph. In the same way, the increase of effort required to move the larger sand grains and the increase of effort to move the particles smaller than 63 μ due to the electrostatic forces can be observed in the shape of the curve. Later authors have slightly modified the graphic expression of Shields' threshold through empirical experimentation [31]. In this way, a band of uncertainty can also be established.

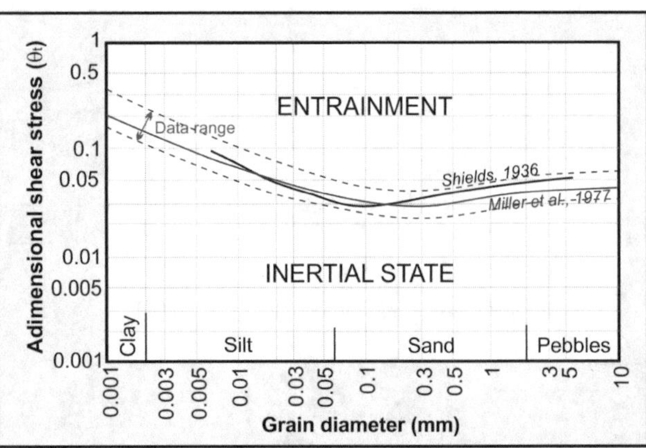

Fig. 14.5 Graph relating grain size to Shields' dimensionless viscous stress (adapted from Miller et al. [31])

The Shields diagram is widely used by sedimentologists because, although less intuitive, its application is more general than the Hjulström diagram. However, it should be used in the transport processes of coastal environments, where changes in salinity and the nature of particles introduce frequent changes in space and time.

14.2.2 Sediment Transport

Once the movement has started after an increase in speed, the grains acquire different types of transport according to the dynamic relationship between inertial forces and kinetic forces (Fig. 14.6).

For larger grains of sand and also *pebbles* and *cobbles*, if the kinetic forces very slightly exceed the inertial ones, the particles are transported without losing contact with the bed. This type of transport is called **traction**. Transport by traction implies that the horizontal kinetic forces are able to overcome the friction with the bed; however, the vertical extraction forces are not able to overcome gravity to lift the clast from the bed. Traction actually groups three ways in which the grains can move: **rolling**, **sliding** or **creeping by impacts**. In this case, the cobbles are dragged or rolled along the bed according to their morphology in such a way that only the most spherical ones can roll, while the flattened ones are preferably dragged.

If, on the other hand, the kinetic forces greatly exceed the inertial forces, **transport in suspension** occurs. In this case, the vertical extraction force is able to largely overcome gravity. In this situation, the friction forces with the bed

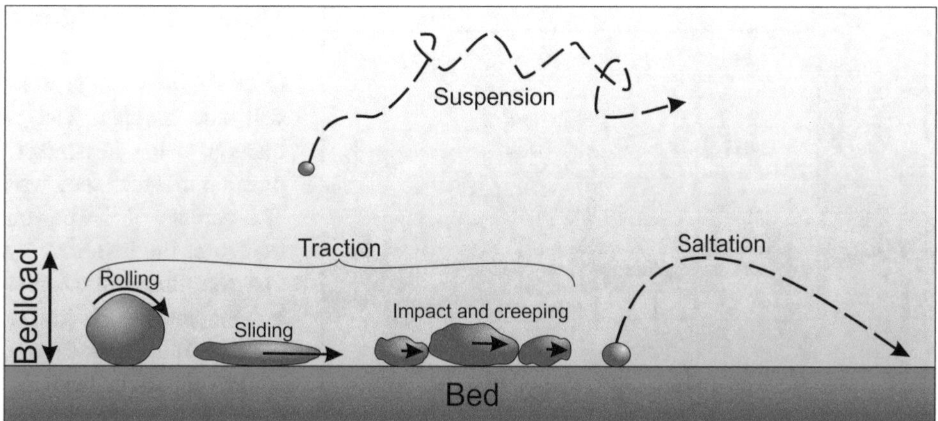

Fig. 14.6 Types of particle transport by fluids

disappear. The permanence in suspension is favored in very fine particles where the electrostatic forces of the grains help to keep them suspended in the fluid no matter how small the speed.

Under intermediate conditions, especially in turbulent flows, conditions occur in which the kinetic forces are greater than the inertial forces. The presence of vortices generates points or moments in which the inertial forces are again greater. In this case, a type of transport called **saltation** occurs. In this type of transport, the grains of sand, or even micro gravels, move by jumping and impacting on the bed. Each impact with the bottom can help other resting particles to obtain the energy needed to overcome the entrainment threshold. These impacts are also responsible for the form of transport called impact **creeping**, named above as a form of traction. In this case, the energy of the impacts is what produces the advance of some clasts without them losing contact with the bed.

The transport that takes place in relation to the bed is called **bed load transport**. This transport groups any of the traction types as well as transport by saltation.

14.2.3 Grain Settling Threshold

Once set in motion, the particles stop moving when the current slows down. It is intuitively understood that, when the speed decreases, there must come a time when the energy of the flow does not have sufficient force to keep the particles in motion. One might imagine that the same threshold that had to be crossed to set them in motion would now set the conditions for sedimentation, but this is not the case. Normally, the clasts continue to be transported when this threshold is crossed to lower speed conditions. This implies that the threshold that selects the particles to stop moving will occur at lower speeds than those that managed to set them in motion. This occurs because the forces acting on the moving particles are not the same as those acting when they are at the bottom [7]. Perhaps, in fact, it is more obvious in particles transported in saltation and suspension, since the friction with the bottom has disappeared while the transport is taking place. On the other hand, the electrostatic forces of the finest particles are not acting to keep them attached to the bed, but to keep them in equilibrium with the fluid. The last factor implies that the angle and surface of application of the viscous forces of the fluid are full when the particle is away from the bed.

Due to all these factors, it is obvious that there is a different threshold for the deposition of particles. This threshold is called the **settling threshold**. The Hjulström diagram also represents this threshold, in this case by dividing into two fields (Fig. 14.6). Above the threshold, the particles are kept in an inertial state, although in this case the inertia is that the particles are still transported. Below the threshold, the particles will be deposited. This threshold does not have an inflection, as was the case with the extraction threshold, but is an ascending line indicating that larger sizes will always be deposited at higher speeds than finer grains.

One aspect of this curve is that it cuts the horizontal axis at the size of 0.016 mm. This means that particles smaller than 16 μ could continue to be transported even at speeds of less than 1 mm/s—i.e., almost zero speed. In this case, the presence of electrostatic forces acts to keep the particles suspended in the fluid. However, it is precisely the presence of these electrostatic forces that can cause fine particles to be attracted and form agglomerates that will behave like larger particles and eventually decant.

Actually, the Hjulström diagram does not show both thresholds separately, as we have seen in Figs. 14.4 and 14.7, but groups both thresholds in the same graph (Fig. 14.8).

This diagram is widely known and reproduced. It clearly shows that the settling threshold is below the entrainment

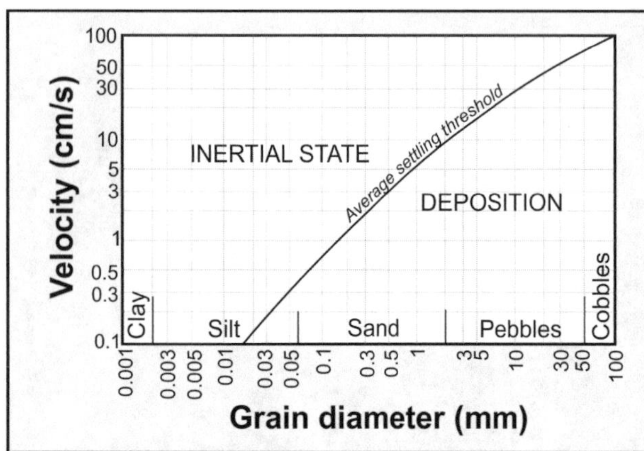

Fig. 14.7 Settling threshold in the Hjulström diagram

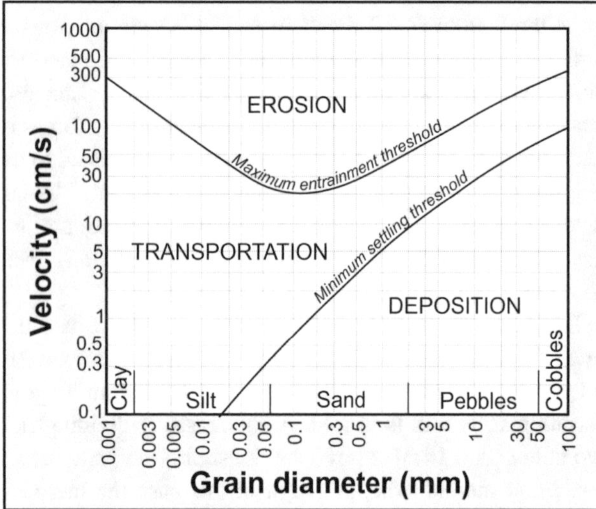

Fig. 14.8 Hjulström diagram showing both entrainment and settling thresholds

threshold. The area above the extraction threshold is commonly indicated as erosion conditions, the area below the selection threshold is indicated as deposition conditions and the field between the two thresholds is indicated as transport conditions. This label in the diagram could confuse anyone who does not know how thresholds really work, as it could be thought that particles will be transported in this field under any conditions, when in fact they would only be transported under these conditions if the particles have been put into motion under more energetic conditions beforehand. The field located between both thresholds is more correctly a zone of inertial conditions, since the particles will behave in this zone as they did in the previous energetic state. Those that were at rest will continue to be at rest when the speed is rising, while, when the speed is falling, the particles that were being transported will continue to be transported.

14.2.4 Sedimentation Velocity

Once the threshold is crossed, the particles that were moving will stop moving. The particles that were transported by traction, being in contact with the bed, will remain in the position in which they were when the threshold was crossed. The particles that were transported by saltation will simply remain at the bed after a last jump. However, the particles that were transported in suspension were very far from the bottom and will begin to settle. The speed at which the particles fall vertically to the bed is known as the sedimentation velocity, as well as the settling or decanting velocity (Fig. 14.9).

The sedimentation velocity is a function of the grain size in relation to the viscosity of the fluid. The temperature changes the settling velocity curves, as it influences the viscosity of the fluid. In general terms, at higher temperatures the particles will settle more quickly as the viscosity decreases. Conversely, lower temperatures will decrease the settling speed. The graph in Fig. 14.9 shows the speed curves in fresh water at 10 and 30 °C. In parallel, salinity has an influence, since higher salinities increase viscosity and vice versa.

This curve shows how the pebbles decant almost immediately, at speeds above 30 cm/s. The sand grains decant at speeds ranging from 2 mm for very fine sand to 20 cm for

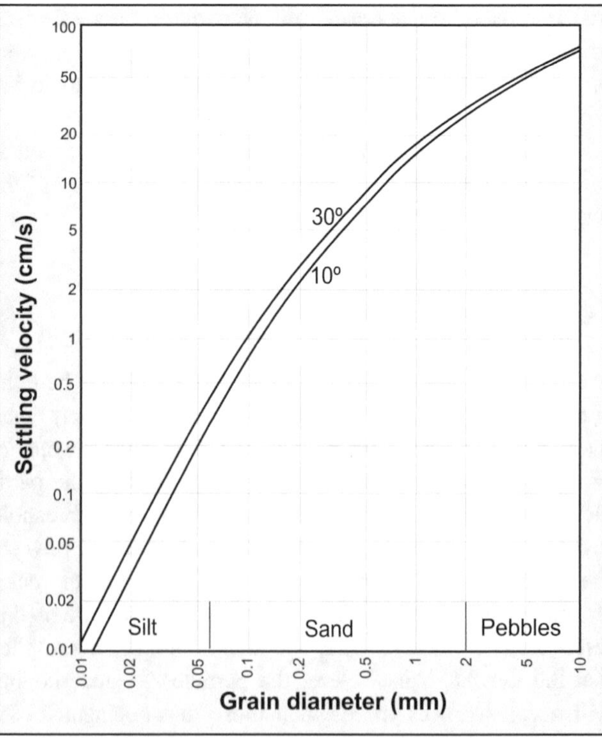

Fig. 14.9 Diagram showing the curves of sedimentation velocity as a function of grain size. Curves are for fresh water at 10 and 30 °C

very coarse sand. Again, the problem with the silt particles lies in their low sedimentation rate, which will be less than 1 mm/s. This means that it takes more than a quarter of an hour for silt particles to pass through 1 m of water. If we add this information to what was observed in the Hjulström diagram—i.e., that these particles require almost zero velocity to decant—we now add the fact that these still water conditions must be maintained for long periods of time for the suspended particles to reach the bottom. Actually, it may be that those grains located in the lower layers of the flow, near the bottom, manage to cross that distance, but in channels several meters deep, it is difficult for the conditions to occur for the particles that travel in the fluid to reach the bed.

Advanced Box 14.1: Calculating the Transport Rate By Currents

Among the most frequently addressed problems in the studies of coastal system dynamics are those related to sedimentary balance. This type of problem always requires a volumetric quantification of the sediment. Thus, it is interesting to be able to calculate the amounts of sediment (in weight or volume) transported by any of the currents acting on the coast.

Many authors have proposed equations to calculate the amounts of sediment transported. All these equations are proposed for different modes of transport. Thus, there are separate equations for the calculation of bed load and suspended load.

AB14.1.1 Potential Bed Load Transport Rate

Bed load means the material transported in contact with the bottom. Normally, the concept of bed load includes material transported by traction (in any form) and saltation. What the equations actually calculate is the capacity of the fluid to transport material in bed load, so the parameter used is called the **potential bed load rate** (Qb). Most formulas express the rate of material transported in units of weight per unit length of cross-section to flow and per unit time (e.g., in grams per centimeter per second).

The most frequently cited formulas in the calculation of potential bed load are those of Einstein [13], Meyer-Peter and Müller [27], Bagnold [6], Yalin [46] and Van Rijn [45]. These formulas are represented by Eqs. 14.4–14.8.

- Einstein [13]

$$Qb = \frac{P}{1-P}\frac{1}{43.5}\left\{\frac{[(\delta_s-\delta_f)gD]^3}{\delta_f}\right\}^{1/2} \tag{14.4}$$

- Meyer-Peter and Müller [27]

$$Qb = \delta_f U_*^3 8\left[1-\frac{0.047gD(\delta_s-\delta_f)}{\delta_f U_*^2}\right]^{3/2} \tag{14.5}$$

- Bagnold [6]

$$Qb = \frac{K1\delta_s U_*^3}{\frac{g(\delta_s-\delta_f)}{\delta_f}} \tag{14.6}$$

- Yalin [46]

$$Qb = 0.635\delta_s DU_* S\left(1-\frac{1}{aS}\right)\ln(1+aS) \tag{14.7}$$

- Van Rijn [45]

$$Qb = \frac{0.053\delta_s(dg)^{1/2}D^{1.5}-T^{2.1}}{\left[D\left(*\frac{g\delta_f}{\mu}\right)^{1/3}\right]^{0.3}} \tag{14.8}$$

In these equations Qb is the potential background charge, g is the acceleration of gravity, D is the mean grain diameter, δs is the density of the solid and δf is the density of the fluid, and μ is the viscosity of the fluid. U_* is the shear rate, or velocity at the water–bed interface. When data for this value are not available, it can be calculated from velocity values (Uz) taken at a depth at a distance from the bottom (z), as long as this distance is less than 1 m, according to Eq. 14.9; [9].

$$U_* = \frac{0.4\ Uz}{\ln\ln\left(\frac{30Z}{ks}\right)} \tag{14.9}$$

In this equation Ks corresponds to the bed roughness (Nikuradse roughness) and is estimated at three times the size representing 10% of the largest sample sizes (Ks = 3D10).

In Einstein's equation (14.3) P is the probability of transport and is calculated through Eq. 14.10.

$$P = 1-\left[\frac{\frac{0.143gD\left(\frac{\delta_s-\delta_f}{\delta_f}\right)}{U_*^2}-2}{\sqrt{\pi}}\right] \tag{14.10}$$

In Bagnold's equation (14.6) k1 is a coefficient that is calculated through Eq. 14.11.

$$k1 = 0.10 \exp(0.17/D) \tag{14.11}$$

In the Yalin equation (14.7) a is a complex relationship between grain and fluid densities, grain size and critical shear rate (14.12) and S is a relationship between bottom shear rate and critical shear rate (14.12).

$$a = 2.45\left(\frac{\delta_f}{\delta_s}\right)^{0.4}\left[\frac{\delta_f U_{*cr}^2}{gD\,(\delta_s - \delta_f)}\right]^{1/2} \tag{14.12}$$

$$S = \frac{U_*^2}{U_{*cr}^2} - 1 \tag{14.13}$$

In these equations U_{*cr} is the critical threshold speed. It is understood that this is the speed at the water–bed interface required to set a particle of a given size in motion (D).

Most of these formulas have in common the variables used in the flow calculation. All of them consider the movement of the grains as the result of the viscous effort of the fluid on the bed and, consequently, all of them use as criteria the thresholds of movement of the particles. Therefore, the contrast of densities between the grains and the fluid, as well as the size of the particles, the bed roughness and the near-bottom flow velocities, are always considered as input parameters in the equations.

The formula proposed by Hans A. Einstein was designed for beds formed by unimodal sands, where all particles have the same diameter, geometry and density and are transported in a one-dimensional flow. In contrast, the equation proposed by Meyer-Peter and Müller, although it also assumes a one-dimensional flow, works for beds with slopes greater than 20% and well-sorted fine gravel grain sizes of up to 3 cm. The Yalin equation, on the other hand, assumes a constant flow of particles transported mostly in saltation. The Bagnold equation is based on criteria of mechanical equilibrium between water flow and bottom clasts, taking into account their size and density, in relation to the roughness of the bed. This equation is considered to be adapted to non-cohesive beds consisting of multimodal sediments of different diameters between sand and fine gravel. All of these equations were designed to be applied to natural flows in river systems; however, the equation suggested by Van Rijn is designed for use in experimental pools and introduces some parameters that are difficult to measure in experiments in natural environments, such as the height of the jumping particles transported. There are more recent equations that are valid for sediments with different grain sizes between sand and gravel, although they present the problem of the great complexity of dealing with the different thresholds of movement that exist for each of the grain sizes that make up the sediment.

Although most of these formulas have been designed for calculations based on river systems, their use may also be appropriate in channels subject to tidal currents, since both types of current work on the same principles. It is sufficient to recalculate by introducing the densities corresponding to coastal waters. As can be seen, each of these formulas has been designed for limited conditions and, therefore, when making calculations in natural coastal systems it is necessary to choose the equation that should be applied according to the characteristics of the bed and the usual flows of the environment to be studied.

The use of any of these formulas in coastal systems requires action on the bed. In any case, to calculate the potential transport through a channel, the following should be measured: (1) the flow section; (2) the flow velocities at less than 1 m from the bed; and (3) water and sediment samples. Different requirements are demanded for samples taken in the field. For water samples, the density must be determined. In the case of sediment samples, the density of the grains must be measured and a particle size analysis must be carried out to determine the mean grain size and 10th decile (in order to determine the bed roughness).

AB14.1.2 Suspended Load Transport Rate

The analysis of suspended transport rates requires three types of baseline data: (1) the flow section; (2) the distribution of flow rates in that section; and (3) the distribution of suspended matter concentrations in the flow section. In natural systems, the flow section and velocity profiles can be determined through field measurements. Suspended sediment or suspended transport concentrations can be estimated analytically or numerically.

The analytical calculation is complex because there are too many variables that determine lateral, vertical and temporal changes of the concentrations of suspended matter at a given moment. The use of equations to calculate the transport capacity of the flow (potential transport rate) always assumes stable and uniform conditions. The equations are always based on different assumptions that facilitate the calculation [26]. These are:

- That the decanting rate does not vary in time and space,
- That the concentrations of suspended matter are so low that interactions between grains can be ignored,
- That the viscosity of the vortices depends on the shear rate and a length scale,
- Other possible changes to the turbulent mixture due to effects such as saline or thermal stratification by temperature, salinity or suspended sediments to buffer or enhance the turbulent mixture are not considered.

One of the most common characterizations of the concentration of suspended sediments in a flow is given by the Rouse profile [20]. Rouse's profile characterizes any differences in suspended matter concentrations in the water column, including turbulent mixing of areas with different suspended particle concentrations and vertical gradients due to particle weight.

The turbulent mixing starts from the idea of the presence of vortices that produce a net displacement of grains from areas with higher concentrations to areas with lower concentrations. On the other hand, it assumes that all particles will tend to move downwards due to their weight, and thus there will be a gradient of increasing particle concentration downwards in the water column. The Rouse profile assumes an equilibrium between the vertical gradient due to decantation and the turbulent mixture that will tend to move the particles upwards.

Another problem that arises in the use of empirical equations is the treatment of suspended sediments containing a variety of grain sizes. The reference concentration must take into account the concentration of each grain size and the shear stress required to keep that grain class moving.

The calculations required for the use of empirical equations in the potential transport of suspended sediment are very inaccurate if they include all those assumptions so far from reality. On the other hand, the inclusion of all these variables in one equation makes their use too complex. The end result is that the easiest and most widely used method for calculating the rate of suspended material transported by a flow is the direct measurement of concentrations by taking water samples at the same time as the direct measurement of velocities.

Equation 14.14 proposed by Bagnold [7] can be applied directly in this case.

$$\mathrm{Qs} = \int_{a}^{h} c(z)U(z)dz \tag{14.14}$$

In this equation the suspended load (Qs) is calculated by integrating the load values at different depths (z) between the fluid–bottom interface (a) and the surface (h). In this case, c (z) marks the concentration at a given depth (determined by the value of z), while U(z) is the flow velocity at that depth, with dz being the depth differential.

14.3 Transport of Particles by Tides

The action of tidal currents in the channels of coastal systems represents a clear example of a continuous increase in speed from zero to reach a maximum speed that is maintained and then decreases again to zero. Throughout this process, the entrainment threshold of each grain size present on the bed will be reached to keep it moving as long as the transport speed is above that threshold. Finally, the speed will decrease until it progressively reaches the settling thresholds, and the particles will be deposited. Then, the same process occurs in the opposite direction.

In accordance with what has been explained in previous paragraphs, there is different behavior between the grains of non-cohesive sediments and those of cohesive sediments, due to the differences between their entrainment and settling thresholds.

In the case of non-cohesive grains, the difference between these thresholds is not very high and marks the delay in depositing the particles once the tide reaches the speed required to set them in motion. In this way, the grains will remain in motion almost until the moment of the tidal inversion. This case can be well-illustrated by the behavior of the average sand grains (Fig. 14.10).

In contrast, for cohesive particles, the difference between these thresholds is very marked. To understand the behavior of cohesive grains we will use the graph of 20-μ particles at the same tide (Fig. 14.11). In this case, the tide must advance for almost an hour until it reaches the necessary speed to set these coarse silt grains in motion. Once these grains are pulled from the bottom, they will remain in motion until the tide rises.

With regard to these fine particles, it must be taken into account that the time the current is stopped during the upturn is very short. Due to their slow settling speed, only those that travel closest to the bed will be deposited there. This means that most of the suspended charge would continue to be transported almost perpetually if the processes of electrostatic agglomeration of the particles did not exist.

The amount of sedimentary material transported in each of the directions depends on the asymmetry of the tidal current curve as a function of the available grain sizes. This asymmetry can be manifested in the velocity reached by the ebb and flow currents, as well as in the action time of both

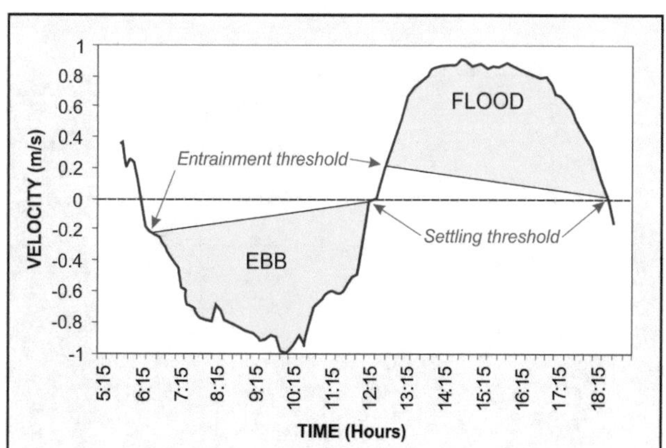

Fig. 14.10 Curve of velocity over time in a tidal cycle, showing the entrainment and settling thresholds. The fields highlighted in yellow correspond to the conditions of transport of medium sand grains

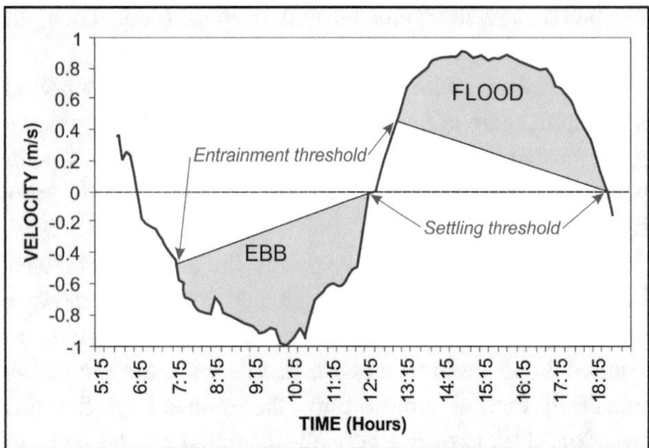

Fig. 14.11 Curve of velocity over time in a tidal cycle, showing when the entrainment and settling thresholds of coarse silts are surpassed (shown in blue)

currents. The result is always a residual balance acting in one of the two senses.

Advanced Box 14.2: Calculating the Transport Rate By Tidal Currents

The potential sediment load (Qb) carried by tidal currents during a tidal cycle can be calculated using the generally proposed equations for currents (Advanced Box 14.1). The particularity of the application of these equations to tides is that the measurements of Qb are usually given in weight per length of flow section and unit of time, and when tidal currents are analyzed these measurements are integrated to cover the whole flow section and all the time of each of the tidal cycles.

In other words, in sedimentary balance studies, it is not enough to know how many grams pass through a centimeter in a second, but rather to seek to know how much sediment transits through a channel during complete ebb and flow cycles in order to calculate the net balance of sediment that is transported in a residual way to one of the two directions.

Firstly, the integration in time is considered. Starting with the data of a time curve of tidal currents taken at a distance of less than 1 m from the bed, calculations are made by dividing the curve into set time intervals, calculating the average speed for each of these intervals. These values are then entered into the equations along with the rest of the measurements (grain density, fluid density and grain size). The final result is a potential load curve with respect to time (Fig. 14.12).

To obtain the total values of sediment transported by ebb and flow, it is sufficient to sum the values of Qb for all of the time intervals in which there was transport, taking into account the direction of the current. Finally, the total values of the flow cycle will be compared with those of the ebb to obtain the value of the net transport.

Once these values are obtained, the total quantities through the flow section are calculated by multiplying them by the length of the bed measured across the channel profile. It should be noted that the values measured at a single point may not be extrapolated to the entire channel because there may be lateral variations in the flow. In this case, the channel should be divided into different sections and measurements of the flow in each of these sections should be made. The calculations obtained from these curves may then be applied to the length of the section for which these measurements are considered valid. A clear example could be the presence of a central bar that divides the section of the channel into two well-differentiated sections. In this case, three different speed curves should be obtained, one in each deep zone and one on the bar. Then, the Qb data of each tidal cycle would have to be multiplied by the length of the bed of each of these sections to finally add up the three sections and thus obtain the total for the whole channel.

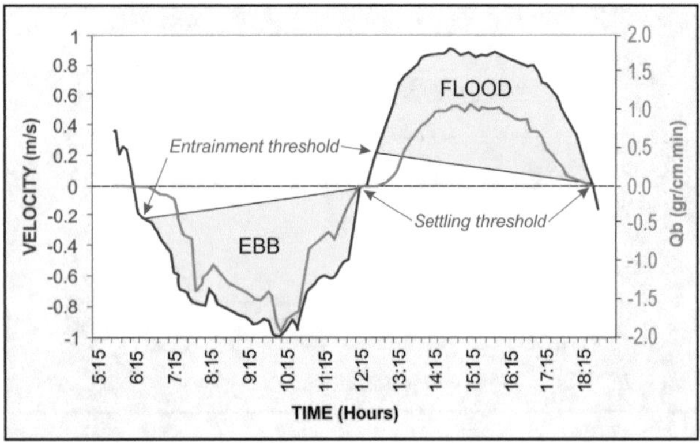

Fig. 14.12 Curve of velocity over time in a tidal cycle (blue) and curve of potential bed load (Qb) at intervals of 1 min. The entrainment and settling thresholds are indicated. Observe that between settling and entrainment thresholds of successive cycles the bed load is 0

14.4 Transport of Particles by Waves

A wave, like a tide, also represents in its movement an increase in the speed of the current followed by a decrease in the same, an increase in the opposite direction and a further decrease. The difference is that the time at which these speed changes occur is measured in seconds and not in hours, as in the case of tides. As a result, greater accelerations occur.

The flow behavior of a wave applied to the bed material differs depending on where the wave acts (Fig. 14.13). In areas of the *shoreface* above the base level of the wave, the water flow at the bed responds to the smaller orbits away from the surface. In this case, the velocity curve is quite symmetrical, generating a sway where only the movement thresholds of very fine and thin sands can be reached (Fig. 14.13a).

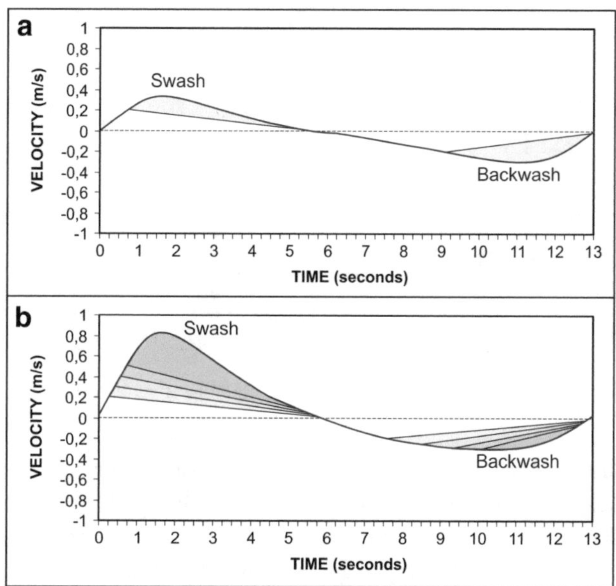

Fig. 14.13 Curves of velocity over time during the passage of a wave. **a** In a shoreface bed. **b** In a surf zone. The thresholds for very fine, fine, medium, coarse and very coarse sands are shown in orange tones

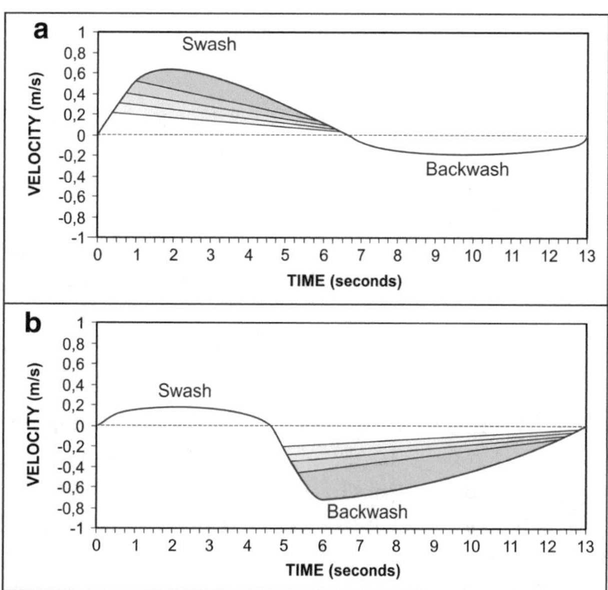

Fig. 14.14 Curves of velocity over time for breaking waves. **a** Spilling type. **b** Plunging type. The thresholds for very fine, fine, medium, coarse and very coarse sands are shown in orange tones

In contrast, in the surfing areas the flow depth is much lower and it is the larger orbits that act directly on the bed. In this case (Fig. 14.13b), there is no longer so much symmetry. The example shows the passage of a wave where the surge develops faster than the undertow, although in both cases the movement thresholds of all classes of sand (in progressively darker shades of orange for larger sizes) are progressively exceeded.

In the swash areas, the type of breaker has a great influence on the development of sediment transport (Fig. 14.14). The most dissipative breakers, such as spilling types (Fig. 14.14a), show a significant increase in velocity during the swash, where all the thresholds of movement of the sand grains are crossed, while the backwash develops lower velocities that do not manage to cross the thresholds of movement even for very fine sands. This implies that the spilling breakers only move the sediment towards the coast, assuming a sedimentary accumulation. Conversely, more energetic breakers such as the plunging type develop higher velocities during the backwash (Fig. 14.14b). In this case, it is the speed of the swash that fails to reach the movement thresholds, while the backwash is able to move sand of all sizes, and even fine gravel.

This type of calculation of transport transversal to the coast is useful to know if the sediment has a displacement that goes mostly landwards or seawards. It can be complemented with calculations of longitudinal transport if we take into account the angles of wave action with respect to the coastline and decompose the velocity vector by extracting the longshore vector and its relationship with the entrainment and settling thresholds for the beach grains.

Advanced Box 14.3: Calculating the Longshore Transport Rate

The calculation of the amount of sediment potentially transported along the coast can be done through empirical equations similar to those described for currents. In this field, the work carried out by different authors in the second half of the twentieth century is essential (e.g., Inman and Bagnold [22]; Komar [25]; Galvin [14]). These works were excellently summarized by Carter [10], who also makes some personal contributions.

In the case of grain transport by waves, it is necessary to introduce the concept of longitudinal flow of wave energy by length of coast (P_L). This parameter relates the wave energy with its propagation speed in relation to the angle of arrival at the coast line (Eq. 14.15) and is given in joules per meter of beach length if we work in SI units of measurement.

$$P_L = 0.0625 \delta f g Hs^2 C \sin 2\alpha \qquad (14.15)$$

where δf is the fluid density, g is the gravity acceleration, Hs is the significant wave height, C is the propagation velocity and α is the angle between the wave crest and the shoreline.

Using this parameter, Van der Graaf and Van Overeem [15] proposed the use of a formula for the calculation of the longitudinal background load (Qb) carried by waves (Eq. 14.16).

$$Qb = \frac{\text{K } P_L}{ga(\delta s - \delta f)} \quad (14.16)$$

where K is a constant that varies depending on the type of surf, g is the acceleration of gravity and a is the degree of packing of the sediment, with δs and δf being the densities of the grains and water, respectively.

In more practical terms, Kamphuis [23] proposed another equation valid for quartz-grain sand beaches (Eq. 14.17).

$$Qb = 1.28 \frac{\tan \beta Hs^2}{D} \sin 2\alpha \quad (14.17)$$

where β is the slope of the beach, D is the average grain size, Hs is the significant height in the breaker zone and α is the angle between the crest of the wave and the coastline.

Subsequent work has proposed certain adjustments to this equation and optimized it for use in mathematical models (e.g., Schoonees and Theron [39]; Damgaard and Soulsby [11]).

14.5 Transport of Particles by Wind

Air is a low-density fluid and therefore wind can be treated like any other fluid flow. Under the action of wind, the grains also present thresholds of movement that can be determined using Shields' dimensionless viscous stress. In this case, the thresholds of grain entrainment by air are two orders of magnitude higher than those needed to put the same particles in motion by water.

Not many authors have devoted their efforts to experimentally determining the thresholds of grain movement by wind. One of these authors [5] has determined the entrainment threshold of grains with a density of 2.6 g/cm^3 as a function of the shear velocity (Fig. 14.15). This density corresponds to the most abundant minerals in sediments such as quartz and feldspars. The threshold was determined experimentally, similar to the Hjulström diagram. This curve also shows the increase in velocity that is necessary to suspend particles smaller than 100 μ, although in this case this difficulty is not attributed to electrostatic forces but to the low roughness of the bed which reduces the air friction on the particles.

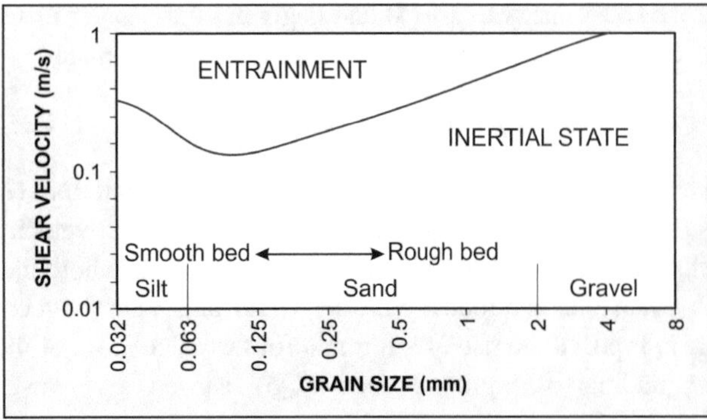

Fig. 14.15 Entrainment threshold of grains under wind action in Bagnold's diagram (1941)

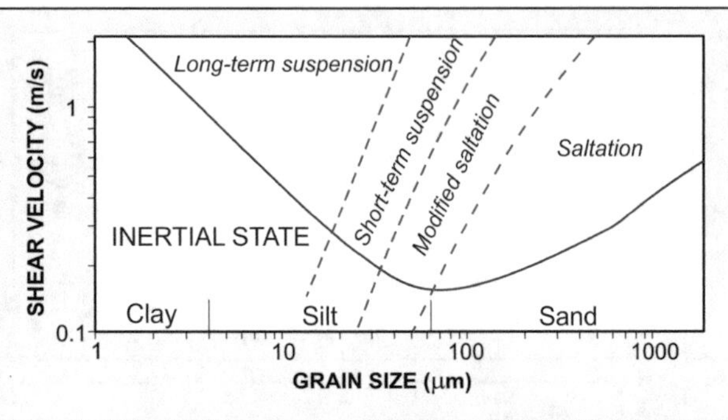

Fig. 14.16 Wind transport modes as a function of grain size [16]

Once in motion, the grains will be transported via saltation or suspension according to their density and size. In this regard, Greeley and Iversen [16] suggested two intermediate forms between continuous saltation and long-term suspension. These are the modes of transport of modified saltation and short-term suspension (Fig. 14.16). In both intermediate cases the particles are partly jumped and partly suspended.

Finally, when the air speed drops, the grains can no longer be transported and fall when they cross the settling threshold. In this case, the grains fall with less resistance from the fluid because air has a lower density than water. This falling speed has also been determined for grains with a density of 2.6 g/cm^3 [5] and is shown in Fig. 14.17.

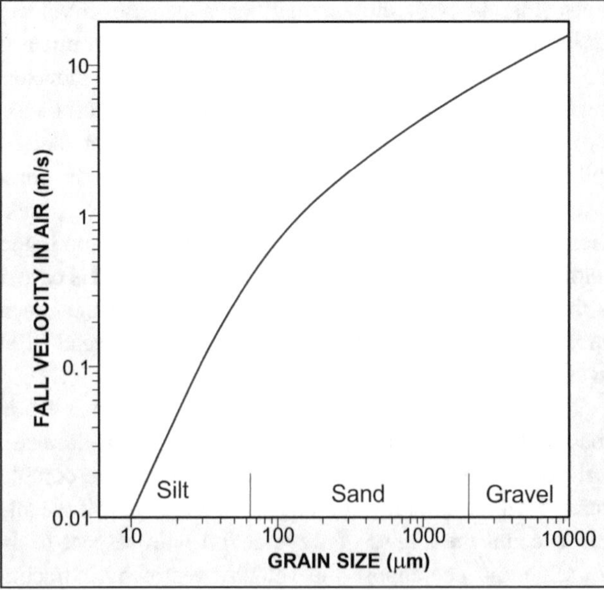

Fig. 14.17 Fall velocity as a function of grain size [5]

14.6 Development and Dynamics of Bedforms

A certain flow velocity higher than the entrainment threshold produces a particular morphological configuration of the bed. These configurations are called **bedforms**. The migration of bedforms over a sedimentary bed also generates a particular internal order associated with the surficial shape. Both surface shape and internal structure have been studied in stratigraphy as sedimentary structures of the top and the interior of the strata, respectively. In this chapter, when referring to bedforms, we do so with respect to their dynamics and will associate them with both their external geometry and their internal structure.

There are different types of bedforms generated by unidirectional flows, bidirectional flows and combined flows.

14.6.1 Unidirectional Bedforms (Current Bedforms)

The most common classification of current-generated bedforms is according to the **flow regime** under which they develop. The flow regime is a state of flow related to the amount of friction of the fluid with the bottom and which controls the amount of energy transferred to the movement of the sediment [41]. It is clear that this regime is directly dependent on the same conditions that have been established to determine the thresholds of particle movement, including flow velocity and grain size. In general terms, for the development of bedforms there are two flow regimes determined by the turbulence of the fluid, the rate of sediment transported and the possible relationships between the shape of the bottom and the shape of the fluid surface. These regimes are low flow regime and high flow regime.

In the **low flow regime**, there is a laminar flow with a low transport rate dominated by the bed load and there is no relationship between the shape of the bottom and the shape of the fluid surface. On the contrary, in the **high flow regime**, there is a turbulent flow that generates a high transport rate, with the undulations of the bed and the surface being in phase.

Thus, the bedforms are developed according to the regime [18]. Those associated with the low flow regime are the forms of **lower plane bed**, **current ripples** and **dunes** (Fig. 14.18).

- *Lower plane bed*: This is a bed configuration that describes a completely flat surface that responds to a very low transport rate where fine particle traction normally dominates. The internal structure that corresponds to this surface configuration is the low regime parallel lamination.
- *Current ripples*: These are asymmetrical centimeter-scale bottom undulations, which are due to a low flow regime with saltation transport dominance. The asymmetry is manifested in the angle and length of the two faces of the ripple. The flow-oriented face (*stoss side*) has a smaller angle, while the opposite face (*lee side*) has a larger angle that coincides with the sediment rest angle according to its grain size (Fig. 14.19a). According to the crest geometry, the ripples can be differentiated into: straight-crested ripples, sinuous ripples and linguoid ripples. Ripples migrate along the bed in the sense of flow, generating an internal sedimentary structure called **cross-lamination**. Each of these ripple geometries corresponds to a type of internal lamination (Fig. 14.19b). Straight-crested ripple migration generates planar cross-lamination (2D). Sinuous-crest ripples generate curved-base cross-lamination, while linguoid ripples generate trough cross-lamination. The latter two have a 3D geometry.
- *Dunes*: These are also asymmetrical bottom undulations, but on a metric scale. Their size is greater than that of the ripples, which is why many authors call them mega ripples or *large ripples*. They also originate during a low flow regime with a dominance of saltation transport, but they are normally made up of coarser grains than ripples. They can also be classified according to ridge geometry, distinguishing between straight ridges, sinuous ridges and linguoid or crescent ridges. Like the ripples, the dunes migrate along the bed in the sense of the flow, generating an internal sedimentary structure which is called **cross-bedding,** of a larger scale but identical geometry to the cross-lamination. Each dune geometry corresponds to a type of cross-bedding (Fig. 14.20). Straight-crested dunes (also called sand waves and 2D dunes) have associated planar cross-bedding. Sinuous-crested dunes are associated with curved-base cross-bedding, while the migration of linguoid and semi-lunar dunes generates trough cross-bedding. The latter two are also called 3D dunes.

The nomenclature of ripples and dunes has changed over time. A magnificent synthesis was made by Ashley et al. [4]. Similarly, the dimensions used to differentiate the bedforms vary from one author to another, although almost all agree that the transition between ripples and dunes occurs at wavelengths of 60 cm and heights of 6 cm.

The bedforms associated with the high flow regime are **upper plane bed** and **antidunes**.

Fig. 14.18 Photographs showing some low regime bedforms. **a** Straight-crested ripples. **b** Sinuous ripples. **c** Linguoid ripples. **d** Sinuous small dunes. **e** Sinuous dunes with superimposed ripples. **f** Dunes totally reworked by linguoid ripples

- *Upper plane bed*: This is a configuration of the bottom that describes a flat surface without relief, but in this case it responds to a very high transport rate. Such a high flow velocity generates a bottom leveling by razing, in which the existence of any relief would be unfeasible. The internal structure that corresponds to this type of bottom is the parallel lamination of the high flow regime (Fig. 14.21).
- *Antidunes*: These are low-relief undulations with a very low ratio between height and wavelength, which differentiates them geometrically from ripples. As high regime bedforms, the antidunes are in phase with the undulations on the water surface. Depending on the dynamics of these forms there are two possibilities: (1) they are static (*standing waves*) or (2) they move against the current. The internal structure of both types of antidunes is different. The standing antidunes develop wavy parallel lamination, while the antidunes develop upstream cross-lamination.

As with the movement thresholds, the bed configuration and the stable shape type is a function of the flow velocity. Harms and Fahnestock [18] showed that, above the movement threshold, there is a transition of bedforms from low to high regime forms with increasing flow velocity (Fig. 14.22).

Through experiments in laboratory flumes, Harms et al. [19] proposed the use of a diagram where fields of stability of the bedforms are established in relation to the grain size and the flow velocity (Fig. 14.23a). This diagram can be interpreted in the same terms as the Hjulström diagram, so that, if we know the grain size of the sediment and its structure, we can determine the range of flow velocities that were able to generate its deposit. Inversely, we can predict the type of morphological configuration that a bed of a given grain size will acquire under a known flow velocity. Years later, new experiments led Southard and Boguchwal [43] to propose a modification of the diagram, establishing new limits for the fields of stability (Fig. 14.23b).

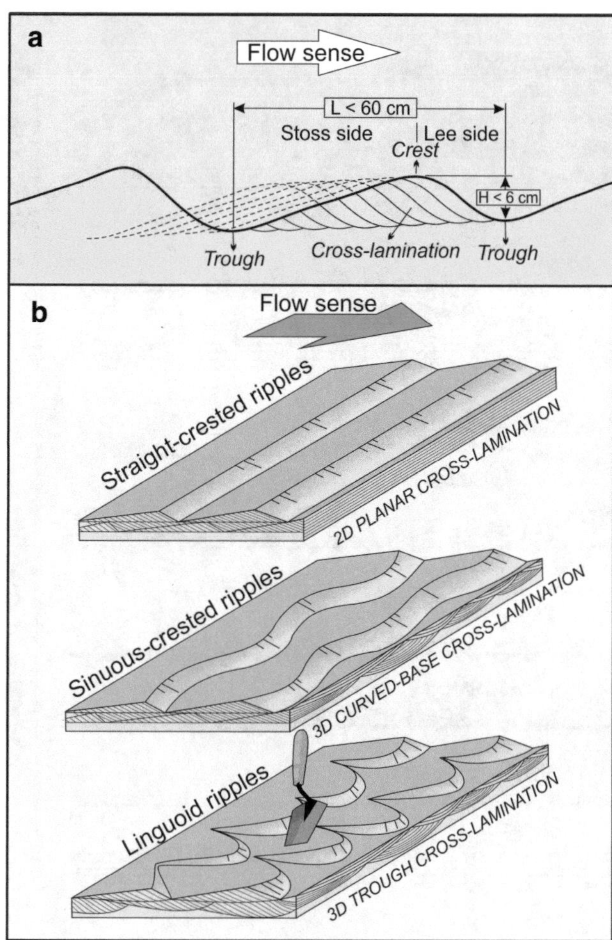

Fig. 14.19 Current ripples. **a** Genesis and dimensions. **b** Different crest morphologies in relation to their internal structure

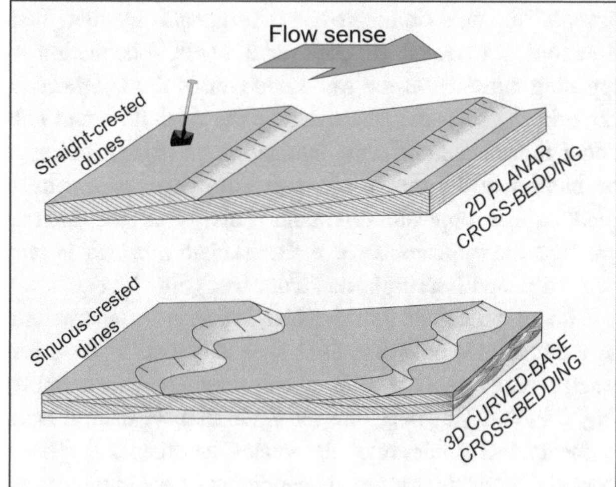

Fig. 14.20 Genesis of dunes and different morphologies in relation to their internal structure

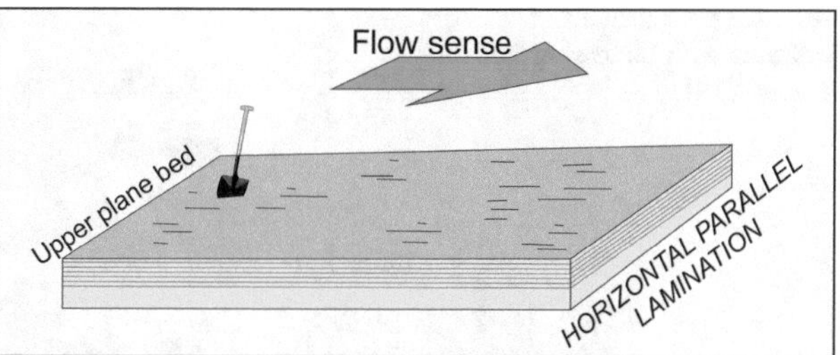

Fig. 14.21 Geometry and internal structure of the upper plane bed

It follows from these diagrams that several series of bedforms can be produced by increasing the speed depending on the size of the available sand. For example, in fine sand sizes, an increase in speed to high regime would lead from the development of ripples to the upper plane bed and antidunes, without lower plane bed and dunes developing. On the contrary, very coarse sand sizes would progressively lead to the development of a lower plane bed, 2D and 3D dunes and antidunes, without the development of ripples or upper plane bed. Only the sizes of medium sands or a progressive increase in grain size parallel to the increase in flow speed would be capable of developing the whole series proposed by Harms and Fahnestock [18], from the lower plane bed to the antidunes, passing through ripples, dunes and the upper plane bed.

It is not only flow speed and grain size that are the control factors of bedform stability. The flow depth also has a great influence on bedform development. Flow depths of less than 1 m narrow the space for a good shear band development, increasing the viscous stress of the fluid on the bed sediment grains. Rubin and McCulloch [38] established diagrams showing the relationship between parameters such as grain size and depth based on experimental data obtained in flumes by previous authors (e.g., Guy et al. [17]). In these diagrams (Fig. 14.24), it can be seen that a decrease in depth lowers the velocity level required to increase the flow regime, thus reducing the boundaries between shape stability fields. These same authors illustrated the relationships between flow velocity, depth and grain size in a three-dimensional diagram (Fig. 14.25).

Changes of bed configurations when the flow velocity increases do not occur immediately. Many of them occur progressively, as it takes time for the bed to transform and reach a form stable with the flow conditions. This time is known as the **equilibrium time**. In the case of ripples, changes between different crest morphologies have always been associated with changes in the flow rate. Thus, many authors identified the presence of straight-crested ripples with very weak currents and, conversely, linguoid ripples with high flow velocities. However, authors such as Oost

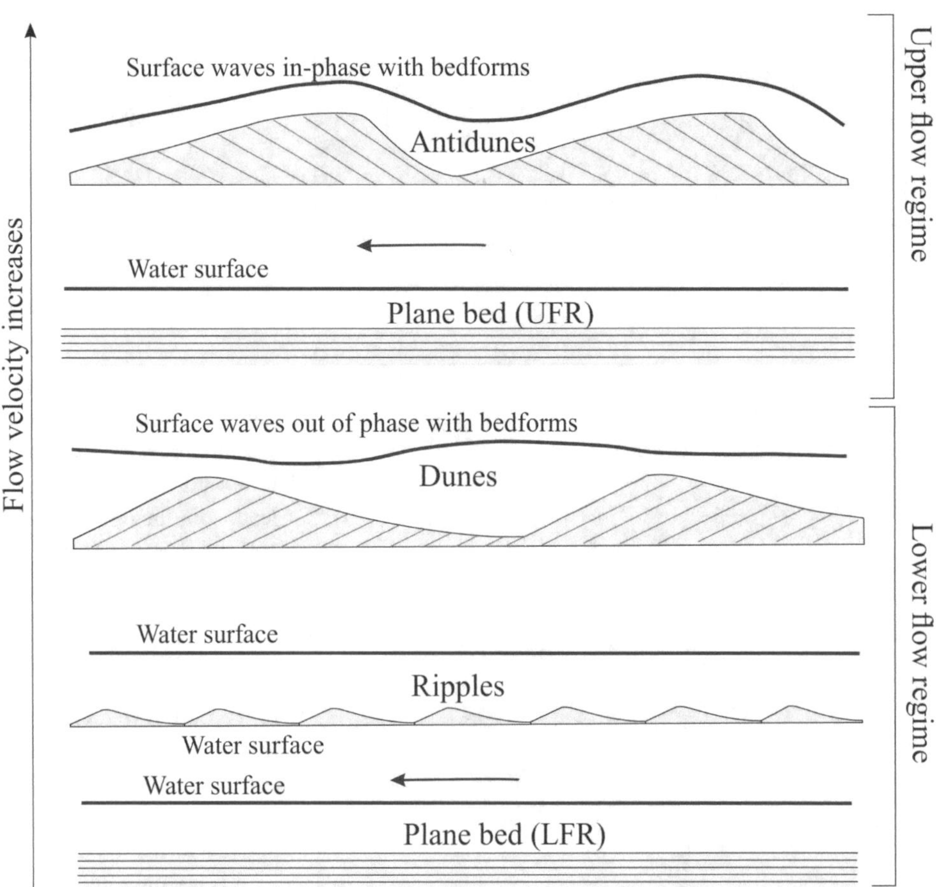

Fig. 14.22 Transition of bedforms with increasing flow velocity [18]

and Baas [37] established that, in the case of ripples, the stable form is always the linguoid morphology. What happens is that the time to reach the stable form varies with the flow velocity (Fig. 14.26). Thus, when the flow velocity is low, the equilibrium time is so high that nature does not offer such long actuation times of the current and, consequently, the stable form is not achieved and straight-crested and sinuous ripples, which are the intermediate metastable states, are preserved.

Another effect of time is seen in the transition between large and small forms. In theory, a regime of smaller flow is not capable of reworking larger forms into smaller ones. Thus, the larger forms would become fossilized and the smaller forms would be superimposed on them. However, in situations where there is not much sediment availability, the sediment of the larger bedforms is used by the smaller ones and, with enough time, the top of the larger bedforms is incorporated into the flow of the superimposed smaller forms. This is the reason why in many cases the surficial morphology of these large forms is not preserved but their internal structure is.

14.6.2 Bidirectional Bedforms (Tide-Generated Bedforms)

The migration of bedforms due to tidal currents is governed by the same parameters described above. The particularity of currents in tidal-dominated environments is that their direction is reversed in each tidal cycle. The action of opposing currents on the same bed causes the bedforms to move in the opposite sense during the tidal flow and ebb. Under these conditions, the lamination or cross-layering of the interior of the sets of these structures will appear to be tilted in the opposite direction. This gives the internal structure the appearance of a fish skeleton, which is why they are called **herringbone structures** (Fig. 14.27).

Although the most intuitive idea is that in these structures the inclination in both directions is equally distributed, this is usually only the case with small shapes such as ripples (Fig. 14.28a). However, the reality is that, in natural environments, one stream often dominates the other [24]. This is most clearly evident in the development of the larger forms. When this occurs, one of the two currents is able to make the

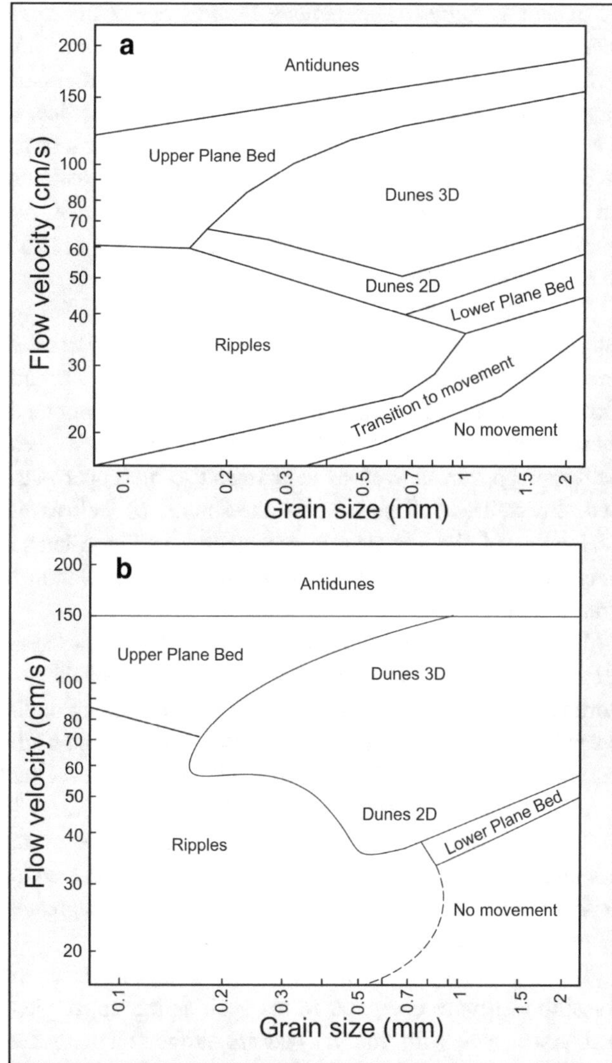

Fig. 14.23 Stability diagrams of bedforms in relation to grain size and flow rate. **a** According to Harms et al. [19]. **b** According to Southard and Boguchwal [43]

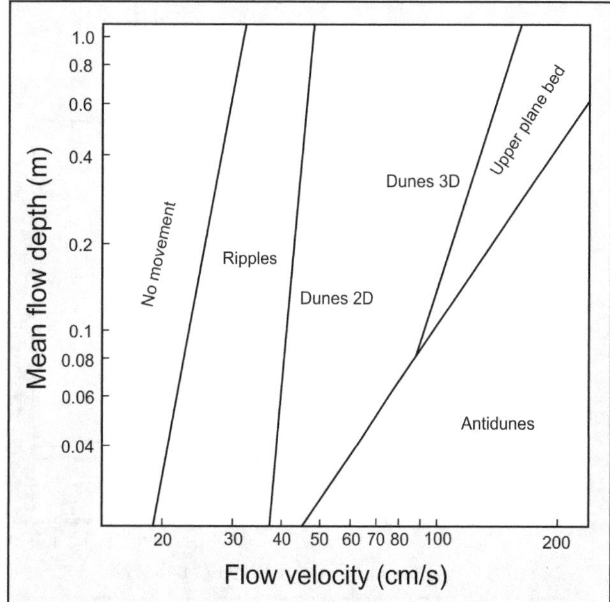

Fig. 14.24 Stability diagram of bedforms in relation to flow velocity and depth [38] The graph is designed for average sand grains between 0.4 and 0.5 mm and with fresh water at 10 °C

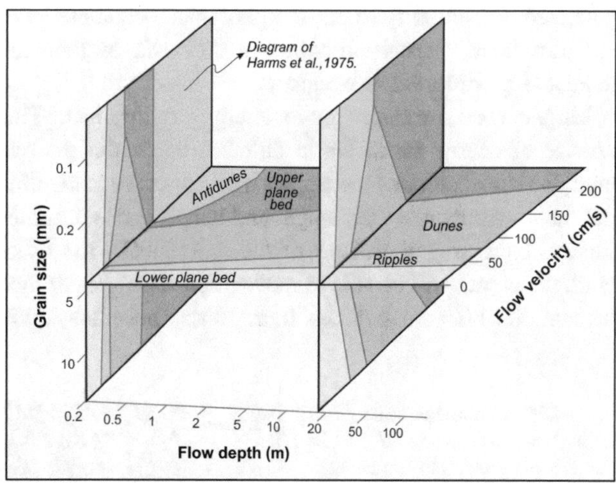

Fig. 14.25 Three-dimensional diagram of bedform stability fields as a function of velocity, sediment size and depth, generalized from natural system data and flume data (adapted from Rubin and McCulloch [38])

bedforms migrate, while the other only retouches the ridges creating reactivation surfaces of the form (Fig. 14.28b). Sometimes the residual current is able to generate bedforms that move in the opposite direction, although with smaller dimensions. In these cases, the dominant current subsequently destroys these shapes totally or partially. In this way, one of the two inclinations is often better preserved in the record of a tidal formation.

14.6.3 Oscillatory Bedforms (Wave Bedforms)

In wave-dominated coastal fronts, the orbital movement of the water is able to transfer energy to the bed through viscous effort. This type of flow generates bedforms on a range of scales depending on the orbital velocity (w), but also on where the shapes develop in relation to the flow depth. These forms play an important role in the transfer of sediments on the waterfront, since their presence increases the viscous friction generating vortices in the near-bed layer and favoring the transport processes [36].

The forms that first come to mind when we talk about oscillation forms are the symmetrical or sinuous straight-crested ripples. These are the forms generated in the beds of the *shoreface* zone, where the action of the waves usually manifests itself in a swaying of the water at the

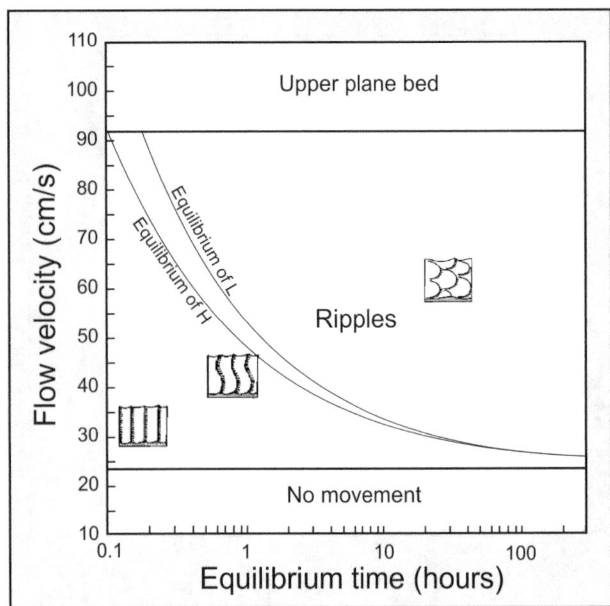

Fig. 14.26 Flow velocity as a function of equilibrium time for ripple height and wavelength in very fine sand of 0.095 mm [37]

bottom. Although these ripples look very similar to those generated by currents, their morphological characteristics and also their internal structure are different, as they are generated by different mechanisms.

Larger waves transfer more energy to the bed. This transfer of energy manifests in different ways. On the one hand, the dimensions of the ripples increase at the same time that the ratio between wavelength and height (L/H) becomes greater, increasing the slope of their faces. On the other hand, the waves will be able to move grain sizes greater than the sand, forming small dunes. In any case, these dunes will be treated as ripples. The increase in the slope of the ripple faces has a clear effect on the behavior of the flow at the limit with the bed. The steep slopes give the ripple a trochoidal shape that generates a vortex in each direction of the flow of a wave [42]. This vortex is responsible for the erosion of the sediment within the ripple and the accumulation of sediment on the opposite side (Fig. 14.29). The existence of two vortices, one in each direction, produces an internal structure in the form of a spike.

There are empirical formulas that attempt to predict the height and wavelength of shapes as a function of parameters similar to those that have been analyzed in the currents. However, there are notable differences in the predictions of these formulas [35]. The conditions under which these bedforms appear, as well as their transit to the upper plane bed, are well established. These thresholds, as well as the L/H ratios of the shapes, are represented in the Allen [1] diagram, as a function of orbital velocity (w) in relation to grain size (Fig. 14.30).

It should be noted that not all wave bedforms are symmetrical. In the areas close to the breakers, one of the two vortices normally dominates over the other depending on the behavior of the wave with respect to the general slope of the beach. In this case, one of the faces will be more developed and the shape will be asymmetrical. Normally, the face with the greatest slope is the one that marks the sense of the dominant vortex. Inside the shape, the spike will also be asymmetrical, so that the sheets of the face with the greatest slope will be thicker.

In the breaking zone, the thin water depth influences the transition to more energetic forms such as the upper plane bed. Associated with this flat bed are surface marks generated by flow deviations due to the presence of coarser

Fig. 14.27 Centimeter-scale herringbone structure in the Phanerozoic Dindefelo formation (Senegal)

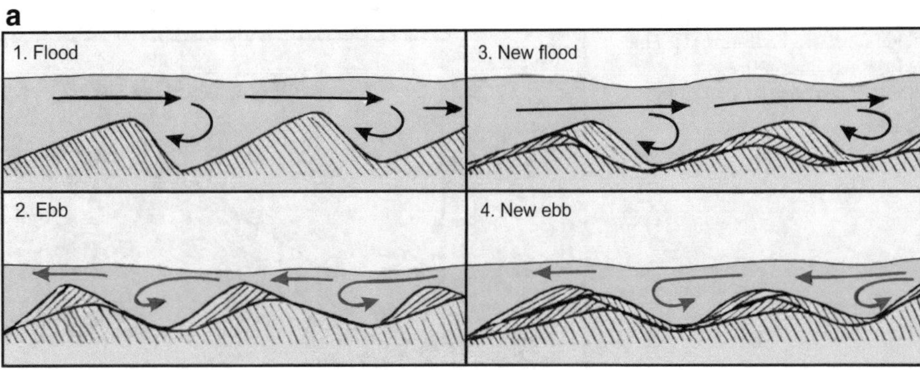

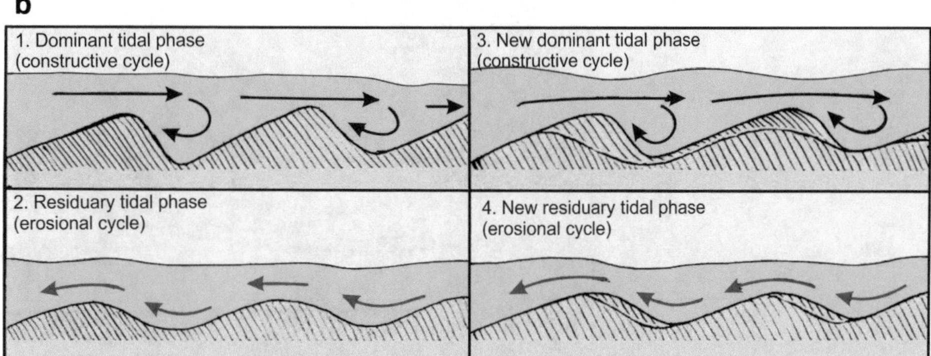

Fig. 14.28 Schemes of generation of herringbones by tidal currents. A: With two equilibrated currents. B: With a dominant depositional current and another residuary erosional one (adapted from Klein [24])

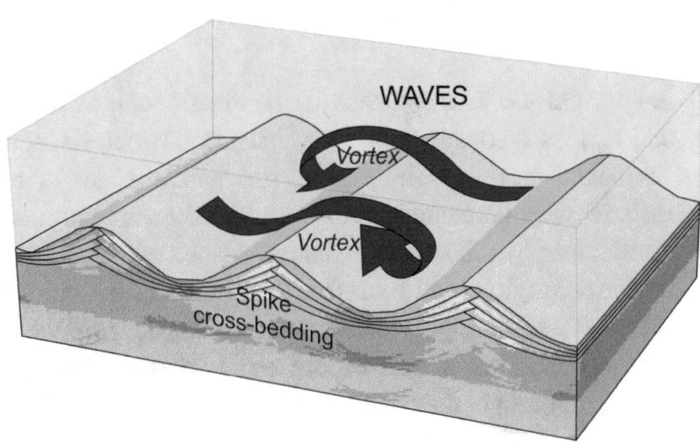

Fig. 14.29 Generation of spike cross-bedding by wave vortices

elements (i.e., shells), such as crescents (Fig. 14.31a) and rhomboidal structures (Fig. 14.31b).

14.6.4 Combined Flow Bedforms

The combination of unidirectional flow and oscillatory flow results in combined flow. In this type of flow, a strong unidirectional current is usually added to an oscillatory flow. The combination of this type of flow is typical of storm action, where high velocity storm relaxation currents are superimposed on the wave oscillation. The result is the acceleration and deceleration of the unidirectional current, generating what are known as ***hummocky structures***.

A hummocky structure consists of a surface on a metric scale similar to an egg box, where depressed areas (*swales*) and elevated areas (*hummocks*) alternate, both with circular to elliptical shapes in plan. The internal structure consists of erosive base sets characterized by low angle corrugated sheets that thicken in the furrows, while becoming thinner or even truncated in the ridges.

The geometry of the internal structure suggests that the hummocky structures behave like stream antidunes; the particularity is that the oscillation of the wave makes it go cyclically from low angle dunes to static antidunes and upward dunes and vice versa, preserving these movements in the internal lamination. This same effect can also be attributed to the influence of internal waves between the turbulent fluid that moves along the bottom and the surface water mass [32].

Due to their origin, these types of structures are typical of *shoreface* areas of those coasts where storms have influence (Fig. 14.32).

14.6.5 Macro-scale Bedforms

Unidirectional flows acting for a very long time can form macroforms of dimensions of tens or hundreds of meters.

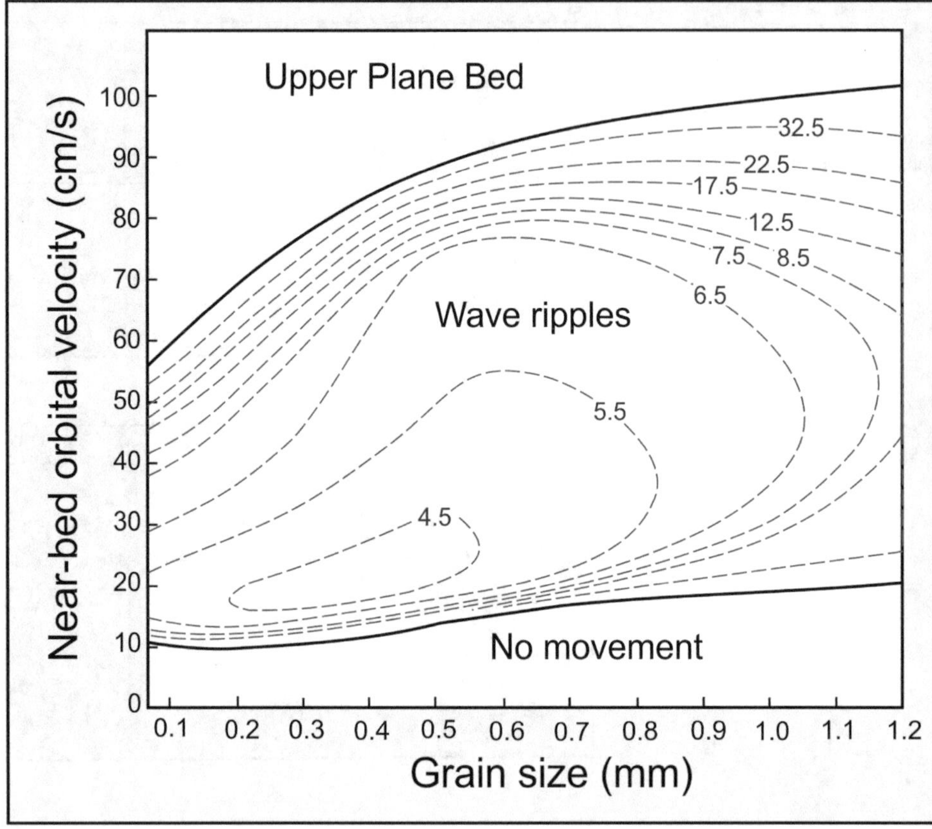

Fig. 14.30 Diagram of stability fields for wave bedforms [1]. The red dashed lines represent different relations of L/H in the ripples

Belderson et al. [8] classified and described these forms on continental shelves, however, they can also be present in *shoreface* zones, as well as in the interior of semi-enclosed bays where there is sufficient space to generate forms of these dimensions, and where persistent streams develop with the necessary velocity. These macro-shapes may be subject to different input regimes. This is how they are defined (Fig. 14.33):

- *Sand patches*: Large *sand patches* deposited on low-mobility bottoms (rocky or cohesive beds) and limited by these same materials. On these sand patches, smaller bedforms develop (ripples and dunes).
- *Sand waves*: Large shapes with wavelengths of up to hundreds of meters that can reach several kilometers in length. On top of these, smaller bedforms such as dunes and ripples also develop. They can develop crests with sinusoidal morphologies and be grouped in trains or be isolated with barchanoid crescent-shaped morphologies.
- *Sand ribbons*: These are macro-shaped *ribbons* that develop longitudinally in the direction of the current action. They are highly evolved forms that develop when they separate from one of the arms in some barchan shapes, although they can also form due to a dynamic shadow behind an obstacle.
- *Gravel furrows*: These are longitudinal forms that are very similar to the sand *furrows* but with a coarser sediment size, which require a high speed to form.

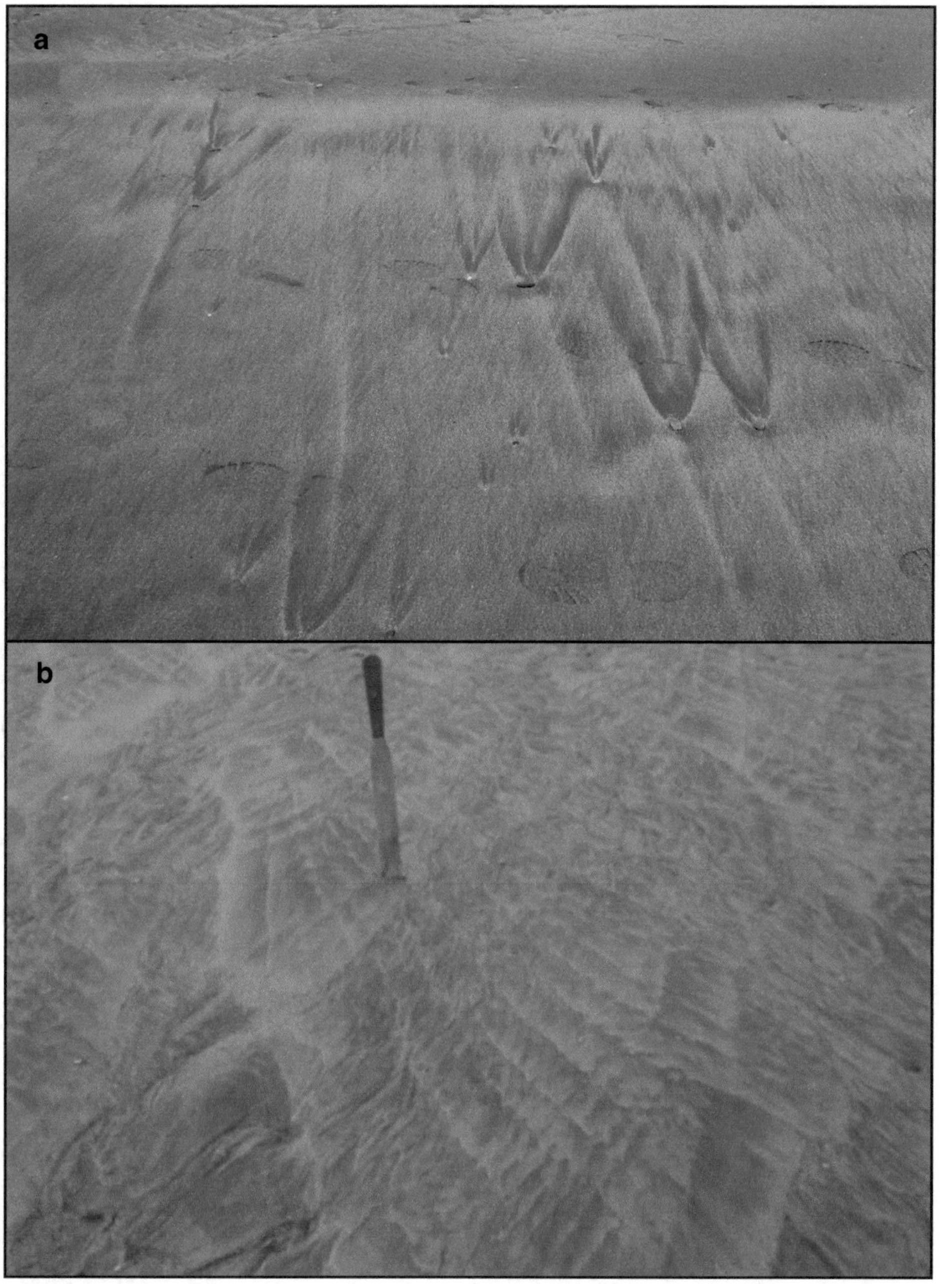

Fig. 14.31 Photographs showing surficial structures associated with the upper plane bed. **a** Crescents caused by action of the backwash on shells. **b** Rhomboidal structures

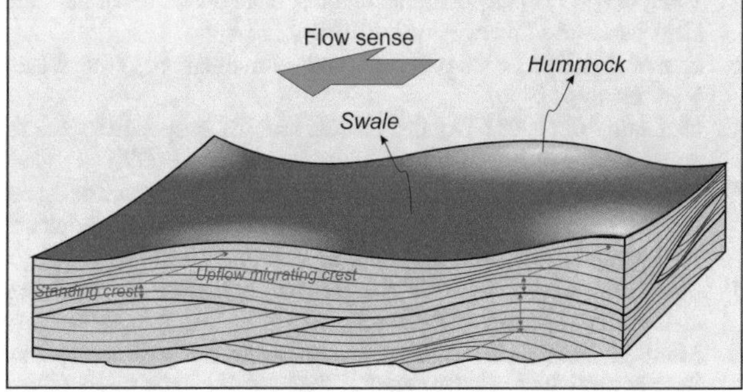

Fig. 14.32 Scheme showing the surface morphology and the internal structure of a hummocky bedform

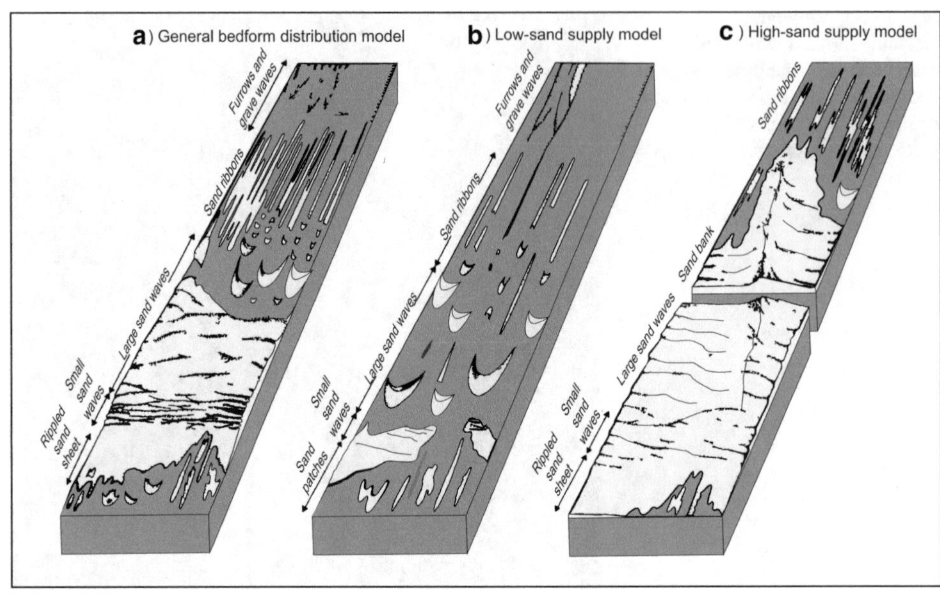

Fig. 14.33 Models of macro-scale bedforms generated by tidal currents in sandy facies on cohesive beds. Each model corresponds to a different sediment supply regime. The average speed of the tidal current at the surface during a spring tide appears in parallel to each zone [8]

References

1. Allen JRL (1984) Sedimentary structures: their character and physical basis. Elsevier, Amsterdam, 592pp
2. Allen JRL (1985) Principles of physical sedimentology. George Allen & Unwin, London, 272pp
3. Allen JRL (1994) Fundamental properties of fluids and their relation to sediment transport processes. In: Pye K (ed) Sediment transport and depositional processes. Blackwell Scientific Publications, Oxford, pp 25–60
4. Ashley GM, Boothroyd JC, Bridge JS et al (1990) Classification of large-scale subaqueous bedforms: a new look at an old problem. J Sediment Petrol 60(1):160–172
5. Bagnold RA (1941) The physics of blown sand and desert dunes. Chapman and Hall, London, 265pp
6. Bagnold RA (1963) Mechanics of marine sedimentation. The sea: ideas and observations. Interscience 3:507–528
7. Bagnold RA (1968) Deposition in the process of hydraulic transport. Sedimentology 10:45–56
8. Belderson RH, Johnson MA, Kenyon NH (1982) Bedforms. In: Stride AH (ed) Offshore tidal sands process and deposits. Chapman and Hall, London, pp 27–57
9. Van Den Berg JH (1987) Bedform migration and bed-load transport of sediment particles in the shallow marine environment. In: Swift JP, Douane DB, Pilkey OH (eds) Shelf sediment transport: processes and pattern. Dowden, Hutchinson and Ross, Stroudsburg, pp 61–82
10. Carter RWG (1988) Coastal environments: an introduction to the physical, ecological, and cultural systems of coastlines. Academic Press, London, 617pp
11. Damgaard JS, Soulsby RL (1996) Longshore bed-load transport. In: Proceedings of the 25th international conference on coastal engineering. Orlando, FL, pp 3614–3627
12. Dyer KR (1985) Coastal and Estuarine sediment dynamics. Wiley Interscience, New York, 342pp
13. Einstein HA (1942) Formulas for the transportation of bed load. Trans Am Soc Civ Eng 107:561–577
14. Galvin CJ (1987) The continuity equation for longshore current velocity with breaker angle adjusted for a wave-current interaction. Coast Eng 11(2):115–129
15. Van der Graaf J, Van Overeem J (1979) Evaluation of sediment transport formulae in coastal engineering practice. Coast Eng 2 (3):1–32
16. Greeley R, Iversen JD (1985) Wind as a geological process on Earth, Mars, Venus and Titan. Cambridge planetary science series 4, Cambridge, 333pp
17. Guy HP, Simons DB, Richardson EV (1966) Summary of alluvial channel data from flume experiments, 1956–1961. US geological survey professional paper 462-I, 96pp
18. Harms JC, Fahnestock RK (1965) Stratification, bed forms, and flow phenomena (with an example from the Rio Grande). SEPM 12:84–115
19. Harms JC, Southard JB, Spearing DR, Walker RG (1975) Depositional environments as interpreted from primary sedimentary structures and stratification sequences. Short Course 2:161
20. Harris CK (2003) Lecture 9: suspended sediment transport II. Sediment transport processes in coastal environments. Virginia Institute of Marine Science
21. Hjulström F (1935) Studies of the morphological activity of rivers as illustrated by the River Fyris. Bull Geol Inst Upsala 25:221–527
22. Inman DL, Bagnold RA (1963) Beach and nearshore processes part II: littoral processes. In: Hill MN (ed) The sea. Wiley Interscience, New York, pp 529–553
23. Kamphuis JW (1991) Alongshores sediment transport rate. J Waterw Port Coast Ocean Eng 117(6):624–640
24. Klein GV (1970) Depositional and dispersal dynamics of intertidal sand bars. J Sediment Petrol 40:195–1127
25. Komar P (1976) Beach processes and sedimentation. Prentice Hall, New Jersey, 429pp
26. McLean SR (1992) On the calculations of suspended load for non-cohesive sediments. J Geophys Res 97:5759–5770
27. Meyer-Peter E, Müller R (1948) Formulas for bed-load transport. In: Proceedings of the 2nd meeting of the international association for hydraulic structures research, pp 39–64
28. Middleton GV (1976) Hydraulic interpretation of sand size distributions. J Geol 84:405–426
29. Middleton GV (ed) (1965) Primary sedimentary structures and their hydrodynamic interpretation. Society for Sedimentary Geology Special Publication 12, Tulsa, Oklahoma, 265pp

30. Middleton GV, Southard JB (1978/1984) Mechanics of sediment movement (1st & 2nd edn). Short course 3. Society for Sedimentary Geology, Tulsa, Oklahoma
31. Miller MC, McCave IN, Komar PD (1977) Threshold of sediment motion under unidirectional currents. Sedimentology 24:507–527
32. Morsilli M, Pomar L (2012) Internal waves vs. surface storm waves: a review on the origin of hummocky cross-stratification. Terra Nova 24:273–282
33. Moss A (1972) Bed-load sediments. Sedimentology 18:159–219
34. Moss AJ, Walker PH, Hutka J (1980) Movement of loose, sandy detritus by shallow water flows: an experimental study. Sed Geol 25:43–66
35. Nelson TR, Voulgaris G, Traykovski P (2013) Predicting wave-induced ripple equilibrium geometry. J Geophys Res Oceans 118:3202–3220
36. Nielsen P (1981) Dynamics and geometry of wave-generated ripples. J Geophys Res Oceans 86:6467–6472
37. Oost AP, Baas JH (1994) The development of small scale bedforms in tidal environments: an empirical model and its applications. Sedimentology 41:883–903
38. Rubin DM, McCulloch DS (1980) Single and superimposed bedforms: a synthesis of San Francisco Bay and flume observations. Sed Geol 26:207–231
39. Schoonees JS, Theron AK (1995) Evaluation of 10 cross-shore sediment transport/morphological models. Coast Eng 25:1–41
40. Shields AF (1936) Application of similarity principles and turbulence research to bed-load movement. Mitteilungen der Preussischen Versuchsanstalt für Wasserbau und Schiffbau, vol 26, 25pp
41. Simons EB, Richardson EV (1961) Forms of bed roughness in alluvial channels. Proc Am Soc Civ Eng 87:87–105
42. Sleath JFA (1984) Sea bed mechanics. Wiley, New York, 335pp
43. Southard JB, Boguchwal A (1990) Bed configurations in steady unidirectional water flows. Part 2. Synthesis of flume data. J Sediment Petrol 60(5):658–679
44. Sternberg RW (1971) Measurements of incipient motion of sediment particles in the marine environment. Mar Geol 10:113–120
45. Van Rijn LC (1981) Computation of bed-load concentration and bed-load transport. Delft Hydraulics Laboratory Research Report S487-L; Delft Hydraulics Laboratory: Delft, The Netherlands, p 43
46. Yalin MS (1963) An expression for bed-load transportation. J Hydraul Div 89:221–250

Part III

Coastal Systems: Dynamics, Facies and Sedimentary Models

Tread the surface level
wearing answers in layers.
Sediments of sediments unfold
to write new cave scratchings.

"Great News From the South Pole"
Das Oath

15 Geologically Controlled Coastal Systems: Rocky Coasts, Bluffs, Cliffs and Shore Platforms

15.1 Introduction

Most of the world's coasts develop in hard or cohesive materials typical of mountainous continental margins. Indeed, since the work of Emery and Kuhn [3], the literature has commonly reported that 80% of the world's coasts can be categorized as rocky or cliffs. However, these coasts do not usually appear in our minds when we imagine a coast. This happens because humans normally settle their resorts on low coasts associated with depositional systems and usually keep them away from steep coasts. The scarcity of research work on these rocky coastal systems does not help to favor them as well-known coasts either. In recent decades, these environments have attracted less attention than other systems located on coastal plains that are more vulnerable to problems associated with global change. However, rocky coasts offer very beautiful panoramas and spectacular views of processes, especially during storms, and play an important role as a supplier of sediments to depositional environments.

The resistance of the materials typical of **rocky coasts** (Fig. 15.1a) makes them slow to dismantle. At the same time, a high wave transport capacity gives this type of coast a distinctly erosive character. Erosion rates in this type of coastline are highly variable, ranging from a few millimeters per year to a few meters per year, depending on the consistency of the rocks in relation to the wave dynamics. Despite these erosive characteristics, associated deposits can also appear on rocky coasts. Normally, these deposits either correspond to debris that the wave is not able to displace over long distances, or are made up of smaller materials that the wave deposits in those places where the wave energy is dissipated.

In most discussions, the concepts of rocky coasts and cliffs are used interchangeably, although there are nuances between these terms. In general, a rocky coast is understood to be a coast formed by rocks of hard and resistant lithology. There is a variety of lithologies among igneous (plutonic and volcanic), metamorphic and consolidated sedimentary rocks (sandstones, siltstones, limestones, dolomites), with a wide range of ages, from Precambrian to Cenozoic. This type of lithology offers greater resistance to erosion, which is manifested in retreat rates measured in millimeters or centimeters.

Very similar erosive morphologies can be generated in lithologies of lower resistance and higher regression rates, although these are normally developed in more recent ages. In these cases, the term **bluff coasts** is used (Fig. 15.1b). Normally, these coasts develop in sedimentary rocks of diverse origin, among marine deposits, tillites, eolianites and beachrocks. These lithological formations correspond to sea levels and climatic stages different from the current one, but they can also develop in non-coherent volcanic lithologies (pyroclastic). The term bluff is much less used, and for many authors the term rocky coast also includes bluff coasts.

The term **cliff** refers to both rocky and bluff coasts that are steep and sloped at an angle of more than 40° [7] (Fig. 15.2a). A cliff is never surpassed by the wave breakers, so it is understood that its height is always greater than the maximum height of the maximum swell acting during the moments of maximum surge (maximum flooding level). Cliffs are the most common feature of both rocky and bluff coasts, where they take the name of *soft cliffs*, so the term cliff coast is also used as a synonym. However, rocky coasts and also bluffs can have slopes less than 40° or levels below the maximum floodable level. In these cases, reliefs are formed that can be surpassed (*overtopped*) by the waves under certain circumstances. These reliefs of lesser slope and height are called **banks** or **benches**.

Regardless of the emergent morphology and the type of materials that control the development of this type of coast, a platform that extends in depth from the coastline and is subject to the action of the waves usually develops on the submerged front. These platforms may have different characteristics of slope, extension, morphology and location with respect to wave and tidal action, and are commonly referred to as **shore platforms** (Fig. 15.2b).

Rocky coasts are widely distributed throughout the world and can be found in different climatic and tectonic

J. A. Morales, *Coastal Geology*, Springer Textbooks in Earth Sciences, Geography and Environment, https://doi.org/10.1007/978-3-030-96121-3_15

Fig. 15.1 **a** Rocky coast in Pliocene basalt lavas (Ngor Island, Senegal). **b** Bluff coast in Cretaceous marls and grainstones (Praia da Luz, Portugal)

environments, although they have classically been associated with tectonically active coasts subject to an uplift process. Their high slope and height may mean that sea level movements do not produce large horizontal movements in the form of transgressions and regressions in the long term. The current rocky coasts have been developing since the arrival of the sea level to its current position after the Flandrian transgression. However, although the dynamics over these environments are very energetic, the resistance of the lithologies means that a large part of the characteristics of these coasts can be inherited from previous geological times.

The possibilities of the combinations of lithology, arrangement of lithological bodies and fractures, as well as differing slopes in relation to processes, lead to a wide variety of erosive morphological features. At the same time, coastal erosion systems are important suppliers of material to other nearby coastal sedimentary environments. In this

Fig. 15.2 **a** Near-vertical cliffs developed in Cretaceous limestones (Sagres, Portugal). **b** Shore platform in a cliff front of Pleistocene sands (Larache, Morocco)

chapter, the variables that control the whole range of processes and forms will be analyzed, as well as the deposits associated with the main erosive features.

15.2 Control Factors

Most of the factors that control the dynamics of rocky coasts are those that affect the intensity of the weathering and erosion processes which contribute to their dismantling. Thus, we will find among them wave energy, tidal amplitude, climate and relative sea level movements. However, the most important control factor is geology, since it is the geological factors that determine the threshold over which the rest of the dynamic factors must act [13]. This is why this type of system is being classified here as geologically controlled.

15.2.1 Geology

There are several geological characteristics that determine the degree of a coast's resistance to erosion. Obviously, lithology is a fundamental factor, but so are the structural characteristics, such as the degree of fracturing and the arrangement of materials with respect to the coastline, in determining to what extent dynamic processes will act on geological formations.

Regarding the lithology, this is the factor that determines the intrinsic strength of the rock (Fig. 15.3). Igneous rocks, such as granites, or coherent volcanic rocks, such as basalts, present high resistance that can be maintained over wide coastal fringes. The presence of plutonic rocks minimizes the importance of the structural factor (Fig. 15.3a). On the contrary, the presence of metamorphic or sedimentary rocks promotes varying resistance by lithological sections, favoring differential erosion. In this case, there is greater structural control and the result is a coast with irregular morphology (Fig. 15.3b). Massive limestones tend to give very vertical cliffs associated with different forms undergoing dissolution processes (Fig. 15.3c). Lithological alterations of undeformed sedimentary rocks with different strengths result in very irregular coasts with numerous collapsed blocks at the base of the cliffs (Fig. 15.3d). Finally, recent sediments and unconsolidated sedimentary rocks offer little resistance to dynamic processes, which are accelerated in these lithologies. The result is the presence of long slopes with many deposits at the cliff toe (Fig. 15.3e).

In certain lithologies, the degree of fracture plays a fundamental role, since erosion and weathering tend to be concentrated in the rock sections with the highest density of fractures. On coasts with homogeneous lithology, the areas with a higher concentration of fractures will be easier to erode and bays will form, while the less fractured areas will be more resistant and will constitute the coastal overhangs.

Finally, the arrangement of materials is a factor to be taken into account, since different arrangements can generate different coastal morphologies. On the one hand, the lithological disposition plays a determining role in the general morphology of the coast. In this respect, authors such as Johnson [9] distinguished between concordant and discordant coasts according to the orientation of the lithological guidelines in relation to the wave front. Furthermore, it is not only the angle between the planes of the lithological units and the waterfront that must be considered, but also the angle between the waterfront and the fractures in three dimensions. This arrangement determines the possibility of different slope processes occurring as part of coastal retreat. It is important to note that it is the disposition of the fractures with respect to the action of the agents that induces the formation of secondary features such as caves, arches, and *stumps* and *stacks*.

15.2.2 Wave Energy

Wave energy exerts the main control over the physical processes of erosion, since it conditions the erosive capacity of the rocks on the waterfront, as well as the potential for transporting the material to other coastal sections.

The energy does not depend only on the dimensions of the wave, but also on the type of interaction with the coast

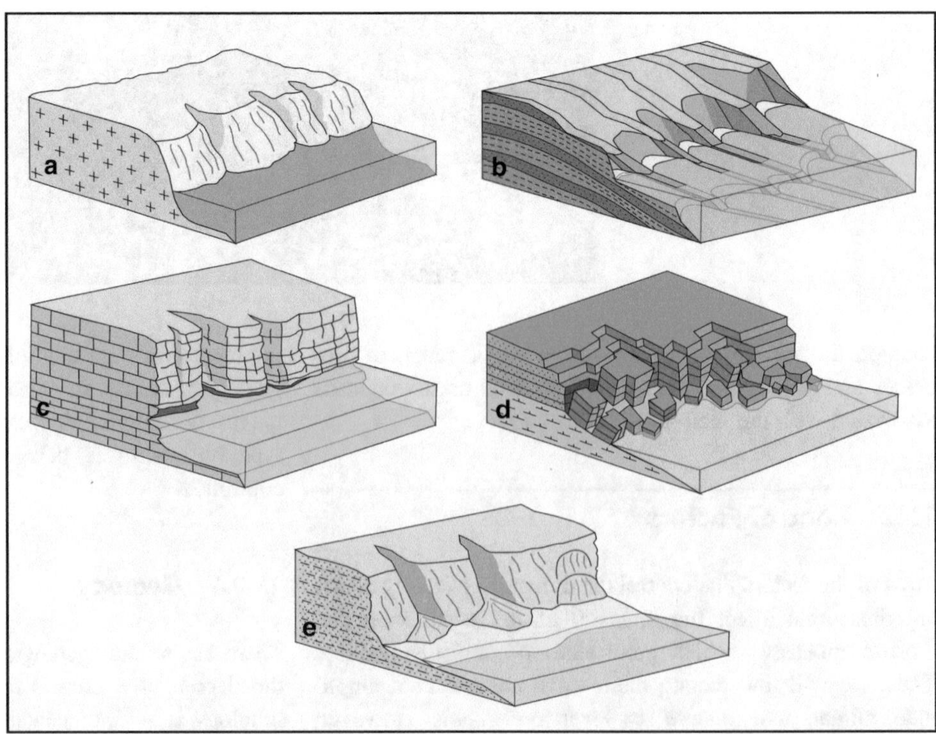

Fig. 15.3 Examples of coast morphologies of different lithologies. **a** Granitic coast. **b** Alternating quartzite–schist coast. **c** Cliff in massive limestone. **d** Cliff in sandstone with a marly base. **e** Cliff in silty sand

that is reflected in the breaker type. In this aspect, the extension and position of the coastal platform and the general slope of the *shoaling* area have a notable influence. If the platform is wide and near the surface, some of the wave energy can be dissipated by friction with the bottom of this platform. If, on the other hand, the platform does not exist, or is located in very deep areas, the waves will reach the coast with all their energy, which will be used entirely in the breaker and will have a greater effect on the physical disintegration of the rock.

Wave energy is also a factor in transport capacity. It is important to measure the transport capacity in relation to its ability to produce material accumulations at the base of the rock face by other processes associated with the slope. If the wave is able to rapidly transport the deposits generated at the foot of the cliff, then the face will remain active and erosion rates will be higher. If the rate of formation of the slope deposits is very high, and only a part of the deposit is able to be transported, then the deposits will accumulate at the base, protecting the waterfront and delaying the retreat.

15.2.3 Tidal Range

The tide controls the topographic level of wave action and the time it acts at each level of a rocky coast. The development of these systems is not related to the tidal range; therefore, rocky coasts can develop in any tidal range. When the tidal range is small, the wave action is restricted to a very narrow area of the rocky front, while on macro-tidal coasts the tide displaces the wave action vertically, dispersing the energy in a wider strip.

15.2.4 Relative Sea Level Movements

The relative movements of sea level combine eustatic movements with the action of tectonics on the ground. The relative rises and falls of sea level can place the action of waves in different positions in the long term, generating shore platforms at different topographic heights. Higher positions of sea level will generate emerged coastal terraces, while lower positions will generate submerged platforms at different heights and depths. The rhythm of the sea level movements in relation to the erosion rate control the mean slope of the coast and the development of some particular morphologies, such as inclined shore platforms.

15.2.5 Climate

The weather controls the wind regime and is therefore a primary factor in wave energy. In this sense, the climate determines the average wave energy throughout the year, but also the number of storms and the time they occur. In addition, the climate also exercises the main control over physical and chemical weathering processes, selecting the type of processes and their intensity.

This influence of the climate causes a latitudinal zoning in the processes that are associated with rocky fronts. At high latitudes, weathering processes related to freezing and the action of large waves in strong storms predominate. In contrast, in tropical latitudes, chemical weathering processes dominate and the overall energy of the waves is lower. At mid-latitudes, global wave energy is higher and there is a dominance of physical weathering processes.

15.3 Processes

Numerous processes act on the rocky coasts, which are grouped according to their nature into physical, chemical and biological; these combine to result in a recessionary effect of the coastline. The coastal morphology that we observe today is the result of several thousand years of continuous action of these processes acting in a combined way. The relative importance of each of these processes in the global erosion of the coast depends on the way they affect the geological nature of the rocks and their structural arrangement. The effect of these processes has been analyzed in detail in the first section of Alan Trenhaile's volume (1987), as well as in the first chapters of Tsuguo Sunamura's work (1992), and is summarized in this section.

15.3.1 Wave Erosion

Direct erosion by waves is the most immediate and important process in the retreat of cliffs. Wave action takes place in a narrow vertical strip of the rocky front that is located in areas close to the average water level and whose position varies according to the magnitude of the waves and the tidal range [24]. The waves actually act on the cliff via three different mechanisms: (1) direct hydraulic strike; (2) abrasion using clastic elements; and (3) action of air compressed by the waves.

- *Direct hydraulic action*: This is the action of direct water impact on the surface of exposed rocks on the waterfront (Fig. 15.4a). In this impact, all of the energy of the wave is instantly converted into deformation and heat. This effect is called *wave pounding* or the "*sledge hammer effect*" by some authors [2]. Since the energy depends quadratically on the height of the waves, the impacts will be more energetic the larger the dimensions of the wave.

The effect of these impacts on the rock will depend on this dissipated energy in relation to the geological nature of the rock. In this respect, it should be borne in mind that the hammering action is not the result of a single wave, but of thousands of waves acting continuously at the same topographic level over millennia.

- *Abrasion*: In this case it is the erosional effect of clastic fragments displaced by the waves on the rock face (grains,

Fig. 15.4 **a** Direct hydraulic action of waves on a cliff base (Llanes, Spain) and effect of air pressure venting through open fractures (photograph courtesy of G. Flor-Blanco). **b** Clastic elements used by waves as tools for abrasion in a cliff base (Larache, Morocco). **c** Potholes in a calcareous shore platform (Menorca, Spain)

pebbles, cobbles and even blocks). This action is more important on surfaces with a low slope, because the dragging effect of these lithic elements can also occur on these surfaces, causing wear and tear on the rock. On cliff faces, the effect of projected clast impacts is greater (Fig. 15.4b). In this case, the effectiveness of the abrasion process depends not only on the geological nature of the rock and the wave energy, but also on the availability of abrasive elements (type and quantity). The effect of this action on the rocky outcrop is generally a polishing of the surface. On other occasions, one or more of the clastic elements may be trapped in a cavity, turning inside it with the arrival of each wave. In these cases, the result will be the genesis of *potholes* of different dimensions (Fig. 15.4c).

- *Action of compressed air*: The arrival of the wave crests to the rocky front usually implies that different volumes of air can be trapped in the rock cavities, increasing their pressure when the breaker occurs. These volumes can be of different dimensions depending on the character of the empty space, ranging from the millimetric volume of the pores, the centimeters in the fractures to several meters in larger cavities of the rock face. The result of the pressure increase varies depending on the scale at which it occurs. The greatest effects usually occur at the fracture level, with the increase in pressure acting as one of the causes inducing fracture growth; this can eventually lead to the opening of cavities to relieve pressure towards the surface (Fig. 15.4a) and may also contribute to the detachment of rock fragments.

In general terms it can be considered that the combined effect of these three processes results in the removal of rock fragments from the surface [22]. Although waves act continuously at these topographic levels, it is during major storms that the effects of their action become more visible. Different works based on the continuous photography of the scars left by the removal of blocks on cliff fronts have shown that storm waves are capable of removing blocks of up to several tons and displacing them even to the top of cliffs more than a dozen meters high (e.g., Trenhaile and Kanyaya 25]; Naylor and Stephenson [17]).

15.3.2 Mechanical Weathering

Physical weathering occurs when internal forces act on the pores or fractures of the rocks towards the outside; these contribute to the appearance or increase of fracturing at different scales, ending with the disintegration of the rock. The three processes that are usually associated with rock disintegration in the coastal strip are: (1) thermal stress; (2) expansion by salt crystallization; and (3) freeze–thaw cycles in periglacial climates.

- *Thermal stress*: This is the effect of the expansion and contraction of the rock due to thermal changes that occur throughout the day (day–night) or the year (summer–winter). In general terms, as the crystals that make up the rock heat up they expand, putting pressure on the rest of the crystals. The different minerals behave differently, giving rise to a differential expansion that involves the appearance of internal forces that end up disintegrating the rock (thermal fatigue). This is not an exclusive effect of rocks located on the coast, but in coastal environments, especially in arid and semi-arid climates, it combines with other processes. This process normally occurs at a microscopic level, between grains; however, it can also occur at a metric scale when rocks of different thermal behavior alternate in an outcrop. In this case, fractures with greater separation may appear and larger blocks may be detached.
- *Expansion by salt crystallization*: In coastal environments, it is common for seawater to enter small rock cavities such as pores or fractures. The entry into the seawater cavities is related to tidal oscillations, but also to the *spray* effect associated with wave breakers. This seawater is, after all, a saline solution that can later evaporate and cause the crystallization of salts inside these cavities. Sodium sulfate, calcium chloride, magnesium sulfate and, above all, sodium chloride, are the salts that most commonly crystallize in these environments. The appearance of these salts inside the cavities means an increase in volume that generates the forces necessary for the disintegration of rock fragments.
- *Freeze–thaw cycles*: This is a process of weathering associated with high latitudes. In this case, the internal forces that break up the rock are generated by the expansion and contraction of water inside the pores and the fractures that develop from changing from liquid to solid state and vice versa. In this case, the expansion of the water when it freezes (volume increase of 9%) is a sufficient force to break up the rocks. In this respect, it should be taken into account that water with a high salt content is more difficult to freeze and, consequently, generates lesser forces. However, this process can occur with fresh water at higher topographic levels of the rock profile, contributing to the disintegration of the rock where the waves cannot act.

15.3.3 Chemical Weathering

Chemical processes take place in all rocky coastal strips, from the underwater areas to the emerged zones, including the intertidal fringe. They consist of mineral transformations due to the interaction with the water and the ions dissolved in it. They are processes that are concentrated on the surface of the rock or at shallow depth, sticking to the area where surface water exerts its influence. Consequently, their speed of action is much slower and less visible than physical processes. This type of weathering mainly comprises four types of chemical process: (1) dissolution; (2) hydrolysis; (3) hydration; and (4) oxidation. These all contribute to the granular disintegration of the rock, facilitating the erosive processes that act with them. The type and intensity of each of these processes has a strong lithological and climatic control. These processes can act together in an effect called **corrosion**.

- *Dissolution*: This implies the wear of the rock due to the chemical disintegration of the minerals, passing the elements that compose the crystals to an ionic state inside the water.
- *Hydrolysis*: This is a process that acts preferentially in feldspars and carbonates, in which the reaction with water molecules breaks the crystalline structure producing its transformation into clay phyllosilicates. It must be taken into account that feldspars are part of the composition of many rocks of different origins, whether igneous, metamorphic or sedimentary, and the carbonates are the base of all calcareous rocks. The action of this process implies a transformation into minerals that pass immediately to the suspended phase, to be easily removed by the waves.
- *Hydration*: This involves the entry of the water molecule into the crystalline structure of the minerals, meaning a transformation that normally involves an increase in volume. The passage from oxides to hydroxides and the hydration of anhydrite into gypsum are the most common examples, but these processes are present only in certain lithologies. When this occurs, the chemical process also acts in combination with the physical process of expansion that this increase in volume implies.
- *Oxidation*: This process involves the loss of electrons from any mineral compound, but the most common reaction in geological environments is associated with that between the iron contained in any crystal structure and the oxygen dissolved in the water. These reactions normally result in iron oxides and hydroxides (goethite, hematite and limonite) which are mobilized through the pores to concentrate in the form of crusts on the surface of the rock or in the more porous areas, controlled by lithology or the presence of joints.

In all of these processes, there is a strong latitudinal control from the temperature of the water. In general terms, chemical processes run more easily at higher temperatures, since, as discussed in Chap. 11, higher temperatures increase the ionic capacity.

15.3.4 Bioerosion

Biological degradation of rocks occurs when organisms drill their surface for shelter and/or food. These organisms, both animal and plant, perform a double biochemical and biomechanical action, since they are capable of producing a physical breakdown effect on the rock grains as well as chemical changes that lead to the disintegration of their mineral structure.

The most effective bioerosion is the colonization of the rock surface by algae in the sub- and intertidal zones. The case of blue-green algae is the best example (Fig. 15.5a). These microalgae are capable of rapidly colonizing the surface of carbonate rocks by penetrating more than 1 mm below the surface. Their activity contributes to the bio-dissolution of large amounts of carbonate, breaking down the rock fabric and undissolved grains, thus facilitating the subsequent action of the waves. A similar action is effected by other types of organisms such as fungi and lichens in the inter- and supratidal zones. The presence of these microorganisms on the surface of the rock causes the appearance of other macroorganisms that feed on them. Among these are some mollusks such as gastropods (snails and limpets, Fig. 15.5b), bivalves (the most common at present are those of the genus *Pholas)* and polyplacophores (chitons). The same activity can be performed by other invertebrates such as echinoderms (Fig. 15.5c), polychaetes, bryozoans, sponges and crustaceans.

The combined activity of bioerosion and chemical weathering processes gives the rock surface an irregular cell texture, made up of holes and partitions known as *honeycomb* (Fig. 15.5d).

15.3.5 Slope Processes

The processes associated with the direct pounding of the waves can only transform the lower zone of the rocky shores. For a coastal system with a high retreating slope, other processes occur that affect the part where the waves cannot reach. The processes of weathering, although they affect the entire profile of the rocky coastline, are slow and, in any case, only contribute to the release of particles to facilitate erosion; once these particles are released, they must be set in motion by other physical processes. In these high areas, high-slope continental processes are responsible for this task. In these processes, gravity plays a very important role, although surface water runoff is also significant. Both contribute to vertically transporting materials from the upper parts of the rock profile to bring them down to a height

Fig. 15.5 Some examples of bioerosion. **a** Blue-green algal cover of a shore platform (Larache, Morocco). **b** Some fractures in a granite enlarged by limpet activity (Pointe de Trévignon, France). **c** Erosional action of an urchin in the base of a cliff (Samaná, Dominican Republic). **d** Honeycomb structure on a shore platform (Algeciras, Spain)

where the waves can exert their action on them. There are four fundamental processes: (1) rockfall; (2) landslides; (3) mass flows; and (4) the action of surface water runoff.

- *Rockfall*: This involves a vertical displacement of rock masses due to the direct action of gravity on consolidated rock faces (Fig. 15.6a). This movement may individually affect rock fragments of different sizes released by mechanical weathering processes or, conversely, involve large rock masses by large-scale destabilization. This destabilization is usually related to the occurrence of decompression fractures parallel to the cliff surface and to the destabilization of the base due to wave action. This massive drop (Fig. 15.6b) can occur in the form of *overturns* or collapses (*toppling*).
- *Slides*: Landslides tend to occur in less consistent lithologies, especially in *bluff coasts*. They occur when there is a displacement of masses in favor of a net surface because the resistance of the base is exceeded by the load it supports. Depending on the geometry of this surface,

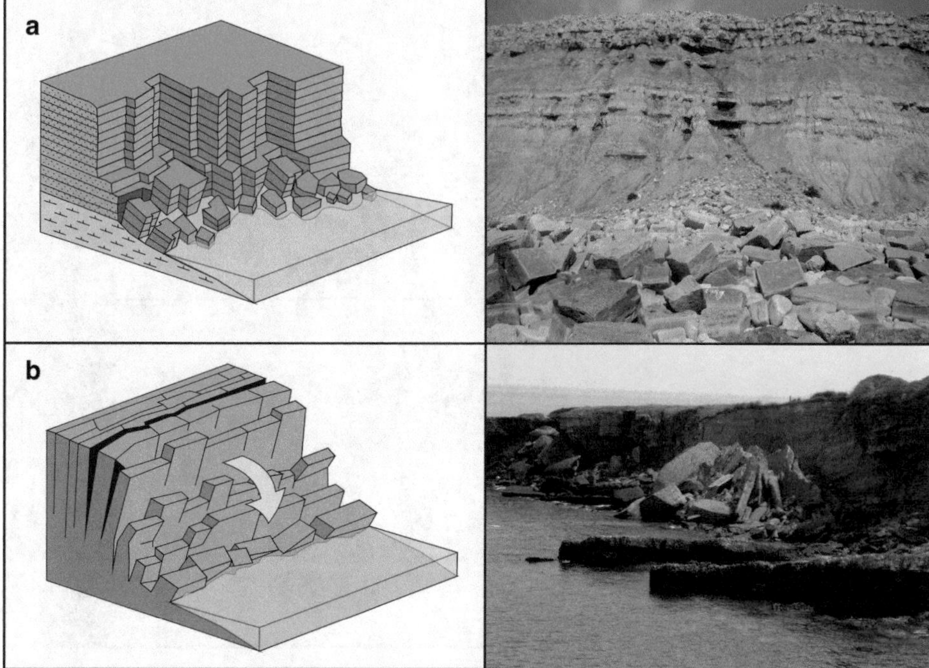

Fig. 15.6 Processes of rockfall associated with the high slope. **a** Small rockfalls, with an example in southern Portugal. **b** Toppling, with an example in northern Spain

displacing slides may occur when the surface is flat (Fig. 15.7a) or rotational slides if the surface is curved (Fig. 15.7b). This geometry is structurally controlled. Another type of landslide occurs due to the presence of fracturing in conjugate families arranged at an oblique angle to the rock surface, which release large wedge-shaped masses of rock (Fig. 15.7c). As with the massive landslides described above, the onset of the landslide is related to the destabilization of the base due to the action of the waves, although it can also begin when very intense or very continuous rains cause a decrease in grain friction by increasing water pressure in the pores.

- *Mass flows*: *Mudflow*, *debris flow* or *grain flow* phenomena may also occur in certain unconsolidated lithologies. In these phenomena, there is a movement of the ground of lower density and more deformation than in the case of landslides. Interstitial water also plays a fundamental role in lubricating the soil and reducing the density of the material and the friction between grains.
- *Surface runoff*: The effect of water currents on the rock surface, especially in lower slope systems, can cause erosion and material transport by fluid flow. This transport manifests in the upper part of the profile in the form of large erosive grooves where water flow is concentrated, while in the lower part of the profile, small-scale alluvial fans can form with the materials transported from the upper part (Fig. 15.8).

15.4 General Morphology and Morphological Features

Taking into account all the factors and processes that influence the development and evolution of a rocky coast as detailed in the previous sections, it can be seen that these present a wide variety of morphologies and morphological features. In general terms, the active morphologies can be

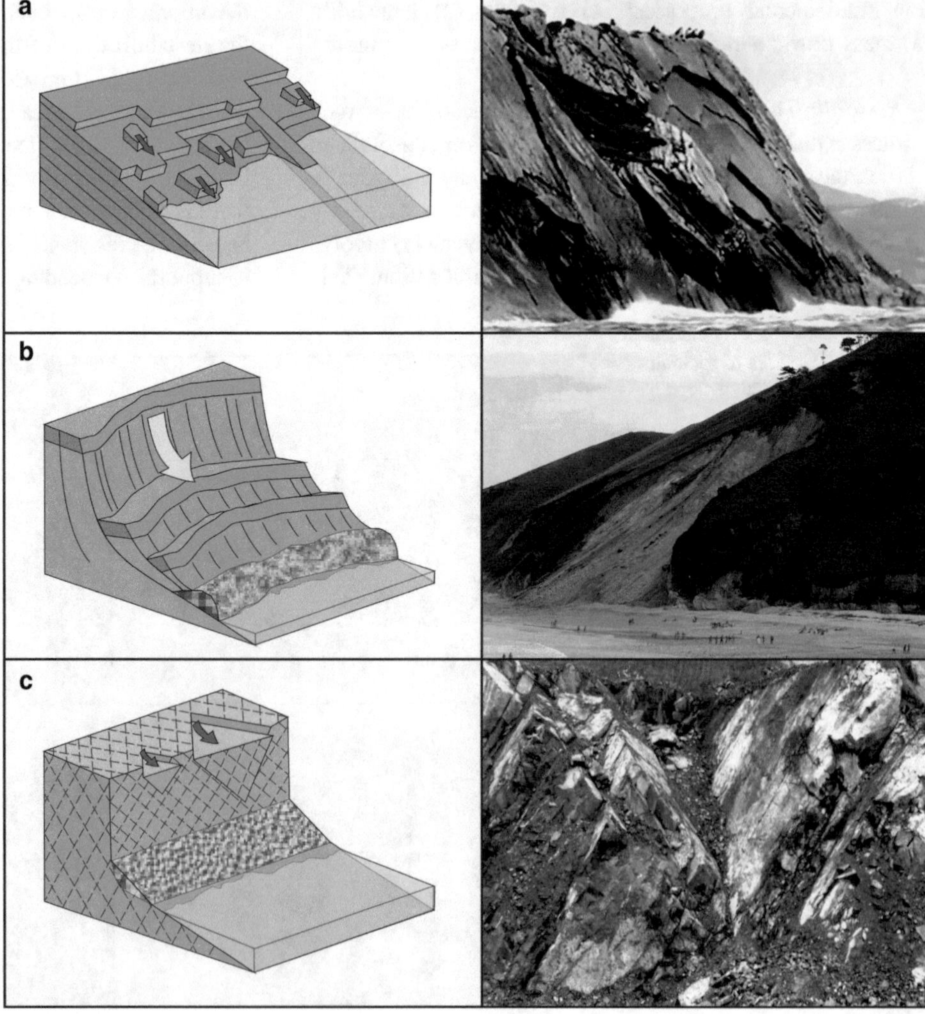

Fig. 15.7 Slide processes associated with the high slope. **a** Planar slides, with an example in northern Spain. **b** Rotational slides, with an example in Northern Spain. **c** Wedge slides by orthogonal fractures

Fig. 15.8 Grooves and alluvial fans caused by flow runoff, with an example in southwestern Spain

grouped into three main features: rocky ramps, shore platforms and cliffs. Each of them can present different features and secondary elements and follow different evolutionary paths according to the combination of the factors and processes mentioned above, as well as their relative importance.

15.4.1 Rocky Ramps

Although this term is not very widespread, it is used to refer to rocky coasts that have a slope of less than 40° (Fig. 15.9). Other terms such as gentle cliffs, benches or uniformly sloping rocky coast are also used to name this concept. Apart from the lesser slope, the processes that affect it, its morphological characteristics and its evolution are very similar to those of cliffs, so many authors include them as a subtype of cliffs. Rocky coasts acquire this morphology when the materials are not very resistant in relation to the continental processes that act in the upper part of the same, resulting in a very fast dismantling rate.

15.4.2 Shore Platforms

Shore platforms are low-slope erosive rock surfaces excavated from the front of some rocky shores. The origin of the platforms is mainly erosive and is due to the recession effect of the rocky front. They can be completely horizontal or have a gentle slope towards the sea that can reach 10° and their scale in a transverse direction to the coastline varies from tens to hundreds of meters. Towards the sea, the edge of the platforms presents a sharp increase in slope that results in a steeper slope that connects the platform with the deeper areas. The slope of the platform is closely related to the dynamics of the erosion processes that generate it in relation to the resistance of the rock. Thus, more resistant rocks tend

Fig. 15.9 Ramp-shaped rocky coast (Algarve, Portugal)

to generate horizontal platforms, while less consolidated rocks generate a greater slope [23]. Within the processes, the tidal range also exerts an important control, in such a way that, the wider the tidal range, the greater the slope tends to be [24]. On the other hand, although there is agreement that platforms constitute a relict form of shoreline retreat, there is still debate about the process that is actually responsible for the development of the shoreline. Classically, they have been referred to as abrasion platforms or wave-cut platforms, however, these terms have recently been questioned by some authors, as they refer to processes that may not be responsible for their development. Specifically, Masselink et al. [13] cite the work of Stephenson and Kirk [21] on the northeast coast of the South Island of New Zealand, in which they determined that the energy developed by the waves was insufficient, even during storms, to cut the rock on the scale of the platform. Thus, they attributed the fundamental role as breakers of the rock in certain cases to the weathering processes.

The presence or absence of the platforms, as well as their inclination and position with respect to the tide levels, is an element to be taken into account in the dynamics of rocky coasts, since these factors condition the action of the waves on the coastal front. In any case, most of the research includes these platforms as one of the elements of the cliff system, incorporating their classification, as well as their dynamics and evolution, in that wider study.

15.4.3 Cliff Systems

The cliffs are very steep coastal forms which hang over 40°. The presence of an almost vertical escarpment divides these coasts morphologically into several elements (Fig. 15.10).

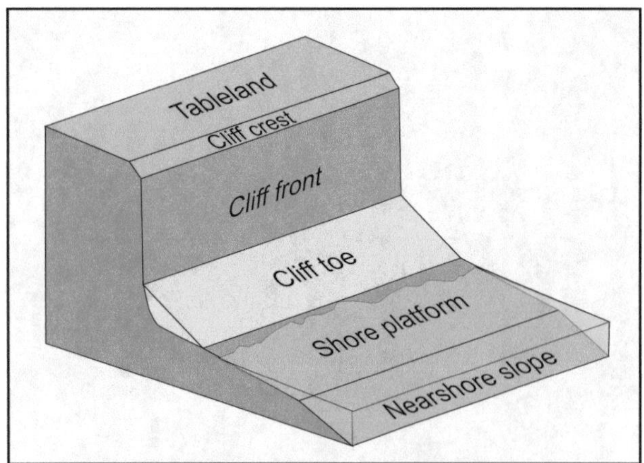

Fig. 15.10 General morphology of a cliff system

The upper part (*cliff top*) has a low-slope morphology, usually called **tableland**. At the marine boundary of this flat surface, the **cliff face** may develop directly, although sometimes there may be a transitional zone of intermediate slope (**cliff crest**). This cliff face is the area that is subject to the most intense erosion and its main tendency is to retreat.

The base of the cliff face can link directly to the **shore platform,** which is a very low-slope area that means the transition to the subtidal zones, where it links to the **nearshore slope**. Between the cliff face and the coastal shelf, a glacier may develop where debris from continental processes accumulates on the cliff face before the waves eventually remove it. This debris accumulation area is called a **cliff toe**. Various types of cliffs can be distinguished according to the presence of these elements of the system and the location of the rocky platform [23].

The **plunging cliffs** have a vertical drop to a platform in the underwater area at several meters' depth (Fig. 15.11). This is typically the case for the more resistant lithologies and the erosion processes are usually very slow, giving the cliff a very low rate of retreat. In these cases, the coastal front is generally quite regular and any irregularities are related to the density of fracturing. The waves break directly onto the cliff face and the direct hydraulic impact is usually accompanied by phenomena related to the action of compressed air. When material collapses from the upper parts, it tends to accumulate below sea level, allowing the waves to continue their action on the cliff face. This continuous action of the waves at the same level develops a linear erosive feature called a **notch**, which is typical of this type of cliff.

The **cliffs with a sloping coastal platform** appear in less resistant lithologies and form coasts with less general slopes. In these cases, a platform develops at the foot of the cliff face, although it is not very long and slopes to the sea by several degrees. In this type of cliff, there is a stronger recession controlled by the slope processes acting on the cliff face. These processes contribute to the development of a cliff toe made up of debris in the intertidal zone (Fig. 15.12).

Wave erosion is based on the destruction and remobilization of these residual deposits. Cliff retreat is controlled by the balance between the capacity of the slope phenomena to generate these deposits and the capacity of the waves to erode them. It is common for erosion to go through a cyclical phenomenon that starts with the collapse of a segment of the front and continues with a long period that the waves need to fragment the blocks and transport them to other areas of the coast.

Although erosion is a process that dominates horizontally, causing the cliff face to recede, erosion also occurs vertically on the platform that is being submerged. Together, the physical processes related to the action of the waves, and

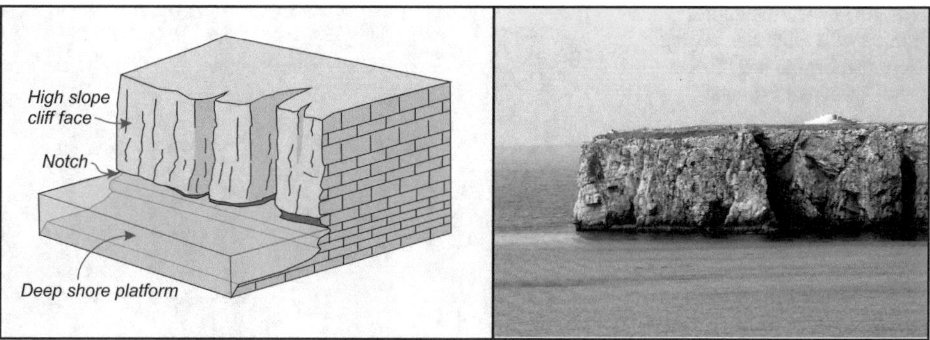

Fig. 15.11 Morphology of a typical plunging cliff, illustrated with the example of Cape São Vincente (Portugal)

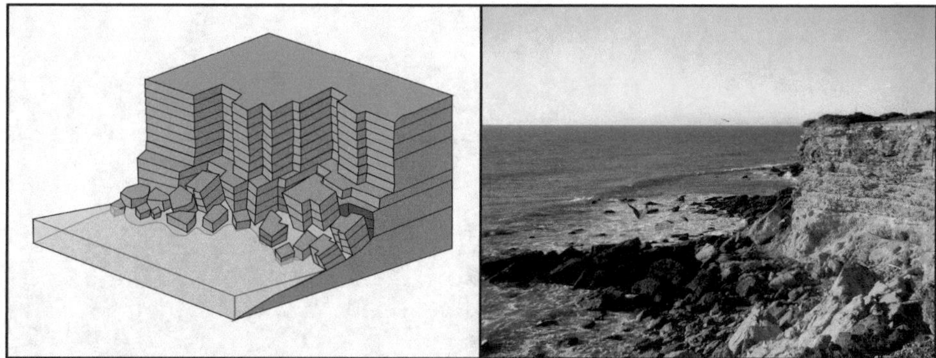

Fig. 15.12 Morphology of a typical cliff with sloping shore platform, illustrated with an example of Chiclana (southern Spain)

the processes of chemical weathering and bio-erosion, make the submerged platform deeper, especially in the sectors closest to the sea. As a result, this platform gains in slope as the cliff retreats.

This type of cliff presents a greater heterogeneity and morphological variety. Lithological alternation is typical, as well as differing sections with different degrees of fracture. Thus, all the features described in Sect. 4.2.1 of this book and illustrated in Fig. 4.2 are associated with these cliffs. These features include *stacks*, *stumps*, *cove beaches*, *caves*, *arches and blowholes* (Fig. 15.13).

Cliffs with a horizontal shore platform develop in even less resistant lithologies and in situations where the retreat of the cliff face is very rapid in relation to the vertical erosion of the shore platform. In these cases, the platform may develop widely in the intertidal zone (Fig. 15.14) to end abruptly in a slope break connecting this platform with the shoreface through a slope. Depending on the wave efficiency and tidal range with respect to the strength of the material, the height of this platform may vary from one system to another depending on its proximity to high or low tide. The presence of an increasingly wide platform causes the waves to dissipate before they reach the front of the cliff, so this type of system usually develops deposits at the base of the cliff. However, during storms, the action of the waves is intensified and these materials can be used as abrasive agents at the base of the cliff.

15.5 Dynamics and Evolution

It is the combination of the processes acting under control factors such as climate in relation to the resistance of the rocks that determines the dynamics and short-term evolution of the rocky coasts and, in particular, of the cliff systems. In addition, other factors such as relative sea level movements control this evolution in the longer term. All attempts by some authors to directly relate the morphology and dynamics of these systems to a single variable have been unsuccessful. For example, attempts to correlate higher wave energy with larger rocky shelf dimensions are not sustained since a large platform will dissipate wave energy before breaking at the base of the cliff (e.g., Stephenson and Kirk [21]). Consequently, any approach to understanding the dynamics of these systems must be made from the point of view of considering multiple variables together.

In any case, the relationship between the short-term dynamics of the cliff system and the following four variables is clear: the resistance of the rocks, the rhythm of the weathering processes, the efficiency of the slope processes and the erosion capacity of the waves, both at the base of the cliff face itself and in the possible deposits generated at the cliff toe (Fig. 15.15).

On the one hand, the strength of the rocks is the result of the combination of lithology, arrangement of the materials

Fig. 15.13 Morphological features of a cliff with sloping shore platform in a cliff near Lagos (southern Portugal)

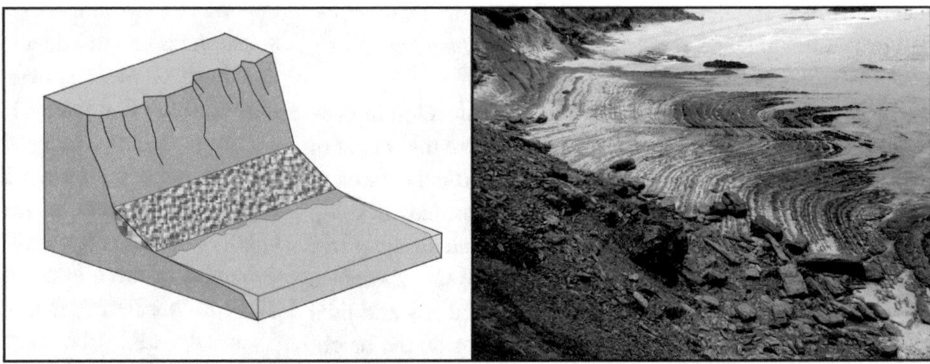

Fig. 15.14 Morphology of a typical cliff with horizontal shore platform, illustrated with the example of Aljezur (southwestern Portugal)

and degree of fracturing. This resistance is what clearly marks the degree of efficiency of the processes, determining which of them can contribute most to the dismantling of the rock at the different possible scales. The result is a different response of each lithology to erosion, resulting in highly variable erosion rates of several orders of magnitude. These rates were summarized by Sunamura [23] and are shown in Table 15.1.

On a millimetric scale, the processes of weathering contribute to the weakening and granular disintegration of the rock, both in the subaerial and submerged parts, including the intertidal strip, which is periodically exposed and

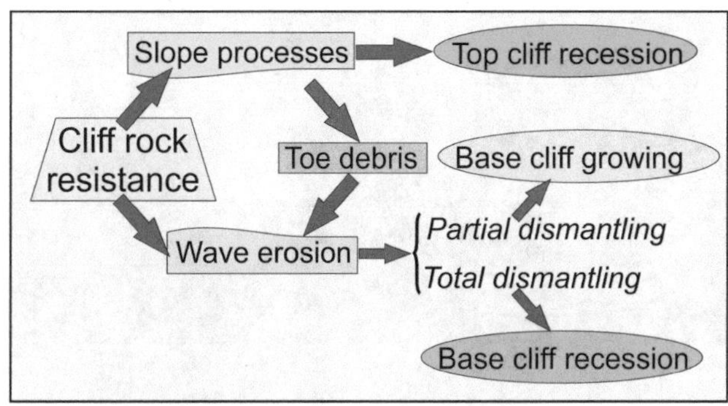

Fig. 15.15 Slope and wave process interactions in cliff systems

submerged. These processes result in micro-erosion that acts on such long time scales that it would be impossible to visualize without the application of precision instruments (e.g., Gómez-Pujol et al. [5]; Moses et al. [15]). In the more resistant lithologies, these seem to be the dominant processes. The granular material generated by these processes is easily removed by the waves. In this case the result is *plunging* cliffs.

On a larger scale, in less resistant lithologies, slope processes combine with wave efficiency to take advantage of the weakening caused by weathering. Slope processes and waves work in different ways. On the one hand, in most cases the slope processes act instantaneously, displacing large volumes of rock in a few minutes, while the rest of the time they remain latent. In between these processes, return periods can be established. The action of the waves is much more continuous, although it is subject to the annual cycles of fairweather waves and storm surges. The volumes of material worked by the slope and wave processes must refer to the same period of time.

This joint action of both types of processes operating at different timescales results in a cyclical model of cliff retreat where three different phases follow each other (Fig. 15.16). In the first phase, the attack of the waves can destabilize the base of the cliff, activating the slope processes. Depending on the nature of the rock, its degree of fracturing and the slope of the rock profile, the slope processes involved may be different, with avalanches, landslides and mass flows acting in the case of unconsolidated lithologies, and rockfall or planar and rotational landslides in the case of coherent rocks. In the second phase, the action of the slope processes produces metric-scale setbacks on the cliff crest and brings down large volumes of rock from the upper reaches to a level where the waves can act on them. In the final phase, the waves are responsible for breaking up and transporting the accumulations generated by the slope processes at the cliff base. The definitive dismantling of the deposits at the foot of the cliff allows the waves to attack the base of the cliff again, thus starting a new cycle of retreat. The duration of each of these cycles depends on the relationships between the nature of the rock and the capacity to transport the waves; the final result is the net retreat of the cliff.

In reality, this conceptual model does not always work this way since, on the one hand, not all slope processes are related to wave destabilization at the base of the cliff and, on the other hand, as shown in the flowchart in Fig. 15.15, there are different possibilities of combination between the rate of action of the slope processes and the rates of wave erosion on the toe deposits. Taking into account the latter consideration, there are three possibilities in terms of the rate of erosion of the base deposits by the waves. These rates affect both the general dynamics of cliff retreat and its slope and morphology [26] (Fig. 15.17).

Under intense erosion, the model expressed in the previous paragraph would be fulfilled. These conditions mean the deposits in the foot are completely dismantled and the retreat is moved upwards through the slope processes. As a consequence, the retreat generates parallel profiles with a uniform slope (Fig. 15.17a). At moderate erosion rates, the foot deposits are only partially dismantled, so that the waves do not cut through and destabilize the base of the cliff. In this case, the slope processes that destabilize the upper part of the cliff are not related to the action of the waves, but to the dynamics of the gravitational processes themselves. Although the upper part of the cliff develops a retreat in the form of parallel profiles, in the lower part there is always a debris deposit (Fig. 15.17b). The third case is that in which the erosion of the base deposits is minimal. As in the second case, slope phenomena occur unrelated to wave action at the base. In this way, the upper part of the cliff suffers a retreat,

Table 15.1 Erosion rates in different lithologies (from Sunamura [23])

Material	Erosion rate (m/year)
Igneous rocks	<0.001
Limestone	0.001–0.01
Flysch and shale	0.01–0.1
Chalk and Cenozoic sedimentary rocks	0.1–1
Pleistocene sedimentary rocks	1–10
Holocene sediments and pyroclastic materials	>10

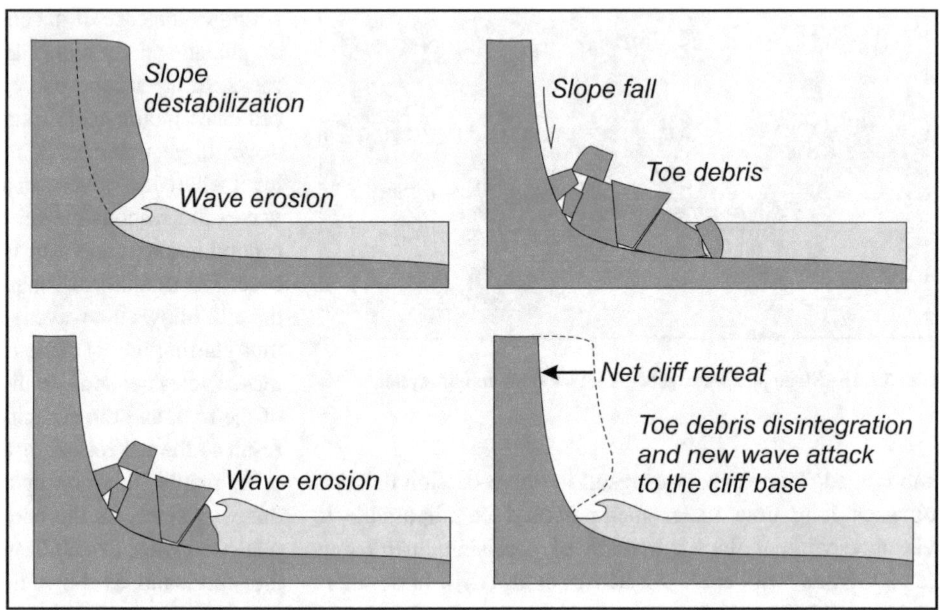

Fig. 15.16 Conceptual model of a cycle of cliff retreat

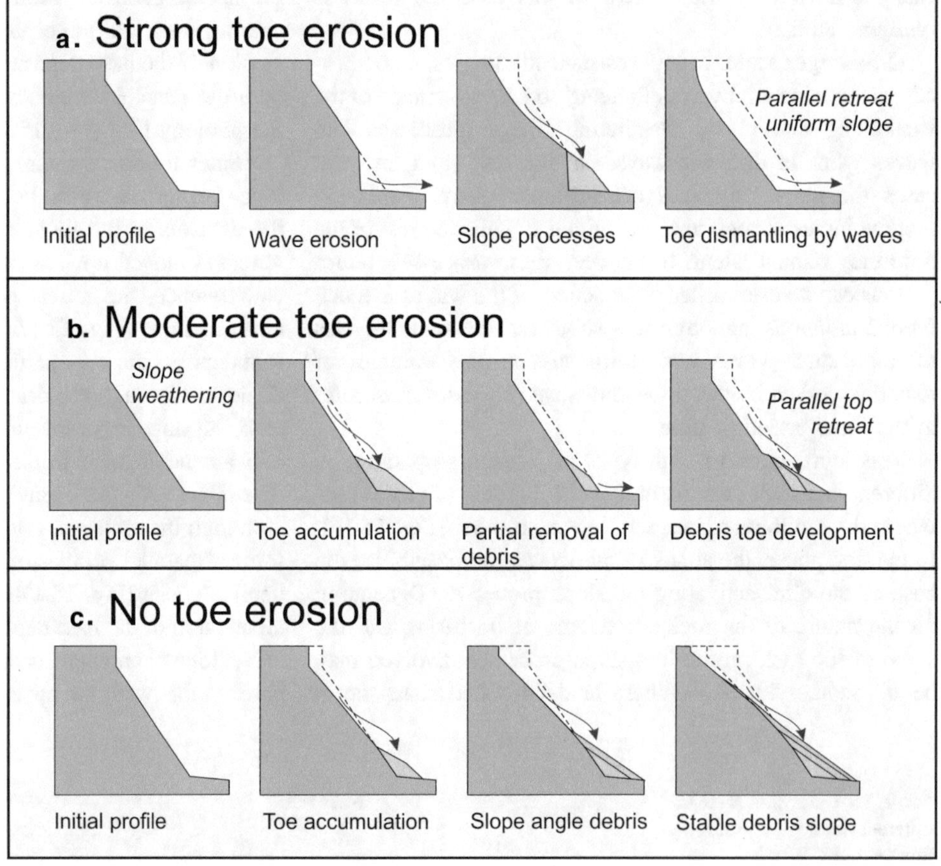

Fig. 15.17 Response models of cliff retreat to toe erosion rate (adapted from Woodroffe [27] and based on Vallejo and Degroot [26])

while at the base the debris accumulates, causing the toe to grow rapidly. The profile of the cliff thus undergoes a rotation until a stable slope is reached (Fig. 15.17c).

In parallel with these erosional processes that act horizontally by causing the cliff face to retreat, others occur that act vertically on the shore platforms [11]. These processes also include weathering (mechanical and chemical), bio-erosion and physical erosion by waves (hydraulic action and abrasion). As on the cliff face, these processes act at different scales and their effect is the loss of height of the platforms. Obviously, the process of vertical erosion on the platforms (*lowering*) is not uniform, but is subject to the same lithological conditioning factors as the cliff face: i.e., the resistance of the material and the factors that determine it (lithology, arrangement of the materials and degree of fracturing).

Among the most visible processes that work on these platforms are the formation and movement of large blocks. Most of the time the blocks are delimited by the presence of discontinuity planes, such as the stratification surface and different families of joints. These blocks can be formed both on the upper surface of the platforms as well as on the submerged front, and can be mobilized during the action of high-energy events such as tsunamis and extreme storms.

The movement of large blocks by tsunamis has been widely documented over the past 20 years (e.g., Schefers and Kelletat [20]; Gracia et al. [8]; Goto et al. [6]; Roig-Munar et al. [19]). According to this work, the blocks can be distributed along the platforms (Fig. 15.18a), be interlocked at the base of the cliff (Fig. 15.18b) or even climb the cliff face and be deposited on the ridge. The movement of blocks due to large storm waves has also been mentioned in the section on processes. These dynamics have been studied, especially in tropical cyclone contexts (e.g., Morton et al. [14]), but also in high latitudes (e.g., Hall et al. [10]; Etienne and Paris [4]). The location of blocks due to storms is so similar to that of tsunamis that there is currently a strong controversy in the literature about the origin of these blocks in some places, often attributed to a mixed origin.

The movement of these blocks on the platforms is episodic. Work by authors such as Nott [18] have tried to establish through formulas the energy and dimensions of the wave (distinguishing between tsunamis and storms) necessary to move the blocks according to their size. Large blocks of several tons can only be displaced in decades or every hundred years. However, smaller blocks can be pulled out and displaced almost every year. The dragging of these blocks has an abrasive effect on the platforms. On many occasions, the displacement followed by these blocks can

Fig. 15.18 Large boulders generated in a cliff front. **a** Boulders displaced over a shore platform (Laghdira, Morocco). **b** Imbricated boulders (Menorca, Spain)

even be distinguished due to the existence of linear scars known as *bruises* [13].

In the long term, the dynamics of the cliff system begins when the sea is positioned at a certain level, remains in that position for a long time and the system begins to retreat. However, the retreat of the cliff face leads to the growth of rocky platforms, and the dissipation of wave energy that takes place on increasingly long platforms means that the slope processes gain importance against the waves in the more evolved stages of the cliff. This trend means that the crest of the cliff begins to retreat more quickly than the base, where deposits that the waves can no longer erode begin to be preserved. The cycle ends when the cliff profile becomes more regular and the base deposits begin to protect the base from extreme wave attack. At this point the cliff becomes inactive.

A continuous movement of sea level rise rejuvenates the erosion processes at the base of the cliff and gives the waves more importance. The result is a much greater retreat, but accompanied by the development of a sloping coastal platform as a consequence of the rising effect of sea level.

Rapid upward or downward movements of sea level interrupted by periods of stability produce a staggered coastline where each stable period leads to the development of a different front and platform at a different level. These steps are known as **coastal terraces** and they can be found both in the emerged area and in the submerged strip of the system [16].

15.6 Associated Deposits

In cliff systems with well-developed shore platforms, there are usually two types of deposits that can be preserved at the base of the system [1]: **cliff-toe deposits** and **cliff-beaches**. These deposits are especially important in cliffs formed in unconsolidated rocks (bluffs).

Toe deposits are of a diverse nature. In consolidated rocks they are colluvial deposits resulting from rockfall and sometimes include large blocks resulting from landslides. In unconsolidated rocks, the slides may be preserved at the base (Fig. 15.19a) and the colluviums may be constituted by sand (Fig. 15.19b); in these lithologies, the cliff base deposits may also develop alluvial fans of different scale (Fig. 15.19c). These deposits, when preserved, usually present important erosive scars from frontal attacks by storm surges.

In the advanced stages of evolution of cliff systems, beaches tend to develop on shore platforms due to increased wave dissipation as these platforms gain width. The volume of material accumulated on these beaches depends directly on the contribution of sediments from the cliff system and on the changes in dynamic conditions generated by factors acting in the long term. Initially, the accumulation of sand on the platforms influences their evolution by accelerating the erosion process, since the beach clasts set in motion by the waves contribute to abrasion. However, when the deposits become thick enough not to move completely during storms,

Fig. 15.19 Examples of toe deposits in a soft cliff. **a** Slide. **b** Colluvial sand deposits. **c** Meter-scale alluvial fan

Fig. 15.20 Examples of cliff-foot beaches. **a** Pebble beach (photograph courtesy of G. Flor-Blanco). **b** Sandy beach in a soft cliff front. **c** Beach of sand covering boulders

they end up protecting the platforms, which eventually stop being eroded. Cliff beaches are governed by the same dynamics that will be studied in Chap. 17 of this book and their deposits have a great similarity; however, there are certain particularities that are present in this type of beach.

Firstly, it should be borne in mind that the supply of material to cliff-front beaches is episodic, as these beaches are normally isolated in circulation cells separated by headlands from the longitudinal transport of the coast. Because of this, these beaches are commonly known as pocket beaches or cove beaches. Due to this isolation, material only reaches these beaches after the action of slope processes or during the dynamic of storms when sediment can be transferred between cells beyond the headlands. Consequently, these beaches suffer from strong variations in sediment volume [12]. The behavior of these beaches has an intimate circular relationship with the retreat of the cliff face. When the cliff face suffers an erosion process, the beach volume suddenly increases, increasing the dissipation of wave energy and protecting the cliff face from further recessions. The material of this beach can be slowly eroded in the subsequent period.

With respect to the sedimentary characteristics of these deposits, it should be considered that they can be either pebble beaches (Fig. 15.20a) or sandy beaches (Fig. 15.20 b). The dynamics differ slightly in both cases, since the energy required to move the pebbles is greater than in the case of sand. The slope of the beach and the internal structure of the sediment are also different. The pebbles are usually very rounded and, if they are discoidal, they usually show overlaps towards the sea without any other type of internal arrangement being observed. On the other hand, the sand usually shows characteristic cross-layering, although this internal structure is different if the beach is dissipative or reflective. These differences are identical to the beaches of other locations and will be studied in the chapter dedicated to beaches. A particularity is the presence of blocks inside the sediments (Fig. 15.20c). These blocks are the same as those studied in shore platforms and are encompassed by beach deposits when they develop there.

References

1. Anthony EJ (2009) Shore processes and their palaeoenvironmental applications. Elsevier, Amsterdam, 519pp
2. Davis RA Jr, FitzGerald DM (2004) Beaches and coasts. Blackwell Publishing, Oxford, 419pp
3. Emery KO, Kuhn GG (1982) Sea cliffs: their processes, profiles and classification. Geol Soc Am Bull 93:644–654
4. Etienne S, Paris R (2010) Boulder accumulation related to storms on the south coast of the Reykjanes Peninsula (Iceland). Geomorphology 114:55–70
5. Gómez-Pujol L, Fornós JJ, Swantesson JOH (2006) Rock surface millimetre-scale roughness and weathering of supratidal Mallorcan carbonate coasts (Balearic Islands). Earth Surf Proc Land 31:1792–1801
6. Goto K, Kawana T, Inamura F (2010) Historical and geological evidence of boulders deposited by tsunamis, southern Ryukyu Islands, Japan. Earth-Sci Rev 102:77–99
7. Goudie A (2004) Encyclopedia of geomorphology. Routledge, London, 1200pp
8. Gracia FJ, Alonso C, Benavente J, Anfuso G, Del Río L (2006) The different coastal records of the 1755 tsunami waves along the South Atlantic Spanish coast. Zeitschrift für Geomorphologie Suppl 146:195–220
9. Hall AM, Hansom JD, Williams DM, Jarvis J (2006) Distribution, geomorphology, and lithofacies of cliff-top storm deposits: examples from the high-energy coasts of Scotland and Ireland. Mar Geol 232:131–155
10. Johnson DW (1919) Shore processes and Shoreline development. Wiley and Sons, New York, 584pp
11. Lange WP, Moon VG (2005) Estimating long-term cliff recession rates from shore platform widths. Eng Geol 80:292–301
12. Lee EM, Brunsden D (2001) Sediment budget analysis for coastal management, West Dorset. In: Griffiths JS (ed) Land surface evaluation for engineering practice, vol 18. Geological Society, Engineering Group Special Publication, pp 181–187
13. Masselink G, Hughes MG, Knight J (2003) Introduction to coastal processes and geomorphology. Routledge, London, 416pp
14. Morton RA, Richmond BM, Jaffe BE, Gelfenbaum G (2006) Reconnaissance investigation of Caribbean extreme wave deposits: preliminary observations, interpretations, and research direction. Open-File Report 1293, USGS, 46pp
15. Moses C, Robinson D, Barlow J (2014) Methods for measuring rock surface weathering and erosion: a critical review. Earth Sci Rev 135:141–161
16. Muhs DR, Rockwell TK, Kennedy GL (1992) Late quaternary uplift rates of marine terraces on the Pacific coast of North America, southern Oregon to Baja California Sur. Quatern Int 15:121–133
17. Naylor NA, Stephenson WJ (2010) On the role of discontinuities in mediating shore platform erosion. Geomorphology 14:89–100
18. Nott J (2003) Waves, coastal boulder deposits and the importance of the pretransport setting. Earth Planet Sci Lett 210:269–276
19. Roig-Munar FX, Rodríguez-Perea A, Vilaplana JM, Martín-Prieto JA, Gelabert B (2019) Tsunami boulders in Majorca Island (Balearic Islands, Spain). Geomorphology 334:76–90
20. Schefers A, Kelletat D (2003) Sedimentologic and geomorphic tsunami imprints worldwide: a review. Earth Sci Rev 63:83–92
21. Stephenson WJ, Kirk RM (2000) Development of shore platforms on Kaikoura Peninsula, South Island, New Zealand: part one: the role of waves. Geomorphology 32:21–41
22. Stephenson WJ, Dickson ME, Trenhaile AS (2013) Chapter 10.11, Rock coasts. In: Sherman DJ (ed) Treatise on geomorphology, vol 10: coastal geomorphology. Academic Press, Cambridge, pp 289–307
23. Sunamura T (1992) Geomorphology of rocky coasts. Wiley, New York, 302pp
24. Trenhaile AS (1987) The geomorphology of rock coasts. Oxford University Press, Oxford, 384pp
25. Trenhaile AS, Kanyaya JI (2007) The role of wave erosion on sloping and horizontal shore platforms in macro- and mesotidal environments. J Coastal Res 23:298–309
26. Vallejo LE, Degroot R (1988) Bluff response to wave action. Eng Geol 26:1–16
27. Woodroffe CD (2002) Coasts: form, process and evolution. Cambridge University Press, Cambridge, 623pp

16 Wave-Dominated Systems I: Barriers and Barrier Islands

16.1 Introduction

Barriers in general, and barrier islands in particular, constitute a significant system between coastal environments. They are sedimentary bodies made up mainly of sand, built by the action of the waves on the front of the mainland and separated from it by a body of water called a lagoon (e.g., [16, 29]). The sedimentary building of the barriers can take place on the rocky substrate of the mainland or on other older coastal formations. The name barrier is based on the fact that the presence of these sand bodies protects the mainland from the direct action of sea waves (Fig. 16.1).

Barrier systems are distributed along the world's coasts and are not limited to a particular climatic or tectonic context. Since Davies [4], most research has stated that these systems constitute about 15% of the coasts at a global level, with barrier islands being the most frequent systems among them, occupying 10% of the world's coasts [27]. They are usually associated with wave-dominated coasts where the budget of sediment is abundant, giving the coast a depositional character. These conditions often appear in passive tectonic contexts and during periods in which the sea level is in a stable position for a more or less prolonged time. Thus, barriers usually appear in the form of chains of islands that develop parallel to the coast, but are also associated with other well-fed environments such as the front of deltas and estuaries.

The dimensions of these systems are highly variable. There are barriers of hundreds of kilometers in length, while others do not exceed hundreds of meters. Something similar occurs with the width. We can find islands of only tens of meters between the coastline and the lagoon, and others with widths of kilometers. The barrier building may be completely flat or rise a few meters above the average water level when the action of the wind develops dune systems on the island. The best-known examples of longer and wider barriers are located in the 600 km-long sandbanks of the Patos-Mirim system, Brazil [28]. Smaller barriers are commonly found by closing bays located on rocky coasts. Examples of such barriers exist along the coasts of all mainlands (e.g., Dobrovolsky and Zalogin [2, 8, 25, 26]).

Due to the not very pronounced relief, the beaches in the front zone and the attractive landscape make these systems very busy. Much of the world's tourist industry is located on barriers and important port facilities are often located in the rear area, so that the land usually has a high price for owners on these barrier islands. On the other hand, those barriers that have preserved their natural character constitute ecosystems of high ecological value. However, the highly dynamic regime imposed by geological processes makes them extremely changeable environments, which can migrate or be eroded continuously or in moments. In addition, the lack of significant relief makes them susceptible to high-energy events such as storms and tsunamis. Therefore, the occupation of these systems always carries a high risk, especially in a realm of global climate change. In order to minimize the impacts of the processes that occur in these systems on the human activities that are based on them, it is important to understand their dynamic functioning.

From a geological point of view, the sedimentary sequences of the barriers as preserved in the stratigraphic record are of particular interest, because they often constitute the storage rock of hydrocarbon deposits [11].

16.2 Control Factors

Barriers are open systems in which there is a clear dominance of waves; however, there is also this dominance on the rocky shores. Why, then, this difference? The answer lies in the influence of other factors. In general terms, the main controls on the origin and dynamics of wave-dominated environments are found in the interrelationships of wave action with the morphology of the seabed, the geometry of the coastline and, above all, sedimentary input [1, 24]. In this case, it is a positive sedimentary supply, at least in the origin of the sedimentary bodies that constitute the barrier. In the long term, other factors such as tectonic context, relative

J. A. Morales, *Coastal Geology*, Springer Textbooks in Earth Sciences, Geography and Environment,
https://doi.org/10.1007/978-3-030-96121-3_16

Fig. 16.1 Aerial view of a sandy barrier in New Jersey (USA)

sea level movements and climate also have a significant influence. Let's discuss what these factors are and how they intervene in the sedimentary dynamics.

16.2.1 Wave Action

The arrival of the waves at the coast is characterized by an interaction with the bed from a depth equal to half the wavelength. This depth was defined in Chap. 7 as the wave base level. At shallower depths, waves interact with the bed, where they are not only deformed in the process described as shoaling, but some of their energy is transferred to the bed and used to transport grains of sedimentary material. The principles governing the transport of sedimentary material by waves were described in Chap. 14. On the one hand, the transverse component of the wave approach vector describes a movement of particles from and to the shore. On the other hand, if the wave arrival is oblique there will also be a longshore transport component which is also known as littoral drift.

From a transverse approach to the shoreline, the orbital speed of each wave generates a movement of the particles on the bed. However, as a wave is an oscillatory movement, the net transport of particles to and from the shore is minimized, except in certain situations. One such situation occurs when the wave orbits are asymmetrical. We have seen in Chap. 14 that the thresholds of grain movement are exceeded in different ways in the swash and backwash movements depending on the type of wave breaker. According to these data, the less energetic breakers will produce a net transport of the sediment landwards, while the more energetic breakers will take it in the opposite direction. Clearly, cycles will occur over time in which one or other direction of net transport will dominate, but, in general, for a barrier to originate the landward transport must dominate. Once a barrier is formed, if the long-term net transport is seawards, it will be eroded and, if this period is long enough, it can be completely dismantled.

The littoral drift transport component acts longitudinally to the coast. This component is fundamental in providing material to the barrier from other stretches of coast connected to the cell where the barrier originates. Littoral drift works like a current and acts in the long term to generate a volume of sediment transport that can be calculated (see Advanced box 14.3). It should be noted that most barriers have been formed by longitudinal growth, which gives an idea of the importance of longshore transport in the genesis and dynamics of barriers. In addition, in the case of barrier islands, the existence of littoral drift is usually responsible for the lateral migration of the channels, which ultimately means migration of the islands themselves.

16.2.2 Tides

The tidal range determines the amplitude of the oscillation of the wave action in the coastal fringe. However, since barriers separate an inland body of water from the open coast, the tidal range, together with the surface of that water body, also determines the volume of water that this restricted water body exchanges with the open coastal waters. This volume is known as the tidal prism and is responsible for keeping an open communication between the two water bodies. The speed of the currents that develop in the passages between islands, and the dimensions of the tidal deltas that form at both ends of these inlets, depend on the tidal prism.

In the frontal part of the system, the tidal currents have less influence. The currents tend to be parallel to the coast in these sectors and have very little capacity to transport sediments. Although the tides also have an oscillatory character, the ebb and flow cycles are usually asymmetric, resulting in a net transport in one of the two directions. The longitudinal action of these tidal currents usually combines with coastal drift and other currents to give a resultant transport parallel to the coast.

16.2.3 Other Currents

The action of wind-induced currents in shoreface areas has been extensively described in Chap. 10. These currents have a weak character and their sediment transport capacity is not very high. However, their combined action with tidal currents has a significant influence on the transverse and longitudinal components of waves. Under these currents, the oscillation of the waves can become asymmetric, generating a net transport component from or to the barrier front. This component can contribute to the construction of the barrier or, conversely, to its dismantling.

16.2.4 Sediment Supply

In the introduction to this chapter, it was noted that a barrier system corresponds to a depositional coast, and therefore the sedimentary balance must be positive, at least in the origin of the barrier. Most of the barriers that exist today are made up of sand, although a small but significant number have developed in gravel. These gravel barriers are concentrated in regions of high clast availability, generally in areas that were covered by glaciers during the Pleistocene [21]. Most of this sediment comes from adjacent mainland areas. The longest barrier island chains in the world developed along the coasts of land areas whose geological nature allows a rapid dismantling, generating high rates of supply [6]. At other times, it is the relict coastal sediments that are reworked in continuous sedimentary cycles. Thus, in situations of limited input, it is the destruction of old barriers that provides the sediment to build new ones.

Although a positive sedimentary balance underlies the origin of the barriers, throughout the life of the barrier this balance can be altered and modified by the factors that have been described in the previous sections. There can be moments of negative sedimentary balance with the consequent erosion of the barrier or part of it. In these variations of the sedimentary balance during the life of the barrier, it is obvious that variations in the wave energy will play an important role; however, modifications in the sedimentary supply that occur at the mesoscale also exert a remarkable influence, although this is frequently ignored in many studies [3].

16.2.5 Mainland Physiography

The physiography of the mainland plays a crucial role in the development of the first moments of the barriers, as well as in the dynamics that develop during their lifetime. Perhaps the clearest example visually is the contrast between the presence of an initially straight coast and an intricate coastline. Both physiographies would generate barriers of different length, orientation and dynamics. However, coastal physiography also shows its influence at other levels, such as the type of interaction that coastal morphology has with the arrival of wave trains. From this point of view, a distinction can be made between coasts that are oriented parallel to the wave ridges and coasts where the arrival is oblique. In the former, the genesis of the barrier would be related to the action of the waves in a transverse direction to the coast, while in the latter the coastal drift would be the factor that plays the most important role in the development of the barrier. Davies [4] called these cases **swash-aligned barriers** and **drift-aligned barriers**. Particular cases of these situations can be generated in irregular coasts in relation to the phenomena of refraction and diffraction.

Another effect that can be understood more intuitively is the variation in depth on the waterfront. Bathymetric differences influence the way wave energy is dissipated. Thus, the slope of the nearshore plays a major role in the type of breaker, along with the dimensions of the wave. There are numerous studies that show the relationship between the position of the barriers and the existence of banks or steps in the waterfront (e.g., Davis and Clifton [5]). This relationship comes about because the genesis of the barrier is related to the descent of energy that occurs at this step, since it is there that the wave loses its transport capacity and deposits the sand, generating the nucleus of the barrier.

16.2.6 Sea Level Movements

The current barriers were established in the period of *highstand* that followed the Flandrian transgression. In general terms, stillstand periods favor the formation of coastal sedimentary systems in optimum supply situations, as they give the systems the necessary time for the dynamic processes to accumulate the material and build up the sedimentary bodies, even in limited input situations. If the budget is sufficient, once the barrier is stabilized there will be a progression of growth that will lead to a shift of the coastline towards the sea, which means a regression.

In the event of a drop in sea level, the model would also be regressive, although the trend would be for coastal systems to be turned inland. Normally, mainland processes would lead to the total or partial dismantling of these formations. Only early consolidation would contribute to the preservation of the sedimentary record of these coastal sequences.

In any case, many of the current studies of barrier systems show that many of them were generated before the time of sea level stabilization and that they migrated to the mainland in the last centuries of slow sea level rise (e.g., [22, 23]). In this context, the process of construction of the barrier island depends on the slope of the coastline in relation to the rate of sea level rise, the input regime and the capacity of the waves to rework sediment. Depending on the combination of these factors, either a part of the sediment of the old barrier or the whole of it will be used to build the new barrier, giving rise to the processes of translation (*roll-over*) or excess (*overstepping)*. This fact influences the preservation potential of the sedimentary sequences of these systems.

16.2.7 Climate

The weather exerts control over some processes that act on the short-term dynamics of the barrier, such as the wave regime and the frequency of storms. In addition, the climatic influence is manifested in the amount of contribution that the rivers are capable of making to the sedimentary systems of the coast. At the same time, the climate also influences some of the processes that occur in the protected environments associated with the barrier. In this sense, it is the climate that marks whether salt marshes (temperate climates), mangroves (tropical climates) or coastal sabkhas (arid climates) develop in supratidal areas.

A less obvious effect of climate is the influence on the nature of the sedimentary input. For example, in tropical climates the content of carbonate clasts generated in the sea basin itself may be more abundant than the terrestrial clasts contributed from the mainland. Similarly, in high latitude barriers the presence of clasts from ancient glacial activity is much higher [6].

In the longer term, and at a global level, changes in the climate will determine the dynamics of the barriers. A change in the climate regime will have an influence on the number and intensity of storms, having a direct consequence on the erosional–depositional character of the system. In the even longer term, it is changes in climate that induce the eustatic movements that control the relative movements of the sea level described in the previous section.

16.3 General Morphology and Associated Environments

The barriers can be classified into three types according to their connection to the mainland (Fig. 16.2). If they are connected to the mainland at two points and completely isolate the lake, they are called **welded barriers** or **bay barriers**. If they are connected to the mainland at one end only, they are called **barrier spits**. In this case, there is a communication between the lagoon and the open sea through a passage channel. Finally, if they appear totally isolated from the mainland, they are called **barrier islands**. Barrier islands usually form continuous chains in which each island is separated from the next by a tidal channel. The action of the waves is concentrated only at the front, while the rear area is dominated by the tide.

The barrier is a sedimentary body that generates a system composed of different sedimentary environments (Fig. 16.3). The presence of the barrier divides the system into three domains that differ in the presence or absence of wave action: the barrier body, the frontal zone (**barrier front**) and the **back barrier** area.

In each of these areas, there is are associated sedimentary environments:

- The front area is made up of **beaches**.
- The **body of the barrier** is usually constituted by sand ridges developed on ancient beaches.
- **Dunes** can develop over the body of the barrier. Occasionally dunes are absent.
- The **lagoon** is located in the area of back barrier, which is a body of shallow water with a restricted connection to the sea. The term is applied to a subtidal zone semi-enclosed by the barrier. The intertidal areas of the lagoon, both attached to the mainland and to the barrier, develop **tidal flats** and the supratidal areas may be occupied by **salt marshes**, **mangroves** or **coastal sabkhas**, depending on the climate in which the barrier is framed.
- Other very frequent structures in the back barrier area are **washover fans**. These are sandy fan-shaped formations with their convexity pointed towards the interior of the lagoon. They are produced during storms which erode the

Fig. 16.2 Different type of barriers. **a**: Welded barrier (Norton Sound, Alaska). **b**: Barrier spit (Mutadua, Mozambique). **c**: Barrier island (Olga Bay, Russia). (Images Landsat/Copernicus from Google Earth.)

front sandy body by opening a channel between the dunes and depositing the eroded sediment at the back.

- In the case of spits and barrier islands, the system is completed with an **inlet** channel. This is a narrow environment subjected to very high tidal energy. This means there is a continuous remobilization of the sedimentary material, which is deposited at its end in both directions of the tidal current forming delta-shaped sedimentary bodies. The external zone (**ebb-tidal delta**) has a very active dynamic due to the interaction of the current with the waves, while the internal zone (**flood-tidal delta**) is fundamentally subject to the tides.

The type of barrier, the relationship between its width and length, and the number of inlets in the case of barrier islands, are regulated by the balance between wave and tidal energy. The classification proposed by Hayes [13], which was described in Chap. 4, differentiates three types of coasts: wave-dominated, tide-dominated and mixed-energy. Of these three types, barrier island systems may appear associated

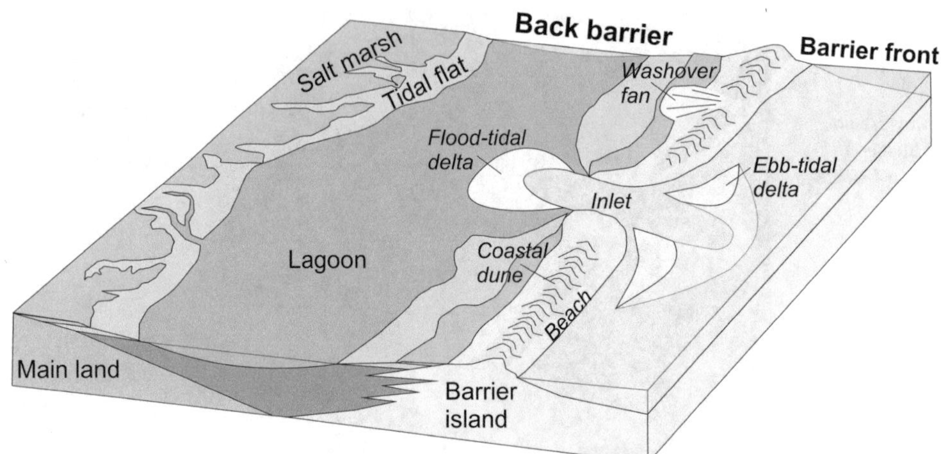

Fig. 16.3 Sedimentary environments constituting a typical barrier system. (adapted from [20])

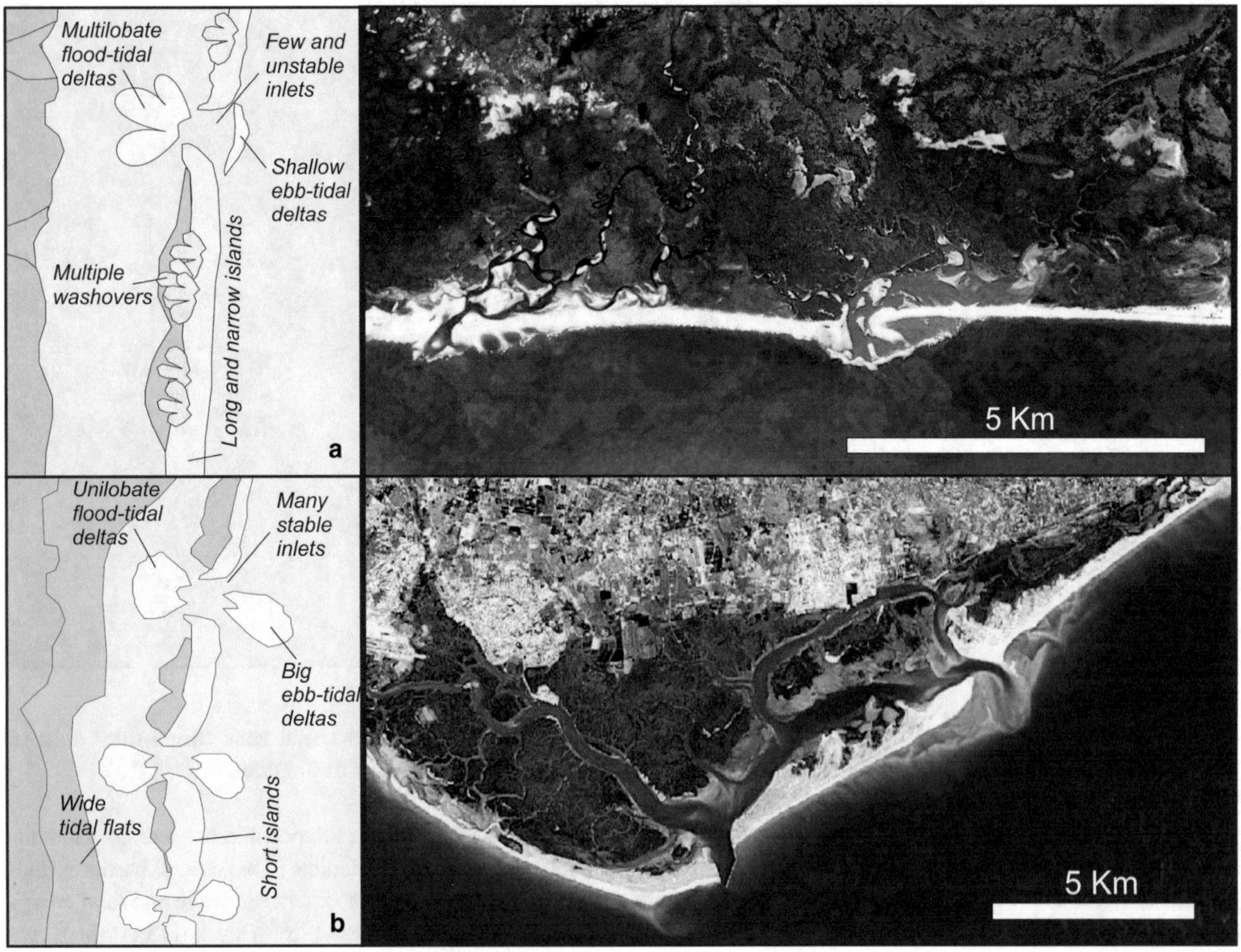

Fig. 16.4 Different types of barrier island systems. (adapted from [13]). **a** Barrier island system of a wave-dominated coast, with the example of Machese Island (Mozambique). **b** Barrier island system of a mixed energy coast, with the example of Ria Formosa (Portugal). (Images Landsat/Copernicus from Google Earth.)

with wave-dominated and mixed-energy, although the barriers would have different characteristics (Fig. 16.4).

- *Wave-dominated coastal barriers* (Fig. 16.4a): These are long, narrow barrier islands, part of large island chains with a low number of *inlets*. The position of the inlets is unstable, as they are affected by frequent migrations and avlusions. Ebb-tidal deltas tend to be small, while flood-tidal deltas are better developed and multilobed. Washover fans are abundant. Smaller-scale estuaries are common in the back-barrier lagoon. These morphologies are related to high wave energy in the face of a low tidal range or in systems where the lagoon is highly clogged and the tidal prism is small.
- *Mixed-energy coastal barriers* (Fig. 16.4b): These are shorter and wider barrier islands separated by a large number of *inlets*. The channels are more stable and tidal reflux deltas with large frontal lobes develop at the marine end. The flood-tidal deltas are unilobed and have large ramps. In the back barrier areas, the lagoons usually have a high rate of clogging, with the tidal flats and supratidal areas being very well developed. Barrier islands acquire this morphology when the energy of the waves is compensated for by a greater tidal range, or where the lagoon is large and a major tidal prism develops. They are also usually associated with the frontal area of estuaries and deltas located on coasts with strong wave influence.

16.4 Genesis, Dynamics and Evolution

The morpho-sedimentary dynamics of the barriers depends largely on the dimensions of the system in relation to wave energy and tidal currents. However, the processes are also strongly related to the type of barrier, so that, broadly speaking, each type of barrier has its own genetic and dynamic characteristics. These characteristics will be described for each type.

16.4.1 Spits

On coasts with abundant input and a strong component of transport by coastal drift, coastal spits can develop. When the coast is very irregular, some spits are connected to coastal headlands where wave diffraction occurs (Fig. 16.5 a). On the other hand, any coastal inlet also tends to develop a spit that partially closes it (Fig. 16.5b). Spits functioning as closure systems are usually associated with large bays, estuaries or estuarine lagoons. In both cases, the tidal prism is important enough to generate currents that keep the inlet generated at the end of the spit open. The action of the tide when passing through these passages usually develops tidal deltas, too.

The genesis and dynamics of spits are completely controlled by coastal drift, but a good sedimentary budget is also necessary to maintain their longitudinal growth rate. Growth is carried out by attaching hook-shaped bars to the curved end of the spit. These bars constitute the continuation of the bars that the waves make as they migrate along the foreshore of the open barrier face (Fig. 16.6). In a spit, the attached bars leave growth marks in the form of berms, so that each berm marks old positions at the spit apex.

In semi-enclosed systems, at the apex of the spit there is usually an inlet that feeds seawater to the lagoon located inside. This inlet is normally associated with a system of tidal deltas whose operation is similar to that located in any inlet between barrier islands. When the inlet and the tidal deltas are present, the curved bars make the spit grow over the facies of this deeper system [18].

16.4.2 Welded Barriers

The presence of welded barriers is linked to systems with small tidal prisms but a very energetic wave regime, although they are also present in systems with a larger tidal prism. In any case, in this type of system, the energy of the waves always dominates over the tides. Many of these barriers were built by the waves closing rocky coastal bays with the sediment eroded in the coastal headlands. Therefore, these barriers are usually short and their sediment is continuously recycled by being inside coastal transport cells that have a limited connection with adjacent cells. The sediment that constitutes them is normally sandy, although this type of barrier has occasionally presented with gravel. The small tidal prism is not capable of generating tidal currents strong enough to keep a permanently open inlet. The isolation of the lake at the back of the barrier from marine waters means that it usually has fresh or brackish water, as it receives water directly from the mainland.

Being in a closed cell, the welded barrier actually dominates the transverse dynamics of the sediment versus the longitudinal component. Due to this fact, the genesis of the barrier is not usually linked to longshore transport, but to the front attachment of wave bars (Fig. 16.7a). In this growth model, under some circumstances there may be a certain longitudinal transport of the sediment that causes it to move from one end of the barrier to the other. If the bay occurs on a coast with a strong coastal drift, then a spit may form at one end of the bay that may end up closing the bay

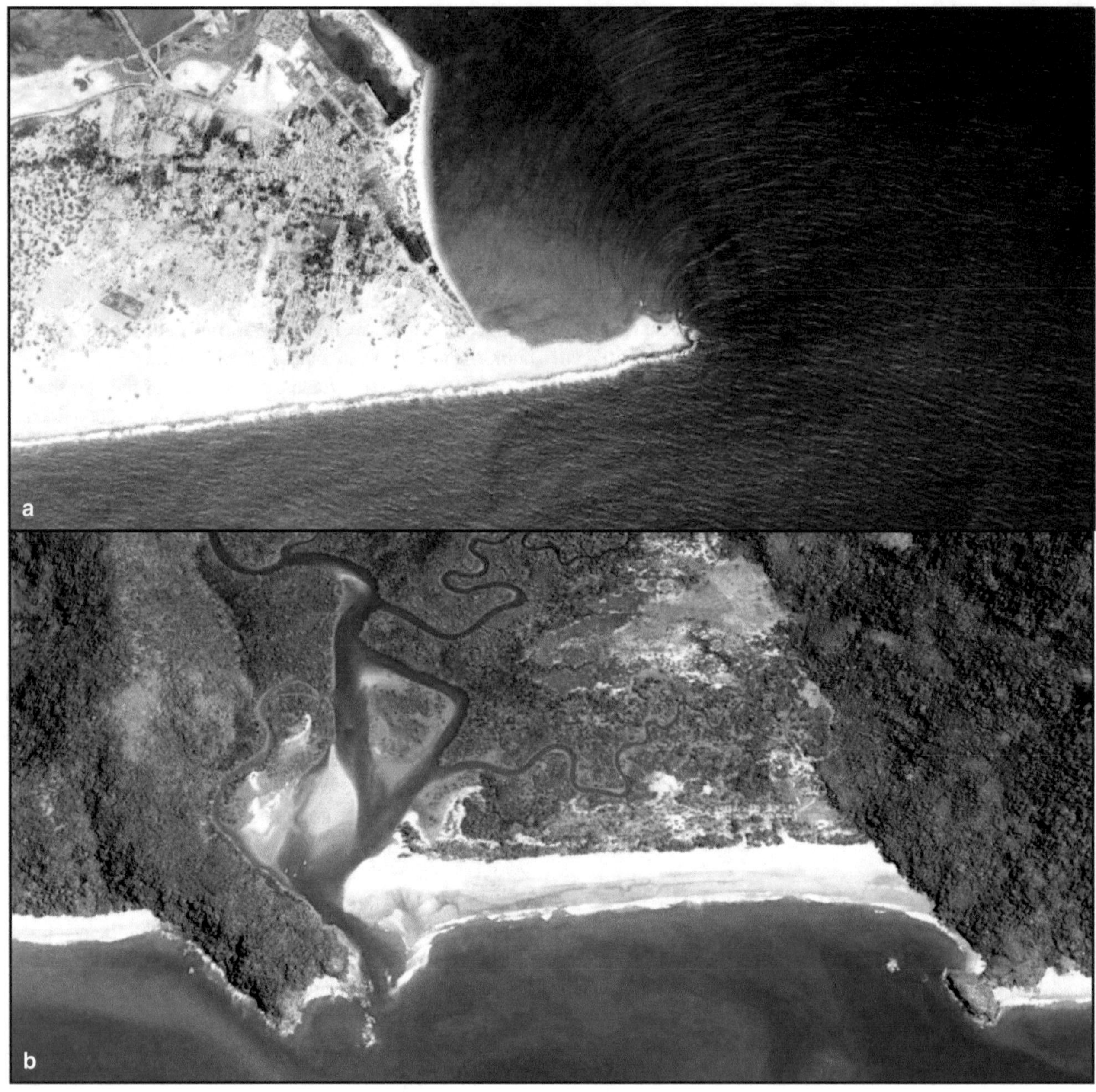

Fig. 16.5 Different locations of spits. **a** Spit formed by diffracting waves in the apex of a headland in Ervadi, (India). **b** Spit closing the Mayan estuary (Myanmar). (Images Landsat/Copernicus from Google Earth.)

completely (Fig. 16.7b). In closed cells, if the arrival of the waves is frontal, the barrier can be subject to a double drift. In that case, two spits will begin to grow at both ends of the bay in opposite directions. Then, the barrier is formed when both spits meet in the center (Fig. 16.7c).

Once the barrier is formed, if the lithology is sandy, sometimes storms can break it completely or create a breach in some sectors. On other occasions, it is the flow of the river system connected to the lagoon that breaks the barrier from the inside out. In both cases, during these moments a connection is formed between the lagoon and the open coast (Fig. 16.8). When this happens, the post-storm situation tends to rebuild the barrier quickly, since most of the sediment is retained within the cell. In situations where there is an ephemeral inlet, before it closes it can migrate in either direction of the barrier if there is a certain longitudinal component of the wave.

In barriers made of gravel, storms are not usually erosional, but it is during the moments of greatest energy when bars can be built that can make the barrier progress.

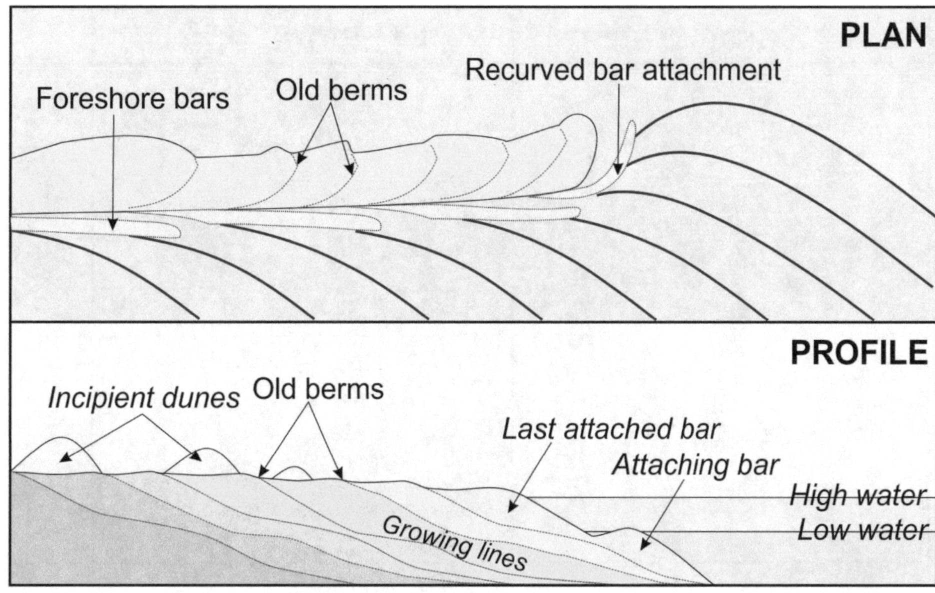

Fig. 16.6 Scheme showing the process of apical growth of a spit

16.4.3 Tombolos

The genesis of a tombolo is related to the presence of a hard element in the coastal front. This element is normally natural and constituted by a rocky outcrop, but it can also be artificial. The interaction of the emerged hard element with the dominant wave trains induces the accumulation of one or two sediment bodies which connect the coastline with this element. There can be three different processes in building these sediment bodies (Fig. 16.9). Some of these are similar to those described in welded barriers. They can occur at different scales.

- *Double drift*: The presence of the hard element refracts the wave trains, inversing the littoral drift. Two converging longshore transports tend to accumulate sediment after the obstacle, creating a prograding band which finally reaches the hard element (Fig. 16.9a).
- *Bar migration*: In a well-supplied coast, the deformation of the wave trains tends to build bars migrating landwards. These bars eventually emerge, forming barriers on both sides of the obstacle. Usually, this process builds a double tombolo (Fig. 16.9b).
- *Spit growing*: On an irregular coast with a strong littoral drift, the apical growth of a spit can link the mainland with the hard element (Fig. 16.9c)

16.4.4 Barrier Islands

More than a century of research has contributed to today's good understanding of barrier island systems. Throughout the twentieth century, great coastal geologists from Douglas Johnson to Richard Davis built their science on the studies of the second half of the previous century, making the genesis and evolution of these systems well known. From the middle to the end of the nineteenth century, three theories for their genesis had been proposed. Each of them tried to explain the origin of all the barrier islands. However, all these theories were put forward in a conceptual way and without the support of experimental data. Later, throughout the twentieth century, some of these hypotheses were demonstrated by other researchers in individual study cases. Today it is accepted that none of the three theories is valid for all systems; however, there are examples of barrier island systems around the world that were generated in all three ways. These theories are: (1) construction of the island on an offshore bar; (2) breaching of a previous coastal spit; and (3) separation from the mainland by flooding due to sea level rise. These genesis models will be described below.

16.4.4.1 Building on an Offshore bar

This theory was initially stated by de Beaumont in 1845 [7] and was supported by Douglas Johnson's field data in 1919 [15]. According to this concept, the shoaling of waves on an initial coastal slope could generate an underwater bar at the place where the waves begin to break. This process would be favored by the presence of a submerged step or relief that would facilitate the dissipation of energy. Once the bar has formed, its presence would make the waves lose energy, depositing more sediment and making the bar grow. Finally, this bar would emerge, preventing the passage of the wave towards the mainland and protecting its rear zone (Fig. 16.10).

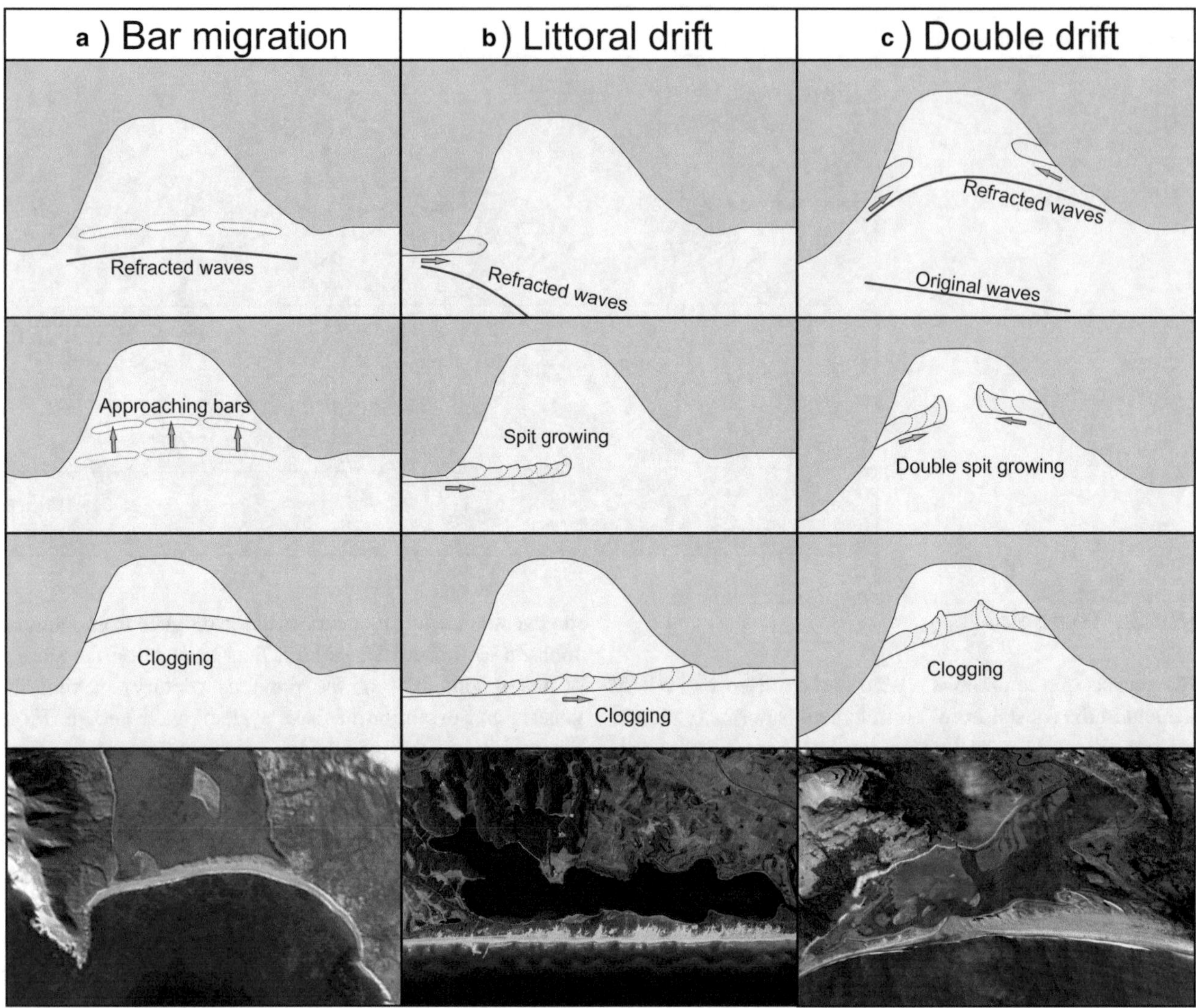

Fig. 16.7 Scheme showing the different possibilities of welded barrier genesis. **a** Transversal bar migration. **b** Littoral drift (spit elongation). **c** Double littoral drift (spit and counterspit attachment)

16.4.4.2 Breaching of a Spit

As a counterpoint to the theory of offshore bars, in 1885 Grove Gilbert [12] suggested that island chains would have generated through the formation of spits. Originally, longitudinal accretion would have occurred through the action of a major littoral drift. In a second stage, a long spit would have fragmented during storms, opening inlets that would become permanent and separate the different islands (Fig. 16.11). In 1968, John Fisher's sedimentological studies in North Carolina made him a great advocate of this theory [9].

16.4.4.3 Detachment from Mainland by Submergence

This theory was conceived by William McGee in 1890 [17]. According to his reasoning, the rise in sea level that occurred during the Holocene transgression would invade some valleys parallel to the coast, isolating coastal sandy formations that were previously attached to the mainland (Fig. 16.12). In 1967, John Hoyt [14] developed this theory, extensively basing his arguments on the fact that, in some coastal systems, there was no sedimentary record under the lagoon that would demonstrate the existence of an open-water situation, such as that expected at a stage prior to the existence of the barrier island. For Hoyt, the width of the lagoon was simply the result of the general slope of the coast, so that on steeper slopes the sand ridge would develop nearer the mainland, leaving no space for a wide lagoon at the time of the marine invasion. Conversely, a coastline with a low slope would allow a wide marine invasion, with the consequent development of a wide lagoon.

Regardless of the way in which the barrier island systems have emerged, the dynamics of all of them are very similar,

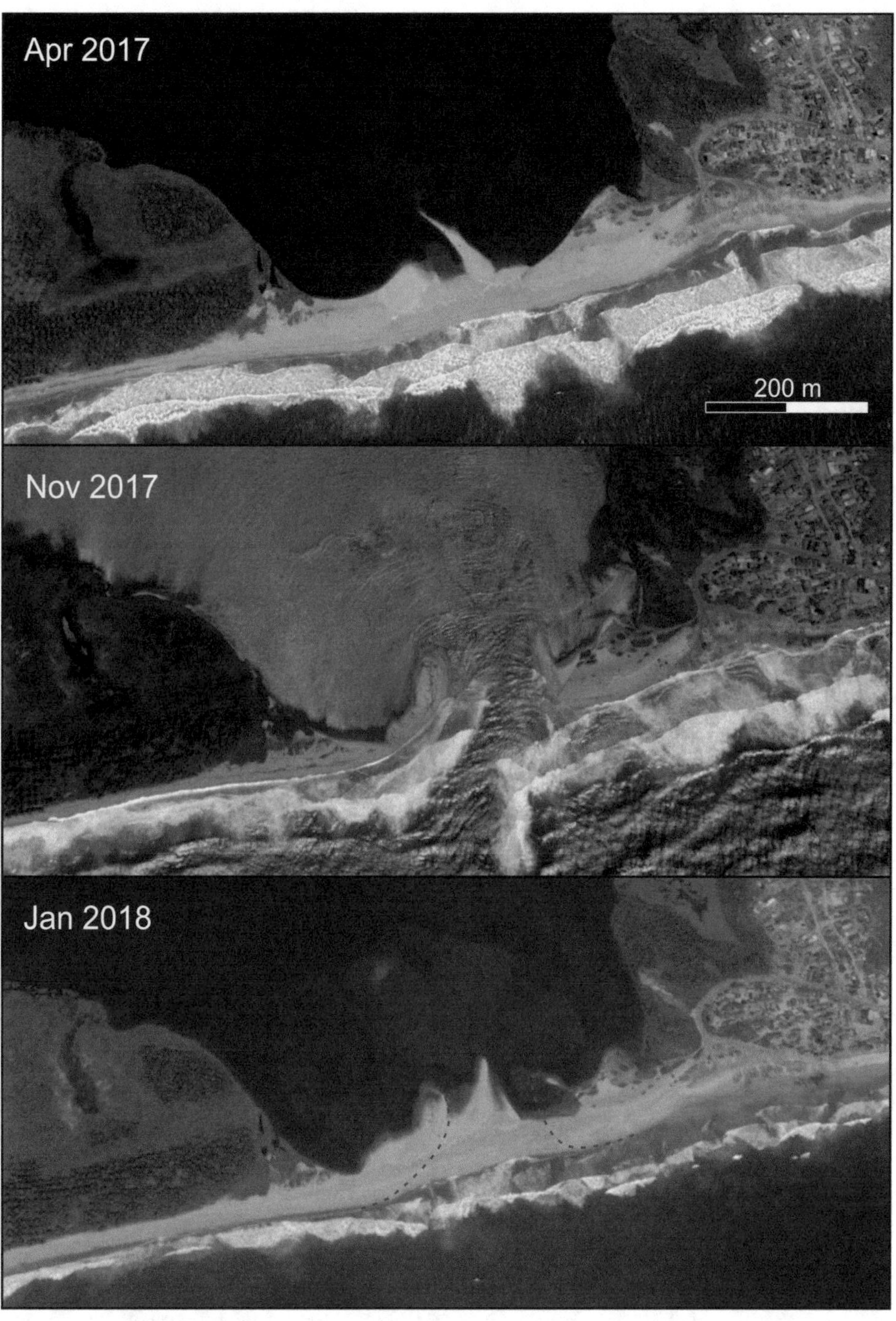

Fig. 16.8 Breaching and rebuilding of a welded barrier (example of the barrier closing Lagune Digboué, Ivory Coast in 2017). (Images Landsat/Copernicus from Google Earth.)

as they are constantly modified by the action of waves (including littoral drift), and variations in the contribution of sediments and the relative movements of sea level have a notable influence on their evolution.

The influence of littoral drift is manifested in the migration of the inlets. Often the islands grow apically at one end with the attachment of curved bars in the shape of a hook, while eroding at the other. The combined effect of these processes leads to the migration of inlets that are located between the ends of two barrier islands (Fig. 16.13, years 2006–2015). The result is a constant renewal of the barrier island sediment while it undergoes a change in shape and position.

Sometimes a strong storm is capable of generating a washover important enough to stabilize and create a new inlet (Fig. 16.13, year 2017). Often the tidal prism is not sufficient to generate currents to keep both inlets (old and new) open and the swell closes one of the inlets. Frequently it is the old one that closes, while the new one gets bigger,

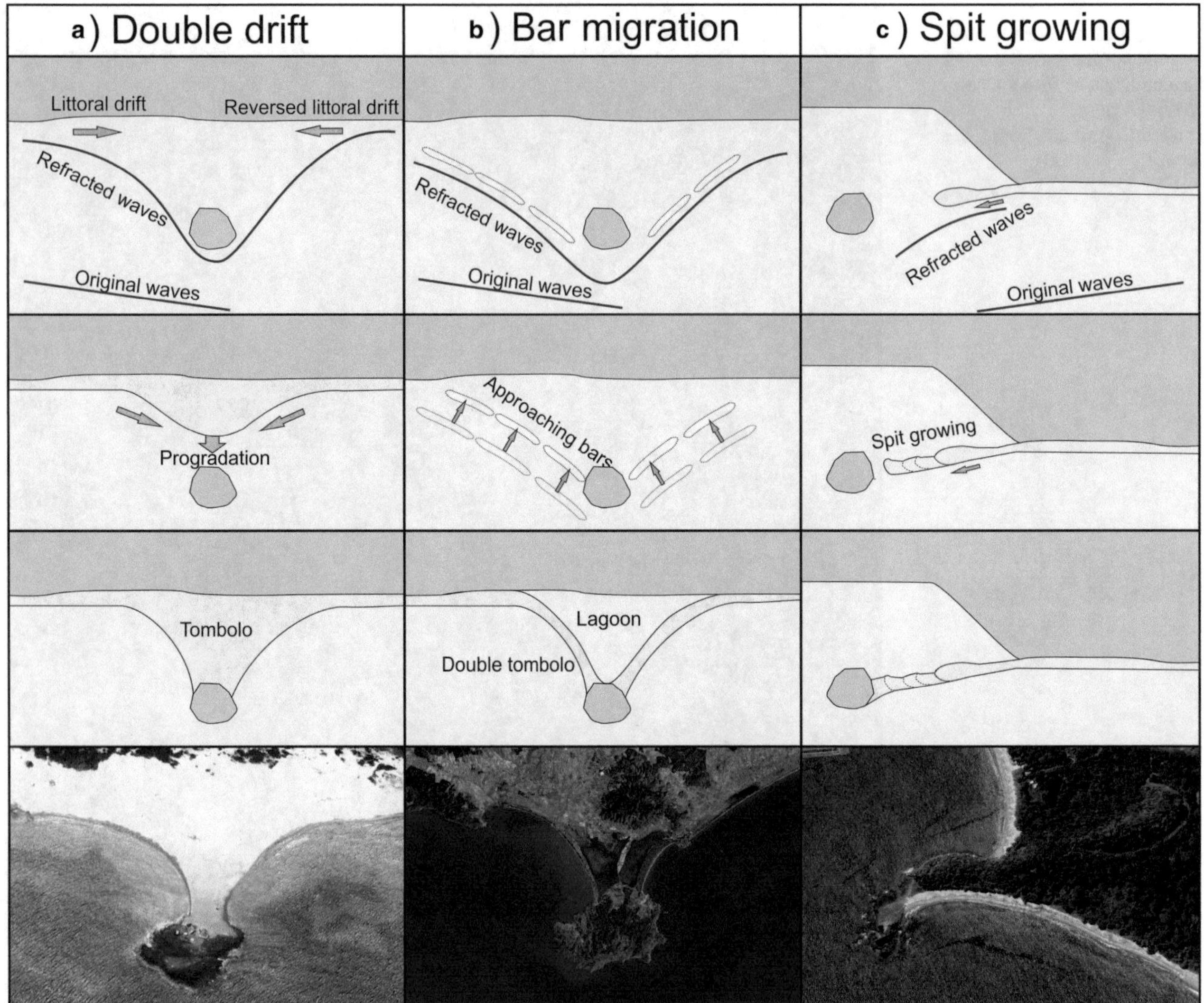

Fig. 16.9 Scheme showing the different possibilities of tombolo genesis. **a** Double drift (example of Cap Serrat, Tunisia). **b** Double tombolo built by bar migration (example of Orbetello, Italy). **c** Spit growing (example of Puerto Viejo, Costa Rica). (Images Landsat/Copernicus from Google Earth.)

starting a new migration cycle. The migration of the inlets also involves the migration of the ebb- and flood-tidal deltas associated with their ends. The ebb-tidal delta is continuously reworked by the waves and is partially destroyed as it moves, yet some of its facies can be preserved among the new sediments at the front of the island. In contrast, the flood-tidal deltas are completely preserved. In this way, the successive lobes that develop during migration remain in the lagoon. This can be seen very clearly in the 2013 frame of Fig. 16.13. These continuous changes in the position of the inlets were responsible for many shipwrecks during the seventeenth and eighteenth centuries, when ships tried to enter the lagoon through the inlets to protect themselves from storms. Clearly, it was impossible for the nautical charts used at that time to reflect these rapid changes.

A common element to all barrier island systems is the sedimentary filling of the rear lagoon at a greater or lesser speed. In the long term, the filling of the lagoon leads to the development of tidal flats and the sediment takes up space that previously had to be filled and emptied by tides. This means a progressive loss of tidal prism. As less water enters and leaves the inlets, the tidal currents that develop there will also be less. Finally, the currents become so weak that the waves end up closing many of the inlets. In this way, a system of mixed tidal–wave dominance, formed by short islands and a large number of inlets, can end up becoming wave-dominated, with very long islands separated by few inlets. If the sedimentary input is abundant, this process is accelerated and the final result may be the complete filling of the lagoon and the final closure of all inlets.

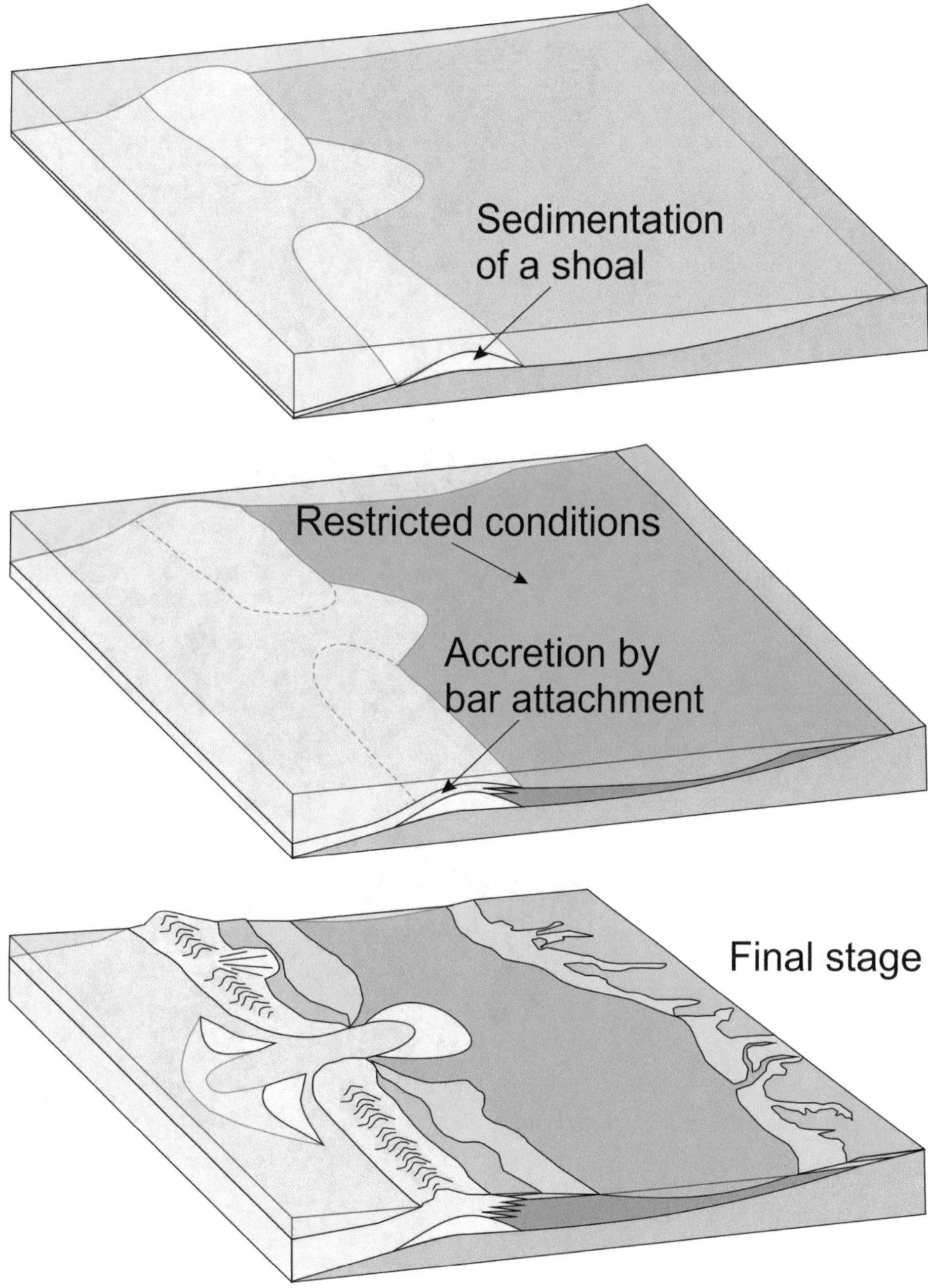

Fig. 16.10 Conceptual model to illustrate the theory of barrier island genesis by building on an onshore bar

16.5 Facies Models

The barriers do not develop their own facies, since these are not one sedimentary environment, but a system composed of numerous environments each of which develops its own facies and sequences of facies. The specific facies of all environments associated with the barrier (beaches, dunes, inlets, tidal deltas, washovers) and the back barrier zone (lagoon, tidal flats, marshes or coastal sabkhas) will be described in detail in later chapters. What really characterizes barrier systems is their facies architecture. The 3D facies architecture is the result of the evolutive history of the barrier. This architecture is also variable depending on the factors acting in the long term, such as sedimentary input and relative sea level movements. These are the factors that also determine the potential for preservation of the facies of the different environments that build up the system. The combination of these contributions results in three different facies architecture models [11]: (1) a retrograding barrier model in a transgressive situation, (2) a prograding barrier model in a regressive situation; and (3) an aggrading barrier model in a sedimentary equilibrium situation. Each of these models will be analyzed separately.

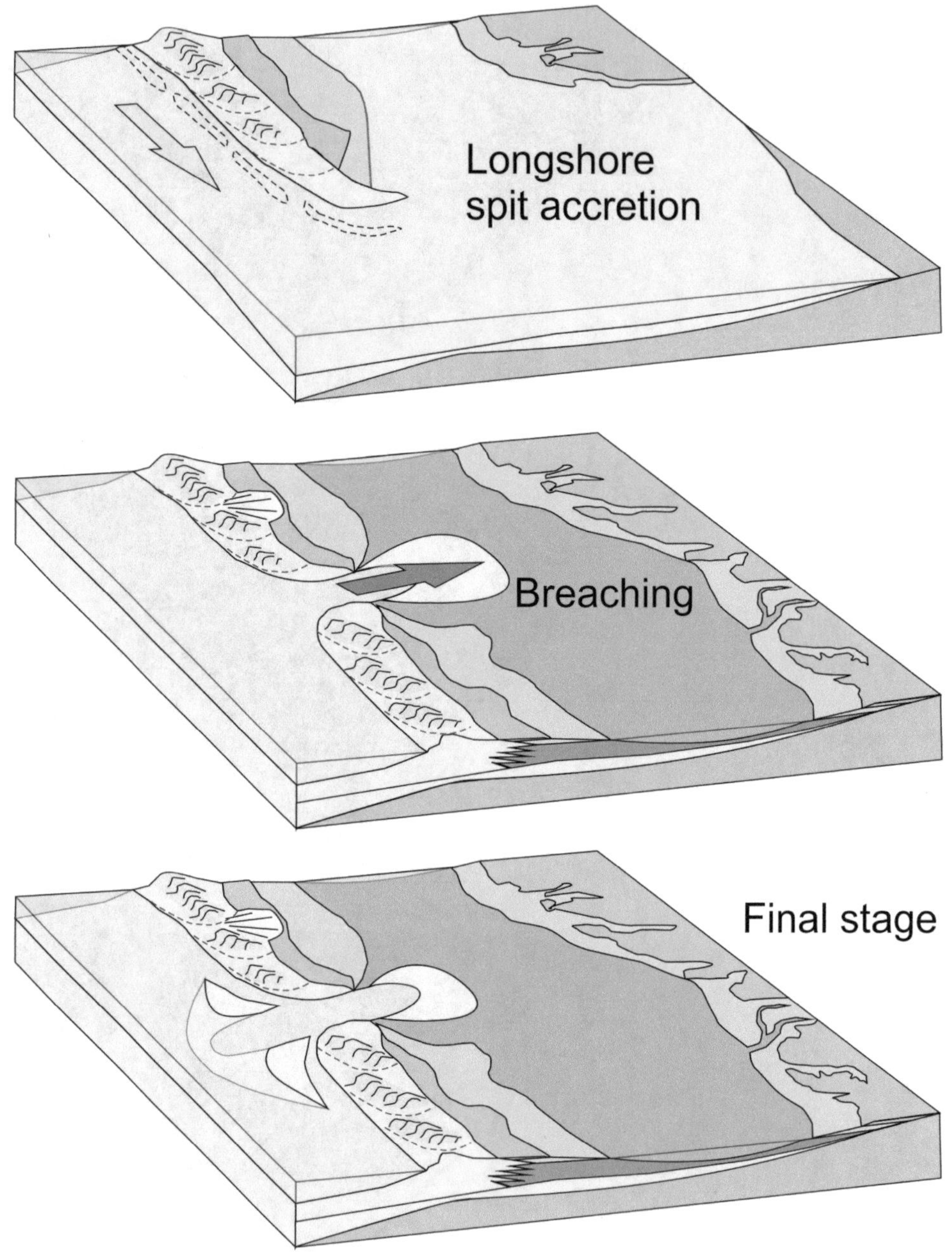

Fig. 16.11 Conceptual model to illustrate the theory of barrier island genesis by breaching of a spit

16.5.1 Retrograding (Transgressive) Barriers

Retrograding barriers are formed under conditions of rising sea level, so that the barriers are forced to migrate towards the mainland. In these conditions, the rate of sea level rise is combined with the rate of sedimentary input, and this combination determines the preservation of the facies of the different environments that make up the barrier during transgression [19]. In many cases, the supply of sedimentary material is scarce and material from older barriers has to be continuously recycled to build new ones. In facies models of transgressive barriers, deposition rates in the associated environments are also important. Normally, the sedimentation rate of lagoons and tidal flats developed in the rear area have very high rates of filling. This means that they are easily preserved under the barrier facies in their migration process towards the mainland.

When the rise is slow and the contribution is scarce, the totality of the sediment of the old barriers is reworked to form the new ones. These barriers are transformed by a process of continuous landwards movement called **roll-over**. In this process, the overwash phenomena that occur during storms play a fundamental role. These frequently break the dune ridge, moving the sand in the form of fans towards the roll-over. In these conditions of slow rise, the dune systems can have enough time to rebuild themselves. In this case, the

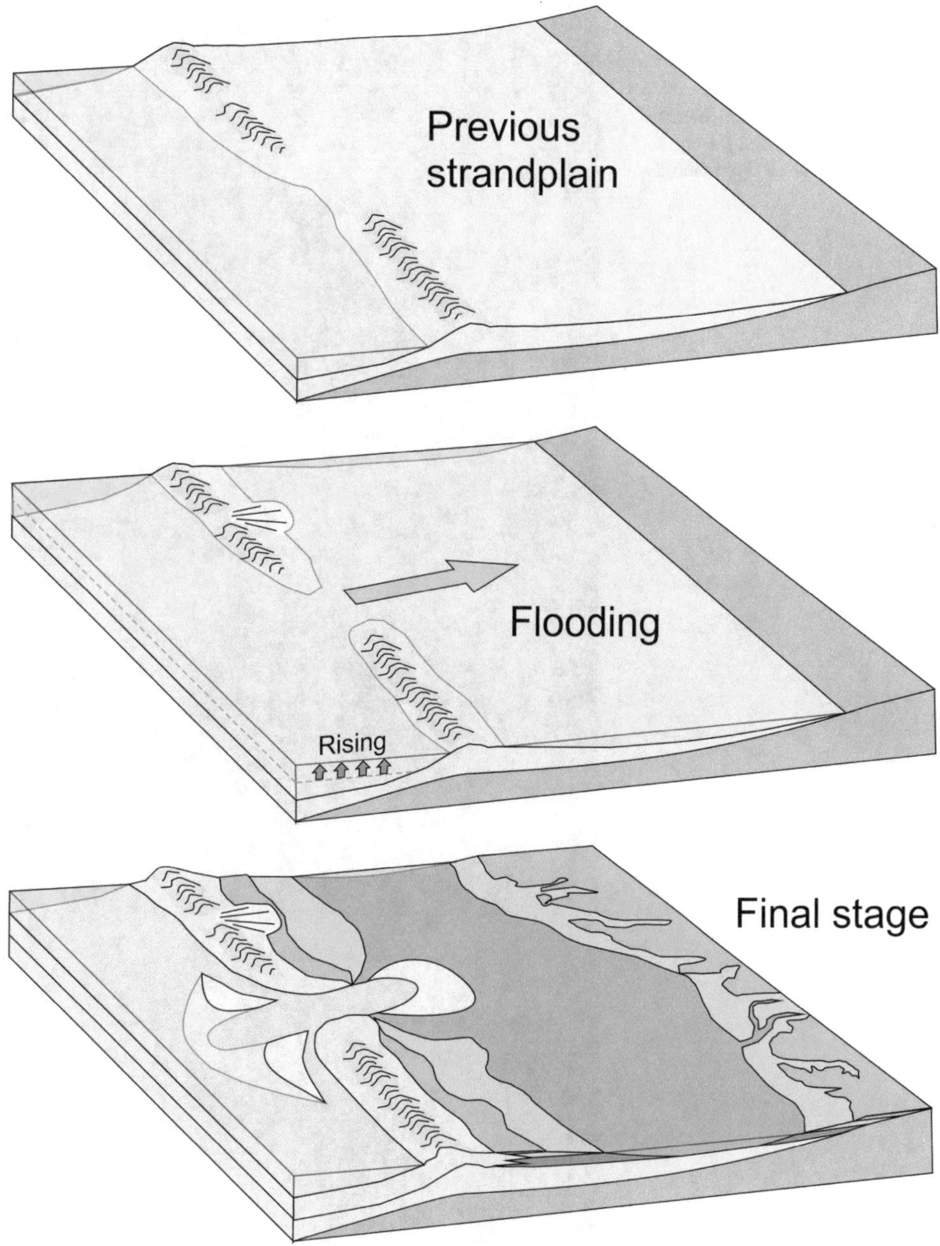

Fig. 16.12 Conceptual model to illustrate the theory of barrier island genesis by detachment from the mainland by submergence

dune building makes the overwash processes difficult and slows down the landward migration of the barrier.

In situations of rapid sea level rise, the barrier is quickly exceeded, being submerged in a situation of less energy to be below the wave breakers. In this case, only a part of the material of the old barrier is reworked to build the new one, located in a position closer to the mainland. This process is known as **overstepping**. The fast-rising conditions allow the preservation of the facies of some environments of the barrier body. However, in these conditions, the rapid rise in sea level prevents the reconstruction of the dunes. It is the material of these dunes that is most easily reworked to build the new barrier.

Thus, the facies architecture model of the retrograding barriers shows a retreat of the barrier facies (Fig. 16.14). In this sequence, the lagoon facies overlap the tidal flat facies and the tidal flat facies overlap the marsh facies in a transgressive sequence. Over these rear facies, the barrier facies are superimposed. In the case of minimal sedimentary input, the barrier facies are thin and rarely exceed 3–4 m [6]. In this model, the importance of the washover facies is noteworthy, interleaved with the muddy lagoon, tidal flat and marshes facies, reflecting the process of displacement of the barrier towards the mainland. In the position of the inlets, it is the characteristic facies of the flood-tidal delta and the inlets that overlap the facies of the barrier environments.

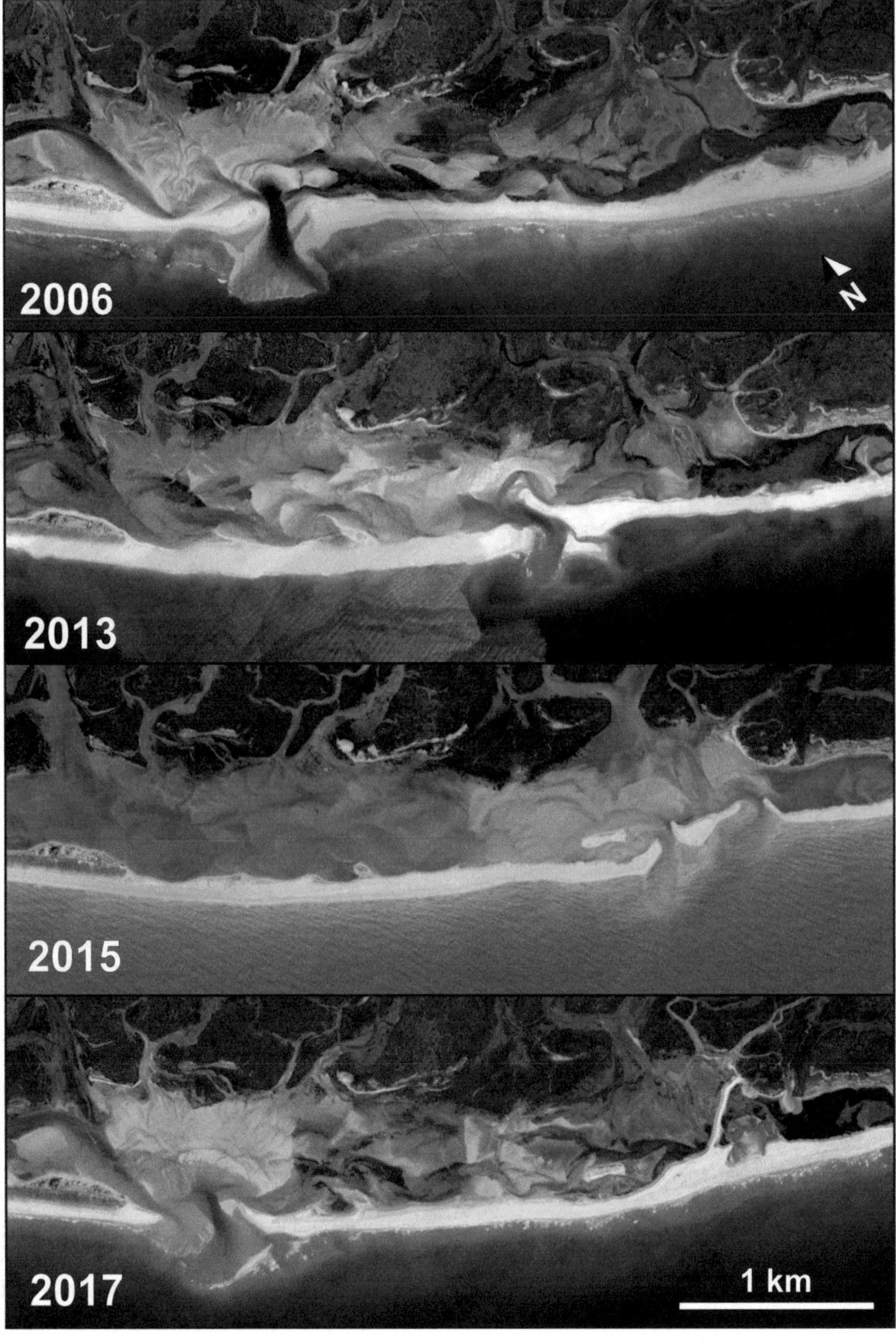

Fig. 16.13 Process of inlet migration (2006–2015) and genesis of a new inlet by stabilization of a washover (2017). Example of Faro Island (South Portugal). (Images Landsat/Copernicus from Google Earth.)

However, the facies of ebb-tidal deltas are usually not preserved, since their sands are easily reworked under transgressive situations to build wave bars. The facies model also includes all the record of the changes in position of the inlet under the body of the barrier island, as well as the presence in the lagoon of old bodies of flood-tidal deltas related to these previous positions of the inlets [24]. There is also a close relationship between ancient flood-tidal deltas and washovers, since the scars corresponding to ancient inlets constitute areas of weakness that are more easily transgressed during storms.

The first barrier island systems that were studied mostly featured this mechanism, as they were built in the early part of the Holocene, during the last stages of the Flandrian transgression when the rate of sea level rise had slowed down.

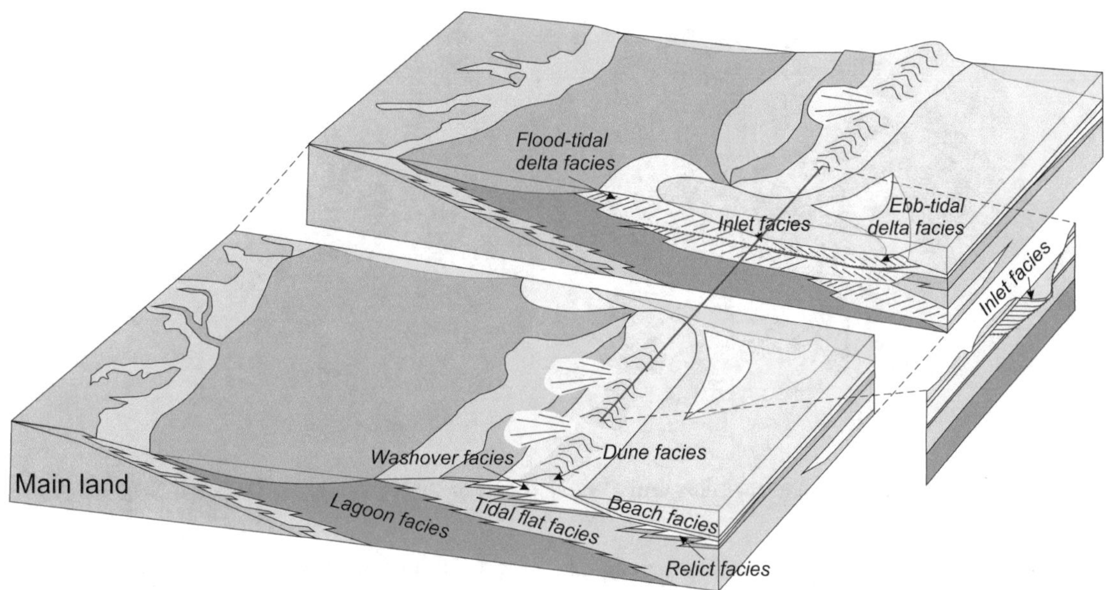

Fig. 16.14 Scheme of facies architecture of a retrograding barrier system (based on Roy et al. [24] and Fitzgerald et al. [10])

16.5.2 Aggrading Barriers

When relative sea level rise is accompanied by a good input regime, barriers are not transgressed except in specific situations. In this case, a vertical growth of the barrier is produced by piling up the sediment of the environments that make up the system (Fig. 16.15). Small variations in the equilibrium between the rates of rise and the volume of sedimentary input can cause oscillations in the coastline that moves towards land or towards the sea in different cycles.

In the frontal area of the island, these oscillations are manifested in stages of erosion or progradation, which accompany the general process of aggradation. On the other hand, in the back barrier zone, slight movements towards the land can originate washovers, which are reflected in an interleaving of its sediments with the lagoon facies, tidal flats or marshes. The rise in sea level makes the depth conditions in the center of the lagoon stable within an order. However, pulses in the rate of rise can cause the margins to produce rhythmic sequences between sub-, inter- and supratidal

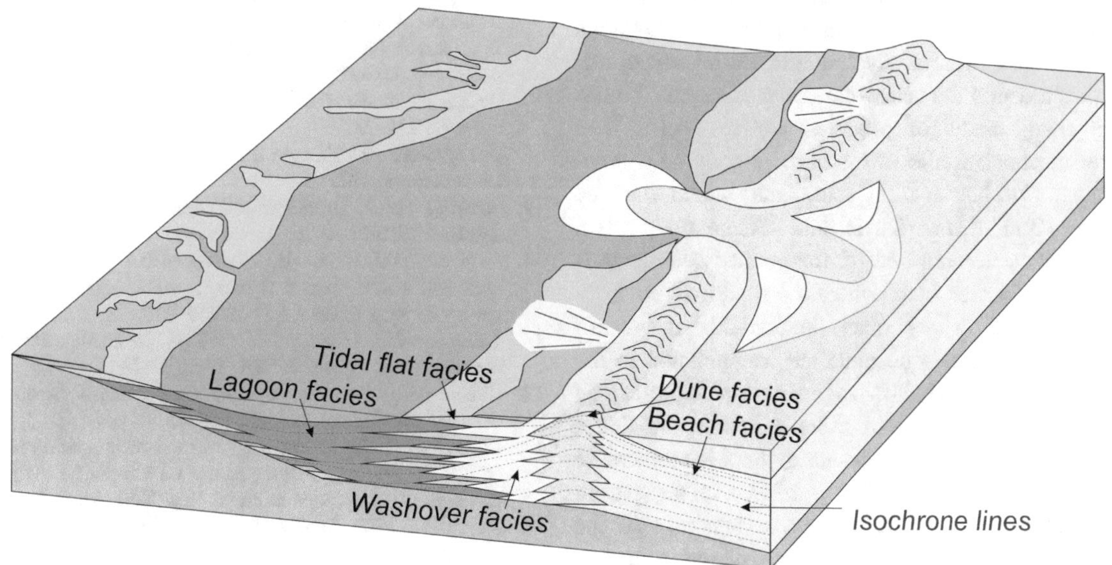

Fig. 16.15 Scheme of facies architecture of an aggrading barrier system (Based on Davis and Fitzgerald [6])

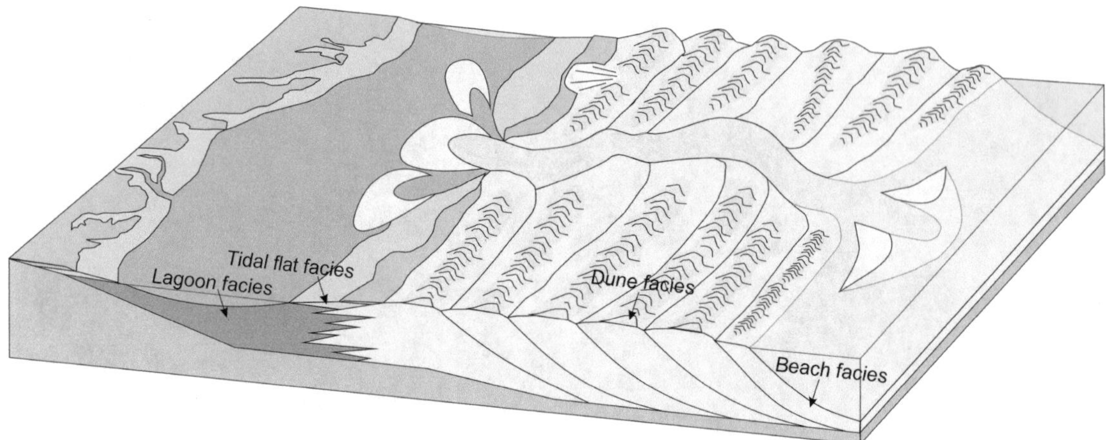

Fig. 16.16 Scheme of facies architecture of a prograding barrier system (Based on Roy et al. [24] and Davis and Fitzgerald [6])

facies. This maintenance of the depth in the lagoon causes the tidal prism to be maintained over time, and the inlets remain open, although they retain their typical migration processes. The aggradation process does not generate vertical sequences of a higher order, since there is no real overlap of sedimentary environments as there is no net displacement of the barrier. This process, on the other hand, is capable of generating sedimentary bodies several tens of meters thick.

The equilibrium conditions necessary for these systems to develop are difficult to achieve, which makes examples of such barriers very exceptional.

16.5.3 Prograding (Regressive) Barriers

The barriers adopt a regressive model under conditions of relative stability of the sea level and good conditions of sedimentary contribution, or under conditions of relative decrease of the sea level. In any case, there is an advance of the coastline towards the sea while the sedimentary bodies acquire a characteristic of progression (Fig. 16.16). The prograding mechanism allows the accumulation of a powerful sequence of barrier-front sediments, which may be more than 10 m thick (Davis and Fitzgerald, 2010). A backfill sequence composed of fine-grained sediments is generated in the barrier front, where the shallower environments overlap the deeper ones (lagoon–tidal flat–marsh sequence). Similarly, oblique units are accumulated in the barrier front, where the shallow sediments overlap the deep ones constituting a sequence of shoreface–foreshore–dunes built on the offshore facies. In this case, there is an increase in grain size towards the top that is related to the greater discharge of energy associated with the breaking of the waves; the exception is in the upper part of the sequence, which would be composed of fine sand associated with the action of the wind. In this context, washover fans are not very well developed, since in most case the dune building prevents the first line of coast from being over-washed by storm waves.

References

1. Anthony EJ (2009) Shore processes and their palaeoenvironmental applications. Elsevier, Amsterdam, p 519
2. Cooper JAG (2001) Geomorphological variability among microtidal estuaries from the wave-dominated south African coast. Geomorphology 40:99–122
3. Cooper JAG, Green AN, Loureiro C (2018) Geological constraints on mesoscale coastal barrier behavior. Global Planet Change 168:15–34
4. Davies JL (1980) Geographic variation in coastal development. Longman, New York, p 212
5. Davis RA, Clifton HE (1987) Sea-level change and the preservation potential of wave-dominated and tide-dominated coastal sequences. J Sediment Petrol 49:167–178
6. Davis RA Jr, Fitzgerald DM (2004) Beaches and coasts. Blackwell Publishing, Oxford, p 419
7. De Beaumont JBE (1845) Lessons of practical geology. Bertrand, Paris, pp 221–252
8. Dobrovolsky AD, Zalogin BS (1982) Seas of the USSR. Moscow University. pp 192
9. Fisher JJ (1968) Barrier island formation: discussion. Geol Soc Am Bull 79:1421–1426
10. Fitzgerald DM, Buynevich IV, Argow BA (2006) Model of tidal inlet and barrier island dynamics in a regime of accelerated sea-level rise. J Coastal Res 39:789–795
11. Galloway WE, Hobday DK (1980) Terrigenous clastic depositional systems. Springer, New York, p 423
12. Gilbert GK (1885) The topographic features of lake shores. US Geol Survey Annual Report 5:69–123
13. Hayes MO (1975) Morphology of sand accumulation in estuaries: an introduction to the symposium. In: Cronin LE (ed) Estuarine research, vol 2. Academic Press, New York, pp 3–22
14. Hoyt JH (1967) Barrier island formation. Geol Soc Am Bull 78:1125–1136
15. Johnson DW (1919) Shore processes and shoreline development. Wiley, New York, p 584

16. Masselink G, Hughes MG, Knight J (2003) Introduction to coastal processes and geomorphology. Routledge, London, p 416
17. McGee WD (1890) Encroachments of the sea. Forum 9:7–449
18. Morales JA, Borrego J, Jiménez I, Monterde JR, Gil N (2001) Morphostratigraphy of an ebb-tidal delta system associated with a large spit in the Piedras Estuary mouth (Huelva Coast, Southwestern Spain). Mar Geol 172:225–241
19. Moore LJ, List JH, Williams SJ, Stolper D (2010) Complexities in barrier island response to sea level rise: insights from numerical model experiments, North Carolina outer banks. J Geophys Res 115:F03004
20. Oertel GF (1985) The barrier island system. Mar Geol 63:1–18
21. Orford JD, Carter RWG, Jennings SC (1991) Coarse clastic barrier environments: evolution and implications for Quaternary Sea level interpretations. Quatern Int 9:87–104
22. Pilkey OH, Cooper JAG, Lewis DA (2009) Global distribution and geomorphology of fetch-limited barrier islands. J Coastal Res 25:819–837
23. Reinson GE (1992) Transgressive barrier island and estuarine systems. In: Walker RG (ed) Facies models: response to sea level change. Geological Association of Canada, pp 179–194
24. Roy PS, Cowell PJ, Ferland MA, Thom BG (1994) Wave-dominated coasts. In: Carter RWG, Woodroffe CD (eds) Coastal evolution: late quaternary shoreline morphodynamics. University Press, Cambridge, pp 121–186
25. Roy PS, Williams RJ, Jones AR, Yassin I, Gibbs PJ, Coaters B, West RJ, Scanes PR, Hudson JP, Nichol S (2001) Structure and function of south-east Australian estuaries. Estuar Coast Shelf Sci 53:351–384
26. Sloss CR, Jones BG, Mcclennen C, de Carli J, Price DM (2006) The geomorphological evolution of a wave dominated barrier estuary: Burrill Lake, New South Wales, Australia. Sed Geol 187:229–249
27. Stutz ML, Pilkey OH (2011) Open-ocean barrier islands: global influence of climatic, oceanographic, and depositional settings. J Coastal Res 272:207–222
28. Tomazelli LJ, Dillenburg SR, Villwock JA (2000) Late quaternary geological history of Rio Grande do Sul coastal plain. Southern Brazil. Revista Brasileira de Geociências 30(3):474–476
29. Woodroffe CD (2003) Coasts: form. Cambridge University Press, Cambridge, Process and Evolution, p 623

17 Wave-Dominated Systems II: Beaches

17.1 Introduction

Beaches are widely represented sedimentary coastal environments, occupying about 20% of the world's coastline [15]. Like the rocky coasts, these are environments dominated by the action of waves, although in this case the genesis of beaches involves their cumulative effect. This type of sedimentary accumulation can be associated with different environments (Fig. 17.1); thus, they can appear in rocky coastline bays (Fig. 17.1a), along cliff faces (Fig. 17.1b) and on barrier fronts.

Vertically, beaches develop from the base level of storm waves, at the bottom of the waterfront, to the height of the storm breakers, where they border the wind-dominated environments. It is a very dynamic environment, since the action of the waves subjects it to continuous short-term morphological change (e.g., Komar [38]). Over time, there is an alternation between moments when the sediment moves landwards and others when it moves seawards. This alternation subjects beaches to cyclical changes that manifest themselves in gains and losses of sediment in the most visible part of the beach. In a general way, the sediment moves transversely between the shallowest and deepest areas, which leads to changes in the topographic profile of the beach.

This very active system of wave dynamics generates an environment where fine particles cannot decant. Thus, the most common sediments on beaches are sand, although fine gravel, pebble and cobble beaches are also frequent. Whatever the grain size of the sediment, beaches are characterized by the development of bars of different morphology. In fact, it is the movement of the bars that causes the change in the shape of the beaches.

The action of the waves at an oblique angle to the beach also induces the longitudinal component of transport on the beachfront. This littoral drift (or longshore current) is cut by the presence of elements such as headlands, inlets, river mouths or human infrastructures. These obstacles divide transport cells that determine the length of the beaches. In this way, there are beaches of very variable lengths, from tens of meters to hundreds of kms, that depend on the morphology of the coast. In general, irregular coasts generate very short beaches limited to small bays, while linear and very regular coasts can generate kilometer-long beaches.

In addition to the waves, beaches can be affected by other dynamic agents. Above all, tides are important. The tide is responsible for the vertical displacement of the direct action zones of breaking waves. The higher the tidal range, the more complex the beaches are and the greater the surface area of the beach affected by the action of the waves.

Another acting agent is the wind. Its origin can influence the direction of the waves and, above all, the air exerts an effect of deflation on the sand to form the dunes or, conversely, introduces sand from the dunes, feeding the beach system.

17.2 Control Factors

17.2.1 Wave Energy

The wave energy is the main dynamic engine of beaches. Waves control the mobility of the sedimentary material on the beachfront, both transversely and along the coastline. Wave energy is used to transport sediment during the shoaling process, but also when the wave reaches the shore and breakers occur. In the shoaling zone, the waves produce a swaying in the transverse direction to the beach that generates symmetrical bedforms such as ripples, while the breakers can produce asymmetrical forms that result from a net movement of the sediment in one direction or another. The breaker type controls the direction of the movement of the material from or to the continent and thus exerts a great influence on the general slope of the beach. The longitudinal transport capacity on all the fringes of the beach not only depends on the wave energy, but also on their angle of incidence on the shoreline.

J. A. Morales, *Coastal Geology*, Springer Textbooks in Earth Sciences, Geography and Environment,
https://doi.org/10.1007/978-3-030-96121-3_17

Fig. 17.1 Photographs showing beaches associated with different coastal systems. **a**: Pocket beach on a rocky coast (Costa Brava, NE Spain). **b**: Beach in front of a bluff coast (Matalascañas, SW Spain). **c**: Beach at the front of a barrier (Nueva Umbría, SW Spain)

17.2.2 Beach Slope

Variations in depth on the beachfront influence the process of wave energy dissipation. On the one hand, the nearshore slope—along with the dimensions of the wave—plays a major role in the breaker type. On the other hand, the beach slope also plays a role in other phenomena related to the interaction between the waves and the coast, such as wave reflection. Thus, smaller slopes will tend to have the wave dissipating most of its energy in material movement during the approach to the coast, while larger slopes will cause an interaction with the bed closer to the shoreline and will cause some of the wave energy to be reflected on the coast and return to the center of the ocean in the form of a reflected wave train. In this way, the processes that take place along the different depths of the beachfront vary with the slope, giving rise to morphological changes due to the accumulation or erosion of material. These morphological changes are manifested in slope modifications, entering a circular model in which the slope, the breaker type and the erosive/accumulative conditions of the beach influence each other.

17.2.3 Grain Size

The grain size of the beach material is determined by the size of the waves and the availability of sediment. These are the same factors that determine the transport capacity of the sedimentary material. The grain size greatly influences the slope of the beach (Fig. 17.2). Smaller grain sizes tend to generate beaches with low slopes, while an increase in grain size usually implies an increase in slope, such that pebble beaches are usually very steep [36, 54].

The grain size also exerts a strong influence on the magnitude of the evolutionary changes experienced by beaches, both those of a cyclical nature and those that occur over a longer period of time. Thus, beaches with finer grains tend to be very dynamic places that are affected by small changes in wave dynamics, while pebble and cobble beaches tend to be more stable.

17.2.4 Tidal Range

The tide controls the topographical level of wave action and the time it acts on each level of the beach profile. On microtidal coasts, the action of the waves always occurs at the same height of the profile, while on coasts with a wide tidal range the tide widely displaces the level of action of the waves, which will disperse the energy by transporting sediments in a broader strip. The tide is responsible for the existence of a specific area delimited by the high and low tidal levels (the foreshore), but it also displaces the limits of the rest of the areas that are located in deeper waters of the subtidal zone. This displacement can mean a significant change in the way energy is transferred from the waves to the sediment, since the high and low areas of the beach usually have different slopes.

Moreover, the tide also influences the number of bars that the waves can develop on the nearshore. On macrotidal beaches, the number of bars that simultaneously migrate landwards can be numerous.

17.2.5 Nearshore Currents

The influence on the beachfront of tidal and wind-induced currents has been described in detail in Chap. 10. From a general perspective, these currents have a limited sediment transport capacity compared with the waves, especially in the breaker zone. It is in the shoreface where the combined action of these currents is able to modify the transport components of the waves. For example, the landward component of these currents generates an asymmetry in the wave oscillation. In general terms, this asymmetry displaces material landwards on the shoreface, although in some cases the effect may be the opposite.

17.2.6 Climate

Climate exerts an important control over all the variables described above. On the one hand, it is the weather that controls the wind regime and, therefore, the wave energy. In this sense, the climate will determine the average energy of the waves that affect the beach and also their distribution throughout the year. Ultimately, it is the climate regime that imposes the number of storms that affect the beach each year, their magnitude, their length of time and the moment at which they occur.

In the short term, the annual climatic cycles mark those rhythmic pulses of accretion and erosion that are so typical of all beaches. Over periods of several years, solar cycles are responsible for the El Niño–Southern Oscillation (ENSO) and North Atlantic Oscillation (NAO). These climatic oscillations are manifested in other interannual cycles of erosion and reconstruction.

A perhaps less evident effect of the climate is the influence on the volume and nature of the sedimentary input. Since most of the sediment on the beaches comes from the continent and reaches the coast through the rivers, and since the flow of the rivers depends directly on the rainfall regime, the connection with the climate becomes immediately obvious.

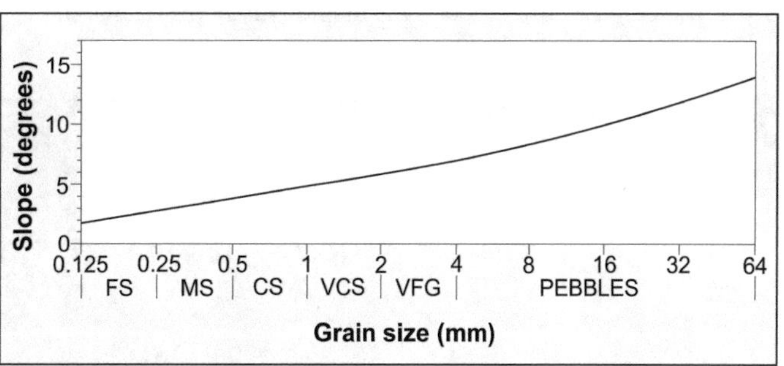

Fig. 17.2 Example of relationships between beach slope and grain size (based on the data of Shepard [54] and Jennings and Shulmeister [36]

17.3 Zonation and Morphology

The zoning of the beach system is determined by the dynamic factors established by the different action of the waves. Using this dynamic criterion, the zones will always occupy strips longitudinally to the coast, succeeding each other transversally.

17.3.1 Fringes of Wave Action

Dynamic zoning is therefore established by the action of the waves on the bed. This action combines, on the one hand, the different effects that the waves produce on the bed and, on the other hand, the deformation that the bed produces on the wave in terms of energy dissipation. The main division is determined by the breaker (Fig. 17.3). The interaction with the bed that has been described in Chap. 7 as the shoaling process [34] begins in areas shallower than the wave base level. Hence, the zone where it takes place is called the **shoaling zone**. In this zone, the action of the wave on the bed is manifested in an oscillatory movement.

The place where the breaker starts is called the **breaker zone** and it is a zone that usually coincides with a change of slope, and most often with the presence of a bar. The breaker zone is the fringe in which the transition from oscillation waves to translational waves takes place. Here, there is a whole strip in which translation waves are produced, called the **surf zone**, which is characterized by a more or less uniform transport of sediments to the shore. The arrival of the wave to the shoreline is a last stroke and is called the **swash zone**. There, a laminar dynamic will be established, represented by *swash* and *backwash* processes.

The surf area may be absent on steep beaches where the wave practically breaks on the beach. These beaches go directly from the breaker zone to the swash zone.

17.3.2 General Zoning

On tidal beaches, the zones of wave action move up and down the slope of the beach. This displacement generates the presence of fringes with different dynamics and divided by wave levels related to the linked action of waves and tides (Fig. 17.4).

The most dynamic area is the one where the wave energy dissipation bands described in the previous paragraphs are most frequently found. This zone is called the **nearshore**. Its lower limit corresponds to the place where the significant waves begin to interact with the bed—i.e., the wave base level—while its upper limit is the middle level of the high tides. This nearshore zone is divided into two zones with different behavior. In the subtidal zone, the strip where the shoaling process takes place is called the **shoreface**. Here, there is also a greater influence of the tidal currents and the wind-induced currents. The shoreface also includes the breaker and surfing zones during low tide. Above it, the intertidal fringe includes the beach between high and low tides. This area, called the **foreshore,** is where the breaker, surfing and swash areas are located during high tide, emerging during low tide.

The subtidal zoning is completed with another deeper band called the **offshore**, which is only affected by storm surges and is delimited by the base level of the extreme storm waves. Above the foreshore there is another supratidal strip dominated by the wind and which precedes the first ridge of foredunes. This area is called the dry beach or **backshore** and its upper limit is the maximum level of swash of extreme storm waves.

17.3.3 Beach Types

The general typology of beaches is established according to the balance between the processes of dissipation or reflection of the energy of the waves, which is in turn established according to the slope of the beach. In the previous sections discussing the influence of the slope, we saw that this balance defines two different situations [60]:

(1) *Steep beaches.* These are characterized by the absence of the surfing area and the breaking and swash areas being located right on the shoreline. Because of this, the swell suffers an important reflection process. They are

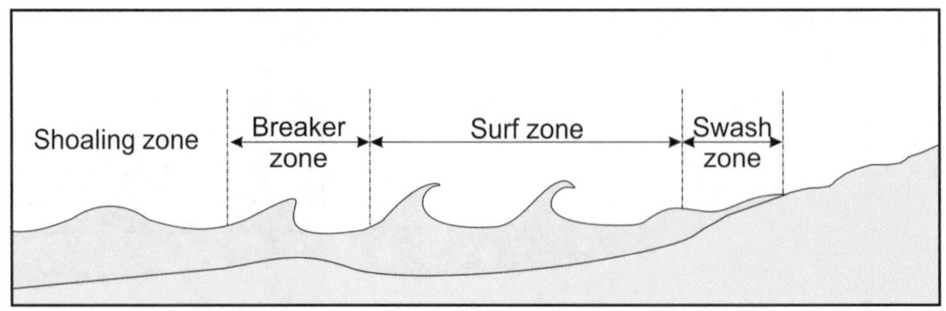

Fig. 17.3 Zoning of the beach strip with respect to the wave breaker [34]

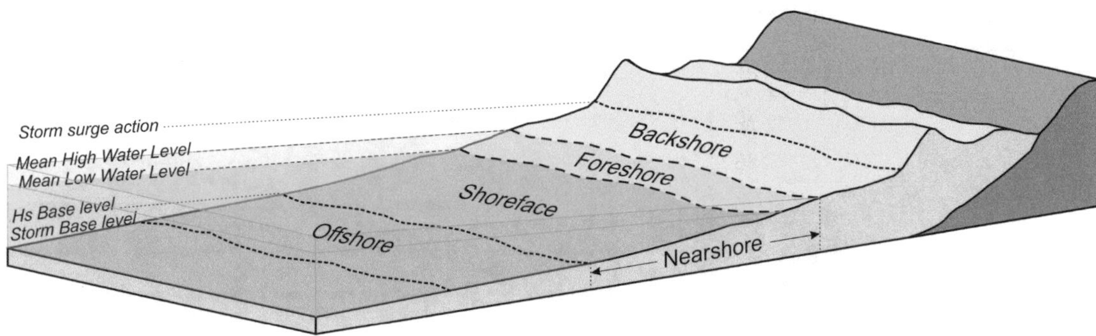

Fig. 17.4 Complete zoning of beaches with tides. (adapted from CERC [12])

dominated by the backwash and have an erosive behavior. These are called **reflective beaches**.

(2) *Gently sloping beaches.* In these, the wave energy dissipates in a gradual way along the surf zone. These are beaches where the landwards component of the wave dominates and they usually have a cumulative character. These are called **dissipative beaches**.

However, these two situations evidently describe the extremes of a wide spectrum of possibilities, with an infinite number of cases in between. In fact, the same authors described an intermediate type. In this type, the slope is not very gentle, but neither is it very steep. So, the wave suffers some dissipation, but there is also reflection, since not all the energy of the wave is dissipated. In this type of beach, there is a complex model of circulation. Wright and Short named this type **intermediate beaches**.

This spectrum of combinations between energy dissipation and reflection depends on the dimensions and frequency of the waves in relation to the beach slope. All this is very much linked to the width of the surf band and the breaker type. These relationships were expressed by Guza and Inman [26] in equation 17.1.

$$\varepsilon = \frac{2\pi^2 H_b}{gT^2 so^2 \beta} \tag{17.1}$$

In this equation, ε is a dimensionless parameter known as **surf scaling**, *Hb* is the height of the wave in the break, *g* is the acceleration of gravity, *T* is the period of the wave and β is the slope angle of the beach.

The surf scaling parameter was actually intended to predict the width of the surf zone, but was later used by Wright and Short to differentiate the type of beach as a function of the dissipation/reflection processes. Thus, dissipative conditions would remain above values of 20, while reflective conditions would occur below values of 2.5. Then, intermediate conditions would occur for values between 2.5 and 20.

The same authors found a relationship between the capacity of mobilization of the sediment by the wave and the presence of certain beach morphologies related to the phenomena of dissipation or reflection. The mobilization capacity was also determined through a dimensionless parameter previously established by Gourlay [25], calculated through an equation (17.2).

$$\Omega = \frac{H_0}{w_s T} \tag{17.2}$$

In this equation, Ω is the non-dimensional fall velocity, H_0 is the original wave height, *ws* is the settling velocity of the average grain size of the beach and *T* is the wave period. With fall velocity, the authors refer to the particle sedimentation velocity when the flow loses the capacity to transport it, and it is calculated for grains of volumetric equivalence to a sphere through the equation 17.3.

$$Ws = \sqrt{\frac{4gD\delta s}{3Cd\delta l} - 1} \tag{17.3}$$

In this equation, Ws is the rate of fall, g is the acceleration of gravity, D is the diameter of the mean grain, δs is the density of the solid grains, δl is the density of seawater and Cd is a hydrodynamic coefficient that depends on the Reynolds number. For quartz grains, the rate of fall velocity in relation to grain size is represented in the graph in Fig. 17.5.

The name used by the authors could lead to confusion, since it does not have dimensions and the word speed could give an idea that it is a vector, when in fact it is a matter of observing the space–time relationship of the breaking wave in relation to the movement of the particles. Davidson-Arnott [15] proposes the use of the term **surf similarity** to designate this parameter.

In any case, according to this parameter Wright and Short defined six types of beach. In addition to the dissipative and reflective extremes, they divided the intermediate beaches into five types. This classification is shown in Table 17.1.

A decade later, Masselink and Short [40] proposed a new classification, this time taking into account not only the parameter, but the relationship between the tidal range and the wave dimensions. They quantified the relationship

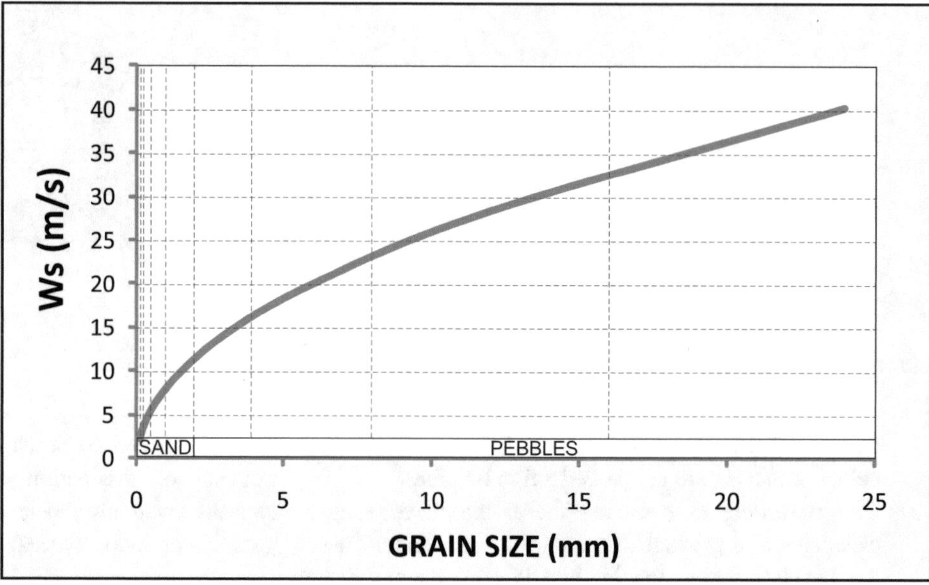

Fig. 17.5 Graph showing the relationship between grain size and orbital velocity

Table 17.1 Beach types according to non-dimensional fall velocity (surf similarity)

Ω value	Beach type
< 1.5	Reflective
1.5–2.5	Low tide terrace
2.5–3.5	Transverse bars and rips
3.5–4.5	Rhythmic bars and troughs
4.5–5.5	Longshore bars and troughs
> 5.5	Dissipative

between tide and waves using the parameter RTR (relative tidal range). RTR actually quantifies the importance of the maximum tidal range (STR: spring tidal range) versus the height of the waves in the breaker (RTR = STR/Hb). According to the authors, for RTR values lower than 3, the influence of the tide would be negligible. Conversely, values higher than 15 would be typical of tidal plains where the tide totally dominates the processes. In the case of intermediate values between 3 and 15, tide and waves combine to control the transport processes. The values of Ω and RTR can be combined to obtain six types of beach (Fig. 17.6). Among these types, the three corresponding to RTR values lower than 3 would be the three initially differentiated by Wright and Short [60], distinguishing three new types of RTR values higher than 3: low tide terrace beaches, low tide bar and rip beaches and ultradissipative beaches.

17.3.4 Bar Morphology

A large fraction of the beaches described in the previous section is characterized by the presence of bars. Classic works differentiate between two distinct types of bars: the transversal bars and the longitudinal bars (e.g., [22, 27, 47, 53]).

The transversal bars characteristic of reflective beaches are commonly known as **rip-and-cusp** systems. The rips and cusps form a system of repetitive forms with regular spacing ranging from a few meters to over a hundred. Actually, these forms constitute the most visible element of the system, since they develop in the emerged zone; however, the transversal bars that are associated with the submerged part develop more widely. The morphology of these systems gives the beach a saw blade appearance (Fig. 17.7).

The longitudinal bar systems characteristic of dissipative beaches are known as **ridge-and-runnel** or **bar-and-trough** systems. Bars or ridges are sandy crests parallel to the beach with a clear tendency to migrate landwards. Behind them, furrows (runnels or troughs) develop, which are elongated depressions in the same direction (Fig. 17.8).

The most common systems have only one bar in the foreshore zone, although there may be a few more in the nearshore zone. However, in macrotidal zones more than one bar can be emerged during low tide (Fig. 17.9a). Under favorable conditions, multiple small bar systems can develop even in microtidal systems (Fig. 17.9b).

There is much debate about the causes that condition the appearance of multiple bars on the beaches under different tidal regimes. Early on, authors who worked on bars (e.g.,

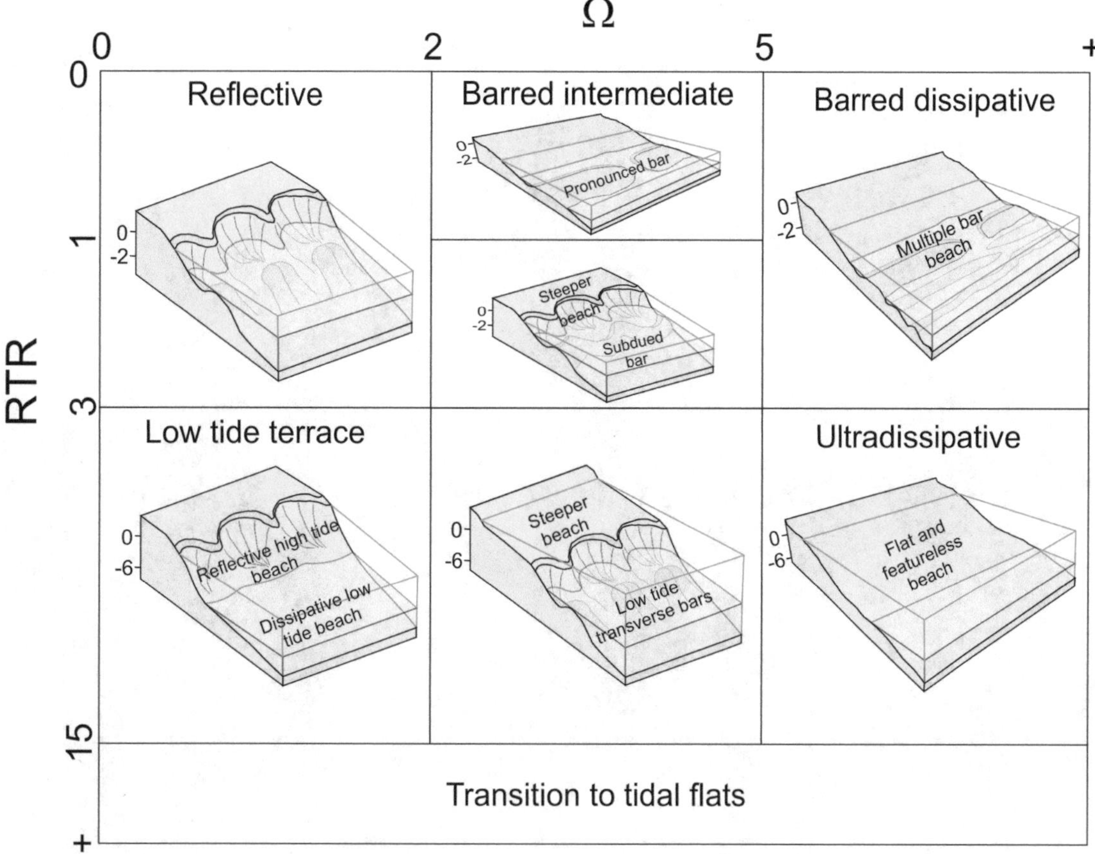

Fig. 17.6 Classification of beach morphology according to the values of Ω and RTR. (adapted from Masselink and Short [40])

[22]) stated that one of the main factors involved in the genesis of a high number of bars, along with wave dimensions and grain size, is the shoreface slope. Thus, the slope conditions the distance to the coast of the first breaks and thus determines the width of the surf zone. In a general way, very wide surf areas subjected to small waves moving fine sediment allow the development of a greater number of bars.

The same studies also established that the dimensions of the bars tend to become larger as they move landward, while their spacing becomes smaller. This is evident from the change in water column above the bars and the deformation that this decrease in depth generates in the wave orbits during the shoaling process. In this regard, it should be noted that the deeper bars do not migrate during the arrival of smaller waves.

In order to try to quantify these factors, Short and Aagard [55] established the parameter B* (17.4). This parameter relates the width of the surf zone to the wave period and the nearshore slope.

$$\mathrm{B}_{*} = \frac{X_s}{gT^2 \tan\beta} \tag{17.4}$$

In this equation, *Xs* is the width of the surf zone, *g* is the acceleration of gravity, *T* is the wave period and β is the nearshore slope. In this case, the grain size is absent from the equation, but the authors understand that it is the grain size that conditions the beach slope.

For values of B_* that are lower than 20, no longitudinal bars will be generated; for values higher than 20, an increase in value will allow a greater number of bars to be developed.

Although two types of bar (transversal and longitudinal) were described from the beginning, in fact these two types are once again the extremes of a broad spectrum (Fig. 17.10). The spectrum includes the crescentic bars as intermediate morphology and a good number of transition morphologies that take into account the orientation of the bars, whether they are attached to the beach or whether they have grooves that separate them [42].

Fig. 17.7 Aerial view of a rip-and-cusp system at Mazagon Beach, SW Spain

17.4 Dynamics and Evolution

17.4.1 Movement of Sediments by Waves

The sediment present on beaches, whether sand or pebbles, is easily transported by the forces that the waves exert on the bed. However, these forces go through cycles, alternating fairweather waves with those developed in periods of storm. When faced with the forces of the waves, the clasts can be transported in two different ways, depending on the relationship between their size and the force of the wave: suspension and bed load. From a morphological point of view, this transport manifests in the development of an upper plane bed or in the migration of bars.

A dynamic concept directly linked to sedimentary environments is the **closure depth**. The closure depth is the bathymetric level at which the waves can start their transport to land. In terms of sedimentary balance, it can be defined as the lower level to which the sand can descend during storms without losing its ability to return to the upper zone of the beach. In morphological terms, it is the shallowest depth that delimits the most marine zone in which there are no significant changes in the bathymetry. Consequently, under this level there is no significant net exchange of sediments, in contrast with the beach zone, which is dynamically active and morphologically changing.

To calculate this parameter, the most energetic conditions of the wave must be taken into account. This depth is established through the relationship between the entrainment threshold of the grains present on the beach and the force of the dominant wave. This strip can be determined from bathymetric profiles and the grain size of the sediment, and represents the area where the processes of remobilization and circulation of sediments take place.

Hallermeier [28] established an equation for the calculation of the closure depth (17.5), which he later modified in 1983 (17.6).

Fig. 17.8 Panoramic view of a ridge-and-runnel system at Agate Beach, Oregon, USA

$$Dc = HsTs\sqrt{g/5000D} \quad (17.5)$$

$$Dc = 0.018HsTs\sqrt{g/D(S-1)} \quad (17.6)$$

Where *Dc* is the closure depth, *Hs* and *Ts* are the significant wave height and period that are exceeded 12 *h* per year (storm surges), *g* is the acceleration of gravity and S is the relationship between densities of sand grains and seawater.

In 1985 this equation was simplified so that the granulometric parameters did not have to be used, and was thus established solely as a function of the wave dimensions (17.7).

$$Da = 1.75Hs - 57.9\left(\frac{Hs^2}{gT^2}\right) \quad (17.7)$$

Above the active depth, the breaker is the key for separating areas with different transport models. At the moment a wave breaks, a sudden discharge of energy is produced, which puts the finest particles in suspension and displaces the coarser clasts in the form of bed load. Depending on the type of breaker, the moving particles can be transported landwards or seawards through the surf zone. On dissipative beaches, the spilling type breakers dominate and the main transport is landwards, while on the characteristic breakers of reflective beaches (plunging, collapsing or surging) it is the undertow currents that cause a dominant transport towards the sea and erosion in the higher parts of the beach.

With waves oblique to the coast, undertow currents are generated at the entrance of each new wave. Then, the transversal vectors are compensated for, but longitudinal vectors appear and a drift current is generated. This component is responsible for the longitudinal transport of sediments along the beach.

17.4.2 Dynamics and Genesis of Transversal Bars

In order to understand the dynamics of a reflective beach with a system of transversal bars, it is necessary to know how the energy discharge occurs on the surfaces of high slopes during the wave-swash process. According to studies by Hughes and Turner [33], the arrival of a wave on a highly sloped surface occurs in several phases (Fig. 17.11).

The first phase involves the concentration of a landward current in a diminishing section. This current increases significantly from the presence of the beach step, which is located at the front of the rip-and-cusp system. In the second phase, the swash is produced when the step is overcome. This means the entry landwards of a sheet of water of a few centimeters, which acts by transporting particles over the beaten surface. In the third phase, the sheet of water moves in favor of the slope as a high-regime flow. This flow is of much greater energy than the previous one and it displaces a great quantity of particles towards the sea. These particles

Fig. 17.9 Panoramic view of multiple bar systems. **a**: Macrotidal beach at Formby, UK. **b**: Microtidal beach of Mississippi Sound, USA

are deposited at the front of the step, when the increase of the flow section decreases the velocities. With the arrival of a new wave, the undertow flow is abruptly slowed down by forming a hydraulic jump at the front of the step. There, the encounter with the new wave takes place. The presence of this jump implies a settling of the particles that the undertow transported down the slope. Finally, at the front of the step, a vortex is produced between the undertow current of the previous wave and the pounding of the new wave. This vortex can remove the particles deposited under the step and transport them uphill when the next wave is produced.

This model takes only a two-dimensional view of the swash. However, the morphology of a reflective beach includes the development of rips and cusps, so the process must be observed three-dimensionally [48]. This perspective allows us to delimit a zone of divergence of the swash towards two successive bays during the arrival of the wave crest at the front of the cusps (Fig. 17.12a).

At same time, a convergence of the crests occurs at the center of the bays. This convergence results in an undertow being channeled into the center of the bays, generating a current stronger than the simple undertow. This rip current is capable of reaching and maintaining the supercritical flow (Fig. 17.12b). Each rip current deforms the crest of the next wave, causing the arriving wave advancing to the cusp zone to give rise to a new zone of divergence (Fig. 17.12c). The encounter between the rip current and the crest of the new wave is what produces both the convergent upturn and the undertow vortex, which in this case are located in the front of the bay areas.

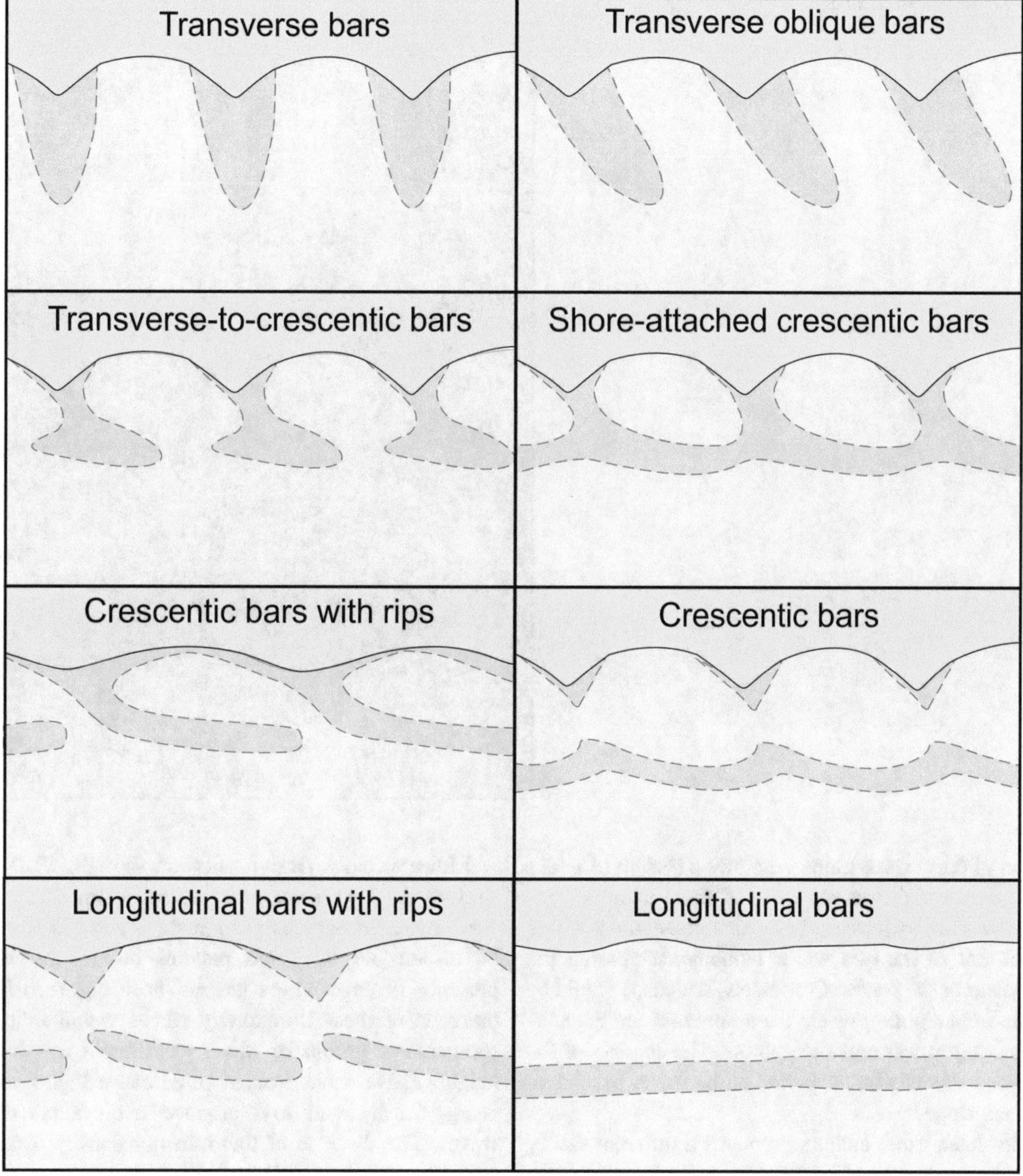

Fig. 17.10 Different beach bar morphologies. (adapted from Masselink et al. [42])

Taking into account the fluid dynamics and the movement of sediments in a rip-and-cusp system, two zones can be differentiated in terms of sediment movement:

(1) *Feeding zone:* Located above the step. This is the zone where the swash and undertow of the waves act. It is where the undertow currents dominate and the energy is concentrated. The flow of particles is produced towards the sea, cutting the breaker.
(2) *Deposit zone:* In this zone, the undertow current faces the pounding of the next wave, cancelling out its forces and depositing the bedload. The sedimentation of transversal bars takes place in these deposit zones.

One of the questions that has interested researchers since the last quarter of the twentieth century has been the cause of the regular spacing of the cusps. A possible origin was proposed by Guza and Inman [26], based on the interaction of the dominant waves with a stationary wave train that usually forms on reflective beaches running parallel to the coast. These waves are called edge waves. If the period of the edge waves is double that of the incident waves, the

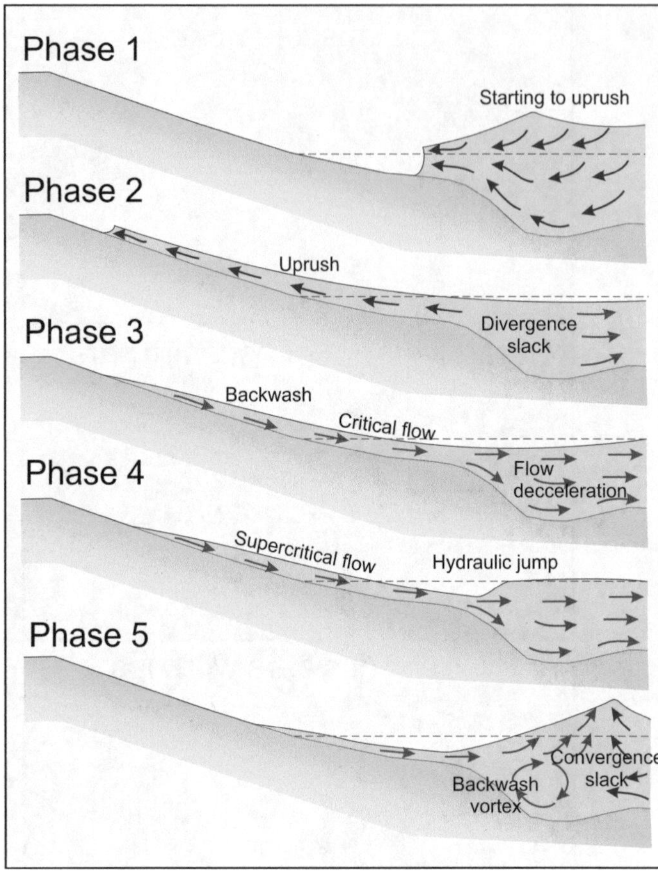

Fig. 17.11 Flow patterns during the different phases of a swash–backwash process. (adapted from Hughes and Turner [33]

interaction of these wave trains generates a system of nodes and antinodes, which was the basis of these authors' proposal. In this way, bays would be formed in the nodes where the amplitudes of the two waves combine, increasing the erosive power of the waves. Conversely, the cusps would be formed in the antinodes, where the amplitudes are counteracted, resulting in less erosional power. The spacing of the cusps depends directly on the period of the waves in relation to the beach slope.

Decades later, other authors proposed a different theory for the origin of the cusps [14, 58]. This is the theory known as self-organization. In this case, the proposal is based on the feedback that exists between the presence of irregularities in the coastline and the arrival of the wave trains. If we start from an irregular coastline, the coastal inlets would generate undertow currents that would deform the waves. This deformation of the waves would cause disturbances that would have a direct effect on increasing the erosion of these coastal inlets. The end result is the regular arrangement of the cusps and bays, whose spacing in this case would depend on the length of the sheet of water in the breaker (swash excursion). These two theories are not contradictory. If there are irregularities originated by a wave interaction, these could initiate a later self-organization that would adapt the resulting forms to the characteristics of the dominant wave.

More recently, the results of the GLOBEX project showed the importance of infragravity waves in non-linear wave transport patterns in high slope beaches. These could be responsible for linear patterns in systems where the presence of edge waves has not been observed [43]. The presence of these infragravity waves would influence the formation of groups of waves of different wavelength and height. These wave groups would cause higher and lower energy density bands to be produced every certain number of waves. The distance of these energy density bands would coincide with the spacing of the cusps.

17.4.3 Dynamics of Longitudinal Bars

There are two mechanisms proposed for the genesis of bars: the mechanism of the breaking point and the mechanism of standing waves. The **breaking point mechanism** was proposed by Aagaard et al. [1] and attributes the deposition of the bar to the convergence between the land transport in the shallow zone and the sea transport in the surf zone (Fig. 17.13a). These differences in transport are due to the orbital asymmetry of the waves in both zones, especially during periods of high wave energy. The **standing wave mechanism** is based on the presence of standing waves in

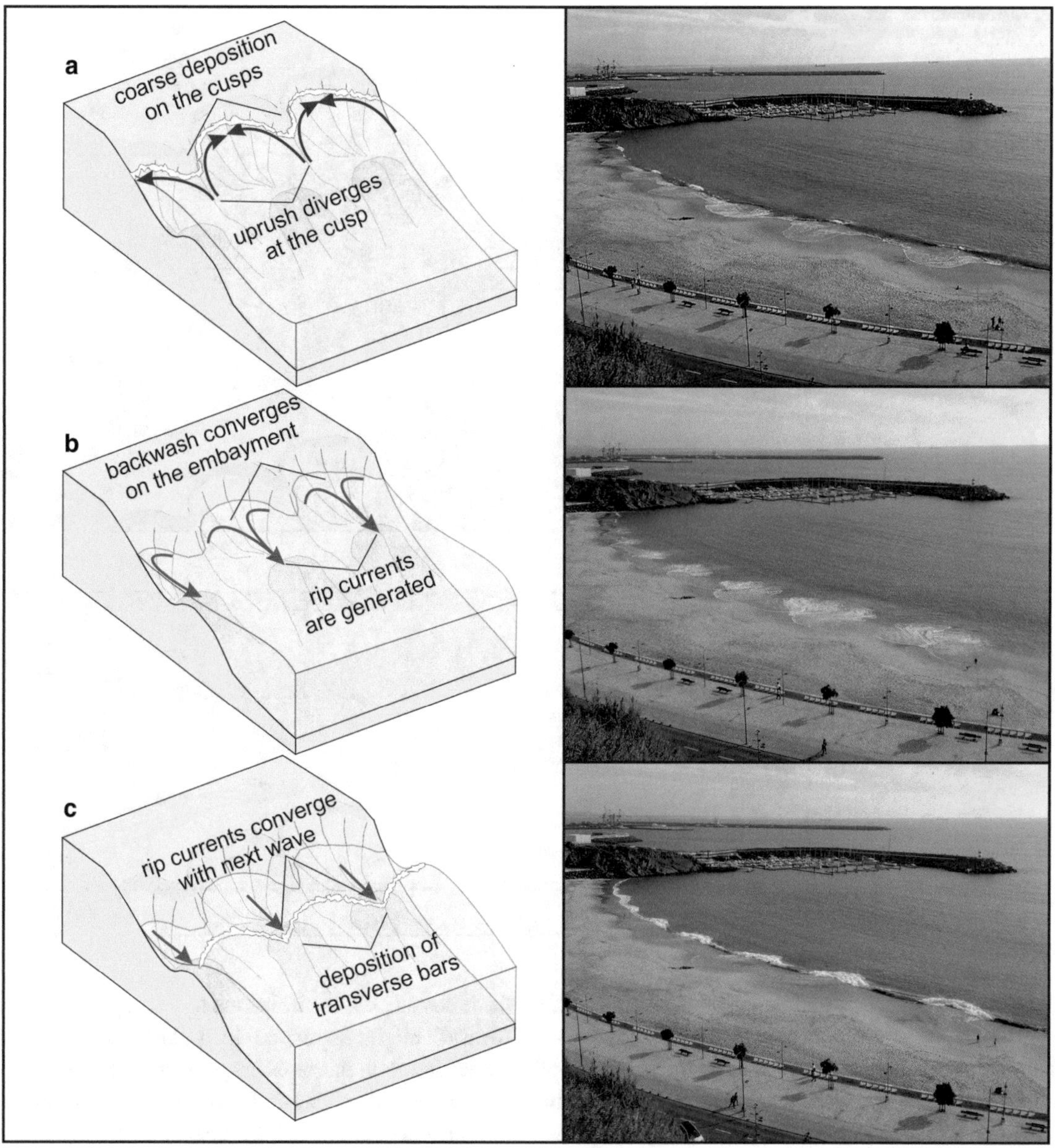

Fig. 17.12 Flow patterns in a rip-and-cusp system. (adapted from the ideas of Pethick [48])

the waterfront area acting parallel to the coast [11, 32]. Similar to what has already been observed in the formation of rips and cusps, the presence of these waves results in the formation of nodes and antinodes. This mechanism would give rise to a system of multiple bars generated in the antinodes, where the currents of different rotation orbits converge (Fig. 17.13b).

Once formed, the bars tend to migrate due to the dissipation of wave energy that is produced over them. Under significant wave conditions, there is usually a migration towards land through the surf zone, especially under the action of spilling type breaks [3]. This implies that the bars cross the shoreface to the foreshore and end up on the beach (Fig. 17.14). The migration rates of these bars are usually greater than 1 m per day, but have in some cases exceeded 30 m per day [21]. In this way, a bar can take less than a week to climb up the foreshore and be fully attached to the upper zone of the beach [52].

The mechanism of migration is simple in the passage of each wave. The swash generates a millimetric sheet inclined towards the land. To do this, it uses sand that is eroded by the swash on the face inclined towards the sea.

There may also be migration of bars towards the sea. This type of migration usually occurs during very energetic wave conditions, in which the breaker is either plunging, surging or collapsing. These types of breaker are characterized by a dominance of the undertow, in which erosion concentrates in the highest areas of the beach and sands are transported

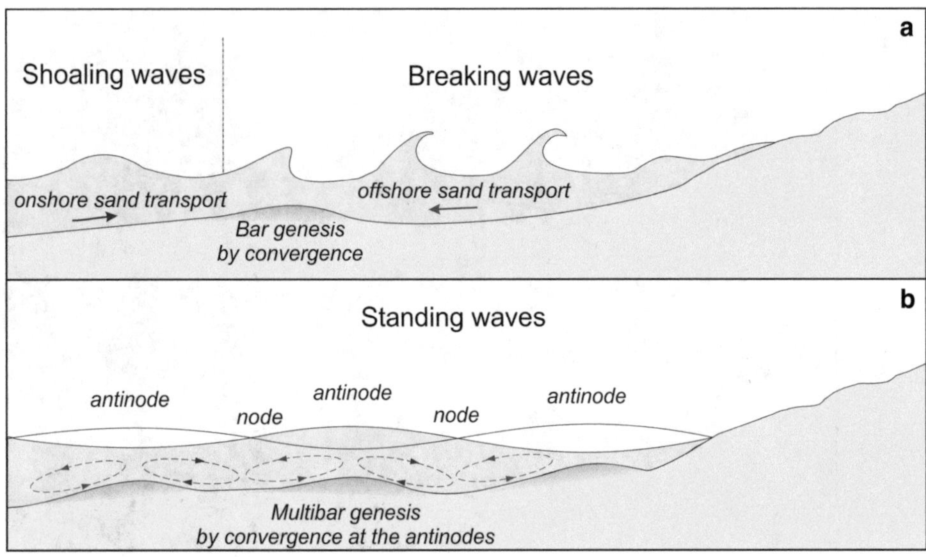

Fig. 17.13 Mechanisms of bar genesis. **a** Breaking point mechanism by Aagaard et al. [1]. **b** Standing wave mechanism by Carter et al. [11]. Figure adapted from Masselink et al. [42]

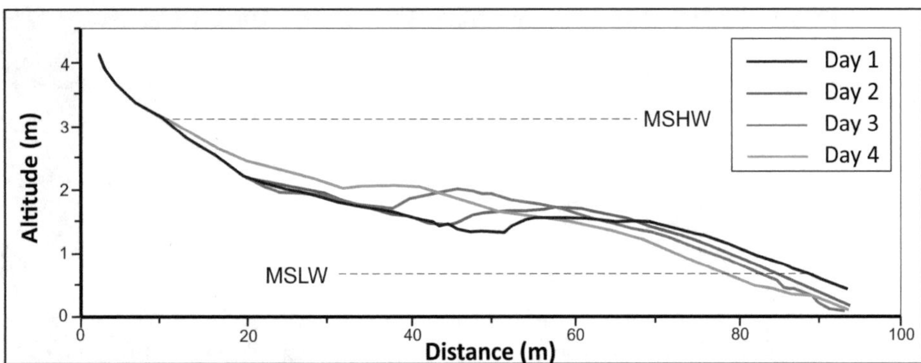

Fig. 17.14 Example of bar migration on the foreshore to be attached to the beach in only four days [52]

towards the sea. These conditions usually occur during storms.

Another characteristic of these systems is the existence of longitudinal currents channeled into the runnels above the bars. These currents increase in magnitude when the bar reaches higher levels and enters the foreshore. When the swash exceeds the bar, entering the runnel, the water cannot return downstream and is forced through the trough. The water thus channeled into the runnel seeks exit to the sea through undertow channels that cross the bar. In these channels, true rip currents can form with an almost continuous character.

17.4.4 Dynamics of Cobble Beaches

The profile corresponding to the size of these clasts is always reflective, so longitudinal bars never appear, but instead there are almost always cusps (Fig. 17.15). This type of high-slope profile is characterized by a coincidence between the breaker and the swash zones, so that the dynamics are dominated by the swash and undertow processes [61]. In these conditions, the influence of the edge waves is minimal. Instead, all the dynamic load falls on the energy flows associated with the breakers, which are usually of plunging, surging or even collapsing type. In these processes, and with such a coarse grain size, the characteristic mode of transport is traction, although in some vortices and for finer sizes, saltation may also occur. Thus, most of the particle transport is concentrated in the vortices that are generated around the break point, where the undertow currents of one wave are confronted with the pounding of the next wave. The most influential studies on these processes have been summarized by Buscombe and Masselink [9].

One of the big differences with sandy beaches is that the porosity reduces the speed of the undertow currents when a part of the water filters through the clasts. This effect also produces a decrease in the wave reflection process, so that the beaches with edges produce less reflection than sandy beaches with the same slope [49]. In this case, sorting plays an important role in permeability, since the presence of different clast sizes reduces pore volume. Another effect of the large grain size is created by the bed roughness, which has a significant influence on the shear stress of the fluid on

Fig. 17.15 Examples of cobble beaches in Asturias, N Spain, showing reflective profiles and rip-and-cusp systems. **a**: Panoramic view of Arreas beach. **b**: Shorefront view of Portizuelo beach. (Photographs courtesy of G. Flor-Blanco.)

the clasts by increasing friction. This roughness also affects the dissipation of the fluid energy in the formation of small vortices near the bed [41].

One of the characteristics of the ridge beaches is that the dynamics during storms introduce most of the changes in the profile, while the fairweather swell hardly causes any movement of the clasts. In this regard, it should be noted that many gravel beaches correspond to relict situations and the clasts are in clear imbalance with current dynamic conditions [46].

17.4.5 Equilibrium Profile

The beach profile is a topographic profile made from the highest areas (usually the dune ridge) to the front of the shoreface, including the berm and the intertidal beach (*foreshore*). It is usually studied during extreme low tides, which expose the entire intertidal zone and a part of the subtidal front. Theoretical studies by engineers compare the beach profile with an equilibrium profile to predict the erosive or cumulative behavior of the beach. The equilibrium profile is the theoretical topographic outline that a beach with a certain granulometry adopts when it is subjected to constant wave characteristics for a long time.

It is already established in the definition that it is a theoretical profile that is rarely reached, since the wave conditions are not constant, but highly variable. However, the use of this profile allows the prediction of behavioral trends between different wave situations. Other profile models take into account only the characteristics of the sediment, considering that the wave regime only introduces profile

Table 17.2 Application equations for the calculation of parameter A as a function of the average grain size (D50), according to the proposal of Dean and Maurmeyer [20]

D50	A
D50 < 0.4	A = 0.41 $D50^{0.94}$
0.4 < D50 < 10	A = 0.23 $D50^{0.32}$
10 < D50 < 40	A = 0.23 $D50^{0.38}$
D50 ≥ 40	A = 0.46 $D50^{0.11}$

oscillations, with the equilibrium profile being a central position of these oscillations.

The theoretical profile is established through models of wave transport calculation. The most widely used method is that proposed by Bruun [8] and modified by Dean [19], which takes into account only the characteristics of the sediment, because of the latter consideration mentioned. The expression for calculating the point-to-point height of the profile is given by equation 17.8.

$$y = Ax^{2/3} \tag{17.8}$$

where y is the beach elevation at a given point, x is its distance from the coastline and A is a grain size parameter related to the sedimentation rate and established from the mean grain size (D50). The calculation of the parameter A depends on the range of grain sizes and is reflected in Table 17.2 [20].

The choice of the 2/3 value for the coefficient is based on the Gaussian distribution of energy dissipation in the profile; however, other authors [7] noted that, on lower energy beaches like the Caribbean ones, the profiles fit better to a 1/2 coefficient.

The balance profile takes a concave shape and fits quite well to beaches without bars, but it departs from the typical profile of dissipative beaches with bars.

One of the drawbacks of an equilibrium profile as conceived by Brunn and Dean is that it is unable to predict changes in response to different wave dimensions. Vellinga [56] proposes the calculation of the equilibrium profile for different waves through a more complex equation (17.9), but with easily determinable input parameters, such as significant wave height and mean grain size (through their fall velocity).

$$y = \frac{Hs}{7.6}\left\{\left[0.47\sqrt{x\frac{7.6}{Hs}^{1.28}\frac{Ws}{0.0268}^{1.28}}\right] - 2\right\} \tag{17.9}$$

In this equation, y is the beach elevation at a given point less than the value 0.75Hs, x is its distance from the shoreline, Hs is the significant wave height and Ws is the value of the average grain size fall velocity of the beach (in m/s).

This equation is applied to obtain the equilibrium profile in the upper section of the beach, between the level reached by the surfacing up to a depth equal to the value of 0.75 Hs. For the lower stretch of beach, between 0.75 Hs and the closure depth is considered a stretch of constant slope (y = 1.25 x).

Using this equation, it is possible to predict, for example, the erosion profile that a beach will have in the face of a storm with a determined wave size, or the response that it would have in the face of a variation in the grain size of the sand that feeds the beach.

17.4.6 Evolution of the Beach Profile

The beach profile responds to variations in the wave regime as a result of the different transverse transport trajectories that occur under waves of different dimensions. Through the observation of the beach profiles, it is possible to monitor the evolution of beaches. Knowing and understanding the evolutionary tendencies of beaches is of notable social interest, since they are locations of important economic activity. However, it is also of interest from a purely sedimentary perspective, since this evolution influences the sedimentary sequences that are preserved in the geological record.

There are variations of the beach profile at different time intervals. Short-term variations are daily modifications, which respond to sediment transport trends under the action of wave trains with similar characteristics. Seasonal variations respond to annual cyclical changes that alternate between clearly constructive and clearly erosive periods. The variations of decades mark the trends of the beach in periods higher than interannual climatic cycles, such as ENSO or NOA, and respond to variations in the intensity and number of storms over the years.

The best-established changes are the seasonal ones (Fig. 17.16). This type of process was established by Shepard [53] and later verified by numerous authors through records in successive profiles. All authors who have addressed the subject agree on a seasonal alternation of the two characteristic processes of erosion and sedimentation. The variations can be summarized in two different moments that follow each other in a cyclical way. The morphology of the profile after calm periods is different from the morphology after the storms.

(1) *Calm conditions*: During these periods, a transport of sediment towards the highest areas of the beach takes place over days, weeks and months. In profiles with longitudinal bars, at the beginning of the process these are located in the lower part of the profile and progressively migrate towards the higher part. In the reflective and intermediate profiles, the calm period is

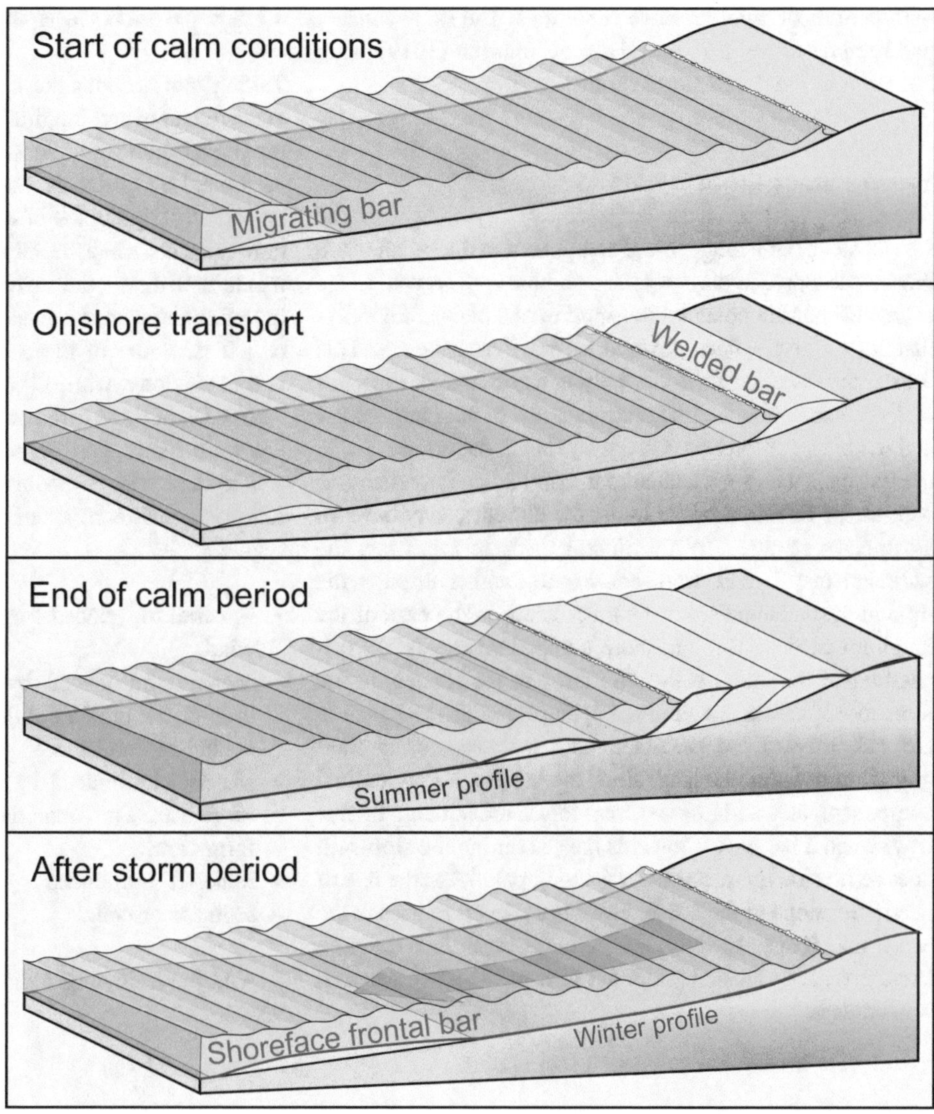

Fig. 17.16 Typical cycle of seasonal variations of a beach profile

the moment of greatest approach to land of cross bars. At the end of the period, the process of attaching the bars to the front of the berm achieves a convex profile at the top and concave at the base.

(2) *Storm conditions*: During these periods, large waves act with a peak of energy on the beach. These storm waves generate erosion processes in the high areas of the beach, accompanied by the transport of sediment to the submerged areas. It is precisely during these processes that the genesis of bars in the lower part of the shoreface begins, due to the mechanism of the point of rupture when the storm surge faces the waves in the process of shoaling. This type of event takes place during short time intervals that, once they stop, are replaced by successive moments of calm. On dissipative beaches, the process of migration of bars towards land is interrupted and they migrate towards the sea. On the other hand, in the reflective profiles, the transversal bars experience a greater development in the submerged part. The characteristic profile after a period of storms is concave in the highest part, where the erosive phenomena are concentrated, and convex in the lowest part, where the undertow accumulates the eroded material.

It is common that, on occasions, small storms interrupt the process of accretion of calm waves; however, the calm period is usually prolonged and fairweather waves quickly rebuild the action of a small storm. It is during large storms, or periods where successive storms occur, that the change in trend restarts the cycle.

These two periods give rise to the characteristic *summer and winter profiles* described by Shepard as the end result of both periods of different wave conditions. The same concept had been described earlier by Johnson Johnson (1919) using the terms *normal and storm* profiles.

17.4.7 Longshore Dynamics

The existence of a longitudinal component to the beach when the waves approach the coast in an oblique way, as well as the calculation of the potential transport of sediments, are topics that have already been covered in previous chapters. This littoral drift is manifested not only in a continuous transport of sediments along the surf zone, but also in the morphology and dynamics of the bars. On the one hand, the orientation of the transversal bars associated to the rip-and-cusp systems, as well as the direction of the undertow currents, may have an asymmetry imposed by the littoral drift. In this case, the sediment may bypass from one bar to another through the rip-and-cusp channels. On the other hand, in the case of the longitudinal bars, the longshore component marks the orientation of the currents that run through the runnels in the foreshore area. The longitudinal dynamics is also reflected in the trajectory of the bar's sediment particles during their migration towards the land, since the bar has a longitudinal component that adds to the transversal movement. In this way, when a bar moves towards the sea during the storms, it does so in a totally transverse trajectory, parallel to the line of maximum slope of the beach; however, when it returns to the beach line, it does so at an oblique angle, reaching a different stretch of beach located under the flow of the place where it was eroded.

17.4.8 Beach Cells and Sedimentary Balance

Taking into account the existence on the beach of a transverse dynamic and longitudinal transport, the beaches would be divided into semi-closed transport cells. Each cell would be limited by obstacles that interrupt the coastal drift totally or partially. Examples of obstacles that serve as boundaries between cells may be capes, river mouths, shoreface elements that divert the arrival of the waves to the coast by inverting the drift, or even artificial elements such as groins or jetties. Some of these cells can be tens or hundreds of kilometers long, while others barely reach a hundred meters.

In each of these cells, or even in a certain stretch of beach, there are inputs and outputs of sedimentary material from and to the systems or bordering the cells (Fig. 17.17). The following sediment inputs can be counted in any coastal segment:

- Material introduced from the land (by the rivers and the wind).
- Material introduced by the waves from the continent through the erosion of dune areas or rocky shores located behind the backshore.
- Material introduced by the waves from the front of the shoreface. This material can be brought in by relict formations.
- Material introduced by coastal drift from the updrift segment or cell.

And the following sediment outflows are also accounted for:

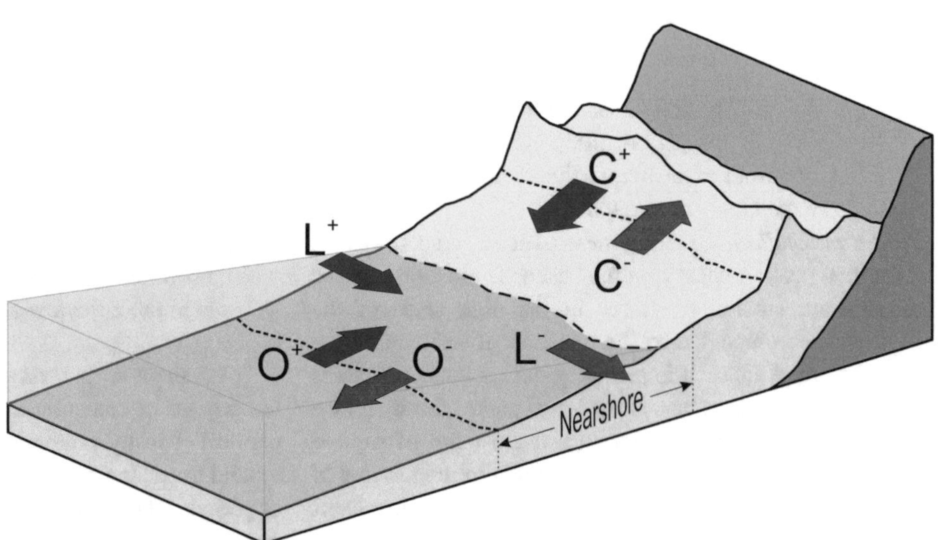

Fig. 17.17 Inputs and outputs in a beach sediment budget. Inputs: C^+ = from the continent. O^+ = transversal transport onshore. L^+ = longshore transport from updrift cells. Outputs: C^- = to the continent. O^- = transversal transport offshore. L^- = longshore transport to downdrift cells

- Material extracted by the waves towards deeper areas.
- Material extracted by drifting towards the segment or cell located downdrift.
- Material extracted by the wind towards the dunes.

The sediment budget will result from the algebraic sum of the sediment volume inputs and outputs, counting the inputs with a positive sign and the outputs with a negative sign.

It is important to precisely determine the time intervals to compare the amounts of sediment from each of these inputs and outputs, since the beach undergoes continuous changes and there are situations that are not comparable. One type of balance that is often established is the annual balance. This balance would reflect the synoptic situation throughout a complete year, taking into account that this would be the probable result of a period with positive balance in the calm period and another with negative balance during the period of storms. Another type of balance is made with longer intervals to evaluate erosive/accumulative tendencies. Given that there are multi-year cycles, such as those associated with climatic oscillations, it is important that the initial and final situations of the balance be located at similar times in the cycle.

17.5 Facies and Facies Sequences

17.5.1 Beach Sediment Characteristics

The sediment on the beaches is the product of the interaction between the beach slope and the dimensions of waves. These interactions determine the process of energy dissipation that is transferred to the transport of sediments. In general terms, the grain size of the sediment is a function of the intensity of the breaker, which depends on both factors. However, there are variations in the grain size across the transverse beach profile, so that the larger clasts are located right in the break zone, with a gradient of size loss towards the higher areas of the beach and also towards the deeper areas [45]. There is also a zone of larger grain size in the zone of the first breakers, where the dissipation of energy begins to be transferred to the movement of particles. This transfer of energy to the movement of particles can be understood in terms of sediment dynamics subjected to the orbital velocity of waves until they cross their entrainment and selection thresholds, as seen in Sect. 14.4. On tidal beaches, these larger grain size zones move vertically, causing a size variation over time in a particular area of the beach profile, which results in an increase in grain size dispersion that coincides with a loss of sorting [5, 6]. In this way, the highest grain size and the lowest sorting coincide precisely where the highest values of energy dissipation are reached. Such variations in sediment size and classification have been observed both on sandy beaches [23] and on gravel and pebble beaches [39].

The differences in grain size are even more apparent between the different layers that make up the beach sediment. These variations between layers reflect temporary variations in energy over longer periods of time. Thus, there is a clear relationship between the average grain size in a layer and the wave energy that generated it. This means the alternation between layers of different grain size reflects the variations in time of the wave dimensions and the changes in the way the wave energy is dissipated. It is very frequent to find intercalations of gravel (siliciclastic or bioclastic) and heavy minerals that correspond to the action of more energetic waves at certain moments. This alternation of layers of different grain size can even become rhythmic, responding to seasonal changes between periods of fairweather and storms (Bascom 1951). Gravel beaches present patterns very similar to sand beaches; in fact, many gravel beaches also present interspersed sand levels that correspond to moments of lower wave energy.

Despite the clear influence of wave energy on grain size in relation to transport capacity, there is a primary control exercised by the availability of sediment. Thus, although in some beaches there is a capacity to transport coarser grains, these will not exist if these grain sizes are not available at the source feeding the system.

Edge beaches usually have less variation in grain size, precisely because of the availability of sediment. Some pebble beaches show no temporal variation in size, despite the existence of clear alternations between waves of different dimensions (e.g., Carr [10]). This lack of variation is due to the absence of clasts of other sizes and in many cases only the largest waves reach the movement threshold of these clasts.

17.5.2 Sandy Beach Facies

The facies on sandy beaches are distributed according to the depth, since this factor, together with the dimensions of the waves, conditions the type of interaction of the wave with the bed. Thus, each of the distinct areas has its own facies. Moreover, in beaches with longitudinal bars, the bars and troughs have different facies, giving rise to characteristic sequences of the ridge-and-runnel systems (Fig. 17.18).

The characteristic facies of each of these areas have been established in successive works by Edward Clifton, and summarized in a synthesis paper [13]. In the case of reflective beaches, the facies are less studied, since the upper part of this type of beaches is erosive and does not usually generate preservable facies. However, facies associated with the transversal bars can be generated. This type of facies has been described by Isla et al. [35].

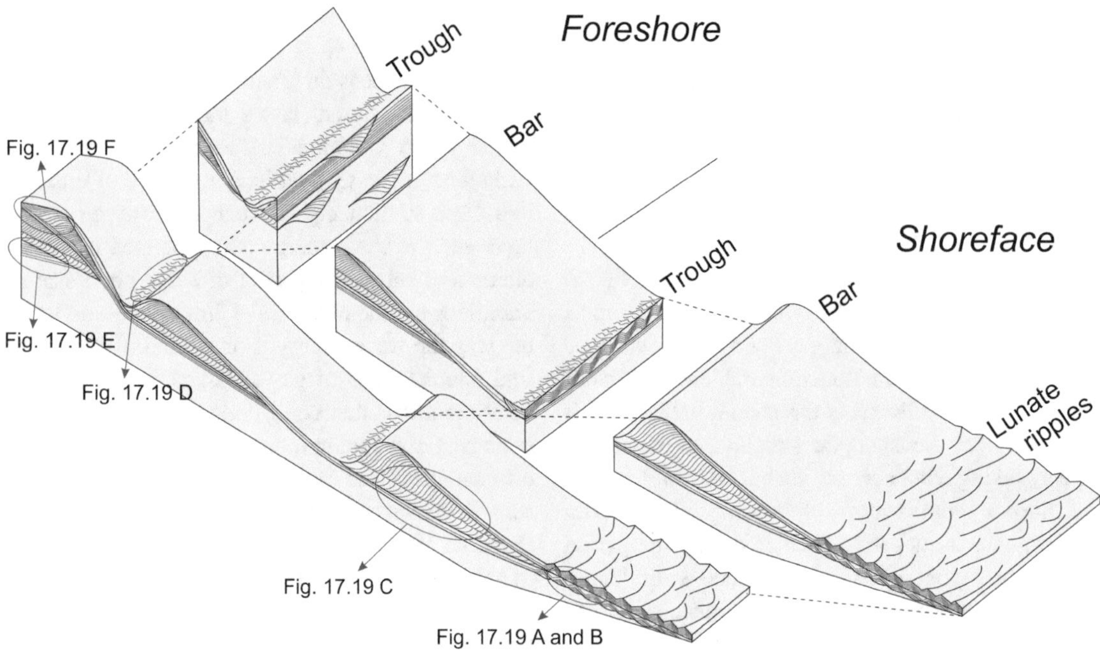

Fig. 17.18 Facies distribution scheme of a ridge-and-runnel system (Based on Davidson-Arnott and Greenwood [16])

The following describes the characteristic facies of each of the distinct areas in the beach systems.

- *Offshore*: Alternating silts, very fine and fine sand with oscillation ripples that can be asymmetrical. These alternations may contain coarser grain size intercalations and *hummocky* stratification corresponding to storm levels. In all cases, the bioturbation usually alters the original structure.
- *Shoreface in the shoaling zone:* In this zone, fine sand is usually sedimented with asymmetrical ripples of lunar morphology (Fig. 17.19a and b). Sometimes, bars with vergent cross-bedding are formed towards the sea, product of the erosion of the storms in the highest zones of the beach. Although these bars are subsequently reworked during fairweather periods, parts can be preserved.
- *Shoreface in the breaker and surf zones*: They are usually composed of medium-sized sand grains. The facies developed in zones of smaller depth to the breaker depend on the type of beach. In beaches with bars, there is also a difference in the position of the facies with respect to the bar. In the outer zone of the bars, a parallel lamination inclined towards the sea is developed, as a result of the asymmetric swash of the waves and the transport of the material towards land. At the crest of the bars, there is cross-bedding inclined towards land, as a result of the migration of the bar in this direction (Fig. 17.19c). In the troughs, asymmetric ripples are usually formed, which migrate in the direction of the littoral drift. Although 3D megaripples may also be present, migrating landwards like the bars.

On reflective beaches there is usually a coarser grain size of the sand. From the point of view of the structures, this zone is characterized by the monotonous development of parallel lamination inclined towards the sea, which passes at deeper levels to parallel sub-horizontal bedding [35] or cross-stratification inclined towards the sea, typical of the transversal or crescentic bars.

- *Foreshore*: The facies developed in the foreshore area also depend on the type of beach. In beaches with bars, the distribution of facies is very similar to the one described in the shoreface zone; however, the inclination of the lamination is usually greater, both in the parallel laminations inclined towards the sea, and in the cross-stratification inclined towards the land. This occurs because the foreshore bars usually have a greater relief than the submerged bars. In profiles parallel to the beach line, the migration and filling structures of the exit channels of the undertow currents are also usually observed. In the troughs, the asymmetric ripples are usually well developed and even acquire linguoid morphologies (Fig. 17.19d). Their migration is always oriented in the direction of the drift. On beaches with multiple bars, these facies repeat as much as the number of bars.

  On reflective beaches, this zone is usually erosive (Fig. 17.19e), although parallel laminations may develop in an ephemeral way, strongly inclined towards the sea. The three-dimensional geometry of these laminations shows the typical concavity of rip-and-cusp systems.

  In both types of beaches, the development of coarser grain layers formed by siliciclastic or bioclastic material

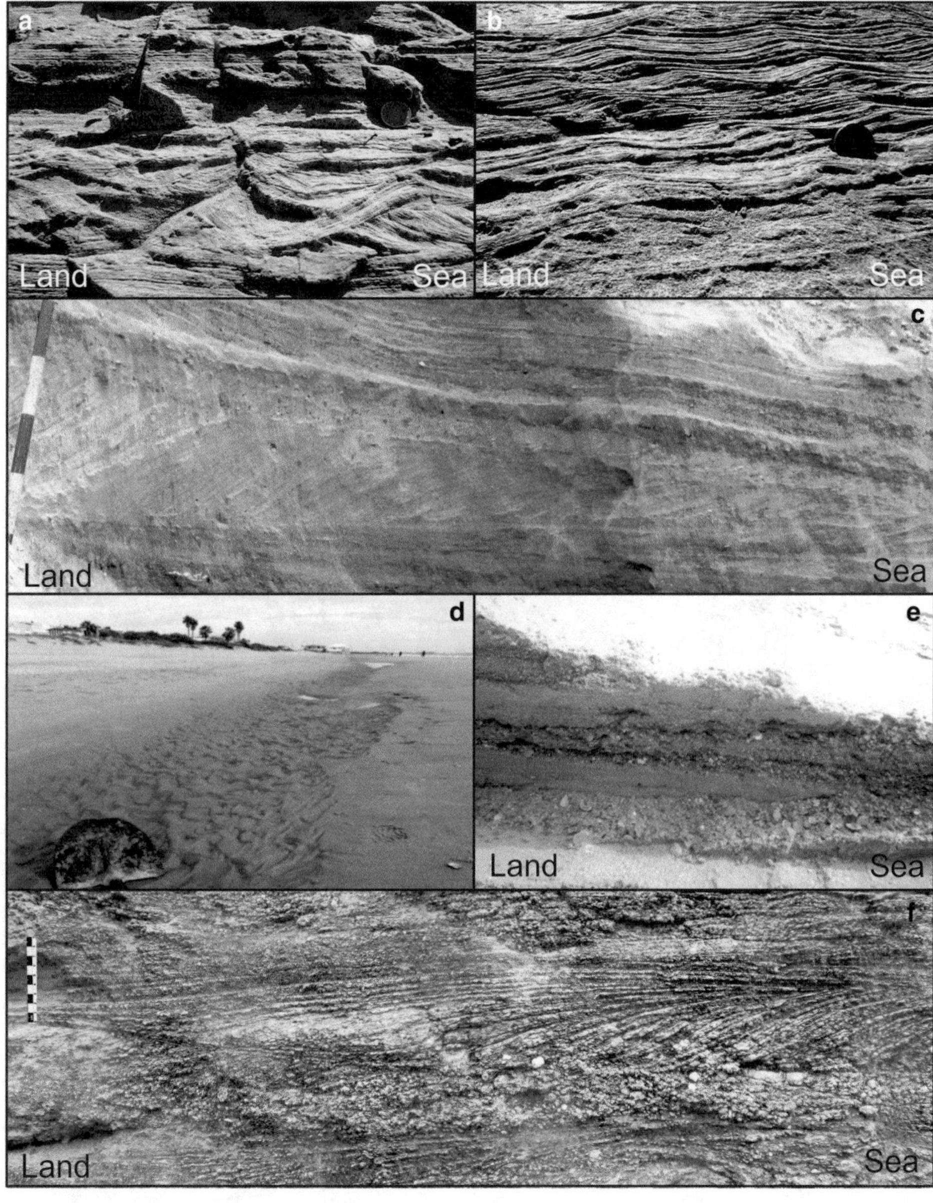

Fig. 17.19 Different facies in beach systems. **a** Wave ripples with erosive curved base. **b** Asymmetric wave ripples. **c** Landward-inclined cross-bedding and seaward-inclined parallel bedding corresponding with the record of a migrating bar. **d** Linguoid ripples in a foreshore trough. **e** Erosive surface cutting a parallel lamination in a reflective beach. **f** Structure of clogging of a backshore trough by a welding bar

and also of heavy mineral sheets is typical of the foreshore. These layers have a residual character and correspond to the deposit of storm waves.

- *Backshore*: This is usually characterized morphologically by the berm. This morphology responds to the final attachment of bars, so its characteristic facies show the development of sands with sets of cross-stratification inclined towards the land, which culminate in concave beds that mean the final filling of the runnels (Fig. 17.19f). On these, parallel lamination can be developed that corresponds to the remobilization of the finest fractions by the wind.

The facies models for this type of environment do not really reflect a three-dimensional distribution of the facies, but, given the tabular character of the beaches, they are limited instead to establishing a sequence of facies that follows the guidelines of Walther's Law under coastal progradation [13, 44]. The different sequences proposed [24, 50, 57] do not present many conceptual or descriptive differences, although the consideration of the thickness of the series and the grain size range are different (Fig. 17.20). Both characteristics are the reflection of the dimensions of the wave, which translates not only into a difference in the energy put into play, but also in the bathymetry in which it can exercise its action and deposit sediments.

In this way, all the proposed sequences mark a coarsening-upwards character, while, at the same time, the sedimentary structures also manifest an increase of the energy towards the top of the sequence as we approach the facies originated in the swash area (Fig. 17.20).

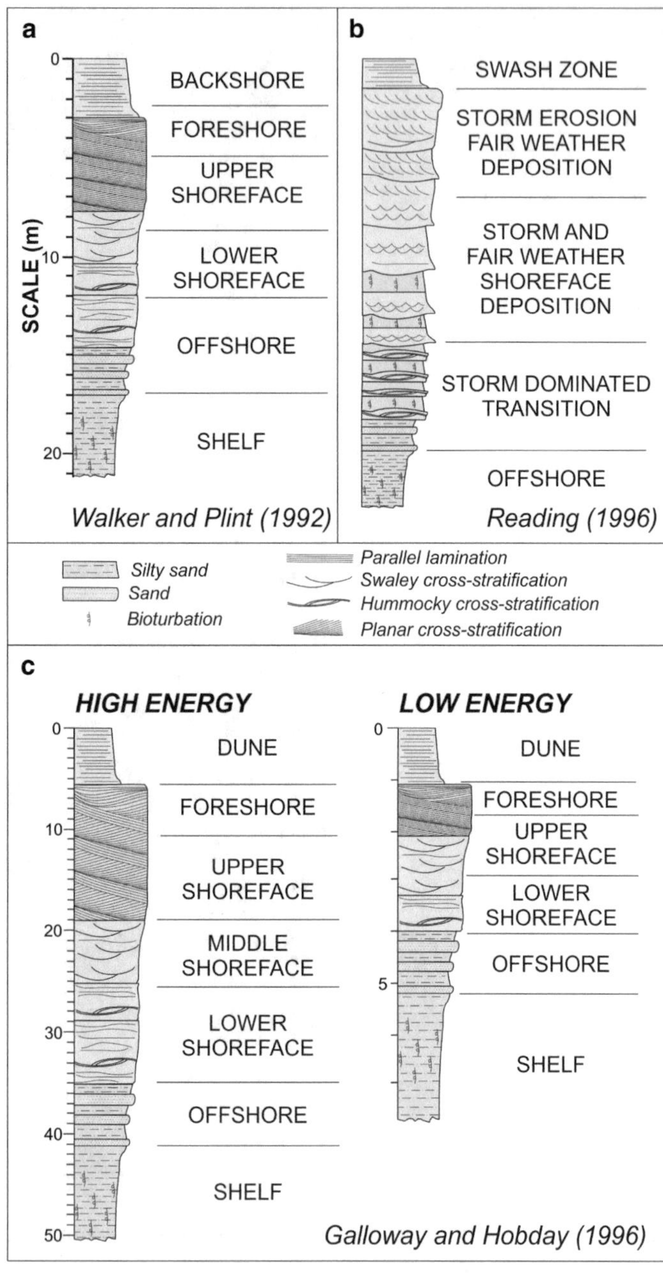

Fig. 17.20 Facies sequences suggested by different authors, based on ancient beach sequences. **a**: Walker and Plint Walker and Plint (1992). **b**: Reading [50]. **c**: High and low energy sequences suggested by Galloway and Hobday [24]

17.5.3 Gravel Beach Facies

Gravel beaches are much less documented than sandy beaches. Excellent synthesis papers are included in Hart and Plint's monograph [30] and the same authors have also published a review article on facies sequences [31]. Gravel beaches are known to be those with grain sizes ranging from 2 to 64 mm, including clasts of granules (2–4 mm) and pebbles (4–64 mm). In general terms, it can be stated that gravel beaches present a dynamic and morphology similar to reflective-type sandy beaches, with concave profiles of high slope. However, their larger grain size requires higher wave energy to transport the larger-sized clasts (2 mm). At the base of the sequence, corresponding to the deposits of the lower shoreface, there is usually a finer grain size with development of wave ripples, often asymmetrical and oriented towards the sea.

In the high part, most of the gravel beaches have reflective profiles, with development of transverse bars. However, in swell conditions, intermediate beaches can develop crescentic bars. One of the characteristics of gravel beach facies is the presence of fining-upward sequences inside the bars.

In the long term, the periods in which they move seawards alternate with others in which they move landward. In the sequence of sediments, this is manifested in a different orientation in the slope of the cross-bedding (Fig. 17.21). In

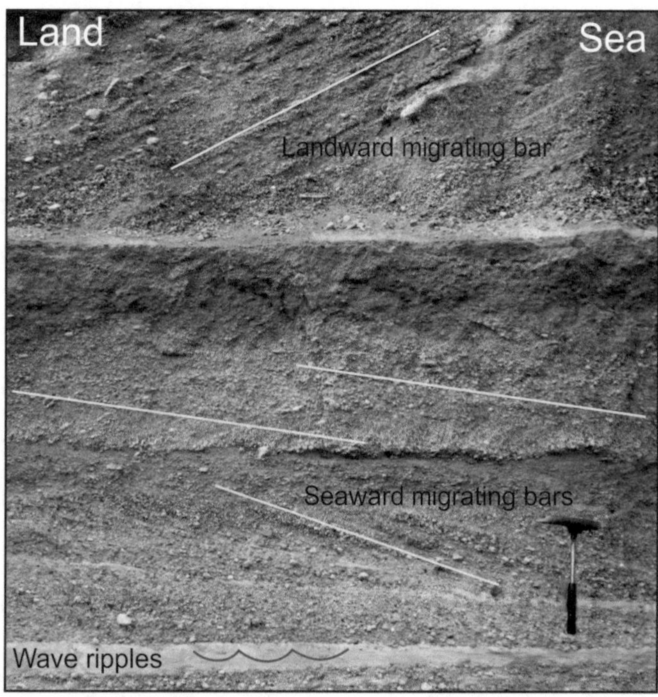

Fig. 17.21 Facies sequence typical of a gravel beach

these sequences, the seaward-inclined stratification typical of the transverse bars originated by the undertow always presents smaller inclinations than the landward-inclined cross-stratifications typical of the crescentic bars.

The larger grain size also influences the inclination of the cross-layering, which represents the slope of the bars. In this case, the bars formed by gravel present a greater angle of rest than the sand bars, with greater inclination in the crossed stratifications, especially in the bars that migrate landward—these can reach angles superior to 25°, as opposed to the 21° typical of sand.

17.5.4 Cobble Beach Facies

Cobble beaches have grain sizes greater than 64 mm, but less than 256 mm. The presence of these clasts is associated with sedimentary availability, as well as with the energetic characteristics of the waves. In most cases, the basis of the sequence is an erosive unconformity over rocky outcrops, since these beaches tend to have limited development on rocky shores. Normally, such large clast sizes prevent the clear observation of the internal structure, although the presence of fining-upward and coarsening-upward sequences is observed, whose base is usually an erosive surface. Occasionally, if the clasts are flattened and interlocked, cross-lamination can be observed, inclined towards the sea and coinciding with the paleoslope. This structure is characteristic of the reflective character of this type of beach.

References

1. Aagaard T, Nielsen J, Greenwood B (1998) Suspended sediment transport and nearshore bar formation on a shallow intermediate state beach. Mar Geol 148:203–225
2. Anthony EJ (2009) Shore processes and their paleoenvironmental applications. Elsevier, Amsterdam, p 519
3. Aubrey D (1979) Seasonal patterns of onshore/offshore sediment movement. J Geophys Res 84:6347–6354
4. Bascom WN (1960) Beaches. Sci Am 203:80–94
5. Birkemeier WA (1985) Field data on seaward limit of profile change. J Waterw Port Coast Ocean Eng 111:598–602
6. Birkemeier WA, Miller HC, Wilhelm SD, Dewall, AE and Gorbics CS (1985) A user's guide to the coastal engineering research center's (CERC) field research facility. US Army Corps of Engineers, Waterway Experiment Station, Vicksburg, Mississippi, USA
7. Bonn JD and Green MO (1988) Caribbean beach-face slopes and beach equilibrium profiles. In: Proceedings of the 21st coastal engineering conference. American Society of Civil Engineers. pp 1618–1630
8. Bruun P (1954) Coastal erosion and the development of beach profiles. US Army Beach Erosion board. Technical Memorandum, pp 44
9. Buscombe D, Masselink G (2006) Concepts in gravel beach dynamics. Earth Sci Rev 79:33–52
10. Carr AP (1969) Size grading along a pebble beach: Chesil Beach, England. J Sediment Petrol 39:297–311
11. Carter TG, Liu PL, Mei CC (1973) Mass transport by waves and offshore sand bedforms. J Waterway Harbors Coastal Eng 99 (2):165–184
12. CERC (1984) Shore protection manual department of the army. US Army Corps of Engineers, Washington. vol 2. pp 559
13. Clifton HE (2006) A reexamination of facies models for clastic shorelines. In: Posamentier HW and Walker RG (Eds.) Facies Models Revisited. SEPM Special Publication, vol 84. pp 293–337

14. Coco G, Huntley DA, O'Hare TJ (2000) Investigation of a self-organisation model for beach cusp formation and development. J Geophys Res 105:21991–22002
15. Davidson-Arnott RGD (2010) Introduction to coastal processes and geomorphology. Cambridge University Press, Cambridge, p 442
16. Davidson-Arnott RGD and Greenwood B (1976) Facies relationships on a barred coast. Kouchibouguac Bay, New Brunswick, Canada. In: Davis Jr RA, Ethington RI (eds) Beach and Nearshore sedimentation. Society of Economic Paleontologists and Mineralogists Special publication, vol 24. pp 149–168
17. Davies JL (1980) Geographic variation in coastal development. Longman, New York, p 212
18. Davis RA Jr, Fitzgerald DM (2004) Beaches and coasts. Blackwell Publishing, Oxford, p 419
19. Dean RG (1977) Equilibrium beach profiles: US Atlantic and Gulf Coasts. Ocean Engineering Technical Report 12
20. Dean RG, Maurmeyer EM (1983) Models of beach profile response. In: Komar PD, Moore J (eds) CRC handbook of coastal processes and erosion. CRC Press, Boca Raton, pp 151–165
21. Elgar S, Gallagher EL, Guza RT (2001) Nearshore sandbar migration. J Geophys Res 106:11623–11627
22. Evans OF (1940) The low and ball of the East Shore of Lake Michigan. J Geol 48:476–511
23. Fox WT, Ladd JW, Martin MK (1966) A profile of the four movement measures perpendicular to the shoreline, South Haven, Michigan. J Sediment Petrol 36:1126–1130
24. Galloway WE and Hobday DK (1996) Terrigenous clastic depositional systems: applications to fossil fuel and groundwater resources, 2nd edn. Springer, Heidelberg, pp 489
25. Gourlay MR (1968) Beach and dune erosion tests. Delft Hydraulics Laboratory Report M935-M936
26. Guza RT, Inman DL (1975) Edge waves and beach cusps. J Geophys Res 80:2997–3012
27. Hagen GHL (1863) Handbook of hydraulic engineering: 3 (in German). Teil das Meer, Berlin, p 918
28. Hallermeier RJ (1981) A profile zonation for seasonal sand beaches from wave climate. Coast Eng 4:253–277
29. Hallermeier RJ (1983) Sand transport limits in coastal structure design. In: Proceedings, coastal structures '83. American Society of Civil Engineers, pp 703–716
30. Hart BS, Plint AG (1995) Gravelly shoreface and beachface deposits. Special Publication of the International Association of Sedimentologists, vol 22
31. Hart BS, Plint AG (2002) Gravelly shoreface deposits: a comparison of modern and ancient facies sequences. Sedimentology 36 (4):551–557
32. Holman RA, Bowen AJ (1982) Bars, bumps and holes: models for the generation of complex beach topography. J Geophys Res 87:457–468
33. Hughes MG, Turner IL (1999) The beachface. In: Short AD (ed) Handbook of beach and shoreface morphodynamics. Wiley and Sons, Chichester, pp 119–144
34. Iversen HW (1952) Waves and breakers in shoaling water. In: Proceedings of the 3rd coastal engineering conference. American Society of Civil Engineers, pp 1–12
35. Isla MF, Schwarz E, Veiga GD (2020) Record of a nonbarred clastic shoreline. Geology 48(4):338–342
36. Jennings R, Shulmeister J (2002) A field based classification scheme for gravel beaches. Mar Geol 182:211–228
37. Johnson DW (1919) Shore processes and shoreline development. Wiley, New York, p 584
38. Komar PD (1998) Beach processes and sedimentation. Prentice-Hall, Englewood Cliffs, New Jersey, p 429
39. Krumbein WC, Griffith JS (1938) Beach environment in Little Sister Bay, Wisconsin. Geol Soc Am Bull 49:629–652
40. Masselink G, Short AD (1993) The effect of tide range on beach morphodynamics, a conceptual model. J Coastal Res 9:785–800
41. Masselink G, Li L (2001) The role of swash infiltration in determining the beachface gradient: a numerical study. Mar Geol 176:139–176
42. Masselink G, Hughes MG, Knight J (2003) Introduction to coastal processes and geomorphology. Routledge, London, p 416
43. Michallet H, Ruessink BG, da Rocha MVLM, de Bakker A, van der A D, Ruju A, Silva PA, Sénéchal N, Marieu V, Tissier M, Almar R, Abreu T, Birrien F, Vignal L, Barthélemy E, Mouazé D, Cienfuegos R, Wellens P (2014) GLOBEX: wave dynamics on a shallow sloping beach. In: Proceedings of the HYDRALAB IV joint user meeting. Lisbon, Portugal. pp 12. HAL open files id: hal-01084718
44. Middleton GV (1973) Johannes Walther's Law of the correlation of facies. Geol Soc Am Bull 84:979–988
45. Miller RL, Zeigler JM (1964) A study of sediment distribution in the zone of shoaling waves over complicated bottom topography. In: Miller RL (ed) Papers in marine geology. MacMillan, New York, pp 133–153
46. Orford JD, Carter RWC, Jennings SC (1991) Coarse clastic barrier environments: evolution and implications for Quaternary sea level interpretation. Quatern Int 9:87–104
47. Otto T (1912) The Darß and Zingst: a contribution to the history of the development of the West Pomerania (in German). Annual Report of the Geograph Soc Greifswald 13:393–403
48. Pethick J (1984) An introduction to coastal geomorphology. Edward Arnold, London, p 260
49. Powell KA (1990) Predicting short term profile response for shingle beaches. Hydraulics Research. Wallingford, Report SR 219
50. Reading HG (1996) In: Sedimentary environments: processes, facies, and stratigraphy. 3rd edn. Blackwell Scientific Ltd., Boston, pp 688
51. Roy PS, Cowell PJ, Ferland MA, Thom BG (1994) Wave-dominated coasts. In: Carter RWG, Woodroffe CD (eds) Coastal evolution: late quaternary shoreline morphodynamics. University Press, Cambridge, pp 121–186
52. Sedrati M, Morales JA (2017) Effect of transversal groins on dissipative barred beach dynamics: a study case in Matalascañas beach, Huelva (in Spanish). Geo-Temas 17:239–242
53. Shepard FP (1950) Longshore bars, longshore troughs. In: Beach erosion board, technical memorandum. vol 20. pp 38
54. Shepard FP (1963) Submarine geology. Harper and Row, New York, p 557
55. Short AD, Aagard T (1993) Single and multi-bar beach change models. J Coastal Res 15:141–157
56. Vellinga P (1984) A tentative description of a universal erosion profile for sandy beaches and rock beaches. Coast Eng 8:177–188
57. Walker RG, Plint AG (1992) Wave- and storm-dominated shallow marine systems. In: Walker RG, James NP (eds) Facies models: response to sea level change. Geological Association of Canada, pp 219–238
58. Werner BT, Fink TM (1993) Beach cusps as self-organised patterns. Science 260:968–971
59. Woodroffe CD (2003) Coasts: form. Cambridge University Press, Cambridge, Process and Evolution, p 623
60. Wright LD, Short AD (1983) Morphodynamics of beaches and surf zones in Australia. In: Komar PD (ed) Handbook of coastal processes and erosion. CRC Press, Boca Raton, pp 35–64
61. Wright LD, Thom BG (1977) Coastal depositional landforms and morphodynamic approach. Progress in Phys Oceanography 1:412–459

18 Wind-Dominated Systems: Coastal Dunes

18.1 Introduction

The action of wind as a transport agent on the coast was analyzed in Sect. 9.3. Its role can be summarized by three different effects: (1) contribution of sand from the continent to coastal systems; (2) erosion by deflation on the surface of beaches; and (3) accumulation in coastal areas protected from waves. In this last sense, deposits of sand from wind deflation can reach the back-barrier systems in various forms. On the one hand, the material transported by wind can feed the coastal lagoons and the tidal flats, contributing to their aggradation. On the other hand, *overblow* processes can build sand fans very similar to washovers but with a purely wind-based origin. Eolian sandy plains can also develop in the supratidal areas of the *strandplains*. However, the most common forms of deposit associated with wind transport are coastal dunes.

A dune can be defined as a mound made up of sand accumulated by the action of the wind. Coastal dunes are those that develop in coastal environments. The origin of the sand that constitutes them is normally linked to the wind deflation on the beaches that are situated at their front [20]. In one way or another, dunes are part of almost all depositional coasts, especially if they are wave-dominated, so they are widely distributed across the world. Many coastlines develop small chains of dunes associated with the upper part of the beaches,however, where sand input is abundant, large areas may be covered by extended dune fields that can even migrate inland [21]. In sand barriers, the dunes occupy the highest topographic level and are what really prevent the action of the waves at the back of the system. There is actually a wide typology of dune forms in relation to the combination of factors involved in their genesis and evolution (Fig. 18.1).

Due to their location and origin, the dunes and the beach are intimately linked and their dynamics act together. For instance, while the deflation on the beach is the main contribution in building the dunes, the dunes themselves form a sand store whose presence is vital in the preservation of the beaches, constituting a supply of sand that feeds the beaches during storms. Similarly, dune systems can be the source of sedimentary contribution to other coastal environments such as washovers.

From an ecological point of view, the dunes constitute one of the most notable environments of the coastal ecosystem, where a wide variety of animal and plant species are developed [17]. Perhaps a geologically more outstanding aspect of the dunes is their capacity to recharge fresh water. Indeed, dune aquifers are sometimes the only source of fresh water in barrier systems.

This chapter will analyze all the dynamic processes that contribute to the genesis and evolution of coastal dunes, as well as their relationships with the adjacent coastal environments and their most characteristic facies.

18.2 Control Factors

Three conditions are needed for dunes to be created: a sufficient supply of sand, a wind with enough energy to transport that sand, and a large surface area on which it can accumulate. There are also other conditions that can influence the development of coastal dunes, such as the climate, the degree of humidity of the land and the vegetation cover [1].

18.2.1 Wind Regime

Section 14.5 discussed how wind transport is a function of the viscous stress of the wind on the sand grains. The viscous stress is a direct function of wind velocity. The formation of dunes requires the existence of winds of intermediate speed between the entrainment and selection thresholds. Velocities below the entrainment threshold will not produce grain movement, while velocities much higher

J. A. Morales, *Coastal Geology*, Springer Textbooks in Earth Sciences, Geography and Environment,
https://doi.org/10.1007/978-3-030-96121-3_18

Fig. 18.1 Coastal dune developed in the backshore area of a sandy beach

than this value will tend to transport so much that they will prevent the genesis of dune forms or even end up eroding the pre-existing dunes [7].

In addition to velocity, a very important issue is the directional regime of the wind. The genesis of dune systems usually requires the presence of persistent winds with a dominant transport component. The persistence of the wind is an even more important factor than its velocity in the formation of dune chains. In coastal systems, the differential heating between the continent and the seawater mass is responsible for the presence of breezes that blow coastwards a large part of the time, ensuring a wind circulation pattern that favors the formation of coastal dunes. It is evident that the orientation of the coast with respect to these trends influences the wind efficiency in both the process of wind deflation on the beaches and the piling up of sand in the dunes.

The existence of winds with different vectors from the dominant direction influences the growth and possible migration of the dunes in these directions. This process then influences the geometry of the dunes, as well as the presence of complex sets of internal stratification.

18.2.2 Sediment Budget

The development of coastal dunes requires a constant supply of sand from the beaches. The dimensions and geometry of the first ridge of dunes depend directly on the volume of input. In conditions of insufficient supply, the dunes do not form. This need is further evidence of the important connection between the beaches and the dunes, which form a circular feedback system of coastal dynamics [19, 22].

The relationships between the volume of sand input, the wind transport capacity and the coastal dynamics are shown in the development of successive chains of coastal dunes. On prograding coasts, the lines of coastal progradation are often marked by the presence of successive ridges of foredunes. However, on stable coasts with sufficient input, those that are generated as foredunes can begin to migrate inland, while new foredunes are built in the original position.

18.2.3 Development Surface

By now it has become clear that coastal dunes develop in connection with the beaches that supply the sediment to build them. But the beaches can appear in different coastal systems: at the front of sand barriers, on the coastal platform of cliff systems or on the strandplains. Thus, the dune chains can reach greater or lesser dimensions depending on where these beaches are located, and whether dune migration landwards will be possible or not.

On sandy barriers, the width of the dune field is limited to the dimensions of the barrier. On spits or drumstick-type barrier islands, the foredune ridges mark the growth lines of the barriers. Where some of these ridges are formed by mobile dunes, the migration landwards is limited by the lagoon presence. In some cases, the dune may migrate into the lagoon, where its sediments may be reworked by the tides and re-sedimented as tidal facies. In other cases, the dunes may migrate over supratidal systems such as tidal flats or coastal sabkhas.

In dunes that develop in the backshore of frontal beaches linked to cliff systems, the wind turbulence caused by the presence of the rocky front favors the development of eolian accumulation at their base. However, the cliff prevents the migration of these dunes towards land. If the cliff face is not completely vertical, some dunes can pile up and climb over it, reaching considerable heights. The dunes may even manage to exceed the height of the cliff and migrate over its crest.

On beaches developed in coastal plains, the dunes have enough space to migrate to land. In these cases, extensive mobile dune fields can develop over the continental systems.

18.2.4 Climate

Coastal dunes can develop in any climate. There are dune systems distributed latitudinally from the poles to the Equator, in different degrees of humidity, ranging from dry deserts to tropical systems [9, 16, 26–27, 15]. With this widespread distribution, climate cannot be used as a diagnostic criterion for coastal wind systems. However, climate determines the type of vegetation and the degree of vegetation cover, as well as the sand moisture level and wind regime. Thus, it can be considered that, although climate is not a determining factor in dune formation, it does exert a significant influence on the development and dynamics of dune systems.

18.2.5 Moisture Content

The degree of humidity of the sand is a factor in the mobility of the grains, since the presence of water in the pores exerts a viscous friction that acts as an inertial force. On a beach surface that is used as a wind deflation platform, there is usually a high degree of humidity, so it is necessary for the air flow to dry the ground before the particles can be moved from it.

The dune system itself may also contain a significant degree of porewater during and after rain, as dunes often act as a recharging element for coastal aquifers. The presence of this moisture in the dunes makes the particles remain longer in the formation and contributes to their preservation.

In the interdune areas, the water table can be exposed by dune migration. In this case, these zones act as a deflation surface, whereby the wet zone is preserved and the zones where the pores are full of air are removed and eroded. The surfaces thus formed can be covered by the migration of a new dune; in these cases, the positions of the phreatic levels can be preserved as ravinement surfaces.

18.2.6 Vegetation Cover

A unique feature of coastal dunes is that they arise from the interaction of physical processes generated by the wind with biological processes influenced by the presence of vegetation. The first studies considered that plants only colonized the dune surface by adapting to it; however, today the relationships between vegetation and dune dynamics are considered to be a system of mutual interaction [12]. Thus, it is now known that biological factors play a role as important as sand transport conditions in the geometry and dynamics of many coastal dune chains.

To start with, plants can play an important role in the genesis of the incipient dunes. Once the dunes are formed, the plant roots contribute to their stabilization. Dune systems well fixed by vegetation can achieve a notable increase in height without the dune changing its position. These dunes are characterized by a complex internal structure, where sets of different inclination angles alternate [5, 8]. Hence, the vegetation cover can be a primary factor in determining the dimensions of the dunes [6].

One consequence of this fact is that the vegetation begins to exert a direct influence on the volume of sand stored in the coastal system to be used during high wave energy conditions. The presence of this sand storage is a controlling factor on the beach profile during storms. In this way, the existence of the vegetated dune not only ensures the survival of the beach, but is a guarantee of self-protection.

18.3 Morphology and Sub-environments

The morphology and dimensions of dunes depend on a complex relationship of all the factors analyzed in the previous section. There is a wide morphological variety of dunes, although the most-used classification is based on genetic criteria, since it is the genesis that imposes the morphology. Firstly, one needs to differentiate between primary and secondary dunes [4, 25]. **Primary dunes** are generated directly on the upper part of the beaches using the sediment from the deflation processes that occur on the foreshore surface. **Secondary dunes** come from the reworking of the sandy material of the primary dunes or from the migration of the same towards the land.

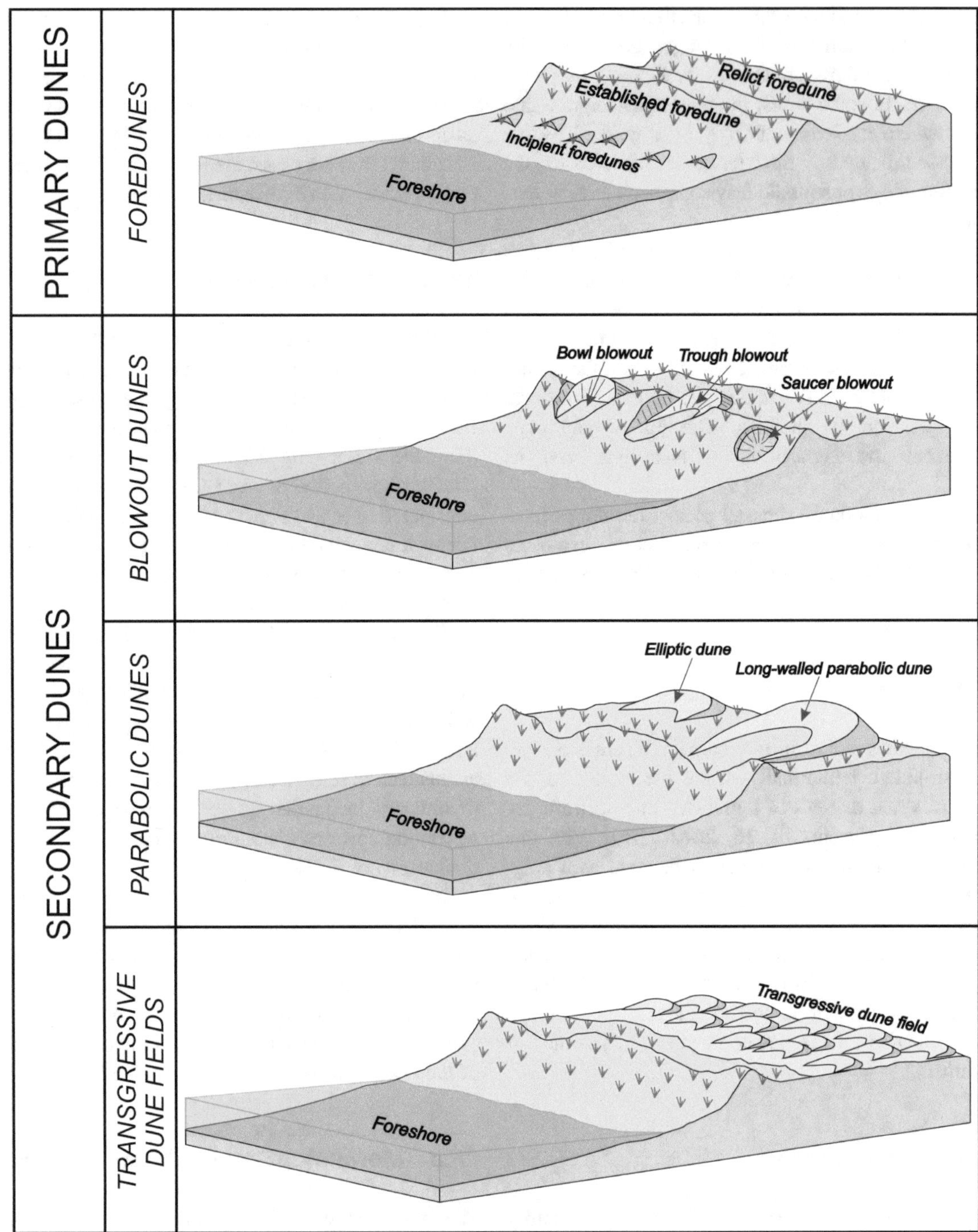

Fig. 18.2 Morphogenetic classification of dunes (based on Davies [4, Short and Hesp 10, Hesp 24], and Tsoar [25])

The primary dunes are known as frontal dunes or **foredunes,** while the secondary dunes can be divided by their morphology and are linked to their degree of evolution into **blowout dunes**, **parabolic dunes** and **transgressive dune fields** (Fig. 18.2).

When there are several chains of coastal dunes, depressed areas known as **interdune zones** may be established between them. These zones can function as wind deflation surfaces, although, depending on the climate and the position of the water table in them, **inter-dune lakes** may also be present.

18.3.1 Foredunes

These are primary dunes and thus formed with the contribution of sand directly from the beach. The name derives from their location at the front of the coast—that is, from their position close to the beach, as they are located in the area immediately behind the backshore. From a physiographic point of view, they are transverse dunes that generally form a continuous ridge parallel to the beach line. Morphologically, they have a convex profile that can be symmetrical or asymmetrical and the height/width ratio depends on their degree of evolution, with those that are more evolved having greater height. According to their evolutionary development [10, 24], the foredunes can be classified as incipient, established or relict (Fig. 18.3).

The **incipient dunes**, also called embryonic dunes, are small discontinuous or continuous mounds formed in the shade of objects or vegetation (Fig. 18.3). They usually have more width than height and their dimensions do not go beyond the metric scale.

The **established dunes** present a greater degree of evolution and are formed by vertical growth of the incipient dunes. They are already continuous chains of convex shapes that correspond to the initial morphological description (Fig. 18.3). They are normally stabilized by vegetation, so the effect of vertical accretion dominates.

Relict dunes are characteristic of prograding coasts. They were once established dunes, but the progradation of the beach has allowed the birth of a new established dune chain in front of them, resulting in these dunes taking up a rear position (Fig. 18.3). They present the same morphology as established dunes, although their growth is less, as they do not receive the direct contribution of sand from the beach due to the presence of a new foredune.

18.3.2 Blowout Dunes

These are dunes that have developed from deflation holes that arise in the foredunes. As they are associated with an erosional phenomenon, they have two different zones: a deflation area and an accumulation lobe. Depending on the geometry of both zones, they can be distinguished as saucer **blowouts**, bowl **blowouts** or trough **blowouts**.

Saucer blowouts have a semicircular deflation surface without a depositional lobe (Fig. 18.4a) or with a low relief lobe. The bowl blowouts present a similar shape to the saucers, but with a greater degree of incision and also a greater development of the depositional lobe (Fig. 18.4b). The trough blowouts present an elongated geometry with a corridor-shaped deflation surface with steep lateral margins and a parabolic-shaped depositional lobe with much vertical development (Fig. 18.4c).

Fig. 18.3 Panoramic view showing the different types of foredunes at Punta de San Jacinto, Doñana National Park (SW Spain)

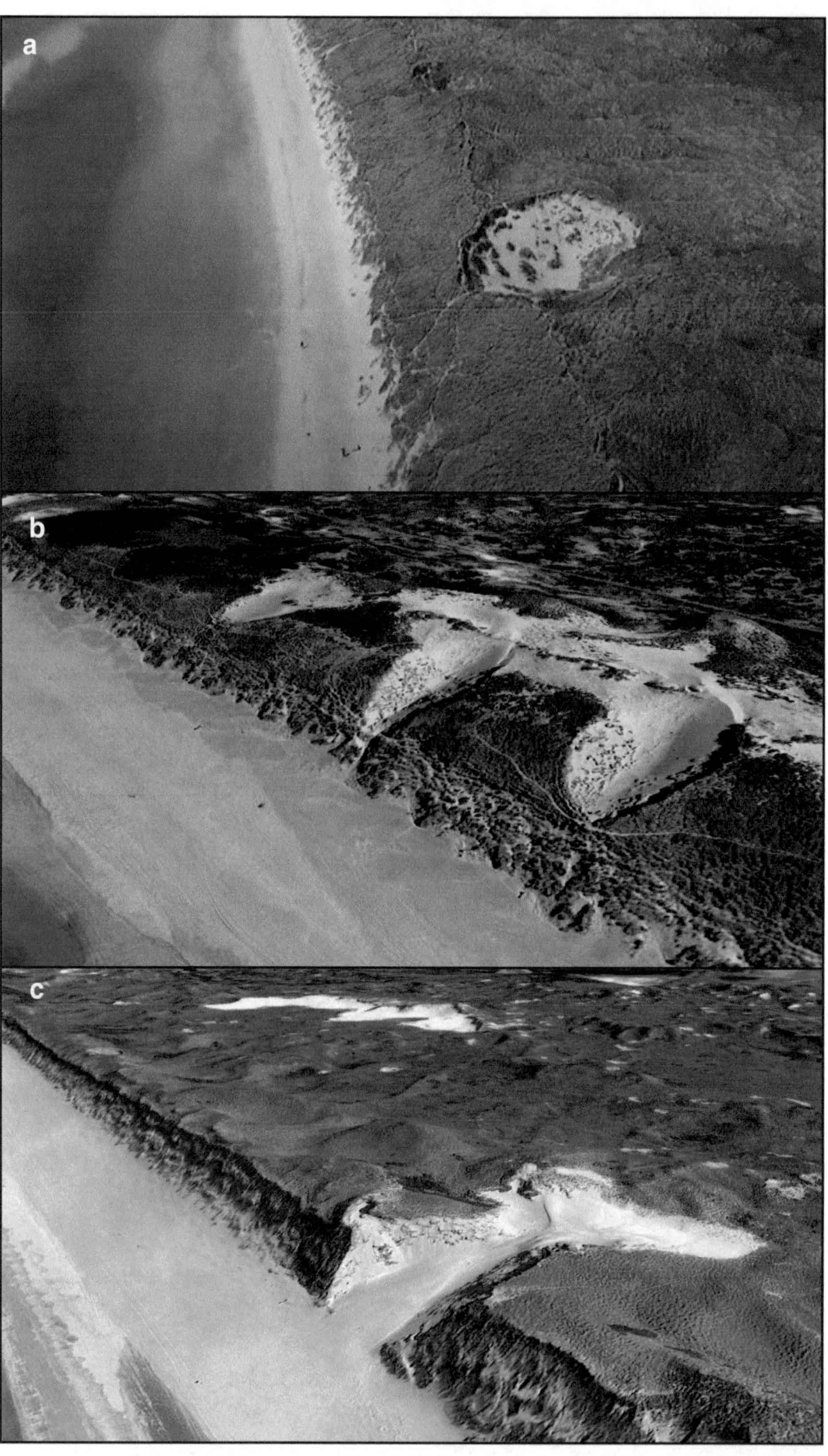

Fig. 18.4 Different types of blowout dunes. **a** Saucer blowout in Cruden Bay dunes (Scotland). **b** Bowl blowout in Haasvelder dunes (the Netherlands). **c** Trough blowout in Heemskerker dunes (the Netherlands)

18.3.3 Parabolic Dunes

These are more evolved dunes that have undergone a process of migration towards land from an original frontal position. The name refers to a morphological criterion, since the dune crest is in the form of a parabola due to the migration of the central part of the dune exceeding that of the extremes, acquiring the shape of a linguoid ripple but of decametric dimensions (Fig. 18.5). Since they are mobile dunes, one of their faces is erosive (stoss side), while the other face is accumulative (lee side). As the erosive face has a parabolic shape, it can be considered that a deflation basin is established in its center, whose orientation is usually parallel to the direction of the dominant wind.

Morphologically, two types of parabolic dunes can be distinguished: the elliptical dunes and the long-walled dunes.

The **elliptical dunes** are not elongated, presenting more semicircular shapes (Fig. 18.5a). They acquire dimensions of hundreds of meters and usually appear in fields where dunes overlap. They are characteristic of areas with a large sandy input but without strong winds, so their migration speed is low.

Fig. 18.5 Different types of parabolic dune. **a** Elliptical dunes in Castilla Beach (SW Spain). **b** Long-walled parabolic dunes in Lagoinha (NE Brazil)

The **long-walled parabolic dunes** have large central deflation basins and very elongated arms (Fig. 18.5a). They tend to develop greater dimensions and can reach the kilometer scale. They are characterized by large areas of flat terrain, large sedimentary inputs and strong and persistent winds. They acquire very high migration speeds.

18.3.4 Transgressive Dune Fields

These are sand deposits that cover large areas and are formed by the landward movement of sediments of wind origin (Fig. 18.6). If the sand is displaced in a laminar fashion due to strong winds, an **eolian mantle** can develop, where the dunes take on the form of low domes. They can also develop due to an extensive migration of dune fields towards land. These sandy formations can reach scales of tens of square kilometers. Within them, dune forms can be as complex as in desert areas, including transverse dunes, parabolic dunes, barchans and barchanoids.

18.3.5 Interdunes

When there are several chains of dunes, or between individual dunes in a field of transgressive dunes, depressed areas develop. These are known as interdune depressions or simply interdunes. In addition to their depressed morphology, the interdune depressions present other characteristics that can be different depending on the nature of these depressions. In areas where dune piling occurs, the character of the interdune is depositional and in these we can find evidence of the dunes that developed in the lower level. On other occasions, dune migration gives these depressions an erosive nature and deflation processes dominate them. A special case of interdune depression that is dominated by deflation occurs when the dunes migrate on a cohesive substrate. In this case, the deflation exposes the surface of the lower formations in the interdune. In many cases, this surface is covered by clasts with a clear residual character (Fig. 18.7a). On other occasions, the deflation process digs down to reach the water table. There, it meets a layer of sand in which the pores are filled with water and the moisture prevents the threshold of particle movement being reached [14]. The interdune is then characterized by a completely flat erosion surface. In situations of aquifer recharge, the water table can rise, flooding the interdune depressions. In that case, interdune lakes are formed (Fig. 18.7b). If the flooding period of these lakes is prolonged, they may have a depositional character, generating characteristic deposits that will depend on the climate where they have developed.

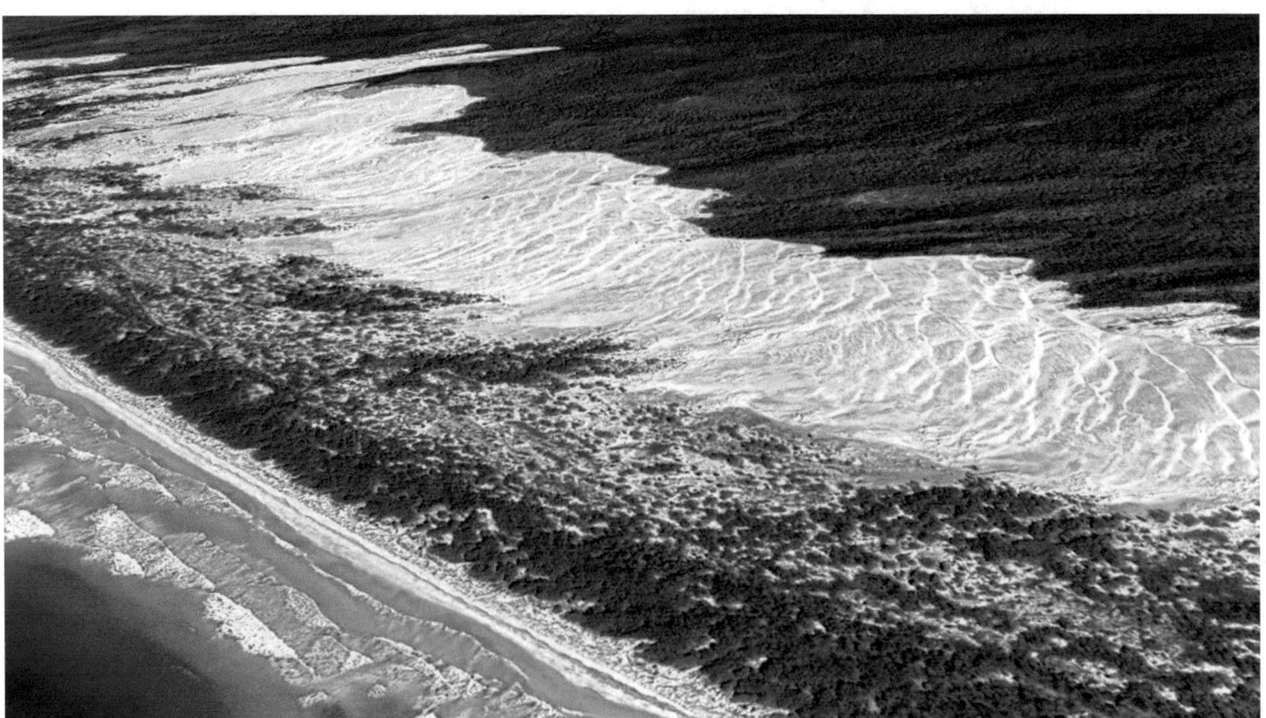

Fig. 18.6 Transgressive dune field in NE Fraser Island (Australia)

Fig. 18.7 Different types of interdune depression. **a** Coarse deflation lag in St Fergus (Scotland). **b** Lagoonal interdunes in the Lençóis Maranhenses National Park (Brazil)

18.4 Dynamics and Evolution

In Sect. 14.5, the mechanisms of entrainment and selection of the particles by the wind were described in detail. The mechanism of entrainment acts preferably in the backshore, as well as in the foreshore during low tide. It is on that surface where this mechanism generates a deflation process, which obtains all the sand particles that are transported towards the coastal supratidal fringe. In this strip, other mechanisms that favor the decrease in wind speed occur, leading to the formation of coastal dunes. In this section we will analyze the mesoscale processes that give rise to the genesis, dynamics and evolution of all the types of dunes that can be present in coastal areas.

18.4.1 Genesis and Evolution of Primary Dunes

The sea winds loaded with sand grains on the surface of the beach can lose speed when entering the supratidal fringe. The main effect of the loss of speed is that the threshold of

Fig. 18.8 Genesis of incipient dunes. **a** Sand sheets formed on the backshore by wind deceleration. **b** Sand mounds formed in the dynamic shadows of plants

particle selection can be crossed, and the particles will then stop being transported and settle on the surface of the coastal land. This deceleration mechanism may simply be due to friction with the ground when the wind enters an area with a slope greater than the slope of the deflation surface. In this case, the continuous accumulation of sand on this slope can give rise to incipient or embryonic dunes (Fig. 18.8a).

Via this mechanism, the presence of any element introducing roughness in the terrain can generate turbulence in the wind. Normally, after this turbulence, a shaded area where particles tend to deposit is generated. These irregularities in the terrain can be rocks, driftwood, accumulations of algae or dead animals, even garbage, but very often it is the dune vegetation that causes this effect. Thus, it is very common for sand piles to form in the shade of plants (Fig. 18.8b). These mounds tend to grow, and in them the influence of saline water decreases as the spray effect diminishes and fresh water is recharged during rainfall. At the same time that the dimensions of the incipient dune increase, the levels of nutrients also increase, giving rise to an ideal place that can be colonized by new plants, which in turn will tend to increase the accumulation of sand. The final result is the generation of a dune that is more and more stable.

As the size of the dune increases, the sand tends to be deposited more and more on the marine face rather than in the shaded area. The presence of vegetation covering the whole dune also causes a large fraction of the sand to be deposited on the crest of the dune, increasing its height. In the absence of erosional processes, the growth of these mounds can in a short time give rise to a continuous ridge of sand that will acquire the characteristics of an established foredune. The dimensions that the foredune can reach depend on factors such as the rate of transport, the extension of the deflation surface and the time that the dune continues to receive sand from the beach front without suffering erosional processes. Due to these factors, it is understandable that dissipative beaches develop the largest dunes. There are cases of foredunes that reach heights of over 20 m [12].

The genesis of a new chain of dunes in front of the old foredune makes it a rear form. The roughness of the new foredune generates greater wind turbulence, making it less effective in transporting sand to the old foredune, which has thus become detached from the deflation surface. In this way, the dune is fixed in its growth. At this point we can speak of a relict foredune [11]. Successive chains of relict dunes are characteristic of prograding beaches, since on these beaches the advance of the coastline towards the sea generates space and guarantees the supply of sand for the development of new foredunes. A prograding coast can show dozens of chains of relict foredunes. These lines mark the position of old shorelines.

18.4.2 Effects of the Waves on the Dunes

Both the incipient dunes and the established foredunes can be subject to the action of waves during storms. The overlift caused by surges and the swash of large storm waves usually attacks the front dunes once or several times throughout the year. These conditions subject the dunes to a process of erosion, which is especially intense when they coincide with spring high tides.

The incipient dunes are the most sensitive to these conditions and can suffer total destruction during the most severe storms. Often, the incipient dunes are ephemeral forms and never become established foredunes [12].

The wave attacks frequently occur at the front of established dunes. In these cases, the front of the dunes suffers an erosion process that generates a vertical escarpment and the dune loses a significant volume of sand (Fig. 18.9b). This sand is redistributed back to the beach and contributes to its preservation during erosional moments. The return of fair-weather conditions allows the beach to function again as a deflation surface and thus the sand will be transported back to rebuild the dune [19].

The first stage of reconstruction of the dune front begins with the development of small incipient dunes that form with the reflection of the wind on the erosive escarpment of the established dune. Due to this wind flow inversion effect, these dunes are known as echo dunes (Fig. 18.9c). The small echo dunes represent an obstacle to the wind flow and they grow until they deviate from the established dune front. At this point, the inversion of the wind flow concludes and the echo dunes begin to function as incipient dunes that end up adhering to the established dune front (Fig. 18.9d). Thus, the dune profile is reconstructed and the foredune can continue its growth. The old erosive escarpment is preserved inside the dune structure as a reactivation surface.

This process can be repeated with the arrival of successive storms, alternating periods of erosion with periods of reconstruction. The internal structure of the dune will preserve the erosive–accumulative history as numerous sets of cross-stratification separated by reactivation surfaces.

18.4.3 Reworking of the Primary Dunes and Evolution Towards Secondary Dunes

Erosional processes in an established foredune can be caused not only by waves, but also by the wind. One of the most common processes that causes wind erosion of foredunes (established or relict) is the "**out-blow**" process. This process usually starts when there is some cause that triggers a loss of vegetation cover in a part of the dune. The local absence of vegetation facilitates the movement of sand grains during times of high wind speed [3]. Among the natural causes that can initiate the loss of vegetation are erosion by waves and gravitational processes by destabilization of the dune front. In recent times, the causes of the initiation of these processes have usually been related to human activities on the dune system [18]. An out-blow process causes an erosive structure known as a **blowout**, which is recognized as a form of secondary dune.

The geometry and dimensions of a blowout depend on the initial size of the surface that has lost vegetation, but also on the speed and persistence of the wind at the local level and the wind flow conditions over the erosive zone. The type and density of the vegetation in the surrounding area also often has an important influence on the development of the blowout. A blowout does not only consist of an erosive zone, but under favorable conditions the eroded sand can accumulate on the margin giving rise to a dune lobe.

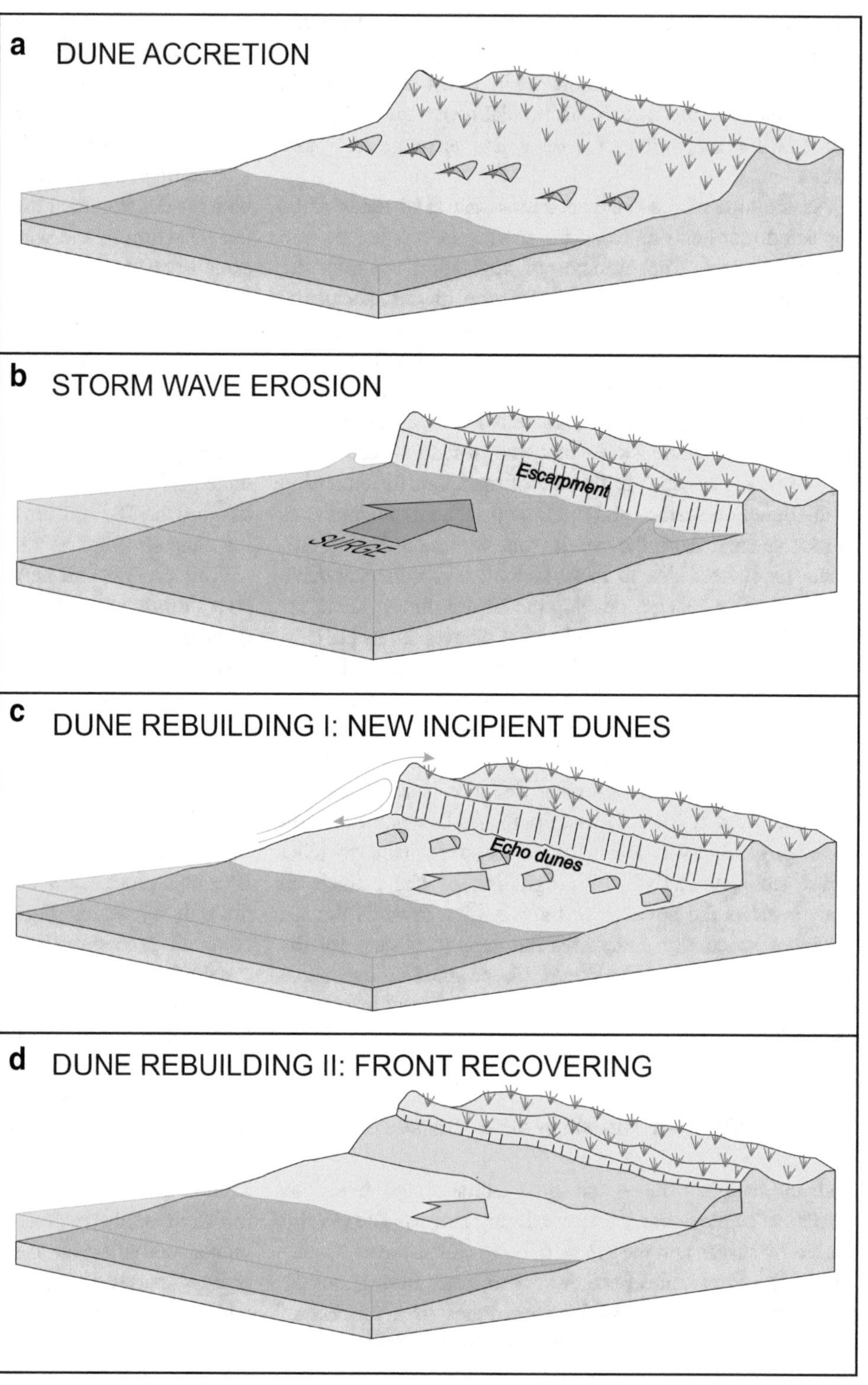

Fig. 18.9 Cycle of wave erosion and wind rebuilding of a foredune

The three types of blowout that have been differentiated morphologically can form an evolutionary sequence. A lack of vegetation of small dimensions can form a saucer blowout (Fig. 18.10a). In this form, the out-blow process usually extracts the sand beyond the dune system without the formation of a cumulative lobe. However, in a more advanced stage of evolution, this initial blowout can enlarge and develop the dune lobe by accumulation of sand on the leeward side [13]. The interaction of wind with a blowout is complex, since the presence of this lobe produces turbulence in the flow that tends to increase erosion. The morphological response is an increase in the size of the blowout. If the process continues, the erosive depression can deepen by developing near-vertical walls. The exit of the wind in the leeward side of the blowout normally implies a decrease in the speed and a tendency to accumulate sand, developing a dune lobe of greater dimensions. In this case, the complete transformation will have taken place to form a bowl blowout. The most advanced process continues with a deepening of the depression and a migration of the lobe, which is fed by the deflation that occurs in the depression. In this case, the lobe migration is usually accompanied by an elongation of the erosive depression. This lengthening usually continues until the depression takes the form of a groove or corridor (Fig. 18.10b). The geometry of this corridor tends to increase the wind speed when crossing it. The process culminates when this corridor completely crosses the profile of the foredune and the dune lobe moves to the rear zone, where the wind acquires a radial disposition when it leaves the corridor.

The dune lobe of a trough blowout can continue to move until it is disconnected from the primary dune. In this case, it acquires a curved shape and can start to be called a parabolic dune (Fig. 18.10c). At first, the dune shape is elliptical and continues to be linked to the deflation corridor located in the primary dune, without the existence of a core in the elliptical dune shape that functions as a deflation basin. If the process of advancement of this dune continues, it will acquire a shape closer and closer to a parabola, developing in its center a deflation basin that is flanked by the two elongated arms of the parabola. This is the type called a long-walled parabolic dune.

It must be taken into account that both types of parabolic dune can also develop during the total dismantling of the primary dune, when different sections of the primary dune begin to migrate landwards at different rates. The final result is the development of parabolic dunes not linked to an existing primary dune. This process is usually associated with episodes of massive destruction of the vegetation cover. The natural causes that can induce the disappearance of the vegetation cover are usually associated with changes in the climate, although there can also be other causes of an anthropic nature.

In very well-fed coastal areas, significant migration of dunes towards land can occur. Then, successive lines of parabolic or even barchanoid dunes develop, spreading and overlapping in an increasingly extensive field. These are known as transgressive dune fields. The source of sand can be limited and the migration of the dunes can lead them to leave the area where they originally developed. The only remainder of them in this area will then be a low-relief eolian mantle. Eolian mantles and transgressive dune fields are part of the same erosive process and are usually related laterally.

18.4.4 Genesis of Minor Structures on the Dune Surface

The action of the wind on the dune surface generates small-scale structures. In most cases these structures are ephemeral and are destroyed by the very evolution of the dune, but sometimes they can be preserved inside. The minor structures that often form at the front are usually trains of small ripples that migrate towards the ridge (Fig. 18.11a). During low-speed winds they tend to concentrate in the most depressed areas, while during moments of high wind intensity they form in the shaded areas, under the protection of plants or the dune crest.

The back side is usually dominated by avalanche processes that form sheets tilted towards the land that tend to achieve an angle of repose. These avalanches originate on the dune crest by destabilization and generally first generate an erosive groove through which the sand slides. The friction

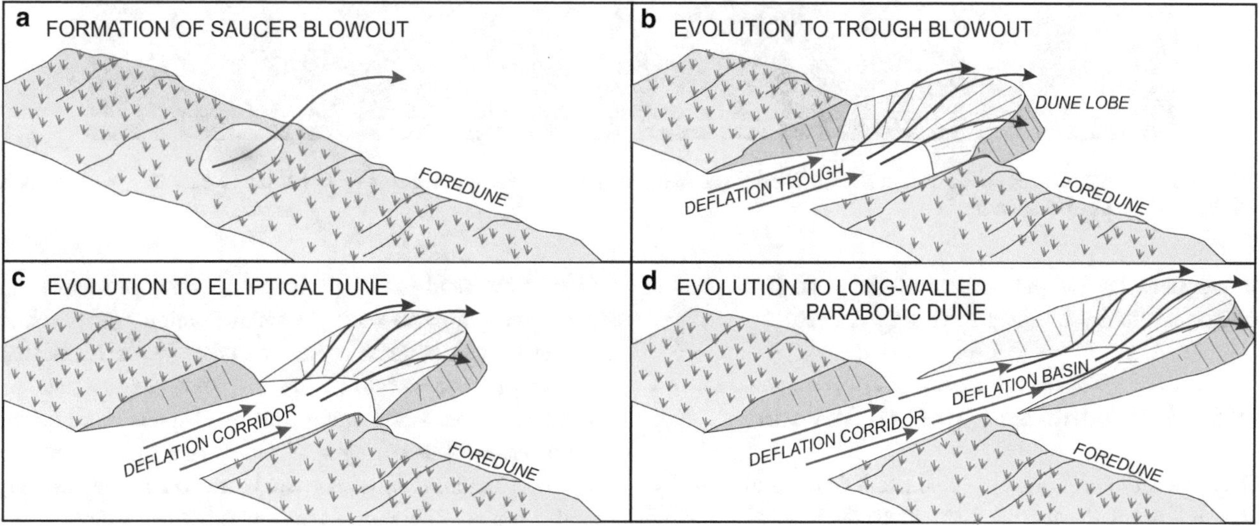

Fig. 18.10 Evolution from a saucer blowout in a primary dune to a secondary parabolic dune (Based on [2])

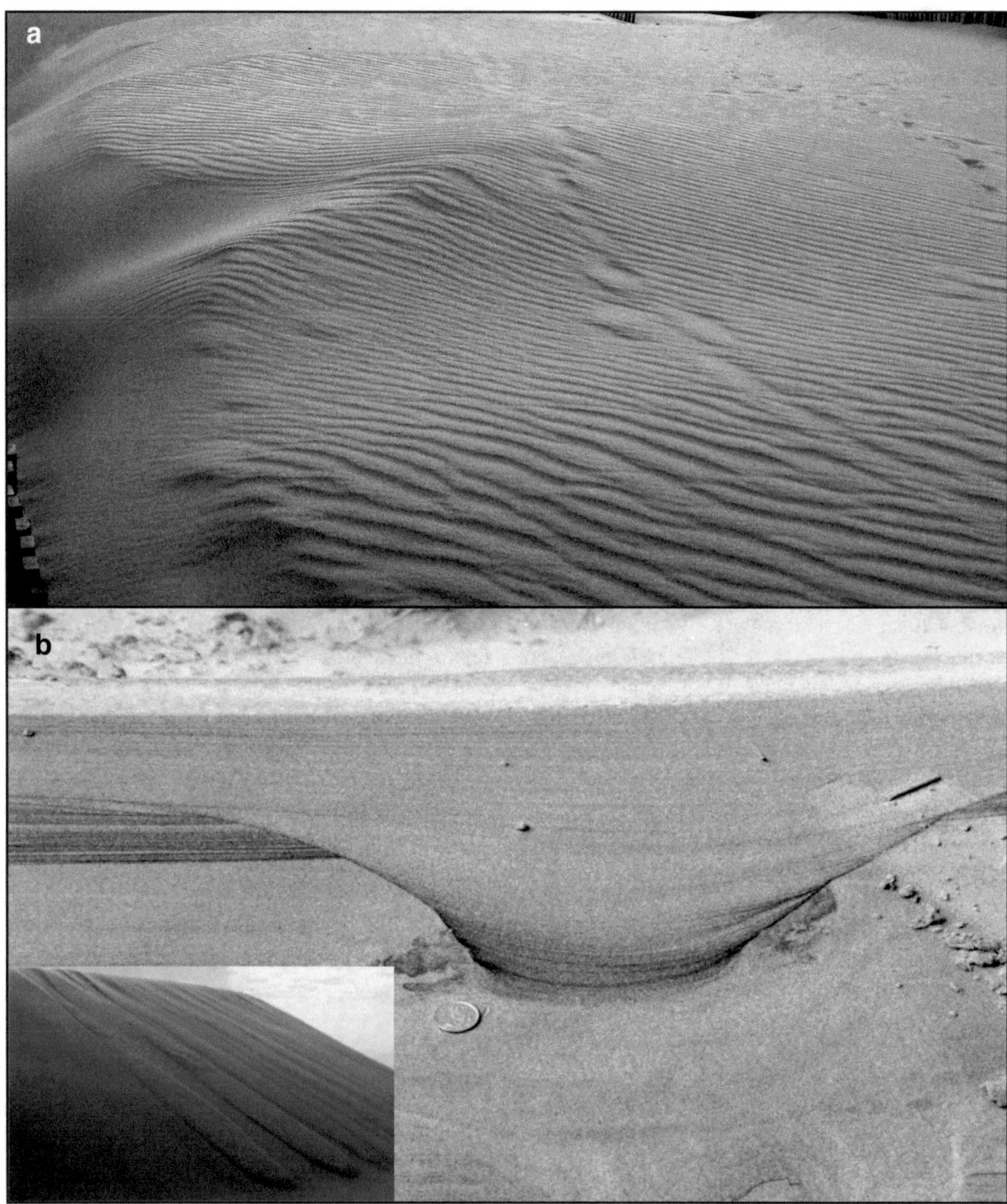

Fig. 18.11 Minor structures on the dune surface. **a** Wind ripple trains on the crest of a foredune. **b** Avalanches on the back of a parabolic dune and detail of the internal structure

of the sand in this groove causes the deposit of parallel sheets that finally fill the groove (Fig. 18.11b).

18.5 Sediments and Internal Structure

The dune sediments present a characteristic textural maturity. They are normally very well-sorted sands, because the wind is the agent that best classifies grains. In coastal dunes, the origin of the sand is usually linked to wave processes that subject the grains to multiple sedimentation cycles which wear away their surface. It must be taken into account that, once the grains pass into the dune system, they can still be reworked by the waves again or several times more during the processes of storm erosion and fairweather reconstruction. This explains why the sands are composed of very rounded grains. This process of reworking in multiple cycles also means that, from a compositional point of view, the dunes are normally composed of the most stable minerals such as quartz, although heavy minerals are also generally concentrated in the dunes. In coastal carbonate systems, some dunes are mainly composed of allochemical elements such as bioclasts, oolites or intraclasts. In these cases, the grain size is usually somewhat larger, since the specific weight of calcite is lower than the more stable silicates. In evaporative coasts, there are also dunes composed mainly of gypsum grains.

The dunes developed in lithologies susceptible to chemical processes can produce cementation phenomena that contribute to the consolidation and lithification of the dune deposits. Lithified dunes are known as **eolianites**. The term eolianite was proposed by Sayles [23] to refer to the limestone dunes of the Bermuda Islands, which have a carbonate nature. For this reason, some authors tend to identify the term eolianite with the calcareous composition of the dunes.

As for the internal structure, in general terms it could be said that the entire internal structure is composed of medium- to large-scale cross-bedded sets that tilt in the direction of wind transport. However, this simple description hides a much more complex reality. Coastal dunes usually present a greater diversity and complexity than desert dunes. The simple fact that the internal structure of coastal dunes is influenced by vegetation already makes a significant difference. This description also does not include such singularities.

A first distinction could be made between the internal structures of primary and secondary dunes, since the foredunes are characterized by the process of accretion, while the secondary dunes are characterized by migration; these processes are marked in the geometry of their respective sets of cross-bedding.

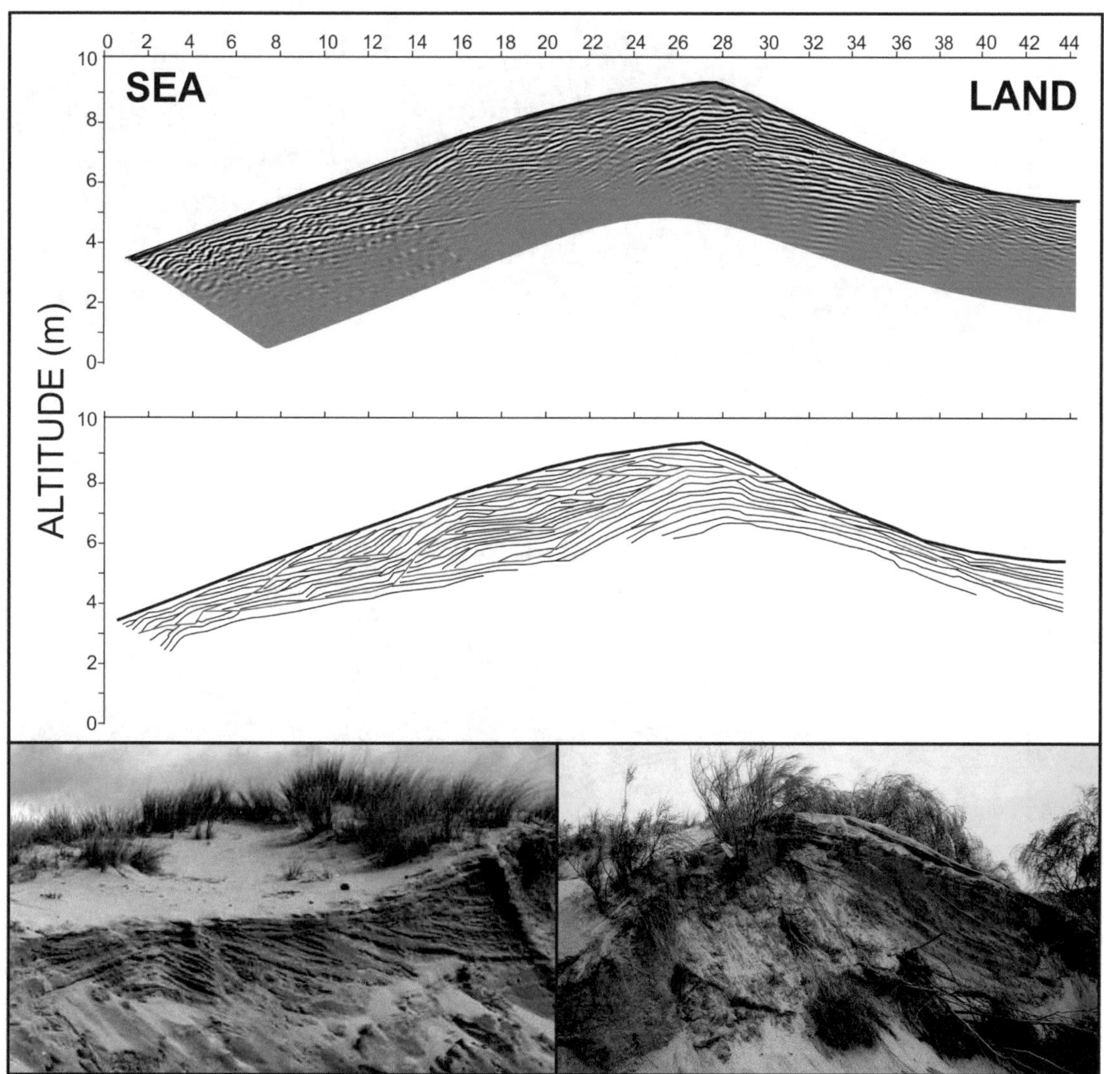

Fig. 18.12 Internal structure of an established foredune. The upper part of the figure shows a GPR record and its interpretation. The lower photographs show the internal structure discovered during the formation of washover channels

18.5.1 Foredunes

The internal structure of the foredunes is characterized by the presence of sets of trough cross-bedding, separated by reactivation surfaces (Fig. 18.12). The internal stratification presents a greater continuity in the core of the dune, with a wedge shape towards both front and rear areas. Some sets appear truncated by others, showing that among the delimiting surfaces are erosive disconformities. The complete structure corresponds to a stack of sets. In humid climates, the presence of vegetation influences the internal structure of the dune. In addition to contributing to the growth of the dune, the vegetation is encompassed by the sandy sediment in such a way that not only the roots but also the entire plants disturb the cross-stratification, especially in the central area of the dune.

In detail, there are two minor structures that can be observed in the sheets that make up the sets of cross-layers: the cross-lamination corresponding to the migration of small ripples and the filled grooves that correspond to the avalanches which occur on the leeward side.

18.5.2 Blowout Lobes

The dune lobes of the blowouts have an even more complex structure than the foredunes. The core of these dunes may present a relict structure made up of remains from the rear of

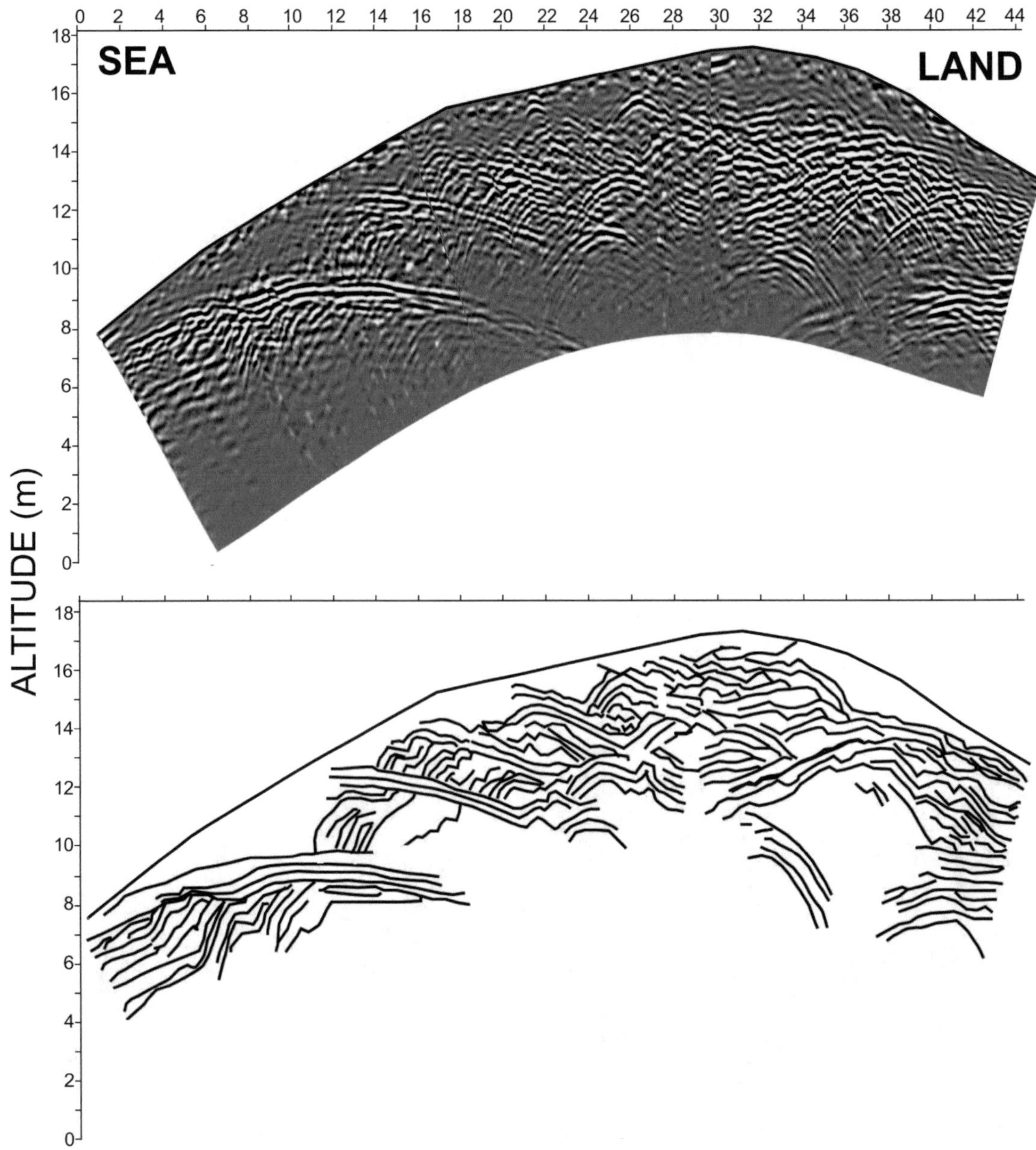

Fig. 18.13 Internal structure of a blowout lobe. The upper part of the figure shows a GPR record and the lower part shows its interpretation. The numerous parabolas correspond to the presence of vegetation

the foredune. Above this, sets of curved cross-stratification are developed as a result of vertical growth, on which sets of plates inclined towards the ground are arranged. The internal structure becomes more complex towards the top, since, as the dune lobe grows, it can include existing vegetation at the back of the dune. Meanwhile, pioneer plants may try to colonize the blowout lobe during periods of slower growth. All these plants are encompassed by the growth of the dune lobe and incorporated into the internal structure, increasing its complexity (Fig. 18.13).

18.5.3 Parabolic Dunes

The internal structure of the parabolic dunes is composed of sets of planar cross-beds or cross-beds with a slightly curved base, separated by reactivation surfaces between sets of different slope (Fig. 18.14). The inclination of the cross-beds ranges from 20º to 40º. Some internal levels present minor cross-laminations correlated to the migration of wind ripple trains. In section, parallel to the crest, there are filling grooves corresponding to the leeward face avalanches. The sets on the windward side are clearly cut by the topographic surface as a result of the migration.

18.5.4 Interdune Sediments

The deflective interdunes do not present sediment except for the residual lags and some laminar sand deposits where ripples develop. These interdunes are preserved as unconformity surfaces. However, depositional interdunes may present characteristic sediments whose nature will depend on the relationships of the interdune with the water level. The characteristic common to all interdune deposits is the sedimentation in sub-horizontal layers since its genesis is not linked to dune migration.

The most characteristic interdune deposits are those that occur in relation to the existence of a sheet of water that covers all or part of the interdune. Under these conditions, swampy or lacustrine sediments can be deposited. These sediments are obviously very much influenced by the sandy

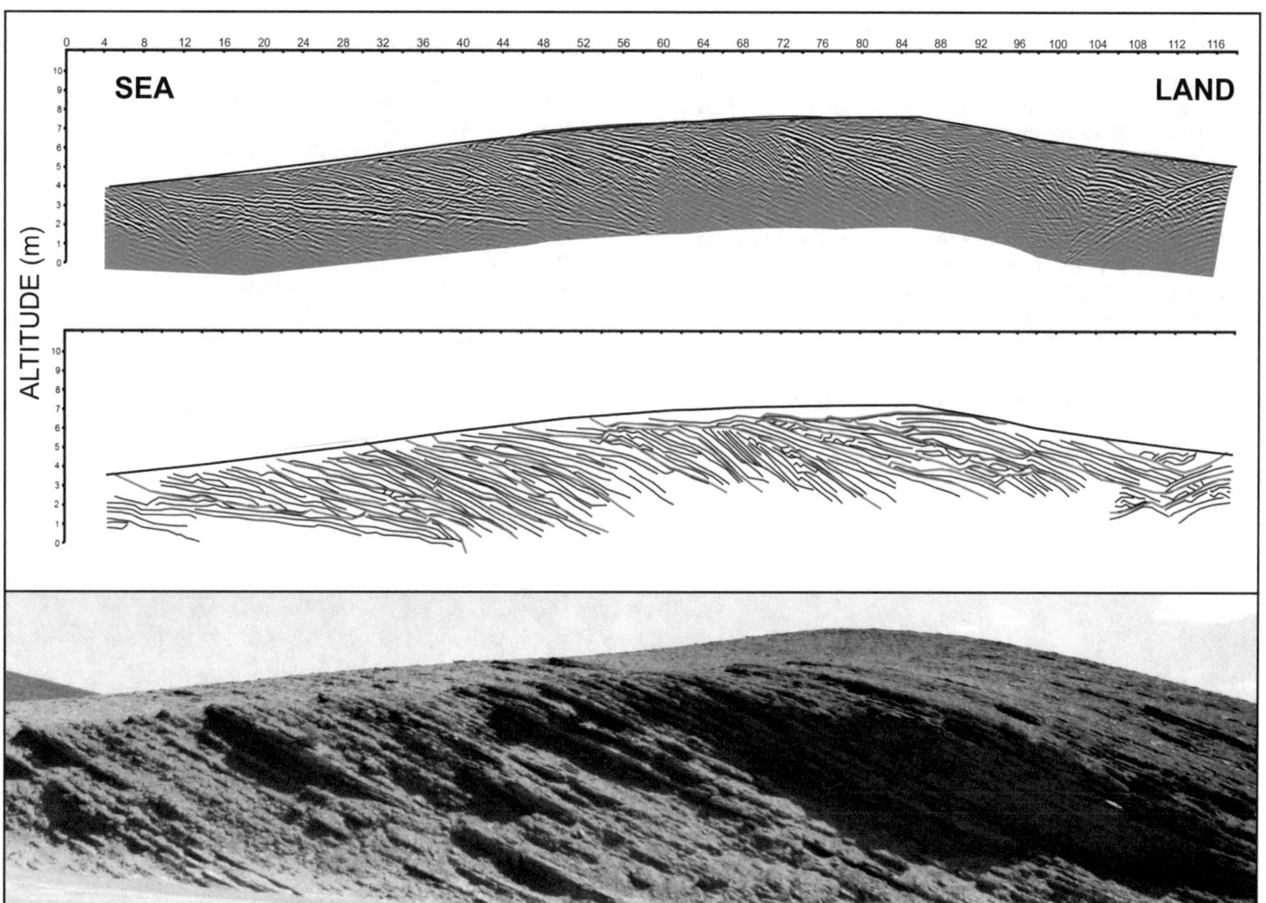

Fig. 18.14 Internal structure of a parabolic dune. The upper part of the figure shows a GPR record and its interpretation. The lower photograph shows the internal structure of an Eolianite

material introduced by the wind, although they are predominantly composed of fine material (clay or silt). Interdune lake sediments may have abundant freshwater microfauna such as foraminifers, ostracods and diatoms, as well as macrofauna of bivalves, gastropods and worms. All these organisms are responsible for a high degree of bioturbation. Swamp sediments are strongly influenced by plants, and peat formation is frequent in these environments. Often, the water level undergoes oscillations that give the flooded interdune environments a short-lived character. In humid climates, periods of drying out can be accompanied by wind deflation of these sediments and during these moments surfaces of unconformity can be generated. In arid climates, the drying of the interdune lagoons is accompanied by evaporation, so it is common to find evaporites associated with these interdune spaces.

References

1. Bauer BO, Sherman DJ (1999) Coastal dune dynamics: problems and prospects. In: Goudie AS, Livingstone I, Stokes S (eds) Aeolian environments, sediments and landforms. Wiley, Chichester, pp 71–104
2. Bird E (2000) In: Coastal geomorphology: an introduction. Wiley, Brisbane
3. Carter RWG, Hesp PA, Nordstrom NF (1990) Erosional landforms in coastal dunes. In: Nordstrom KF, Psuty N, Carter B (eds) Coastal dunes: form and process. Wiley, Brisbane, pp 217–252
4. Davies JL (1980) Geographical variation in coastal development. Longman, London, p 212
5. Doing H (1985) Coastal fore-dunes zonation and succession in various parts of the world. Vegetatio 61:65–75
6. Duran O, Moore LJ (2013) Vegetation controls on the maximum size of coastal dunes. Proc Natl Acad Sci 110:17217–17222
7. Fryberger SG (1979) Dune forms and wind regimes. In: McKee ED (ed) A study of global Sand Seas. United States Geological Survey Professional Paper 1052, pp 137–140
8. Goldsmith V (1973) Internal geometry and origin of vegetated coastal sand dunes. J Sediment Petrol 43:1128–1143
9. Goldsmith V (1985) Coastal dunes. In: Davis RA (ed) Coastal sedimentary environments. Springer, New York, pp 171–236
10. Hesp PA (1988) Foredune morphology, dynamics and structures. J Sediment Geol Special Issue: Aeolian Sediments 55:17–41
11. Hesp PA (1999) The beach backshore and beyond. In: Short AD (ed) Handbook of beach and shoreface morphodynamics. Wiley, Brisbane, pp 392
12. Hesp PA (2002) Foredunes and blowouts: initiation, geomorphology and dynamics. Geomorphology 48:245–268
13. Hesp PA, Hyde R (1996) Flow dynamics and geomorphology of a trough blowout. Sedimentology 43:505–525
14. Hotta S, Kubota S, Katori S, Horikawa K (1984) Sand transport by wind on a wet sand surface. Coast Eng 19:1265–1281
15. Kelletat D (1995) Atlas of coastal geomorphology and zonality. J Coastal Res Special Issue 13:286
16. Klijn JA (1990) Dune forming factors in a geographical context. In: Bakker TWM, Jungerius PD, Klijn JA (eds) Dunes of the European coasts, Catena, Supplement vol 18. pp 1–14
17. Martínez ML, Psuty N (eds) (2004) Coastal dunes: ecology and conservation. Springer, Heidelberg
18. Nordstrom KF, Arens SM (1998) The role of human actions in evolution and management of foredunes in the Netherlands and New Jersey, USA. J Coast Conserv 4:169–180
19. Psuty NP (Ed) (1988) Dune/beach interaction. J Coastal Res Special Issue 3:1–136
20. Pye K (1983) Coastal dunes. Prog Phys Geogr 7:531–557
21. Pye K, Tsoar H (1990) Aeolian sand and sand dunes. Unwin Hyman, London, p 396
22. Sherman DJ, Bauer BO (1993) Dynamics of beach dune systems. Prog Phys Geogr 17:413–447
23. Sayles HW (1931) Bermuda during the ice age. Proc American Acad Arts Sci 66:382–467
24. Short AD, Hesp PA (1982) Wave, beach and dune interactions in southeastern Australia. Mar Geol 48:259–284
25. Tsoar H (2001) Types of aeolian sand dunes and their formation. In: Balmforth NJ, Provenzale A (eds) Geomorphological fluid mechanics, 582. Springer, Berlin, pp 403–429
26. van der Maarel E (1993a) Dry coastal ecosystems: polar regions and Europe. Elsevier, Amsterdam
27. van der Maarel E (1993b) In: Dry coastal ecosystems: Africa, America, Asia and Oceania. Elsevier, Amsterdam

Tide-Dominated Systems I: Inlets and Tidal Deltas

19

19.1 Introduction

Tidal inlets are channels that separate different barrier islands or are located between a barrier and the mainland. It is through these channels that the mechanism of tidal circulation between the open marine environment and the environments protected by the barrier takes place. **Tidal deltas** are sandy bodies located at the ends of the tidal inlets in both directions of the tidal currents. **Ebb-tidal deltas** are those located on the sea front, in contact with the nearshore area, while **flood-tidal deltas** are developed at the back barrier area, in contact with the semi-restricted body of water represented by the lagoon.

Conceptually, tidal inlets and tidal deltas are defined as part of the barrier island systems (Fig. 19.1a), where the barrier island constitutes the confining element and the tidal inlet is the channel that separates two different islands. However, later tidal deltas have come to be defined as possible elements of other coastal systems [19], such as estuaries confined by barriers (Fig. 19.1b) or barrier reef systems (Fig. 19.1c). In all cases, the absence of confining conditions, both seaward and landward, allows the building of extensive sandy bodies that are essentially very similar morphologically and genetically to the underwater portion of river deltas. Tidal deltas can theoretically arise on any depositional coast where the sandy input is adequate and where the substrate physiography and relative sea level fluctuations allow bodies with this geometry to develop [6].

Because they are subjected to the action of tidal currents and waves, these environments have a high dynamic mobility and their position and morphology is in constant evolution. As they correspond to passage environments between open and restricted waters, navigation through these systems has always historically been a factor to be taken into account. In the sixteenth to eighteenth centuries, many shipwrecks took place in these passages when vessels tried to take refuge in the restricted systems using charts that did not reflect these changes. This is why, today, many of these environments have been artificially modified by stabilizing the position of their channels with jetties and maintaining their depths by dredging.

Keeping these inlet channels open is not only important for the development of ports that are located in lagoonal waters, but also to preserve the ecology of the natural environments that are associated with these systems, since these places are used as breeding grounds for many species of marine fauna.

There are several systems in the world that are very well studied, such as those on the East Coast of the United States and the Wadden Sea [3], but there are also well-documented cases on the coasts of Iceland, Portugal and Spain, as well as other less-studied coasts such as Brazil, Senegal, Ivory Coast and Mozambique.

19.2 Control Factors

Although factors such as sediment input and the initial coast slope significantly determine the dimensions of deltas, other factors such as tectonic and eustatic conditions may even inhibit their development. The main factors that control the genesis of tidal deltas and their short-term dynamics, however, are tidal currents and wave energy [3].

19.2.1 Tidal Currents

The action of the tidal currents is the main physical process acting on the deltas. The speed of the currents depends directly on the volume of water that circulates through the tidal inlet draining the protected part of the barrier system. This volume of water drained by the tide is known as the **tidal prism** and it depends on the tidal range in relation to the dimensions and morphology of this restricted area.

The dimensions, shape, dynamics, distribution and abundance of tidal deltas were originally associated with the magnitude of the tidal range [16]. For this author, the abundance of tidal deltas would be at a maximum on coasts

J. A. Morales, *Coastal Geology*, Springer Textbooks in Earth Sciences, Geography and Environment,
https://doi.org/10.1007/978-3-030-96121-3_19

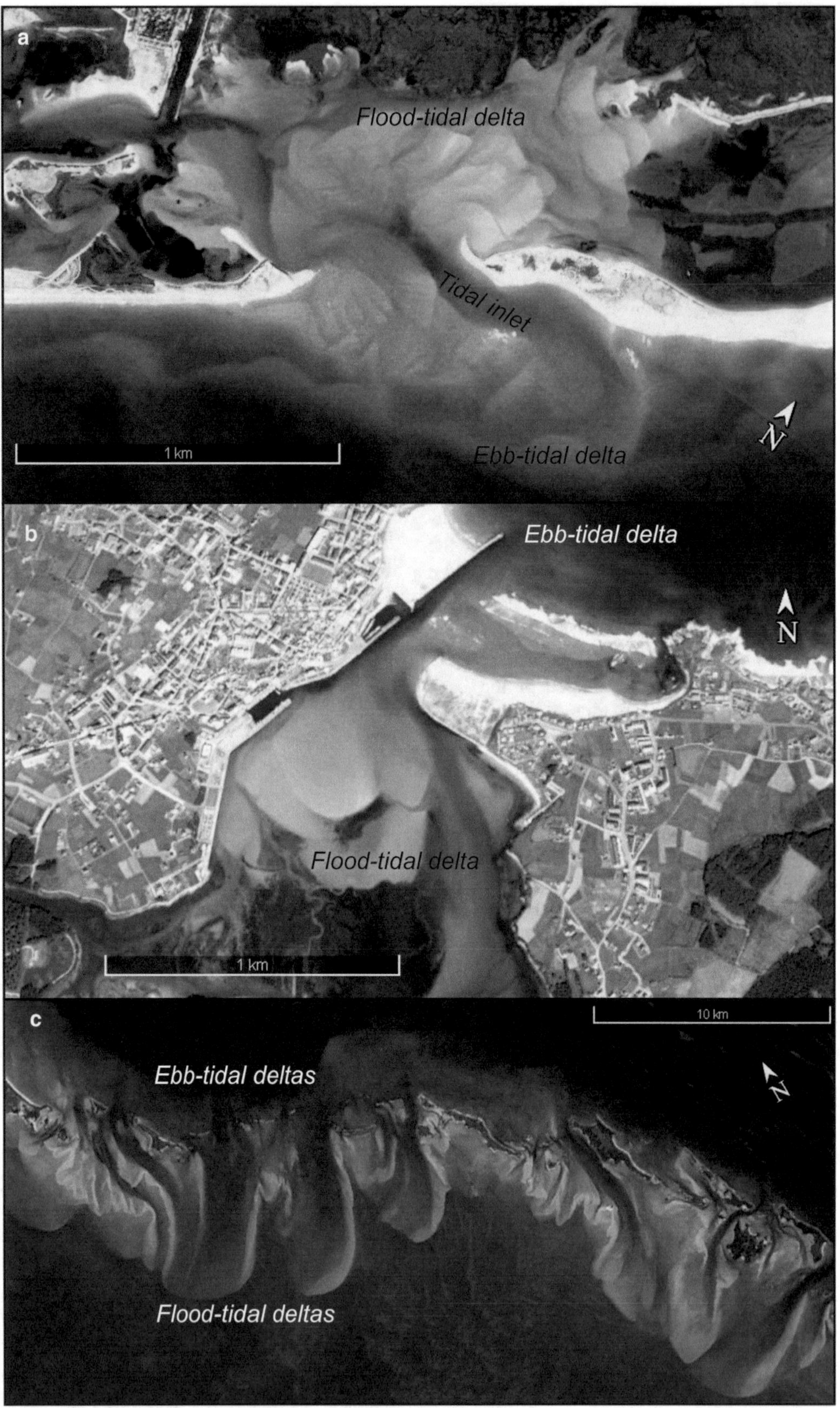

Fig. 19.1 Tidal deltas in different coastal systems. **a** Associated to a barrier island system (Fuseta, Portugal). **b** Associated to barrier estuaries (Foz, N Spain). **c** Associated to reef keys (Nassau, Bahamas). (Images by Landsat, Google Earth.)

with tidal ranges between 2 and 2.5 m, then decreasing sharply towards greater amplitudes so that they would no longer form on coasts with tidal ranges greater than 5 m. This probability would also decrease towards smaller ranges, although not so abruptly, so that tidal deltas would appear, though less frequently, even on microtidal coasts with almost no tidal influence. Thus, according to this criterion, the distribution of tidal deltas along the coasts of the world would be determined by the tidal range.

While flood-tidal deltas are subject only to tidal currents in and out of the lagoon, ebb-tidal deltas are further influenced by tidal currents that run parallel to the shoreline. These currents are usually of lesser magnitude than the currents that circulate through the tidal inlets. Nevertheless, they have an important effect on the deviation of these ebb-tidal deltas, generating an asymmetry factor on the orientation of the inlets and on their morphology.

19.2.2 Wave Energy

The genesis of the barriers is related to the action of the waves, so this factor must also be taken into account as a dynamic agent in the inlets that separate them. The waves developed during storms can even be a determining factor in the creation of new inlets from overwash processes, facilitating the exit of the tide through them. In addition, the swell is the main agent that takes charge of closing some inlets by means of the attachment of parallel bars to the coast when these lose functionality. For this reason, the waves represent a crucial factor, both in the number of inlets in barrier island systems, and in their immediate evolution.

Ebb-tidal deltas are permanently subjected to the action of waves. At the front of these deltas there is a conflict between the capacity of sediment transport by tidal currents and the capacity of sand remobilization by waves. While ebb currents tend to generate sand bodies perpendicular to the coastline, waves rework the sand to try to place it in parallel. In this way, the influence of waves is unequivocally manifested in the existence of wave bars in the frontal zone of the ebb deltas.

19.2.3 Combined Action of Waves and Tidal Currents

The conflict between the sediment mobilization capacity of tides and waves would suggest that the final morphology of the tidal inlets and, above all, of the ebb-tidal deltas, depends on a balance of forces. Although Hayes [16] had associated the morphology and distribution of tidal deltas with the tidal range, some years later the same author studied some particular cases in more detail, and linked these factors to the energy balance between the tidal currents and the dominant waves [16]. With this new criterion, tidal deltas would be typical of mixed-energy coasts, decreasing their frequency when one agent dominates the other (Fig. 19.2). This decrease in frequency when the tide becomes more important is related to the absence of barriers on the coasts where tides dominate. The number and size of inlets also decrease when the waves become dominant, as the wave bars generally tend to close them.

The energy balance between tides and waves influences not only the abundance of tidal deltas on the coasts, but also the morphology of tidal inlets and associated tidal deltas. [9] distinguished four types of inlet morphology depending on the dominance of one or another agent (Fig. 19.3). The presence of two different typologies corresponding to mixed energy conditions highlights the significance of these systems.

19.3 Morphology and Sub-environments

There are differences in the distribution of processes and facies within flood- and ebb-tidal deltas, as well as their morphological variations, so they will be considered separately here.

19.3.1 Ebb-Tidal Deltas

The morphological model of an ebb-tidal delta (Fig. 19.4) consists of several sub-environments:

- The body starts from a wide and deep channel dominated by ebb currents. This is known as the **main ebb channel**.
- This channel is flanked by elongated **channel margin linear bars** in a perpendicular or oblique direction to the coast.
- These bars separate the main ebb channel from two **marginal flood channels.**
- Towards the sea, the ebb channel connects with a **terminal lobe,** which is the most distal part of the delta and the most important sedimentary body. The terminal lobe originates due to the loss of speed of the tidal ebb when it meets the waves and deposits its sedimentary load, mainly sand.
- On the sides of the terminal lobe and on the channel margin linear bars, the waves cause swash bars to migrate, which can be attached to the barrier island or fall into the marginal flood channel to be reworked by the tide [14].

The different relative energy between tidal currents and waves that has been previously mentioned gives rise to

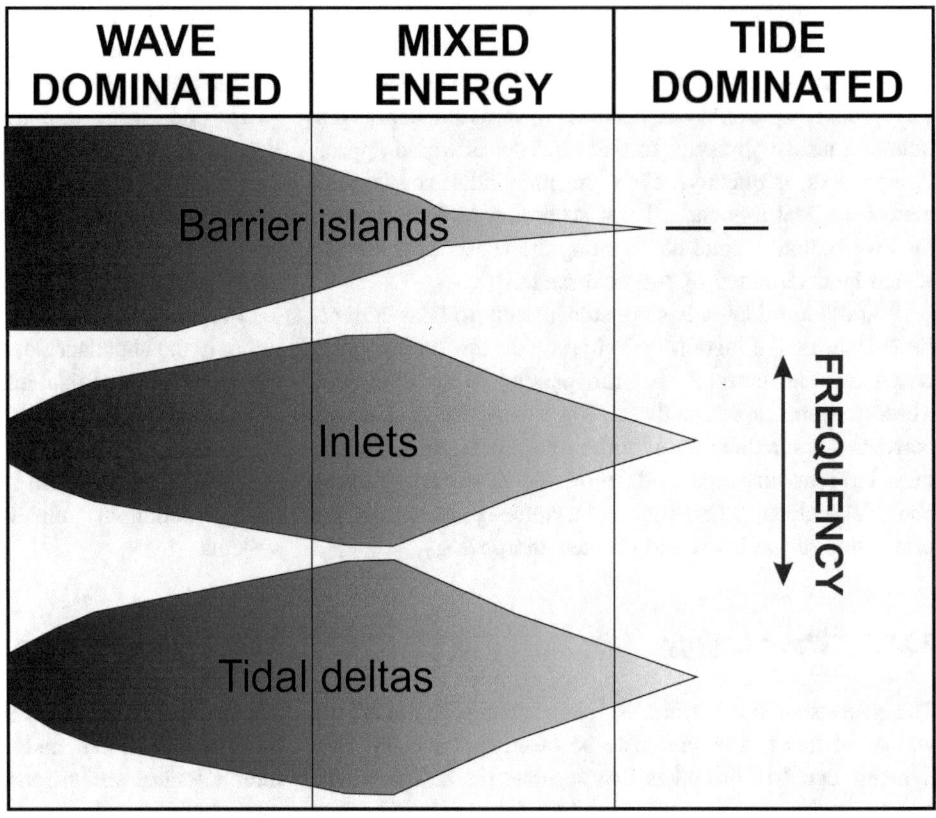

Fig. 19.2 Frequency of barrier islands, inlets and tidal deltas according to the dominance of tide and wave energy, adapted from Hayes [16]

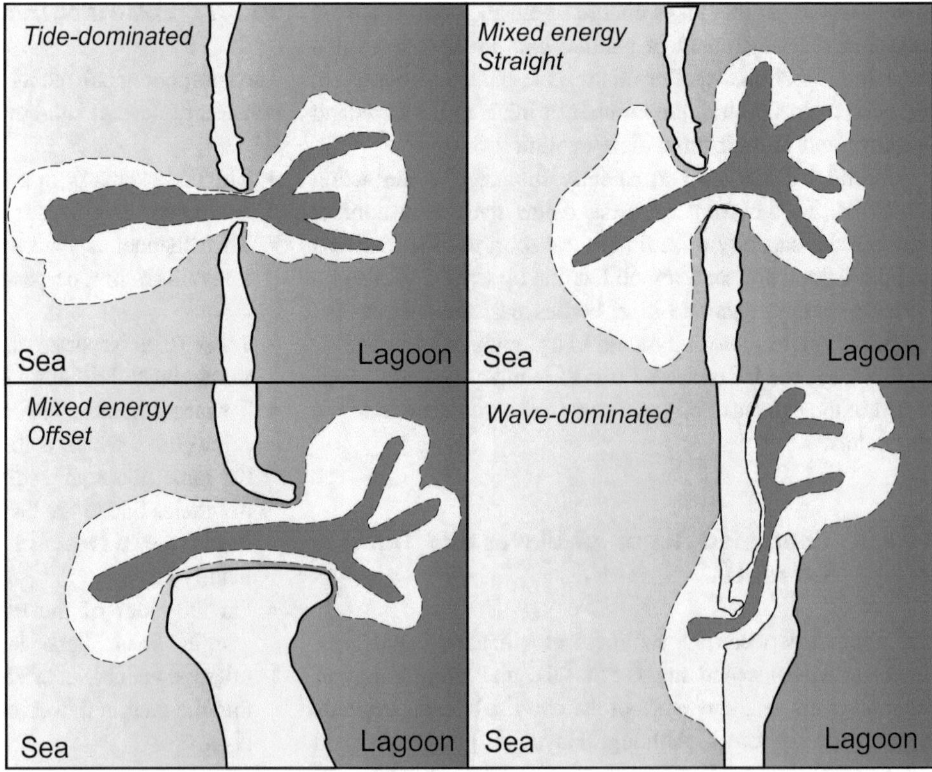

Fig. 19.3 Scheme showing the different types of tidal inlets according to the balance of energy between tides and waves [9]

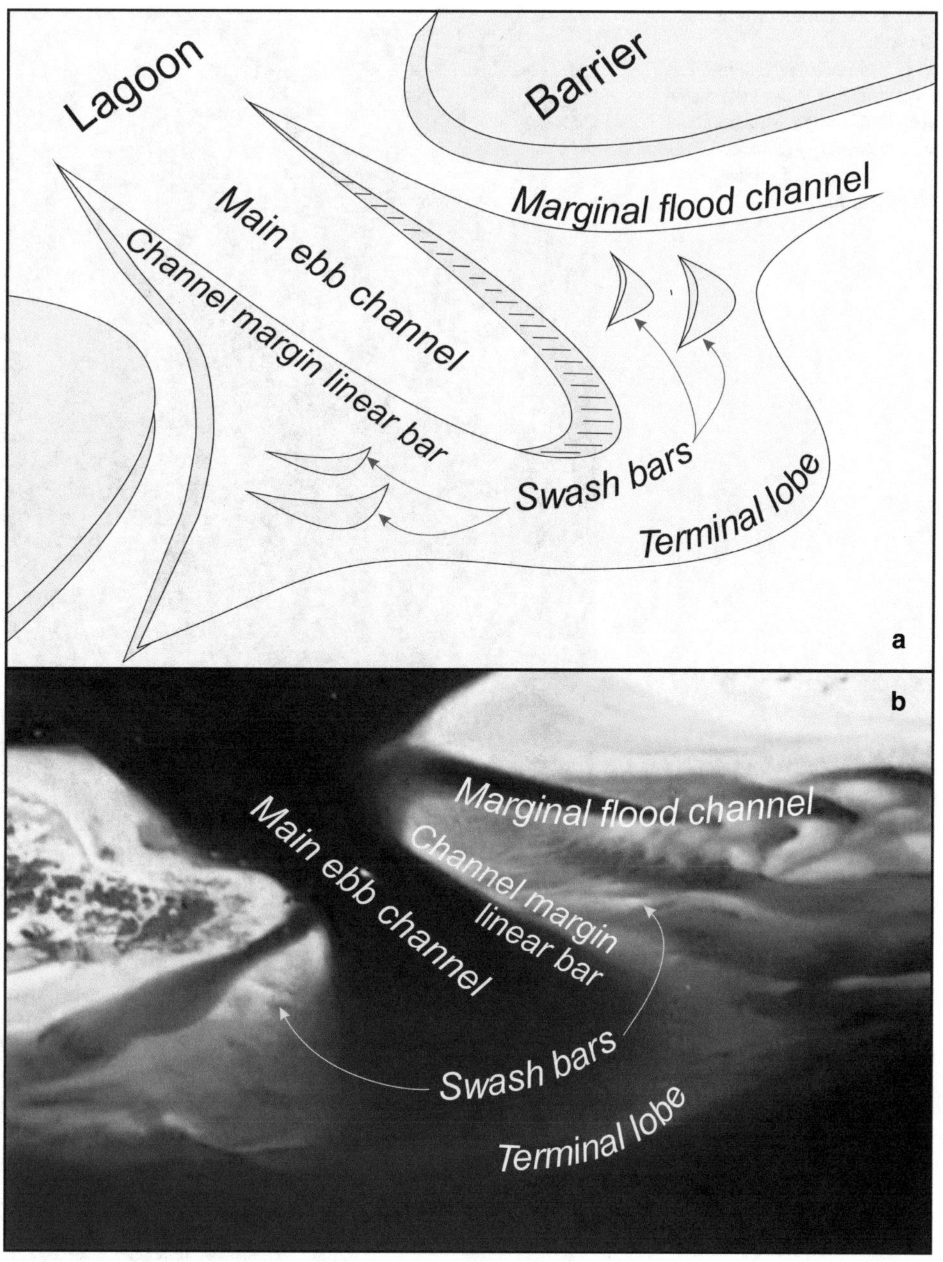

Fig. 19.4 **a** General morphological model for an ebb-tidal delta as suggested by Hayes [16]. **b** Example showing the elements of the general model

important morphological variations in the ebb deltas. This variability can be summarized by distinguishing the three types of ebb-tidal delta (Fig. 19.5).

(1) Tide-dominated ebb deltas, which are characterized by a perpendicular arrangement to the coast (Fig. 19.5a). In this type of delta, the tidal inlet is erosive and very deep, sometimes even reaching the substrate. This depth reduces its mobility, so that it does not usually present migration. Its sandy sedimentary body develops with greater length than width, reaching deeper and further areas from the coast.

(2) Wave-dominated ebb deltas, on the other hand, are usually smaller in length as well as wider, with a form dominantly parallel to the coast (Fig. 19.5b). In this case, the inlet is usually shallow and highly unstable with constant changes in position and morphology. The marginal flood channels are generally absent. The terminal lobe of the delta takes the form of a crescent and is often found in the intertidal fringe. This produces a number of swash bars that attempt to completely close the inlet because tidal currents are lower. They tend to have a marked asymmetry and a high rate of migration.

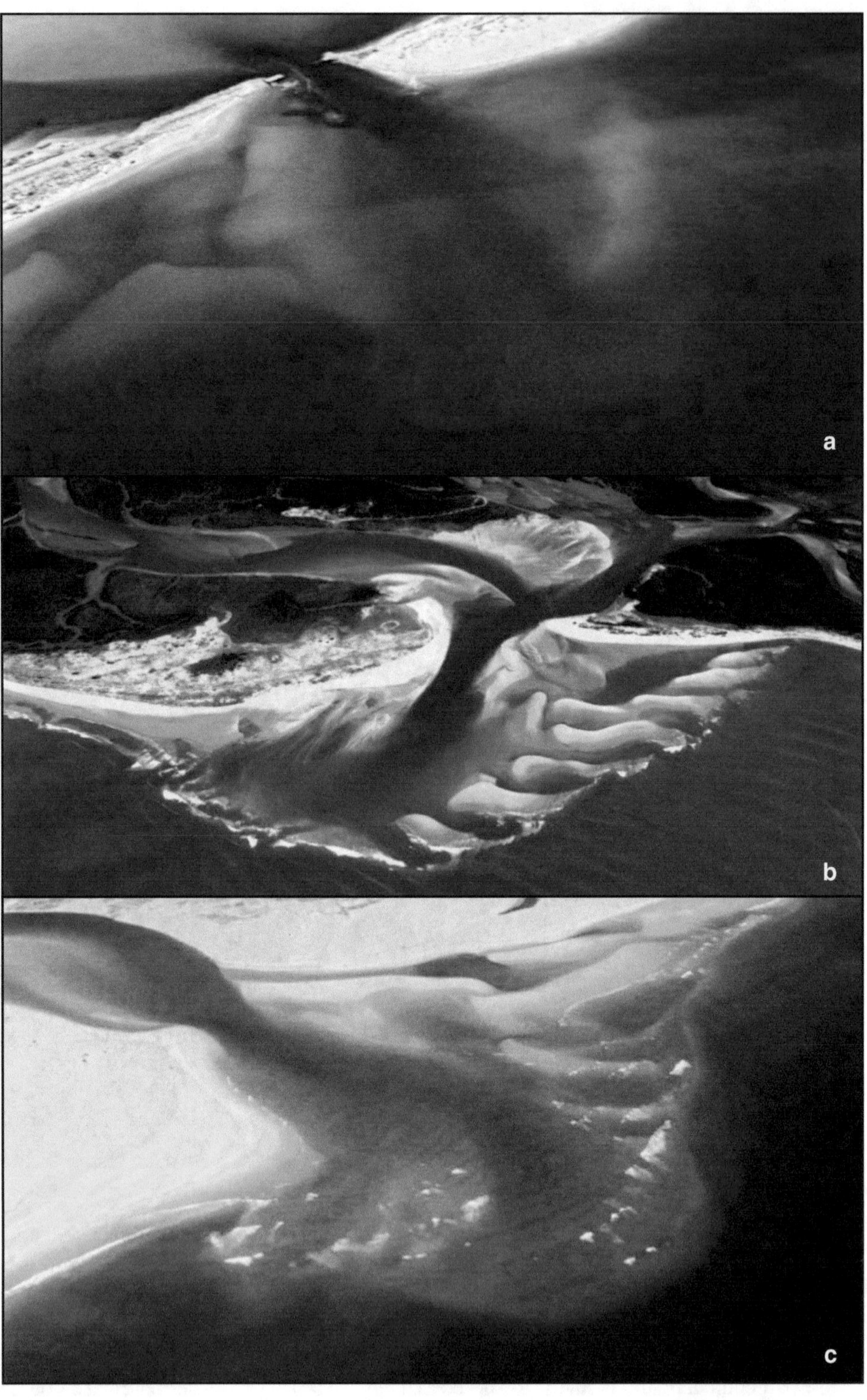

Fig. 19.5 Different types of ebb-tidal deltas. **a** Tide-dominated (Nabule inlet, Myanmar). **b** Wave-dominated (Machesse inlet, Mozambique). **c** Mixed energy (Saharabedi inlet, India). (Images: Landsat from Google Earth.)

(3) The third type of ebb-tidal delta is the mixed type. It has a smooth and rounded morphology, in response to the combined action of tides and waves. Of the three types, this is the one with the most complex dynamics and sedimentology, presenting an intermediate character between the other two types (Fig. 19.5c). Here, again, a wide and deep inlet appears, connecting directly to a main ebb channel that becomes progressively shallower towards the terminal lobe. The marginal flood channels may be absent or present only at one of the margins. Marginal bars tend to be triangular rather than linear in shape and develop abundant swash bars and other

bedforms due to tidal action. A similarity to tide-dominated deltas is that the terminal lobe is subtidal. However, this is usually very shallow.

The presence of wave bars is very significant in wave-dominated and mixed-energy deltas, where these migrate over wide marginal bars. Then, these behave like swash platforms. However, this is a minor process in tide-dominated deltas, where the marginal bars take on a more linear character and are arranged perpendicular to the coast. Another important difference is that in tide-dominated deltas the terminal lobe is further away from the coast and the wave action on it is less intense [3]. In general terms, deltas tend to be smaller when the wave influence increases. A common characteristic of all ebb deltas is the presence of bedforms in all environments, although these bedforms are mostly oriented according to the dominant current.

Both wave-dominated and mixed-energy deltas are usually asymmetrical. Oertel [28] researched studies of the deltas of the East Coast of the United States and attributed the causes of this asymmetry to the direction of the ebb currents that circulate in the open coastal zone and divert the ebb that comes out of the inlet. According to the relative importance of these currents, he differentiated four different types of morphology (Fig. 19.6). When the outgoing ebb current has little importance with respect to the external currents, the terminal lobe will have a longitudinal development to the coastline and a position closer to land (Fig. 19.6a). Conversely, when the outgoing currents are much greater than the currents in the open zone, the terminal lobe will have a greater development being located in areas far from the inlet (Fig. 19.6d). These two situations will tend to present a greater symmetry. When the currents of the open zone are asymmetrical and one of the currents is greater than the other, it induces an asymmetry in the orientation of the deltas (Fig. 19.6b, c).

Later, [32] added to these tidal current interactions the incidence of wave trains on the shoreline (Fig. 19.7). For this author, the relative importance between waves and outgoing ebb currents is the factor that really controls the geometry and dimensions of ebb deltas. In this case, the tidal prism controls the speed of the outgoing currents. The presence of a frontal wave brings the terminal lobe closer to land (Fig. 19.7a), while oblique waves add or subtract tidal currents from the open zone to generate a coastal drift component that deflects the outflow backwash (Fig. 19.7b–d). A well-defined littoral drift increases the asymmetry. This asymmetry of the ebb deltas also induces an inlet instability that manifests in the migration of the entire system associated with the tidal channel.

19.3.2 Flood-Tidal Deltas

The shape and dimensions of the flood deltas are not related to the balance between tides and waves, as the waves do not act on this environment. There is, however, a variation in their shape depending on the tidal amplitude. Broadly speaking, it can be said that, on microtidal coasts, the flood deltas tend to be multilobate, since the reworking by the ebb is minimal. These are the narrowest deltas, rarely reaching 3 m in thickness. In contrast, in areas where the tidal range is

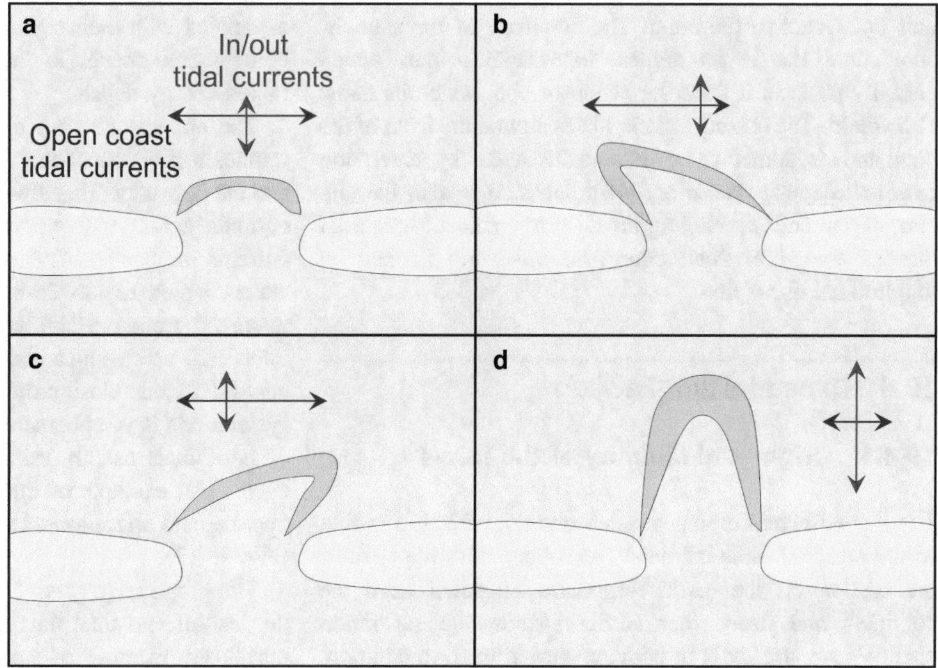

Fig. 19.6 Different combinations of in/out tidal currents with outer tidal currents to obtain different ebb-tidal delta morphologies, adapted from Oertel [28]

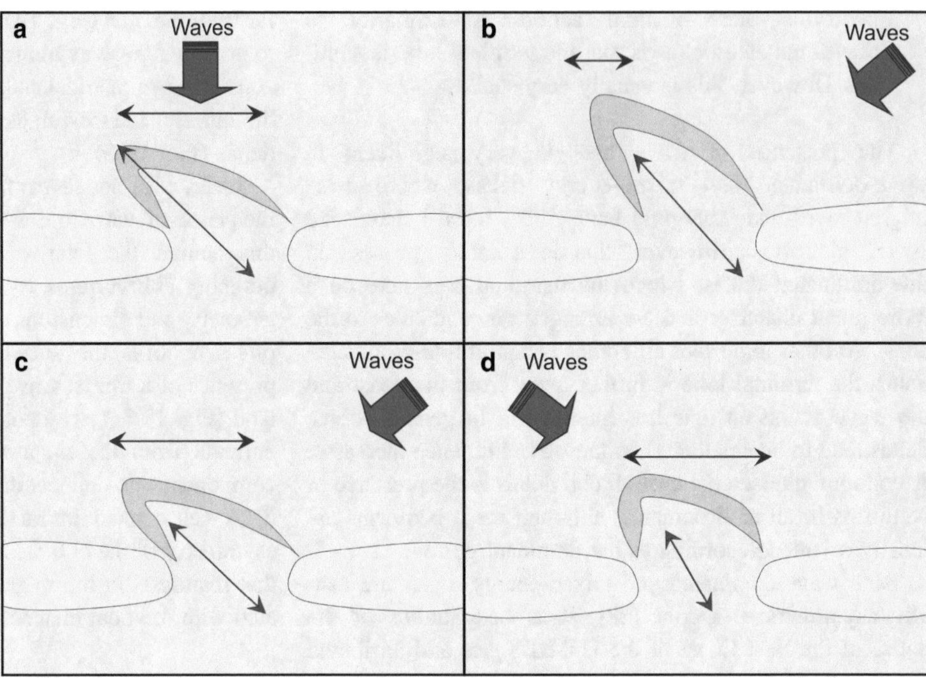

Fig. 19.7 Different combinations of in/out tidal currents with outer tidal currents and oblique wave trains to obtain different ebb-tidal delta morphologies, adapted from Sha [32]

greater than 1.5 m, the velocity of the flood currents is high enough to transport a large amount of material. Then, the flood currents rework the sediment, creating their own morphology, and the delta usually acquires a horseshoe shape.

The general model reflects this second situation (Fig. 19.8). It consists of a **flood ramp** connected to the tidal inlet. This flood ramp loses depth and thus allows bedforms to migrate in the flood direction. The flow ramp can develop a **tidal flat** in its shallowest area. This tidal flat is shaped like a horseshoe and is surrounded by very shallow flood channels connected to the ramp. The boundary of the semicircular tidal flat is its highest topographical part which partially protects it from the action of ebb, hence its name **ebb shield**. The ebb current circulates around the front of the delta and can sometimes overcome this screen by generating overflow deposit named **spillover lobes**. It is also the ebb current that ends up circling the delta to connect to the inlet through **marginal flood channels**, which are flanked by longitudinal **ebb spits**.

19.4 Dynamics and Facies

19.4.1 Origin and Mobility of the Inlets

There are different origins of inlets and tidal deltas. The first of the possibilities is intimately associated with the birth of the barriers. If the barrier originates separated from the continent and from other barrier islands, the separation channels are already born with the barrier itself. In addition, the associated tidal deltas develop when the barrier stabilizes. The second possibility is that the inlets originate from a process of breaching of the barrier. Normally, this process of rupture is associated with the dynamics of storms and the incidence of overwash processes. These processes originate new inlets in previously stabilized barriers, being able to generate islands from spits or to divide previously existing islands.

Once formed, an inlet can tend to migrate when the incidence of the waves is oblique and is accompanied by littoral drift phenomena. Normally, the migration of inlets is associated with their asymmetry and the dominating agent, being much greater in the cases of wave-dominated and mixed-energy deltas.

The appearance of a new inlet by a breaching process implies a division of the tidal prism between the previous and the new inlet. This always induces a decrease of the tidal currents across the two inlets. The decrease of the tidal currents implies an increase of the relative influence of the waves, which tries to close the inlet by attaching frontal bars reworked from the ebb-tidal deltas. Generally, one of the inlets ends up closing and the system chooses to maintain the new inlet while closing the old one. This means that barrier systems can have ephemeral inlets, but also that the opening of new inlets usually starts migration and barrier renewal cycles. An example of this type of cycle generated by the opening and migration of an inlet was analyzed in Chap. 16 (Fig. 16.12).

The progressive loss of tidal prism due to the infilling of the lagoon into tidal flats and marsh environments usually means the increase of wave dominance in the ebb deltas

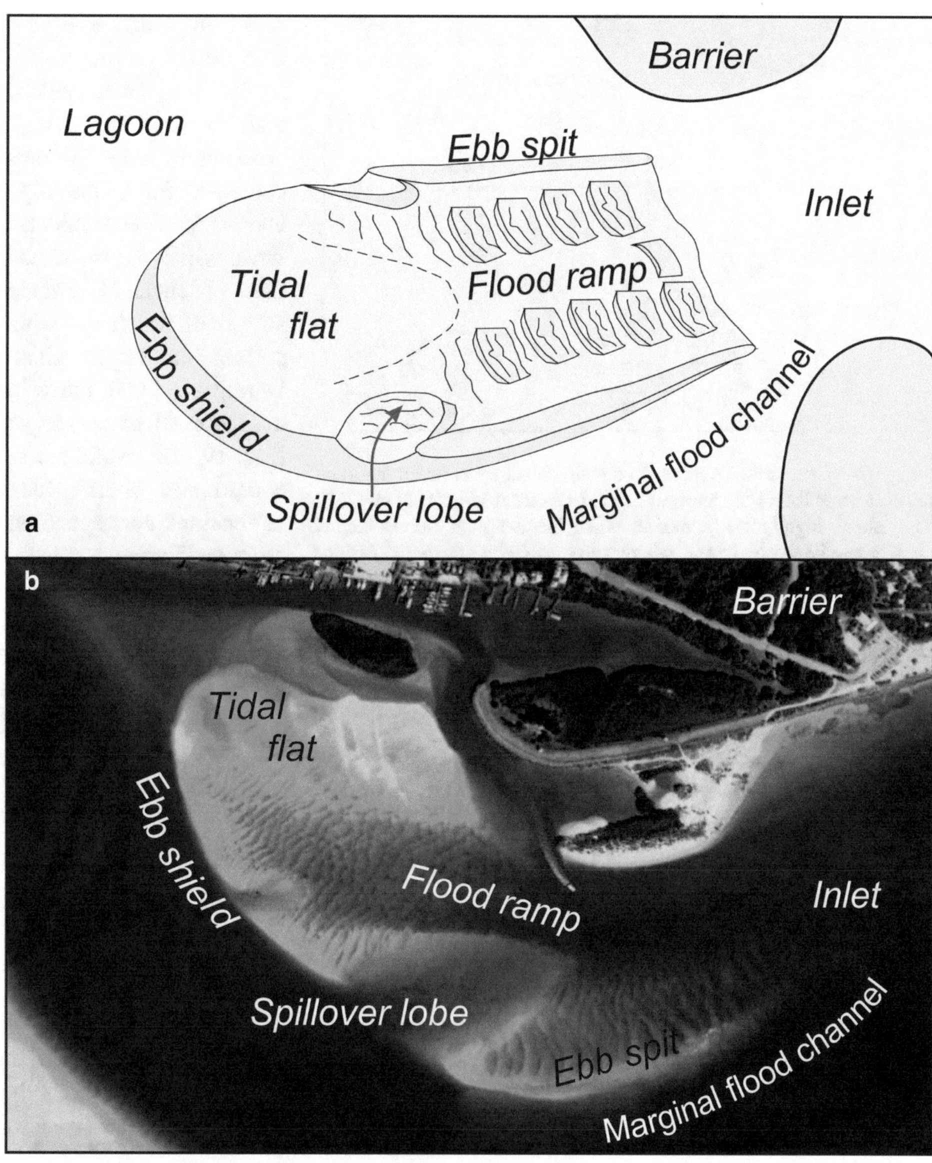

Fig. 19.8 **a** General morphological model for a flood-tidal delta, adapted from Hayes [16]. **b** Example showing the elements of the general model

associated with the inlet. The final stage of the cycle corresponds to the total filling of the back barrier system and the final closure of all the inlets.

19.4.2 Dynamics and Facies of Ebb-Tidal Deltas

Ebb-tidal deltas develop large differences between the main ebb channel and the marginal flood channels in terms of the asymmetry of velocities and duration of tidal currents (Fig. 19.9). The main channel develops a time/velocity curve in which a clear asymmetry is observed in the direction of the ebb, in both the developed velocities and duration of these currents. On the other hand, in the marginal flood channels the asymmetry is opposite, with greater flood velocities and longer duration being observed.

This diagram shows how the inversion of the currents to a flood situation occurs earlier in the marginal flood channels than in the main ebb channel (situation marked in Fig. 19.9 with an ellipse). Thus, in the last moments of the cycle, the flood begins to penetrate the lagoon through the marginal channels, while the ebb conditions in the main channel are still maintained. This time interval results in the current circulation model (Fig. 19.10) presenting a transitional situation between the general ebb and flood conditions (Fig. 19.10b).

Regarding the sediment transport and the generation of bedforms, in the main ebb channel the velocity conditions necessary for the migration of ripples are reached only half an hour after the beginning of the ebb current. About another half an hour later, the velocity necessary for the migration of dunes is reached. Under these conditions, the dunes remain

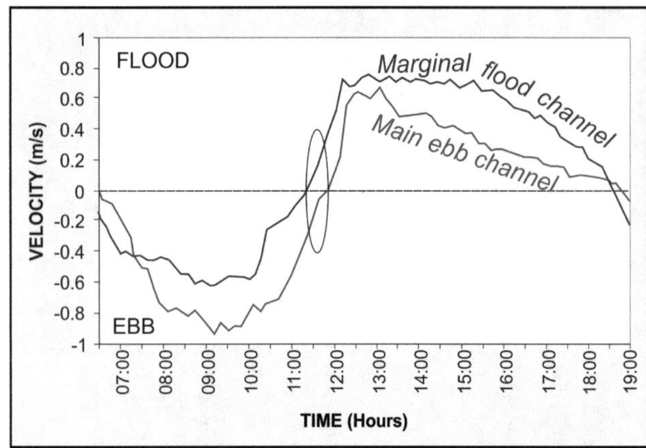

Fig. 19.9 Time/velocity curves in a main ebb channel and a marginal flood channel for a tidal range of 2.5 m (data from Morales et al. [25]). The ellipse highlights a situation when the marginal flood channel already has flood conditions, whereas the main ebb channel maintains the ebb current

under migration for more than three and a half hours. Conversely, during the flood semicycles, the speed required for the dune movement is maintained for only a little more than 75 min (Fig. 19.11a). Thus, in the main ebb channel, cross-laminations and cross-bedding are usually preserved, corresponding to the migration of ripples and dunes in the ebb sense. These calculations are made for medium sand sizes, which are the ones that preferentially circulate along these channels. Nevertheless, larger sand levels and gravel-size shell fragments with 2D cross-bedding may also be present. In the marginal channels, the opposite occurs. During the flood, the velocity conditions necessary for the migration of dunes are only maintained for about 40 min (Fig. 19.11b), while during the ebb, these conditions barely exceed two hours. Due to this fact, small-scale cross-laminations corresponding to the migration of ripples and cross-bedding generated by dune migration, both in the

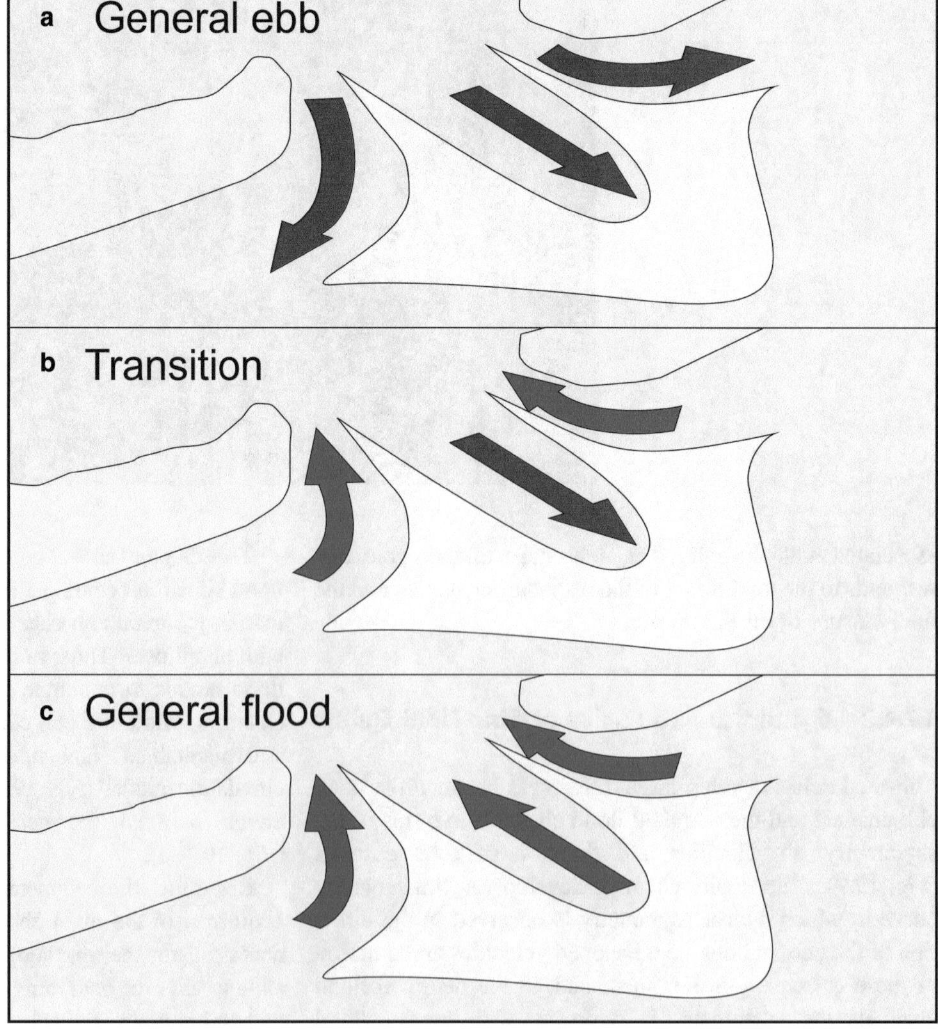

Fig. 19.10 Conceptual model showing the general tidal circulation in an ebb-tidal delta. **a** During ebb conditions. **b** Interval of flood inversion in the marginal channels. **c** During flood conditions, adapted from Hubbard and Barwis [21]

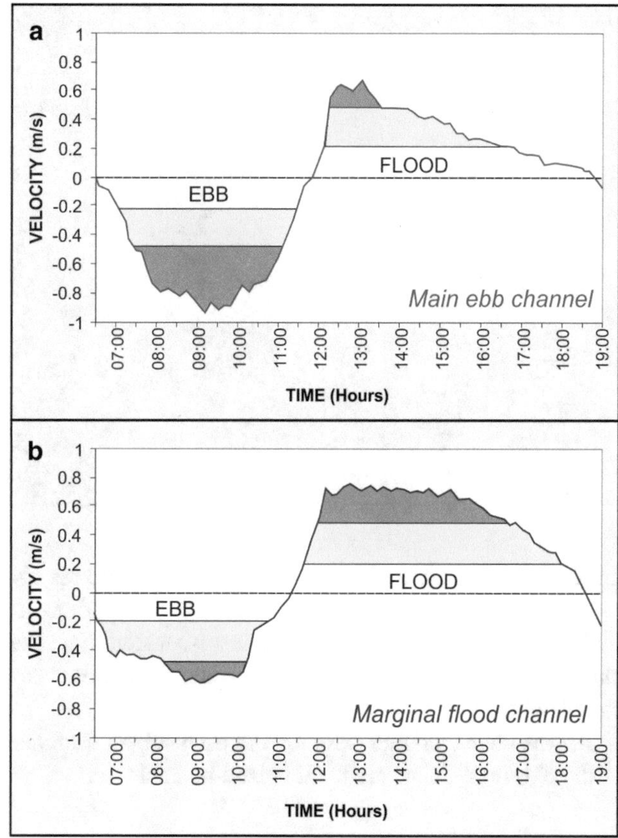

Fig. 19.11 Time/velocity curves of Fig. 19.9, where the thresholds for development of ripples (yellow) and dunes (orange) are marked. The curves correspond with the main ebb channel **a** and a marginal flood channel **b** (data from Morales et al. [25])

direction of the flood, are mainly preserved [25]. This dynamic is consistent with that described by other authors (e.g., [3, 21]).

In channel margin linear bars, the current pattern is much more complicated, as they do not stick to a channel and have rotational rather than bidirectional patterns. In addition, it should be noted that these bars are emerged during the last moments of the ebb and first moments of the flood. In general, ebb currents dominate. However, flood currents from the marginal channel to the main one can also play a role [21].

Due to this distribution of currents, there is a wide variability of bedforms with different dimensions and orientations. Thus, in the areas that flank the main ebb channel, 3D dunes of metric dimensions usually develop. Sometimes, these shapes have superimposed linguoid ripples (Fig. 19.12a) and sinuous ripples with perpendicular directions may be installed in the troughs between ridges of the larger forms (Fig. 19.12b). At the seaward edge of the swash platform, swash bars are developed and start to migrate inward (Fig. 19.12c). Extensions between different bars develop sinuous wave ripple fields with a general direction more or less parallel to these bars (Fig. 19.12d).

In the frontal lobe, there is a pattern of currents similar to the main ebb channel, although with lower speeds, since the section is larger. Therefore, ebb bedforms and structures dominate, although the grain size is finer.

The action of waves also affects the frontal lobes, since they are pounded for quite some time due to their shallower

Fig. 19.12 Different bedforms developed in the marginal linear bars (swash platforms). **a** Medium 3D dunes with superimposed linguoid ripples located in the border of the main ebb channel. **b** Medium 3D dunes with sinuous crests and perpendicular ripples developed on the troughs. **c** Intertidal swash bars developed on the front of the platform. **d** Sinuous crested wave ripples located on the troughs between intertidal swash bars

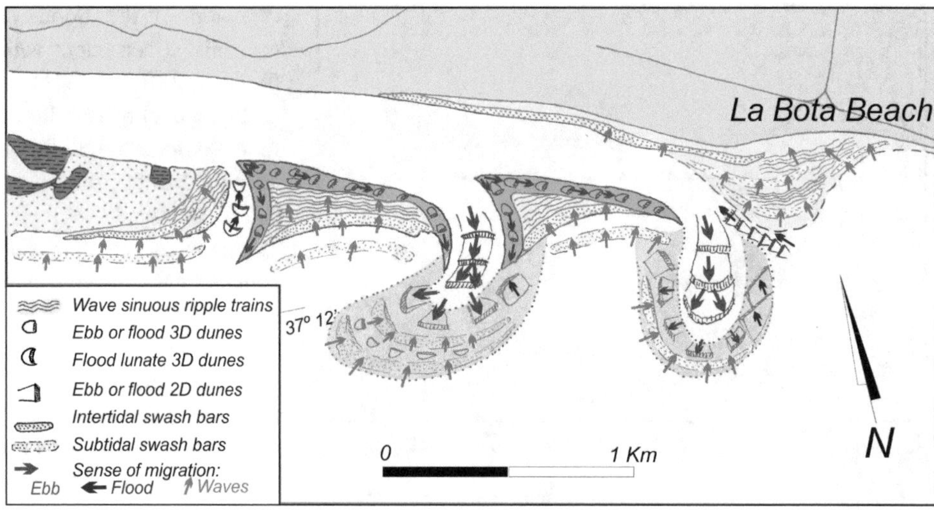

Fig. 19.13 Bedform distribution and sand transport patterns in a mixed-energy ebb-tidal delta system (Piedras inlet, Spain). This scheme is based on the type and direction of the main bedform lee facies [24]

depth. This swash that takes place in the frontal lobe has a special character. On the one hand, the waves are refracted by changing direction, losing energy and depositing the material they transport. On the other hand, these are waves without a runoff, so that the process also produces an inward migration of bedforms, especially during the moments of flood. These bedforms are usually swash bars and 2D dunes, as well as 3D dunes and linguoid ripples.

The forms are not generally preserved in the frontal lobes, although they are preserved in the channel margin bars [20]. A good scheme of the distribution of bedforms and their preferential directions of migration in the environments associated to a mixed energy ebb-tidal delta system was proposed in the work of Morales et al. [24] based on the Piedras inlet in SW Spain (Fig. 19.13).

In the shallower areas of the delta and, above all, in those places protected from the action of waves (such as the less energetic areas of the main channel, adjacent to the channel margin bars and the marginal flood channels), there are also usually interleaved sheets of mud that correspond to the settling of the material transported in suspension that takes place during the moments of slacks, when the currents stop completely.

All the characteristics described in each sub-environment of the ebb-tidal deltas are summarized in Table 19.1.

19.4.3 Flood-Tidal Deltas

From a dynamic point of view, there are great differences between the different elements of the delta. As with the ebb-tidal deltas, the velocities and durations of the tidal currents are clearly asymmetric. There are, however, different types of asymmetries in the different sub-environments of the delta (Fig. 19.14). In order to better understand the dynamic functioning of the flood-tidal delta, it is necessary

Table 19.1 Summary of the main characteristics of the different sedimentary sub-environments within an ebb-tidal delta

Sub-environment	Processes	Bedforms	Sedimentary structures	Grain size
Main ebb channel	Tidal ebb and flood reworking	Ripples and dunes in the ebb direction	Metric and decimeter scale cross-layers inclined towards the sea	Fine and medium sand. Shell lag deposits
Marginal flood channel	Tidal flood and ebb reworking	Ripples in the flood direction	Decimeter scale curved base cross-lamination towards the land. Flaser beds	Fine and medium sand. Silt
Channel margin bars	Tidal currents with rotational patterns. Wave swash	Ripples with directional dispersion. Swash bars	Decimeter scale cross-lamination	Medium sand. Shell fragments
Frontal lobe	Tidal ebb. Wave oscillation	Ripples in the direction of the ebb reworked by waves	Metric scale cross-bedding inclined towards the land. Cross-lamination of decimeter scale. Flaser beds	Very fine sand. Silt

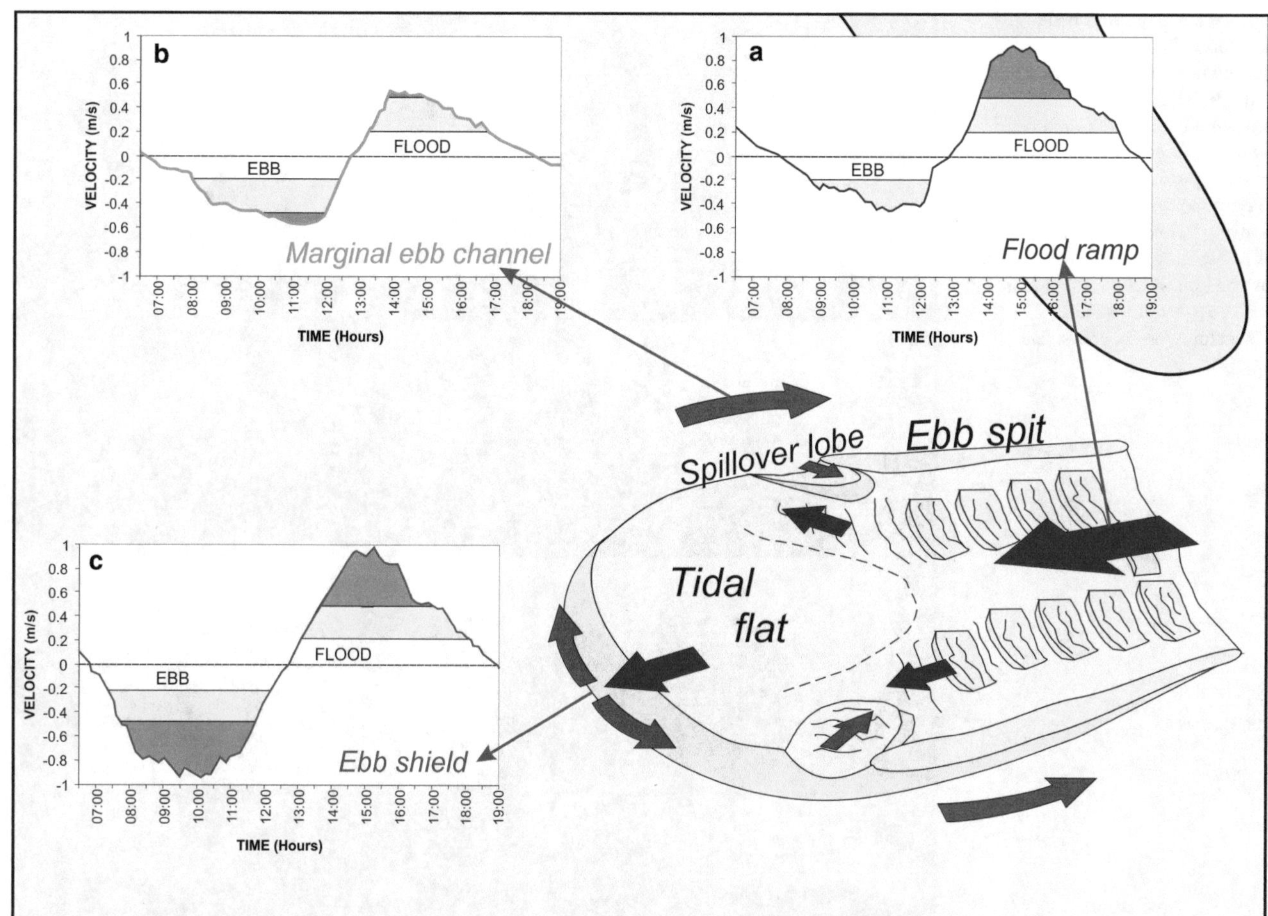

Fig. 19.14 Morphology and flow patterns in a flood-tidal delta. **a** Current velocity/time in the flood ramp. **b** Current velocity/time in the marginal ebb channels. **c** Current velocity/time in the ebb shield. In all curves, the thresholds for migrating bedforms are indicated: ripples (yellow) and dunes (orange)

to consider the velocity/time curves in three different zones: (1) the flood ramp; (2) the marginal ebb channels flanking the ramp; and (3) the tidal flat front (ebb shield).

Similar patterns of tidal currents and bedform migration are present in the flood ramp. The curve shows a clear asymmetry in the flood direction (Fig. 19.14b), so that once the flood current starts, the minimum velocity required for the start of ripple migration is reached almost immediately. About one hour later, the 2D and 3D dunes begin to migrate, and these conditions remain for at least three hours. Conversely, during the ebb, only the velocity required for the start of the ripple migration is reached. These conditions are maintained for about three and a half hours. Consequently, the dominant bedforms are dunes and ripples oriented in the flood direction (Fig. 19.15a). The migration of these forms generates cross-stratification and cross-laminations with mostly landward-sloping sheets (Fig. 19.15c).

A situation contrary to that described occurs in the marginal ebb channels which flank the delta as well as in the spillover lobes. In these environments, ebb currents dominate (Fig. 19.14b). Most of time they are self-currents, able to make ripples migrate. Nevertheless, dune migration conditions can be reached during the ebb cycle for about an hour (Fig. 19.15b). The typical structures of these environments will then be complex small-scale cross-laminations sloping in the direction of the ebb. However, the same structures sloping in the opposite direction are also less abundantly present, and together they form the typical herringbone structures (Fig. 19.15d). In all the sectors, more or less thick layers of fine material usually appear interspersed with the sands. These layers correspond to the settling of material transported in suspension during the moments of the current lapses. These constitute flaser and wavy bedding structures, depending on the average energy of the place where they appear (Fig. 19.15e).

One situation that differs from those previously described is that which occurs on the ebb shield, which is a shallower area where higher speeds are reached (Fig. 19.14c). In this case, there is a near symmetry of tidal currents. Consequently, the conditions for the migration of

Fig. 19.15 Different bedforms developed in the sub-environments of a flood tidal delta. **a** 3D dunes with superimposed ripples on a flood ramp. **b** 3D dunes and ripples on an ebb spit. **c** Cross-bedding of a flood dune with superimposed ripples. **d** Herringbone structure of a reversing form developed on an ebb spit. **e** Interbedded mud and cross-bedded sands. **f** Structures on an ebb shield

3D dunes are reached during both ebb and flood (Fig. 19.15f). The particularity is that the ebb and flood circulate in perpendicular directions. In this area, 2D and 3D dunes are dominantly developed, both oriented in flood and ebb directions.

Frequently, the tidal flats of the central zone of the delta quickly aggrade until they are colonized by halophytic vegetation and transformed into marshes. At this point, the flood stops going beyond this zone and sticks to the flood channels that surround it. In this situation, this environment increases its depth because the currents increase. From then on, all the water exchange between the lagoon and the sea takes place through these channels and the flood delta stops working as such. The evolution continues when new flow deltas are built at the end of these channels. All of the characteristics described in each sub-environment of the flood deltas are summarized in Table 19.2.

19.5 Facies Models

The depositional models proposed for these environments are built on ideal sequences. Thus, their theoretical spatial variations have been obtained mostly in existing tidal deltas. This means that they are comparable with the geological record under conditions of maximum preservation potential.

It is evident that the general models suffer variations in thickness and geometry of the sedimentary bodies, depending on the type of delta as stated in the introduction.

19.5.1 Ebb-Tidal Deltas

In general terms, it can be said that the facies of the ebb-tidal deltas are arranged over other external facies of the barrier

Table 19.2 Summary of dynamic and sedimentary characteristics of the different sub-environments linked to the flood-tidal deltas

Sub-environment	Processes	Bedforms	Sedimentary structures	Grain size
Main channel and flood ramp	Tidal flood and ebb reworking	Ripples and dunes in the flood direction	Cross-bedding of decimeter and metric scale inclined towards the land	Fine and medium sand. Shell lag deposits
Tidal flat	Tidal flood and ebb reworking	Ripples in the flood direction	Decimeter scale curved-base cross-bedding inclined towards the land. Flaser, wavy and lenticular bedding	Fine and medium sand. Silt
Marginal ebb channels	Tidal ebb and flood reworking	Ripples in the ebb direction	Decimeter scale curved base cross bedding inclined towards the sea. Herringbone and flaser beds	Medium and fine sand. Silt
Ebb shield	Tidal ebb and flood	Ripples and dunes in the flood direction reworked at the sides by the ebb	Decimeter scale cross-bedding. Herringbone beds	Fine sand

island system, such as the nearshore facies. Assuming a stable sea level, the internal disposition of the facies generated in the different components of the tidal delta corresponds to a regressive model of progradation. In this scenario, both ebb and flow channels fill up when they are abandoned (by breaching), when they migrate or when they lose an effective section due to the decrease of the tidal prism. An ideal sequence was proposed by Sha and De Boer [34] for Frisian Island ebb deltas. This is consistent with the sequence proposed by Imperato et al. [23] for the US East Coast.

In the proximal sector (Fig. 19.16a), the sequence from base to top would consist of:

(a) Erosive contact with nearshore facies.
(b) Lag deposit of shells and shell fragments at the base of the marginal flood channel, on which are arranged sets of medium to coarse sands with flood-oriented trough

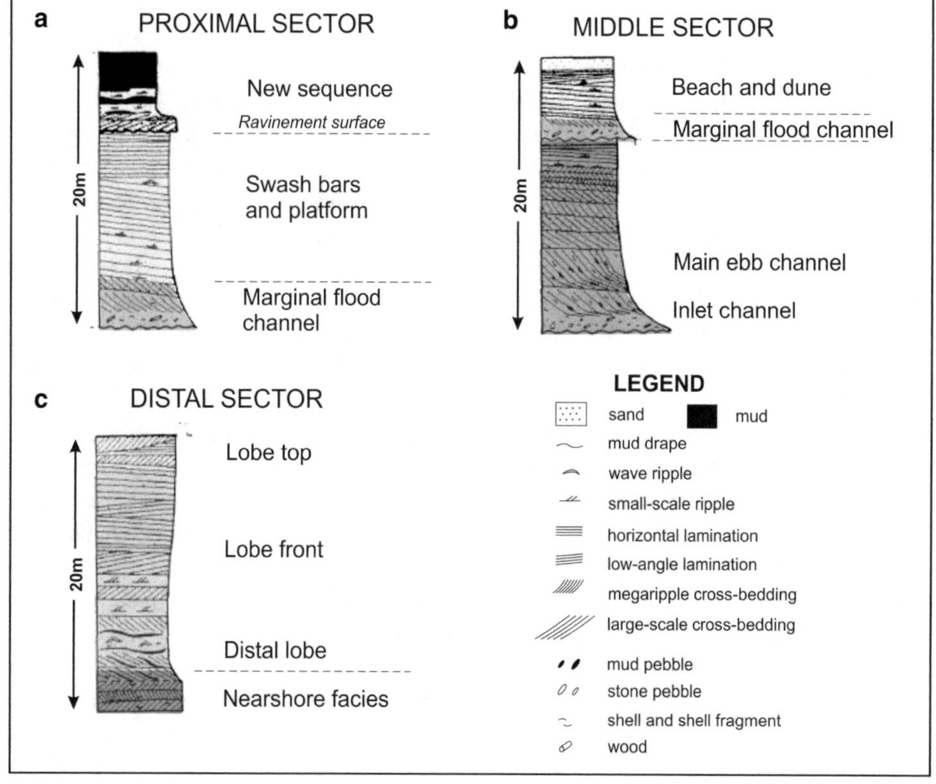

Fig. 19.16 Ideal sequences for an ebb-tidal delta, adapted from Sha and De Boer [34]

cross-bedding of decimeter scale. These facies are not very thick (less than 1 m) due to the fact that they are generated during the operation of the channel, when the dominant process is the sedimentary bypassing of sands towards the tidal inlet.

(c) Deposit of more than 2 m of fine and very fine sands with ripple-type cross-lamination. Dominant bedforms are oriented in the flood sense and less abundantly in the ebb sense. These facies are interlaminated with muds in variable proportion forming flaser and wavy bedding structures. It is a very bioturbed body that corresponds to the infilling stage of the marginal flood channel when losing functionality.

(d) These facies, oriented laterally towards the sea, are interleaved with parallel laminated fine sands with abundant lenses of shell fragments and inland-oriented planar cross-bedding. This level corresponds to the marginal linear bars and to the swash bars that migrate over them. The lamination is planar due to the action on these of spilling-type breakers, without undertow that could redistribute the sediment on the surface.

(e) The sequence is completed with sets of planar cross-bedded fine sands, corresponding to wave-generated 2D bars that migrate towards the land.

In the middle sector (Fig. 19.16b) are the facies of the main ebb channel, which are embedded through erosive contact with the shoreface facies. These channel facies are the thicker of the system, being able to exceed 15 m, and consist of:

(a) Body of more than 5 m composed by sets of medium to thick sands with intercalations of shells and shell fragments with ebb-oriented trough cross-bedding. These correspond to the migration of 3D dunes through the bed of a main ebb channel that migrates laterally and allows the preservation of these forms in the less energetic areas.

(b) Above this level there is about 1 m of fine and very fine sand with flaser bedding structures and abundant bioturbation. These are interpreted as the less active sectors of the ebb channel.

(c) Finally, the top culminates the sequence with about 5 m of fine sand with parallel lamination, corresponding to the already described swash bar facies.

In the distal sector (Fig. 19.16c), fine and very fine sand bodies are developed in which sets of cross-bedding originate from both the tidal ebb and the wave action. These correspond to the frontal lobe facies, which laterally transform into parallel lamination associated with tidal bundles, where spring and neap tidal cycles are recognizable. These facies are arranged over the shoreface facies or even over the offshore facies if the delta is large enough.

19.5.2 Flood-Tidal Deltas

The sequence that results from the process of a flood-tidal delta advancing into the interior of the lagoon is a regressive sequence of shallowing. This sequence presents at the base the sediments corresponding to the deepest environments, which will be progressively transformed until reaching the supratidal zone. In general terms, the facies of the flood deltas are arranged over the facies of the lagoon, and they are interleaved both longitudinally and laterally. Like ebb deltas, flood deltas can undergo variations in thickness and geometry. The theoretical sequence is composed of five lithological units [3]:

(a) Sedimentary continuity with the fine lagoon sediments.

(b) On the lateral faces of the delta there are 2–3 m of fine sand at the base of the series. These present sets of small-scale trough cross-bedding and cross-lamination. These structures correspond to the migration of 3D dunes and ripples. Normally, (more abundantly) dunes are oriented in the ebb direction. These facies correspond to the typical sediments of the marginal ebb channels and ebb spits.

(c) Over these facies, the most important sedimentary body of the delta by its size and thickness is arranged. This is the one corresponding to the flood ramp and the shallow flood channels. These consist of metric sets of flood-oriented planar cross-bedded fine and medium sands. The sets are separated by laminations of coarser sediments or by sets of ripple-type cross-laminated sands. This sedimentary body can exceed 6 m in thickness depending on the tidal range.

(d) Limiting the facies of the flood ramp towards the lagoon, a horseshoe-shaped body of fine sand is located. This presents ripple-type lamination oriented in a very dispersed way, although those directions that coincide with the orientation of the limit of the sedimentary body tend to dominate. These are the facies corresponding to the ebb shield.

(e) On the top of these facies are the sediments of the tidal flat. These are very fine sands with mud intercalations forming flaser and wavy bedding structures. In any case, these are affected by abundant bioturbation. The sands usually present flood-oriented ripple-type lamination.

This sequence could be covered by saltmarsh mud bioturbed by roots when the delta has ceased to be functional. On the other hand, the sequence can also be covered by a body of land-oriented cross-bedded fine sands corresponding to the installation of a washover fan. This situation occurs when the tidal inlet is totally closed, but its traces remain as a depressed area of the dune system. This depression can be over-washed by waves during storms. This has been observed in the geological record by Murakoshi and Masuda [27] in the Pleistocene infilling of the Paleo-Tokyo Bay (Fig. 19.17) and also by Boersma [2] in a Miocene flood-tidal delta of Rhine Bay (Germany). In the geological record, some terms of the sequence can be repeated when there are pulses of relative sea level rise.

19.5.3 Architectural 3D Facies Model in the Barrier System Framework

The morphological variability of the sandy bodies of tidal inlets and tidal deltas associated with barrier island systems has been explained as a response to the action of tides and waves. Nevertheless, other factors such as the tidal prism, freshwater input, the nature of the substrate, the initial slope, the sedimentary input and relative sea level movements also exert influence. All these controls make the relationship between the sandy bodies of the ebb- and flood-tidal deltas, and other sandy bodies corresponding to the rest of elements of the barrier island system, respond to different architectural schemes.

These relationships have been studied by Hubbard et al. [22], who reduced the geometry of barrier island systems to a three-case model according to whether they were dominated by tidal currents or wave energy or a combination of the two (Fig. 19.18). These models have been proposed assuming a 100% preservation potential and moderate upward movement of sea level.

Thus, in the tide-dominated model the main sandy bodies are located outside the island, so that in this case the ebb-tidal delta is the environment with the greatest thickness and surface extension. Here, the flood-tidal delta is usually absent. This is the model that presents the greatest stability of the tidal inlet.

Conversely, the wave-dominated model shows a significant development of multilobate flood-tidal deltas. At the same time, the ebb-tidal deltas are usually smaller and crescent-shaped in the direction of the coastline. In this case, the swash platforms are poorly developed. More than one tidal inlet and a high migration rate may occur, contributing to the presence of more than one flood delta in the geological record, as described by Boersma [2] in the German Miocene.

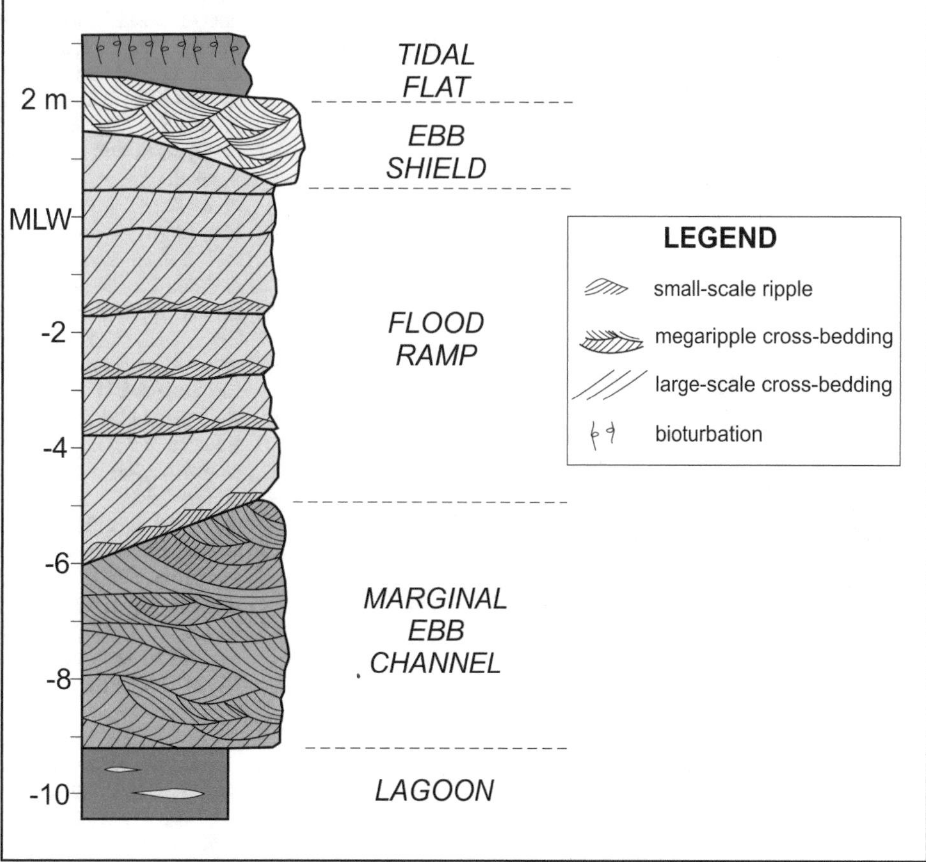

Fig. 19.17 Stratigraphic sequence of the Pleistocene infilling of the Paleo-Tokyo Bay, adapted from Murakoshi and Masuda [27]

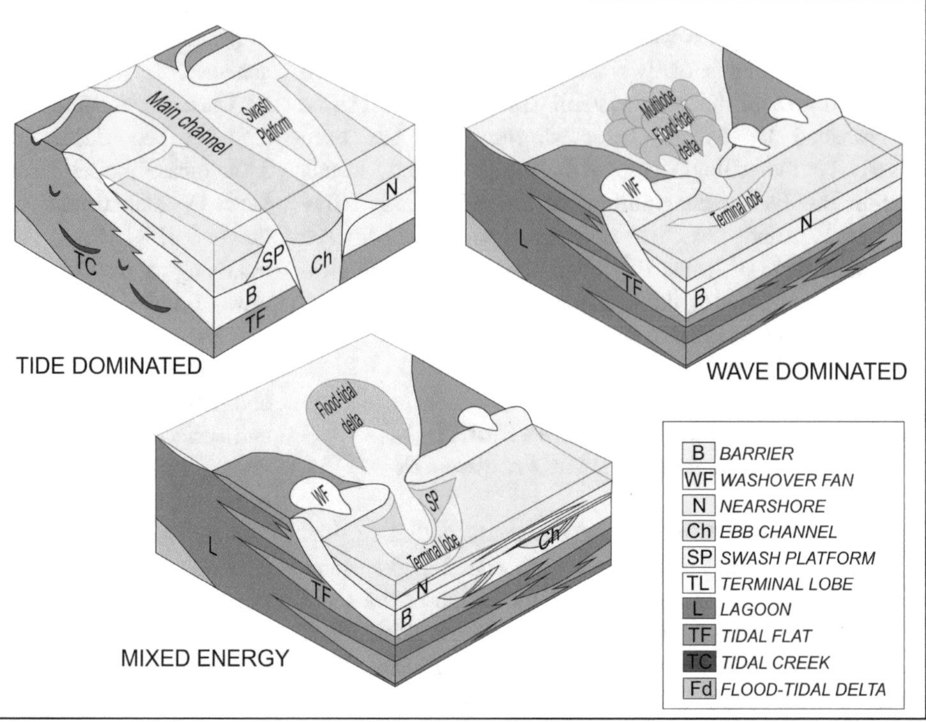

Fig. 19.18 Conceptual block diagrams showing the geometry and 3D relationships of sedimentary bodies generated by the three types of tidal inlets, adapted from Hubbard et al. [22]

The mixed or transitional model presents intermediate characteristics. Here, the extension of the ebb- and flood-tidal deltas is usually more equilibrated. However, it is the model that can present more variations in one or other direction, depending on the relative importance of the energy of tidal currents and waves. Usually, there is only one main channel and several marginal channels, which are generally unstable and have a high migration rate. So, channel filler sediments are well represented in the open zone sequence.

References

1. Aubrey DG, Weisher L (eds) (1988) Hydrodynamics and sediment dynamics of tidal inlets: lecture notes on coastal and estuarine studies. Springer, Heidelberg, p 454
2. Boersma JR (1991) A large flood tidal delta and its successive spill-over apron: detailed proximal-distal facies relationships (Miocene Lignite Suite, Lower Rhine Embayment, Germany). In: Smith DG, Reinson GE, Zaitlin BA, Rahmani RA (eds) Clastic tidal sedimentology, vol 16. Canadian Society of Petroleum Geologists Memoir, pp 227–254
3. Boothroyd JC (1985) Tidal inlets and tidal deltas. In: Davis RA (ed) Coastal sedimentary environments. Springer, Heidelberg, pp 445–532
4. Boothroyd JC, Hubbard DK (1974) Bedform development and distribution pattern, Parker and Essex Estuaries, Massachusetts. Miscellanea Paper 1–74. Coastal Engineers Research Center, 39pp
5. De Boer PL, Nio S (eds) (1988) Tide influenced sedimentary environments and facies. Reidel Publishing Co., Boston, p 530
6. Davies JL (1980) Geographical variation in coastal development. Longman, New York, p 212
7. Davis RA, Clifton HE (1987) Sea-level change and the preservation potential of wave-dominated and tide-dominated coastal sequences. J Sedim Petrol 41:167–178
8. Davis RA, Fox WT (1981) Interaction between wave- and tide-generated processes at the mouth of a microtidal estuary: Matanzas River, Florida. Mar Geol 40:49–68
9. Davis RA, Gibeaut JC (1990) Historical morphodynamics of inlets in Florida: models for coastal zone planning. Technical paper 55, Florida Sea Grant College Program, 81pp
10. Finley RJ (1978) Ebb-tidal delta morphology and sediment supply in relation to seasonal wave energy flux, North Inlet, SC. J Sedim Petrol 48:227–238
11. Fitzgerald DM, Nummedal D (1983) Response characteristics of an ebb-dominated tidal inlet channel. J Sedim Petrol 53:833–845
12. Gallivan LB, Davis RA (1981) Sediment transport in a microtidal estuary: Matanzas River, Florida. Mar Geol 40:69–84
13. Gibeaut JC, Davis RA (1993) Statistical geomorphic classification of ebb-tidal deltas along the West-Central Florida Coast. J Coast Res 18:165–184
14. King CAM (1972) Beaches and coasts, 2nd ed. St. Martin's Press, New York, 570pp
15. Kumar N, Sanders JE (1975) Inlet sequence formed by migration of Fire Island Inlet, Long Island, New York. In: Ginsburg RN (ed) Tidal deposits. Springer, Heidelberg, pp 75–83
16. Hayes MO (1975) Morphology of sand accumulation in estuaries: an introduction to the symposium. In: Cronin LE (ed) Estuarine research, vol 2. Academic Press, New York, pp 3–22
17. Hayes MO (1980) General morphology and sediment patterns in tidal inlets. Sed Geol 26:139–156
18. Hayes MO (1991) Geomorphology and sedimentation patterns of tidal inlets: a review. In: Coastal sediments, vol 91, American Society of Civil Engineers, pp 1343–1355
19. Hayes MO, Fitzgerald DM (2013) Origin, evolution, and classification of tidal inlets. J Coast Res 69:14–33
20. Hine AC (1975) Bedform distribution and migration pattern on tidal deltas in the Chatman Harbor Estuary. Cape Cod,

Massachusetts. In: Cronin LE (ed) Estuarine research, vol 2. Academic Press, New York, pp 235–252

21. Hubbard DK, Barwis JN (1976) Discussion of tidal inlets sand deposits: example from the South Carolina Coast. In: Hayes MO, Kana TW (eds) Terrigenous clastic depositional environments. Technical Report 11. CRD Univ. of South Carolina, Columbia, pp 158–171
22. Hubbard DK, Oertel G, Nummedal D (1979) The role of waves and tidal currents in the development of tidal inlets sedimentary structures and sand body geometry: examples from North Carolina, South Carolina and Georgia. J Sedim Petrol 49:1073–1092
23. Imperato DP, Sexton WJ, Hayes MO (1988) Stratigraphy and sediment characteristics of a mesotidal ebb-tidal delta, North Edisto Inlet, South Carolina. J Sedim Petrol 58:950–958
24. Morales JA, Borrego J, Jiménez I (2001) Morphostratigraphy of an ebb-tidal delta system associated to a large spit in the Piedras Estuary mouth (Huelva Coast, S.W. Spain). Mar Geol 172:225–241
25. Morales JA, Delgado I, Gutiérrez-Mas JM (2015) Bedform variability and flow regime in a barrier-inlet system. The mesotidal Piedras mouth (Huelva, SW Spain). Geol Acta 13:137–153
26. Moslow TF, Tye RS (1985) Recognition and characterization of Holocene tidal inlet sequences. Mar Geol 63:129–152
27. Murakoshi N, Masuda F (1991) A depositional model for a flood-tidal delta and washover sands in the late Pleistocene Paleo-Tokyo Bay, Japan. In: Smith DG, Reinson GE, Zaitlin BA, Rahmani RA (eds) Clastic tidal sedimentology, vol 16. Canadian Society of Petroleum Geologists Memoir, pp 219–226
28. Oertel GF (1975) Geomorphic cycles in ebb-deltas and related patterns of shore erosion and accretion. J Sedim Petrol 47:1121–1131
29. Oertel GF (1988) Processes of sediment exchange between tidal inlets, ebb deltas and barrier islands. In: Aubrey DG, Weisher L (eds) Hydrodynamics and sediment dynamics of tidal inlets. Springer, Heidelberg, pp 297–318
30. Okazaki H, Masuda F (1995) Sequence stratigraphy of the late Pleistocene Palaeo-Tokyo Bay: barrier islands and associated tidal delta and inlet. In: Flemming BW, Bartholoma A (eds) Tidal signatures in modern and ancient sediments, vol 24. IAS Special Publication. Blackwell Science, Oxford, pp 275–288
31. Oost A (1995) Sedimentological implications of morphodynamic changes in the ebb-tidal delta, the inlet and the drainage basin of the Zoutkamperlaag tidal inlet (Dutch Wadden sea), induced by a sudden decrease in the tidal prism. In: Flemming BW, Bartholoma A (eds) Tidal signatures in modern and ancient sediments, vol 24. IAS Special Publication. Blackwell Science, Oxford, pp 101–120
32. Sha LP (1989) Variation in ebb-delta morphologies along the West and East Frisian Island, The Netherlands and Germany. Mar Geol 89:11–28
33. Sha LP (1990) Sedimentological studies of the ebb-tidal deltas along the West-Frisian Islands, The Netherlands. PhD thesis, Utrecht. Geol. Ultraiectina 64. 160pp
34. Sha LP, De Boer PL (1991) Ebb-tidal delta deposits along the West-Frisian Islands (The Netherlands): processes, facies architecture and preservation. In: Smith DG, Reinson GE, Zaitlin BA, Rahmani RA (eds) Clastic tidal sedimentology, vol 16. Canadian Society of Petroleum Geologists Memoir, pp 199–218

Tide-Dominated Systems II: Tidal Flats and Wetlands

20

20.1 Introduction

Tidal flats are depositional surfaces where sedimentation of mud and sand occurs and that are located in the space limited by sea level between high and low waters. Because of their location, they are completely covered by water during high tide and are exposed to air at low tide (Fig. 20.1). Obviously, these are tide-dominated environments and therefore have minimal wave influence. Thus, they develop along open coasts with low relief and affected by low energy waves (open-coast tidal flats) or on coasts with higher wave energy but in areas protected from direct wave action, behind barriers, spits or reefs (back-barrier tidal flats).

In the first case, when they have developed on low wave energy coasts or macrotidal coasts, tidal flats are born as independent environments and are usually located inside bays that amplify the effect of tidal currents [15]. The development of massive deposits of fine sediments results in this type of coastline being generally known as muddy coasts. The best-studied examples at a global level are found in Mont Saint-Michel in France, Inchon in Korea, the Wash in England, the Bay of Fundy in Canada, the Bay of San Sebastian in Tierra del Fuego, Argentina and the Gulf of California in Mexico. When they appear in protected areas of high wave energy coasts, tidal flats can also appear associated with other sedimentary environments (Fig. 20.2). In this context, they can develop as sub-environments of estuaries, deltas or barrier island systems [17].

At present, the widest tidal flat is the Yellow Sea in Korea, which reaches a width of 25 km [6]. Despite its large extent, the distribution of its sedimentary texture is similar to those described for the smaller tidal flats, such as those of the North Sea [57].

Most of the current tidal flats are places where mainly siliciclastic sedimentation takes place. This is the case for the best-known examples. Nevertheless, there are also significant examples of tidal flats consisting of carbonate and evaporite sediments, such as the flats associated with Laguna Madre in Texas, the northern zone of Andros Island in the Bahamas or the Trucial Coast in the Persian Gulf. Much of what we know about the tidal flats preserved in the geological record comes from research carried out in the existing ones through the application of the principle of uniformitarianism.

The scientific interest in tidal flats as a sedimentary environment is reflected in the many monographs devoted to their study [12, 20, 37, 25, 26, 42, 43, 45]. In addition to appearing as a chapter in a large fraction of the works dedicated to the general study of sedimentary environments [27, 41, 50, 56, 31], among others), there are also numerous articles that are constantly published in scientific journals focused on the dissemination of this topic. In addition to the purely scientific engagement with tidal flats, they have also been found to be of great economic interest due to the deposits of uranium, oil and natural gas that can be found in these systems.

20.2 Control Factors and Global Distribution

20.2.1 Tidal Regime

The distribution of tidal flats on the world's coasts has been classically associated with tidal range. Thus, this type of system was at first described as part of the macrotidal coasts (for tidal oscillations greater than 4 m). Later, tidal flats were also described on mesotidal and even microtidal coasts. However, tidal flats associated with tidal ranges below 4 m were described as exceptional cases, since their abundance is considered much lower than on macrotidal coasts.

In any case, tidal dynamics are responsible for the sediment dynamics in these environments. It is the tidal current that controls the transport, deposition and distribution of sedimentary material in the extension of the plain. In turn, the temporal variation of the currents is regulated by the periodic oscillations of the tidal range. These variations result in modifications to the grain size that can be transported and the bedforms that can be generated. The tidal range is also responsible for the presence of a marked

J. A. Morales, *Coastal Geology*, Springer Textbooks in Earth Sciences, Geography and Environment,
https://doi.org/10.1007/978-3-030-96121-3_20

Fig. 20.1 Panoramic view of a tidal flat comparing the moments of high and low waters (Wadden Sea, the Netherlands)

vertical zoning, both by the levels of exposure and submergence, as well as by the velocities that the current develops at each level of the flat.

All these features are recorded in the sedimentary sequence in the form of variations in the internal lamination of the sediment that is generated in each of these zones.

20.2.2 Wave Regime

Following the ideas introduced by Hayes [22], it is not the tidal range, but the balance between the tidal range and the wave dimensions, which determines the possibility of a tidal flat developing on a given coast. For this author, the distribution of tidal flats is at its maximum on tide-dominated coasts and decreases on mixed-energy coasts, being minimum or absent on wave-dominated coasts. It is understood, then, that the influence of the waves for the development of tidal flats must be minimal or null. However, as already mentioned, tidal flats can appear on wave-dominated coasts as long as the existence of a morphological element generated by the waves itself inhibits the action of the waves—so, for instance, the creation of an area of dynamic shade for the waves in which only the tides act. This explains the presence of tidal flats on microtidal coasts and wave-dominated coasts, as is the case of the Bahamas Islands or the mud flats of the East Coast of Florida.

20.2.3 Sediment Supply

The development of a tidal flat requires a significant supply of sedimentary material. In general terms, it can be asserted that the high sedimentation rate of these environments is responsible for their characteristic low slope, influencing the loss of wave energy by dissipation while favoring the acceleration of tidal currents. In most cases, it is fine-grained material, although in some areas coarser material can also be transported. The origin of this material is generally linked to the presence of river mouths when the tide is associated with estuaries or deltas, although it can also be of marine or coastal origin when the sediments come from other nearby coastal environments.

There are situations in which the absence of a continuous terrigenous contribution determines a domain of reworking processes of precipitated materials in the subtidal areas of this same environment. Then, material generated by biological activity can also be deposited dominantly. In these cases, the nature of the most abundant sediments is carbonate or evaporite.

20.2.4 Organic Activity

The activity of the organisms on the tidal flats is of vital importance. As they are tranquil environments, they are

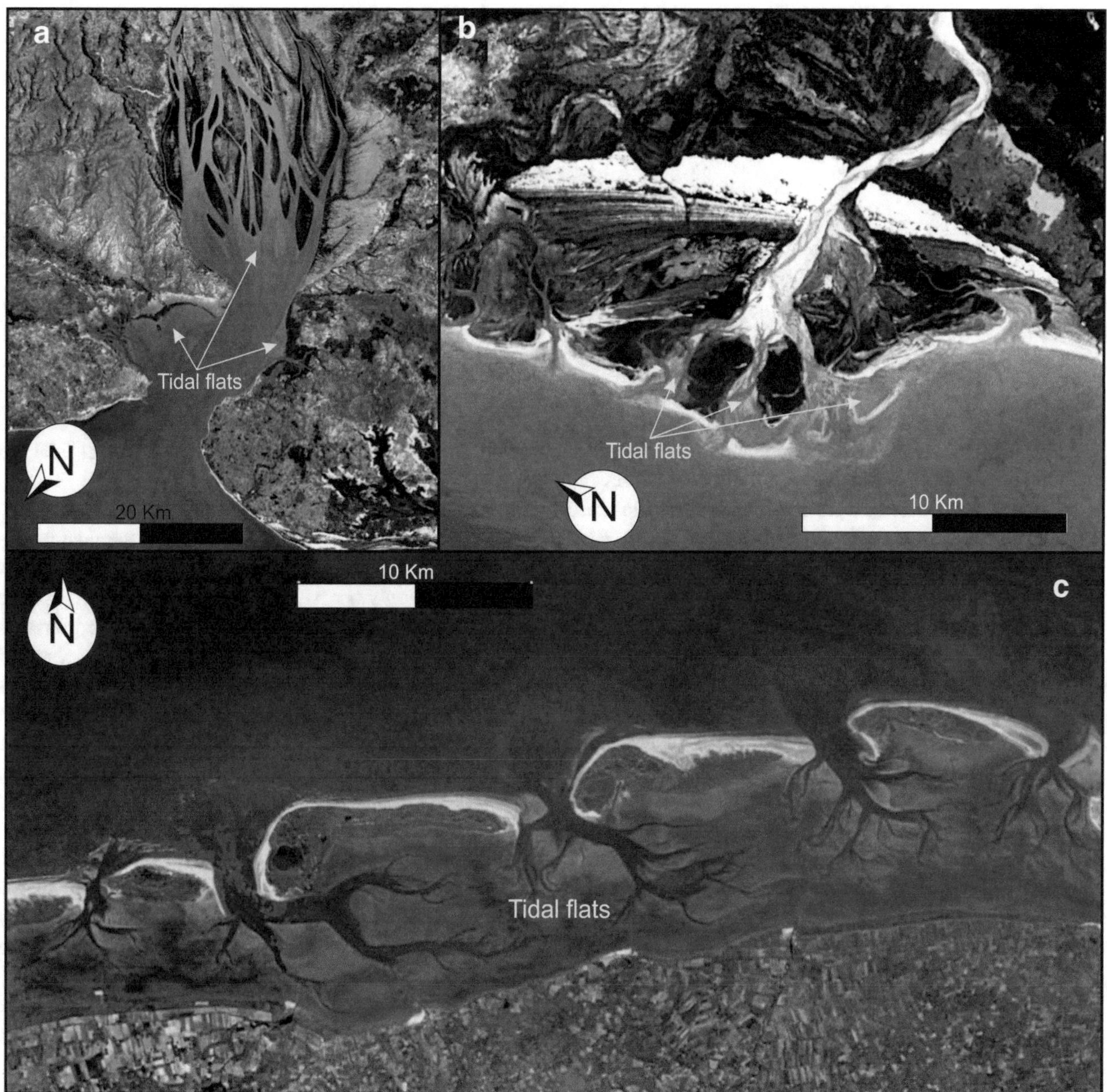

Fig. 20.2 Development of tidal flats linked to other major environments. **a** Estuaries (example of a tide-dominated estuary in Amborovy, Madagascar). **b** Deltas (example of a mixed-energy delta in Ankatrafay, Madagascar). **c** Barrier island systems (example of the Wadden Sea, Germany). (Images: Landsat/Copernicus from Google Earth.)

home to ecosystems with a wide profusion of life, both animal and plant. This influence of the organisms on the sedimentary environment, which has already been described in Chap. 12, has a special influence in the case of tidal flats, where four types of processes related to organic activity usually occur:

- Firstly, the accumulation of soft parts, shells or pellets can lead to the formation of **biogenic sediment**.
- Secondly, the organisms (specially plants) can act as a screen for the currents by slowing them down and causing decantation. Many plants can also trap sediment particles by means of cohesive mucous membranes (**biomechanical sedimentation**).
- In addition, some organisms are able to modify the chemical conditions of their surrounding environment by inducing precipitation or flocculation processes (**biochemical sedimentation**).

- Finally, the activity of the organisms on the sedimentary substrate causes a significant alteration of the internal structure of the sediment (**bioturbation**).

20.2.5 Climate

In addition to dynamic and biological factors, climate introduces notable variations in processes, reflected in the nature of the sediment. In principle, tidal flats can develop in any climate, although there are particularities in the flats depending on the climate in which they develop.

- In regions of temperate climate, tidal flats composed of siliciclastic sediment are predominant. The tidal flats usually cover the highest parts of the flat, while in the central part a mixture of mud and sand accumulates, and finally in the lower areas and the channels mostly sand accumulates.
- In sub-arctic areas, they are characterized by a huge fraction of edges dragged by the ice blocks that in turn leave small grooves on the surface of the plain.
- In the Tropics, the composition of the sediment depends on the sediment supply. If the terrigenous contribution is abundant, the differences between the flats of temperate climate are minimal. In this case, the marshes of the supratidal zones are replaced by mangroves. If, on the other hand, the siliciclastic contribution is minimal, clear waters and high temperatures will allow the development of flats where carbonate sedimentation predominates.
- In the conditions of arid climates, areas of occasional flooding and extreme evaporation known as coastal Sabkhas occur in the high part of the tidal flats. These areas are frequently affected by drying processes, so it is common to find desiccation cracks and also growth in the pores of the anhydrite, gypsum and halite crystal mud.

20.3 Morphology and Sub-environments

The processes of transport and deposition of sediments, as well as the colonization of organisms, lead to the existence of several sub-environments or elongated zones parallel to the coast. These zones have a variable width that depends on the slope of the tidal flat in relation to the tidal range. In general terms, there are three perfectly distinct zones: the supratidal zone, the intertidal zone and the subtidal zone. In turn, the intertidal fringe is compartmentalized into several different zones.

In most cases, three intertidal units can be distinguished, which are named according to their height or dominant sediment [2]. This is the type of zoning described in the coastal flats of the Wadden Sea in Germany and the Netherlands [47, 48, 49, 35, 36] and also in the Bay of San Sebastian in Tierra del Fuego, Argentina [51, 52, 53]. However, there are cases where there is more complex zoning, such as the Wash on the east coast of England, where up to five intertidal zones can be distinguished [13]. Each of these zones has a different distribution of dominant processes and preserved sediment. Most of the large tidal flats described coincide with the existence of a decrease in grain size from the subtidal to the supratidal areas.

However, in small tidal flats where the aggradation is so rapid that no areas are distinguished, the decrease in grain size takes place over time. This is the case of some tidal flats that have developed on mesotidal coasts and are associated with back-barrier areas [29].

Following the most general model, in a typical tidal flat the areas that can be distinguished are (Fig. 20.3):

(a) *The subtidal area*: The subtidal zone of the tidal flats is located below the mean spring low waters. So, its upper zone will only occasionally be exposed during extreme tides. This area may have different characteristics depending on its location (open-coast or back-barrier) or be associated with another type of tidal system. Its identification is important, since this is the area with the greatest preservation potential. Thus, the subtidal zone of tidal flats can be constituted by a typical nearshore, by a bay, a lagoon or by different types of channels in cases where they are associated with deltas or estuaries.

(b) *The intertidal flat*: This is found between the mean spring low waters and the mean high waters in such a way that it is subject to continuous alternation between periods of exposure and submersion. However, the number and duration of these exposures and submergences do not affect the whole flat equally. So, the topographically lower part of it presents higher rates of submergence and, conversely, its higher part presents higher rates of exposure. Obviously, this transition is gradual, although it results in the existence of three different zones within the intertidal flat: low or sandy intertidal flat, mid or mixed intertidal flat and high or muddy intertidal flat. The names of these zones are related to the most frequent sediment in them.

(c) *The supratidal area*: This appears above the mean high waters so that it is only flooded during the high tides and also during periods of storm. In warm and temperate climates, it is abundantly colonized by vegetation; for that reason, these are known as **coastal wetlands**. The kind of vegetation varies according to the type of climate, so marshes or mangroves can develop. Within the supratidal zone itself there may be another topographic zone marked by a succession of

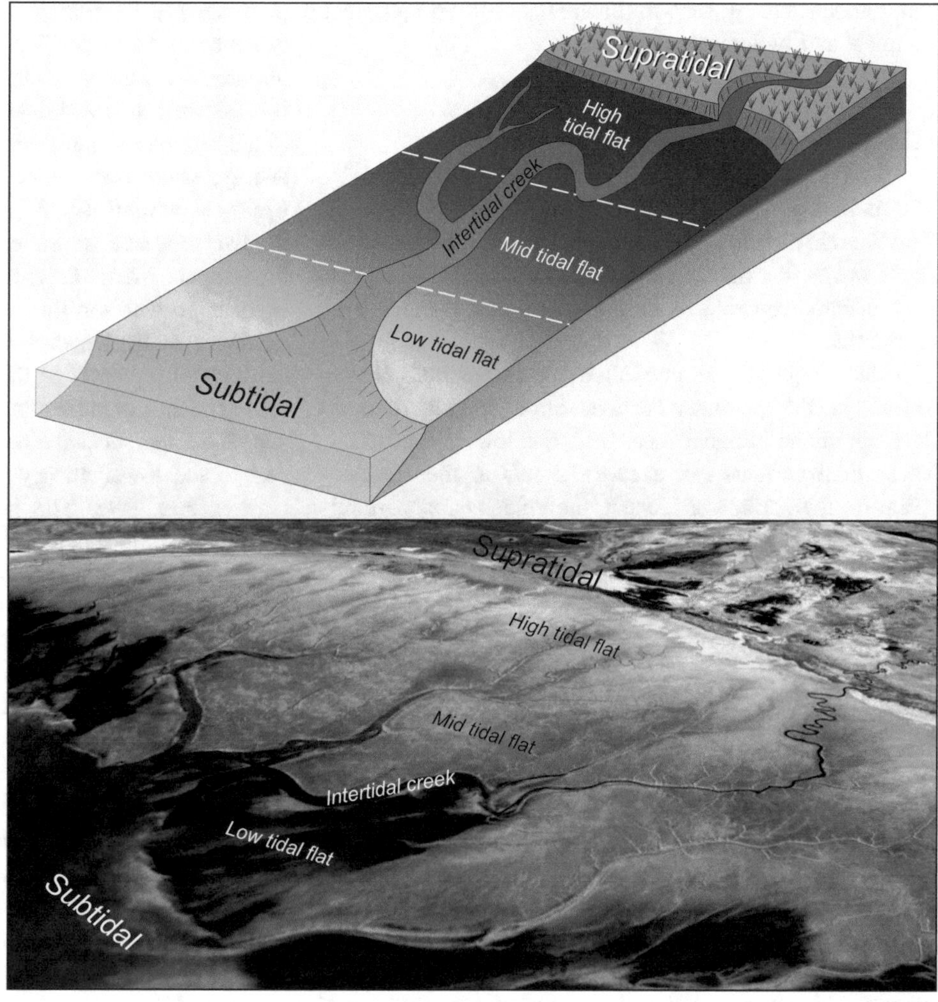

Fig. 20.3 Vertical zoning in the tidal flats with an example from Tierra del Fuego (Chile)

plants and animals that colonize it, although this is not always present.

In arid climates, the vegetation of this area does not develop and the sediment that reaches it during extreme tides is subjected most of the time to the action of the wind. This is how the **wind tidal flats** develop.

Both the intertidal and supratidal zones are crossed by a complex network of meandering channels, which constitute the tidal drainage system. This drainage system presents a hierarchy in terms of its functionality and topographic location [23, 59]. Thus, there are feeder and distributor channels (subtidal) and terminal channels (intertidal).

In the open-coast tidal flats developed inside a bay, the subtidal zone does not present overly strong currents. This allows the development of wide muddy deltas in the area where the intertidal channels flow into the subtidal area. This is a different type of ebb delta from those described in barrier island systems that have been identified in the tidal flats of the Bay of Cadiz, for example.

The distribution of the tidal drainage system of tidal flats represents, in a way, the early pattern of the tidal network later consolidated on the marsh or mangrove areas that, consequently, may be inherited from the older intertidal flats [32]. Some depressions not yet filled, such as ponds or salt pans, may even be relict forms of the initial stages of the drainage network [13, 34].

20.4 Processes and Dynamics

Sedimentary dynamics in tidal flats are controlled by a series of processes related to: (a) tidal dynamics (water levels and currents); (b) the contribution of land sediments; (c) the action of waves on the flat; (d) the processes of chemical precipitation and flocculation; and (e) the organic activity. All these processes are in turn influenced by the climate, which controls: the occurrence of surges, the wind regime, the sedimentation rates, the physicochemical conditions of the environment and the type of colonizing organisms. For a better understanding, we will differentiate between physical, chemical and biological processes.

20.4.1 Physical Processes

The most important physical (mechanical) processes on the tidal flats are obviously related to tidal activity. However, the small waves that agitate the waters flooding the plain also play an important role, as do the storms that eventually act on the flat.

Tidal energy is distributed unevenly in the areas characterized in the previous section. Stronger tidal currents develop in the subtidal areas and the low intertidal flat. These tidal currents can exceed 1.5 m/s in the channels, while on the surface of the flat, the velocity ranges rarely exceed 0.5 m/s [38]. However, these velocities are sufficient in both areas to transport sand and generate bedforms. Tidal currents introduce sandy sediment into the tidal flat through subtidal channels and transport it to higher topographic areas. There, it is trapped, taking into account that as we move in height not only the currents are increasingly smaller, but they are also of shorter duration. On the other hand, it often happens that the tidal currents are asymmetrical. This situation is synthesized in the diagram proposed by Postma [33] (Fig. 20.4).

This diagram shows how the flood currents raise the grains to levels where the ebb currents remove them for a shorter time but are not able to return them to the initial point. From this point, a new flood current would transport them to a higher point and, in turn, a new ebb current would not be able to return them to the initial point. In this way, successive currents would make the particles go back to places where the currents would no longer be able to set them in motion.

This demonstrates that the distribution of the energy of the currents in the different zones of the tidal flat determines the sedimentary processes and the facies that are deposited in them. Thus, in the subtidal zones, bed load transport predominates, while in the upper levels of the flat, decantation of suspension grains becomes relatively more important. On the intertidal flat, a mixed process takes place between bed load and decanting of suspended matter. Most of the bed load is transported during the moments of greatest energy, while decantation occurs during the slack of the upstream currents [26, 27]. The relative importance between the two processes is gradual, so that, as we move to higher topographic areas, decantation processes become increasingly important. On the other hand, as a consequence, the existence of tidal cycles of different period has a cyclical capacity of transport of the currents that is reflected in the characteristics of the sediment deposited in each cycle. In the supratidal flat, decantation processes clearly predominate, due to the lower energy of the currents when the water reaches this level. The high exposure rate can generate drying cracks in this high zone.

The channels draining the tidal flat usually reach maximum energy during ebb times. The meandering morphology causes currents to concentrate in certain areas. This results in a migration similar to that produced in meandering river channels, with erosion on the concave margin and sedimentation on the convex one (Fig. 20.5). This erosion can exceed millimeters per day and, if it takes place on cohesive sediments, generates abundant soft pebbles in the tidal flats [36]. The action of this process in carbonate tidal flats is responsible for the genesis of intraclasts. In both cases, these lithological elements become part of the material transported through the bed of the channel and are finally deposited in the *point bars* that are generated in its convex margins [21].

The innermost part of the channels, especially if they drain supratidal areas (salt marshes and mangroves), usually presents a cumulative character. These channels are filled with cohesive sediments and the filling process results in a narrowing with an increase in the slope of the margins (Fig. 20.6). This narrowing is often accompanied by a deepening that is achieved by erosion of the bed, so it can accumulate sediments in a residual way [23].

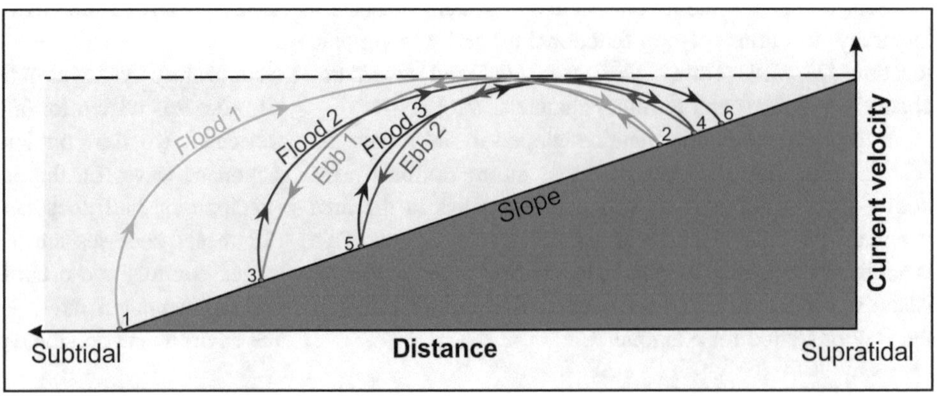

Fig. 20.4 Diagram of the transport of solid grains ascending the tidal flat with decreasing tidal current velocity, adapted from Postma [33]

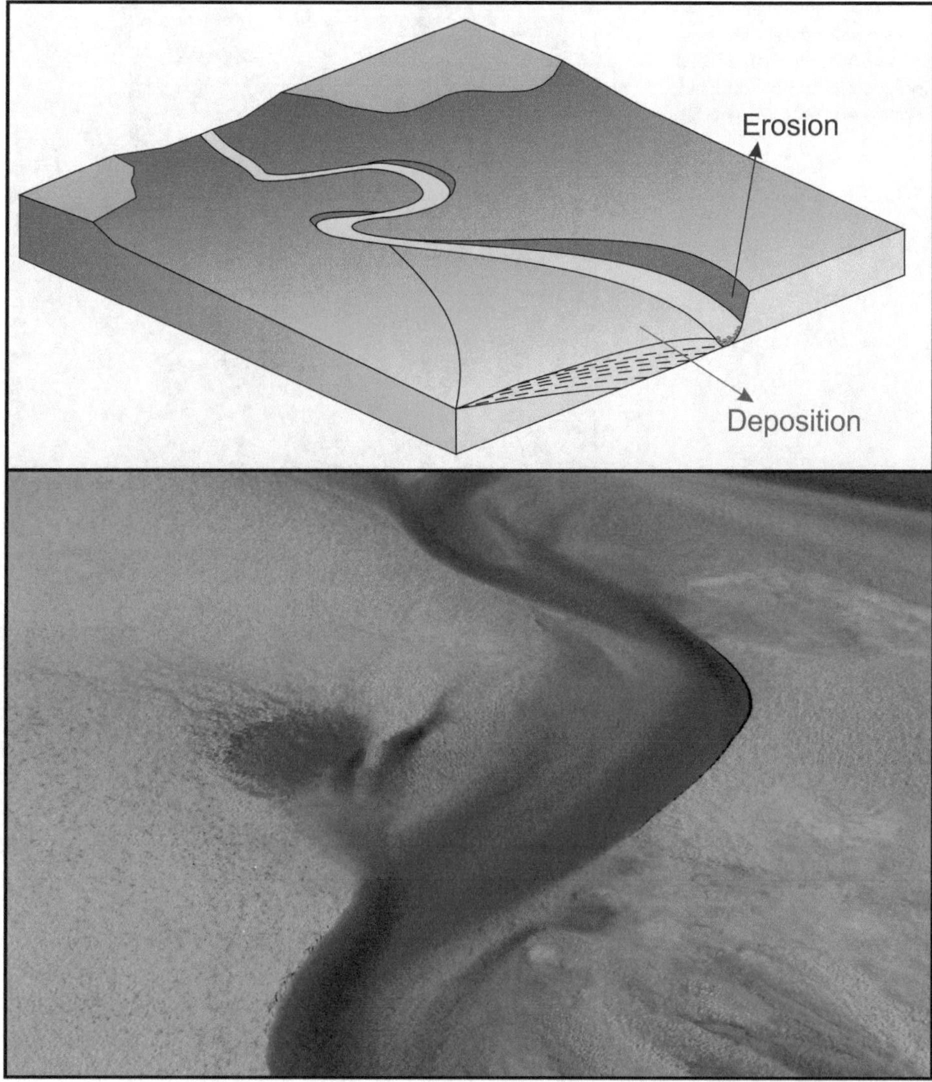

Fig. 20.5 Scheme and view of a meandering tidal creek, with erosion in the concave margin and development of a tidal point bar by deposition in the convex margin

In the highest areas of the intertidal zone, the importance of tidal currents in the transport of sandy material is lessened when faced with the action of small waves [7]. These waves can be incoming from open coastal areas or be generated within the tidal flat itself during high tide if there is sufficient *fetch*. This process has a significant influence on the distribution of the sedimentary facies in the flat.

The action of eventual storms can also affect tidal flats in two different ways. On the one hand, in the lower areas there is strong erosion and reworking of previously deposited sediments. On the other hand, it is during storms when large amounts of coarse sediment can be introduced into the higher areas of the flat that accumulates in the form of cheniers [53]. This sediment has a high preservation potential, because the range of energy normally put into play in these areas is much lower and cannot rework this coarse material. In the intertidal zones of carbonate tidal flats, it is precisely in the moments after the storms that most of the deposition takes place. Since these are clean water environments during fairweather conditions, it is during the periods following storms that the water has the greatest amount of suspended matter from the subtidal areas [46].

20.4.2 Chemical Processes

Chemical processes play an important role in tidal flats as suppliers of autogenous sediment. On those tidal flats where terrigenous sedimentation is predominant, the flocculation processes which take place during times of lower tidal currents result in a significant amount of organic sediment being added to the sediment from silting.

However, it is in the carbonate flats where chemical processes play a more important role. In these environments the siliciclastic contribution from land is minimal and almost all of the sediment comes from the action of chemical

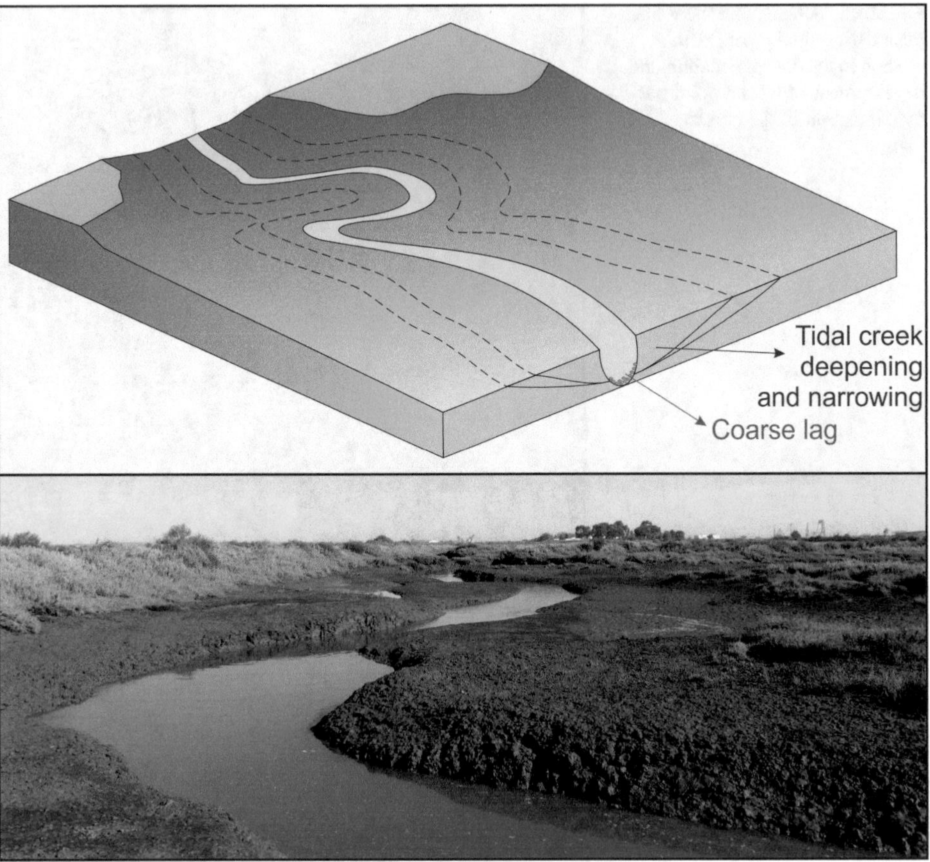

Fig. 20.6 Scheme and view of a depositional tidal creek, with narrowing in the margins and deepening at the bottom, and development of a lag deposit

processes. In Chap. 11, the shallow subtidal zones were characterized as the place where most of the primary precipitation of calcite (*carbonate factory*) occurs. This precipitated calcite is subsequently reworked by physical processes until it is distributed throughout the intertidal zone. Organisms have, in addition, a remarkable influence in the primary precipitation of calcite, catalyzing the process by providing the main part of the elements composing these facies, i.e., the pellets [5].

Other types of chemical processes that take place on the tidal flats occur in their topographically highest area. There, interstitial water that fills the pores during exposure periods circulates through them, resulting in capillary precipitation of aragonite and dolomite that act as cement leading to crust formation [41]. These crusts go on to form intertidal beachrocks [9]. However, the action of energetic physical processes can fragment them into intraclasts [14, 21]. Evaporite crystallization of gypsum, anhydrite and also dolomite can accompany this cementation by capillarity [5].

In the supratidal areas, the capillary circulation of fresh water can produce the first diagenetic transformations, extracting the Ca and replacing it with Mg through a dolomitization process. However, the result can easily be confused with the typical cementation of a *beachrock* [21].

20.4.3 Biological Processes

Perhaps the most typical example of biomechanical sedimentation is the screen action that plants exert when they are immersed in an aqueous flow. The presence of plant formations contributes to generate turbulence, slowing down the flow speed due to friction and contributing to siltation. Plants also exercise chemical control of the environment when they are submerged, contributing to biochemical sedimentation. By remaining submerged, the tissues of some plants are capable of secreting substances that modify their chemical environment to maintain their osmotic balance. These substances cause changes in the factors that determine the ionic stability of certain elements, inducing chemical precipitation and flocculation processes. These biophysical and biochemical sedimentation processes are present in both clastic [18] and carbonate environments [21].

Another type of organic activity that induces sedimentation is exercised by some types of algae that act as a sediment trap. The most characteristic case is that of algae that achieve the growth of geological structures through the adhesion of detrital particles. Thus, blue-green algae (*cyanophyceae*) give rise to the formation of stromatolites (Fig. 20.7a) and oncolites, while some red algae (*rhodophyceae*) by the same

Fig. 20.7 Biologically induced structures. **a** Stromatolites. **b** Rhodolites

process generate spherical structures called rhodolites (Fig. 20.7b). A similar phenomenon occurs in siliciclastic environments, where the activity of microorganisms such as diatoms in the pores of the sediment generates mucous membranes that agglutinate the particles and give them a cohesive character, preventing the currents from remobilizing them [11].

While the processes described above contribute to the accumulation of sediment on the tidal flats, bioturbation processes have just the opposite effect. Bioturbation involves the alteration of a large part of the sediment due to the activity of macroorganisms, which totally or partially destroy the previous internal order. On the other hand, some organisms digest the organic matter that represents the bonding element that keeps the sediment grains together, allowing the currents to put it back in motion.

The arrangement of all of these biological processes along the tidal flats is clearly zonal, as the organisms are distributed vertically according to their degree of tolerance to exposure and submersion levels (Fig. 20.8). Thus, animals such as crustaceans, gastropods, bivalves, annelids, foraminifera, ostracods, or plants such as higher halophytes, green algae, cyanobacteria, and diatoms are located in specific fringes within the intertidal zone [55, 56]. In general terms, the degree of bioturbation increases towards the higher areas, since the downwards migration of bedforms can be inhibitory to benthic activity. However, the presence of certain organisms such as algal mats or eelgrass can

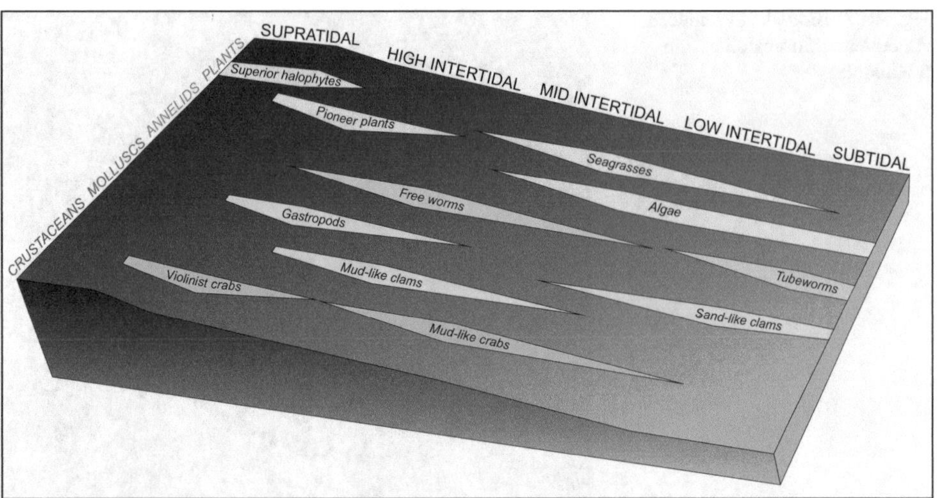

Fig. 20.8 Vertical distribution of benthic organisms in a mid-latitude tidal flat, adapted from Weimer et al. [56]

contribute to the fixation of the sediment, increasing the content of shaly material and inhibiting the migration of bedforms. In short, organic activity can alter the normal sedimentary zoning of tidal flats [44].

20.5 Sediments and Bedforms

The distribution of sediment in the tidal flats is clearly determined by the vertical distribution of sedimentary processes. It should be noted that tidal currents only act on the surface of the tidal flat when the tidal water reaches the level at which that surface is located (Fig. 20.9). The subtidal area is always submerged, so the currents in that zone experience the full range of ebb and flood velocities, from zero to maximum speed. In contrast, the higher elevations are covered by the tide at times when the speed is decreasing. The most direct consequence of this is the presence of a vertical and horizontal gradation of sediment grain size. Thus, the coarser-grained material is found towards the subtidal area, while the finer material increases towards the supratidal zone. This fact means that the lower zone of the flat is generally known as the sand flat and the upper zone as the mudflat, while the intermediate zone is called the mixed flat.

On the other hand, the different associated bedforms are also graduated vertically and horizontally. Figure 20.9 shows the thresholds for the development of different bedforms. In the subtidal areas and low intertidal flats, where higher current velocities are reached and the sediment is thicker, dunes and ripples develop. Towards the mid flat, the most frequent forms are the ripples, while in the upper flat

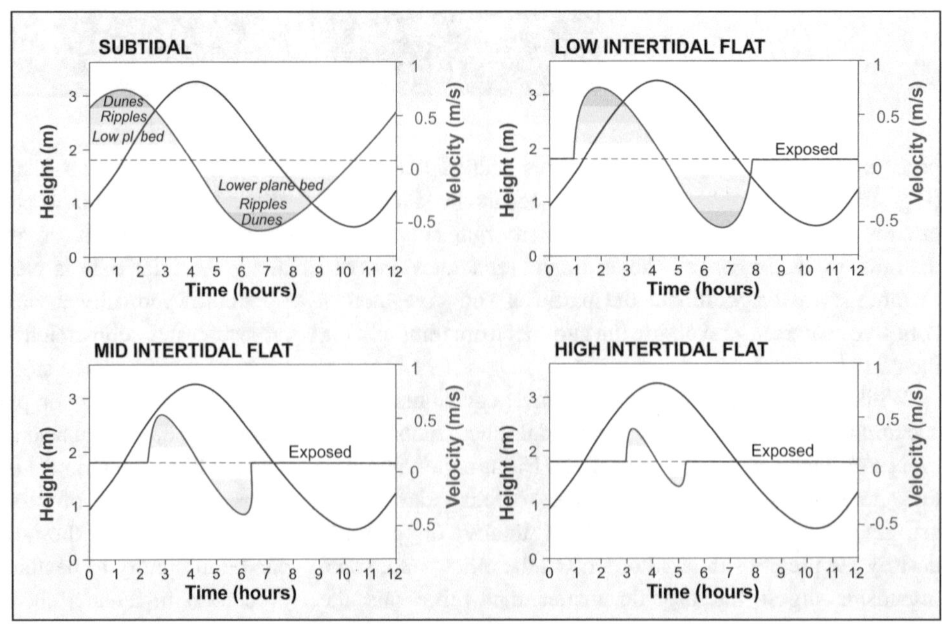

Fig. 20.9 Vertical distribution of tidal currents and development of bedforms in each vertical zone of the tidal flat

the current is only able to develop a low regime plane bed and then parallel lamination dominates.

20.5.1 Sandy Tidal Flat

The most frequent bedforms in this area are 2D and 3D dunes (sand waves and megaripples), often with superimposed ripples (Fig. 20.10a). The internal structure corresponding to these dunes are sets of planar or trough cross-lamination (Fig. 20.10b). The dimensions and orientation of the bedforms and their corresponding cross-layering are controlled by the velocity and the asymmetry of the tidal currents. The alternating cycles of spring and neap tides can generate differences in the dimensions of the bedforms as well as in the thickness of the cross-laminations generated in each cycle [35, 24, 4, 54].

If the velocities in both directions of the current are of the same intensity, and the input is sufficient, the structures can show bipolarity (Fig. 20.10c). In section, it appears as a herringbone cross-layering or cross-bedding [35]. If, on the contrary, the contribution is insufficient, each tidal semicycle reworks the sediments and bedforms generated during the previous semicycle. Then, the preservation of bedforms is minimal [10]. When asymmetry is the dominant feature, the internal structure will present abundant reactivation surfaces. In this case, the most frequent sets of cross-bedding will be those oriented in the direction of the dominant current [8, 24].

The waves can also leave their mark in this low intertidal zone, so that in open-coast tidal flats they can make bars migrate in a very similar way to the beach ridges [28]. The rate of migration of these bars is determined by the combined factors of the tidal range and the wave energy.

20.5.2 Mixed Tidal Flat

The most frequent bedforms in the mid intertidal zone are ripples (Fig. 20.11a) whose typical internal arrangement is cross-lamination in sets of centimeter scale (Fig. 20.11b). These structures also present double vergence (herringbone-type), developed when the migration of the bedforms under ebb and flood currents is reversed.

However, throughout the tidal cycle, there is an alternation between moments of high current and moments of zero speed that occur during tidal slacks. Thus, there are intervals where mechanical sedimentation and moments of flocculation and settling dominate. The migration of small-scale bedforms is fossilized by covering them with muds from settling during high tide surges. This dynamic alternation results in an interleaving between cross-laminated sandy facies and muddy facies. These alternations can be classified into three groups according to the relative proportion of muddy and sandy sediments they contain. Each group, in turn, can be subdivided according to the morphology of the mud and sand lenses (Fig. 20.12).

Conceptually, the structures can be classified into flaser bedding, wavy bedding and lenticular bedding.

(a) *Flaser bedding*: Corresponds to those sets of layers in which the proportion of sandy sediments dominates over the muds (Fig. 20.13a).
(b) *Wavy bedding*: These are structures in which the sandy and muddy sediments are in the same proportion (Fig. 20.13b).
(c) *Lenticular bedding*: These are the deposits in which the proportion of muddy sediments is greater than the sandy lenses. The sand sheets can reflect a symmetrical cross-lamination if they have been generated by wave ripples, since, as mentioned, small waves can act on the area where these structures originate (Fig. 20.13c).

These mixed structures of sand and mud are distributed vertically in the tidal flat, so that in the lower areas the flaser structures dominate, while towards the higher areas lenticular ones take over.

In carbonate tidal flats, the sandy fraction is usually composed mainly of pellets; however, depending on the average energy involved in the subtidal zone, and the type of organisms that inhabit the environment, intraclasts, oolites and bioclasts may also be present. Meanwhile, the fine fraction is usually formed by micrite. The origin of this micrite can be directly from the processes of chemical precipitation, although the micritic grains can also be remobilized from the subtidal zones and deposited by settling in the intertidal areas.

Settling is the most important process in the high intertidal zone and in the supratidal area. Only muddy material is deposited there. In these areas, the presence of tidal cycles is recorded in the lamination that characterizes the tidal flat facies. These structures are generally known as rhythmites (Fig. 20.13d). In this way, the cycles of spring and neap tides are reflected in the grain size of the different laminae (*tidal bedding*), while the cycles of long period (six-monthly) are reflected in the changes of thickness of the plates (*tidal bundles*).

In open-coast tidal flats, wave action can develop a type of macroform called cheniers (Fig. 20.14a, b). These sediment bodies have the geometry of chains consisting of coarse material (very coarse sand and gravels). These are normally generated during periods of storm [40, 51]. However, they can also develop due to the action of fairweather waves under high supply conditions [30]. The material that constitutes the

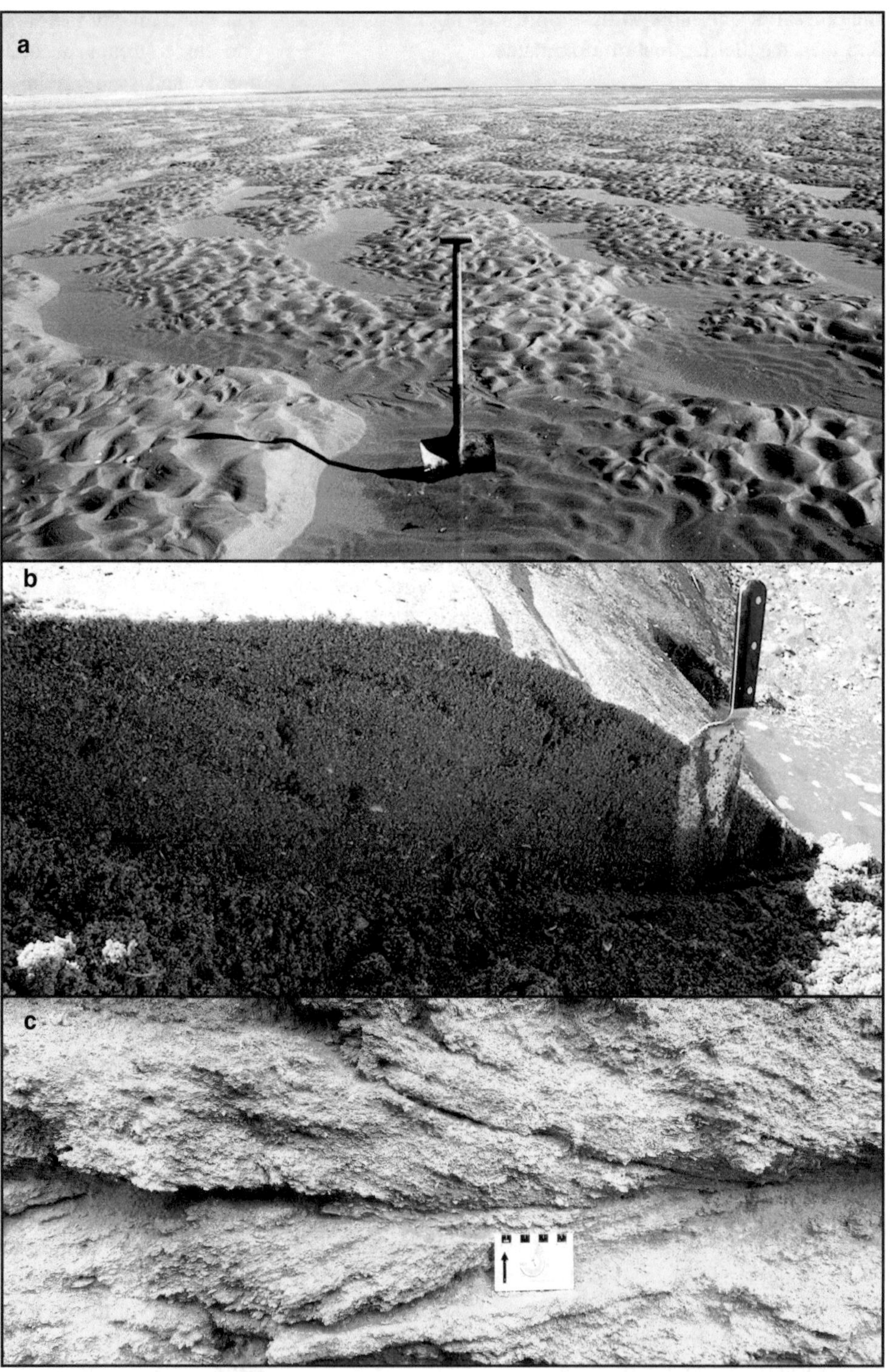

Fig. 20.10 Bedforms typical of low tidal flats. **a** 3D dunes with superimposed ripples. **b** Internal cross-bedding of a 3D dune. **c** Herringbone cross-bedding

cheniers can be contributed by coastal drift or be coarse material from the residual fraction of the reworking of muddy sediment [1]. Its internal structure is usually composed of metric sets of bioclastic sands or gravels with planar cross-bedding inclined landwards (Fig. 20.14c). In the interior of the set, the plates can have different grain-size, forming positive sequences. These sequences are arranged on the muddy facies of the high tidal flat.

20.6 Facies, Facies Sequences and Facies Models

The typical depositional facies of tidal flats are defined by three fundamental factors: lithology, internal ordering, and type and degree of bioturbation [19]. Obviously, all these characteristics are determined by the topographic position

Fig. 20.11 Sinuous ripple field in a mid-tidal flat **a**, showing the internal structure **b**

which, in turn, influences the sedimentary processes generated by the facies. Thus, in most tidal flats we can distinguish: (a) subtidal facies, (b) sandy flat facies; (c) mixed flat facies; (d) muddy flat facies; (e) supratidal facies; and (f) intertidal channel facies. These depositional facies coincide with the zoning described for this type of sedimentary environment.

20.6.1 Subtidal Facies

In back-barrier tidal flats, these are usually represented by the infilling sequences of subtidal channels. Although in current tidal flats these represent less than 50% of the surface of the plain, their migration makes their facies appear below those of the intertidal zone. They tend to be the thickest facies, and their thickness is equal to the depth of the channel that is filled, thus being able to exceed a dozen of meters. The sediment is usually composed of sandy material forming positive sequences. The bedforms present vary in scale from sand waves to megaripples and current ripples. All these bedforms generate cross-laminations and cross-bedding whose crests form open angles with the channel margin. The sheets are inclined in both directions (herringbones), but usually only one of the two is dominant in response to the asymmetry of the currents. The orientation of the bedforms presents a fanning that responds to the meandering character of the channels [56]. Subtidal flasers may also appear in the convex margin of the channels when cohesive mud is deposited during slacks [39].

In open-coast tidal flats, the subtidal facies are usually muds or muddy sands with the most representative structure being the parallel lamination slightly seawards inclined. In addition, the bioturbation of these fine sediments is usually significant. These facies may also include some of the coarse elements available as soft clasts or bioclasts [52].

20.6.2 Sandy Tidal Flat Facies

This remains submerged for most of the tidal cycle. It is almost completely constituted of sands of different grain sizes which are moved as bed load. This sand forms various types of bedforms such as sand waves and megaripples, on

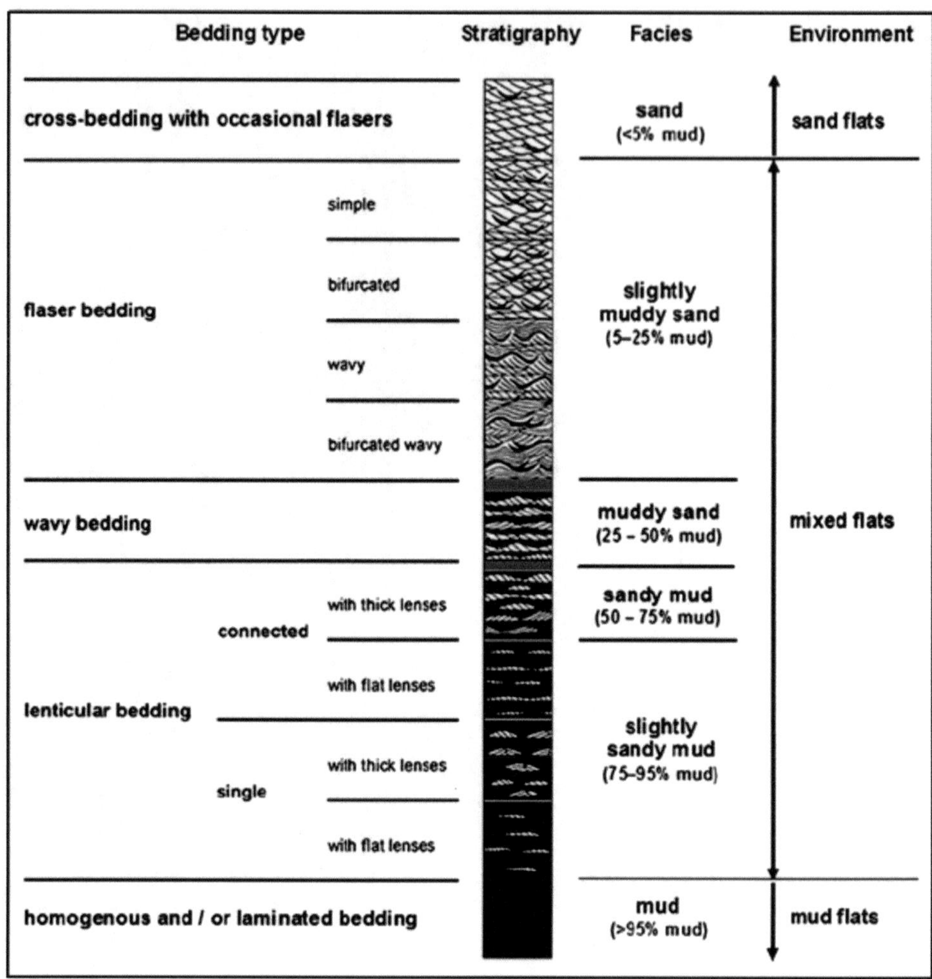

Fig. 20.12 Internal structures characteristic of tidal environments [16, 39, 17]

which current ripples with variable orientations are superimposed. In carbonate tidal flats, most of this sand is made up of pellets, although bioclasts and intraclasts can also be frequent. The internal structure presents cross-bedding with frequent reactivation surfaces, including the presence of herringbone-type structures. Bioturbation is scarce, due to the high degree of instability of the sandy substrate: some organisms such as *Lanice conchilega* or *Arenicola marina* build fixed galleries.

20.6.3 Mixed Tidal Flat Facies

This occupies the mid part of the intertidal flats and receives both suspended sediments and bed load. Thus, it is characterized by a mixed lithology composed of alternating layers of sand and mud. This alternation is due to two types of sediment transport. On the one hand, bed load takes place, both immediately after submersion as well as before emersion. On the other hand, the settling of suspended matter occurs when water covers the flat. This generates plane bed and current ripples. These are covered and fossilized by cohesive sediments. Internally, this facies presents flaser, wavy or lenticular bedding, depending on the volumes of sand or mud contributions and the topographic position.

20.6.4 Muddy Tidal Flat Facies

This is dominated by fine-grained sediments, silts and clays, deposited during the brief periods of submergence by the high water, when current velocities are minimal, thus forming a characteristic tidal lamination. Burrowing bioturbation by the local infauna is abundant, causing partial destruction of the internal structure of the sediments. Superficially, they can also show root bioturbation, which represents the advance of the colonization by pioneer-type vegetation from the supratidal area.

In cool climate intertidal flats, the internal structure of the sediments can also disappear due to the effect of ice and thaw in those sectors of greater air exposure. The channels that constitute the drainage network are poorly developed

Fig. 20.13 Structures of interbedded sands and muds. **a** Flaser bedding. **b** Wavy bedding. **c** Lenticular bedding. **d** Tidal rhythmite

and are frequently abandoned, giving rise to areas of permanent waterlogging until they are completely filled. In carbonate tidal flats, the fine sediment is mostly composed of pellets. Crusting and mud cracks are frequent in arid climate areas. Sediments composed only of a bioclastic sand fraction of microorganism shells are also typical in carbonate flats.

20.6.5 Supratidal Facies

The swamp facies (marshes or mangroves) are composed of the finest sediment on the flat and have a rhythmic parallel lamination similar to that observed in the muddy flat. In this case, the lamination is usually intensely altered by root bioturbation of the only plants capable of colonizing such an extreme environment (usually of the *Espartina* or *Mangal* genera, depending on the climate). The galleries of a few crustaceans that prefer to live on the lower zone of the marsh may also be present (e.g., fiddler crabs of the genus *Uca*). The long periods of air exposure cause the organic matter to oxidize, giving rise to a characteristic reddish color. Abundant mud cracks and evaporite crusts (gypsum, dolomite or aragonite) also occur in areas under an arid climate [56]. Both the burrowing and the holes left by the roots, as well as the mud cracks, are usually filled with sand or iron oxides. This chemical environment gives rise to the neoformation of pyrite [3].

20.6.6 Tidal Creek Facies

Mainly fine sediment from suspension is deposited in the intertidal part of the drainage channels. In the higher areas of tidal flat, and also across the swamps, these channels undergo the processes of lateral migration with erosion in the concave margin and deposition in the convex one, in the form of a characteristic point bar. Rhythmic sheets of fine sediment inclined towards the channel are deposited in the accretion bars. At the base of the channel, there is normally a lag deposit composed of the shells that cannot be transported by the tidal currents. When the channel is totally filled with this coarse material, it is covered by finer sediments [50].

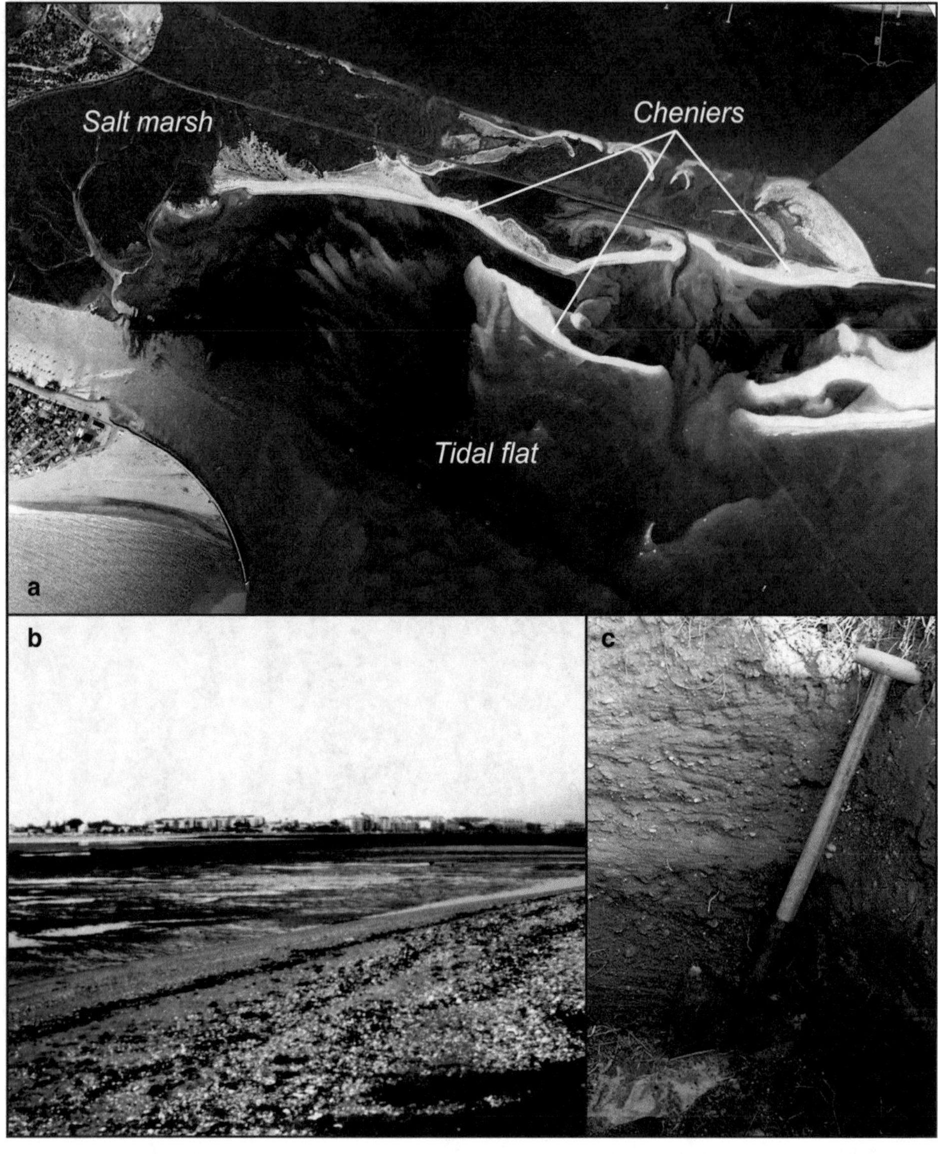

Fig. 20.14 Cheniers developed on a tidal flat. **a** Aerial view. **b** Panoramic view. **c** Chenier facies

20.6.7 Facies Models

The continuous progression of the different depositional facies, according to Walther's Law, generates a vertical shallowing sequence [38]. In this way, the sandy flat facies are superimposed on the subtidal facies, and on these appear the alternation of sands and muds of the mixed flat and on this the muddy flat, finishing the sequence with the swamp facies (Fig. 20.15).

There are numerous examples of this type of sequence of depositional facies in the literature. All of them roughly agree, although there is a wide range of differences in the small details of the facies constituents. This diversity is introduced by the variability of environmental control factors such as tidal range, climate, relative sea level movements, the volume and nature of sedimentary inputs, the mean slope of the flat, its extension and its location with respect to other sedimentary environments.

The most general typical sequence for siliciclastic tidal flats was proposed by Reineck and Singh [38], based on data provided by the Wadden Sea mudflats (Fig. 20.15a). Actually, this sequence is representative of all the clastic tidal flats, simply by varying the vertical scale according to the tidal range. In the case of carbonate tidal flats, a standard sequence was proposed by Wright [58], using the microtidal example of the tropical climate and transgressive framework of Andros Island in the Bahamas (Fig. 20.15b).

These ideal sequences would correspond to the final filling of the fully extended tidal flat. The sequence would only be cut off by the presence of intertidal creeks, which would have their own filling sequence. This infill sequence is developed over the erosive surface created by the channel

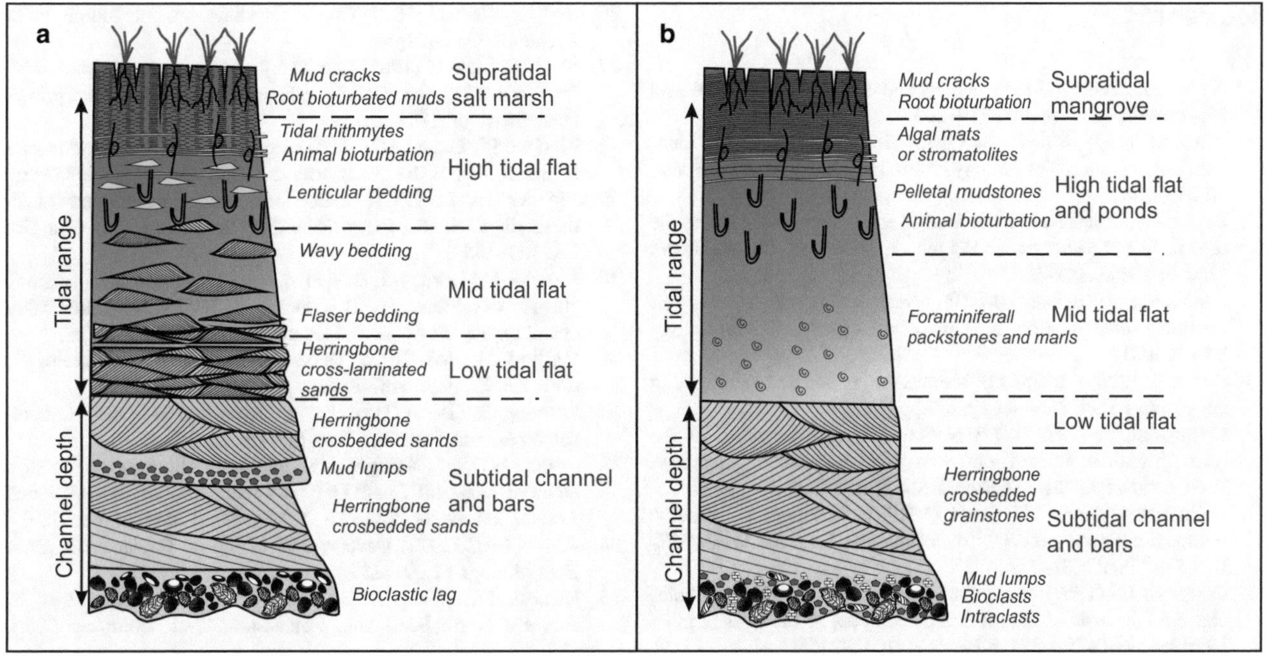

Fig. 20.15 Sequences generated in tidal flats in different conditions. **a** Classical tidal flat [38]. **b** Carbonate tidal flat [58]

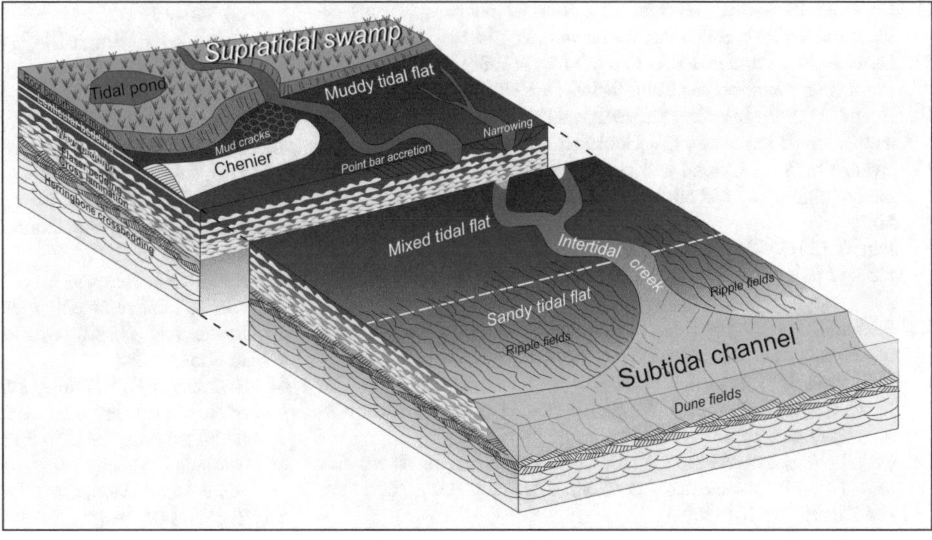

Fig. 20.16 Block diagram showing the facies relationships of the different sub-environments of the tidal flats (based on Boggs [5])

incision and consists of a basal lag that fills the channel bed, and a sequence decreasing in energy that is very similar to that of the adjacent tidal flat, but of lesser thickness [23]. The bodies where the sequence of the intertidal channel develops may have different geometry, depending on whether the channel tends to migrate or simply fills by lateral narrowing. The upper part of the tidal flat may have chenier coarse facies superimposed. The three-dimensional development of these infill sequences is quite simple and is summarized in the block diagram presented by Boggs [5] (Fig. 20.16).

The facies models of carbonate tidal flats, also called peritidal flats, have an equivalent three-dimensional arrangement. However, there may be differences in the detail of the composition of facies from both the lithological point of view and the sedimentary structures, in accordance with the previous paragraphs.

References

1. Augustinus PGEF (1989) Cheniers and cheniers plains: a general introduction. Mar Geol 90:219–229
2. Amos CL (1995) Siliciclastic tidal flats. In: Perillo GM (ed) Geomorphology and sedimentology of estuaries. Elsevier, Amsterdam, pp 273–306
3. Berner RA (1981) Authigenic mineral formation resulting from the organic matter decomposition in modern sediments. Forschritte der Mineralogie 59:117–135
4. Boersma JR, Terwindt JHJ (1981) Neap–spring tide sequences of intertidal shoal deposits in a mesotidal estuary. Sedimentology 28:151–170
5. Boggs S (1995) Principles of sedimentology and stratigraphy, 2nd ed. Prentice Hall, New Jersey, 744pp
6. Chung GS, Park JA (1977) Sedimentological properties of the recent intertidal flat environments. Southern Nam Young Bay, West coast of Korea. J Oceanogr Soc Korea 13:9–18
7. Collins MB, Amos CL, Evans G (1981) Observations of some sediment-transport processes over intertidal flats, the Wash, UK. IAS Spec Publ 5:81–98
8. Collinson JD (1969) The sedimentology of the Grindslow Shales and the Kinderscout Grit: a deltaic complex in the Namurian of Northern England. J Sedim Petrol 39:194–221
9. Davies GR (1970) Algal laminated sediments, Gladstone Embayment, Shark Bay, Western Australia. Mem Am Assoc Petrol Geol 13:169–205
10. Davis Jr RA (1992) Depositional systems, 2nd ed. Prentice Hall, New Jersey, 603pp
11. De Boer PL (1981) Mechanical effects of micro-organisms on intertidal bedform migration. Sedimentology 28:129–132
12. De Boer PL, Van Gelder A, Nio SD (eds) (1988) Tide influenced sedimentary environments and facies. D. Reidel, Dordrecht, p 530
13. Evans G (1965) Intertidal flat sediments and their environments of deposition in the Wash. Q J Geol Soc Lond 121:209–245
14. Evans G (1970) Coastal and nearshore sedimentation: a comparison of clastic and carbonate deposition. Proc Geol Assoc 81:493–508
15. Fan D (2012) Open-coast tidal flats. In: Davis RA Jr, Dalrymple RW (eds) Principles of tidal sedimentology. Springer, pp 187–230
16. Flemming BW (2003) Flaser. In: Middleton GV (ed) Encyclopedia of sediments and sedimentary rocks. Kluwer, Dordrecht, pp 282–283
17. Flemming BW (2012) Siliciclastic back-barrier tidal flats. In: Davis RA Jr, Dalrymple RW (eds) Principles of tidal sedimentology. Springer, pp 231–268
18. Frey RW, Basan PB (1985) Coastal salt marshes. In: Davis RA (ed) Coastal sedimentary environments, 2nd edn. Springer, Heidelberg, pp 101–169
19. Frey RW, Howard JD (1978) Mesotidal estuarine sequences: a perspective from the Georgia Bight. J Sedim Petrol 56:911–924
20. Ginsburg RN (1975) Tidal deposits. Springer, New York, p 428
21. Hardie LA (1986) Stratigraphic models for carbonate tidal flat deposition. Q J Colorado School Mines 81:59–74
22. Hayes MO (1979) Barrier island morphology as a function of tidal and wave regime. In: Leatherman SP (ed) Barrier Island. Academic Press, pp 1–27
23. Hughes (2012) Tidal channels on tidal flats and marshes. In: Davis Jr RA, Dalrymple RW (eds) Principles of tidal sedimentology. Springer, pp 169–300
24. de Vries Klein G (1970) Depositional and dispersal dynamics of intertidal sand bars. J Sedim Petrol 40:195–1127
25. de VriesKlein G (1976) Holocene tidal sedimentation. Dowden, Hutchinson and Ross, Inc., Stroudsburg, PA, 423pp
26. de VriesKlein G (1977) Clastic tidal facies. Cont. Illinois Public Education Co., 327pp
27. de Vries Klein G (1985) Intertidal flats and intertidal sand bodies. In: Davis Jr RA (ed) Coastal sedimentary environments. Springer, Heidelberg, pp 187–224
28. McCave IN, Geiser AC (1978) Megaripples, ridges and runnels on intertidal flats of the Wash, England. Sedimentology 26:353–369
29. Morales JA (1997) Evolution and facies architecture of the mesotidal Guadiana River Delta (SW Spain–Portugal). Mar Geol 138:127–148
30. Morales JA, Borrego J, Davis RA Jr (2014) A new mechanism for chenier development and a facies model of the Saltés Island chenier plain (SW Spain). Geomorphology 204:265–276
31. Perillo GME (ed) (1995) Geomorphology and sedimentology of estuaries. Elsevier, Amsterdam, p 427
32. Pestrong R (1972) Tidal flat sedimentation at Coley Landing, southwest San Francisco Bay. Sed Geol 8:251–288
33. Postma H (1967) Sediment transport and sedimentation in marine environments. In: Lauff GH (ed) Estuaries, vol 83. American Association for the Advance of the Science Memoir, pp 158–179
34. Redfield AC (1972) Development of a New England salt marsh. Ecol Monogr 42:201–237
35. Reineck HE (1963) Sedimentary structure in the area of the southern North Sea. Abh. Depressions. Nat. Company 505:1–138
36. Reineck HE (1967) Layered sediments of tidal flats, beaches and shelf bottoms of the North Sea. In: Lauff GH (ed) Estuaries, vol 83. American Association for the Advance of the Science Memoirs, pp 191–206
37. Reineck HE (1972) Tidal flats, vol 16. SEPM Special Publication, pp 146–159
38. Reineck HE, Singh IB (1980) Depositional sedimentary environments, 2nd ed. Springer Verlag, Heidelberg, 549pp
39. Reineck HE, Wunderlich F (1968) Classification and origin of flaser and lenticular bedding. Sedimentology 11:99–104
40. Russell EJ, Howe HV (1935) Cheniers of Southwestern Louisiana. Geogr Rev 25:449–461
41. Shinn A (1983) Tidal flat environment. In: Scholle, Bebout, Moore (eds) Carbonate depositional environments. Mem Am Assoc Petrol Geol 33:173–210
42. Smith DG, Reinson GE, Zaitlin BA, Rahmani RA (eds) (1991) Clastic tidal sedimentology. Can Soc Petrol Geol Mem 16:307
43. Stride AH (1982) Offshore tidal sands. Chapman and Hall, London, p 222
44. Swinbanks DD, Murray JW (1981) Biosedimentological zonation of Boundary Bay tidal flats, Fraser River Delta, British Columbia. Sedimentology 28:201–237
45. Thompson RW (1968) Tidal flat sedimentation on the Colorado River Delta. Mem Geol Assoc Am 107:133
46. Tucker ME, Wright VP (1992) Carbonate sedimentology. Blackwell Sci. Publ, London, p 482
47. Van Straaten LM (1954) Composition and structure of recent marine sediment in the Netherlands. Leidse Geol Meded 19:1–110
48. Van Straaten LM (1959) Minor structures of some recent littoral and neritic sediments. Geol Mijnbouro 21:197–216
49. Van Straaten LM (1961) Sedimentation in tidal flat areas. J Alberta Soc Petrol Geol 9:203–226
50. Vilas F (1989) Estuaries and intertidal flats. In: Arche A (ed) Sedimentology, vol 1. CSIC Madrid, pp 351–495
51. Vilas F, Arche A (1987) Plain of cheniers in the Bay of San Sebastian, Tierra del Fuego (Argentina). Hisp Geol Rec 21–22:245–251
52. Vilas F, Arche A, González Bonorino G, Isla, FI and Ferrero H (1999a) Intertidal sedimentation in San Sebastian Bay, Tierra del Fuego, Argentina. Hisp Geol Rec 21–22:253–260.

53. Vilas F, Arche A, Ferrero M, Isla FI (1999) Subantarctic macrotidal flats, cheniers and beaches in San Sebastian Bay, Tierra Del Fuego, Argentina. Mar Geol 160:301–326
54. Visser MJ (1980) Neap–spring cycles reflected in Holocene subtidal large-scale bedform deposits: a preliminary note. Geology 8:543–546
55. Weerman EJ, Herman PMJ, van de Koppel J (2011) Macrobenthos abundance and distribution on a spatially patterned intertidal flat. Mar Ecol Prog Ser 440:95–103
56. Weimer RJ, Howard JD, Lindsay DR (1982) Tidal flats and associated tidal channels. In: Scholle PA, Spearing D (eds) Sandstone depositional environments, vol 31. Memoirs of the American Association of Petroleum Geology, pp 191–245
57. Wells JT, Huh OL (1979) Tidal flats muds in the Republic of Korea: Kinhae to Inchon. Office Naval Res Sci Bull 4:21–30
58. Wright VP (1984) Peritidal carbonate facies models: a review. Geol J 19:309–325
59. Zeff ML (1988) Sedimentation in a salt marsh-tidal channel system, southern New Jersey. Mar Geol 82:33–48

Fluvial-Influenced Systems I: Estuaries

21

21.1 Introduction

River mouths can be considered as transitional environments affected by river dynamics and marine processes (tides and waves). In general terms, there are two main types of river mouth: estuaries and deltas. The first of these are established as those systems where the main sedimentation occurs in a coastal valley, while in the second ones the sedimentary body develops as a coastal prominence that grows towards the open coast.

There are numerous definitions of the concept of estuary; however, practically all of them refer to a semi-enclosed basin where fresh river waters meet saline marine waters. There is an excellent summary of the evolution of estuary definitions over time by Perillo [25]. Etymologically, the word estuary comes from the Latin *aestus*, meaning tide, which gives an idea of the importance of tidal processes in these systems. Perhaps the most frequently cited definition in books on estuaries is that given by Pritchard [27]: "An estuary is a semi-enclosed body of coastal water that has a free connection to the open sea and where the sea water is measurably diluted with fresh water coming from a land-based drainage" (Fig. 21.1).

It was also Perillo [25] who highlighted the main shortcoming of this definition—that is, it does not include the influence of the tides, when it is the tide that provides most of the energy that controls the process of mixing waters in every estuary. This is curious, since the same Pritchard who proposed this definition also used the tide as the main criterion for classifying estuaries.

One long ongoing discussion regarding estuarine systems concerns the location of their innermost boundary. In this regard, Dyer [11] proposed the most internal point of influence of the tide as the limit with the river. From this perspective, the action of the tide substantially modifies the river water mechanisms in terms of sediment transport and also in terms of ecology.

In the context of the development of the principles of sequential stratigraphy, Russell [30] related the genesis of estuaries to a relative sea level rise, emphasizing their role in highstand system tracts. Thus, the starting point of today's estuaries would be the Holocene post-glacial transgression (Flandrian transgression). From the marine invasion of the pre-existing valleys, the estuaries are characterized as systems that tend towards sedimentary infilling, which causes a regressive aggradation in their interior.

Thanks to this characteristic infilling, estuaries often develop complex facies models that can be preserved in the sedimentary record. Normally, estuaries are identified by three fundamental characteristics: (1) the presence of tidal facies; (2) the relationship of these facies with bodies of fluvial origin; and (3) the elongated geometry of the bodies developed along a valley.

At the end of the twentieth century, after several decades of advances based on the knowledge of current estuaries, Dalrymple et al. [5] approached these environments from a more geological bias, emphasizing the role of the tidal facies as a signature of the estuary sediments. For Dalrymple and his team, the estuarine facies are those in which tidal activity can be established. One problem addressed by these authors was the identification in the geological record of the last point of tidal influence, defined until then as the boundary between river and estuary. It must be taken into account that, in the innermost areas of the fluvial–marine systems, there is a transition between the estuary and the river in terms of processes and facies. In river-dominated estuarine areas, the erosional action of the river can mask the sedimentary result of tidal activity. With this in mind, these authors proposed that the most internal point where tidal facies are preserved must be considered to be the geological limit of the estuary. In the same way, a transition with the wave-dominated facies is produced towards the sea. Taking into account these considerations, these authors proposed a definition that contemplates these aspects from a geological point of view: "An estuary is the portion towards the sea of a flooded valley system that receives sediments from both fluvial and marine sources, and contains facies influenced by fluvial, tidal and wave processes. The estuary is considered to extend from the

J. A. Morales, *Coastal Geology*, Springer Textbooks in Earth Sciences, Geography and Environment,
https://doi.org/10.1007/978-3-030-96121-3_21

Fig. 21.1 Example of an estuarine system: a semi-enclosed mass of coastal water connected with the sea and where the fresh fluvial water undergoes dilution

inner edge of the tidal facies at its headwaters to the outer edge of the coastal facies at its mouth."

21.2 Control Factors

Given all the considerations set out in the previous section, it is clear that three main factors control the dynamics of estuaries: river, tides and waves. However, there are secondary factors that also influence the geological development of the facies: the climate and the local geology [9] (Fig. 21.2).

On the one hand, the climate exerts a control on the precipitation regime that in turn controls the fluvial contributions, but in the highest latitudes it also has influence over the ice fraction. One only has to consider that one type of estuary, the fjords, has developed on glacial valleys and is strongly influenced by ice dynamics. In addition, it is the climate that controls the wind regime and thus the energy of waves that influence, in a fundamental way, the processes of the marine environment of the estuary.

The geology exerts an important influence on the morphology of the valleys. Weak lithologies allow the rivers to dig wide valleys that, when flooded by the sea, form open estuaries with a typical funnel shape. Conversely, the presence of resistant lithologies forces rivers to run through narrow valleys, developing very confined estuaries when the sea level rises.

21.2.1 River Discharge

It is evident that river discharge exerts one of the main controls over the development of estuarine processes. The aqueous flow of the river determines the speed of the fluvial currents, thus controlling the capacity of sedimentary contribution, both in the form of bed load and of the material transported in suspension. The volume of river water in relation to time is also fundamental in the establishment of the water mixing processes that characterize the interior of these environments. These aspects have been explained in Chap. 9 of this book.

21.2.2 Tidal Action

As mentioned in the introduction to this chapter, the tide is, by definition, one of the factors that control the dynamics of the estuary and allow its facies to be recognized in the geological record. The influence of the tidal regime on water mixing processes has already been demonstrated; however, this is not its only influence. In the central areas of the estuaries, the tidal currents are responsible for the redistribution of sediments brought by the river as bed load, generating bedforms and structures that characterize the tidal facies in these sectors of the estuary. On the other hand, the rise and fall of the tidal level is responsible for a vertical zoning of the margins, similar to that described in the tidal flats, due to the rates of exposure and submergence. However, the tidal regime is also responsible for the temporal variations in river discharges, which are slowed down during the moments when the tide is entering the estuary. In this way, the tidal wave can propagate to the river above the mixing zone without the salt water being able to enter these exclusive freshwater zones.

In short, according to the criteria of Perillo [25], it can be stated that the integral evolution of the estuary is controlled by the tidal dynamics even in the innermost areas of the estuary. This influence is not only fundamental in the distribution of the estuarine facies, but also in the ecology of the estuary, considered as an ecosystem.

21.2.3 Wave Energy

Wave action is not fundamental for the development of estuaries; however, the waves are present in the marine area

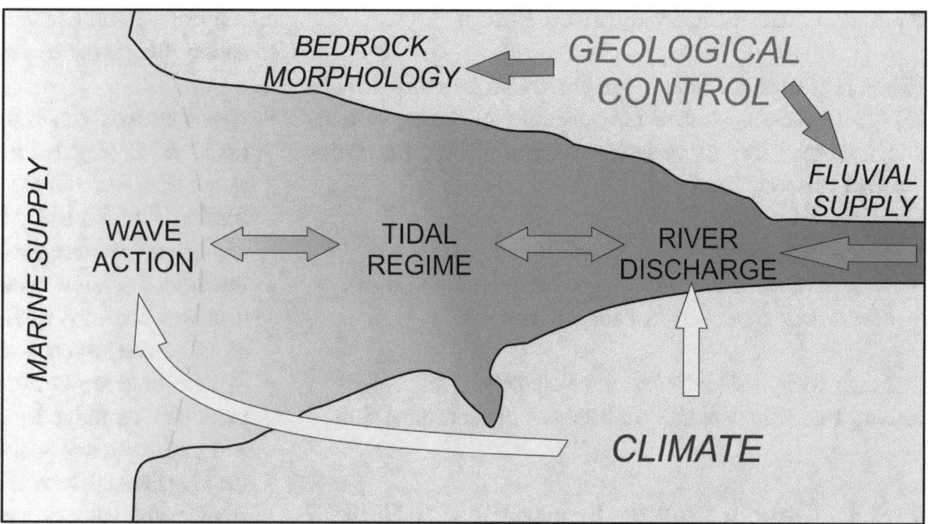

Fig. 21.2 Factors influencing the development of estuaries, adapted from Dionne [9]

of many estuaries and influence the development of facies and coastal sedimentary bodies that tend to develop in these environments. In this way, in some types of estuaries the waves are capable of generating barriers that partially constrict the inland water body, conditioning the passage of the tide towards these areas. In any case, the wave energy, when it is present, usually limits the outermost zone of the system.

Regardless of the river's capacity to provide water, there is no doubt that the dynamics of estuaries, like other coastal systems, are controlled by the balance of tidal and wave energy to redistribute this sediment. As will be seen later, this balance is used as a fundamental criterion when characterizing estuaries.

21.3 Classification and Morphology

Over a century of estuary studies, a large number of classifications have been proposed and numerous criteria have been used to characterize estuaries. Many of these classifications use geographical criteria, while others use purely biological or ecological arguments. In this section, only those with a more direct geological application will be considered.

21.3.1 Hydrological Classification

Since an estuary is the meeting point between a river system and a tidal one, a process of water mixing takes place in the estuary channels whose characteristics allow a classification of estuaries. In this regard, one of the most accepted classifications is that proposed by Pritchard [26]. This classification distinguishes three types of estuaries:

1. *Horizontal or salt wedge estuaries*: In these estuaries, the river flow is much greater than the tidal flow. A layer of fresh water is formed and circulates at the top of the water column, while the salt water introduced by the tide circulates at the bottom. The contact between the two waters of different chemical characteristics is net and very abrupt. This limit is normally named the **halocline**.
2. *Vertically homogeneous or totally mixed estuaries:* In this kind of estuary, the tidal flow is greater than the river flow, and the water becomes totally mixed. The transition between salt and fresh waters occurs progressively, with the waters becoming more saline seawards and fresher landwards. When the estuary is wide enough, the Coriolis effect is important, making the entering salt water concentrate in one margin during floods, while the exiting fresh water during ebbs takes place in the opposite margin. As usual, there is a different direction of circulation in each of the Earth's hemispheres.
3. *Partially stratified estuaries*: In these systems, the river and tidal flows have the same order of magnitude. This type represents a sequence of infinite possibilities between one and the other extremes. Normally, there is a certain vertical distribution of water, but also a horizontal gradient, with different waters on the surface and at the bottom, and an intermediate mixed brackish layer.

This classification has been subsequently adopted by several authors, who have themselves modified the specific thresholds of the relationships between river and tidal flows in order to distinguish the different types of mixtures [10, 16, 19, 31, 32]. However, all of these authors maintain the qualitative relationships expressed in the previous section. The dynamics of these types of mixtures, as well as the cloud of turbidity associated with these processes, have already been extensively studied in Chap. 9 of this book.

21.3.2 Classification by Tidal Range

Hayes [17] used the tidal range criteria proposed by Davies [6] for the classification of the generality of coastal systems and applied these to estuarine systems. Thus, he distinguished between:

- *Microtidal estuaries:* For ranges under 2 m.
- *Mesotidal estuaries:* For ranges between 2 and 4 m.
- *Macrotidal estuaries:* For ranges over 4 m.

Later, Kirby [18] introduced the concept of the hypertidal estuary for those estuaries with ranges of more than 6 m.

21.3.3 Classification by Propagation of Tidal Wave into the Estuary

Another type of criterion used for the classification of estuaries is that proposed by Le Floch [13], who studied the model of tidal wave propagation when entering estuaries. Each of these models is produced according to the greater or lesser relative importance of the processes of convergence and friction with the bed. Convergence is produced by the narrowing of the margins of the estuary, which would cause an amplification of the tidal range. Friction with the bed has the opposite effect of attenuating the tidal range. This classification differentiates three types of estuaries:

- *Hyposynchronous estuaries*: These are those in which the tidal wave progressively decreases in amplitude as it propagates towards the head of the estuary. The effect of friction with the bed dominates.
- *Hypersynchronous estuaries*: In these, the tidal wave increases its range when propagating to the interior, to decrease abruptly when entering the fluvial sector. The effect of convergence of the margins dominates.
- *Synchronous estuaries*: These are where the tidal wave maintains its range by propagating towards the headwaters of the estuary to also decrease sharply when entering the river area. There is a balance between the effects of friction and convergence.

21.3.4 Genetic Classification

Several years after proposing the hydrological classification, Pritchard [27] proposed a new classification of estuaries using morphology as the main criterion, but associating each morphology with genetic conditions. This classification was actually the modification of a previous one made in 1952. It distinguishes four basic types of estuary (Fig. 21.3):

Fig. 21.3 Different types of estuary according to Pritchard [28]. **a** Drowned fluvial valley (Gironde estuary, France). **b** Drowned glacial valley (Erik Harbour fjord, Baffin Island, Canada). **c** Barrier estuary (Bontalakoduru estuary, India). **d** Tectonic estuary (Maitland Bay, NW Australia). (Images Landsat/Copernicus from Google Earth.)

- *Flooded river valley*: These normally develop in coastal plains and present fusiform bar systems that have their axis parallel to the axis of the valley. These estuaries are generated under conditions of intense tidal current speeds.
- *Flooded glacial valley or fjord*: These are estuaries linked to the loss of ice from an ancient glacial valley. They are described as more or less straight and deep valleys, with a high rocky threshold.
- *Barrier estuaries*: This type of estuary is distinguished by the presence of a characteristic barrier mouth generated by wave activity. The system is developed in coasts of very low relief. According to the author, it is the most common type of estuary. The state of sedimentary infilling is important in these.
- *Tectonic estuaries*: These are flooded tectonic depressions. They are usually deep and have a complex physiography. Given their great depth, they are not usually much filled by sediments.

21.3.5 Physiographic Classification

Two decades after Pritchard's last proposal, Fairbridge [12] elaborated a new classification, this time using the physiographic criteria as the main guideline. This classification includes eight types of estuary (Fig. 21.4):

- U-shaped high-relief valley profile estuary (fjord).
- V-shaped moderately high relief valley profile estuary (fjard or firth).
- V-shaped moderate relief valley profile (ria or aber) and karst-incised estuaries (cala).
- Funnel plan-shaped low relief estuary (coastal plain estuary).
- L-plan shaped low relief estuary (bar-built estuary).
- D-plan shaped low relief seasonally blocked estuary (blind estuary).
- Estuaries located in deltaic distributaries (delta front estuary).
- Compound estuary (tectonic estuary).

The shape of each one of these estuaries is determined by the relationships between the factors described above, with a remarkable association to the balance between the relative movements of sea level (combination between eustatism and tectonics) and the volume of sedimentary contribution; other factors such as the relief of the coast are also significant. This last factor is conditioned by the geology of the rocky outcrops.

21.3.6 Dynamic Classification

The most recent classification was made by Dalrymple et al. [5] who used the dynamic criterion as a definition, distinguishing two types of estuary according to the dominant agent. The classification of estuaries is part of a broader consideration that includes all coastal environments as part of a system in which the three fundamental agents (river, tide and waves) intervene in the short term, whereas in the long term there is a control by the relationships between sea level and sedimentary input. In this classification, the estuaries would be systems that are born after a transgression, but tend over time to be transformed into deltas through a process of sedimentary filling. According to this classification, there would be two extreme types of estuary:

1. *Wave-dominated estuaries*: Waves are the dominant agent in the marine zone of the estuary, capable of developing a sand barrier that partially closes the estuary (Fig. 21.5a). In this marine zone, tidal energy is used to keep the inlet open, at whose ends ebb and flood tidal deltas develop. Most of the tidal energy is dissipated across the systems associated with the inlet and is quickly lost to the interior. The central zone of the estuary behaves like a large confined body of water, with a dynamic very similar to that of a lagoon. It is in this central zone where the process of water mixing takes place. The internal zone is dominated by fluvial processes and is characterized by the development of a fluvial delta over the body of water in the intermediate zone. This delta is called the **bayhead delta**. This type of estuary corresponds morphologically with Pritchard's [28] barrier estuary or the so-called bar-built estuary described by Fairbridge [12].
2. *Tide-dominated estuaries*: This type of estuary is characterized by low wave energy. Consequently, the waves are not able to build a confining barrier. For this reason, the tide is the dominant agent, both in the marine part and in the central part of the estuary. The tidal power has the capacity to develop tidal bars longitudinal to the estuary's axis, separating different channels in which the ebb and flood currents are concentrated (Fig. 21.5b). The system thus maintains its funnel shape. In the most continental part, the river is not able to develop a true bayhead delta, because the tide is capable of reworking all the river sediments. This fluvial area is usually limited to a single channel that passes through sections with straight and meandering morphology as an indicator of the dominance of the river or tide in each section.

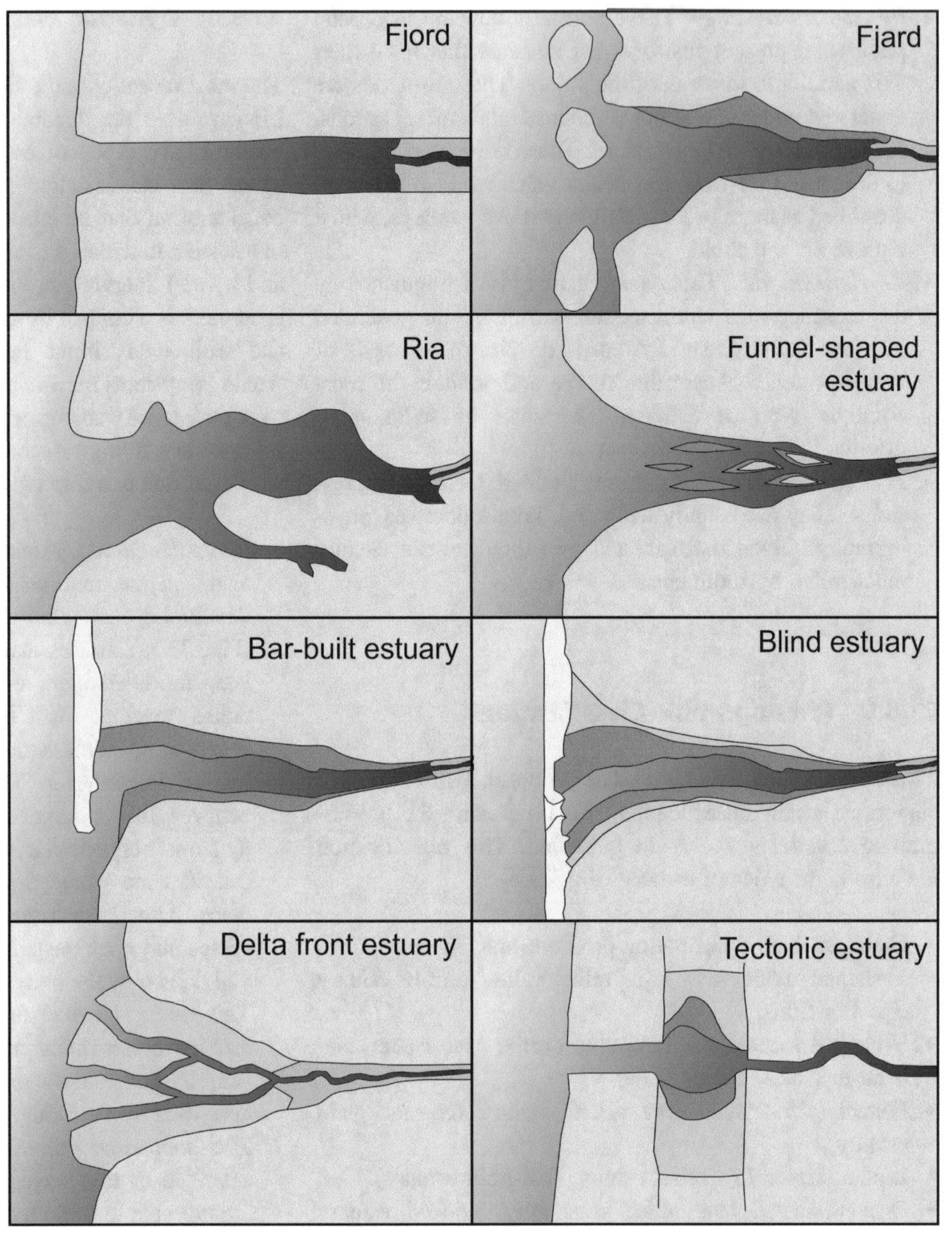

Fig. 21.4 Physiographic classification, adapted from Fairbridge [12]

This classification is not limited to the simple morphological description, but extends to the description of the distribution of processes in each sector of the estuary and is completed by the facies models that result from the total filling of each type of estuary. It is, therefore, the classification with the widest geological application of all those described.

21.3.7 Domains and Sub-environments of Estuaries

Whatever the type of estuary, there is a longitudinal zoning that takes into account the distribution of energy among the different agents that control sedimentation: river, tide and waves (Fig. 21.6). Thus, an estuary is divided into three domains: marine, central and fluvial.

- *Marine estuary*: This is dominated by marine processes (tides and waves). The environments present different morphologies and distribution according to whether waves or tides dominate. In a wave-dominated estuary, the mouth complex develops the **barriers** that confine the estuary. This barrier delimits an **inlet** channel that allows the passage of the tide and at its ends develops **flood and ebb tidal deltas**. In tide-dominated estuaries, **longitudinal tidal bars** are deposited, which in this sector show an incipient development and are generally found in the subtidal **area**.

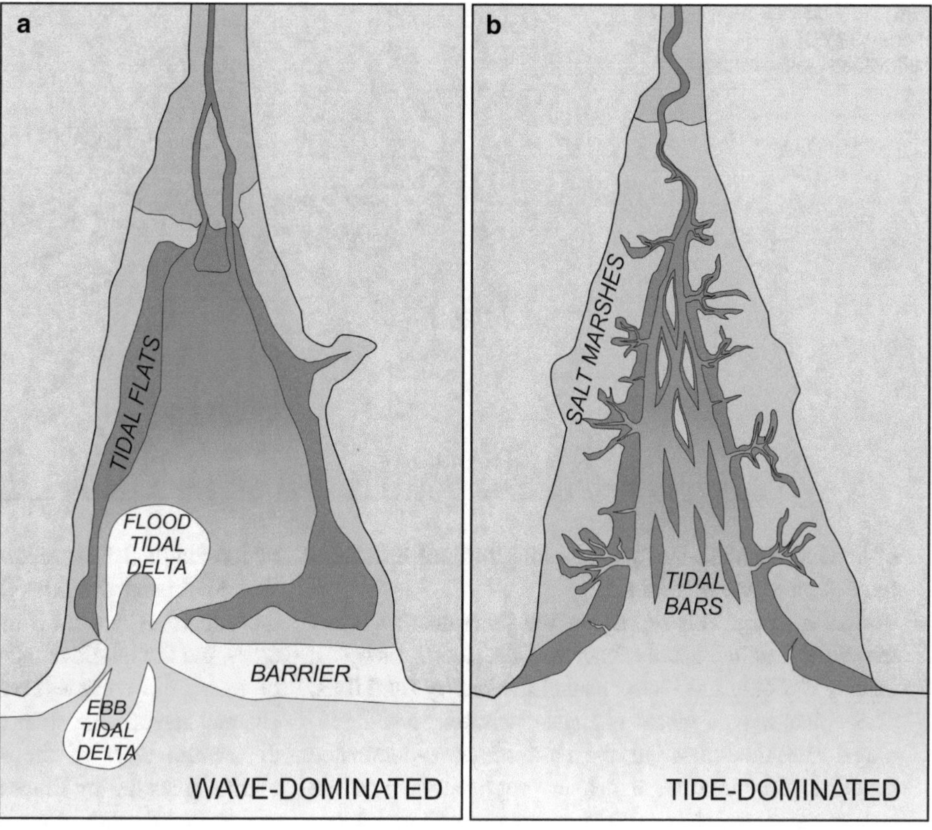

Fig. 21.5 Dynamic classification of estuaries. **a** Wave-dominated estuary. **b** Tide-dominated estuary, adapted from Dalrymple et al. [5]

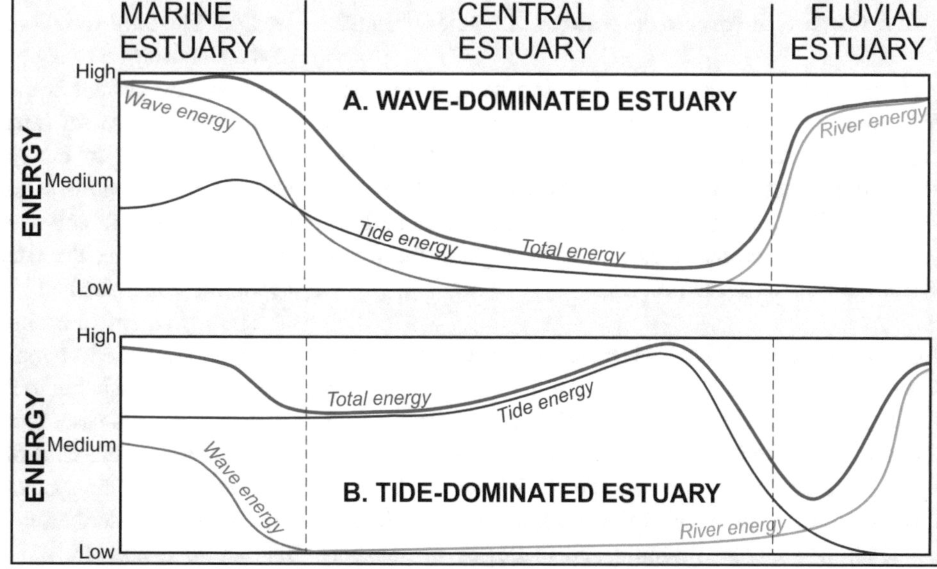

Fig. 21.6 Longitudinal domains of estuaries according to their distribution of energy. **a** Wave-dominated estuary. **b** Tide-dominated estuary, adapted from Dalrymple et al. [5]

- *Middle or central estuary*: This is an area with a net dominance of tides and where the mixture of fresh and salt water occurs. In wave-dominated estuaries with a low degree of sedimentary filling, it corresponds to a wide low-energy area dominated by the processes of flocculation and settling processes—and thus the sedimentation of fine material. In this case, due to its amplitude, it is usually called the **central estuary basin**. Associated with this basin, intertidal environments such as **tidal flats** and **salt marshes** are widely developed. In estuaries with a high degree of clogging, this central basin is replaced by an **estuarine channel** or several tidal drainage channels with anastomosing morphology, which separate inter- or supratidal islands. In the case of tide-dominated estuaries, there is always a dense network

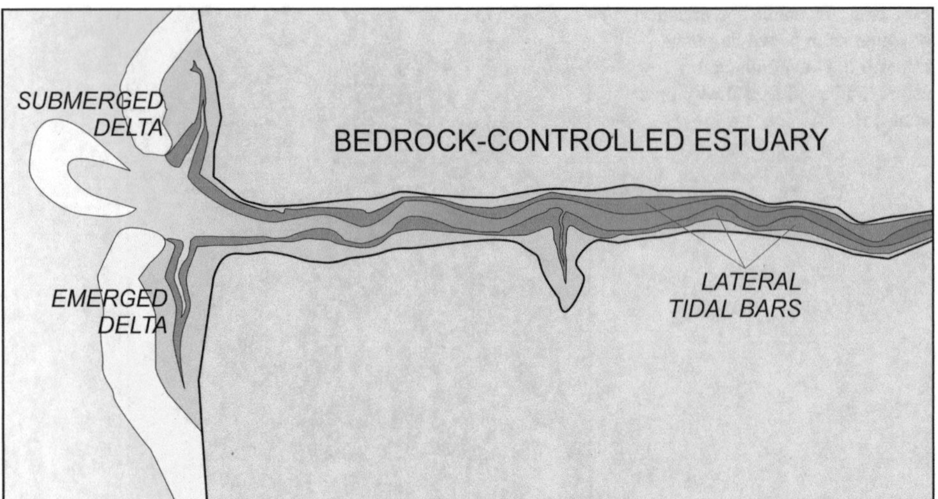

Fig. 21.7 Scheme showing the morphology of a bedrock-controlled estuary

of braided channels that separate **longitudinal intertidal bars** in the shape of a spindle.

- *Fluvial estuary*: This represents the innermost sector of the estuary, with a net dominance of the river's energy. It usually has only one main channel flanked by **tidal flats**. These flats have a mixed character between river floodplains and tidal flats. In the case of wave-dominated estuaries, this area has a deltaic progradation over the central basin, which is called the estuary **bayhead delta**. In tide-dominated estuaries, large marshes develop, which are crossed by a single very sinuous **estuarine channel** that connects the river to the network of braided channels.

In addition to this longitudinal zoning, there is a vertical bio-sedimentary zoning controlled by the tidal levels of exposure and submergence. This is the same vertical distribution of processes and sediments that was described for the tidal flats. This zoning is very evident in the central basin of wave-dominated estuaries, as well as throughout the whole system of tide-dominated estuaries.

Advanced Box 21.1. Narrow Bedrock-controlled Estuaries

Estuaries that are installed in very incised and narrow valleys present a distribution of processes and facies that differs from the general models described by Dalrymple et al. [5]. In this case, the morphology of the substrate exerts an important control over the dynamics and processes of transport and deposition. This fact led Chaumillon et al. [3] to describe these as a different type, called incised bedrock-controlled estuaries, based on examples studied in the NE coast of France.

The main feature of these estuaries is the presence of a single estuarine channel, since the narrowness of the valley does not allow the lateral development of the broad central basins typical of wave-dominated estuaries, nor the presence of longitudinal bars separating several tidal channels as in tide-dominated estuaries (Fig. 21.7). The sedimentation in this estuarine channel is distributed longitudinally into three areas, the fluvial, tidal and wave domains, respectively; in all of them, the energy is very high. The narrowness of the channel means that river currents can be maintained up to the central areas of the estuary, and may even reach the marine areas during times of flooding. The same narrowness means that the tide can move many kilometers inland. This characteristic of strong tidal and river currents makes these estuaries normally behave as well-stratified estuaries (salt wedge). The only variations that may occur laterally are due to the sinuosity that may be present in the incised valley. In this way, the current is distributed laterally in the channel, making it possible for **lateral tidal bars** to develop on the margins in place of the typical longitudinal bars. In these bars, an active sedimentation is produced, mainly composed of intercalations of cohesive sediments with sandy material. At the same time, the deeper areas of the channel act as *bypassing* zones for bed load-transported clasts.

The asymmetrical action of tidal currents makes the river sediment continue to bypass until it reaches the open coast, where a delta can begin to form even from the early stages of evolution. In this case, the deltaic progradation in the open coast is compatible with the presence of the estuary in the confined part of the system. This type of dynamic distribution has been described in the Guadiana estuary in the SW Iberian Peninsula [15, 23].

21.4 Dynamics and Evolution

21.4.1 Dynamics

The sedimentary dynamics of estuaries are controlled by the distribution of energies of the agents that control the transport of sediments, as well as by the processes of water mixing. The water mixing processes are responsible for the generation of new solid particles in the central zone of the estuary through flocculation processes. These flocculated grains have a very fine size (less than 16 microns) and a strongly cohesive character. The new flocculated particles become part of the turbidity maximum and are subsequently redistributed to the other zones of the estuary before being deposited in the less energetic zones. The dynamics of the processes of water mixing and the settling of particles from the cloud of turbidity have been extensively described in Sect. 9.2.

In general terms, the fine particles that settle from the maximum turbidity are usually retained and deposited inside the estuary. Due to their cohesive nature, these grains, once deposited, are not usually resuspended, resulting in very active sedimentation in the internal areas of the estuary. In this way, the estuary tends to fill up with sediment through a process of aggradation that also contributes to the narrowing.

When the tidal wave enters the estuaries, it suffers a deformation. The origin of this deformation is due to the higher speed of movement of the wave during high tides, because the water column is larger and there is less friction with the bed. This causes an asymmetry of the wave, which translates into a longer duration of the ebb semicircle (Fig. 21.8). Similarly, the ebb currents will not only act for a longer time, but will develop stronger currents. Because of this, river sediments transported as bed load tend to circulate downstream until they are deposited in the innermost position of the mixing zone. It is in this zone where the river sedimentary discharge takes place, as the river current meets the flood tidal current at its highest speed. In wave-dominated deltas, it is where the estuary's bayhead delta begins to form. This delta tends to evolve through a progradation of the sedimentary bodies of the delta over the fine sediments of the central basin. In tide-dominated deltas, the bayhead delta does not develop, since sand is recirculated to the marine sectors through the deeper areas of the channels during the tidal ebb. The shallower areas are characterized by the appearance of sandy shoals. It must be taken into account that the course of the fastest currents of ebb and flood do not usually coincide, in the same way that the course of a Formula 1 racing circuit would not coincide if it were to run in the opposite direction. Shoals tend to develop where the lower speed of both currents coincide (Fig. 21.9).

The longitudinal tidal bars begin to develop over these lower areas. In this case, the lower energy sectors are located in the dynamic shadow zones generated by these bars during the moments of maximum speed of the tidal currents. In these areas, there is a sedimentation of cohesive materials that contributes to the vertical growth of the bars, which end up reaching the intertidal zone and further into the supratidal area.

In the marine zone of the wave-dominated estuaries, the action of the waves controls the accumulation of sandy material. The development of barriers restricts the action of the tides in the inner zone of the estuary. However, the sand of marine origin manages to circulate inside the estuary during the action of the flood tidal currents, developing wide flood-tidal deltas that also grow towards the central basin.

21.4.2 Evolution

The tendency of estuaries to accumulate sediments in their interior has as a consequence of loss of depth in the short term (decades), and on the longer temporal scale (centuries to millennia) the total filling of the estuary. The speed of filling of each estuary depends on the relationships between the volumetric capacity of the estuary and the rate of sediment retention within it. In general terms, it can be said that the evolution of an estuary starts from the moment when a valley is flooded and ends when it completes its filling and evolves into a delta [5]. Each type of estuary presents different evolutionary patterns, however, they all have in common the fact that the evolution tends towards the occupation of its entire surface by supratidal environments, while the surfaces drained by the tide diminish and the tidal prism decreases. One consequence of these phenomena is that there is an increase in the degree of canalization of the system while the model of tidal wave propagation towards the interior areas of the estuary is modified. The sedimentation rate in different estuaries of the world gives values ranging from 1 to 12 mm per year [21]. However, data based on the comparison of sedimentary environment mapping, bathymetric maps and hydrodynamic data showed that the interior volume of an estuary changes at a rate of less than 0.1% per year without resulting in dynamic changes in the period studied [20]. This means that, despite the reduction in volume, the hydrodynamic conditions in an estuary can be maintained over periods of time greater than a century.

- *Wave-dominated estuary*: The evolution of wave-dominated estuaries was described by Roy et al. [29]. The filling of the estuary takes place through three different processes: (1) the aggradation of estuarine accretion bodies in the central basin, (2) the seaward progression of the estuary's bayhead delta onto the central basin sediments; and (3) the landward progradation of the seaward-domain flood-tidal deltas also onto the central basin sediments (Fig. 21.10). Thus, the resulting effect of this triple action is the retraction of the central basin. The entrance of the sedimentary bodies of the inner parts of the estuary into the intertidal zone leads to the development of tidal flats over them. These tidal flats are distributed in the form of islands separated by tidal

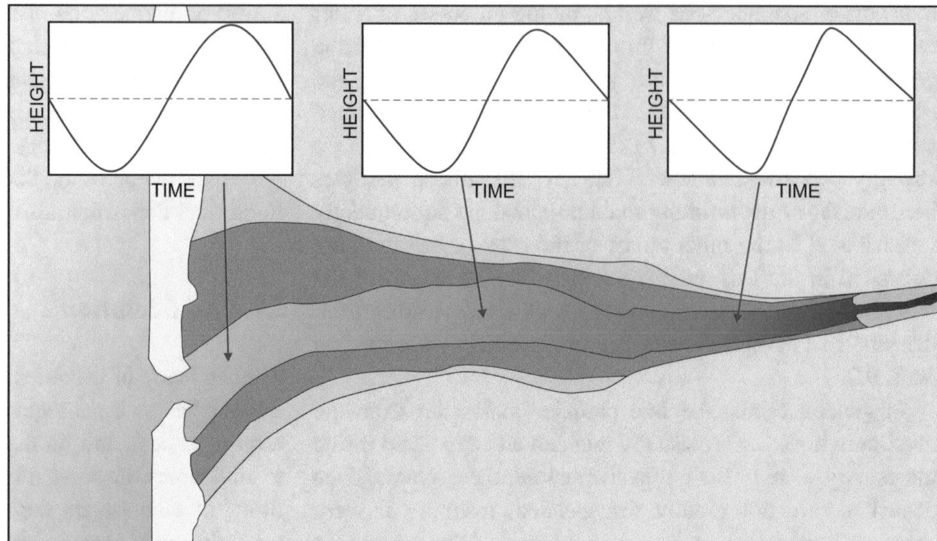

Fig. 21.8 Deformation of the tidal wave when propagating into an estuary

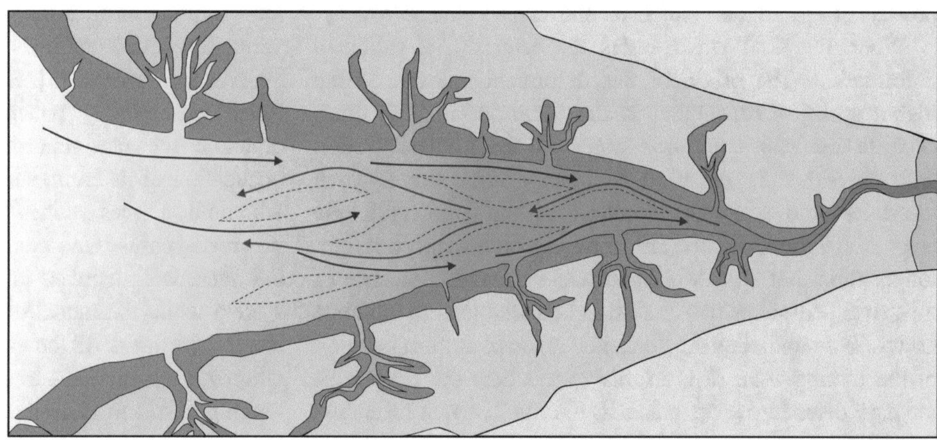

Fig. 21.9 Scheme showing a pattern of tidal circulation of flood (blue) and ebb (red) currents. Shoals (dashed lines) are formed in the areas of less energy, adapted from Dalrymple et al. [4]

channels. These islands tend to transform into salt marshes as a result of sedimentary aggradation (Fig. 21.10b). As the islands grow vertically, the channels that separate them begin to deepen. The filling culminates when the flood delta channels migrating towards the interior manage to connect with the channels of the bayhead delta, constituting what from that moment on will be estuarine channels, with a morphology very similar to that of an anastomosing river (Fig. 21.10c).

The loss of tidal prism also implies dynamic changes in the environments outside the estuary, as tidal currents will be less and the ebb delta will begin to be dominated by waves. At the same time, the front of the barriers will begin to be reworked by the waves, resulting in an erosional phenomenon (Fig. 21.10b). The total filling of the estuary implies that the sands of the ebb-tidal delta can be used by waves to build bars that reattach to the barrier front beaches, causing them to prograde (Fig. 21.10c). At this point, the tidal energy in the final infill can be so low that the estuary can behave like a blind estuary.

- *Tidal-dominated estuary*: Perhaps the best description of tidal-dominated estuaries is provided by Dalrymple et al. [4]. When the tide is the dominant energy, tidal currents are responsible for all sediment transport within the estuary. The development of longitudinal tidal bars is the process that controls the evolution of the entire estuary (Fig. 21.11). The bars develop from sandy shoals that sediment in the lower energy areas of the currents (Fig. 21.11a). As the innermost bars aggrade until they reach the intertidal zone and further toward the supratidal zone, new subtidal bars appear on their front, closest to the sea domain (Fig. 21.11b). The channels trace braided trajectories between the emerged bars. In more advanced stages of evolution, some channels are more efficient than others from a hydrodynamic point of view. The rest of the channels lose their functionality and are thus abandoned until they are filled with sediments. In a parallel way, several bars can be merged, thus expanding the marsh areas. The innermost zone is reduced to a single meandering estuarine channel (Fig. 21.11c). This channel

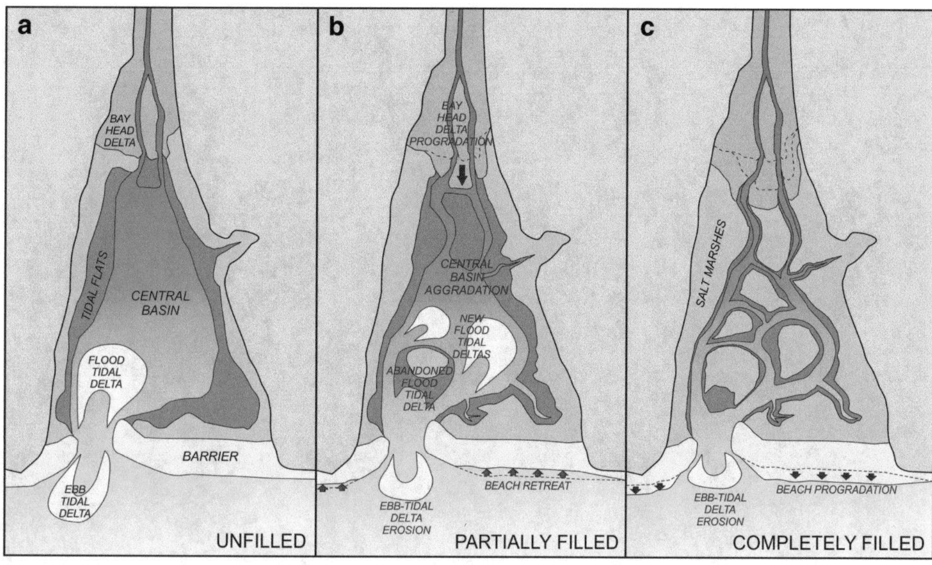

Fig. 21.10 Typical evolution of a wave-dominated estuary, adapted from Roy et al. [29]. **a** Initial stage with a low rate of infilling. **b** Intermediate stage with a partial infilling. **c** Stage of total infilling

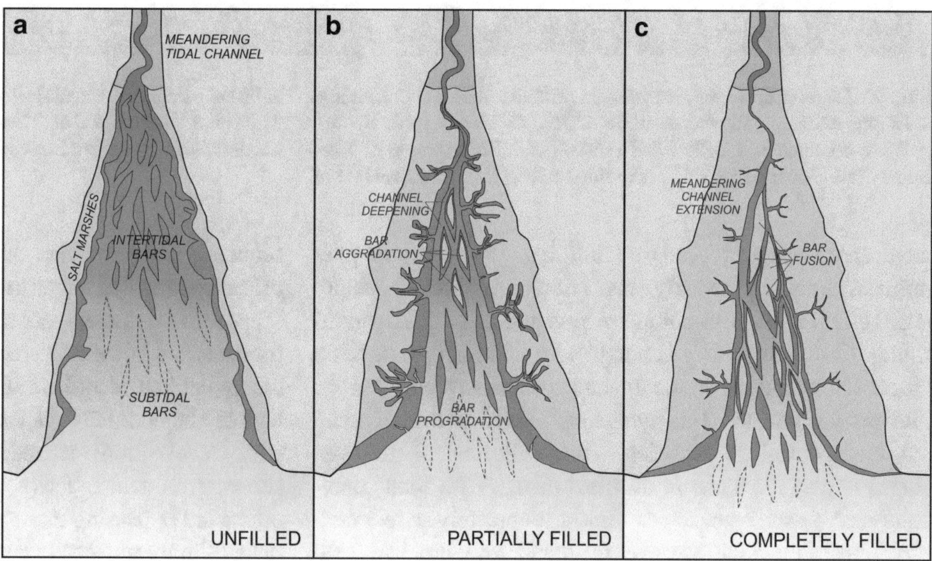

Fig. 21.11 Typical evolution of a tide-dominated estuary, adapted from Dalrymple et al. [4]. **a** Initial stage with a low rate of infilling. **b** Intermediate stage with a partial infilling. **c** Stage of total infilling

progressively changes in response to the filling of the estuary. It usually tends to deepen as it narrows, since all tidal drainage must take place through it.

Today, estuaries appear on coasts all over the world in any of the filling stages, since each estuary evolves at a different speed depending on the volume of sediment available. For that reason, some of the intermediate evolutionary stages of estuaries dominated by waves and tides appeared in some of the first morphological classifications as characteristic types. Thus, the initial unfilled stage could be classified as a drowned fluvial valley or ria-type estuary (Fig. 21.12a). Wave-dominated estuaries in intermediate stages are clearly barrier estuaries from a morphological point of view (Fig. 21.12b), while in their final stage these estuaries can be classified as blind estuaries (Fig. 21.12c). Finally, the intermediate stages of tidal-dominated estuaries would be classified physiographically as funnel-shaped estuaries (Fig. 21.12d).

As the sedimentation process fills the estuaries, the boundaries between the different areas change position in a progressive migration towards the sea. This has a clear influence on the distribution of the facies. Another effect of the filling of the estuaries is the increase in the deformation of the tidal wave. The relationships between the effect of channel narrowing and depth loss can be modified over time, with an effect on the model of wave propagation into the estuary. However, it is common for the tidal wave to become more asymmetrical in the sense of the ebb. This means an increase in the circulation rate of the bed load to the sea in

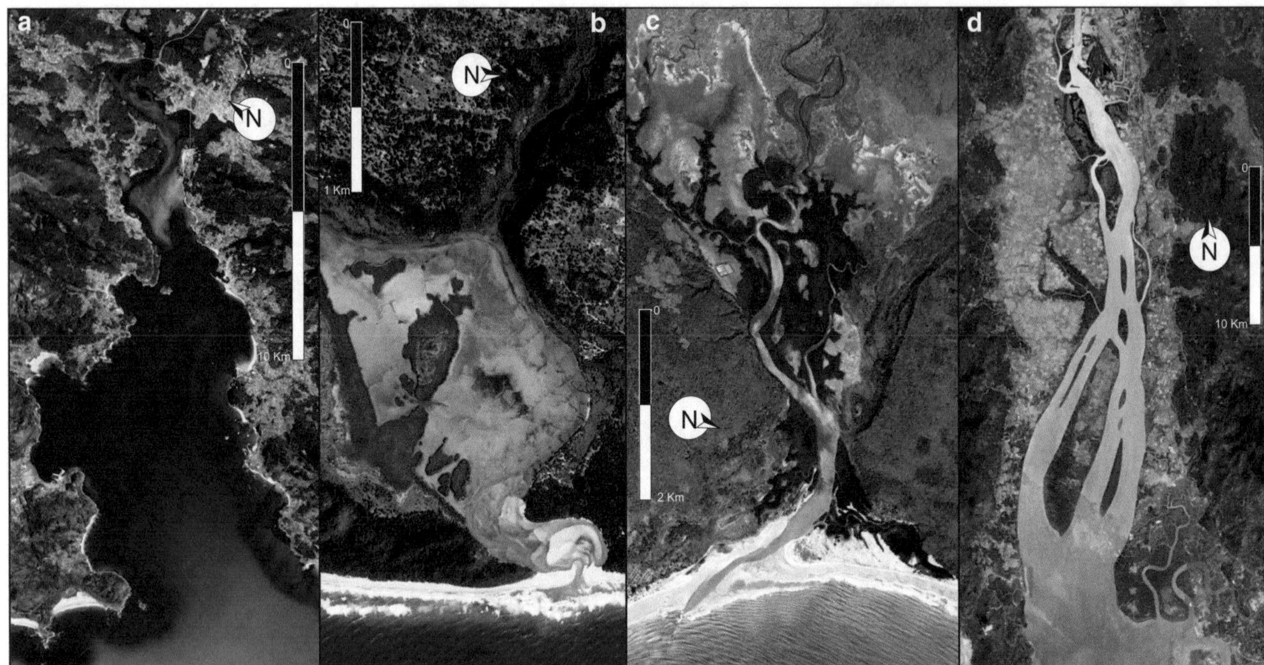

Fig. 21.12 Examples of estuaries in different states of infilling. **a** Empty estuary, with the example of Ría de Muros (NW Spain). **b** Wave-dominated partially filled estuary, with the example of Kosi estuary (NE South Africa). **c** Totally filled estuary, with the example of Kijiweni estuary (Tanzania). **d** Tide-dominated partially filled estuary, with the example of Dawei estuary (Myanmar). (Images Landsat/Copernicus from Google Earth.)

advanced stages of estuarine infilling. These same phenomena appear in the early stages of tide-dominated estuary channels, but are also common in estuaries with advanced filling. Once the estuary is completely filled and becomes a single channel, there are no dynamic differences between the two types of estuary. This dynamic is similar to that of rocky substrate-controlled estuaries. The evolution of the last stages is also the same in all three cases. With total canalization, the estuary becomes almost completely a sedimentary bypassing zone towards the coast and the bed load sediment of fluvial origin can reach the marine zone after several cycles of tidal transport. From this moment on, the river mouth system will have finished its stage as an estuary and will begin a new stage as a delta by starting a process of progradation towards the sea.

21.5 Depositional Facies

As can be seen in the previous sections, an estuary is a complex environment, whose distribution of sedimentary sub-environments is different depending on the morphology, the stage of evolution and the type of dominant agent. Obviously, there is a wide variability of facies depending on the factors that control sedimentation; however, there are a number of common characteristics that can be synthesized. Each of the environments that build the estuary has a sequence of lithofacies that depends on the variability of processes that can occur in it. These sequences of lithofacies (depositional facies) characterize each sedimentary environment and allow us to recognize them when they appear in the geological record. In this regard, it should be noted that part of the sediments is trapped in estuaries, while another part travels until it reaches the open coast. In some sub-environments, active sedimentation dominates, while others are bypassing environments [24]. Thus, the preserved facies simply reflect the nature of the sediment that manages to remain in the estuary. As early as 1957, Moore and Scruton made a magnificent summary of the facies and bedforms present in the various environments associated with the estuaries. This information has been supplemented by numerous papers over the last few decades. The following paragraphs summarize the details.

21.5.1 Depositional Facies Characteristic of Wave-Dominated Estuaries

Wave-dominated estuaries are the systems with the most pronounced bipolarity, because the sub-environments associated with the central basin produce the typical black muddy facies. These are associated with the turbidity maximum and the lowest energy of the whole estuary. The sandy facies are arranged both towards the river area and towards

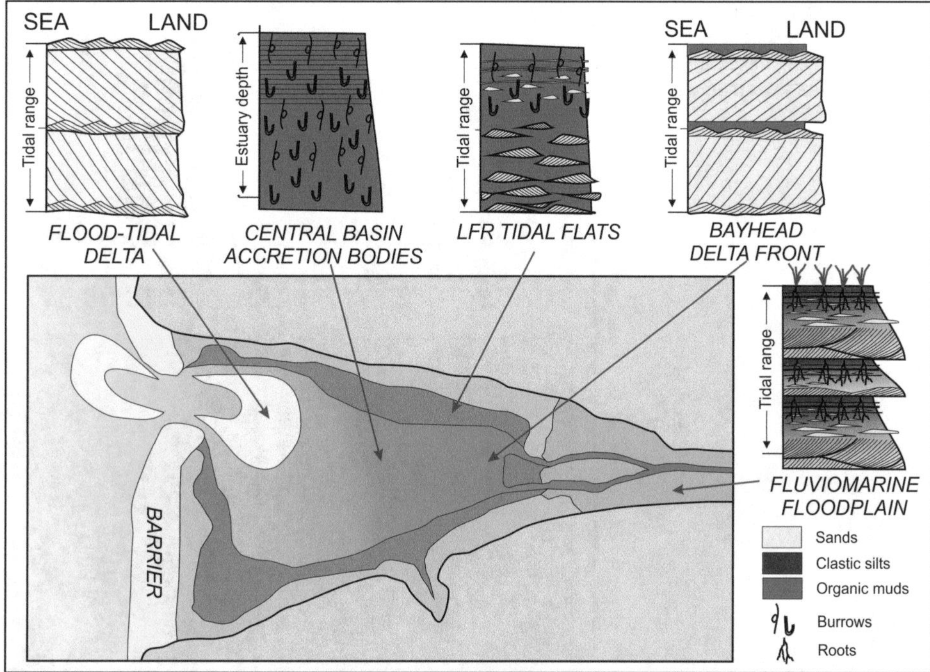

Fig. 21.13 Depositional facies generated on the different sub-environments linked to the early stage of a wave-dominated estuary

the marine domain. The characteristic lithofacies sequences (depositional facies) of this type of estuary are the following (Fig. 21.13).

- *Bayhead delta front*: The front of the bayhead delta develops prograding bars of a sandy and gravelly nature. These are sands with planar cross-bedding on a metric scale. When there is a wide central basin, these bars develop lobe geometries; however, in more advanced stages, the geometry becomes longer until it is reduced to the interior of the channels.
- *Fluviomarine floodplain:* These are fine-grained deposits with alternating centimeter-scale layers of silts of fluvial origin, black muds with parallel lamination of tidal origin, siliciclastic sands with ripples and immature gravels without internal organization. These alternations respond to energy fluctuations corresponding to the regime of fluvial contributions and extreme tides when they flood this part of the estuary. All the facies are finely interspersed in the form of fining-upwards sequences. The sequences are often bioturbated by the activity of plants.
- *Estuarine accretion bodies*: These are subtidal bodies that fill the central basin of the estuaries. They are constituted fundamentally by massive or laminated organic muds, strongly bioturbated, with some fine intercalations of sands that correspond to the distal contributions from the bayhead delta or from the marine system.
- *Tidal flats*: These develop on the intertidal margins of the central basin and could be called the lower flow regime (LFR) tidal flats. The LFR have mainly muddy facies or gray sandy muds, with strong bioturbation by burrows of bivalves and annelids always evident. Only in the vicinity of the marine area and the front of the bayhead delta does the percentage of sandy fraction increase and intercalations of sand and mud appear.
- *Barrier environments*: The wave-dominated environments are considered as a part of the confining barrier. Those located at the front as beaches, dunes and ebb-tidal deltas are not strictly inside the estuary. The only back barrier environments developed inland are flood-tidal deltas. In this case, their facies have been described in Chap. 19 and correspond to horseshoe-shaped sandy bodies whose internal structure are sets of cross-bedding with a dominant landward vergence.

21.5.2 Depositional Facies Characteristic of Tide-Dominated Estuaries

This type of estuary is characterized by a high tidal energy in the central part of the estuary. So, it does not develop a textural bipolarity as pronounced as in the type described above. The depositional facies that characterize these estuaries are the following (Fig. 21.14).

- *Subtidal bars*: These represent the embryonic development of the bars. Their characteristic facies are sands that develop ripple and dune fields. These sandy facies are usually intercalated with mud levels, developing flaser and wavy bedding. The lower contact of each sandy level usually presents erosive characteristics. In relation to

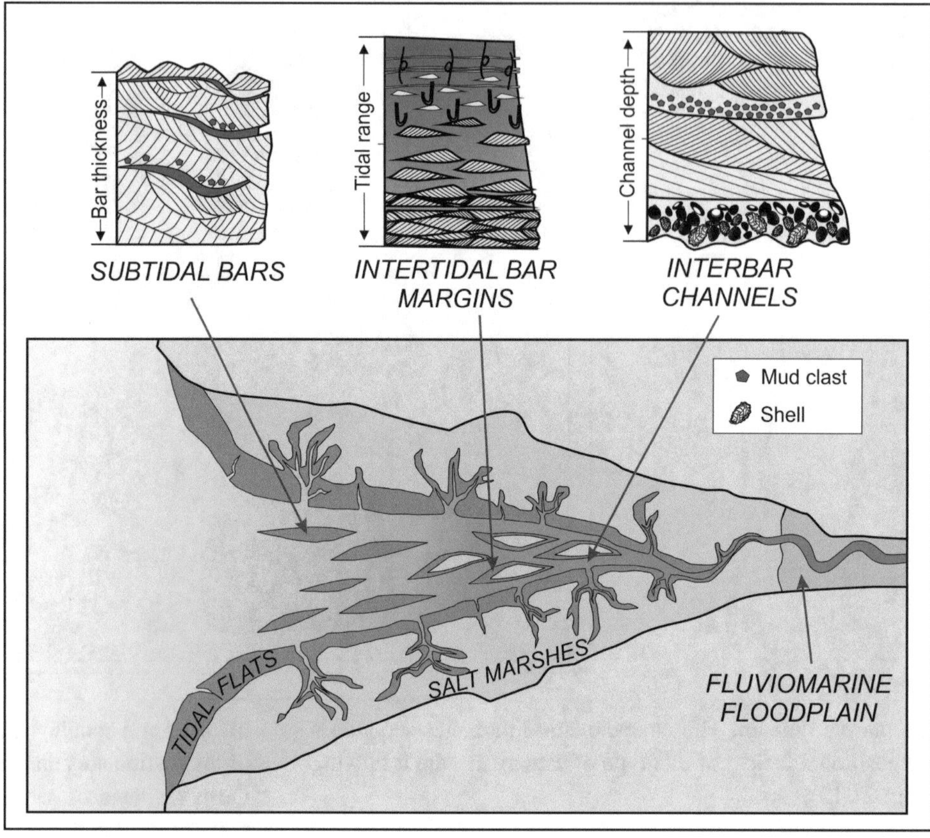

Fig. 21.14 Depositional facies generated on the different sub-environments linked to the early stage of a tide-dominated estuary

these moments of erosion, the facies also tend to present abundant soft clasts.

- *Intertidal bar margins*: When aggrading to the intertidal zone, the longitudinal bars facies become more diverse. There are differences between the areas closer to very active channels and those located in areas of dynamic shadow. All of them present characteristics of sand–mud interlayering. The intertidal margins of more active channels develop into upper flow regime (UFR) tidal flats, with characteristic flaser structures. Conversely, the channels that are abandoned develop lower energy structures such as wavy and lenticular bedding. Towards the supratidal areas, the muddy facies start to dominate the environment, with characteristic laminated muds that present very visible tidal bundles. Bioturbation by burrowing of organisms such as bivalves, annelids and crustaceans can be intense. This activity can even destroy the primary laminate structure. There are channels that are definitely abandoned and filled with muddy sediments until they reach the level of the marshes and are then colonized by vegetation. This process involves the fusion of previous longitudinal bars to form increasingly extensive supratidal islands.
- *Interbar channels*: The channels between bars concentrate the maximum tidal energy of the system. They usually develop sandy beds with significant mesoform fields (2D and 3D dunes) oriented in both senses of the tidal current. Consequently, their characteristic facies present sets of trough cross-bedding forming metric-scale herringbone structures. The mixing of waters also occurs in these channels, reflected in the existence of black mud levels on the bed. These levels can be partially reworked to produce soft clasts that are preserved inside the sandy facies.

In addition to these, fluviomarine floodplain environments very similar to those already described are also present in the fluvial zone of the wave-dominated estuaries. Also found are LFR lateral tidal flats similar to those described and salt marshes, typical of all estuaries in an advanced state of filling.

21.5.3 Depositional Facies Characteristic of Estuaries in an Advanced State of Infilling and Bedrock-Controlled Estuaries

When the estuaries are completely filled, the distribution of facies is very similar to that developed in estuaries controlled from the beginning by the rocky substrate. This system is reduced to an estuarine channel that can be single or forked and that connects the river directly to the open sea. In both

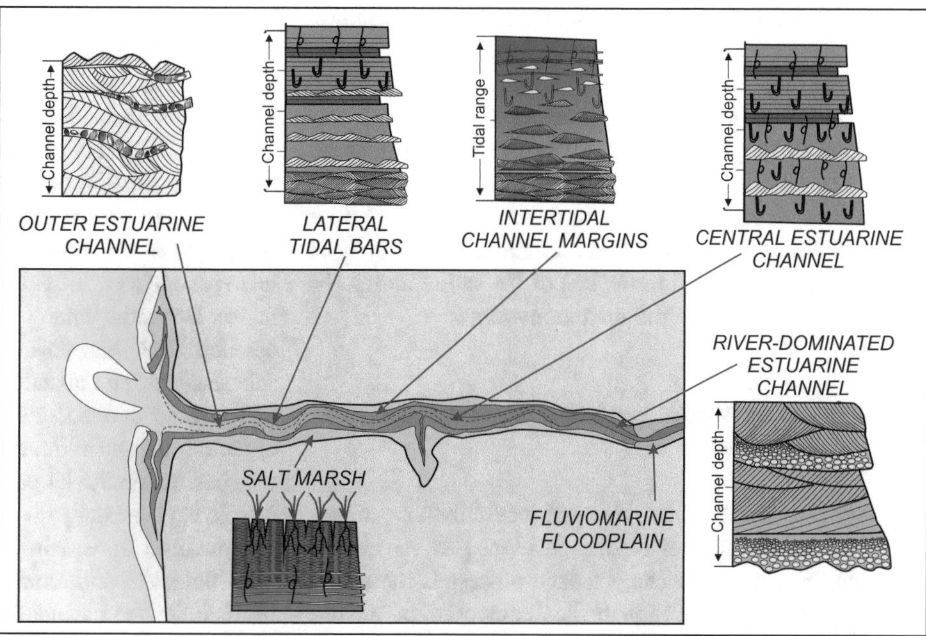

Fig. 21.15 Depositional facies generated in the different sub-environments linked to a rock-bounded estuary (the sequences are also valid for the late stages of infilling of wave-dominated and tide-dominated estuaries)

types of system, the intertidal areas are reduced to narrow fringes located on the estuary channel margins. In all of them, marshes occupy the highest areas in the marine and central zone of the estuary, while the same topographic areas are occupied by the fluviomarine floodplain in the fluvial domain. The difference between the different types of estuary lies in the surface area occupied by the marshes, being much narrower on rock-controlled estuaries. The depositional facies developed in the characteristic environments of these estuaries are as follows (Fig. 21.15).

- *Bypassing estuarine channels*: In the fluvial domain, macro-forms composed of gravel are present. These gravels can alternate with very coarse-to-medium sands in the deeper areas of the channel or even with sandy silt in the shallower areas of the system. This lithofacies alternation is mainly due to seasonal and interannual flow changes, which infer variations in both the type of fluvial contribution and in the relative domain between the river and the tide. In the central domain of the estuary, the deeper zones of the channel mainly present alternating muddy sands and black muds. In the marine estuary, the sequence is made up of monotonous deposits of clean, coarse-grained, non-bioturbated sand with frequent lenses of bioclastic gravels. Towards shallow areas, the grain becomes finer, and can then also present muddy matrix and burrowing bioturbation.
- *Shallow estuarine channels (lateral tidal bars)*: These facies have transitional characteristics with the deeper channel to which they are linked. They are usually composed of meso- and micro-formed sands. Sometimes the bedform fields are covered by black tidal muds, which are usually preserved as thin lenticular layers. The sand–mud association becomes predominantly muddy towards shallower areas of the channel, where monotonous bodies of black muddy sands bioturbated by annelids usually appear.
- *Intertidal channel margins*: These facies are very similar to those of the UFR tidal flats of more incipient estuaries; however, they present a greater slope, reflected in the inclination of the tidal parallel lamination. The dominant facies are muddy sands that pass upwards to bioturbated muds. Some centimetric intercalations of shells of residual origin can be found. Sometimes, millimeter- to centimeter-scale intercalations of muddy sands and muds (wavy bedding) can also appear. These are due to changes in the energy of the environment, produced by the alternation of spring and neap tidal cycles.
- *Salt marshes*: The sequence produced in the marshes consists almost exclusively of plant accumulations (peat) and root-bioturbated muds. The mud usually presents tidal lamination if it is not totally destroyed by the action of the roots. In the zones near the marine sandy contribution, the facies can be constituted by muddy sands or even present intercalations of clean sand reflecting energetic alternations of the environment. The top of the sequence consists of a decimetric level of saline mud with mud cracks.

21.6 Facies Models

The three-dimensional facies models also show very marked differences between wave- and tide-dominated estuaries in terms of the geometry of the sedimentary bodies. The geometric differences are the result of the different evolution of these estuaries. The architectural facies models shown correspond to the final stage of filling of both types of systems after a long period of sea level stability and with sufficient sedimentary availability. These models represent the disposition of sedimentary bodies formed by the depositional facies (lithofacies sequences) described in the previous section. Both models coincide in presenting a basal body of fluvial nature, developed in the bed of the valley during the lowstand period prior to the marine invasion.

21.6.1 Facies Model for Wave-Dominated Estuaries

The sequence that has been classically described as characteristic of these estuaries is that developed in the central basin. In this sector, the characteristic process is the aggradation of estuarine accretion bodies composed of organic black muds coming from the water mixing process. This aggradation leads the sediments to reach the intertidal zone, where they will begin to sediment the tidal flat bodies and finally marshes (Fig. 21.16).

The river zone is characterized in depth by a large body of sand and gravel with a prograding disposition towards the central basin. In this way, the river bodies that constitute the estuarine bayhead delta overlap the estuary accretion bodies until they reach high levels where the fine facies of the fluviomarine floodplain develop in a tabular body. The marine zone shows a facies model similar to the one observed in barrier island lagoon systems. Towards the interior of the estuary, the sandy bodies of the flood-tidal deltas stand out, which present a progradation towards the central basin, migrating over the black muds of the estuary accretion bodies. These sandy bodies are the basis for the development of tidal flats and marshes when the level reaches the intertidal zone. The estuarine channel facies are installed in the last stages of infill and develop longitudinally by cutting into the facies of the estuarine accretion bodies.

21.6.2 Facies Model for Tide-Dominated Estuaries

The facies model developed by the infilling of a tide-dominated estuary presents a much simpler architectural facies model. Early papers on this type of estuary asserted that the river area does not develop a true bayhead delta, due to the absence of a large central basin and the fact that river sediments are reworked upon reaching the tidal channels. However, according to Tessier [33], in the first stage of filling, this type of estuary presents an estuary-type morphology that allows the development of this type of delta. Thus, the base of the sequence presents an estuarine accretion body of a muddy nature that corresponds to the filling of the estuary in the incipient stage. This body is linked landwards with the sand and gravel facies of the initial bayhead delta (Fig. 21.17). Volumetrically, these sediments represent the greater fraction of the estuary's filling.

The upper part of the infilling along the entire estuary is composed of the elongated bodies of the longitudinal bars, interspersed with the interbar channels composed of sandy facies. Upwards, when reaching the intertidal zone, the bars develop UFR tidal flats, which are finally transformed into salt marshes. The abandoned channels are transformed into tidal creeks that become blind when they are filled with organic mud and ultimately they are also transformed into marshes. In the fluvial part of the system, the aggradation of the old bayhead delta gives rise to the development of a wide fluviomarine floodplain drained by a channel that passes from fluvial to tidal domain with a reflection in the change from straight to meandering morphology.

The described architectural model presents some variations in different estuaries [33], especially in the lower part of the model. These variations correspond to the history of the infill, since the depth of this type of estuary allows that some bodies developed during the transgressive stages prior to the stabilization of the sea level can also be recorded.

Advanced Box 21.2. Mixed Models of Dynamic Changes

The sedimentary infilling of the estuaries necessarily implies a decrease in the tidal prism. This decrease in water volume of tidal drainage results in a loss of tidal current velocity. The lower tidal energy increases the relative capacity of waves to build a barrier in the marine area of the estuary. Thus, an estuary that begins its filling with longitudinal tidal bars due to tidal dominance can end its evolution as a wave-dominated estuary if the waves build a barrier closing the estuary mouth.

This change in the dynamic conditions brings with it a change in the facies architectural model. Thus, the internal sector and the base of the estuary sequences can be filled according to the tide-dominated estuary model, while the surficial part of the marine zone can be filled according to the wave-dominated model (Fig. 21.18). In this way, the innermost sector will show the typical architecture of longitudinal bars and UFR tidal flat, separated by tidal channels. It is common to find in this sector marsh islands that result from the fusion of several bars by filling of the channels that separated them. At the same time, the marine sector will be composed of the sedimentary bodies of the barrier and the ebb- and flood-tidal deltas. The intermediate sector, in this case, may correspond to the filling of a central basin located between the longitudinal bars and the bodies closing the estuary.

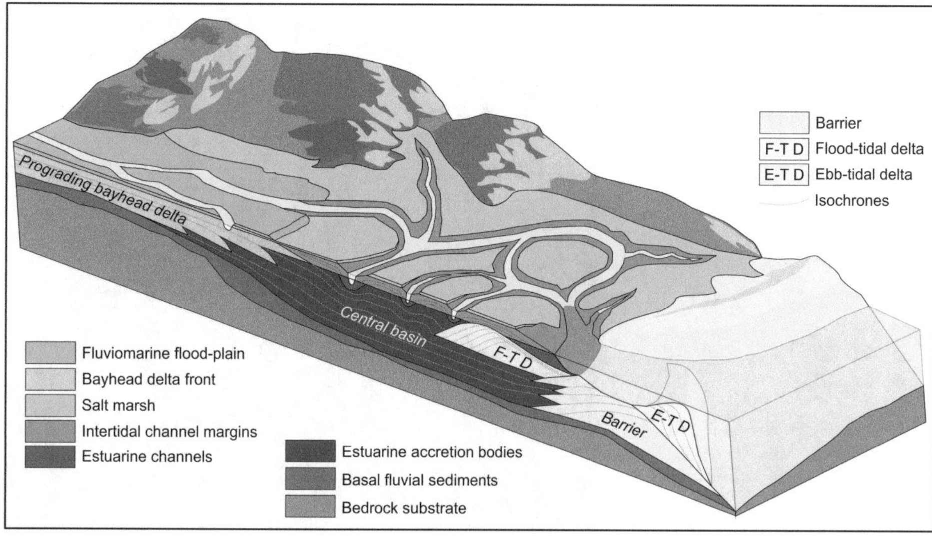

Fig. 21.16 Facies model for a wave-dominated estuary (based on Roy et al. [29] and Dalrymple et al. [5])

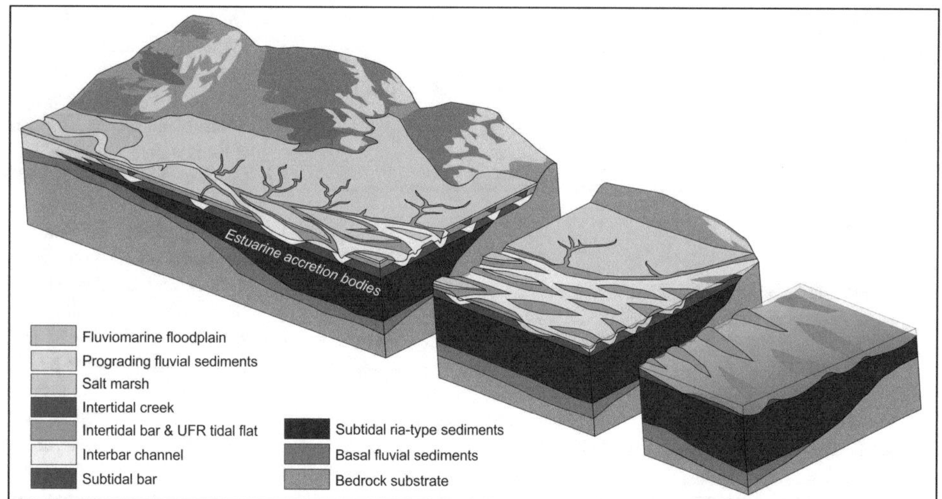

Fig. 21.17 Facies model for a tide-dominated estuary (based on Tessier [33])

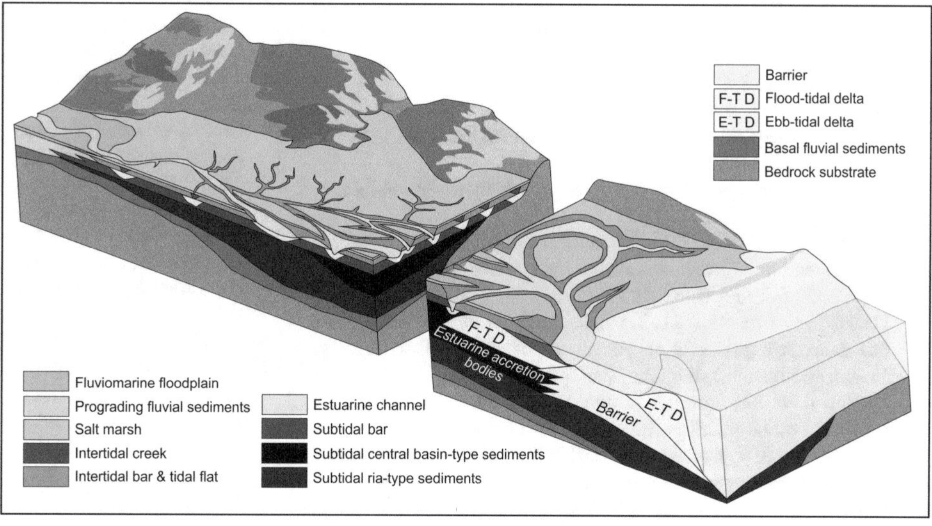

Fig. 21.18 Mixed facies model for an estuary under a change from tide-to-wave domination (based on Aguilar et al. [1])

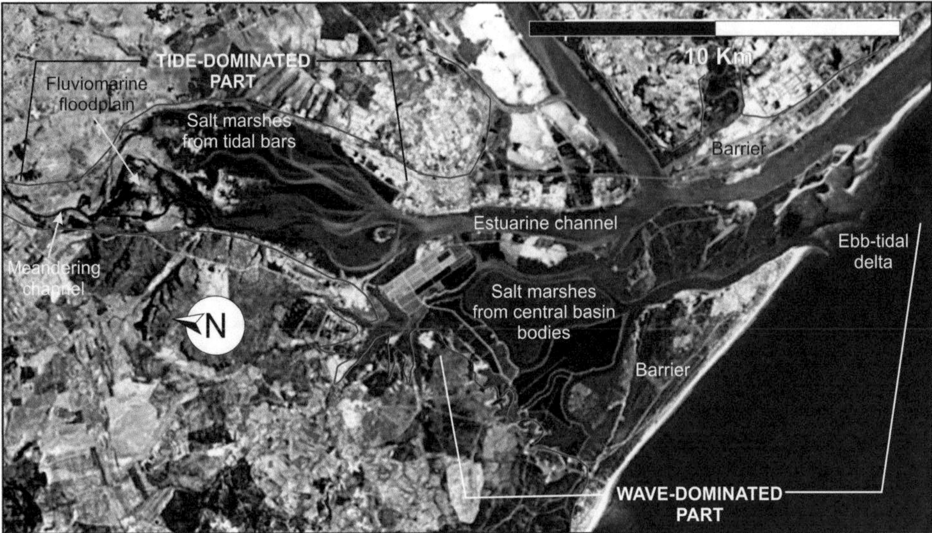

Fig. 21.19 Aerial photograph of the Odiel Estuary (SW Spain), displaying a morphology that reflects a dynamic change from tide- to wave-domination

Within this architectural scheme, the estuaries have four distinct morphological zones. The mesotidal estuary of the Odiel River in SW Spain (Fig. 21.19) fits this model of filling [1].

References

1. Aguilar ME, Morales JA, Morales-Mateo R, Feria MC, González MA, González-Batanero D (2019) Geometry of the upper units of the Holocene infilling of the estuarine channel of Odiel River del Odiel (Huelva, SW Spain) (in Spanish). J Geol Soc Spain 32 (1):127–142
2. De Boer PL, Nio S (eds) (1988) Tide influenced sedimentary environments and facies. Reidel Publishing Co, Boston, p 530
3. Chaumillon E, Proust JN, Menier D, Weber N (2008) Incised-valley morphologies and sedimentary-fills within the inner shelf of the Bay of Biscay (France): a synthesis. J Mar Syst 72:383–396
4. Dalrymple RW, Mackay DA, Ichaso AA, Choi KS (2012) Processes, morphodynamics, and facies of tide-dominated estuaries. In: Davis RA, Dalrymple RW (eds) Principles of tidal sedimentology. Springer, Heidelberg, pp 79–107
5. Dalrymple RW, Zaitlin BA, Boyd R (1992) Estuarine facies models: conceptual basis and stratigraphic implications. J Sedim Petrol 62:1130–1146
6. Davies JL (1964) A morphogenic approach to world shorelines. Geomorphology 8:27–42
7. Davis RA, Clifton HE (1987) Sea-level change and the preservation potential of wave-dominated and tide-dominated coastal sequences. J Sedim Petrol 41:167–178
8. Davis RA, Fox WT (1981) Interaction between wave- and tide-generated processes at the mouth of a microtidal estuary: Matanzas River, Florida. Mar Geol 40:49–68
9. Dionne JC (1963) Towards a more adequate definition of the St. Lawrence estuary. Zeitschrift für Geomorphologie 7:36–44
10. Dolgopolova EN, Isupova MV (2010) Classification of estuaries by hydrodynamic processes. Water Resour 37(3):268–284
11. Dyer KR (1997) Estuaries: a physical introduction, 2nd ed. Wiley, New York, 195pp
12. Fairbridge RW (1980) The estuary: its definition and geodynamic cycle. In: Olausson E, Cato I (eds) Chemistry and biogeochemistry of estuaries. Wiley, New York, pp 1–36
13. Le Floch P (1961) ropagation de la Marée dans l'Estuaire de la Seine et en Seine Maritime. Université de Paris, Thèse Doctorat d'État, p 507
14. Gallivan LB, Davis RA (1981) Sediment transport in a microtidal estuary: Matanzas River, Florida. Mar Geol 40:69–84
15. Garel E, Pinto L, Santos A, Ferreira O (2009) Tidal and river discharge forcing upon water and sediment circulation at a rock-bound estuary (Guadiana estuary, Portugal). Estuar Coast Shelf Sci 84:269–281
16. Hansen DV, Rattray M Jr (1964) Gravitational circulation in straits and estuaries. J Mar Res 23:104–122
17. Hayes MO (1975) Morphology of sand accumulation in estuaries: an introduction to the symposium. In: Cronin LE (ed) Estuarine research, vol 2. Academic Press, New York, pp 3–22
18. Kirby R (1989) Sediment problems arising from barrage construction in high energy regions: an example of the Severn Estuary. In: Telford T (ed) Third conference on tidal power. Institution of Civil Engineers, London, pp 189–200
19. Knauss JA (1997) Introduction to physical oceanography. Prentice Hall, New Jersey, p 309
20. Lane A (2004) Bathymetric evolution of the Mersey Estuary, UK, 1906–1997: causes and effects. Estuar Coast Shelf Sci 59(2):249–263
21. McManus J (1998) Temporal and spatial variations in estuarine sedimentation. Estuaries 21(4A):622–634
22. Moore DG, Scruton PC (1957) Minor internal structures of some recent unconsolidated sediments. Bull Am Assoc Pet Geol 41:2723–2751
23. Morales JA, Ruiz F, Jiménez I (1997) The role of the estuarine deposition in the sedimentary exchange between the continent and the littoral. The Guadiana Estuary (SW Spain–Portugal) (in Spanish). Revista de la Sociedad Geologica de España 10:309–325
24. Nichols MM, Biggs RB (1985) Estuaries. In: Davis RA (ed) Coastal sedimentary environments, 2nd edn. Springer, Heidelberg, pp 77–186
25. Perillo GME (1995) Definitions and geomorphologic classifications of estuaries. In: Perillo GME (ed) geomorphology and sedimentology of estuaries, vol 53. Developments in Sedimentology. Elsevier, Amsterdam, pp 17–47

26. Pritchard DW (1955) Estuarine circulation patterns. Proc Am Soc Civil Eng 81:717/1–717/11
27. Pritchard DW (1967) What is an estuary? Physical viewpoint. In: Lauff GH (ed) Estuaries, vol 83. American Association for the Advancement of Science Publications, pp 3–5
28. Pritchard DW (1960) Lectures on estuarine oceanography. In: Kinsman B (ed) JHopkins Univ pp 154
29. Roy PS, Thom BG, Wrigt LD (1980) Holocene sequences on an embayed high energy coast: an evolutionary model. Sed Geol 26:1–19
30. Russell RJ (1967) Origin of estuaries. In: Lauff GH (ed) Estuaries, vol 83. American Association for the Advance of Science Memoirs, pp 93–99
31. Silvester R (1974) Coastal engineering II: sedimentation, estuaries, tides, effluents and modeling. Development of Geotechnical Engineering. Elsevier, Amsterdam, 338pp
32. Simmons HB (1955) Some effects of upland discharge on estuarine hydraulics. Proc Am Soc Civil Eng 81:729/1–729/20
33. Tessier B (2012) Stratigraphy of tide-dominated estuaries. In: Davis RA, Dalrymple RW (eds) Principles of tidal sedimentology. Springer, Heidelberg, pp 109–149

22 Fluvial–Influenced Systems II: Deltas

22.1 Introduction

Like estuaries, deltas represent another kind of river mouth. In this case, the river has been able to accumulate such a significant amount of sediment that it has been able to build a large sedimentary structure, thus forming a coastal ledge (Fig. 22.1). Therefore, a delta is defined as a prograding body built with the sediments contributed by a river at its mouth and which protrudes from the coast into the sea [1]. According to [2], a delta usually evolves from a simple sedimentary lobe located at a river mouth to a complex three-dimensional entity that could be defined as a sedimentary system. Authors such as [3] understand that, although the primary source of materials is the river, tides and waves can also act on these sediments. In this case, marine processes are responsible for the total or partial redistribution of sediments of fluvial origin, although waves can also act as a supplier of sediments to the delta front by contributing sands from adjacent marine areas [4].

According to this definition, deltas can also develop at the mouths of rivers that flow into lake waters—in this case. Their sediments will not be redistributed by tides, although in large lakes they can be reworked by waves. On the other hand, small deltas can be built in the inland water bodies of estuaries in their early stages of evolution. These are the estuarine bayhead deltas studied in Chap. 21.

The distribution of deltas on present-day coasts has been analyzed by Inman and Nordstrom [5], who estimate that most deltaic systems are located on passive margin coasts, although there are also a significant number of deltas developed in protected areas of active margins, such as the back zone of island arcs. In addition to the deltas currently developed on the coasts of the world, there are numerous examples of deltas in the geological record that have been very well studied.

From an ecological point of view, deltas are diverse and prolific ecosystems. They are highly sensitive to changes in sea level and river basins, and are greatly impacted by human activities. Throughout history, they have been a common site for important cultural settlements. One only needs to recall that early civilizations such as Egypt and Mesopotamia settled on delta plains. The proper name of this environment was coined by these civilizations, enduring to our times from the first definition of a delta as made by the geographer Herodotus of Halicarnassus in the fifth century BC. Nowadays, large cities and important industrial and port facilities are often developed in deltas. Due to the nature of their sediments, they also present great economic interest, since these sedimentary formations are likely environments for the accumulation of substantial amounts of energy resources, such as coal, oil, natural gas and other elements of value including uranium [2].

22.2 Control Factors

The dynamics of deltas and, by extension, their morphology are related to a variety of factors (Fig. 22.2). On the one hand, the characteristics of the drainage area of the river system that feeds it determines the volume and nature of the water and sediment inputs, as well as their distribution over time. On the other hand, several factors intervene in the basin where the deposition occurs, such as the marine processes that redistribute the sediment and the relative sea level movements. These movements are in turn related to the tectonic regime of the basin and global eustasism. In this regard, it should be noted that the presence of the delta sedimentary prism usually induces strong subsidence in the coastal areas where it develops. Finally, it must be considered that climate is the overall driver, exerting a notable influence on the rest of the factors [6].

22.2.1 Fluvial Discharge

The water inputs that reach the deltaic system from the river are important in terms of the dynamics of water mixing that occurs in the distributary channels or at the marine front. This

J. A. Morales, *Coastal Geology*, Springer Textbooks in Earth Sciences, Geography and Environment,
https://doi.org/10.1007/978-3-030-96121-3_22

Fig. 22.1 Example of a deltaic system (Po River Delta, NE Italy): a prograding sedimentary body built by the river at its mouth. (Image Landsat/Copernicus from Google Earth.)

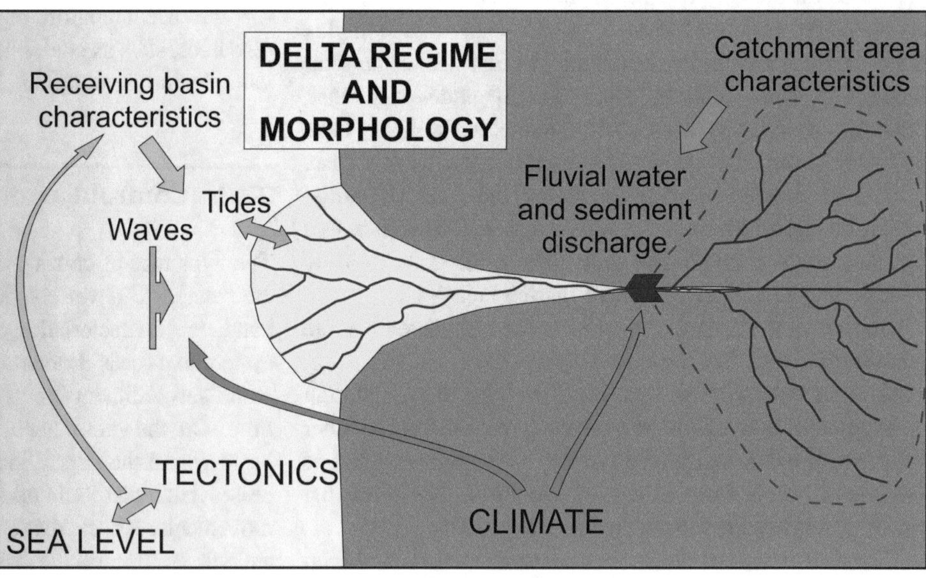

Fig. 22.2 Factors influencing the development of deltas (Based on [6])

mixing involves important flocculation and diffusion processes that have already been discussed in Chap. 9 of this book. However, the fluvial inputs that most influence deltaic development are sedimentary inputs. The fluvial systems that feed deltas usually contribute a significant volume of sediment, both in the form of suspended matter and bed load. These materials are the ones that mostly become part of the deltaic bodies. However, the materials contributed by the river can be redistributed by marine agents such as waves or tides.

22.2.2 Wave Action

Waves are present in most deltaic systems and are absent only in deltas developed in confined seas, on coasts with an orientation contrary to the prevailing winds or in deltas associated with small lakes. The action of the waves is usually manifested in a significant reworking of the sediment that acts in the opposite direction to the prograding trend imposed by the river. When waves are present, the sediments that the river throws onto the open coast are used by the waves to build bars parallel to the delta front. The waves migrate these bars landward to attach them to the areas between the distributaries, building beaches attached to the delta front or barrier island systems. In any case, waves impose their dynamics on the sediment and, being responsible for its final deposition, their action is reflected in the sedimentary sequences.

22.2.3 Tidal Action

The tide is usually present in deltas that develop on the coasts of oceanic areas; however, it is absent in deltas of restricted seas such as the Mediterranean or the Gulf of Mexico and, of course, also in lacustrine deltas. The presence of tides provokes the entry of seawater into the interior of the distributary channels, making them function as small estuarine channels. Thus, the mixing of water and the processes associated with the turbidity maximum take place inside them. These have been classified as delta front estuaries by Fairbridge [7]. The tide is also responsible for the development of tidal flats in the sheltered fringes of the delta. When the tide is so significant that it becomes the dominant force, materials are reworked into longitudinal bars similar to those observed in tide-dominated estuaries. However, in deltaic systems, these bars develop outward from the coastline and not in the interior of the valleys as occurs in the case of estuaries.

22.2.4 Long-Term Factors: Relative Sea Level Movements

Relative sea level movements act over long periods of time, but largely determine the genesis of deltas and their subsequent development. Some authors have associated the development of deltaic systems to moments of relative sea level lowstand, since after a relative fall there is no marine invasion of the river mouths and, therefore, there is no possibility of estuarine formation. Thus, they would form part of the *lowstand system tracts* [8]. However, deltas can also be generated during times of stand following a sea level rise or even during the rise itself. Under these conditions, it is the relationship between the space generated by the sea level rise and the volume of sediment brought by the river that sets the pattern of its development. Thus, a delta will develop if the volume of sediment contributed by the river exceeds the volume of space generated by the rise. To understand delta dynamics, both parameters must be considered with respect to time, so the rate of input and rate of rise must be compared. In the case of systems developed in the context of a sea level rise, sedimentation initially takes place in a coastal valley invaded by the sea. This valley is subsequently filled in and begins to progress towards the outer part of the coast. Thus, deltas developed after or during a sea level rise evolve from a previous estuary once it has been infilled.

The study of ancient deltaic formations shows that their sequences have a clear cyclical character [9]. Large deltas fed from extensive river basins are maintained over very long timescales spanning different cycles of sea level movements. These systems alternate between prograding and retrograding phases of growth. Prograding phases occur when sea level falls or when the rate of sedimentary input exceeds the rates of sea rise. Conversely, retrograding phases are erosional in nature and are associated with times when the rate of sea level rise exceeds the rate of input. The typical example of a large delta where such cycles are observed is the Niger Delta on the west coast of Africa. Such sequences have been studied frequently in deltaic systems preserved in the geologic record.

22.3 Classification and Morphology

Although dictionary definitions of a delta emphasize a triangular plan shape, there is actually a wide morphological variety. In addition to the triangular shape that gives it its name, there are deltas with cuspated (rhomboidal), lobated (arched), elongated (fingered), intricate and other mixed shapes (Fig. 22.3). From the point of view of grain size, there is also great variability, from muddy deltas to gravel deltas, including sandy deltas and deltas of mixed lithology.

22.3.1 Classification of Fluvial Deltas

The morphological variability of deltas is a consequence of their sedimentary dynamics. Ultimately, this is the dominant agent in deltaic processes that imposes the morphology. Gallowayw et al. [3] proposed a classification adopting this criterion that continues to be used today. The practical character of this classification consists of the association of morphology with the hydrodynamic agents that control it. The classification consists of a triangle at whose vertices are placed the three potential agents of sediment transport and reworking: river, waves, and tides (Fig. 22.4).

River-dominated deltas show elongated and fingered morphologies that are commonly referred to as "bird's-foot" shaped, especially when associated with fine sediments. The fingering is due to the presence of numerous distributary channels separated by levees, leaving in the frontal zone concave spaces known as intradistributary bays. The progradation of the system occurs from the sedimentation of lobate bars whose reworking by marine processes (waves

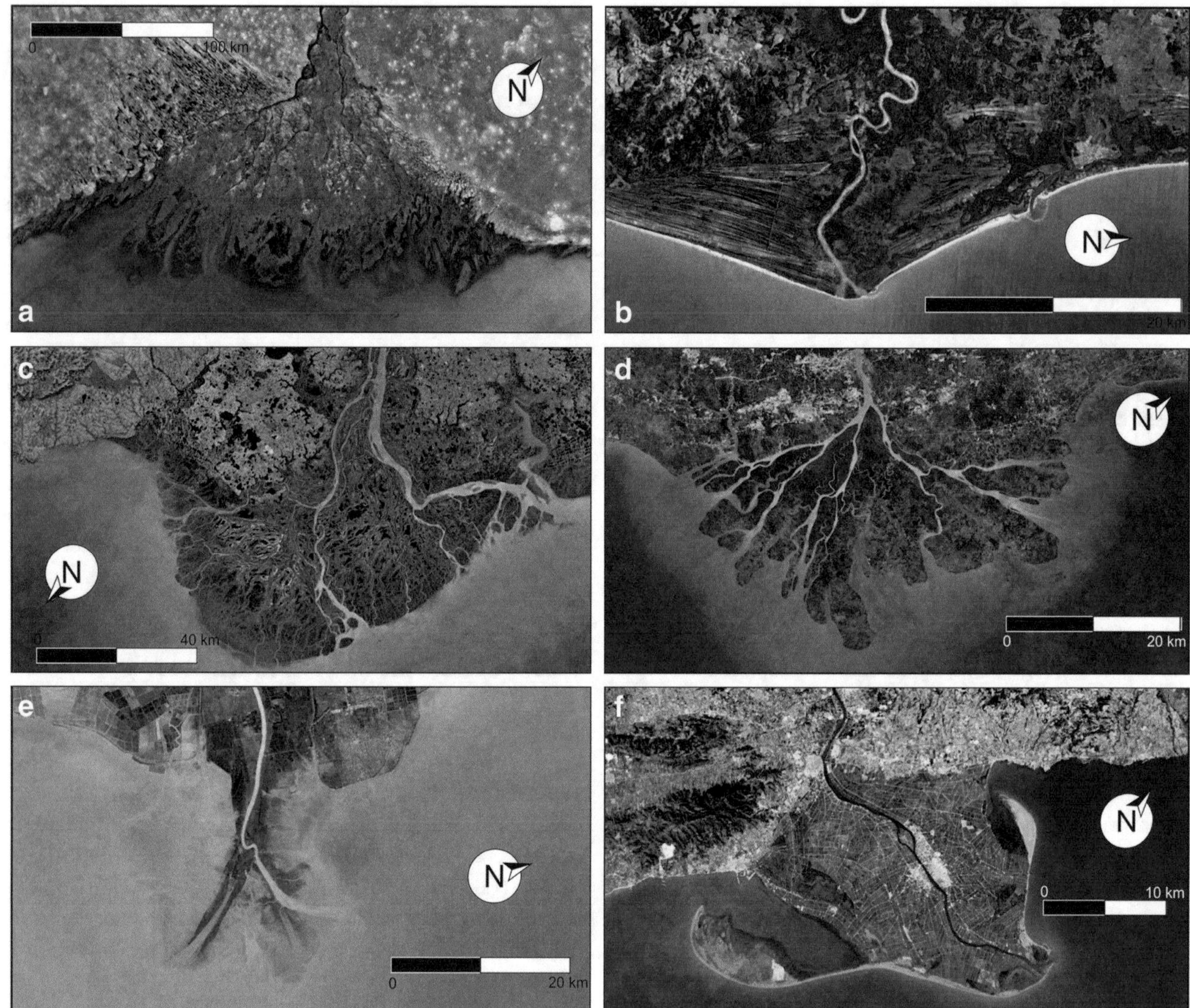

Fig. 22.3 Different morphologies of deltas. **a** Triangular shape (Volga River Delta, Russia). **b** Cuspate shape (Jequitinhonha River Delta, Brazil). **c** Lobate shape (Yukon River Delta, Alaska, USA). **d** Braided shape (Mahakam River Delta, Indonesia). **e** Bird's-foot shape (Yellow River Delta, China). **f** Mixed cuspate/lobate shape (Ebro River Delta, Spain) (Images Landsat/Copernicus from Google Earth)

and tides) is minimal or null. For this reason, the sediments maintain the morphology imposed by the fluvial discharge. They usually occur in rivers that flow into closed, microtidal basins with reduced fetch. The typical example, and also the most studied case, is the Mississippi River Delta.

Wave-dominated deltas present a very different aspect, developing cuspated morphologies. All fluvial sediments tend to be reworked by the waves to build parallel bars that end up adjoining the shore to make the shore prograde into the sides of a reduced number of distributaries or a single channel. The waves can also act as a supplier of marine sediments that add to the volume of fluvial input. Thus, the deltaic physiography is similar to that of a strandplain, with no interdistributary environments. They are usually associated with sandy lithologies. The classic example is the delta of the São Francisco River in Brazil, but similar deltas appear along the Brazilian coast in a number of river mouths.

In **mixed river–wave deltas**, the waves are not able to rework all the sediments brought by the river and a mixed morphology is presented. There are a high number of distributaries, as in river-dominated deltas, but the waves are able to build barrier islands between them, closing the interdistributary bays, which are then converted into lagoons. The delta front is not as cuspate, since the waves displace the sediments, giving the delta a lobated or arched morphology. These are usually deltas that present varied lithologies, including muds and sands distributed in different sub-environments. The most typical example is the Nile, although other deltas located on the Mediterranean microtidal coasts also belong to this type.

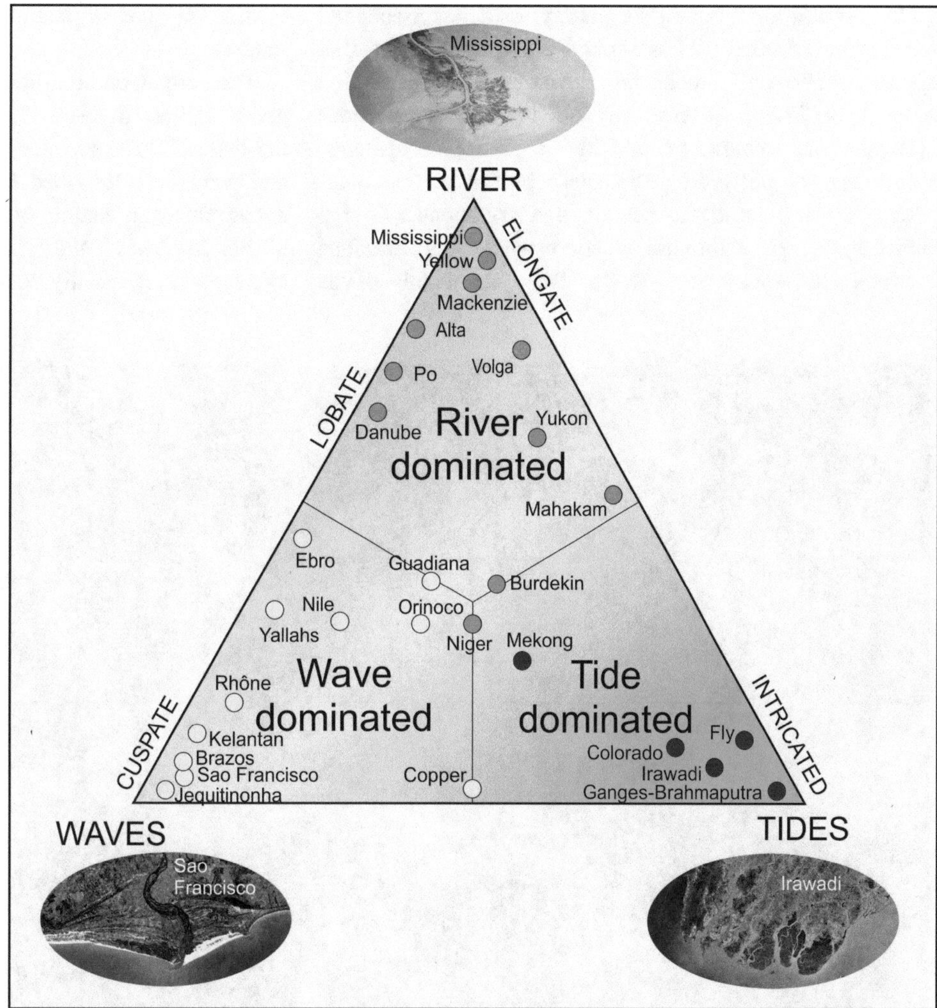

Fig. 22.4 Triangular classification of fluvial deltas after [3], indicating the position of some major deltas (data from Briggs et al. [10])

Tide-dominated deltas present an intricate shape and physiography very similar to the tide-dominated estuaries described in Chap. 21. For this reason, their physiography has been referred to by some authors as estuarine morphology. Morphologically, they consist of a large number of braided or anastomosing channels separated by longitudinal tidal bars. The difference in the case of deltas lies in the location of the bars outside the fluvial bay. With these, it is the tidal energy that is responsible for reworking the sediments of fluvial origin, although much autochthonous sediment can also be generated due to water mixing processes. These are deltas that develop on macrotidal coasts but with reduced fetch or dominant winds acting from land, both circumstances inhibiting wave action. Lithologically, they are deltas made up of fine sediments (muds and fine sands). The best-known example is the delta of the Ganges and Brahmaputra Rivers in India.

In **mixed river–tidal deltas**, the tide is not able to rework all the sediments contributed by the river, with faster rates of progradation than in tide-dominated deltas. As in tide-dominated deltas, there are the characteristic tidal bars and braided channels, but the overall morphology of the delta is much more elongated, and wider interdistributary bays are also present. Sediments tend to be somewhat coarser than in tide-dominated deltas. The classic example is the Mahakam River Delta in Indonesia.

The **mixed wave–tide marine-dominated deltas** present complex morphologies, with an inner part characterized by the presence of typical tidal bars, but with an outer part that presents barrier islands rather than interdistributary bays. They usually occur on macrotidal coasts that also present high wave energies. The most studied example is the Copper River in Alaska, although it is not really a delta, but an estuary.

Intermediate river–wave–tide deltas are the most complex of all. They are usually large deltaic systems whose overall morphology depends on the agent exerting the greatest influence on sediment distribution. The most characteristic example is the Niger River Delta on the west coast of Africa.

In addition to Galloway's typology for deltas associated with river mouths [11] described a type of delta that appears when an alluvial fan develops directly on the coast (Fig. 22.5a). These are called **fan-deltas**. These systems are characterized by high-slope systems. Fan-deltas are powerful sedimentary bodies with a completely alluvial terrestrial section that continues into the submerged region through a zone where the interface between the two systems occurs. They are usually deltas formed by coarse sediments, mainly gravels, but also sands.

The arrival of high-gradient braided rivers at the coast gives rise to a special type of delta closely linked to fan-deltas. These are the **Gilbert-type deltas** (Fig. 22.5b) and were first described by Grove Gilbert [12]. They are deltas consisting mainly of gravels that develop at the mouth of braided rivers and have minimal reworking by other agents, so they usually occur in lacustrine waters.

Fig. 22.5 Coarse-grained deltas. **a** Fan-delta, with the example of Agua del Pueblo Creek, Chile. **b** Gilbert-type delta, with the example of Avellano River Delta, Chile. (Images Landsat/Copernicus from Google Earth.)

22.3.2 Domains and Sub-environments of Fluvial Deltas

Being constituted by a powerful sedimentary body, deltas have a visible subaerial part and a significant submerged part. Classically, the structure of deltas has been divided into three different components (Fig. 22.6). The emerged part is usually flat or of very low slope, constituting the section known as the **delta plain**. In front of this plain there is a zone of greater slope that is constituted by the main bed-loaded sediments that manage to leave the mouths of the distributaries. This region is called the **delta front**. Beyond the front, there is a very low-slope zone consisting of sediments decanted from the suspended matter plumes. This is the section called the **prodelta**.

Different sub-environments are distributed in the delta plain. The inclusion and distribution of sub-environments vary among the different types of deltas, since they are the result of the action of certain processes that work in different ways (Fig. 22.7). In general terms, the environments that may occur in this plain are as follows:

- *Distributary channels*: These are the channels responsible for distributing freshwater flow and sediment load, connecting the river source with the open sea. Their morphology can be varied, ranging from straight to meandering channels and constituting networks of simple, bifurcated or rejoining channels, depending on the dominant processes. In river-dominated deltas, their behavior and dynamics are no more than extensions of typical river channels; however, in tide-dominated deltas, they behave like estuaries, with water mixing and all processes related to the turbidity maximum taking place within them. In wave-dominated deltas, the number of distributaries is usually very small, while in tide-dominated deltas there is usually a high number of channels, which may be even higher than in the case of river-dominated deltas.
- *Levees*: These are established as raised areas on the margins of distributary channels. Their behavior is identical to that of the natural levees of the fluvial systems, marking the overflow level of the channels during floods. They occur as terrains colonized by subaerial vegetation. They are usually better developed in river-dominated deltas, although they can also be present in other types of deltas.
- *Interdistributary swamps*: These are large vegetated areas that extend between the channels in the rear area of the levees, having a somewhat more depressed relief than their surroundings. This topography makes these areas susceptible to flooding during river floods, and also during spring high waters in those deltas affected by tides and by surges in deltas affected by waves. Evidently, they would also be areas flooded by mixed surge and spring high tide processes during extreme meteorological events. The vegetation that occupies these environments depends on the climate of the region where the delta develops. In deltas of tropical areas, mangroves and tree swamps are common, while in mid-latitudes the most common swamps are marshes and in deltas of periglacial areas permafrost may even develop.
- *Delta plain lakes*: These are the depressed areas between different slight distributaries that may not reach the topographic level to become emerged. In this case, these areas are occupied by interdistributary lakes. Oxbow lakes may also form in abandoned distributaries. The dynamics of these environments is similar to that of any lacustrine environment, with endorheic characteristics

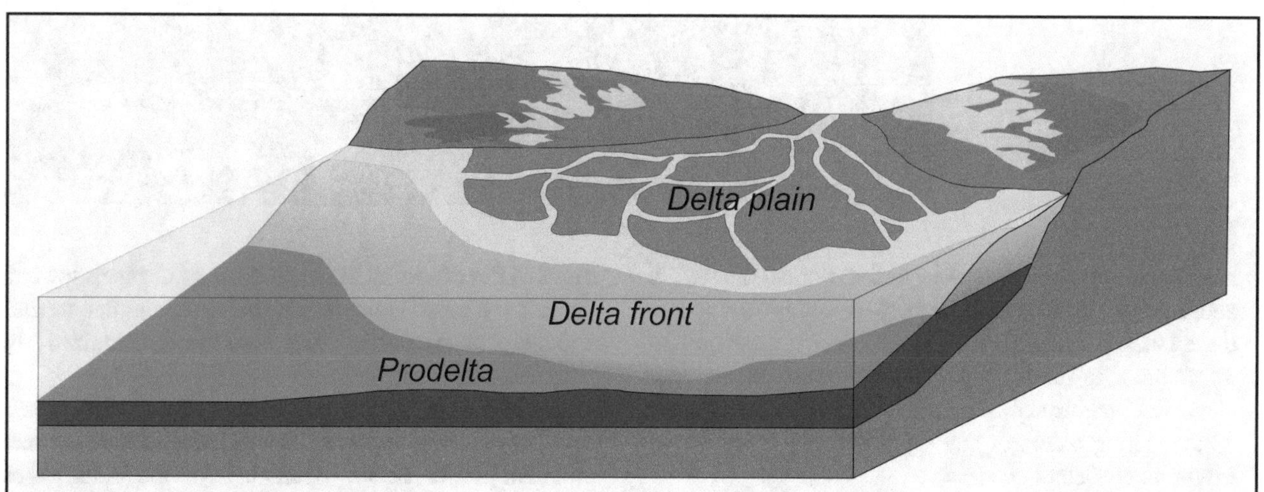

Fig. 22.6 Typical components of a delta in the emerged and submerged areas

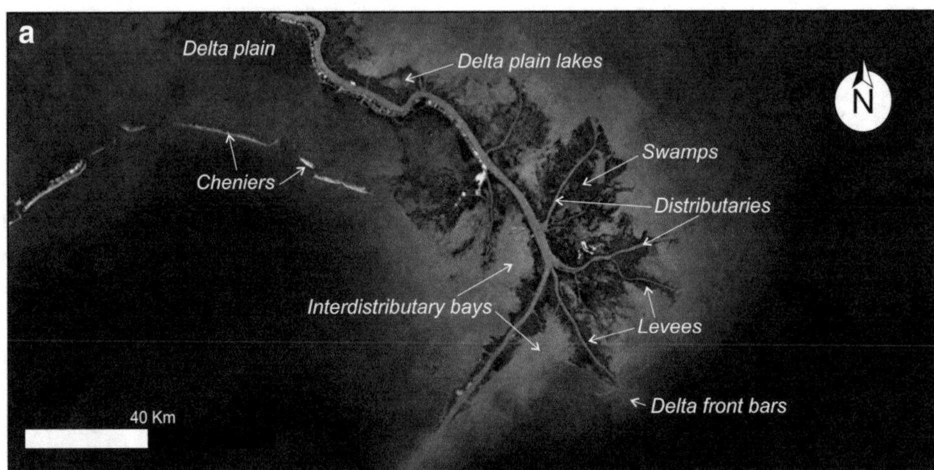

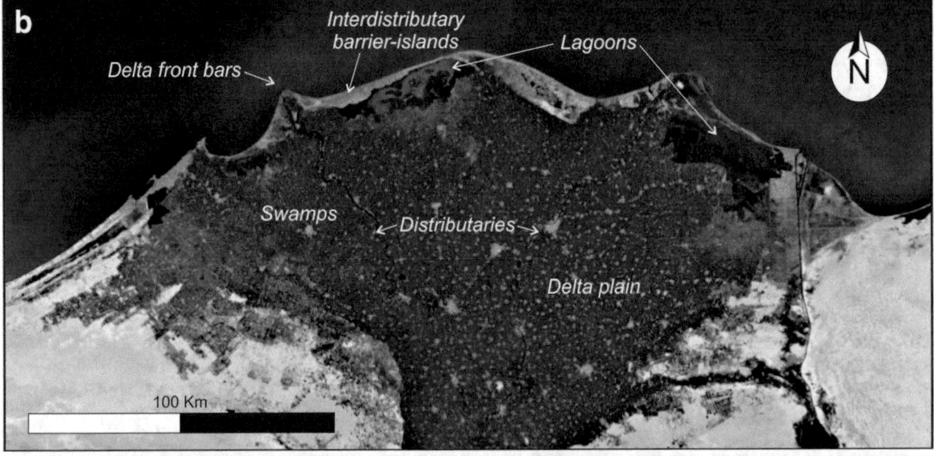

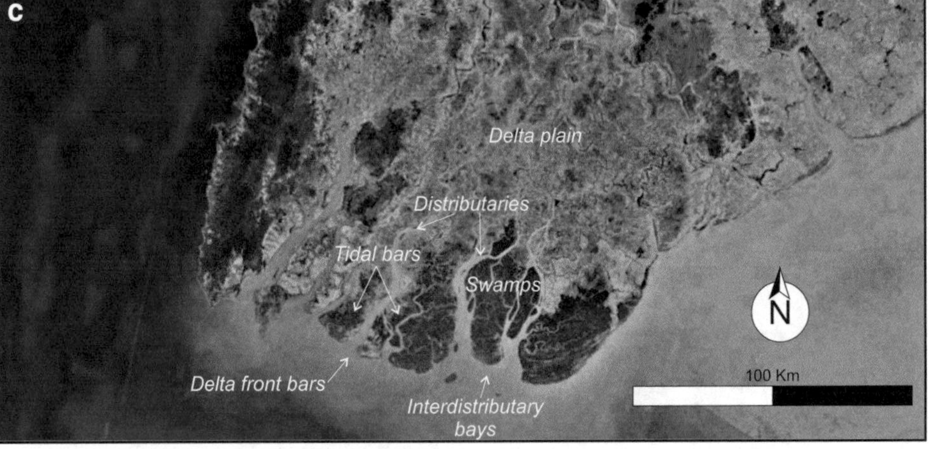

Fig. 22.7 Sub-environments linked to the delta plain of different types of delta. **a** Fluvial-dominated delta, with the example of the Mississippi River Delta, USA. **b** Intermediate river–wave delta, with the example of the Nile River Delta, Egypt. **c** Tide-dominated delta, with the example of the Irrawaddy River Delta, Myanmar. (Images Landsat/Copernicus from Google Earth.)

and stagnant waters. However, during floods, the waters and sediment distributed by the deltaic system can access these lakes.

In addition to these, other different environments may develop in the frontal zone of the delta plain in response to the marine processes that act at the delta front. These environments can include:

- *Submerged levees*: These represent the continuation of the levees of the delta plain towards the submerged zone, limiting the extension of the distributary channels towards the marine zone. However, in this case, and due to their topographic position, they are not colonized by vegetation.
- *Interdistributary bays*: These develop in deltas where waves are not an important factor, being located between different levees in the open area of the delta. Their dynamics are not directly affected by the currents circulating through the distributary channels, but are more

closely related to the movements of the open water masses.

- *Barrier island/lagoon systems*: In deltaic systems where waves are important, a large fraction of the sediments of the delta front is reworked towards the coast in the form of bars. In these cases, the bars can build barrier islands, thus closing the interdistributary bays, whose position will then be occupied by lagoons.
- *Strandplains*: In wave-dominated deltas, the entire deposit of fluvial sediment is reworked by waves to build prograding beaches. In this type of delta, the whole delta plain is made up of old beach strands and successive primary dune systems.

  At the front of the distributaries, just at the apical zone of the submerged levees, the material transported along the channels is deposited in the form of bed load. In that position, the environment that makes the delta front prograde is located:
- *Delta front bars*: These are bars made of the coarsest material of the delta. Their shape depends entirely on the dominant agent that acts on them. In river-dominated deltas, their shape is usually elongated, with an ellipse-shaped plan, since no marine agent reworks the morphology of these bars and they are preserved just as the river deposits them. When the tide is the marine agent that reworks the sediments, the elongation is accentuated, acquiring a spindle shape. Meanwhile, in deltas dominated by waves, they acquire crescentic shapes.

In deltas located in areas affected by severe storms, part of the coarse material deposited on the delta front can be remobilized by storm waves to build sandy ridges that cause them to migrate over the delta plain. These ridges are preserved in the upper part of the plain, since no other agent reaches this area during the normal functioning of the delta. These ridges are called **cheniers** and have already been described as forms that can occur in association with tidal flats.

22.4 Dynamics and Evolution

Deltas are sites of intense interaction between river and marine processes. In this dynamic, the river represents the main supply of sediment, while waves and tidal currents tend to rework all or part of the sediments coming from the river. For a delta to develop, it is necessary that the river sediment supply exceeds the capacity of waves and tides to redistribute this sediment to other coastal environments. When this occurs, the result is a net accumulation of sediment in the form of a large sediment prism. The dynamics of deltas are regulated by this interplay of processes.

22.4.1 Delta Plain Dynamics

In summary, it could be said that the dominant process in the delta plains is aggradation. All the environments linked to this deltaic domain tend to fill in until they reach the level of the maximum level of the fluviomarine floods.

As mentioned above, the sedimentary dynamics of all deltas is strongly conditioned by fluvial dynamics, since the river plays the role of main sediment supplier. It must be taken into account that this supply is subject to seasonal and interannual oscillations in river flow. We can, then, consider that deltaic construction goes through three different phases: periods of regular flow, periods of low flow and flood events. The relative duration of each of these phases is regulated by the climatic regime.

During normal regime periods, two types of process occur in distributary channels, depending on the presence or absence of tides. In non-tidal deltas, distributaries function in a similar way to river channels. Through them, sediment is transported from the river system to the open coastal waters. In these deltas, the main deltaic accretion processes occur at the delta front. In the case of tide-influenced deltas, the distributary channels behave as estuarine channels, and processes associated with water mixing occur within them. Processes such as flocculation and the concentration of suspended matter at the meeting of freshwater and saltwater masses cause the appearance of a turbidity maximum inside these channels. The dynamics of these processes have already been described in Chap. 9. Different tidal cycles cause the transit of sediments to the delta front, although a large part of the fine sediments may be trapped in the marginal areas of the delta plain.

Despite the drama they represent for humans occupying the deltas, floods are very beneficial for delta dynamics, since the highest rates of sediment delivery and the periods of greatest delta accretion occur during these times. During river floods, when the channels overflow their limits, a large part of the sediment load is immediately deposited on the banks. This is caused by the rapid loss of velocity and transport capacity that occurs during overtopping. This causes the sediments to pile up and form levees. The breaching of the levees cuts transversely through these sub-environments, producing the transport towards fans of coarse sediment that are deposited on the swamps or lakes located in their rear zone. The scale of these breaks determines the dimensions of the crevasse splays, which can range from several tens of meters to kilometer scale. The erosive zone in the levee remains depressed after the flood, so that the fan can be reactivated through this same break in successive floods, acquiring larger and larger dimensions. In interdistributary swamps, currents only act during river floods. In these environments, there is a characteristic sheet flow that is also slowed down by the presence of plants. The consequence is a loss of velocity, resulting in the deposition of finer sediments.

22.4.2 Delta Front Dynamics

Sediments that reach the mouth of the distributaries do so in the form of suspended matter and bed load. Once in open water, the path followed by both types of material is different. In both cases, the mouth of the distributary behaves as an effluent. From this effluent outlet, the suspended matter undergoes diffusion processes. These processes depend to a large extent on the vertical distribution of densities of the aqueous flow. Bates [13] described three forms of density distribution in outflows from deltaic effluents: homopycnal, hypopycnal, and hyperpycnal flows (Fig. 22.8).

In **homopycnal flows**, the densities are homogeneously distributed in the water column (Fig. 22.8a), whereas in the other two types of flow there is a vertically heterogeneous distribution. This type of flow is associated with tide-dominated deltas, where mixing of water has occurred within the distributaries and the effluent water has a homogeneous vertical distribution of suspended matter. **Hyperpycnal flows** occur when the denser waters are extruded from the lower part of the water column (Fig. 22.8b). These flows also occur in deltas with tidal influence and mixing processes in the interior of the channels in which there has also been previous settling, although they can also be associated with deltas with a thicker load concentrated in the lower part of the water column. On the contrary, in **hypopycnal flows** most of the suspended matter is located at the surface (Fig. 22.8c). This type of flow occurs when freshwater injected above the seawater flows directly out of the mouth of the channels. The outflow of turbidity above or below the water column influences a factor known as buoyancy.

In reality, sediment diffusion processes and the vertical density distribution model depend on the balance between outflow velocity, turbulent friction with the bed and buoyancy of the outflow with respect to the coastal water mass [14]. Different combinations of these three factors result in the presence of three different dispersion phenomena (Fig. 22.9): inertia-dominated dispersions (axial jets),

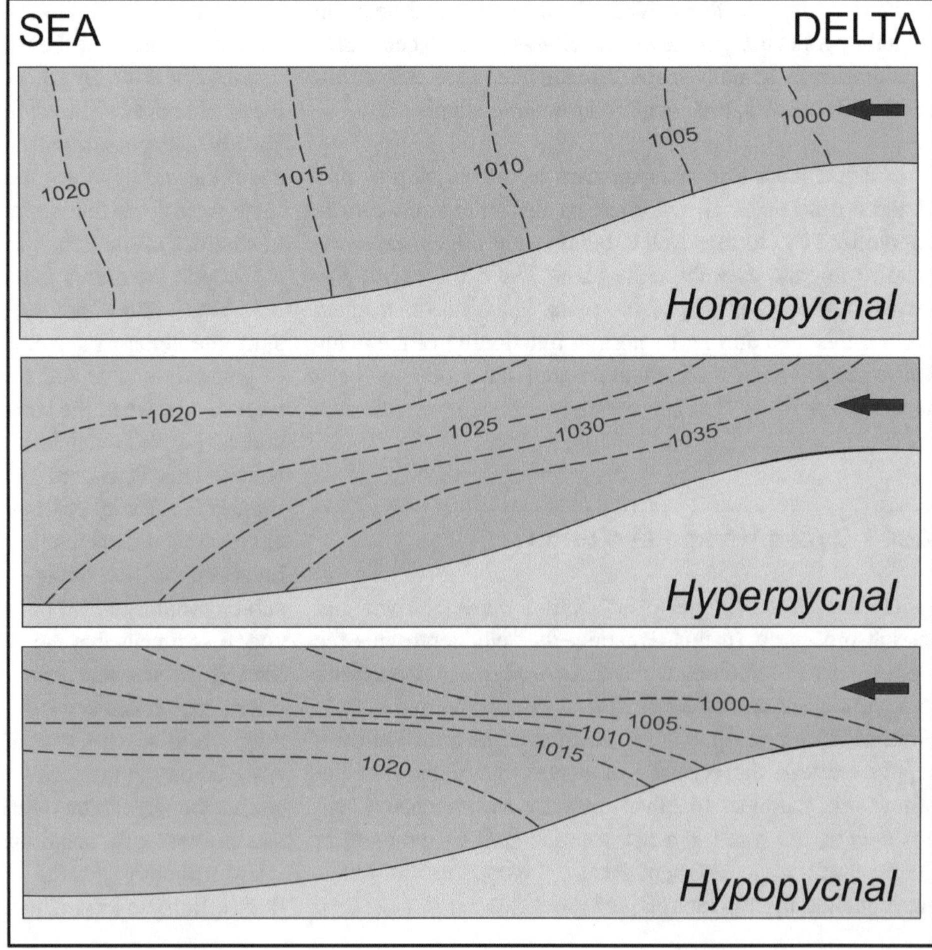

Fig. 22.8 Scheme showing the different vertical distributions of water density. **a** Homopycnal pattern. **b** Hyperpycnal pattern. **c** Hypopycnal pattern. The isopycnal units are expressed in g/cm^3 (Adapted from Bates [13])

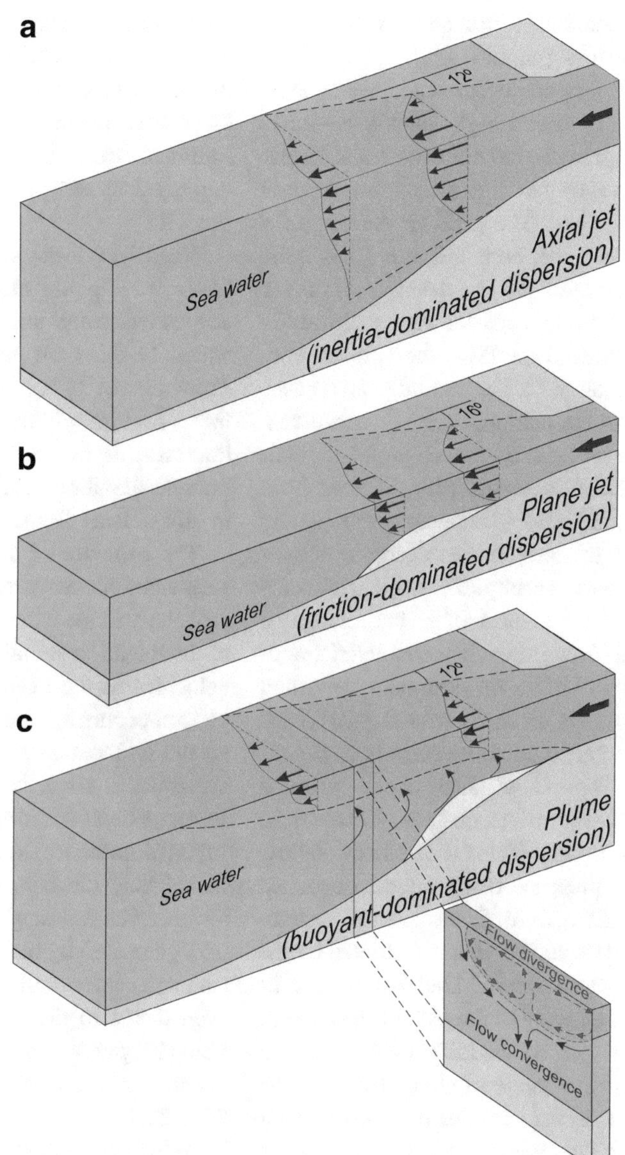

Fig. 22.9 Scheme showing the three models of dispersion described by Wright [14] and by Orton and Reading [15]. **a** Axial jet. **b** Plane jet. **c** Surficial plume

friction-dominated dispersions (plane jets) and buoyant-dominated dispersions (plumes) [15].

Inertia-dominated dispersions occur when water exits the mouth of the distributary with a high-velocity homopycnal flow and enters a deep, high-gradient delta front. Under these conditions, the buoyancy factor is absent and there is little or no friction with the bed. The axial jet consists of a turbulent flow that disperses under homogeneous conditions. In the core of the turbulent flow, eddies develop that favor the exchange between the flow and coastal waters. However, towards the sides and bottom of the flow, the velocity vectors decrease due to friction with the coastal water mass. This decrease in velocity at the margins of the flow causes a lateral extension from the outlet at a reduced angle of about 12º. The mean flow velocity decreases with distance from the effluent until it gradually disappears completely (Fig. 22.9a). The result of this progressive decrease in flow velocity is a loss of transport capacity. Thus, the sediments end up being deposited in the form of a bar elongated towards the sea. Inertial flows are also characteristic of newly created distributaries before the sediment deposited on the outflow bar causes its front to lose depth. These conditions do not usually persist for long periods of time.

Friction-dominated dispersions also occur when a high-velocity homopycnal flow arrives onshore. However, in this case, the friction process dominates as it enters shallow water, usually due to the presence of sediments deposited previously at the same mouth front. Plane jets are characterized by minimal vertical expansion that is restricted by the shallow depth of these coastal waters. Under these conditions, there is a more rapid slowing of the overall flow due to the dominance of friction with the bed. This causes lateral expansion to occur at a greater angle of about 16°. Friction with the bed also manifests in the presence of a strong shear that drastically reduces the flow velocity towards the bottom (Fig. 22.9b). At the surface, energy dissipation also occurs towards the sides, but in a more progressive manner. The result is that plane jets do not extend as far seaward as axial jets. The drastic loss of velocity has an immediate effect on sedimentation, so this type of dispersion generates bars very close to the effluent that contribute to a loss of depth and a further increase in friction.

Buoyancy-dominated dispersions occur when freshwater reaches seawater directly through the distributary with moderate or low velocity, forming a hypopycnal flow. Under these conditions, the salt wedge occurs in the frontal zone of the channel and not in its interior. Thus, turbidity plumes originate in the upper part of the water column and are dispersed at the surface (Fig. 22.9c). The lack of contact of these flows with the bed inhibits the action of friction, so that the loss of velocity of the currents to the sides and bottom of the flow occurs abruptly due to shear with the seawater. Due to this effect, the waters of the plume enter into a double convection, moving outward in a spiral shape. This difference in velocities between the two waters induces turbulence in the seawater that is in contact with the plume, favoring dispersion towards areas further from the distributary. The sediment from these plumes is dispersed seaward and deposited in areas away from the delta front, usually in the prodelta.

Negatively buoyant dispersions can also form when the water leaving the mouth of the distributary has so much suspended matter that it has a higher density than the coastal water. In this case, the hyperpycnal flow will result in the formation of a bottom-transiting plume. The behavior of these bottom plumes is similar to the surficial plumes; however, as they are influenced by friction with the bed, their deceleration occurs more abruptly, depositing their load in areas closer to the delta front.

The behavior of plumes mainly affects the dispersion of suspended matter, while that of jets also influences bed load, especially in the case of plane jets. Obviously, the deposition of bed load material is more closely linked to the flow velocities in the near-bottom shear zone than with the dispersion occurring in the water column. The velocity loss that occurs at the outlet of the distributaries, especially in the near-bottom zone, is responsible in all cases for the deposition of delta front bars. The progressive or abrupt loss of velocity influences the dimensions and morphology of the bars. Thus, the bars generated by axial jets are elongated towards the sea, acquiring a **lunate morphology**, while the bars generated by plane jets have **lobate morphologies**. The bars associated with plume-shaped flows develop in areas even closer to the mouth of the distributary and acquire a less elongated and more laterally extended morphology. These bars are commonly referred to as **bar-finger sands** (Fig. 22.10).

In the previous sections, the behavior of dispersion at the mouth of the effluent has been explained, taking into account only the energy of the river and considering the sea as a static body of water. However, in most instances, tides and waves also exert their influence, both in the dispersion pattern of suspended matter and in the development of delta front bars.

The main effect of the tide is the entry of seawater through the distributaries, contributing to mixing within them. This effect favors the action of homopycnal flows in the delta front during ebb times, which are accompanied by dispersions dominated by inertia and the development of lunate bars. However, the presence of opposing tidal currents reworks these bars during flow cycles, when seawater enters the distributaries. The action of these currents reworks the marine end of the bars and gives them a spindle-shaped morphology (Fig. 22.11). The movement of the sandy sediment in these bars is reversed during a tidal cycle, although the currents may preferentially use certain sides of the bar.

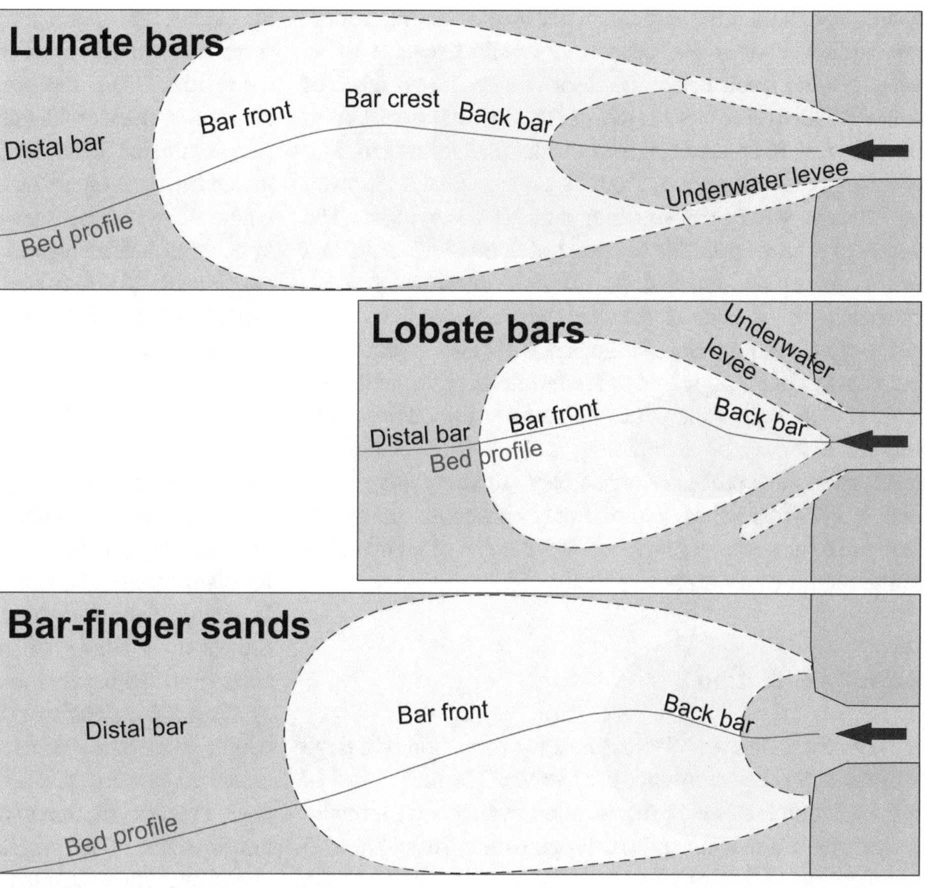

Fig. 22.10 Morphology of different types of delta front bars without wave and tide action

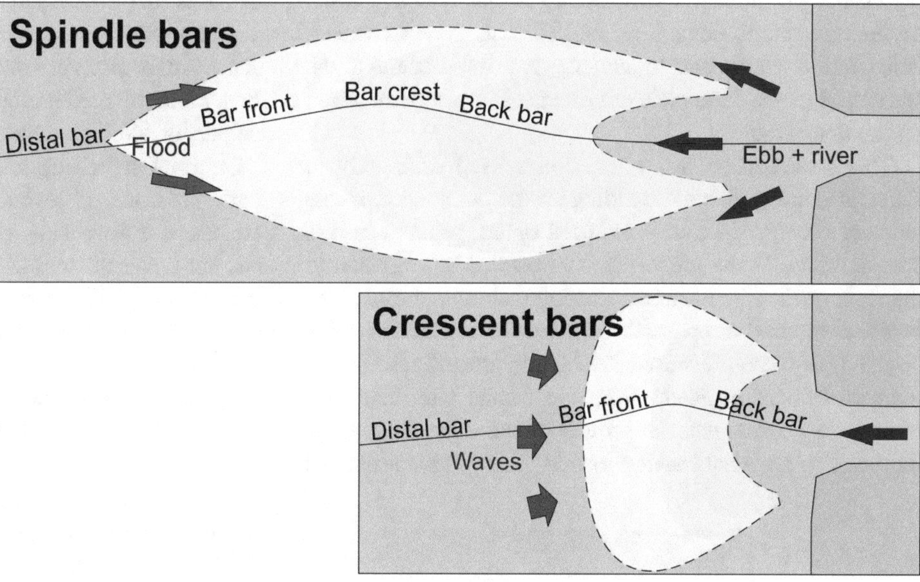

Fig. 22.11 Morphology of delta front bars under tidal and wave action

Thus, it is common for one side of the bar to be used mostly by ebb currents and the other by flood currents, although the net transport is always seaward, consummating the process of progradation.

The action of high-energy waves on distributary front effluents has an effect similar to friction with the bed [16]. On the one hand, waves hinder buoyancy by dispersing suspended matter throughout the water column. On the other hand, they shorten the dispersion distance by acting against the outflow current and causing a drastic decrease in its velocity. The result is that the dispersion in the presence of waves behaves like a plane jet. In this way, the bars tend to be deposited in an area closer to the distributary mouth. The presence of the distributary front bars represents for the swell an element where the shoaling process takes place. The action of the waves on the sediment of these bars results in a reworking of the material of the bars towards the land, especially at the ends of the bars, where the action of the outflow jet is minor. Thus, surge bars will develop parallel to the coast, which migrate towards the delta plain until they end up against its front. Via this process, the outflow bars acquire a crescentic morphology (Fig. 22.11). Being the result of similar processes, these bars are completely identical to ebb-tidal deltas, with the only conceptual difference being that they are not developed at the outlet of an inlet, but at the mouth of a distributary channel.

22.4.3 Evolution

By definition, deltas evolve towards the formation of a large sedimentary prism at the mouth of a river. The main effect of this evolution is **deltaic progradation**, which is what finally forms this sedimentary prism. Progradation, in addition to representing a seaward advance of the coastline, involves the generation of a characteristic internal structure consisting of the frontal attachment of sedimentary bodies with sigmoidal or oblique morphology (Fig. 22.12). This process causes delta front environments to advance over the prodelta at the same time as the delta plain environments are superimposed on the delta front.

This progradation pattern manifests itself differently in different types of delta, depending on the dominance of one or other process. In river-dominated deltas, progradation is best observed in the growth of bars created by diffusion at the front of the distributaries. The internal structure of these bars is composed of prograding foresets that extend over the prodelta. In wave-influenced deltas, this process results in a redistribution of sediment from the frontal bars that contributes to a homogeneous growth of the delta front environments in a typical beach dynamic. The oblique arrival of the waves transports the sediment along the coast, away from the mouth of the distributaries, and forms a continuous delta front. Thus, the progradation of the delta front takes place in a manner identical to that of the strandplains. In tide-dominated deltas, it is the superposition of tidal bars that generates a progradation mechanism. Finally, in deltas where there is a joint action of waves and tides, successive systems of barrier lagoon islands usually develop, and their attachment in the frontal zone generates a progradation–aggradation mechanism.

It may seem that a continuous flow of sediment from the river, coupled with a constant reworking of the sediment by marine agents, would generate an uninterrupted progradation of the delta. However, the progradation process is modified by the delta dynamics and the preferential action of different sectors of the distributary network may shift progradation from some segments of the delta front to others. This is the process described by Coleman Prior [18] as **delta switching**. In reality, delta switching is an effect that can be the consequence of three different causes: channel extension, channel switching and lobe switching.

Channel extension occurs as a consequence of progradation and seaward advance of the delta plain. When extension occurs, the orientation of the channels as they reach the delta front may change, causing the delta to change the direction of its progradation. When there is more than one distributary channel, the relative proportion of load between the two channels may be different. The consequence is that the deltaic sector at the front of that channel will prograde faster. This load ratio between different channels may also vary over time with the extension process. Thus, areas of preferential progradation may migrate from one part of the delta to another.

Channel switching is another mechanism that is linked to the presence of several distributaries. However, in this case, there is no change of relative load between the channels, but rather the loss of functionality of one of them. This phenomenon shifts the preferential transport to another channel, causing the progradation to migrate to the area in front of it. Normally, channel switching occurs due to internal processes of the channels in the delta plain linked to fluvial dynamics, such as plugging or capture.

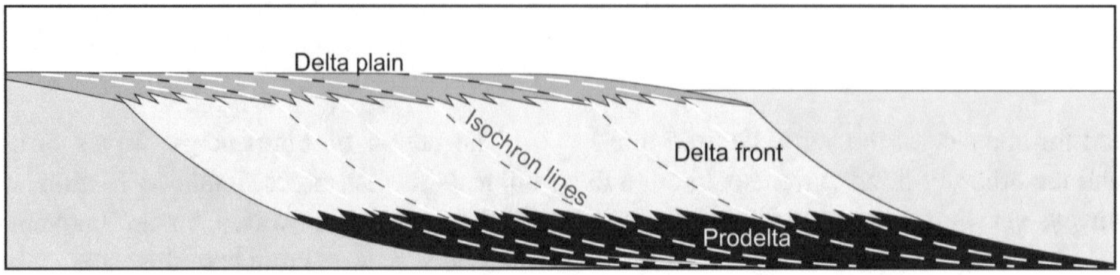

Fig. 22.12 Conceptual model of the progradational structure of a delta (Adapted from Gani and Bhattacharya [17])

Lobe switching involves the loss of functionality of complete branches of distributive channels. This process deactivates the progradation in complete areas of the delta front, activating other different areas and making the deltaic progradation occur by complete lobes.

The areas abandoned by channel switching and lobe switching processes are usually reworked and eroded by marine processes. This erosion is preserved in the record as a ravinement surface. The presence of successive processes of progradation and abandonment of deltaic lobes gives rise to the cyclicity observed in many deltas preserved in the geological record.

22.4.4 Post-depositional Processes in the Delta Front

The delta front reaches high slopes in some deltas, especially in the case of coarse sediment deltas such as fan-deltas and Gilbert deltas. In these environments, the possibility of **gravitational processes** is high. These re-sedimentation processes require three conditions: (a) a high slope; (b) low cohesion between sediment grains; and (c) a trigger for the initiation of movement. The low cohesion of the sediment can be found in a low degree of packing that may be related to a high rate of sedimentation. Movement is initiated by destabilization of sediments previously deposited at the delta front. The causes of destabilization can be diverse. They are usually associated with the action of high regime flows induced by fluvial activity, but can also be caused by the action of storms or seismic movements. The processes involved may be of different nature, depending on the density of the material that is set in motion downslope. These vary in a wide range between landslides, slumps and debris flow, which involve higher density, and other flows of lower density, such as turbidity currents.

Turbidity currents deserve special attention, as they may be responsible for the deposition of significant volumes of sandy sediment in the prodelta. A turbidity current is a dense, highly concentrated flow with negative buoyancy. This causes it to move downslope in a near-bottom fringe reaching a high velocity and high shear rate. The flow may be initiated directly at the mouth of an effluent during a negative buoyancy plume, or it may be initiated by transformation of other types of gravitational movements from sediments already existing at the delta front. In the latter case, it is common for the flow to start from the erosion of incised channels that feed the sediment supplying the flow.

The upper part of the delta front acts as a feeder for these processes, while the base of the delta front and the prodelta receive the materials eroded there. In the feeder zone, the occurrence of these processes is preserved as depressions, erosional scars and incised channels. Higher density processes do not move the material very far from the source, while lower density processes can move the material long distances. This material, especially in the case of turbidites, may be deposited as lobes in the prodelta.

Another post-depositional process that is linked to deltaic sedimentation is **subsidence**. Subsidence of the delta base due to the weight of the significant sedimentary prism generated in these systems is a common process. Sometimes subsidence is due to increased packing of the lower materials due to compaction. At other times, however, the weight may affect the bedrock, which yields and sinks under the weight. Subsidence manifests mainly in the upper part of the sequence, where different subsidence pulses can cause a transgressive effect. On the other hand, differential subsidence in some parts of the delta can cause the appearance of adaptive faults, which cut the most recent sedimentary formations.

A phenomenon related to subsidence is **diapirism**. In this case, the weight of the upper materials on the fluidized muds of the prodelta can cause them to rise and intrude diapirically into the sands of the delta front. These diapirs may even reach the surface and extrude in the form of mud volcanoes.

22.5 Depositional Facies

The depositional facies developed in each of the sub-environments present in the deltas depend to a great extent on the variability of the processes occurring in them. It is evident that these facies will depend on the type of delta, since in each of them the sub-environments present are different and so the combinations of processes in them will also be different [19].

22.5.1 Depositional Facies Characteristic of River-Dominated Deltas

In deltas dominated by river action, many of the features observed in the sub-environments are similar to those described in purely fluvial environments (Fig. 22.13).

The **distributary channels** are dominated by sandy facies that are organized in sets of seaward-sloping cross-stratifications. There are also sets of ripple-type cross-laminations, as well as scour-and-fill structures, gravel intercalations and silt lenses.

The lithofacies sequences of **delta plain swamps** are similar to those of meandering river floodplains. They consist of the superposition of fining-upwards sequences generated by the weak currents that flow across this plain during flood periods. The base of each sequence may consist of a level of medium to fine sands with parallel lamination, followed by finely laminated silts and ending with clays

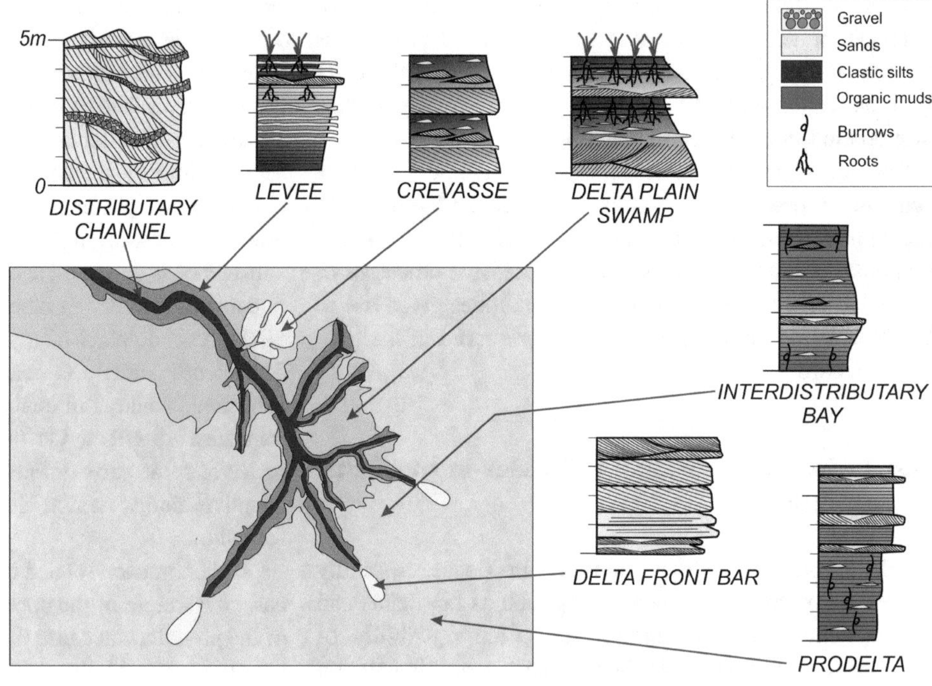

Fig. 22.13 Depositional facies generated on the different sub-environments linked to a river-dominated delta

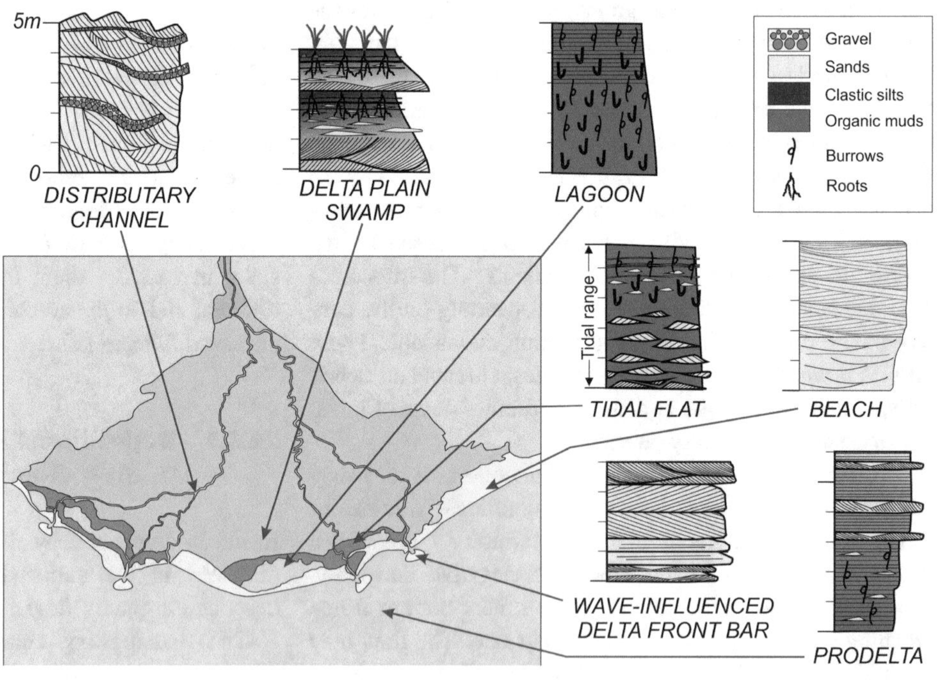

Fig. 22.14 Depositional facies generated in the different sub-environments linked to an ideal intermediate delta

marked with drying cracks. The whole sequence is usually intensely bioturbated by plant activity. Also typical is the accumulation of plant tissues that can lead to the formation of peat layers, which will be transformed into coal.

At the margins of distributary channels, environments such as levees and crevasse splays are also very similar to those present in meandering rivers. Their facies are similar as they are developed by the same processes. The facies of the **levees** are characterized by fine alternations between sands and silts, where parallel lamination dominates, although sometimes the sands may show ripple development. These sequences are also usually highly bioturbated by roots. The **crevasse splays** are organized as fining-upwards sequences that begin with a cross-bedded, fan-geometric body of sands intruded from the river. Each sequence ends with the development of a silt body that may have sand lenses in the form of flaser structures. These sand lenses disappear towards the top of the sequence, where parallel-laminated silts dominate.

At the mouth of the distributaries, **delta front bars** develop. These bars are the result of the deposition of sedimentary material due to diffusion processes. In river-dominated deltas, the bars are composed of cross-stratified sand bodies that progress towards the sea. At the base of the sequence, represented by the most distal part of the bars, there are intercalations of sand and mud.

In the area located in the delta front between the distributaries are the **interdistributary bays**. The facies developed in these bays are characteristically fine, although some levels may be interspersed with ripples developed when these environments are crossed by flows during floods. These muddy sediments usually have a high organic matter content and are greatly bioturbated by invertebrate activity.

The sediments of the **prodelta** are very similar to those of these bays. They are finely laminated silts and clays, often interrupted by sheets of sand that may enter during times when the floods extend the jets or by the action of turbidity currents. It is also common that their internal structure is disturbed by the bioturbation of marine organisms.

22.5.2 Depositional Facies Characteristic of Wave-Dominated Deltas

The deltas dominated by waves are the ones with the least variety of environments, since only distributary channels, beach ridges and eolian dune ridges exist in them. None of these environments is exclusive to deltas. The depositional facies generated in distributary channels are indistinguishable from those generated in river-dominated deltaic channels. Likewise, the facies of the sandy strands that constitute these deltas are identical to the facies characteristic of any beach environment such as those that were studied in Chap. 17 of this book (see Fig. 17.18). For these reasons, some authors consider that they are not true deltas, since they do not present diagnostic facies and their geologic record cannot be differentiated from the record of a strandplain. On the other hand, not even all the sediment accumulated in this type of delta comes from the river, since it is common for part of it to reach the delta transported by littoral drift [20].

22.5.3 Depositional Facies Characteristic of Tide-Dominated Deltas

If a tide-dominated delta is compared with a tide-dominated estuary, it can be seen that the environments present are identical: interbar channels, spindle-shaped subtidal bars and intertidal marginal bars. The only difference is conceptual, and lies in the development of sedimentary bodies inside or outside the funnel bay. In fact, some authors refer to this type of delta as an "estuarine delta" and their morphology as "estuarine form" [10]. Similarly, there is no difference between the depositional facies of these environments. Thus, the facies sequences of the tidally dominated deltas are the same as those shown in Fig. 21.14.

22.5.4 Depositional Facies Characteristic of Intermediate Deltas

There is a wide variety of typology of intermediate deltas, depending on the relative transport capacity among the main agents. The deltas in which there is a greater variety of sub-environments and a greater richness of facies are those found in the central part of [3] triangle. In the intermediate deltas, there are areas dominated by each of the agents that control delta dynamics. Thus, in the innermost zone of the delta plain there are river-dominated environments. In this zone, distributary channels and interdistributary swamps also develop, and levees, crevasses and oxbows may exist in the abandoned channels. These facies do not differ in any way from those already described in river-dominated deltas.

In the frontal zone, there are environments dominated by waves, with a succession of beach sequences that, in this case, constitute barrier island systems together with dune ridges. There are also no differences in the depositional facies with the barrier islands that appear in non-deltaic contexts. Behind these islands are located systems dominated by the tide. In this case, lagoons are installed and tend to fill in until they become tidal flats. The depositional facies characteristic of these sub-environments are also the typical lithofacies sequences that have already been described.

Here, the differences between the different types of intermediate deltas are not so much in the nature of the facies, but in their distribution and amplitude, which are reflected in the geometric scale and thickness of the sedimentary bodies that contain them.

22.6 Facies Models

As can be seen in the previous section, there are no depositional facies exclusive to deltas. The innermost facies may be mistaken for fluvial facies, while other outermost facies may be associated with a variety of coastal systems such as barrier islands, estuaries, bays or strandplains. Therefore, none of these facies can be used individually as diagnostic criteria in the identification of fossil deltas. However, deltas are complex systems and their three-dimensional architecture of depositional facies, studied integrally, can be used to recognize them in the sedimentary record. However, the dynamic complexity of deltaic systems means that there is no single type of facies architecture characteristic of all deltas. Far from it, in fact, as each particular delta seems to present a different geometrical arrangement in its facies. Nevertheless, clear facies models have been established for each of the deltas constituting the vertices of Galloway's triangle, as well as for some intermediate types. These models present a quite different architecture among them and only coincide in the occurrence of prograding mechanisms. For this reason, the geometry of prograding mechanisms has been widely used in the geological literature.

22.6.1 Facies Model for River-Dominated Deltas

The absence of waves and tides that rework the sandy sediment at the delta front, the large number of distributaries and the ease of occurrence of switching phenomena give the facies architecture of river-dominated deltas an elongated geometry and a characteristic bird's-foot shape (Fig. 22.15).

In this model, distributary channel facies appear elongated in divergent directions, flanked on both sides by levee facies. These levee facies extend underwater in the frontal zone of the channels, guiding the currents outward, where dispersion phenomena occur. It is there that the delta front bars are formed.

The facies model shows numerous bar-finger sands. The arrangement of the sandy bodies within these bars clearly shows the prograding mechanism characteristic of these deltas. These bars overlie the thin facies of the prodelta and are laterally bounded also by thin facies, in this case of the interdistributary bays. The fine facies of the bays may be interrupted by sandy bodies that laterally connect with the distributary channels. These are the crevasse splay facies. Over all these facies are superimposed the marsh or mudflat facies characteristic of the delta plains.

22.6.2 Facies Model for Wave-Dominated Deltas

Wave refraction phenomena lead to the reworking of the delta front sands landward, to form prograding beaches on the marine side of the delta. This process gives the delta a characteristic cuspate geometry and a very simple facies architecture consisting of the coalescence of beach ridge bodies. It is this amalgamation of sandy bodies that shows the prograding mechanism in this type of delta (Fig. 22.16).

Each of these beach ridges may be overlain by foredune ridge facies. The presence of numerous foredunes instead of parabolic dune fields is the superficial diagnosis of the progradation. On the other hand, in these deltas the number of distributaries is small, often presenting a single channel that connects the river with the open coast crossing the entire bar system. In this type of delta, the channel switch phenomenon is frequent, which is reflected in the presence of abandoned channel facies that also cross the system. These old channels are observed on the surface as scars that cut the dune system. At the mouth of the distributaries, crescentic bars develop as a result of wave-modified spreading phenomena. However, these bars are not usually preserved and are completely destroyed by waves when a channel is abandoned.

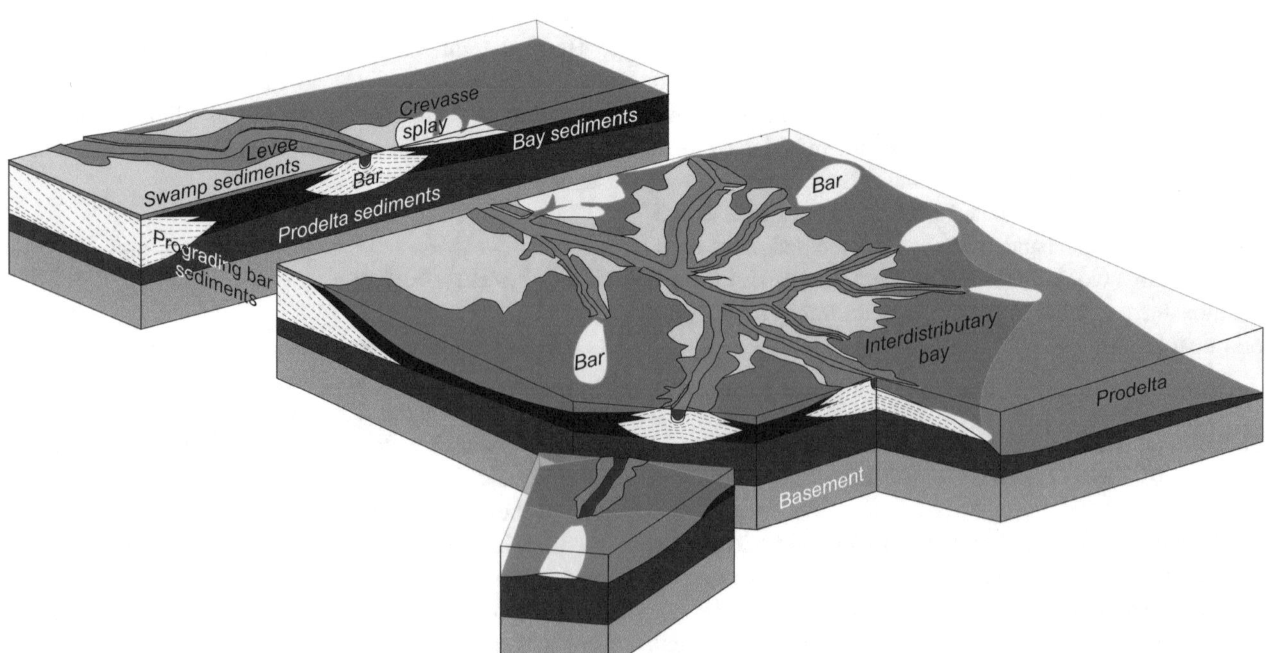

Fig. 22.15 Facies model for a river-dominated delta (based on Reineck [21])

Fig. 22.16 Facies model for a wave-dominated delta (Adapted from Weise [22])

It should be noted that in this model the prodelta facies are absent, with the delta front bars migrating directly over the shoreface facies. The cause lies in the action of the waves on the dispersion phenomena, since the finer particles are washed by the waves and transported to other marine environments.

22.6.3 Facies Model for Tide-Dominated Deltas

Where tidal currents dominate the deltaic system, the distributaries form a network of rejoining channels separated by spindle-shaped bars oriented in the tidal direction. The characteristic sandy facies of these channels are interbedded between the bar facies (Fig. 22.17). Progradation causes these bars to grow above the muddy prodelta sediments. In the innermost part of the delta, tidal flat facies overlying the bar facies are characteristic. These tidal flats may be filling in some of the channels, which are then abandoned and closed, causing the more evolved bars to coalesce. Swamp or salt marsh facies cover the tidal flats when their topographic level reaches the high tide level. Filled channels are usually observed on the surface as geomorphological scars in the marshes. The abandonment of channels in the inner parts of the system gives rise to lobe switching phenomena, which are very common in this type of delta. These phenomena are also reflected in the progradation cycles of the different parts of the system.

The facies model also shows the connection of the deltaic system with the former estuarine system located in the inner part of the river valley. The base of the sedimentary fill in this part of the system is composed of fluvial sediments corresponding to the lowstand period and estuarine accretionary bodies (estuarine type) developed during the estuarine period preceding the development of the delta.

22.6.4 Facies Model for Intermediate Deltas

The intermediate deltas in which the three main processes (rivers, tides and waves) act together present very complex facies architecture (Fig. 22.18). This scheme shows the frontal advance of the delta based on the attachment of sandy facies bodies of beach ridges that form barrier islands separated by distributary channels. Unlike the wave-dominated delta model, each of these ridges is separated from the previous one by tidally dominated depositional facies sequences.

Each of the wave-dominated bodies shows internal evidence of progradation through the attachment of new bars. However, the rear of these bodies show landward fingerings as evidence of sandy body migration in this direction. The upper part of these sandy fingerings corresponds to the development of washover fans. Coastal dune facies are developed over the entire body.

Between different barrier islands, distributary front bars develop due to dispersion phenomena, which are of the plane jet type due to wave action. The bars acquire a crescentic morphology that is very similar to that of ebb-tidal deltas. As in wave-dominated deltas, these bodies are not usually preserved, being reworked by the waves to build the beach ridges that accrete the barrier islands.

A tidally dominated regressive sequence is present in the back zone of each of these sandy bodies. In this sequence, the vertical succession of depositional facies shows how the system progressively somerizes as a result of sediment aggradation. At the base of the sequence are the subtidal lagoon facies, which progressively change to tidal channel as the system narrows. Over these facies, wide tidal plains develop and are covered by marsh or mudflat facies. This entire tidal system is crossed by a dense network of tidal creeks that flow into the main distributary channels of the delta.

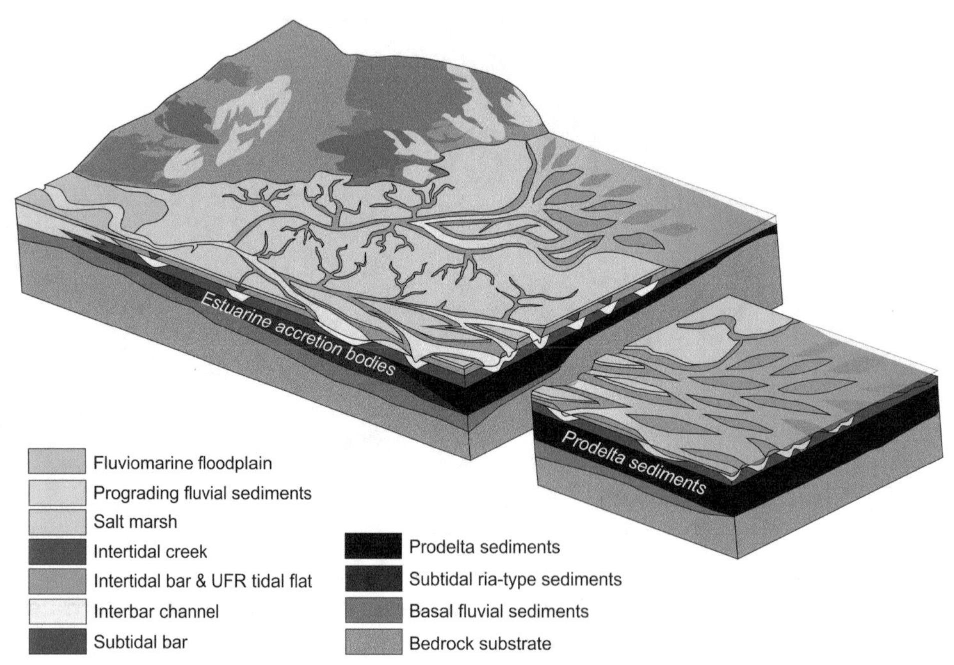

Fig. 22.17 Facies model for a tide-dominated delta (Based on Maguregui Tyler [23])

Fig. 22.18 Facies model for a river–wave–tide intermediate delta (Based on [24]). The pattern shown in this facies model is repetitive in time, which results in a progradational setup of the system

River-dominated facies develop only in the innermost parts of the system, where fluvial sediment fills a smaller number of distributaries, and lateral bars, levees, crevasses and fluvio-marine floodplains similar to those described in elongated deltas may develop.

22.6.5 Facies Model for Coarse-Grained Deltas

Gilbert-type fan-deltas and other coarse-grained deltas do not present a delta plain in the strict sense, since their emerged part is a high-slope zone that connects an alluvial

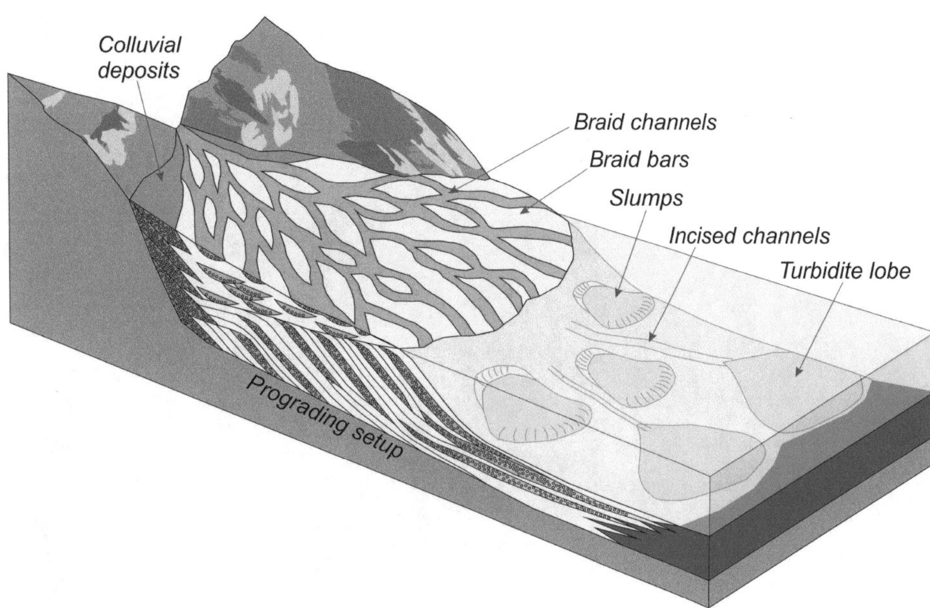

Fig. 22.19 Facies model for Gilbert-type deltas (Adapted from Nemec [25])

fan or a braided river with the coastline (Fig. 22.19). In this emerged part, the typical facies pattern of a braid plain is developed, with interlocking channels separated by bars. These channels have a high capacity for avulsion and are filled with mixed gravel facies during the action of channelized or sheet flows. These delta sectors may also be affected by destabilization processes such as landslides and slumps. Fan-deltas and Gilbert deltas differ in the relative extent of the emerged zone, as well as in the slope. Gilbert deltas have a lower slope and a greater extension in relation to the submerged part.

The delta front usually develops the prograding mechanism characteristic of deltas. In this case, the prograding bodies are made up of alternations of gravels, sands and silts from the jets coming from the emerged part during confined and unconfined flows. This prograding structure is usually interrupted by erosional scars formed by the onset of mass movements such as submarine landslides, slumps and debris flows, as well as by incised channels formed by turbidity currents.

The prodelta usually consists of fine sediments interrupted by lenses of coarser facies generated by turbidity currents. Turbiditic deposits have the characteristic Bouma Sequence within them. This type of deposit occurs less frequently in fan-deltas, whose facies are dominated by debris flow deposits.

Advanced box. 22.1 Isopach distribution

In present-day environments, the different types of delta are easily recognized by their elongate, cuspate, lobed or intricate morphology. However, in the sedimentary record the recognition of delta types is not so simple. The facies architecture can be a clear indication, but in the intermediate types there is such a variety that it is can also be very difficult to identify them. One of the criteria for recognition of delta type was proposed by Coleman and Wright [26] and consists of the study of the thickness arrangement of the sedimentary bodies. These thicknesses are reflected in their isopach maps (Fig. 22.20).

In river-dominated deltas, the isopachs are arranged in a digitate shape, although in this case the design resembles the leg of a gecko more than that of a bird (Fig. 22.20a). A totally opposite form is show by the wave-dominated deltas, where the thicker strips are arranged parallel to the coast (Fig. 22.20b). At the third vertex of Galloway's triangle, tidally dominated deltas present a distribution of isopachs perpendicular to the coastline, with a shape reminiscent of a baseball bat (Fig. 22.20c).

In addition to the three delta types that occupy the extremes of the triangle, Coleman and Wright distinguished three more intermediate types. In this sense, when there is a mixture between fluvial transport and wave reworking, two types are distinguished according to the greater or lesser relative importance of the wave compared with the volume of sediment contributed by the river. When the volume of fluvial sediments cannot be completely reworked by the waves, the innermost part of the system maintains the orientation of maximum thickness parallel to the river, while at the delta front the isopachs become curved with a preferential elongation parallel to the coastline (Fig. 22.20d). When the wave reworking capacity equals the fluvial input, the thicknesses are distributed in a more regular way, and then the isopachs acquire a more lobulate morphology (Fig. 22..22.20e). In the center of the triangle is the mixed model controlled equally by river, tide and waves. In this case, the isopach pattern is distributed by zones, being fingered in the innermost sector and intermediate between types B and C in the central and frontal zones (Fig. 22.20f).

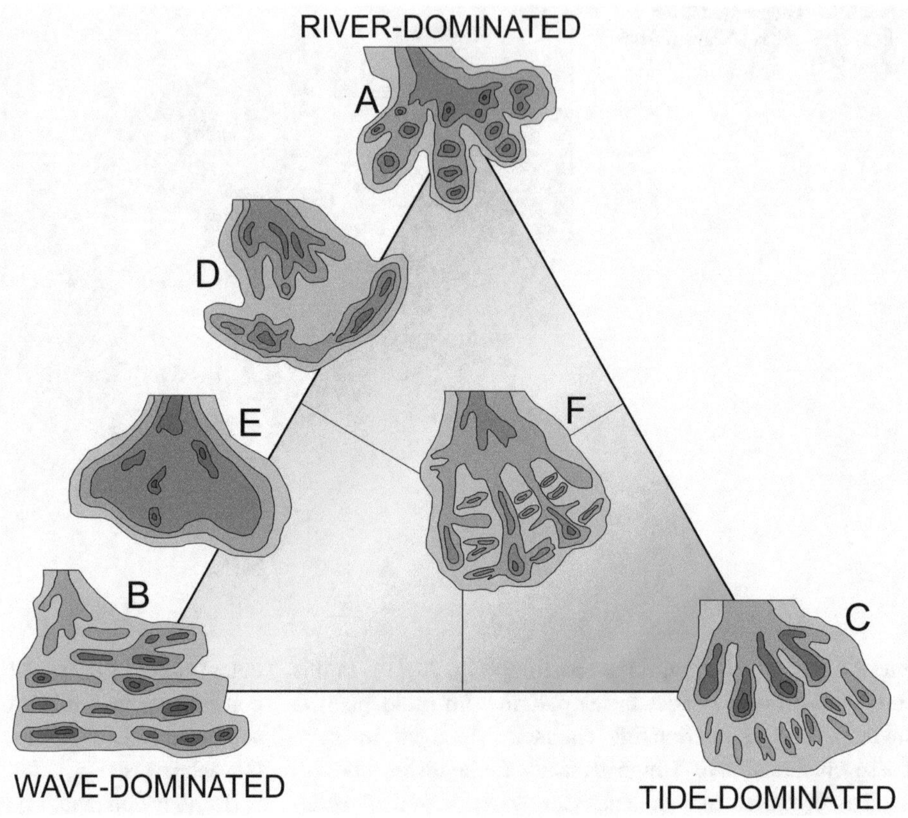

Fig. 22.20 Isopach distributions of different types of delta (Adapted from Bhattacharya [27])

Coleman and Wright's study of the distribution of isopachs actually shows a continuous spectrum, depending on the relative work capacity among the three main agents, in which the six types described are merely identifiable patterns.

22.7 Delta Cycles

All deltas go through successive cycles. From a physiographic point of view, each of these cycles involves the succession of a phase of extension of the delta plain, followed by another phase of reduction of its surface. From a sedimentary point of view, however, they can have different origins. In general terms the cycles can be classified into autocycles and allocycles. Autocycles are controlled by changes in the dynamic conditions that determine sedimentation in the delta, while allocycles are imposed by external conditions such as relative sea level movements caused by the combination of eustatic variations and subsidence.

In the autocycles, the most important factor is the sedimentary contribution in relation to two variables: the accommodation space and the wave reworking capacity. Each cycle is composed of a prograding phase and a retrograding phase. The prograding (regressive) phase occurs when the volume of input exceeds the levels of accommodation and erosion at the delta front (Fig. 22.21a). Conversely, during the retrograding (transgressive) phase, the eroded volume exceeds the sedimentary input and the delta front retreats, causing the delta plain to lose extension (Fig. 22.21b). The result is that the regressive phase develops the prograding mechanism described in the previous sections, while the transgressive phase develops discordant surfaces that separate different prograding units (Fig. 22.21 c). Thus, each unit developed during one of these cycles is composed of a set of obliquely arranged strata, where progradation dominates over aggradation. This type of cycle was described by Frazier [28] in present-day deltas, and identified in numerous fossil deltas.

It should be noted that these cycles can occur on a delta-wide scale when variations in input volume are due to a general cause that affects the entire delta. The scale of these cycles is in the order of thousands of years. Lower order autocycles may occur in specific delta lobes due to the switching mechanism. When this second case occurs, the prograding phase of one lobe coincides with the retrograding phase of another. The timescale of these cycles is in the order of hundreds of years.

Allocycles are more complex, since they are due to phenomena unrelated to the input and processual dynamics of the delta (Fig. 22.22). Thus, allocycles are imposed by subsidence variations or by sea level movements (or a

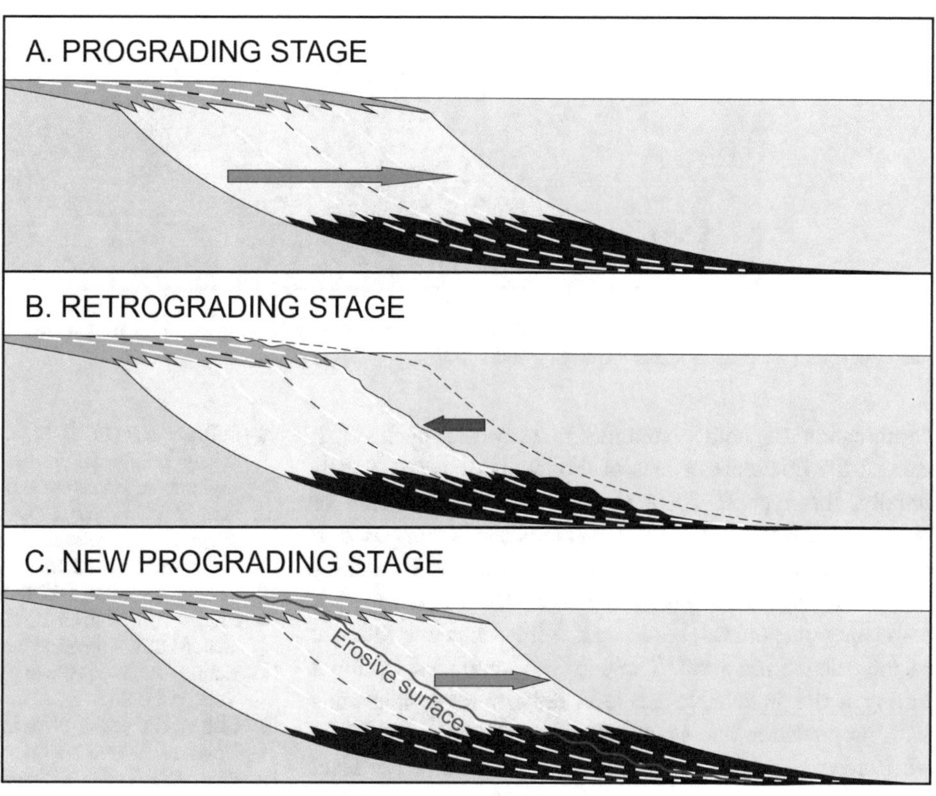

Fig. 22.21 Stages of autocycles in a delta lobe

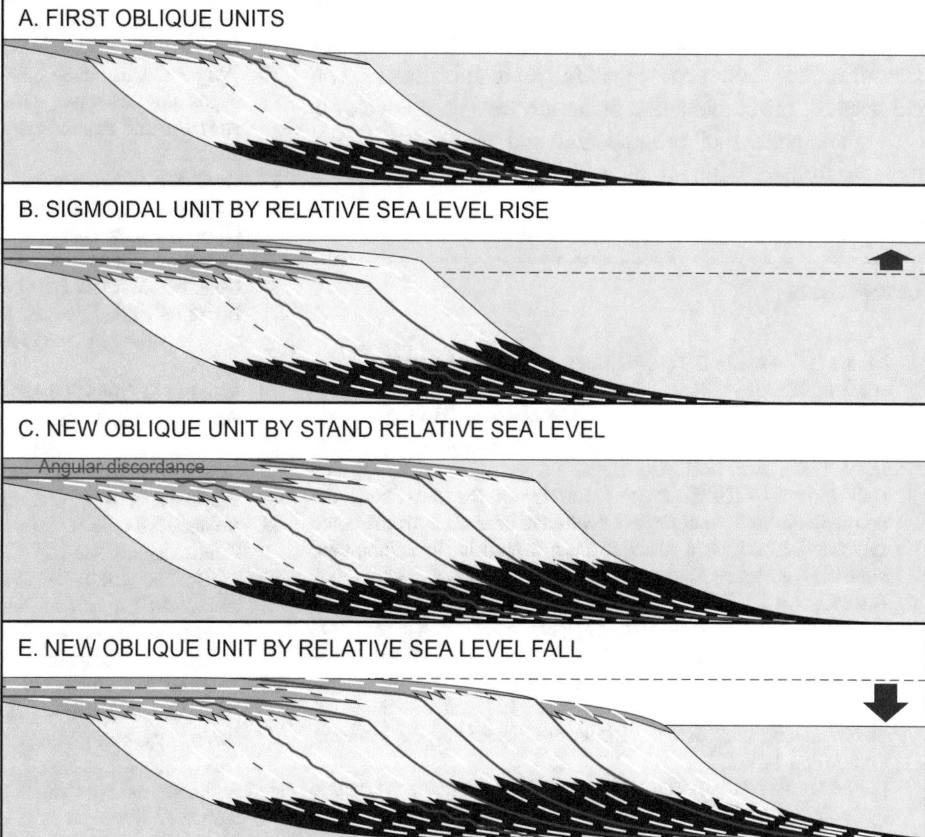

Fig. 22.22 Possible stages of allocycles in a delta lobe

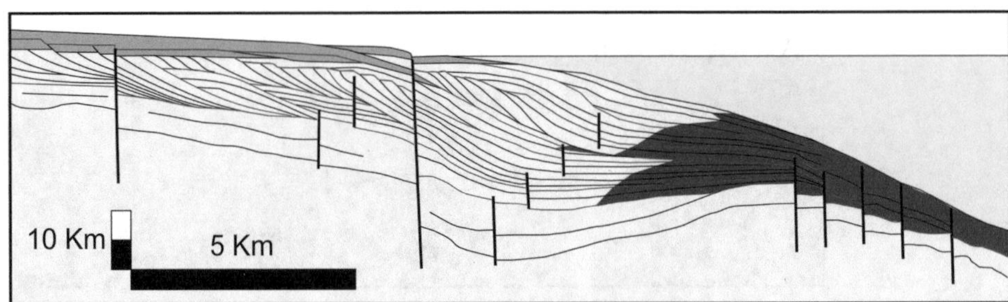

Fig. 22.23 Seismic profile across the Brazos River Delta (Texas, USA) showing the different units resulting from the combination of autocycles and allocycles (Adapted from Bhattacharya Walker [27])

combination of both phenomena). They are cycles that exceed the timescale of tens of thousands of years. Superficially, this type of cycle also involves an extension or reduction of the extent of the delta plain, and is composed of a regressive and a transgressive phase. However, the causes of the transgressive phase are not due to erosion or reworking phenomena, but to marine invasions due to a rise in the relative sea level. Thus, cycles dominated by input during a rise in relative sea level develop sigmoidal units with progradation and aggradation mechanisms (Fig. 22.22 b). Conversely, cycles generated during a relative sea level decline develop oblique units where there is only progradation and the upper boundary of the sedimentary bodies composing the unit is increasingly lowered (Fig. 22.22d). Contrary to what is observed in the autocycles, the units generated by different allocycles are separated by angular unconformities and corresponding paraconformities. The end result of the combination of autocycles and allocycles is a complex pattern of progradation and progradation that may be further modified by subsidence and affected by faulting (Fig. 22.23).

References

1. Moore GT, Asquith DO (1971) Delta: term and concept. Geol Soc Am Bull 82:2563–2568
2. Fisher WL, Brown LF Jr, Scott AJ, McGowen JH (1969) Delta systems in the exploration for oil and gas: a research colloquium: Univ. Texas, Bur. Econ. Geology
3. Galloway WE (1975) Process framework for describing the morphologic and stratigraphic evolution of deltaic depositional systems. In: Broussard ML (ed) Deltas: models for exploration. Houston Geological Society, Houston, Texas, pp 89–98
4. Bhattacharya JP (2006) Deltas. In: Posamentier HW, Walker RG (eds) Facies models revisited. SEPM Special Publication 84:237–292
5. Inman DL, Nordstrom CE (1971) On the tectonic and morphological classification of coasts. J Geol 79:1–21
6. Elliott T (1986) Deltas. In: Reading HG (ed) Sedimentary environments and facies. Blackwell Scientific Publications, Oxford, pp 113–154
7. Fairbridge RW (1980) The estuary: its definition and geodynamic cycle. In: Olausson E, Cato I (eds) Chemistry and biogeochemistry of estuaries. John Wiley and Sons, New York, pp 1–36
8. LeBlanc RJ (1975) Significant studies of modern and ancient deltaic sediments. In: Broussard ML (ed) Deltas: models for exploration. Houston Geological Society, pp 13–85
9. Scruton PC (1960) Delta building and the deltaic sequence. In: Shepard FP, Phleger FB, Andel TH (eds) Recent sediments Northwest Gulf of Mexico. American Association of Petroleum Geologists, Tulsa, Oklahoma, pp 82–102
10. Briggs D, Smithson P, Addison K, Atkinson K (1977) Fundamentals of the physical environment. Routledge, London
11. Holmes A (1965) Principles of physical geology. Ronald, New York, p 1288
12. Gilbert GK (1885) The topographic features of lake shores. US Government Printing Office, Washington, p 123
13. Bates CD (1953) Rational theory of delta formation. Am Asso Petrol Geol Bull 37:2119–2162
14. Wright LD (1977) Sediment transport and deposition at river mouths: a synthesis. Geol Soc Am Bull 88:857–868
15. Orton GJ, Reading HG (1993) Variability of deltaic processes in terms of sediment supply, with particular emphasis on grain size. Sedimentology 40:475–512
16. Wright LD, Thom BG, Higgins R (1980) Sediment transport and deposition at wave dominated river mouths: examples from Australia and Papua New Guinea. Estuar Coast Mar Sci 11:263–277
17. Gani MR, Bhattacharya JP (2005) Bedding correlation versus facies correlation in deltas: lessons for quaternary stratigraphy. In: Giosan L Bhattacharya JP (eds) River deltas: concepts, models and examples. SEPM, Special Publication, vol 83, pp 31–47
18. Coleman JM, Prior DB (1982) Deltaic environments of deposition. In: Scholle PA, Spearing D (eds) Sandstone depositional environments. American Association of Petroleum Geologists, pp 139–178
19. Wright LD (1985) Deltas. In: Davis RA (ed) Coastal sedimentary Environments, 2nd edn. Springer, Heidelberg, pp 1–76
20. Reading HD, Collinson JD (1996) Clastic coasts. In: Reading HG (ed) Sedimentary environments. Processes, facies and stratigraphy. Blackwell Science, Oxford 704p
21. Reineck HE (1970) Marine sandkörper, rezent und fossil. Geologie Rundschau 60:302–321
22. Weise BR (1980) Wave-dominated delta systems of the upper cretaceous San Miguel Formation, Maverick Basin, South Texas. Bureau of Economic Geology, University of Texas at Austin, Report of Investigations, vol 107, 39p
23. Maguregui J, Tyler N (1991) Evolution of Middle Eocene tide-dominated deltaic sandstones, Lagunillas Field, Maracaibo Basin, western Venezuela. In: Miall AD and Tyler N (eds) Three-dimensional facies architecture of terrigenous clastic sediments and its implications for hydrocarbon discovery and recovery. Concepts in Sedimentology and Paleontology, vol 3. SEPM, Tulsa (USA), pp 233–244

24. Morales JA (1997) Evolution and facies architecture of the mesotidal Guadiana River delta (S.W. Spain-Portugal). Mar Geol 138:127–148
25. Nemec W (1990) Aspects of the sediment movement on step delta slopes. In: Colella A, Prior DB (eds) Coarse grained deltas. IAS Special Publication 10. Blackwell Science, Oxford, pp 29–73
26. Coleman JM, Wright LD (1973) Formative mechanisms in a modern depocenter. In: Stratigraphy and petroleum potential of Northern Gulf of Mexico, Part II. New Orleans Geological Society Seminar, pp 90–139
27. Bhattacharya JP, Walker R (1992) Deltas. In: Walker RG, James NP (eds) Facies models: response to sea-level change. Geological Association of Canada, pp 157–177
28. Frazier DE (1974) Depositional episodes: their relationship to the quaternary stratigraphic framework of the Northwestern Portion of the Gulf Basin. Texas Bureau of Economic Geology Geological Circular 74–1, Austin

23 Chemically and Biologically Controlled Systems: Carbonate Coasts and Reefs

23.1 Introduction

The previous chapters have described how chemical and biological processes are active on all coasts, although it is the physical processes that determine the patterns of morpho-sedimentary evolution. However, on many coasts, chemical and biological processes exert the main control of sedimentation and, therefore, are directly responsible for coastal morphology. Carbonate coasts and reefs, as part of these systems, are the best example of coasts controlled by chemical and biological processes [31].

Carbonate shores are located in the warm, clear waters typical of tropical seas. These are coasts that are associated with platforms where there is elevated carbonate production. Chemical processes are rarely solely responsible for this high carbonate production, so, in most cases, organisms are responsible for inducing these chemical processes as part of their biological activity. In particular, reefs often appear as the main element of carbonate coasts and are the clearest example of shorelines built by biological activity (e.g., [3, 21, 33]). This type of coast is not only associated with the shallow areas of continental shelves, but also constitutes the typical shoreline of many oceanic islands in areas located below 30° at both latitudes (Fig. 23.1).

In the past, there were many types of organism capable of building reefs, but today corals are the most common species that make up these formations. Coral reefs are large structures of limestone secreted by these marine invertebrates to build their skeleton. In fact, each of these organisms is not capable of producing more than a few grams of limestone; however, as they are colonial organisms, together they produce large quantities of rock. The coral colony lives in a thin layer on the surface of the structure, settling their skeletons on layers of older dead corals. In this way, the coral structure extends to the surface, out to sea and into deeper water to form large, complex systems. Although it is the corals that form the solid structure of the reef, they would not be able to perform this task without the presence of algae. The tissues of reef-building corals maintain a symbiotic relationship with a special type of algae. In this relationship, the algae supply photosynthetic material to the coral, and in turn the coral provides shelter to the algae and feeds them with nutrients contained in their metabolic wastes.

In addition to coral reefs, there are other biogenic structures that are not related to corals. A particular type of reef is generated by colonies of tubeworms. Serpulids and sabellarids are two types of worms that build important reef structures by attaching themselves to the rigid tubes in which they live. Serpulids build their calcareous tubes directly from secretions, while sabellarids use their secretions to build tubes by agglomerating sand grains and shell fragments. These organisms begin the construction of their structures on a coherent bed (rocky or cohesive) that provides a substrate for anchoring. As in the case of corals, the attachment of the tubes of the new organisms to the abandoned structure of dead worms builds the reef. Worm reefs are also more characteristic of tropical climates.

A third type of reef common today is the oyster reef. These reefs are completely different from the previous two. While oysters prefer turbid and brackish water conditions, they can also form their reefs in temperate waters. These conditions mean that oyster reefs can be associated with siliciclastic environments such as bays, lagoons and estuaries.

In the context of shallow carbonate environments, reefs can occur in different locations. While in many cases reefs are found on the coast itself, in other cases they are found in offshore positions and extend into deeper areas. Thus, we can find reefs on the coastal fringe, but also in the form of offshore barriers (on the outer edge of rimmed platforms), in shallow patches of carbonate platforms, or in oceanic atolls. In this chapter we will deal with reefs in general, including those found in areas that are not strictly coastal.

Although reefs are the primary environments on carbonate shores, there is a suite of environments associated with reefs or that constitute non-reef carbonate shores. Among these environments are lagoons, tidal deltas, tidal flats, mangroves and coastal sabkhas.

J. A. Morales, *Coastal Geology*, Springer Textbooks in Earth Sciences, Geography and Environment,
https://doi.org/10.1007/978-3-030-96121-3_23

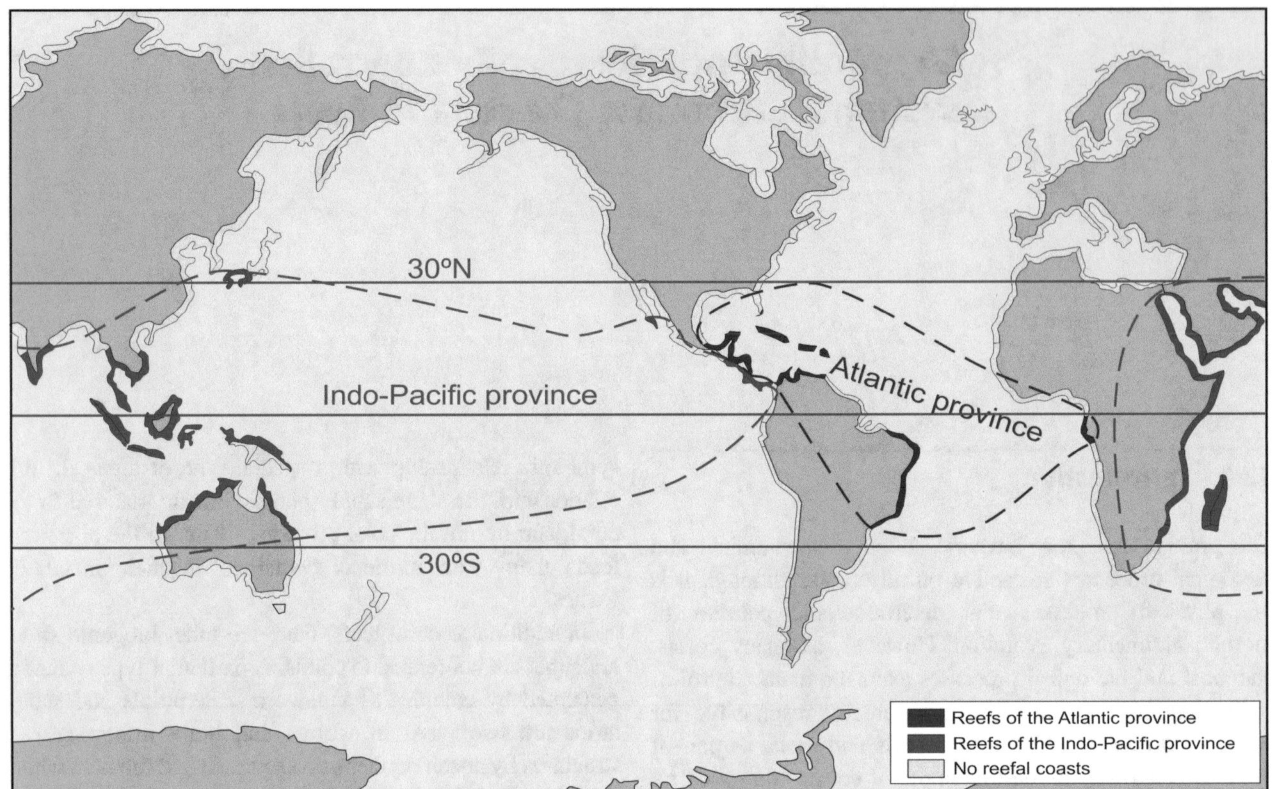

Fig. 23.1 Global location of the main carbonate and reef coasts, showing the biological provinces (Adapted from Davies [4]). Many oceanic islands in both provinces also present carbonate and reef coasts

23.2 Control Factors

The environments present in carbonate environments are controlled by the distribution of organisms capable of generating carbonates; however, this distribution is in turn controlled by the interaction of several physicochemical (temperature, salinity, nutrient concentration and water clarity) and sedimentary (substrate type and volume of terrigenous inputs) factors and is strongly influenced by wave and tidal activity (Fig. 23.2).

23.2.1 Organic Activity

On carbonate coasts, the existence of reef-producing organisms controls the distribution of other physical processes such as tides and waves. Corals and calcareous algae are responsible for building the reef structure. The presence of this structure provides a hard substrate for anchoring other organisms, but also divides the coastal zones, as the reef structure protects the back areas from direct wave action. The skeletons of these organisms are also an important source of sediment for the rest of the carbonate coastal areas. Wave action removes fragments of the coral structure that then become part of the sandy fraction to be redistributed to other environments. Other non-reef-forming carbonate shell macrofauna, such as echinoids, mollusks, barnacles and bryozoans, can also be a source of bioclastic sediment when their remains are fragmented and redistributed by waves. Similarly, the shells of planktonic and benthic microorganisms, such as foraminifera, ostracods and coccolithophorids, can become carbonate grains. In restricted environments, diatoms and radiolarians are a source of siliceous grains.

In the aforementioned cases, organisms contribute to the generation of sedimentary bodies that give the environment a prograding nature, extending the coasts towards the sea. Although some organisms exert an erosional effect by excavating the sediment or perforating the rocks, their action is of minor importance. In this sense, some types of fish also extensively corrode the reef structure; however, this action contributes to the generation of large volumes of carbonated sandy sediment [10].

23.2.2 Physicochemical Controls

There are several physicochemical parameters that control both the primary production of carbonates and the distribution of organisms in carbonate coastal areas. Among them

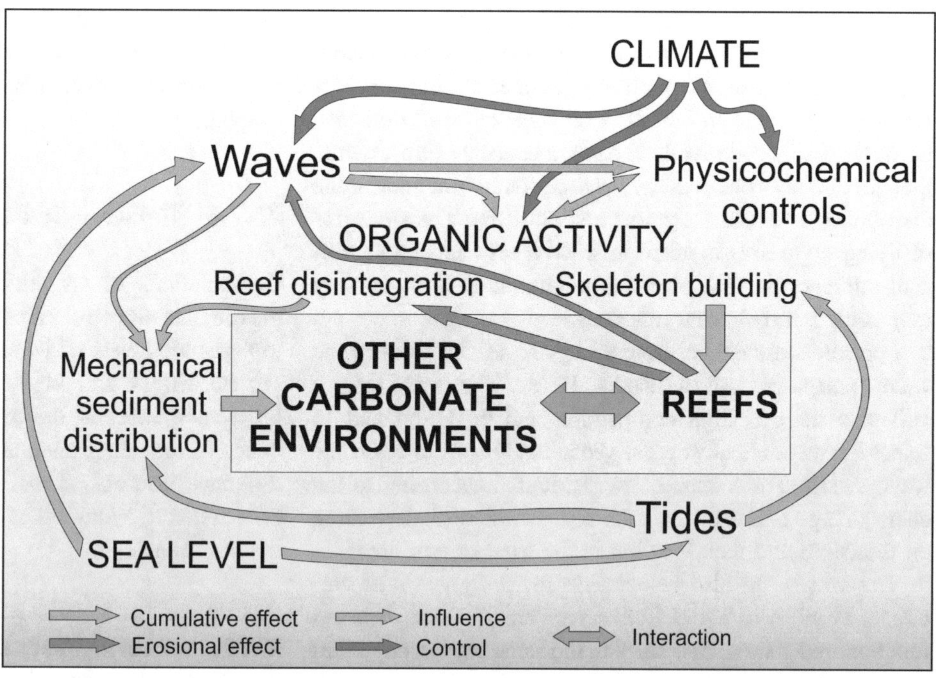

Fig. 23.2 Factors influencing the development of carbonate coasts and reefs

are: temperature, salinity, oxygenation, water clarity and nutrient concentration.

- **Water temperature** is a major parameter in the control of living organisms [32]. Most organisms capable of forming reefs do not tolerate cold water. One of the main reasons is that corals can only reproduce if the water temperature exceeds 20 °C. This is the factor that maintains the reefs. This factor also keeps coral reefs and mangroves restricted to tropical latitudes. The high evaporation rates associated with warm waters are also capable of generating carbonate environments uncontrolled by the activity of organisms. In these high temperature environments, reefs cannot form because the algae associated with corals cannot develop above 35 °C.
- The distribution of many organisms is also limited by **salinity**. In particular, in the case of corals, the formation of carbonate skeletons does not occur when salinity is less than 27%°. Because of this conditioning, reefs do not develop in areas near river mouths where important diffusion phenomena occur [18].
- Corals need **clear water** in order to form reefs. On the one hand, the zooxanthellae algae that nourish the coral need light in order to photosynthesize and synthesize the organic matter on which the coral feeds. On the other hand, both corals and algae need to be free of sediment on their surface to effectively perform their vital functions [25]. Although corals can remove a certain amount of sediment from their surface by secreting mucus, a high sedimentation rate will prevent them from performing their functions and kill them [2].
- The limit of **light** penetration also depends on water clarity and determines the depth to which the coral bodies can develop [18]. Most corals can form carbonate skeletons effectively up to 30 m deep. Although some of them can grow in deeper zones, corals are rarely able to form reefs below a depth of 50 m.
- Reefs are usually associated with good **water oxygenation** [18], but this is not a limiting factor, since zooxanthellae produce oxygen during photosynthesis. However, the concentration of **dissolved CO_2** in the water can be a limiting factor. A high concentration of this gas contributes to an acidification of the water that can be lethal to corals.
- The concentration of **nutrients** in the water also greatly influences reef development [11, 19]. Nutrients are the nitrogen and phosphorus compounds used by photosynthetic organisms to make organic matter. It is evident that a low concentration of nutrients will prevent zooxanthellae from adequately performing photosynthesis. However, excessive nutrient content can lead to the development of other types of algae that will eventually smother the coral [9]. Another harmful effect that can occur is the proliferation of zooplankton to such an extent that it ends up consuming all the oxygen and causing eutrophication of the environment.

23.2.3 Wave Action

The energy range of waves in these environments is several orders of magnitude higher than the energy of the other physical agents acting on them [23]. The way in which waves dissipate their energy on a carbonate coast in general, and on a reef front in particular, defines its erosional or depositional character, but also exerts a significant influence on water clarity and dissolved oxygen concentration. Factors that are directly controlled by wave action in turn control the distribution of sediment grains and nutrients. The strategies of living organisms to adapt to wave energy have a significant influence on their physical design and also on the way they carry out their vital functions.

For corals, moderate wave energy is needed to keep the water transparent and to disperse larvae. This wave energy will also help to limit competition and predation and to select the most effective coral geometric shapes to resist the wave swash. These shapes are distributed according to the energy range that exists in each zone of the reef, depending on the depth and their location in the front or rear areas.

The waves are also responsible for the genesis of carbonate beaches in some frontal reef areas. These beaches function under wave dynamics in the same way as any other beach.

23.2.4 Tide Action

In coral reefs, the tide level controls the upper limit of growth of the reef body; since corals cannot withstand prolonged periods of aerial exposure, they cannot grow vertically above the mean low spring tide. On other reefs, the levels of exposure and submergence imposed by the tide exert another type of control. For example, oyster reefs are concentrated in the central part of the intertidal zone according to the minimum time of inundation and the time when the tide carries the highest concentration of suspended matter filtered by the oysters.

In other coastal carbonate environments, the tide is the main factor controlling the dynamics. In this sense, the tide is responsible for the transport and deposition of carbonate sands in the tidal deltas located between reefs or other sand bodies, as well as for introducing the carbonate muds that are deposited in the lagoons and tidal flats behind them [30].

23.2.5 Climate

Climate exerts control over the physicochemical parameters that influence organic activity [27]. In this sense, water temperature, salinity and nutrient distribution are mainly controlled by climate. On the other hand, climate controls the precipitation regime and insolation, exerting a notable influence on water salinity. Finally, climate regulates the wind regime, and with it the wave regime. In general terms, it can be said that it is the climate that really controls the dynamics and distribution of carbonate shores and reefs. For this reason, this type of environment is found specifically in the intertropical band discussed in the first section of this chapter.

23.2.6 Relative Sea Level Changes

Combinations of subsidence, tectonic uplift and eustatic movements directly influence changes in the location of topographic levels of wave and tidal action, as well as reef growth fringes. Although these changes occur on very slow timescales, most of the present-day reef bodies show evidence of organic growth at times when the sea level was in different positions. Thus, the resulting facies architecture often reflects evolutionary trends controlled by these positional changes.

23.3 General Morphology, Associated Environments and Forms

Carbonate shores are zoned according to the distribution of reef growth, the physical processes affecting the reef and the development of sedimentary bodies resulting from the action of these processes. In general terms, geologists differentiate three zones in carbonate shores: the reef front, the reef body and the back-reef zone.

The **reef front** is the area located in the open part of the reef, exposed to wave action. The **reef body** is the carbonate structure generated by the accumulation of coral skeletons. The **back-reef zone** is the innermost domain of the reef, protected from wave action by the reef body, but subject to tidal action. It is usually a zone dominated by sedimentation based on carbonate grains produced by chemical precipitation (carbonate factory), secreted by planktonic and benthic microorganisms or introduced from the outside by tidal currents.

The area of the reef front usually has high slopes, presenting gradients between 30° and 40°. This slope is usually not constant and is often interrupted by steps and horizontal zones that developed during lower sea levels. In the upper zones, it is common for algal ridges to develop above the low tide level, which corals cannot overcome. These ridges can rise more than 1 m above the body of the reef, which is protected behind them. At the front of the algal ridge, carbonate beaches may also develop, which the waves accumulate from the fragments torn off the corals or from the debris produced by herbivorous and corallivorous fishes. In the deepest areas of the reef front, there is a wide diversity of coral forms that are distributed vertically depending on the light and their tolerance to hydrodynamic stress, subaerial exposure and the amount of suspended matter. In these areas it is also common to find a system of structures called **spurs and grooves**, where longitudinal ridges of coral growth alternate with erosive grooves filled with bioclastic gravels. These structures usually occur in reefs where wave energy is moderate, since grooves are generated by undertow currents. The contact of the spur and groove systems with the open sea can be smooth or abrupt.

Most of the reef body develops behind the algal crest. The upper part of this body is usually flat, as it is vertically limited by exposure to the air, and is therefore often referred to as the **reef flat**. The extent of this reef plain can vary from tens of meters to several kilometers, depending on the factors discussed in the previous section. Most of the reef body is constructed from the skeletons of dead corals, while the reef flat is where the main biological activity of the reef is concentrated. In this area, the wave action is not intense, as it is protected by the algal crest; however, the high exposure of organisms to ultraviolet rays is a significant environmental stress factor.

In the back-reef zone, waves are totally absent, although tidal action is present. The most common environments are lagoons, tidal flats and mangroves, where there is a tendency for sedimentary aggradation. There are usually inlets between the different reef bodies that allow seawater to enter the lagoon, which can be associated with the development of ebb and flood tidal deltas at both ends of the channel.

Not all reefs have the same three-dimensional distribution of these three zones and their associated environments. Classically, several types of reefs have been differentiated according to the geometric arrangement of these three zones [28], distinguishing between: fringing reefs, barrier reefs and atolls. In addition to these three main types, there may also be smaller reef bodies in the form of patches and tabular flats.

23.3.1 Fringing Reefs

This type of reef develops attached to the mainland or separated from it by a very narrow and shallow back-reef zone (Fig. 3a). Its width depends on the overall slope of the coast. The frontal zone usually has a high slope with a good vertical distribution of coral forms that show a high growth rate (Fig. 3b). This frontal zone also usually has pinnacles separated from the coral front (Fig. 3c).

The reef crest may have continuity in the reef body, although it most often has discontinuities in the form of channels that allow the tide to pass into the reef flat. These channels can reach tens of meters in width, and may even extend some hundreds of meters [8]. Coral growth is scarce in the reef flat, being frequently covered by fragments of dead corals (Fig. 3d) or fine sediments, or sometimes colonized by seagrass (Fig. 3e). The reef flat is often attached to land, with no true back-reef zone. The contact with land is often occupied by a carbonate beach (Fig. 3f), constructed of bioclastic sand composed mainly of coral fragments.

23.3.2 Barrier Reefs

Barrier reefs are elongated islands separated from the mainland by a wide shallow-water lagoon (Fig. 4a). As with fringing reefs, the coral bodies of the reef body may be relatively continuous or separated by channels that act as inlets, allowing the passage of open marine waters into the lagoon due to tidal action. At both ends of these inlets, flood- (Fig. 4b) and ebb-tidal deltas develop, although the ebb deltas are usually less developed because the frontal areas have a very high gradient.

The characteristics of the reef front and reef flat can be very similar to those observed in fringing reefs. The reef front is where most of the coral growth occurs (Fig. 4c), while the reef flat may have different environments that vary between the development of live coral fields, dead coral skeletons, or sedimentary sand or carbonate mud flats. The presence of an algal crest just in front of the main body of the reef, as well as spurs and grooves to the fore of the frontal zone, are common in this type of reef (Fig. 4d). Sometimes, waves are able to form carbonate beaches that partially cover the algal crest or the body of the reef front.

What really distinguishes this type of reef from those described above is the existence of a large-scale lagoon that separates it from the land. This lagoon usually has a very active sedimentation of carbonate muds, although, in large-scale systems, the sector of the lagoon closest to the mainland may also have terrigenous material (extraclasts) from land. In the vicinity of the flood-tidal deltas, there is an increase in sediment grain size, due to the greater influence of tidal currents and material from the exterior in these sectors. Patch reefs may appear along the lagoon. These reef patches can be coralline, although they can also be formed by other organisms such as ostreids.

23.3.3 Atolls

Atolls are reefs that form a ring-shaped barrier around a central lagoon (Fig. 5a). They sit on the cusp of submerged volcanic edifices, which may sometimes outcrop in the central zone of the lagoon (Fig. 5b). These structures develop mostly in the Indian and Pacific oceans, with very few examples in the Atlantic [28]. Their shape can vary between circular, elliptical, parabolic or irregular, and their scale varies from hundreds of meters to several tens of kilometers. Geometrically, they are usually asymmetrical, due to the greater persistence of the swell on one of their faces.

Fig. 23.3 Different elements linked to a fringing reef in Puerto Plata (Dominican Republic), **a** Aerial view showing the general aspect of a fringing reef, **b** Branching coral forms in the reef front, **c** Vertical wall of a pinnacle in the reef front, **d** Dead corals in a reef flat, **e** Border between the reef front and a seagrass prairie developed on the reef flat, **f** Carbonate beach in the inner part of a reef flat

The coral building zone shows few differences from a barrier reef and the beaches that usually form in the higher areas of the reef flat. There is also a great similarity in the distribution of environments associated with the lagoon, with coral patches scattered among calcareous mud surfaces, although in this case the facies show a total absence of terrigenous material. Despite this facies similarity, the major difference between atolls and barrier reefs is the architectural

Fig. 23.4 Features of barrier reefs, **a** Aerial view showing the general aspect of a barrier reef in Belize (Landsat image from Google Earth), **b** Aerial view of a flood tidal delta developed between different reef bodies in the Bahamas (Landsat image from Google Earth), **c** Coral forms in a reef front (Belize), **d** Spurs and grooves in a frontal reef area (Florida Keys)

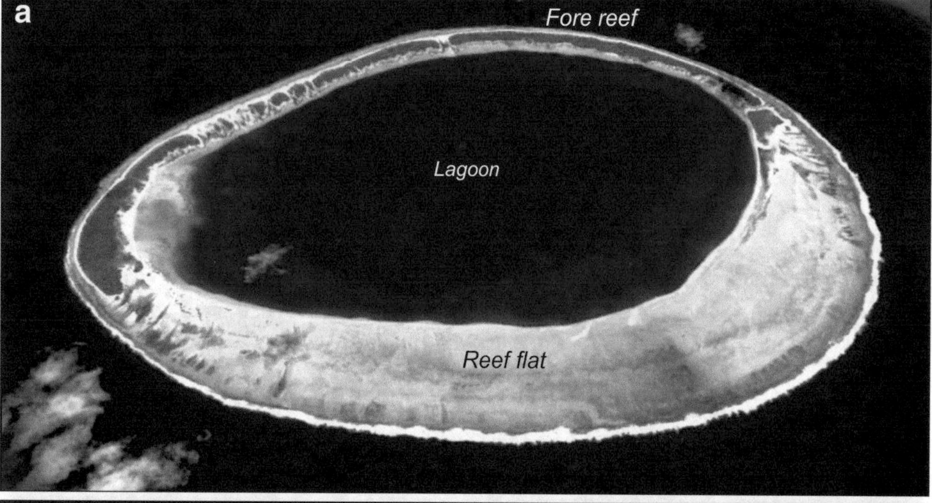

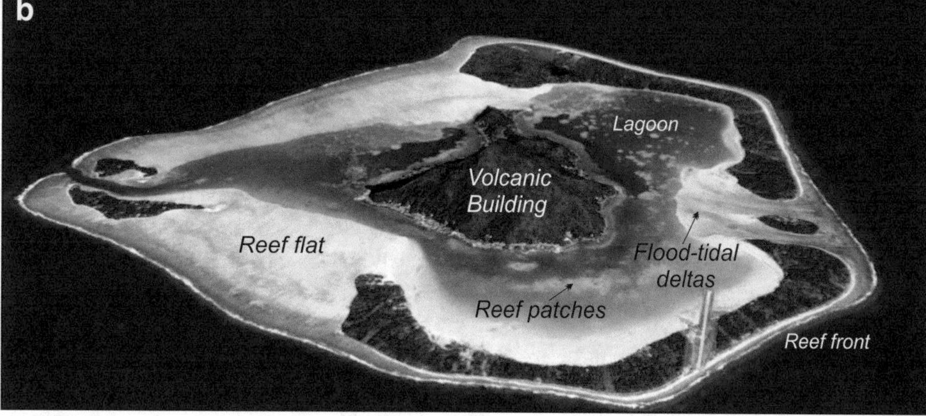

Fig. 23.5 Features of atoll reefs, **a** Asymmetric ring-shaped atoll (Hiti Atoll in the Tuamotu Archipelago, French Polynesia), **b** Atoll around a central volcanic structure (Maupiti Atoll in the Society Islands, French Polynesia) (Images Landsat/Copernicus from Google Earth.)

arrangement of these facies, as will be discussed in Sect. 23.5.

23.4 Dynamics and Evolution

The dynamics and evolution of carbonate shores is closely related to the interrelationship of the factors listed in Sect. 23.2 of this book. In the short term, three factors determine the dynamic functioning of the systems associated with the reef bodies: reef growth, wave action and tidal dynamics. On a longer timescale, climate variations and relative sea level movements influence the vertical and horizontal displacements of the environments associated with these coasts and are determinant in the facies architecture. In the following section, these effects on the dynamics will be discussed individually.

23.4.1 Reef Growth

In coastal reefs, biological rhythms determine the growth rate of the reef body and control the distribution of the other dynamic agents, thus setting the general dynamic patterns of the sedimentary environment. In the reef body, there is an ecological zonation controlled by the vital factors of the corals (water agitation, luminosity, ultraviolet radiation and nutrients) and these factors vary with depth and reef zone (front, top or back). In each of these ecological zones there are corals with different shapes, ranging from encrusting, globous, plate, branching, fragile branching and fan-shaped (Fig. 23.6).

The growth rate is different for each species, where the coralline form is a determining factor [26]. Environmental factors also significantly influence these growth rates [20]. Depending on these determinants, longitudinal growth rates range from 1 to 16 mm per month. Despite these coral growth rates, the volume of the reef body itself develops at a much slower rate. This difference is due to the fact that coral growth leaves numerous gaps in the overall structure. These gaps are filled more slowly with a matrix consisting of secretions from other organisms (mainly calcareous algae, sponges and bryozoans) as well as fragments of corals and other organisms with fragile structures, broken and reworked by the waves. This implies that much of the constructive activity performed by corals is, on the one hand, compensated for by the destructive activity of other organisms (bioerosion) and wave energy, but, on the other hand, the material generated by this destructive activity contributes to complete the massive structure of the reef.

However, longitudinal growth is not the only parameter for determining coral growth, as it results from the biochemical process known as calcification. The calcification rate is the rate of metabolic production of calcium carbonate by living organisms. This rate ranges from 1 to 10 kg per year per square meter of coral surface. Obviously, this calcification will result in different longitudinal growth rates depending on the density of the resulting calcium carbonate crystalline structure, but it will also result in a different value of reef volume depending on the density of the reef structure. In general terms, different authors have shown that, in recent decades, there has been a decrease in the rate of calcification in all reefs worldwide (e.g., De'ath et al. [5]). This decrease has been attributed to the effects of ocean water warming and acidification, due to increased concentrations of dissolved CO_2.

Once the reef body reaches the water surface it acts as a physical barrier to the waves. Thus, the action of the breaker takes place only at the reef front, restricting the environments associated with the lagoon, which are dominated by the action of tides.

23.4.2 Wave-Induced Dynamics

It has been described how wave action directly influences the shape and growth of corals and subsidiarily the growth rate of the reef body. A different orientation of the reefs with respect to the dominant wave trains and the arrival of storm waves will determine ecological differences. On the one hand, the average energy dissipated by the waves controls the type of coral forms that can proliferate on the crest and the reef front. This will determine the morphological differences in these environments according to the different growth rates of each of the coral species capable of resisting this energy [13]. In this sense, it should also be taken into account that waves and wave-induced currents are mainly responsible for the distribution of both larvae and nutrients that feed corals [29]. Due to this fact, seasonal differences in the wave regime induce the presence of ecological cycles [15].

Conversely, the morphology of the reef front conditions the wave energy dissipation gradients and the way in which waves can reflect part of their energy back to the open ocean or, under certain conditions, cross the reef plain and penetrate the back-reef areas. In many reefs, the presence of a reef ridge acts as a bar separating two zones with different behavior. The behavior of these two zones depends on the topographic height of the ridge. At the front of the crest, phenomena associated with wave reflection and backwash are responsible for the appearance of bottom currents that control sediment transport. If the arrival of waves is oblique to the reef front, these currents can circulate longitudinally in the form of littoral drift. However, the morphology of the reef front tends to channel these currents to deeper areas with a strong coarse material transport component (Fig. 7a).

Fig. 23.6 Different coral forms, **a** Encrusting coral, **b** Globous coral, **c** Plate coral, **d** Branching coral, **e** Fragile branching coral, **f** Fan-shaped coral

These currents are responsible for the genesis of spurs and grooves and the filling of the grooves with bioclastic sands and gravels from the dismantling of corals and other organisms on the reef crest.

When the waves exceed the topographic height of the reef crest, the water circulates over the reef flat, transporting significant amounts of sandy bioclastic fragments to the back reef zones (Fig. 23.7). Most of the wave energy in these

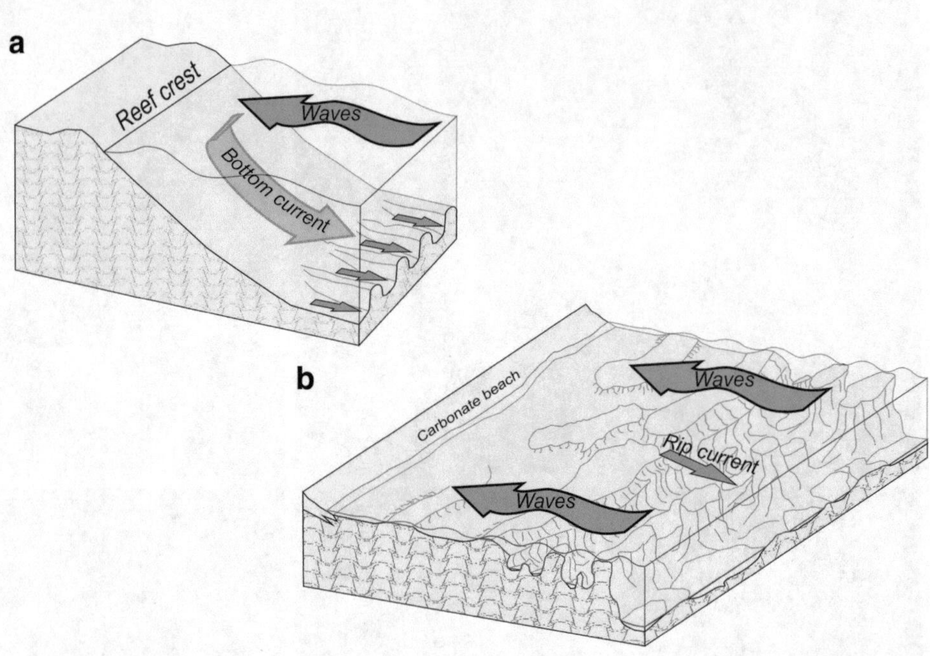

Fig. 23.7 Genesis of wave-induced currents in a reef. **a** In the reef front, **b** Through the reef crest entering the reef flat

shallow areas is dissipated by friction. On fringing reefs, the sandy material transported by the waves is usually trapped on the reef flat, although under favorable conditions it may reach the mainland to form a back beach. In barrier reefs and atolls, these sandy bodies are preserved in the form of washovers at the contact between the reef body and the lagoon. The water that overtops the frontal ridge to enter the flat or lagoon usually returns to the open ocean in the form of rip currents entering channels that pass through the reef body. These currents also transport sediment into the deep areas.

The channels also allow waves to pass into the lagoon, especially in the case of long wavelength waves. Another way in which waves can penetrate the back areas of the reef is by passing through depressed areas of the reef crest when it is irregular. In these cases, the surges associated with the action of persistent winds provide an additional rise in the water level that facilitates this process. When this occurs in barrier reefs, complex refraction and diffraction processes determine the distribution of erosional and depositional zones on the continental shores of the lagoon [14]. Even so, the overwash through the reef crest always exerts an attenuation effect on the waves that act in the rear zone. This attenuation increases with the width and height of the reef body. In many instances, wave attenuation can also build sand or carbonate gravel forms on the reef crest, which grow vertically until they completely close off the wave passage to the reef flat along segments of the barrier. These forms can exceed the height of the spring high tides and begin to behave like islands with frontal beaches.

In atolls, where the reefs restrict the direct connection with the ocean and where the tidal prism is usually very small, the overwash processes become the main mechanism responsible for the environmental renewal of the lagoon water and the introduction of sandy material into the lagoon. In some atolls, it has been observed that the surge that accompanies storms generally raises the sea level, causing sediment-laden seawater to enter the lagoon from one side while the lagoon water exits from the opposite side [1]. This type of process increases the rate of sedimentation in lagoon environments.

23.4.3 Tide-Induced Dynamics

Most of the reefs are found in mesotidal and microtidal coasts. Tidal influence is felt in two different aspects: water levels and tidal currents. The main tidal currents occur in the inlet channels between reef bodies, especially in barrier reefs with extensive lagoon systems. These internal tidal systems allow drainage and renewal of lagoon waters. The more extensive lagoons of mesotidal systems may have large tidal prisms allowing the development of strong currents in the channels that cross the reef body. These channels, unlike inlets in barrier island systems, are morphologically complex. Their bed may be erosional or sedimentary, the latter being filled with very coarse material.

Towards the reef front zone, the channels usually flow into slopes or areas of high gradient, in most cases inhibiting the development of ebb-tidal deltas. However, in the case of lower slopes, ebb deltas tend to be complex and develop between irregular reef bodies. In the area of the reef flat, in contact with the lagoon, flood-tidal deltas usually develop. This type of delta is very frequent and presents a remarkable development, having dynamic and sedimentary characteristics that are identical to the flood-tidal deltas of the barrier island systems.

The filling of lagoons on barrier reefs and atolls can bring sediment to the low tide level. When this occurs, tidal flats develop, although many authors refer to them as peritidal environments. The nature of the sediment in these flats is carbonate and has been studied in detail in Chap. 20. In some of these flats, stromatolites may develop, built by the action of cyanobacteria.

Tide levels are essential to the functioning of carbonate tidal flats, but they are also important in the area of the reef front since they determine the height that the coral body and the algal crest can reach, in addition to controlling the swell action band. The joint action of tides and waves will be analyzed in the following section.

23.4.4 Dynamics of Combined Waves and Tides

The combination of wave action and tidal levels controls the sediment transport from the outer zones to the interior of the lagoon across the tidal flat. The magnitude of the waves combined with the tidal levels, marked by alternating high and low waters and spring and neap tidal cycles, controls the depth of water in the reef flat. In this regard, it should be noted that wave propagation over the flat requires a minimum depth.

Wave inflow across the reef flat at times of high waters drives the transport of nutrients and larvae to the back areas of the bar, which contributes to the ecological renewal of the lagoon water. In parallel, this process also determines the volume of sediment transferred from the reef front areas to the inner areas. It has already been mentioned that the drainage of lagoon waters due to simple tidal action is very important; however, it is necessary to take into account that many reef systems (especially atolls) develop very small tidal prisms. In these cases, and also on microtidal shores, waves may be the main agent responsible for the renewal of lagoon waters. In each stretch of the reef, the relative importance between the volume of water introduced by waves and that introduced by tidal currents is different. In this respect, important differences will be established depending on the spacing and depth of the connections between the lagoon and the open sea.

23.4.5 Long-Term Evolution: Relative Sea Level Movements

The current coral structures are the result of a relative marine stabilization that has been maintained for at least the past 4500 years. However, these reef systems are heirs of the systems that followed the Flandrian transgression that began about 12,000 years ago. In fact, many of them developed on the remains of reefs developed during the Eemian interglacial period, when the sea level was at a position very close to the present one. Studies of ancient reef systems show that reefs can adapt to sea level changes when they occur at a slow rate, since the system can respond by accommodating its growth rate to compensate for these changes. In this context, sea level changes imply modifications in the energy of the environment, as well as in the ecological patterns that control the amount of nutrients. These variations entail changes in the configuration of coastal systems to which organisms must adapt. Whether or not they succeed will depend on the direction of these changes, as well as on the rate at which the changes occur. When changes occur at a slow pace, the communities of reef-building organisms manage to adapt by migrating the reef body landward or seaward by constructing new structures on the substrate of those that are being abandoned.

In the event of a fall in sea level, the reef flats are the most vulnerable environments, as they would be easily exposed to continental processes, constituting the morphology known as **reef terraces**. In this case, these are the parts of the reef that are first degraded when their materials are reworked by surface waters, even becoming karstified. One particular example occurs in carbonate beaches that are exposed to fresh waters, producing a cementation process that turns the carbonate sands into a solid beachrock. The core of some of the barriers that constitute the current reefs is made up of older reefs that have been left hanging after a small fall in sea level. The presence of these cores is an elevated area that prevents waves from passing over it to the back areas. They are often areas where current carbonate beaches can develop due to their action as a dissipator of wave energy.

Conversely, a rise in sea level creates an accommodation space that stimulates vertical growth of the main body of the reef. The relationship between the rate of rise and the rate of reef growth can give rise to three situations [24]. These situations are known as sustained vertical growth (keep-up), forced vertical growth (catch-up) and drowning (give-up). These three cases can be recognized in the evolution of some reefs in response to upwelling during the Holocene transgression, but are also reflected in the evolution of many atolls where relative upwelling occurs by subsidence. In the case of **sustained vertical growth**, the rate of sea level rise can easily be matched by the rate of reef growth and, consequently, there is continued aggradation of the reef body. In many cases, vertical growth does not occur in all areas of the reef and involves a shift of reef environments towards the mainland. In the case of **forced vertical growth**, the rate of upwelling begins at a higher rate than the rate of growth; however, the reef responds by increasing its rate of growth until it finally matches the rate of upwelling and the reef survives. When this occurs, it is even more common for the reef to migrate landward. In the latter case, the rate of upwelling is so high that the reef is unable to compensate for the upwelling by increasing the rate of growth. In this situation, the depth of the reef body increases until a threshold is reached at which the organisms fail to survive and **reef drowning** occurs. Even so, it is possible that the larvae manage to colonize shallower environments and new reef bodies are formed in the newly flooded areas.

The vertical growth model was used by Charles Darwin to construct his theory of atoll formation (Fig. 23.8). Darwin deduced that the reefs of the Pacific archipelagos began by forming fringing reef bodies on the shores of a volcanic island. Subsequently, the subsidence of the volcanic structure would cause a vertical growth of the reef body, which would move away from the emerged part of it until it formed a circular chain of barrier reefs around it with a doughnut-shaped lagoon separating the two elements. Finally, the subsidence of the entire volcanic structure would leave on the surface only the atoll barriers around a central lagoon.

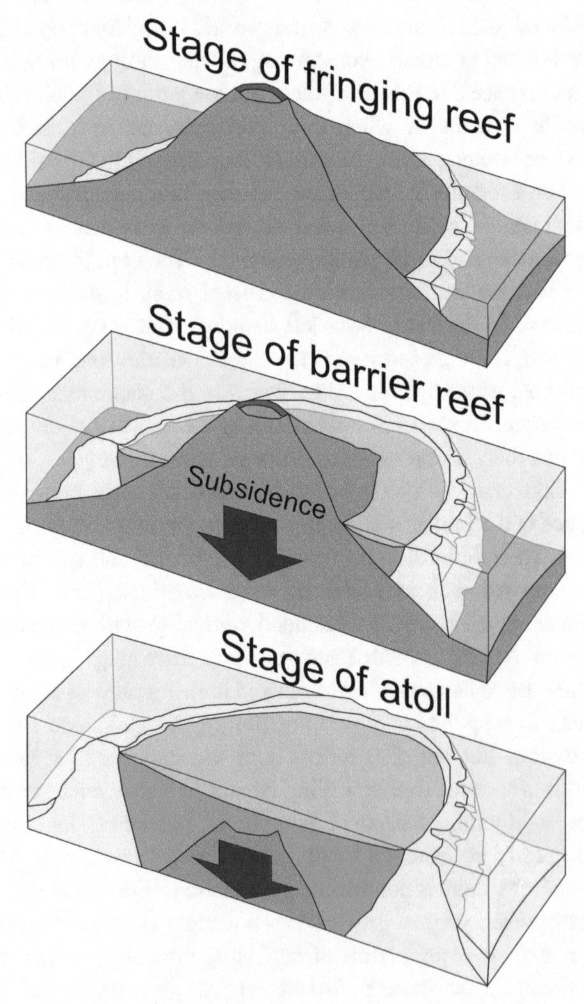

Fig. 23.8 Conceptual model showing Darwin's theory of atoll genesis

In many cases, the rate of sea level rise is not constant, but takes place in pulses. Sometimes a rising pulse drowns a reef that is abandoned at depth while the spread of larvae in the new submerged zones consolidates new reef bodies. The former reef plains are preserved underwater in the form of submerged terraces. Successive horizontal terraces may be separated by more or less vertical slopes several meters high as steps. These steps are visible in the bathymetry of many Caribbean reefs.

23.5 Facies

In principle, the complete study of a carbonate coast makes it easy to distinguish between the reef formation and the sedimentary environments built with transported carbonate material. From this point of view, a first distinction could be made between reciprocal framework facies and detrital facies.

23.5.1 Reef Framework Facies

A quick glance may lead one to think that a reef framework is built as a more or less homogeneous body constituting a single facies. This facies was named boundstone by Dunham [6]. Taking as a criterion the function of the corals in building the carbonate body, [7] were somewhat more precise and distinguished three facies: bafflestone, bindstone and framestone. However, from a sedimentological point of view, there is a much wider variety of facies. To begin with, there is a major structural difference between dome-growing organisms (coralline algae and encrusting and globose corals) and interlacing corals (plate-shaped, branching, fragile branching and fan-shaped corals). The main difference is that the first organisms build a solid structure with very few voids, while the latter build a cavernous structure full of voids that will later be filled with sands or gravels made up of bioclastic fragments or coralline algal growths.

Recent studies [22] have distinguished seven different reef framework facies. The main criterion used is the type of coral growth, as each type is indicative of the environment in which the corals develop. Each of these forms is designed to adapt to environments with higher or lower wave energy, luminosity, dissolved oxygen and nutrient availability, and are therefore indicative of specific locations within the reef structure. The identification of these facies in boreholes has allowed us to interpret the evolution of numerous reefs in response to Holocene sea level variations (e.g., [12, 16]). The facies that have been distinguished are as follows (Fig. 23.9):

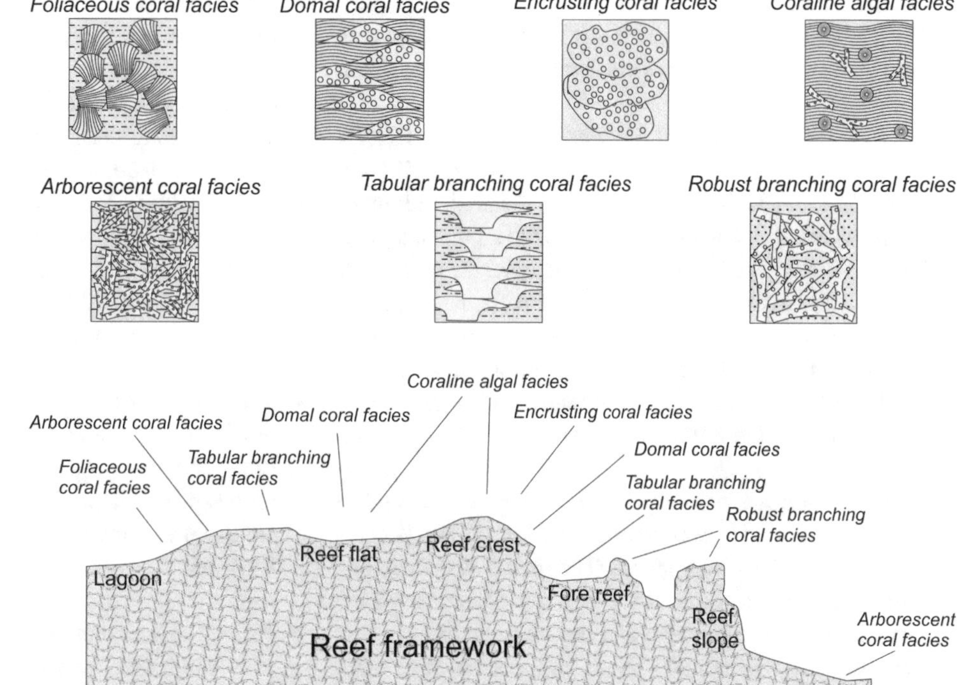

Fig. 23.9 The seven reef framework facies of [22], indicating their common location in a reef-building profile

- *Coralline algal facies*: These are laminated crusts of centimeter or decimeter thickness. They may include in situ corals or fragments of reworked corals. They are located on the flat and the reef crest because their compact growth allows them to resist the indirect action of the breaking waves.
- *Domal coral facies*: Coral bodies intermixed with encrusting coralline algae or scattered colonies separated by bioclastic sands and silts. They are located on the flat and the reef crest, and also on the frontal slope, since these corals are capable of resisting the action of the highest energy waves.
- *Facies of encrusting corals*: Masses developed in situ, sometimes related to coralline algal facies and sometimes with a matrix of detrital material of different sizes. They develop in the reef crests of high wave energy.
- *Robustly branching coral facies*: Bodies formed by a cavernous structure of branching corals, with the cavities filled with a matrix of bioclastic gravels and sands. They develop in the upper part of the reef front, as they resist some wave energy, although not the direct action of the breakers.
- *Tabular coral facies*: Masses with a rigid structure consisting of in situ or reworked tabular corals with a matrix of carbonate sands or silts. These facies are typical of the reef flat and reef front, excluding the ridge, as they are sensitive to wave spray.
- *Arborescent coral facies*: Structures formed by finely branched corals with latticework filled with fragments of the branches or simply accumulated from fragments of reworked branches. They develop in the lower parts of the reef front or in the back zone of the reef, as these corals are fragile and do not support the continuous movement imposed by wave oscillation.
- *Foliaceous coral facies*: These are sediments formed by flat corals and often enveloped by muddy sediment or fine sands. They develop in sheltered areas in the transition between the reef flat and the lagoon, as these corals support higher turbidity environments.

23.5.2 Detrital Facies

Montaggioni [22] studies include three detrital facies associated with the reefs, in addition to the seven framework facies already described. These facies are:

- *Skeletal debris facies*: These are formed by fragments of corals and coralline algae pulled from the reef in a matrix of carbonate sand. The origin of these facies is direct wave action.
- *Carbonate sand facies*: These are sands of different sizes consisting of fragments of carbonate organisms (corals, coralline algae, mollusks, green algae and foraminifera). The origin of the grains may be physical fragmentation due to wave action and currents or biological fragmentation by fish and other vegetarian and corallivorous organisms.
- *Carbonate mud facies*: These are made up of fine particles from direct inorganic precipitation or from the

accumulation of defecation or skeletons of microorganisms. They form the beds of lower energy zones, such as lagoons and seagrass meadows developed in the reef flat.

These facies describe well the sediments that are directly linked to the reef body; however, they do not describe the totality of sediments that can be found associated with other carbonate coastal environments linked to the lagoon and tidal action. Kendall [17] distinguishes a much greater number of detrital facies, considering that the type of carbonate particles which constitute gravels, sands and muds, as well as other biologically controlled facies, provide information on the particular depositional environment within carbonate coasts. Kendall identified the following facies (Fig. 23.10):

- *Gravels and sands of skeletal grains*: These correspond to the carbonate sand facies described by Montaggioni [22] for reefs. These are sands composed of fragments of skeletons of reef-building organisms. They are typical of beaches, grooves between reef bodies and reef slope aprons.
- *Gravels of grapestones*: These are coarse sediments consisting of complex grains that appear as aggregates of sand grains cemented by aragonite. They are revealed in margins of lagoons with moderate wave energy.
- *Pisolite gravels*: Pisolites are subspherical grains formed by regular concentric sheets of micrite around a core. Their size exceeds 2 mm. Spheres of smaller sizes are called oolites. Their formation requires rolling transport on a micritic background. They are typical of inner margins of lagoons with moments of high energy.
- *Oolite sands*: Oolites are laminated subspherical grains that are similar to pisolites but smaller in size. Their most frequent size is between 0.2 and 0.5 mm, although they can reach 2 mm. They characterize the sandy shallows of high tidal energy lagoon margins and almost entirely constitute the sands of carbonate tidal deltas.
- *Intraclast sands*: Intraclasts are irregular grains formed by previously cemented carbonate fragments. Intraclast sands are composed of fragments of carbonate crusts removed from supratidal zones by energetic events. They are usually found accumulated in the form of cheniers in tidal flats or in aprons developed in deeper areas of the reefs.
- *Pelletal sands*: Pellets are grains composed of micrite, with a massive internal structure. Sizes are between 0.1 and 0.5 mm in diameter. Their origin is a product of the fecal activity of organisms. They are typical facies of lagoons.

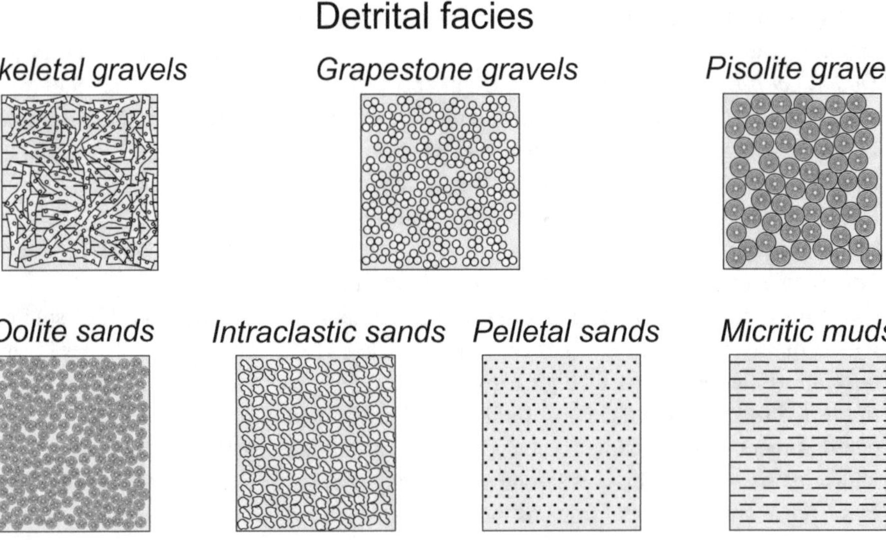

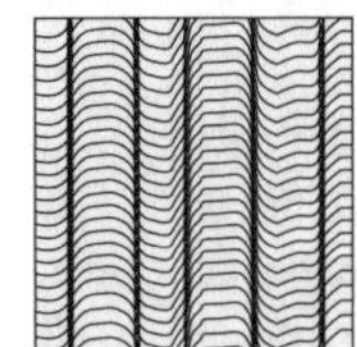

Fig. 23.10 Detrital and biologically controlled facies defined by Kendall [17]

- *Micritic muds*: These are non-organic chemical precipitates. Their deposition requires calm subtidal waters, so they are usually formed in lagoons.
- *Stromatolites*: These are built by the action of blue-green algae on a fixed substrate. They alternate sheets formed by algal filaments and others by carbonate particles trapped by the algae. The lamellae have an irregular shape. They are characteristic of the intertidal zones of the carbonate tidal flats developed in the continental part of the lagoon.
- *Oncolites*: These are rounded but irregular clasts with concentric micritic lamination around a core. Like the stromatolites, the precipitation of the lamellae is the result of the activity of blue-green algae, but in this case on a mobile substrate. They usually develop in very shallow lagoons and very clear waters.
- *Rhodolites*: Very similar in appearance and origin to oncolites, but developed by rhodophycean algae. They also develop in lagoons.

Obviously, the components of the detrital facies may appear mixed, resulting in a much larger number of mixed facies.

23.6 Facies Models

The way · in which these facies intertwine three-dimensionally depends on the type of coastline. This results in a facies model for each reef type: fringing reefs, barrier reefs and atolls.

23.6.1 Facies Architecture of Fringing Reefs

The architectural facies model for a fringing reef is the simplest of all. The reef structure rests directly on a continental substrate. This substrate can be any type of rock, although it is often older reef systems. For example, in many Caribbean reefs, the present development is supported by an Upper Pleistocene structure that developed in the last interglacial period (Eemian), more than 100,000 years ago, when the actual sea level was about 2 m above the present level.

The reef structure consists of the classic distribution of environments described in Sect. 23.3.1, with a reciprocal ridge separating the environments of the reef front and the reef flat (Fig. 23.11). The entire framework is built by boundstone facies, although the types of structural facies described by Montaggioni [22] can be differentiated in the different sectors. The reef ridge is made up of massive coralline algal facies, dome corals and encrusting corals. On the reef front, pinnacles built by robust arborescent corals are often found; between them there may be spaces where the facies of tabular branching corals dominate. In some of these spaces there are detrital facies of gravel and sand, dominated by skeletal elements. In the transition to deep zones, a vertical slope of several meters in height may extend, whose front is built of robust branching corals. This slope is interrupted by furrows where metric-scale aprons composed of skeletal and intraclastic sands and gravels develop. At the base of the slope, within the photic zone, there is usually a lower slope section. This zone of the reef is dominated by arborescent coral facies, although their fragility causes them to frequently fracture, forming fields of skeletal sands whose grains are mainly composed of reworked fragments of this type of coral.

In the reef flat zone, there are large areas dominated by detrital sedimentation (facies of carbonate sands and muds). The composition of the sands is dominated by skeletal fragments and oolites, although grapestones can also be found. These facies may be accompanied by others generated by organisms, such as oncolites, rhodolites and stromatolites. The muds are of micritic origin and have a high organic matter content. On top of these muds, seagrass plains usually develop. In the continental zone, there is usually a beach attached to the mainland. This beach is made up of sands and microgravels with clasts of different origins, although skeletal fragments usually dominate. Between these detrital facies, there are usually patches of reefs composed of arborescent, tabular and foliaceous coral facies.

23.6.2 Facies Architecture of Barrier Reefs

Barrier reefs present a much more complex facies architectural model (Fig. 23.12). As in the fringing reef model, the reef ridge separates the reef front and reef flat environments, although in this model there is an extensive shallow-water lagoon that separates the reef structure from the mainland.

In the distribution of structural facies of the barrier reef, there are practically no differences with the fringing reefs, nor in the distribution of detrital facies of the reef front—although in this sector there may be systems of spurs and grooves, in whose depressions skeletal gravels and sands accumulate. There are, however, differences in the reef crest: in some sectors the waves manage to accumulate significant bodies of skeletal gravels on the crest, developing skeletal sand beaches on its front. Conversely, in other sectors the waves frequently cross the reef crest, developing bodies of oolitic sands on it, and extending towards the reef flat in the form of washovers. There are also numerous passage channels between the reef crests. These channels are tidally dominated and have the same function as the inlets of barrier islands. The bed of these channels can be erosional or depositional. In the latter case, skeletal and oolitic sands with herringbone structures may accumulate on the bed. At the lagoon-facing end of these channels, flood-tidal deltas are frequent, built mainly by oolitic sands arranged in

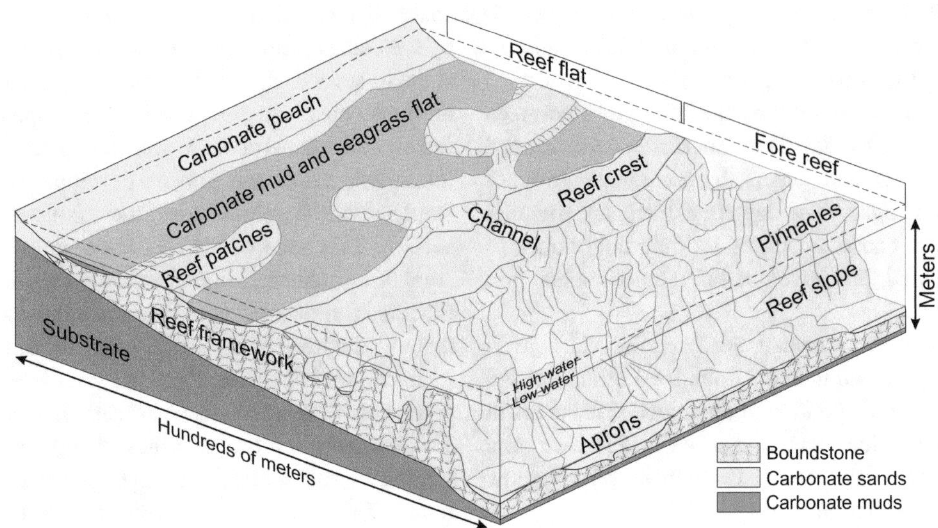

Fig. 23.11 Scheme of facies architecture of a fringing reef

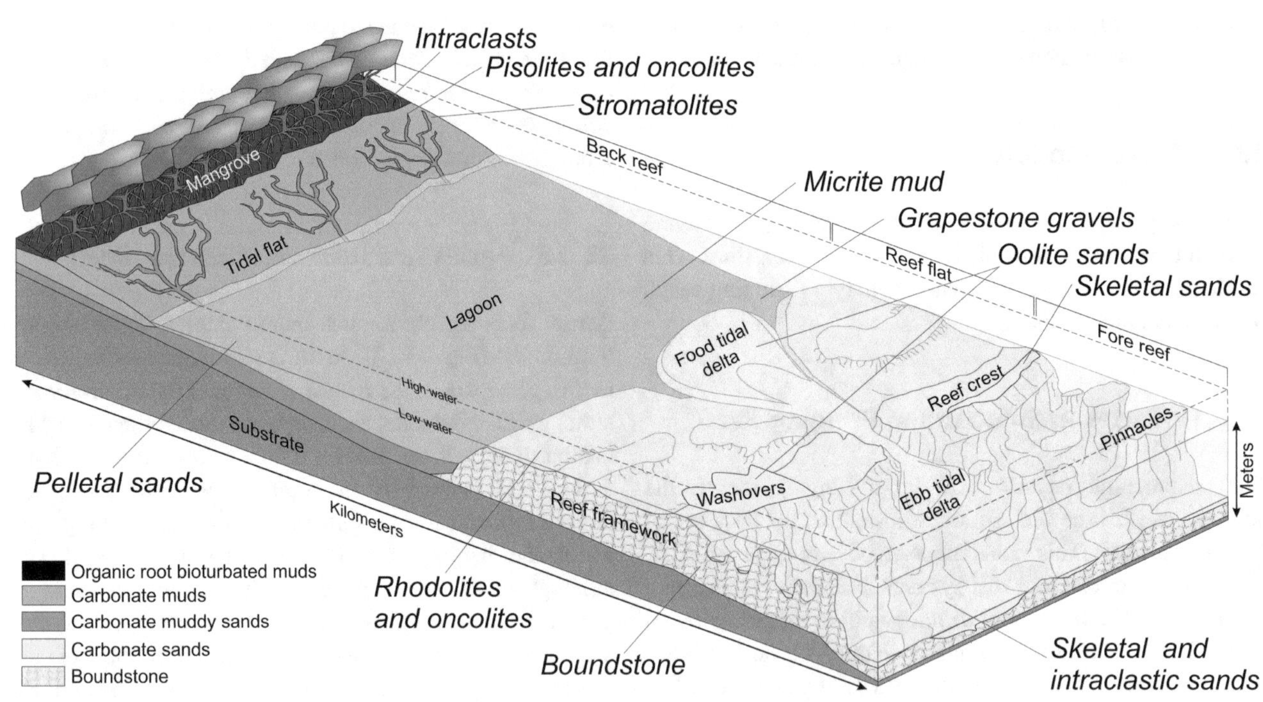

Fig. 23.12 Scheme of facies architecture of a barrier reef

herringbone cross-stratified sets, although sets showing landward migration of carbonate sandy bodies dominate.

In addition to branching, planar and foliaceous coralline facies, other fixed algal constructions such as stromatolites usually develop in the reef flat, although these can also develop on a mobile substrate (oncolites and rhodolites). Towards the lagoon, grapestone sand bodies and pelletal sediments also develop. These sandy sediments change to carbonate mud facies in the central zones of the lagoon. Towards the mainland, the influence of land input is apparent, and the muds may be of a marly type, incorporating the terrigenous fraction.

The intertidal zones adjoining the mainland often develop tidal flats, presenting a facies sequence that was described in Chap. 20 of this book (Fig. 20.15b). The upper part of this sequence may also consist of stromatolites, pisolites and oncolites. The supratidal zones usually develop mangroves, although in arid climates coastal sabkhas may also occur.

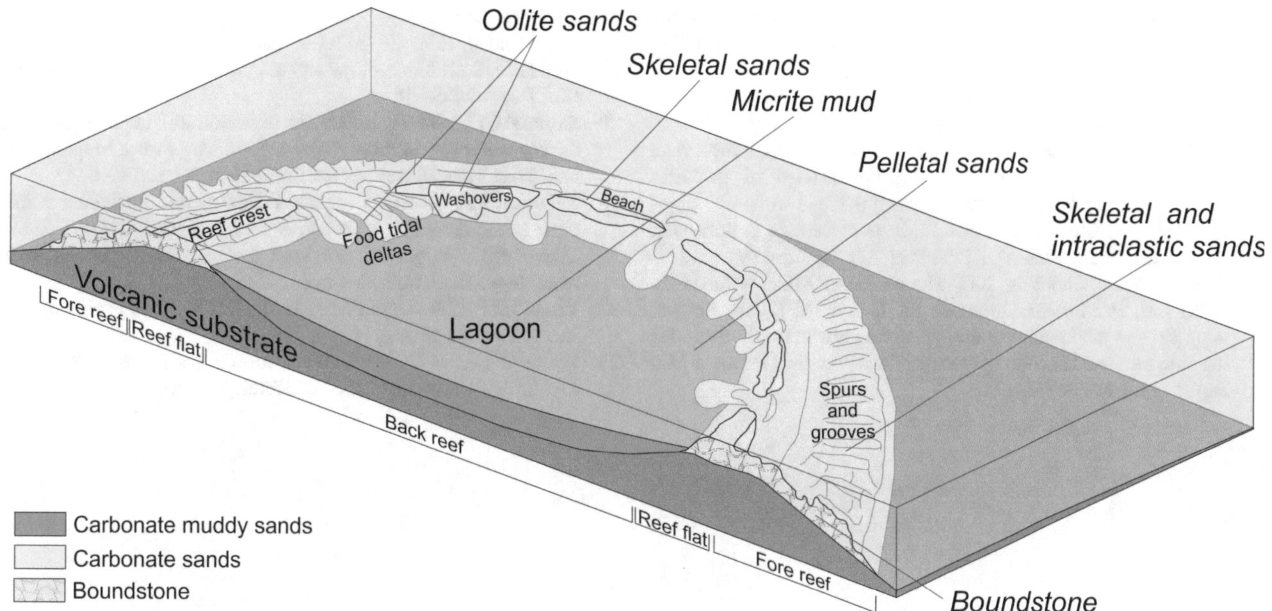

Fig. 23.13 Scheme of facies architecture of an atoll

23.6.3 Facies Architecture of Atoll Reefs

The distribution of the facies with respect to the reef body is very similar to that described for the barrier reefs, except for the fact that the reef structure is annular (Fig. 23.13). Thus, in those atolls that occur in mesotidal environments, there are ridges that develop detrital bodies in their upper part, which are interrupted by channels that develop flow deltas towards the lagoon. The reef front usually also has spur and groove systems, although these end in a slope of much greater depth than in most barrier reefs. Lagoon zones, however, do not usually have the intertidal systems described in the near-continent zone—since there is no continent—although intertidal systems may develop in the final stages of lagoon filling.

References

1. Callaghan DP, Nielsen P, Cartwright N, Gourlay MR, Baldock TE (2006) Atoll lagoon flushing forced by waves. Coast Eng 53:691–704
2. Cortes JN, Risk MJ (1985) A reef under siltation stress, Cahuita, Costa Rica. Bull Mar Sci 36:339–356
3. Davidson-Arnott R (2010) Introduction to coastal processes and geomorphology. Cambridge University Press, Cambridge, 442pp
4. Davies JL (1980) Geographical variation in coastal development. 2nd edn. Longman, New York, 212pp
5. De'ath G, Fabricius K, Lough J (2013) Yes—coral calcification rates have decreased on the last twenty-five years!. Mar Geol 346:400–402
6. Dunham RJ (1962) Classification of carbonate rocks according to depositional texture. In: Ham WE (ed) Classification of carbonate rocks—a symposium, pp 108–121
7. Embry AF, Klovan JE (1971) A Late Devonian reef tract on Northeastern Banks Island, NWT. Can Pet Geol Bull 19:730–781
8. Guilcher A (1988) Coral reef geomorphology. Wiley, Chichester, 228pp
9. Hallock P, Schlager W (1986) Nutrient excess and the demise of coral reefs and carbonate platforms. Palaios 1:389–398
10. Hixon MA (1997) Effects of reef fishes on corals and algae. In: Birkeland C (ed) Life and death of coral reefs. Chapman and Hall, New York, pp 230–248
11. Hopley D (1989) Coral reefs: zonation, zonality and gradients. Essener Geogr Arbeiten 18:79–123
12. Hopley D, Smithers S, Parnell, K (2007) The geomorphology of the great barrier reef, development, diversity and change. Cambridge University Press, Cambridge, 532pp
13. Hubbard DK (1997) Reefs and dynamic systems. In: Birkeland C (ed) Life and death of coral reefs. Chapman and Hall, New York, pp 43–67
14. Kench PS, Brander RW (2006a) Wave processes on coral reef flats: implications for reef geomorphology using Australian case studies. J Coast Res 22:209–223
15. Kench PS, Brander RW (2006b) Response of reef Island shorelines to seasonal climate oscillations: South Maalhosmadulu atoll, Maldives. J Geophys Res 111(F01001):1–12
16. Kench PS, McLean RF, Nichol SL (2005) New model of reef-Island evolution: Maldives, Indian Ocean. Geol 33:145–148
17. Kendall CGStC (2005) Carbonate classification. SEPM Strata: http://www.sepmstrata.org/page.aspx?&pageid=89&4
18. Kinsey DW (1985) Metabolism, calcification and carbon production I. Systems level studies. In: Proceedings of the V international coral reef symposium, vol 4, Tahiti, pp 505–526
19. Kinsey DW (1991) The coral reef: an owner-built, high density, sully serviced, self-sufficient housing estate—or is it? Symbiosis 10:1–22
20. Kleypas JA, McManus JW, Menez LAB (1999) Environmental limits to coral reef development: where do we draw the line? Am Zool 39:146–159
21. Masselink G, Hughes MG, Knight J (2003) Introduction to coastal processes and geomorphology. Routledge, London, 416pp

22. Montaggioni LF (2005) History of Indo-Pacific coral reef systems since the last glaciation: development patterns and controlling factors. Earth Sci Rev 71:1–75
23. Morang A (2004) Coastal geology. US Army Corps of Engineers, University Press of the Pacific, 297pp
24. Neumann AC, Macintyre I (1985) Reef response to sea level rise: keep-up, catch-up and give-up. In: Proceedings of the fifth international coral reef congress, vol 3, pp 105–110
25. Pastorok RA, Bilyard GR (1985) Effects of sewage pollution on coral reef communities. Mar Ecol Prog Ser 21:175–189
26. Pratchett MS, Anderson KD, Hoogenboom MO, Widman E, Baird AH, Pandol JM, Edmunds PJ, Lough JM (2015) Spatial, temporal and taxonomic variation in coral growth: implications for the structure and function of coral reef ecosystems. Oceanogr Mar Biol Annu Rev 53:215–295
27. Stoddart DR (1968) Climatic Geomorphology: review and reassessment. Prog Geogr 1:161–222
28. Stoddart DR (1969) Ecology and morphology of recent coral reefs. Biol Rev 44:433–498
29. Storlazzi CD, Jaffe BE (2008) The relative contribution of processes driving variability in flow, shear, and turbidity over a fringing coral reef: West Maui Hawaii. Estuar Coast Shelf Sci 77:549–564
30. Thomson RE, Wolanski E (1984) Tidal period upwelling within raine Island entrance, great barrier reef. J Mar Res 42:787–808
31. Tucker ME, Wright VP, Dickson JAD (1990) Carbonate sedimentology. Blackwell, London, 482pp
32. Veron JEN (1986) Corals of Australia and the Indo-Pacific. Angus and Robertson, Sydney, 644pp
33. Woodroffe CD (2003) Coasts: form, process and evolution. Cambridge University Press, Cambridge, 623pp

Part IV
Coastal Evolution on a Geological Time Frame

Sleeping geology
on the isolated shore
for millions of years.
Mysterious evolving
problem solving.

"Platypus"
Mr. Bungle

Climate: Climate Variability and Climate Change

24

24.1 What is Climate?

Climate is the set of atmospheric variables existing in a given place over a long period of time [26]. Among the variables that constitute climate are precipitation, atmospheric humidity, temperature, atmospheric pressure, insolation and wind. Climate should not be confused with **weather**, because, although the latter is an expression of the same variables, it is considered at an immediate moment and not over a prolonged period of time. A particular climate is often referred to by reference to the average values of the variables given above. The extreme values of these variables and the probability of reaching them are also often characterized. However, the climate actually involves the characterization of changes in the climatic variables over short periods of time (daily, monthly and annual). In this case, we speak of the **climatic regime** or **climate variability**.

24.2 What is Climate Change?

Climate change involves any change in the statistical values of the aforementioned variables over the long term (at least decades). In fact, when we speak of climate change, we are referring to changes in the climatic regime, ignoring changes occurring over shorter periods of time, which are considered climate variability. In any case, change is an inherent characteristic of the climate whatever the time period considered, so that there has never been a period in the history of the Earth in which all of the variables considered have remained constant. In other words, the term climate change is redundant in the same way that it would be redundant to say *rising upward* or *falling downward*.

Climate changes can be due to natural causes external to the Earth system, such as changes in solar radiation, as well as internal causes such as changes in the way the Earth receives that radiation (changes in the Earth's orbit or changes in atmospheric composition). On the other hand, changes can also be human-induced. Despite the existence of all these possible causes, talk today of climate change at the popular level (and also for many scientists) has become practically synonymous with *global warming caused by humans through the emission of greenhouse gases*.

24.3 Climate Forcing and Climate Mechanisms

24.3.1 Explaining the Global Climate Machine

The 10 km of thickness of the troposphere, the lower layer of the atmosphere, represent a thin sheet with respect to the more than 6000 km of the Earth's radius. However, it is there where most of the air is concentrated and where the climatic phenomena of our planet are forged. In this layer, air circulation takes place in the form of convective cells. The driving force of these are the differences in solar radiation received by the Earth's surface and the thermal exchange between the land or water surface and the air in contact with it. Generally speaking, tropical regions receive more solar energy than polar regions, and this energy is redistributed around the planet thanks to the circulation movements of the atmosphere and the ocean influenced by this energy gradient [26]. The excess of energy received in the intertropical regions causes a warming of the air that decreases its density so that it rises. This ascent creates a permanent belt of low pressure over the Equator called the **equatorial low** or **doldrums**, and causes continuous rainfall at these low latitudes. As this air rises, it is replaced by air arriving from higher latitudes (trade winds). This phenomenon is known as **intertropical convergence** and it generates a convective circulation system called the **Hadley cells** (Fig. 24.1). Between the two Tropics, and on both sides of the Equator, there are two circulation cells that rotate in opposite directions. This phenomenon controls the entire atmospheric circulation, the climatic distribution of the globe and its precipitation patterns. The strip of latitudes influenced by the Hadley cells has a humid and warm climate,

J. A. Morales, *Coastal Geology*, Springer Textbooks in Earth Sciences, Geography and Environment,
https://doi.org/10.1007/978-3-030-96121-3_24

with two distinct seasons (rainy and dry) based on the latitudinal displacement of the cells according to the position of the Earth with respect to the Sun, taking into account the tilt of the Earth's axis.

In the polar circles, the opposite effect occurs. Receiving less solar radiation, the air cools and densifies, circulating downward from the upper parts of the atmosphere. This generates a high-pressure system over the Poles. The excess of downward air generates a convective cell where cold winds at the Earth's surface circulate towards the lower latitudes. These cells are known as the **polar cells**. It is evident that in the polar cells the climate is cold and dry and the lowest temperatures on the planet are produced there.

The surface winds generated in both cells are affected by the Earth's rotation and the Coriolis effect, thus deflecting westward. However, the mid-latitude climate is dominated by the movement of the wind belts which, by way of compensation, circulate eastward in both hemispheres. These are the regions where the cold air masses originating in the polar regions and the warm air rising from the Tropics collide, forming a third type of cell called the **mid-latitude cells**.

In the contact between the mid-latitude cells and the polar cells, masses of air of different temperatures meet, producing an effect called the **polar front**. In the polar front, the warm air from the middle latitudes that circulates towards the higher latitudes rises above the cold air that circulates towards the lower latitudes. This front has a wavy shape, since its position varies depending on the relative wind speed. These undulations of the polar front are known as **Rossby waves**.

At the contact between the mid-latitude cells and the Hadley cells is a belt of high pressure known as the **subtropical high** or **horse latitudes**, located around 30° in both hemispheres. These high pressures cause air to descend from the upper layers of the atmosphere. On reaching the Earth's surface, the air is divided and deflected into each of the cells.

Although it may seem that the distribution of these zones is constant, there is great variability in temperatures and precipitation in each of them throughout the year—as well as between different years, since the polar front and the subtropical high occupy different positions (further north or further south) depending on the oceanic water temperature, which varies from one year to another in relation to solar irradiation.

In the contact of air masses in the horse latitudes, and also in the polar front, horizontal spinning convective cells induce **low pressure centers**. These cells always develop at the contact between the atmosphere and oceanic water due to

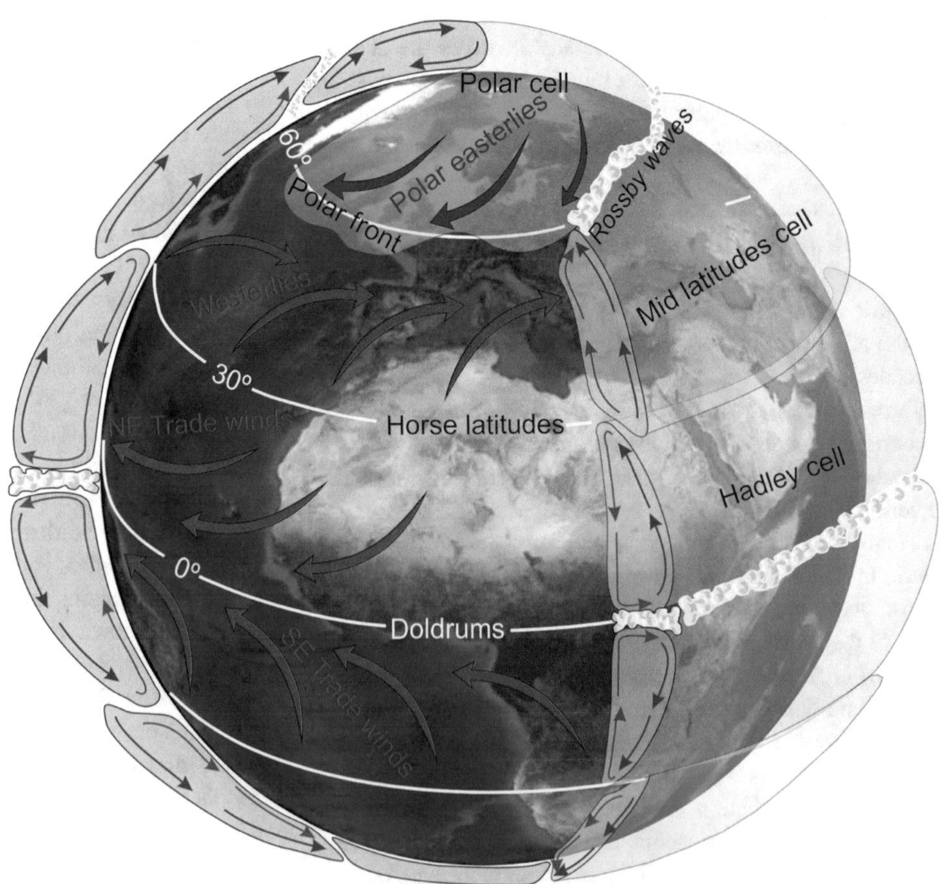

Fig. 24.1 Global convective atmospheric circulation cells. The scale of the atmospheric thickness has been amplified to show the cell winds more clearly

the thermal exchange between the two fluids. The low pressure cells generated in the horse latitudes are called **tropical cyclones** (also hurricanes or typhoons, depending on the ocean in which they originate), while those associated with the polar front are called **squalls** or simply **storms.** Within each of these rotating cells, there is a contact zone between air masses of different temperatures, called **fronts**. In these fronts the cold air mass is introduced under the warm air mass forcing it to rise and generating cloudiness due to adiabatic decompression. The rotating cells have the particularity of moving across the ocean in which they form until they reach the coasts, where they enter the continent and move over it. The climate of mid-latitudes depends entirely on the passage of low pressure cells, which alternate with the establishment of high pressure centers called **anticyclones**.

In mid-latitudes, seasonal climatic differences depend on the position of anticyclones and the trajectory of low pressure gyres. These trajectories depend in turn on the latitudinal position of the polar front and the subtropical highs. The evolution of the coasts of these latitudes is strongly influenced by the passage of these cyclones, which result in strong storms.

24.3.2 The Role of the Sun as the Main Climate Driver

It is clear from the previous sections that the origin of the existence of atmospheric convection cells is insolation. In short, the Sun plays a fundamental role in the Earth's climate, as it is the only source of energy that heats the Earth's surface. Therefore, it can be said that the Sun is the main driver of climate processes [20, 24]. Scientific knowledge about the Sun allows us to state that the star is less stable than it may seem. The energy it emits is far from constant and presents seasons and storms, as well as going through rhythms of activity that are reflected in the existence of irradiation cycles. Today, it is known with certainty that solar irradiation is directly linked to the existence of the phenomenon known as sunspots.

The number of sunspots varies almost daily. A continuous count of sunspot numbers (Fig. 24.2a) shows cycles varying from eight to 15 years. Measurements of the Sun's total irradiance (Fig. 24.2b) meanwhile show a remarkable correlation with the sunspot number curves. While solar irradiance data are not available beyond several decades, there are sunspot number counts from the early seventeenth century (Fig. 24.2c). These data give an accurate picture of the resilience of decadal cycles over centuries, and also allow us to observe how the amplitude of these cycles oscillates secularly, marking other cycles of longer duration.

The decadal cycles of solar irradiance directly influence the temperature distribution of the planet, and cause extensions and retractions of the amplitude of the atmospheric cells described in the previous section. This gives rise to climate oscillations such as the El Niño–Southern Oscillation (ENSO), the North Atlantic Oscillation (NAO) and others located in different oceanic sectors. The differing mean temperature also has a direct influence on the tracks followed by storms. This is true both for storms generated in the horse latitudes and those originating in the polar front.

Longer solar cycles are responsible for longer-term trends of change. For example, the cycle of minimum activity inferred from the near disappearance of sunspots between 1645 and 1715 is known as the Maunder minimum. This period coincided with a global cooling episode known as the "Little Ice Age." A similar episode, although of lesser intensity, occurred between 1798 and 1832 during the Dalton minimum.

In addition to the total solar irradiation, the emission of ultraviolet rays also goes through cycles that directly influence temperatures. As well as direct heating, ultraviolet rays are responsible for the breakdown of ozone molecules in the upper layers of the atmosphere. A third process that significantly influences the Earth's climate is the solar wind. Over time, there are also significant cyclical variations in this process that have indirect effects on climate. An intense solar wind contributes to blocking some of the cosmic rays coming from outer space. The arrival of more or less cosmic rays on Earth causes changes in the production of radioactive atoms in the upper layers of the atmosphere. This phenomenon has consequences in the generation of clouds and, therefore, of cyclical changes in precipitation.

24.3.3 Internal Modes of Climate Variability: Short-Term Climatic Cycles

The term **internal modes of climate variability** refers to changes in climate variables related to the way in which the planet's surface receives solar radiation [8]. Thus, variability related to changes in atmospheric composition, such as those linked to the concentration of greenhouse gases and the entry into the atmosphere of particles from volcanic eruptions, are excluded from this concept. On the contrary, these modes of variability consider the patterns of thermal exchange of the oceans, continents and ice sheets with the atmosphere. The best-known short-period cycles of change are the ENSO and the NAO. Among the most studied cycles of longer duration are the Pacific Decadal Oscillation (PDO) and the Atlantic Multidecadal Oscillation (AMO).

To understand how these cycles work, we can analyze the most studied of them: **ENSO**. This is a climatic pattern that consists of a cyclical change in the climatic parameters of the

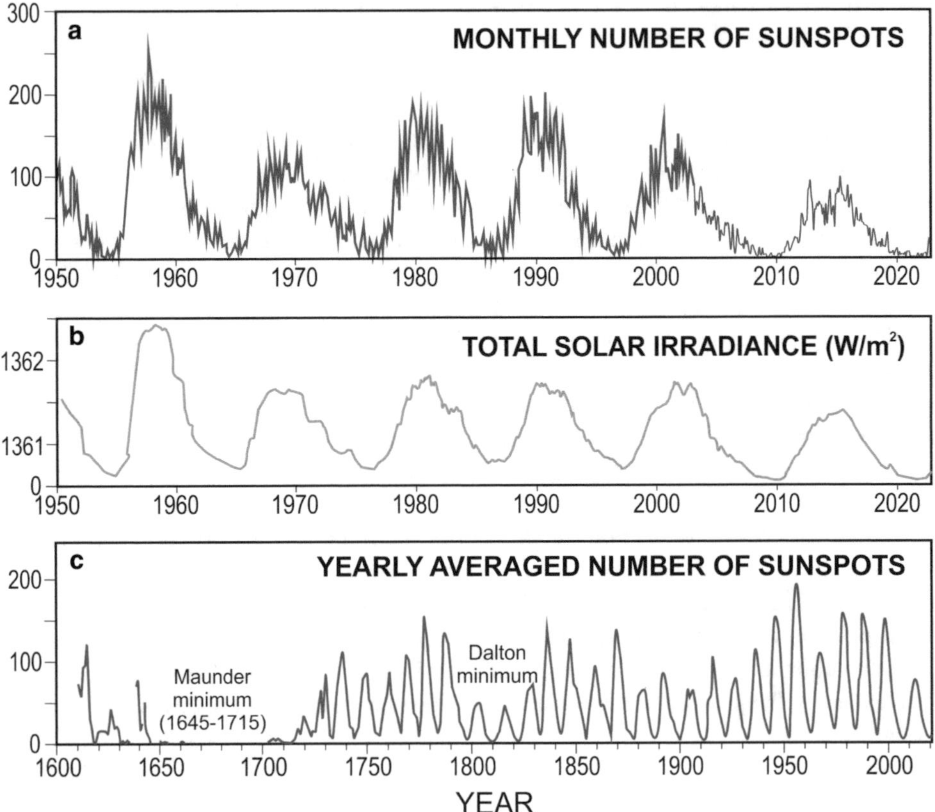

Fig. 24.2 Cycles of solar activity. **a** Daily number of sunspots since 1950. **b** Total solar irradiance (W/m^2). **c** 365 day-averaged number of sunspots since the beginning of records

equatorial Pacific over periods ranging from eight to 15 years. The oscillation consists of a transit between two opposite phases, one of warming of the eastern Pacific surface (known as El Niño) and the other of cooling of the same area (called La Niña). At one extreme, during El Niño, the ocean receives more solar radiation and overheats. The warming of the ocean surface (Fig. 24.3a) is transferred to the atmosphere, which accompanies the increase in temperature with an increase in humidity as it receives masses of evaporated water. Higher temperature and humidity imply an increase in cloudiness and, consequently, in precipitation. This warm and humid air mass expands towards higher latitudes in both hemispheres, shifting the boundaries of all climatic zones in the same direction. At the opposite extreme, during La Niña, the ocean receives less solar radiation and cools (Fig. 24.3b). This cooling causes a temperature transfer from the atmosphere to the water and the air loses moisture as it cools. Consequently, climate cells contract and their boundaries move toward the equatorial zones.

The alternation in time of oceanic water temperature oscillations accompanied by the expansion and contraction of atmospheric cells are responsible for some extreme cyclical changes observed in the contact zones between cells. Thus, El Niño is linked to the occurrence of droughts in some places, while in others there are torrential rains and extreme events such as tropical cyclones. La Niña is responsible for the same phenomena, but in different places [9, 32, 35]. The rest of the known oscillations have similar effects [10, 14, 28, 40] and all of them together cause extreme atmospheric phenomena at the global level (Fig. 24.4).

The study of long-period climatic cycles is approached using different methods: dendrochronology, coral growth phases, rhythmic lamination in sediments and speleothem growth. Any of these methods applied to a particular area shows that decadal alternation of droughts and very wet periods appear recurrently in mid-latitude areas throughout the Holocene (e.g., Cook et al. [5]; Büntgen et al. [7]).

The climatic variability produced by these short-period cycles profoundly affects the dynamics of coastal systems. For example, changes in precipitation lead to a change in the water and sediment input regime to estuarine and deltaic systems, and also affect the sediment input to open coastal environments. Many studies have shown the relationship between these cycles and the change in estuarine mixing type, affecting suspended matter concentration, salinity, dissolved oxygen and nutrient flux.

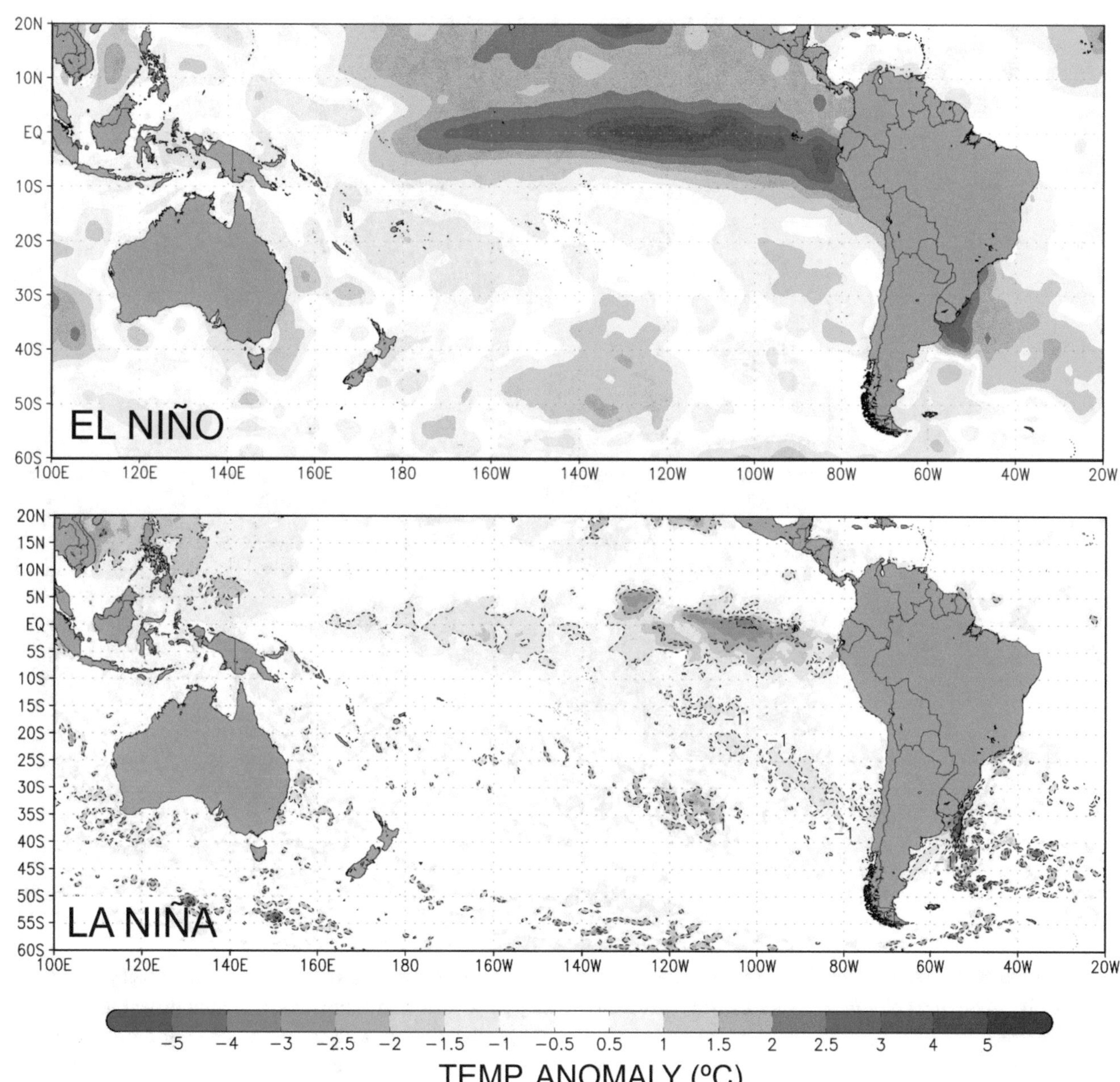

Fig. 24.3 Distribution of temperature deviation during an ENSO cycle. **a** Warming of the equatorial Pacific area during El Niño. **b** Cooling of the same area during La Niña. (Images from the Peruvian National Service of Meteorology and Hydrology.)

24.3.4 Earth's Orbital Influences: Long-Term Climatic Cycles

In addition to the variability produced by solar activity and internal modes, there are other changes that are produced by orbital cycles of the Earth. The repetitive cycles of the Earth's orbit affect the angle of incidence of solar radiation, thereby controlling the effects of incident solar energy. Among these cycles are: the angle of inclination of the Earth's axis, the motion produced by the precession of the equinoxes and the degree of eccentricity of our planet's orbit (Fig. 24.5). The climatic cycles induced by the periodic oscillations of these variables are known as Milankovitch cycles after the Serbian astronomer who discovered them.

From an early age in school, we are taught that the Earth has two motions: one of rotation around itself and the other of translation around the Sun. This apparent simplicity hides orbital variations that occur in both movements. These variations are due to the influence of other bodies present in the solar system. First, the Earth's axis of rotation about itself is displaced with respect to the normal movement of the orbital plane passing through the center of the planet. The angle formed by the two axes is known as **obliquity**, although it is also popularly known as the "tilt of the Earth's

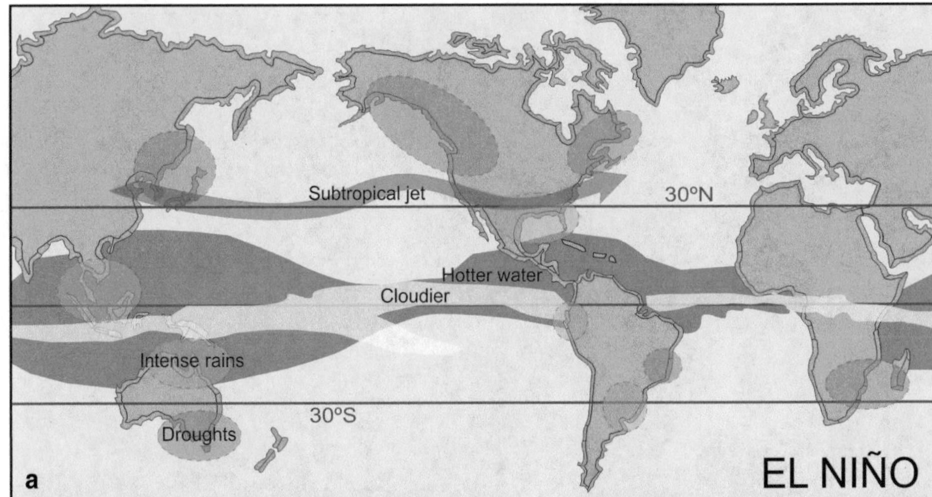

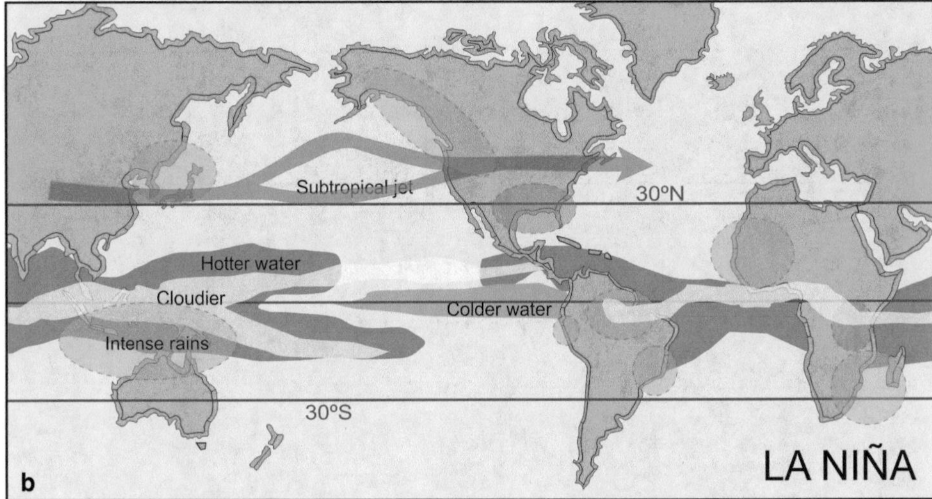

Fig. 24.4 Distribution of anomalous droughts and intense rains associated to internal modes of climatic variability

axis." The obliquity oscillates between 22.1 and 24.3°. The difference between these obliquity values oscillates in cycles of 41,000 years.

Meanwhile, the axis of rotation of our planet changes its position with respect to the orbital plane in such a way that a rotation around the normal position of this plane takes place. This spin is known as axial **precession** and manifests itself in a variation in the orientation towards fixed stars such as the Pole Star (in the Northern Hemisphere) or the Southern Cross (in the Southern Hemisphere). The period of a precession orbit ranges from 19,000 to 24,000 years, with a mean value of 25,771 years. This motion is induced by the gravitational effect of the Moon as it revolves around our planet. The sum of the movements of precession and obliquity variation produce an apparent effect of vibration of the precession trajectory. This vibration is known as **nutation**.

The third orbital cycle occurs in relation to the Earth's translation around the Sun. This motion occurs in a rotation with an elliptical orbital path where the Sun occupies an eccentric position. The degree of similarity of this ellipse to a circle is called **eccentricity**. Over time, the ellipse changes shape between maximum and minimum eccentricity values, completing cycles with a period ranging from 95,000 to 125,000 years, and a cyclic rate of about 400,000 years. These changes are related to the gravitational action of the two major planets of the solar system: Jupiter and Saturn.

The precession, obliquity and eccentricity cycles do not act in phase and combine in a non-linear way. These orbital cycles directly influence the Earth's climate. Obliquity controls the latitudes that receive the most insolation. At the same time, eccentricity controls the length of the seasons and equatorial insolation [4]. Among the combined effects of these cycles are the Pleistocene glaciations, which fit the eccentricity cycles almost perfectly. The influence of Milankovitch cycles on sedimentation has been corroborated by numerous studies in different sedimentary environments. Among them, those carried out in coastal systems are especially interesting, because the sea level movements induced by these climatic cycles are the origin of displacements of the coastline, thus generating coastal sequences.

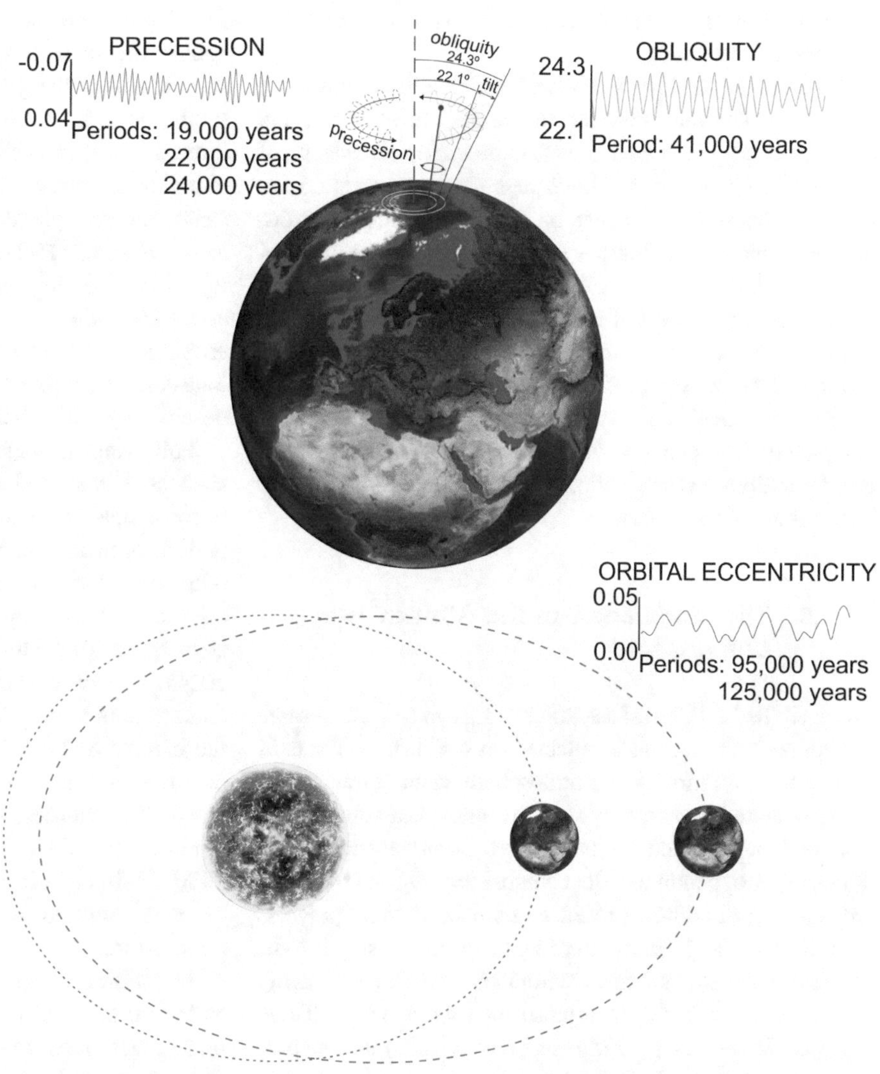

Fig. 24.5 Orbital origin of the Milankovitch cycles

24.3.5 The Processes of Absorption and Reflection of Solar Energy

When solar radiation reaches the Earth's surface, whether continental or ocean surfaces, a transfer of energy takes place. In this process, the terrain of the continents or oceanic water absorbs part of the solar energy. When the energy is absorbed by the oceanic water, overheating occurs, which directly affects the oceanic circulation processes, but also has a response in the atmospheric circulation systems, since this excess energy ends up being transferred to the air. The existence of oscillations such as ENSO that are based on this energy transfer has been discussed in Sect. 24.3.3.

Actually, El Niño's warming of water is only a local example of the general process of energy transfer to the ocean. At the global level, the absorption of solar radiation from the seawater mass is the driver of the ocean circulation that was discussed in Sect. 10.3.1. The surficial ocean circulation cells are the ocean's response to the flow of energy in the atmosphere from the Tropics to the polar regions, which is influenced by the Coriolis force and the presence of the continents. On the other hand, part of this surface water circulates at depth due to the thermohaline circulation. Both currents constitute the so-called **ocean conveyor belt** [23].

As they circulate, ocean currents transport large masses of warm water to latitudes and depths where only cold water would otherwise exist. Warm water currents to deep areas contribute to heat storage in the ocean floor waters. Conversely, currents also transport cold water from higher latitudes and depths to areas of the water surface that would be warmer were it not for the arrival of these water masses. This phenomenon contributes to the existence of a secondary interaction between the oceanic water mass and the atmosphere, so that a heat exchange takes place between them to achieve a thermal balance. This heat exchange contributes to the regulation of the Earth's climate by reducing latitudinal temperature differences.

Much of the energy is reflected and makes its way back into space. The ratio between the amount of absorbed and reflected energy is known as **reflectance**. The reflectance of the surface on which the Sun's rays strike depends on the nature of this surface, but also on its color. In climatic terms, the reflectance of the Earth's surface, expressed in terms of the percentage reflected of the total energy received, is known as the **albedo**. Terrestrial albedo depends on the lithology of the terrain, the degree and type of vegetation cover and the existence of ice cover. This last factor is of great importance, since ice reflects solar radiation almost completely. The presence of large ice-covered surfaces thus greatly inhibits the absorption of energy and contributes to the cooling of the planet.

24.3.6 The Modulation of the Atmospheric Composition

To reach the Earth's surface, solar energy must pass through the atmosphere and, once reflected on the surface, the radiation must pass through the atmosphere again before finally being scattered into outer space. The amount of solar energy absorbed or reflected by our planet is modulated by the atmosphere, depending on its composition (Fig. 24.6). There are certain gases whose presence in the atmosphere prevents the reflected energy from returning to space, causing it to be retained in the atmospheric environment. This excess energy results in an increase in atmospheric temperature. These gases are known as **greenhouse gases** (GHGs) and include common gases such as carbon dioxide (CO_2) and less common gases such as methane (CH_4). Greenhouse gases absorb and release energy more efficiently than other gases in the atmosphere, such as nitrogen and oxygen. Thus, small increases in CO_2 concentrations, and even smaller increases in CH_4 concentrations, have a marked effect on global warming.

The presence of greenhouse gases in the atmosphere is controlled by biogeochemical cycles, as they can be residues of biological activity and can also be stored in the geological record by becoming part of stable organic structures that are fossilized. The abundance of these components in the atmosphere can also be increased by anthropogenic processes such as fuel combustion.

Apart from greenhouse gases, other airborne particles, called aerosols, have a bimodal effect on the Earth's energy balance. On the one hand, they can cause atmospheric cooling by reflecting and dissipating incoming solar energy. On the other hand, they can contribute to warming by absorbing energy arriving from space or reflected at the Earth's surface. The small solid and liquid particles that function as aerosols can be released into the atmosphere by natural processes such as volcanic eruptions or evapotranspiration from the ground, but they can also have a human origin, as in the case of forest fires or emissions generated through industrial activities. Each type of aerosol works in one way or another, cooling or warming the planet. In this sense, for example, the most common aerosol, water vapor, also acts as a greenhouse gas. In the opposite direction, the injection into the stratosphere of enormous quantities of ashes and other aerosols, solid or liquid, due to volcanic eruptions is capable of producing sudden and sometimes long-lasting cooling of the planet. This type of phenomenon deserves special attention.

Following volcanic eruptions with higher explosivity, such as Plinian and Pelean eruptions, the atmosphere can become opaque to incoming solar radiation for prolonged periods of time, even lasting for several years. The ashes and other pyroclastic particles thrown into the air as a consequence of these explosions increase the reflection in the stratosphere of solar energy that does not penetrate the troposphere. Consequently, during these periods, the planet tends to cool down. The effect of some volcanic eruptions on the climate and on the temperature of the planet has been documented on several occasions. In the nineteenth and twentieth centuries, the eruptions of Tambora in 1815, Krakatoa in 1883, Katmai in 1912, Mount St. Helens in 1980, El Chichón in 1982 and Redoubt in 1990 were able to produce significant decreases in the Earth's global temperature.

In addition to volcanoes, other natural phenomena, such as the diffusion of huge dust clouds from deserts, and also human activities can produce effects similar to those of volcanic aerosols.

24.4 A Brief History of Earth's Climate Changes

Throughout its history, the Earth's climate has changed over all timescales due to the multiple factors discussed in the previous sections. Thus, combinations of solar radiation variations (cyclic and episodic), the Earth's orbital cycles, atmospheric composition (and its content of greenhouse gases and aerosols), as well as the different paleogeographic configurations of the planet (due to plate tectonics) have contributed to the existence of constant changes in the climate of our planet. Sometimes, some of these effects have combined to amplify the changes towards cooling or warming. Sometimes, these variables have combined to set in motion unique mechanisms such as the natural release of large quantities of greenhouse gases retained in sediments or the increase in the albedo effect of the surface of ice caps.

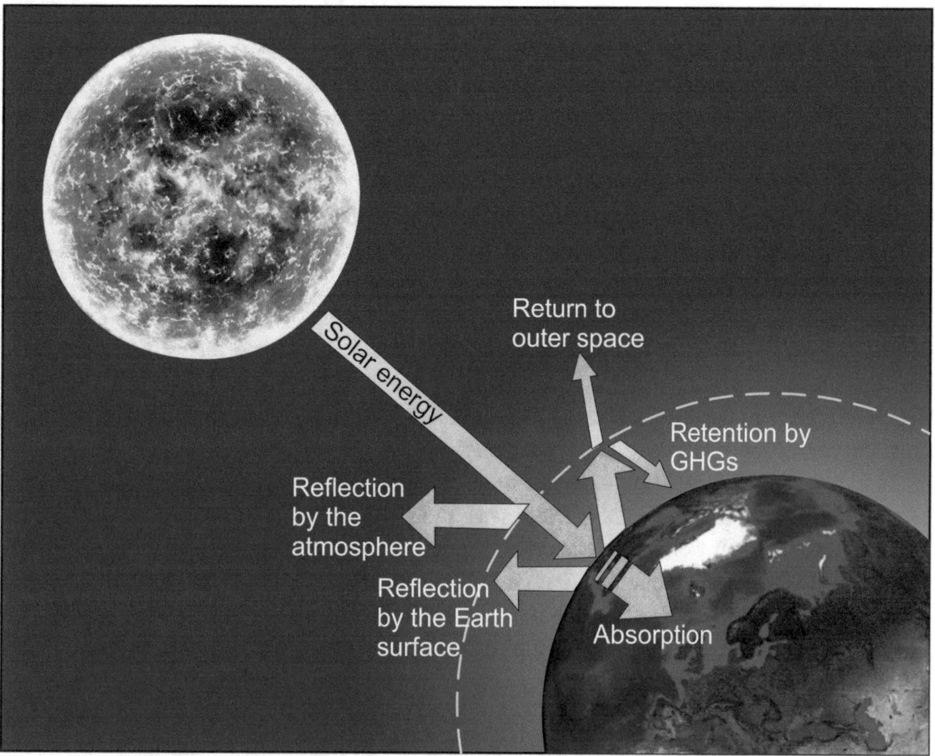

Fig. 24.6 Processes of absorption, reflection and the greenhouse effect

There is ample evidence in the geological record to reconstruct the successive changes in climate that occurred during this history (Fig. 24.7). In this section we will analyze the most significant among them.

Certainly, when speaking of dramatic changes in climate, the first thing that comes to mind are the Pleistocene glaciations (Pt in Fig. 24.7). During this period, there were four glacial periods separated by interglacial stages, with a cyclicity of 100,000 years. It is evident that this value of cyclicity coincides with the periodicity of the cycles of eccentricity of the Earth's orbit, so it follows that the glaciations were mainly related to this phenomenon. However, the Pleistocene glaciations were not the only ones in the history of the Earth, and not even the most dramatic.

Starting from the beginning, but also moving to the most dramatic of climate changes, there was a time in the Precambrian when the entire planet remained frozen for more than 90 million years. This is the period known as the Cryogenian, between 650 and 560 million years ago (Cr in Fig. 24.7). There is geological evidence that the oceanic water mass froze to a depth of 1000 m from the Poles to the Equator. This geological interval is popularly known as "*Snowball Earth*." The mechanisms that led to such a radical glaciation are not yet well understood, but most hypotheses

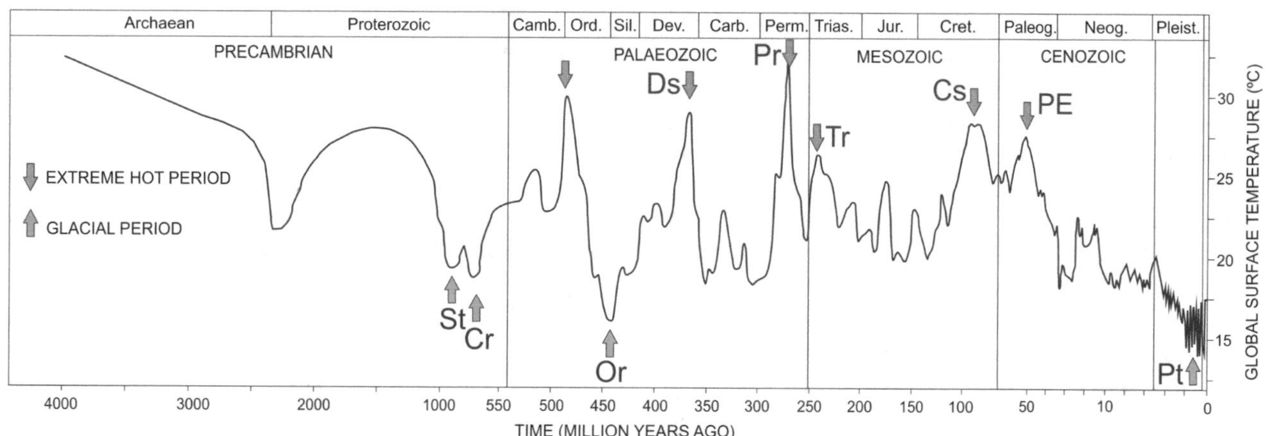

Fig. 24.7 Curve of averaged temperature throughout Earth's history, showing the main glacial periods and the extreme hot events

that attempt to explain them involve a feedback of external and internal effects that contributed to the total cooling. The most plausible hypothesis is related to several initial factors: (1) the Earth was at a time of maximum eccentricity and obliquity; (2) the continents were concentrated in the intertropical zone; and (3) the oceanic circulation was interrupted, doing extreme the thermal contrast. Under these conditions, there was a concentration of ice at the Poles which increased the albedo effect. The coincidence of high albedo and poor thermal redistribution by the ocean currents fed back into the drop in temperatures and the growth of ocean ice masses. Once the glaciers reached tropical latitudes, the greenhouse gases could not counteract the heat loss and the ocean eventually froze completely.

This time interval was the second known major glaciation, since an extreme glacial period was recorded 730 million years ago (the Sturtian glaciation, St in Fig. 24.7). However, it is not proven that this first glaciation was able to freeze the Earth completely. What is well known is that our planet froze almost completely again during the Ordovician, about 447 million years ago (Or in Fig. 24.7). In this case, the glaciation was preceded by a drastic drop in atmospheric CO_2 levels. However, the study of oxygen isotopes shows that the period of glaciation was much shorter, not exceeding 1.5 million years.

Some scientists believe that, with a complete glaciation, our planet could have been frozen forever, since the enormous albedo effect would completely prevent the retention of energy in the atmosphere. So, what were the mechanisms that contributed to the thawing? On the one hand, the CO_2 emitted by the volcanoes would increase its concentration in the atmosphere. On the other hand, the dry environment linked to the presence of ice would prevent the existence of rain and, therefore, no amount of CO_2 could be used in the formation of carbonic acid to be stored geologically through weathering. In addition, the absence of a liquid water body meant that CO_2 could not be stored in the form of carbonates, either. All these mechanisms together led to a thousand-fold increase in atmospheric CO_2 concentration in just 10 million years. As atmospheric warming freed some areas from ice, the decrease in the albedo effect fed back into the system in the opposite direction, increasing thawing. Meanwhile, perhaps the geothermal thawing of the areas near the volcanic foci also played an important role.

In contrast, there have also been times of extreme global warming. The most extreme of these occurred at the end of the Permian, 250 million years ago, when the average temperature of the planet rose by 5° and caused one of the largest mass extinctions in Earth's history (Pr in Fig. 24.7). During this period, the polar ice caps melted completely. This warming was mainly due to causes related to atmospheric composition. In this case, the most plausible hypothesis relates the onset of this rise to an increase in non-explosive basaltic eruptions. The lava flow would have heated the ocean floor to the point of releasing huge amounts of methane hydrate stored there. As discussed in the previous section, methane is a much more potent greenhouse gas than CO_2. Once water saturation in this dissolved gas is reached, enough methane would be released into the atmosphere to explain the temperature rise observed in the oxygen isotopic analysis.

There are also several episodes of global warming caused by increased atmospheric CO_2 concentrations. These occurred in the Upper Devonian (390–359 million years ago; Ds in Fig. 24.7), the Middle–Upper Triassic (245–208 million years ago; Tr in Fig. 24.7), the Upper Cretaceous (100 million years ago; Cs in Fig. 24.7) and the Paleocene–Eocene (60–45 million years ago; PE in Fig. 24.7) periods. None of these episodes reached temperatures as extreme as the Permian, but they were much higher than that observed today. In particular, the last was one of the most rapid global warmings ever recorded in Earth's history. In a period of less than 100,000 years, the global temperature rose several degrees.

In each case, these episodes of extreme climate caused mass extinctions. However, the end of all of them meant an explosion of life and a multitude of new beings appeared, which quickly recolonized all the ecological niches of the planet.

In addition to the climatic changes of natural origin described in the previous paragraphs, there is currently sufficient instrumental and geological evidence to suggest that we are immersed in a global change influenced by human activity. There is a broad scientific consensus on the influence of the increase in atmospheric concentrations of CO_2 of industrial origin on the global increase in temperatures. Other voices affirm that it is possible that solar activity trends are the cause of global warming, although with a notable influence of human activity accelerating the rate of change (e.g., Lüning and Vahrenholt [24]). Apart from the scientific discussion on the degree of influence of solar activity and humans, what is undeniable is that the rate at which this global change is occurring far exceeds that observed in the Paleocene–Eocene, since it is occurring on the scale of centuries and not on the scale of thousands of years as occurred in the case of changes related to natural causes.

24.5 Effects of Climate Changes on the Coast

All coastal systems studied in this book are susceptible to changes in climate—both short-term changes that fall under climate variability and those that occur in the long term. The intensity and direction of wind, the size and origin of waves (fairweather and storm), the magnitude of weather surges, the amount of precipitation, water and sediment discharges from rivers, water temperature, the path and velocity of ocean currents, turbidity and water quality, and the distribution of organisms are all controlling factors in the dynamics of coastal environments, and all of them are controlled by the climate.

Today, there are numerous data instruments collecting information at fixed stations and on board satellites that show changing trends for many of the abovementioned parameters. However, these instrumental records are limited in time and only record changes over a few decades. The geological record can, however, provide data for much longer periods. In this sense, data obtained from boreholes in present-day coastal sediments provide records for thousands of years, while rock sequences can do so for intervals of millions of years. In many cases, this type of data has been used for paleogeographic, paleoecological and paleoclimatic reconstructions that allow us to interpret quite accurately the changes that occurred in coastal systems over periods of the past as a consequence of climate variations (e.g., Cronin and Walker [11]; Bacino et al. [1]; Rice et al. [29]).

Looking ahead, dynamic modeling studies based on climate-dependent changes in hydrological parameters allow us to understand trends in change and to predict the future behavior of coastal environments as a function of current climate change. For example, there are growing efforts to project estuarine responses to thermal change [18], but many other efforts are focused on modeling changes in sediment transport patterns (e.g., Samaras and Koutitas [30]).

The impacts of climate change on coastal systems can affect some variables that are discussed below: temperature (global and regional), regional precipitation, wind and storm regimes, sedimentary processes and sea level rise [8].

24.5.1 Temperature

Temperature variations are the most obvious of all climate changes. Most climate changes have been recorded as thermal changes and are easily analyzed using oxygen isotopes ($\delta^{18}O$). It is true that these variations control the functioning of the coast as an ecosystem. However, the influence of thermal changes on the dynamics of coastal systems is not so obvious. On the one hand, many groups of microorganisms that inhabit coastal areas, such as foraminifera, ostracods, dinoflagellates, diatoms and radiolarians, were affected by the modification of their biogeochemical cycles during the thermal changes that occurred during the Pleistocene [22]. The distribution and relative abundance of these organisms can be used as paleoclimatic indicators in coastal sediments and can contribute to making paleoclimatic reconstructions based on sedimentary records. There are some examples of such studies showing the modifications of estuarine dynamics induced by temperature changes during the Holocene (e.g., Scanes et al. [31]). On the other hand, in terms of macroorganisms, thermal change may also contribute to the modification of the distribution of sediment-disturbing species in sheltered coastal environments such as lagoons and tidal flats. Similarly, the distribution of plants that colonize subtidal environments (mudflats and mangroves) may also be modified by variations in thermal ranges [25].

Of all coastal environments, reefs will certainly be the most directly and indirectly influenced by thermal changes. As these environments are directly related to the growth of organisms, the response of the system to thermal changes is immediate. A warming of the coastal waters has a direct impact on the algae that are symbiotically related to corals. A disappearance of the algae leads to a bleaching of the reef that continues with ecological deterioration and ends with the death of the corals. Indirectly, the increase in water temperature influences its ability to dissolve gases. Warmer water increases its concentration of dissolved CO_2 and thus increases its acidity, preventing coral growth [3].

In more recent times, the sedimentary record shows that many species of organisms inhabiting coastal systems have undergone major shifts in their geographic distribution in the last century, driven by thermal change. Looking ahead, seawater temperatures are expected to increase in the coming decades due to greenhouse gases. This warming could lead to continued changes in the distribution of species of all types of organisms (animals and plants) with low tolerance to temperature changes [17].

24.5.2 Precipitation

Any change in climate implies a modification in rainfall patterns at the regional level. In turn, rainfall patterns have a direct influence on river water and sediment flow. Thus, circulation, water mixing systems and sediment dispersal associated with river mouths such as deltas and estuaries are profoundly modified. Within these systems, the modification of these processes leads to a direct change in the physico-chemical properties of water, such as salinity, dissolved oxygen or nutrients that condition the ecological development of the communities that inhabit these systems [33]. In general terms, it can be stated that the current change presents a clear trend towards the global eutrophication of fluviomarine systems (e.g., Howarth et al. [19]).

Changes in rainfall do not only affect coastal systems associated with river mouths. The modification of sediment inputs brought about by precipitation changes also affects coastal systems open to waves, since these also receive inputs from the mainland via rivers. An increase in precipitation is usually accompanied by an increase in the rate of sedimentation in terrigenous coastal systems, which will be better nourished. However, other coastal systems sensitive to high rates of siliciclastic input will be adversely affected. An example of such systems is reefs, which would be diminished by the increased water turbidity that accompanies increased precipitation.

24.5.3 Wind Regime and Storms

Climate changes imply a variation in the extension of the climatic cells and a displacement of their limits. This necessarily entails an important change in the wind regime and, with it, in the wave regime. There are documented cases of rotations in the direction of the prevailing winds that have influenced coastal dynamics by changing the direction and intensity of coastal drift, or the depositional/erosional regime. On the other hand, a warming or cooling of the ocean surface directly influences the process of storm generation, as well as the intensity that these can reach. This type of phenomena can be preserved in coastal facies in the form of "tempestites," but also in the form of erosive surfaces or changes of direction in the berm lines observed in prograding systems such as wave-dominated deltas or strandplains.

With respect to current climate change, there is talk about an increase in the frequency and intensity of storms [41]. The number of tropical storms (Fig. 24.8) seems to have increased in recent decades [25, 39], but so have storms associated with Rossby waves at the polar front acting at higher latitudes [38]. The increase in these storms is expected to alter the seasonal sediment balance in open coastal systems and increase coastal erosion in these regions.

In addition to the direct effects of waves on the waterfront, the surge that accompanies storms causes flooding, and produces water renewal in coastal systems protected from wave action such as lagoons, estuaries and deltas [13]. In many cases, these events cause beneficial effects from an ecological point of view (e.g., Paerl et al. [27]). Some of these effects can become long-lasting through processes such as the opening of new inlets in barrier island systems or the deepening of river mouth channels [2, 12]. Large storms can also affect coastal wetlands during storm surges, contributing to a short-term increase in sedimentation rates [6] or the deposition of cheniers.

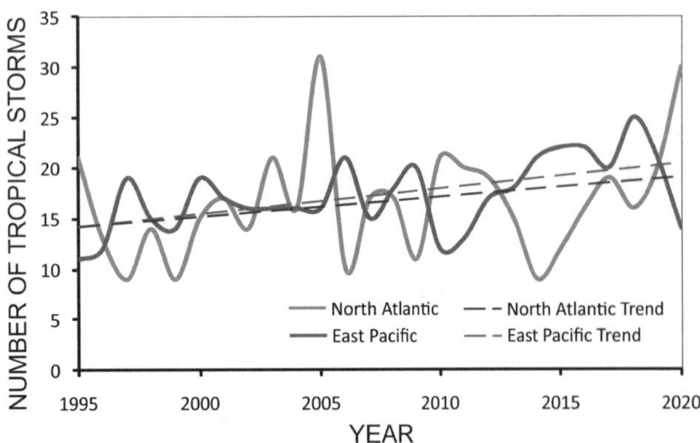

Fig. 24.8 Number of tropical storms per year in the North Atlantic and Eastern Pacific oceanic areas since 1995, showing the increasing trends

24.5.4 Sedimentary Processes

The sedimentary dynamics of coastal systems are profoundly affected by climate. The climate-influenced factors discussed above (temperature, precipitation, and wind and wave regime) are primarily responsible for the dynamic functioning of coastal systems. Thus, the influences of climate on each of these factors have a direct effect on sedimentary processes. The three phases of the sedimentary cycle (erosion, transport and sedimentation) modify their degree of action and their distribution in space and time in response to these changes.

One of the main direct effects is related to variations in precipitation and river flow, since these two variables largely control the volume of sediment input to coastal environments. But there are also other effects on input caused by temperature variations. For example, an increase in temperature and the resulting decrease in glacier ice mass completely modifies the sediment input to fjords [36].

In sheltered systems such as deltas, estuaries, lagoons, tidal flats or marshes and mudflats, the combined variations in temperature and river flow not only influence the volume of sediments, but also their nature. As for siliciclastic sediments, their distribution depends to a large extent on the dynamics of the turbidity maximum. These clouds of suspended matter are controlled by river flow, which in turn is controlled by the precipitation regime. Changes in estuarine mixing conditions imply changes in the dynamics of these turbidity clouds. For example, an increase in freshwater flow is usually accompanied by a change in mixing conditions. The change from saline wedge conditions to well-mixed ones leads to a marked increase in the volume of flocculated material, making the sediments in the central part of estuaries finer and more cohesive [16]. This phenomenon may also influence the type of diffusion that occurs at the mouths of distributary channels at the front of deltaic systems.

Within a closed system, not only clastic sediments are deposited, but also the accumulation of organic matter has a significant influence. This organic matter comes from the remains of organisms (animal and plant) and from flocculation processes. These processes, in turn, depend on climate-influenced parameters such as temperature, salinity, turbidity and light penetration. A large part of this organic matter comes from the productivity of microalgae fed by high concentrations of nutrients and another part from the organisms that feed on these algae, thus sustaining the entire food chain [15, 37].

In open coastal systems such as beaches, dunes and barrier islands, changes in sediment dynamics are more closely linked to variations in wind and wave regimes (including storms). As discussed in the previous section, these factors can modify the littoral drift transport cells, the volume of material transported along the coast, the volumes of seasonal transverse sand movement to and from the coast, and the overall sediment balance. In general terms, many authors agree that current climate change is shifting open systems towards more erosional dynamics [21].

Looking at changes in the trends of current sedimentary processes, it is difficult to differentiate those that are directly related to global warming from those related to the alteration of coastal systems induced by direct human action. On the one hand, the deforestation carried out in many continents in connection with agricultural development (eighteenth century in Europe and nineteenth century in the rest of the world) led to a significant increase in the sediment load contributed by rivers. On the other hand, the retention of sediment in dams in the last century represents a negative effect at the global level [34]. These effects have also had a notable influence on the dynamics of deltas and estuaries, as well as on the nourishment of coastal areas adjacent to them.

24.5.5 Sea Level

It has been seen throughout this chapter that one of the main manifestations of climate change on any timescale are thermal changes. When warming or cooling is large enough and prolonged enough, it is directly reflected in the volume of water stored in glaciers. The extent of ocean ice masses has a connection to the volume of water in the oceans and, by extension, to sea level. The relationship between sea level movements and the last glacial periods is well known. In this case, the expansion and retreat of the ice involved vertical movements of more than 100 m in the sea level.

It is evident that such important movements have an influence on coastal evolution. However, such changes take place at a slow rate and the coast has a long-term response. The magnitude of the importance of sea level in coastal evolution merits a specific chapter in this book.

References

1. Bacino GL, Dragani WC, Codignotto JO (2019) Changes in wave climate and its impact on the coastal erosion in Samborombón Bay, Río de la Plata estuary, Argentina. Estuar Coast Shelf Sci 219:71–80
2. Bailey JF, Patterson JL, Paulhus JLH (1975) Hurricane Agnes rainfall and floods, June–July 1972. United States Government Printing Office. Geological Survey Professional Paper 924. Washington, DC
3. Baker AC, Glynn PW, Riegl B (2008) Climate change and coral reef bleaching: an ecological assessment of long-term impacts, recovery trends and future outlook. Estuar Coast Shelf Sci 80:435–471
4. Berger A, Loutre MF, Mélice JL (2006) Equatorial insolation: from precession harmonics to eccentricity frequencies. Clim Past Discuss 2(4):519–533
5. Büntgen U, Trouet V, Frank D, Leuschner HH, Friedrichs D, Luterbacher J, Esper J (2010) Tree-ring indicators of German summer drought over the last millennium. Quatern Sci Rev 29:1005–1016
6. Cahoon DR (2006) A review of major storm impacts on coastal wetland elevations. Estuaries Coasts 29:889–898
7. Cook BI, Smerdon JE, Seager R, Cook ER (2014) Pan-continental droughts in North America over the last millennium. J Clim 27:383–397
8. Cronin TM (2015) Climate change. In: Kenish MJ (ed) Encyclopedia of Estuaries. Springer, Heidelberg, pp 122–128
9. Cronin TM, Dwyer GS, Schwede SB, Dowsett H (2002) Climate variability from Florida Bay Sedimentary Record: possible teleconnections to ENSO, PNA, and CNP. Climate Res 19:233–245
10. Cronin TM, Thunell R, Dwyer GS, Saenger C, Mann ME, Vann C, Seal RR (2005) Multiproxy evidence of Holocene climate variability from estuarine sediments, Eastern North America. Paleoceanography 20:PA4006
11. Cronin TM, Walker H (2006) Restoring coastal ecosystems and abrupt climate change. Clim Change 74:369–374
12. Davis J, Laird B, Ruzecki EP, Schubel JR, Huggett RJ, Anderson AM, Wass ML, Marasco RJ, Lynch MP (eds) (1977) The effects of Hurricane Agnes on the Chesapeake Bay Estuarine system. The Johns Hopkins University Press. Chesapeake Research Consortium Publication 54. Baltimore
13. Davis RA Jr, Knowles SC, Bland MJ (1989) Role of hurricanes in the Holocene stratigraphy of Estuaries: examples from the Gulf Coast of Florida. J Sediment Petrol 59:1052–1061
14. Enfield DB, Mestas-Nunez AM, Trimble PB (2001) The Atlantic multidecadal oscillation and its relation to rainfall and river flows in the continental US. Geophys Res Lett 28:2077–2080
15. García-Gómez JC, González AR, Maestre MJ, Espinosa F (2020) Detect coastal disturbances and climate change effects in coralligenous community through sentinel stations. PLOS One 15(5): e0231641
16. Geyer WR (1993) The importance of suppression of turbulence by stratification on the estuarine turbidity maximum. Estuaries 16:113–125
17. Helmuth B, Harley CDG, Halpin PM, Oandapos DM, Hofmann GE, Blanchette CA (2002) Climate change and latitudinal patterns of intertidal thermal stress. Science 298:1015–1017
18. Hordoir R, Samuelsson P, Schimanke S, Fransne F (2019) Changes of the overturning of a fjord-type estuary in a warmer climate, a test case in the Northern Baltic sea. Cont Shelf Res 191:104007
19. Howarth R, Chan F, Conley DJ, Garnier J, Doney SC, Marino R, Billen G (2011) Coupled biogeochemical cycles: eutrophication and hypoxia in temperate estuaries and coastal marine ecosystems. Front Ecol Environ 9:18–26
20. Hoyt DV, Schatten KH (1997) The role of the sun in climate change. Oxford University Press, Oxford, 279pp
21. Komar PD (1998) Beach processes and sedimentation. Prentice Hall, New Jersey, 546pp

22. Kucera M, Rosell-Melé A, Schneider R, Waelbroeck C, Weinelt M (2005) Multiproxy approach for the reconstruction of the glacial ocean surface (MARGO). Quatern Sci Rev 24:813–819
23. Kuhlbrodt T, Griesel A, Montoya M, Levermann A, Hofmann M, Rahmstorf S (2007) On the driving processes of the Atlantic meridional overturning circulation. Rev Geophys 45:1–32, RG2001
24. Lüning S, Vahrenholt F (2016) The sun's role in climate. In: Easterbrook EJ (ed) Evidence-based climate science. Elsevier, Oxford, pp 283–305
25. Nicholls RJ, Wong PP, Burkett VR, Codignotto J, Hay J, McLean R, Ragoonaden S, Woodroffe CD (2007) Coastal systems and low-lying areas. In: Parry ML, Canziani OF, Palutikof J, van der Linden PJ, Hanson CE (eds) Climate change 2007: impacts, adaptation and vulnerability. Contribution of Working Group II to the Fourth Assessment Report of the Intergovernmental Panel on Climate Change. Cambridge University Press, Cambridge, pp 315–356
26. NOAA (2009) Climate literacy: the essential principles of climate science. US Global Change Research Program, Washington, 18pp
27. Paerl HW, Bales JD, Ausley LW, Buzzelli CP, Crowder LB, Eby LA, Fear JM, Go M, Peierls BL, Richardson TL, Ramus JS (2001) Ecosystem impacts of three sequential hurricanes (Dennis, Floyd, and Irene) on the United States' largest lagoonal estuary, Pamlico Sound, NC. Proc National Acad Sci 98:5655–5660
28. Prasad M, Sapiano M, Anderson C, Long W, Murtugudde R (2010) Long-term variability of nutrients and chlorophyll in the Chesapeake Bay: a retrospective analysis, 1985–2008. Estuaries Coasts 33:1128–1143
29. Rice JA, Simms AR, Buzas-Stephens P, Steel E, Livsey D, Reynolds LC, Yokoyama Y, Halihand T (2020) Deltaic response to climate change: the Holocene history of the Nueces Delta. Global Planet Change 191:103213
30. Samaras AG, Koutitas CG (2014) Modeling the impact of climate change on sediment transport and morphology in coupled watershed-coast systems: a case study using an integrated approach. Int J Sedim Res 29:304–315
31. Scanes E, Scanes PR, Ross PM (2020) Climate change rapidly warms and acidifies Australian estuaries. Nat Commun 11:1803
32. Schmidt N, Lipp EK, Rose JB, Luther M (2001) Analysis of ENSO related trends in Florida precipitation and streamflow. J Clim 14:615–628
33. Schubel JR, Pritchard DW (1986) Responses of upper Chesapeake Bay to variations in discharge of the Susquehanna River. Estuaries 9:236–249
34. Syvitski JPM, Vörösmarty CJ, Kettner AJ, Green P (2005) Impact of humans on the flux of terrestrial sediment to the global coastal ocean. Science 308:376–380
35. Swart PK, Dodge RE, Hudson HJ (1996) A 240-year stable oxygen and carbon isotopic record in a coral from South Florida: implications for the prediction of precipitation in southern Florida. Palaios 11:362–375
36. Szczucinski W, Zajaczkowski M, Scholten J (2009) Sediment accumulation rates in subpolar fjords—impact of post-Little Ice Age glaciers retreat, Billefjorden, Svalbard. Estuar Coast Shelf Sci 85:345–356
37. Uncles RJ, Stephens JA, Harris C (2015) Estuaries of southwest England: salinity, suspended particulate matter, loss-on-ignition and morphology. Prog Oceanogr 137:385–408
38. Wang S, McGrath R, Hanafin J, Lynch P, Semmler T, Nolan P (2008) The impact of climate change on storm surges over Irish waters. Ocean Model 25:83–94
39. Webster PJ, Holland GJ, Curry JA, Chang HR (2005) Changes in tropical cyclone number, duration, and intensity in a warming environment. Science 16:1844–1846
40. Xu J, Long W, Wiggert JD, Lanerolle LWJ, Brown CW, Murtugudde R, Hood RR (2012) Climate forcing and salinity variability in Chesapeake Bay, USA. Estuaries Coasts 35:237–261
41. Yasuda T, Nakajo S, Kim SY, Mase H, Mori N, Horsburgh K (2014) Evaluation of future storm surge risk in East Asia based on state-of-the-art climate change projection. Coast Eng 83:65–71

25 Relative Sea Level: Eustatism and Tectonics

25.1 Why Sea Level Movements Are Relative

The coast is the place where the sea meets the continent. It is obvious that the location of this meeting place depends on the height of the water level as well as on the height of the terrain. Both of these change over time. Sea level movements in absolute terms can have the same effects on the coast as vertical land movements. In fact, the coast will undergo displacements towards the continent or towards the sea in response to movements in both elements, or the combination between them. When we study a given coast, we always speak of relative movements because what we observe are the results of the net movement, without knowing whether it is the sea or the land (or both) that is moving [28]. Determining the causes of these movements invariably requires more detailed study.

While relative sea level changes are observable at the regional level, the study of global changes always requires a comparison of the results obtained on different coasts. In this way, it is possible to discriminate between changes that are due to land movements (local) and those that are actually due to absolute sea level movements (global). At least five global and four regional processes influence the relative sea level position on any given coast [6]. The contribution of each factor is different for each coast. The causes of these changes, as well as the effects on the coast of these changes, will be discussed separately in this chapter.

25.2 Global Movements: Eustatism

Eustatism is the phenomenon that causes absolute rises and falls of the sea level and, therefore, takes place globally. Eustatic variations can be associated with different geological phenomena, although it is estimated that these phenomena can be attributed to two types of origin: (1) variation of the volume of water in the oceans and (2) deformation of the receiving basin.

25.2.1 Global Changes in the Volume of Ocean Waters

Most eustatic movements are associated with variations in the volume of water in the oceans. Thus, an increase in volume will be in response to a rise in sea level and, conversely, a decrease in sea level will be in response to a loss of volume. Assuming that the total mass of water in the world has remained constant over the last few hundred million years, there are several possibilities that could cause the volume of ocean water to vary.

The most obvious cause for the change in the mass of water contained in the oceans is the change of state of this water. That is, when a significant part of the water is transformed into ice, the volume of water remaining in the liquid state decreases, and vice versa [16]. This phenomenon directly links eustatic changes to global climate changes. Specifically, glaciations are the most frequent phenomenon that originates this type of eustatic movement, which is why this is called **glacioeustatism**.

The best known glacioeustatic variations are those related to the four glacial periods that occurred during the Pleistocene. These variations involved rises and falls of more than 100 m (Fig. 25.1). The present position of the sea level and the development of all today's coastal systems are linked to the last of these rises, as a consequence of the melting of the Würm glaciation and the subsequent stabilization over several thousand years. The total disappearance of polar ice associated with periods of extreme warming, such as those occurring at the end of the Permian or during the Paleocene–Eocene, meant a rise in water levels up to several tens of meters above the present level.

On the other hand, there are causes that can vary the mass of water stored in continental areas. In addition to glaciers, this water may be retained in lakes, aquifers and permafrost. Water retention in continental areas varies with climate. A drier climate transfers water to the oceans, causing the mean sea level to rise. Conversely, a humid climate increases

J. A. Morales, *Coastal Geology*, Springer Textbooks in Earth Sciences, Geography and Environment,
https://doi.org/10.1007/978-3-030-96121-3_25

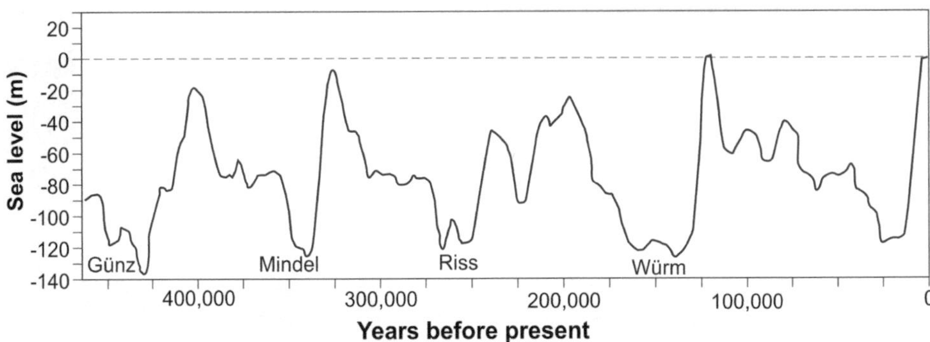

Fig. 25.1 Glacioeustatic changes during the Pleistocene

the level of water retained on the continents and lowers the sea level [20].

Minor variations in the volume of ocean water that are not associated with the oceanic water mass may also be related to climate. For example, the volume of liquid water varies with changes in temperature (thermal expansion). Generally speaking, water density increases as temperature decreases [33] while, on the contrary, a warming implies an increase in volume. As an example, an increase of 1 °C in an ocean 4000 m deep would imply a vertical rise in sea level of 60 cm [21].

These eustatic variations are directly linked to climate variations and their causes were analyzed in detail in Chap. 24 of this book.

25.2.2 Changes of Shape and Size of Oceanic Basins

The dynamics related to plate tectonics introduce horizontal stresses in the oceanic crust that produce changes in the morphological configuration of the seafloor. On the one hand, the opening of some ocean basins while others are narrowing can modify their capacity to store water. On the other hand, horizontal stresses related to the expansion of the ocean floor can be transformed into vertical movements in the vicinity of subduction zones and mid-ocean ridges. The bathymetric position of the ridges may also undergo changes due to the arrival of magmatic masses to the reservoirs located beneath them, or because of periods of more intense volcanic effusion. In this situation, there may also be upwelling of the seafloor related to the increase in mean rock temperature (thermal expansion and contraction).

These deformations in the shape and size of ocean basins produce sea level changes on a global scale (Fig. 25.2). Although in absolute terms these phenomena can produce eustatic movements of several tens of meters, in general the rate at which they occur is very slow, in the order of centimeters per century.

25.2.3 Effects on Coastal Systems

Considering only eustatic movements (that is, assuming local tectonics that do not modify the vertical position of the continents and in the absence of the effects of sedimentation and erosion), a rise in sea level implies a landward displacement of the coast [25]. Under the same assumption, a fall in sea level would imply a seaward displacement of the coast.

An absolute rise in sea level leads to transgressive conditions and results in inundation and landward migration of coastal environments (Fig. 25.3a). If the rise is slow and progressive, the facies superposition will respect Walther's Law with the facies in reverse order, placing the deeper facies on top of the shallower ones. However, if the ascent is abrupt, depth conditions change rapidly and may go from very shallow to deeper environments, lacking intermediate terms, without respecting Walther's Law.

Conversely, an absolute sea level fall leads to regressive conditions and results in emersion of coastal environments and seaward progradation of systems (Fig. 25.3b). In the offshore seafloor, a rise in sea level will lead to a transition to deeper conditions, while a fall will lead to a transition to shallower conditions. Thus, the deeper facies will be located on top of the shallower facies. If subsidence occurs slowly, the facies will be arranged in the order established by Walther's Law; however, if sharp pulses of subsidence occur, intermediate terms may be missing.

Of all the eustatic movements in the history of the Earth, the one that followed the melting of the Würm transgression is the one that gave rise to the starting point for the evolution of all present-day coastal systems. This is the upwelling known as the Holocene transgression (or Flandrian transgression). This rise occurred at a rate that exceeded 2 cm per year at its peak [6].

After a stabilization period lasting more than 4500 years, the mean sea level has started to rise again, reaching a rate of 3.1 mm/year in recent decades (TOPEX website). This is a new eustatic change due to thermal expansion of water and

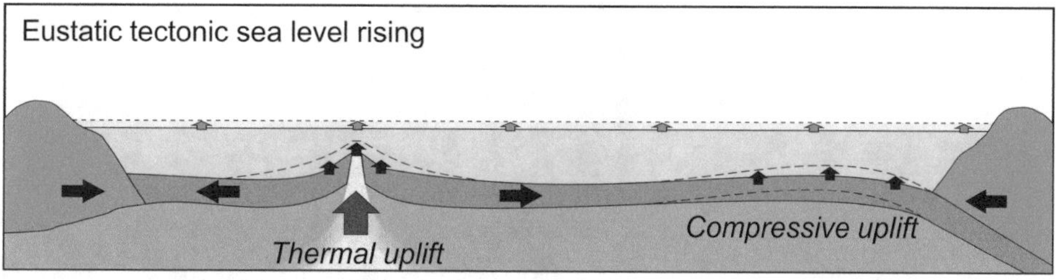

Fig. 25.2 Eustatic changes due to vertical movements of the oceanic bed. The figure indicates two mechanisms of tectonic sea level rising. The inverse movements of sea level fall may be produced by thermal contraction and tectonic relaxation

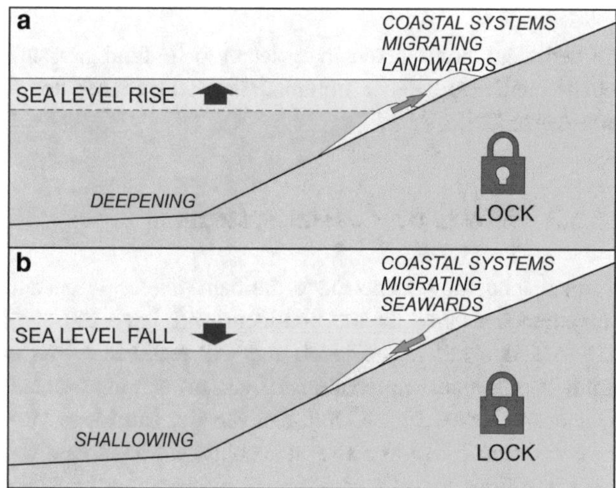

Fig. 25.3 Effects of eustatic sea level movements. **a** Sea level rise. **b** Sea level fall

melting of land glaciers. The effect on current coastal systems is diverse [15]. In barrier island systems, the rise means that dune systems are frequently being overtopped by washover events. Many dune ridges have already disappeared and islands are undergoing landward migration by shifting over sediments of the inner systems [32]. In estuaries, sea level rise is leading to a change in water mixing patterns and a shift of the salt wedge upstream (e.g., Murphy et al. [18]; Hong and Shen [24]. In other coastal wetlands, changes in tidal circulation are modifying some of the parameters controlling organic sedimentation, such as nutrient distribution or dissolved oxygen [4]. Other coastal systems that are highly dependent on organic activity, such as marshes, mangroves or reefs, need to achieve vertical growth rates to compensate for the rate of sea level rise in order to avoid inundation.

Advanced box 25.1 Global eustatic movements: the Exxon curve

Since the development of sequence stratigraphy, one of the main interests of geologists has been the establishment of a global eustatic curve through Earth's history. In the 1970s, oil company geologists had established the principles of interpreting seismic profiles in terms of relative sea level position. During these years, exploration for energy resources had led to the study of several continental margins throughout the world, and by the end of that decade, valuable information was available. The first curve of eustatic cycles was proposed by Vail et al. [31]. This was based on the study of seismic profiles of passive continental margins in a variety of regional tectonic settings. For the establishment of the global eustatic curve, the authors discriminated between sea level movements that appeared in all basins (global cycles) and those that appeared only in some of them (regional tectonic movements). To establish the magnitude of the global changes, they averaged the magnitudes observed in all the basins studied, taking into account their tectonic framework.

During the following decade, many authors made efforts to improve that curve. That task was much easier for more modern sedimentary formations. Chappell and Shackleton [5] produced the first high-resolution global eustatic curve for the last 250,000 years using data from a series of New Guinea rift terraces. It was the same team of authors, this time led by Haq [17], who proposed an improved curve of eustatic cycle for the most recent periods of Earth's history (Fig. 25.4). As these authors worked for an oil company, the curve was named on behalf of their organization as the Exxon Curve. It is a curve in permanent revision, since this group permanently collects information from seismic profiles, boreholes and stratigraphic sections, from the most diverse parts of the world, and maintains an updated data bank to check and modify previous proposals. The most accurate Exxon curve covers the time interval for the last 255 million years, from the Upper Permian to the present day. Data available for earlier time intervals have allowed curves to be proposed, but not to a similar degree of accuracy.

25.3 Regional Movements: Uplift and Subsidence

In addition to the eustatic changes that can be produced by tectonic activity, there are other tectonically induced changes at the regional level that can be much more rapid and also of

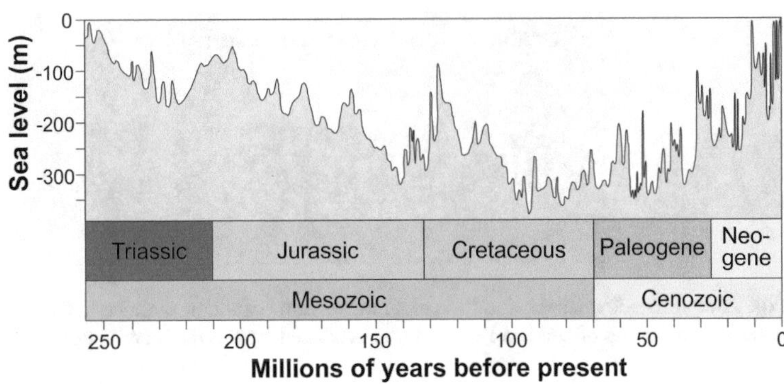

Fig. 25.4 Curve of eustatic sea level movements for the last 255 million years (built with data by Haq et al. [17])

greater magnitude. As a consequence of local tectonic activity, the floor of ocean basins can rise or sink [12].

25.3.1 Causes of Regional Movements

The processes leading to ocean floor uplift or downfall can have several origins. Uplift can be related to: (a) elastic deformation due to compressive tectonic processes; (b) isostatic readjustment; and (c) thermal expansion. The downfall process is known as **subsidence** and can also have several origins: (a) tilting due to extensional tectonic processes; (b) isostatic subsidence; (c) thermal shrinkage subsidence; and (d) compressional subsidence of older sedimentary formations due to the loading of upper sediments. Isostatic upward and downward movements are often regulated by the gain or loss of mass in the ice caps. This phenomenon is known as glacio-isostatic rebound and is due to the viscoelastic response of the Earth's mantle to mass variations in the large ice sheets [13]. Continental uplift and downfall by isostasy can also occur at locations associated with orogenic fronts. An orogeny always implies a crustal thickening that leads to an isostatic subsidence response. Similarly, a crustal thinning by extension or by erosion of a mountain range has as a response an isostatic readjustment of elevation [29].

Loading subsidence is much more common in smaller areas. Some coastal environments characterized by rapid sedimentation and high sedimentary input are capable of generating sedimentary bodies that overlap one another over a short space of time. For this reason, many deltas, for example, have accumulated thick aggradational and progradational bodies during the Holocene. Such rapid sedimentation does not guarantee that the sediments adopt the highest degree of packing during the depositional process. The vertical accumulation of successive sediment bodies implies an increase in weight on the lower bodies, which forces them to optimize their space by increasing their packing and decreasing their volume. This loss of volume of the lower bodies has the effect of sinking the surface [1]. Sometimes, the loss of groundwater, oil or gas contained in the sediment pores leads to a decrease in fluid pressure, which facilitates grain accommodation and contributes to subsidence [26].

25.3.2 Effects on Coastal Systems

Vertical land movements have the same effect as eustatic movements. A rise of the continent will have the same response as a fall in sea level, and will result in a loss of depth and the superimposition of shallower environments over deeper ones (Fig. 25.5a). Conversely, subsidence will have the same effect as a rise in absolute sea level, and will cause the floor to sink, resulting in a deepening and relocation of coastal systems to the continent (Fig. 25.5b).

25.4 Combined Tectonics and Eustatism

Relative sea level changes refer to variations in sea depth at a given point during a specific time interval due to the combined action of regional tectonic movements and eustatism.

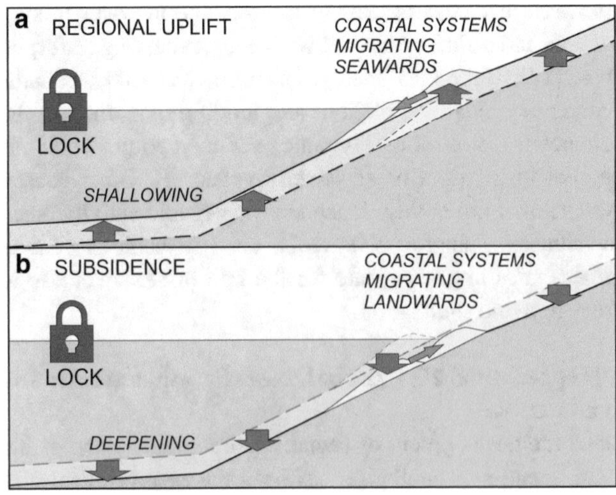

Fig. 25.5 Effects of regional land movements. **a** Uplift. **b** Subsidence

Thus, relative sea level rise may be due to eustatic rise (Fig. 25.6a), subsidence (Fig. 25.6b) or a combination of both. Among the possible combinations for relative sea level rise to occur are: eustatic rise accompanied by subsidence (Fig. 25.6c), rapid subsidence accompanied by slow eustatic subsidence (Fig. 25.6d) and rapid eustatic rise accompanied by slow tectonic rise (Fig. 25.6e). The highest rates and magnitudes of relative uplift will occur in the third case, when eustatic uplift and subsidence coincide.

On the contrary, a relative sea level fall is an observable effect whose cause can be found in a eustatic lowering (Fig. 25.7a), a tectonic uplift (Fig. 25.7b) or a combination of both phenomena. Within these, combinations may occur: eustatic subsidence accompanied by tectonic uplift (Fig. 25.7c), rapid eustatic uplift accompanied by slow eustatic subsidence (Fig. 25.7d) and rapid eustatic fall accompanied by slow subsidence (Fig. 25.7e).

An episode of relative sea level stand will result from absolute stability or when both phenomena counteract each other.

The response to a relative rise in sea level is the process of migration of coastal systems towards the continent, generating a deepening sequence. Conversely, the effect of a relative fall is the process of migration of coastal systems seaward, generating a shallowing sequence. From the point of view of the geological record, these are the effects we see. That is why geologists always speak of relative rise and fall. Some authors have even questioned whether eustatic movements act in isolation without the effect of regional tectonic movements [23].

Advanced Box 25.2 Ocean surface topography

Until the middle of the twentieth century, geologists spoke of sea level as if the ocean had an equipotential surface adapted to the geoid. It was assumed that this novel sea level varied only on very short timescales due to wave action, surges and tides, but that it maintained an average position. It was assumed that the level varied in the long term because of eustatic changes, but when it changed it did so simultaneously all over the world. However, the beginning of the use of topographic instruments on satellites orbiting the Earth completely changed this perception.

Spacecraft in orbit around the planet carry precision instruments that make very close measurements of the Earth's topography and also of the height of the ocean surface. Specially designed radar altimeter systems are used to make extremely accurate measurements of the altitude of the ocean surface. These instruments achieve an accuracy of less than 3 cm. Both land and ocean topography are processed and characterized in the same way. Normally, both are expressed as a positive or negative height relative to the geoid—i.e., above or below this theoretical surface. In the sea, the geoid would be the shape of the sea surface if there were no topography—i.e., what was hitherto called "sea level." It is clear that the marine topography measured by satellite instruments is an instantaneous topography. Therefore, it is necessary to process the values obtained to eliminate the effect of short-term variations such as those from tide and wind [3]. The result obtained is a relief that shows elevated and depressed areas on the ocean water surface (Fig. 25.8).

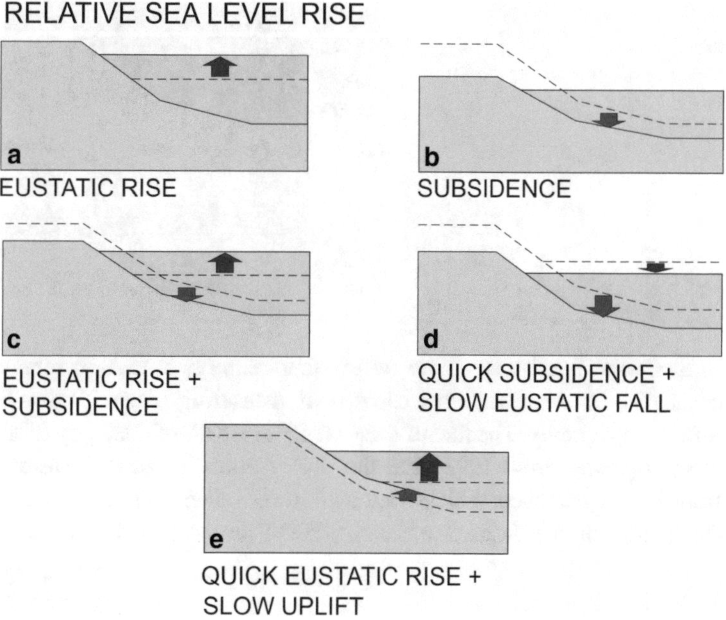

Fig. 25.6 Possible combinations of processes resulting in a relative sea level rise

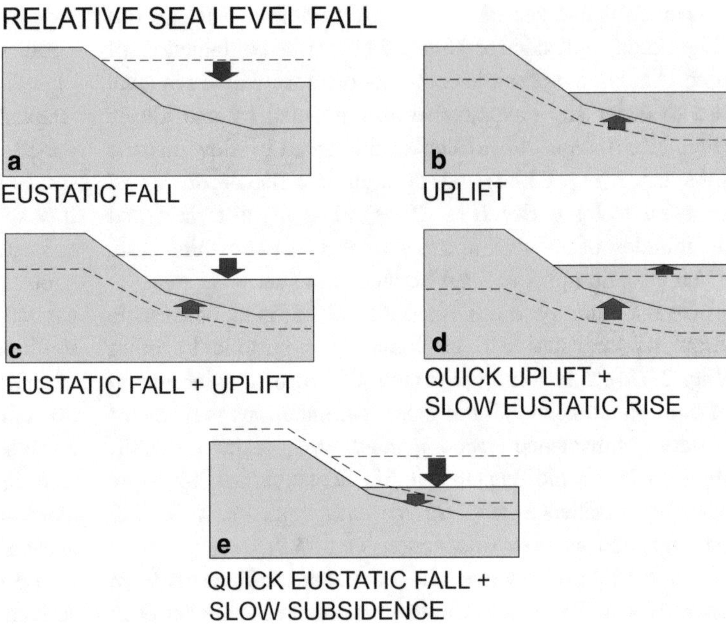

Fig. 25.7 Possible combinations of processes resulting in a relative sea level fall

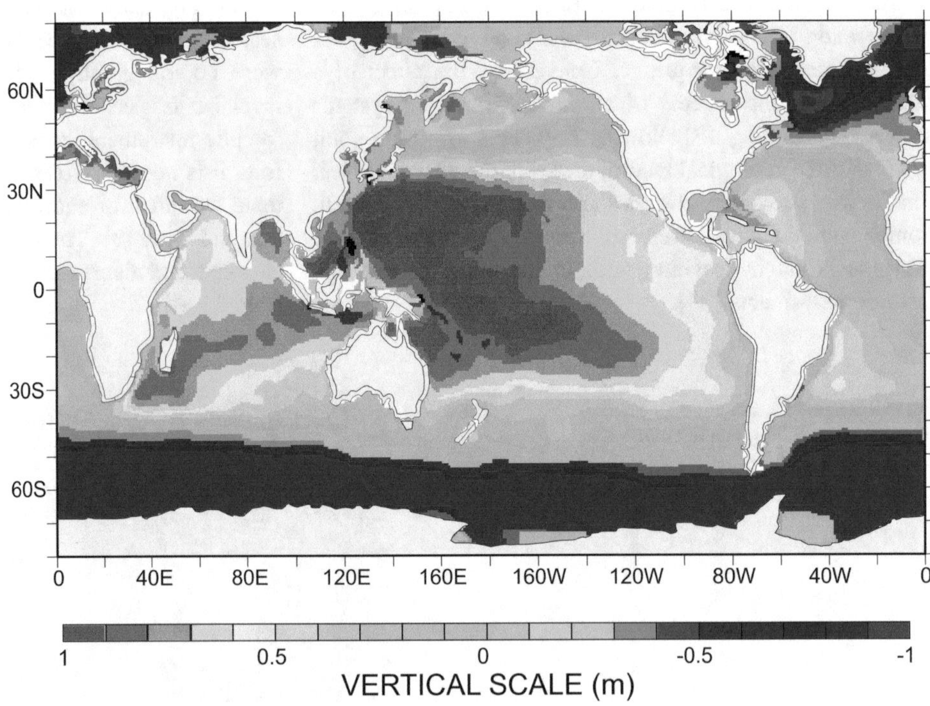

Fig. 25.8 Mean dynamic ocean surface topography obtained for the period 1992–2008 (adapted from Andersen and Knudsen [2])

Since the beginning of the twenty-first century, NASA has launched three satellites capable of measuring ocean surface topography. The first of these was TOPEX/Poseidon, which operated from 1992 until the fall of 2005. In 2001 Jason-1 was launched, which continued measurements until the launch of the Jason-2 mission (OSTM/Jason-2). This probe was launched in 2008 and operated until 2016, when Jason-3 was launched, which continues its work today.

The existence of sea surface relief is due to three types of cause: (1) gravity differences; (2) atmospheric and oceanic circulation; and (3) thermal expansion and contraction of water. The establishment of ocean surface topography is important for the study of ocean circulation and tides. These observations are also used to establish interannual, decadal and longer-term weather patterns. Changes in the position of these highs and lows of ocean surface relief imply that eustatic changes need not occur simultaneously or be of the same magnitude around the world.

25.5 Transgressions and Regressions

Relative movements of sea level are manifested in the coastal fringe as advances or retreats of the coastline that are known as transgressions and regressions. The term **transgression** is applied to an encroachment of the sea onto an emerged zone, implying a landward advance of the coastline. In contrast, a **regression** is defined as a retreat of the sea, so that a previously submerged area becomes part of the continent, with a shift of the shoreline seaward.

The phenomena of transgression and regression are most clearly manifested in coastal environments, where a minimal variation of the coastline means an abrupt modification of sedimentation conditions and a horizontal displacement of sedimentary environments. However, transgressions and regressions can also be deduced from the observation of deposits in deeper locations, such as shelf environments, by studying the variability of sediment bathymetric conditions.

It should be noted that the existence of transgressions and regressions does not depend only on relative sea level movements. The balance between erosional and accretionary processes, and the rates of erosion or accretion with respect to the rates of relative sea level movement, are the real drivers of transgressive and regressive movements. For example, with a stable sea level, we can obtain a transgression by erosional invasion of the sea or a regression by progradation caused by sedimentation. The combination of the rhythms of relative sea level movements and accumulation/erosion rates was analyzed graphically by Curray [8]. According to this author's criteria, transgressions and regressions can both take place under different conditions of sedimentation, erosion or relative sea level movements. Thus, four types of transgression and four types of regression are distinguished (Fig. 25.9).

In the case of regressions, these can occur under erosional conditions as long as there is a decrease in sea level at a rate that exceeds the rate of erosion (RI). More commonly, however, regressions occur under depositional conditions. Depositional regressions usually occur under conditions of relative sea level fall accompanied by sedimentary accumulation (RII and RIII). In these cases, the preservation of regressive depositional sequences depends on the relationships between the rate of fall and the rate of sedimentation. When the rate of relative sea level fall exceeds the rate of sedimentation, mixed depositional regressions (RII) occur, in which new coastal sedimentary bodies can develop, disconnected from the previous ones, which are hanging in the emerged zone. If, on the other hand, the rate of input exceeds the rate of fall, a discontinuous depositional regression (RIII) occurs, in which new coastal bodies develop in the form of an offlap, resting on the previous ones. Regressions can also occur only by sedimentary accumulation under a stable sea level or even with a slightly rising relative sea level. These are depositional regressions (RIV). In this case, the coastal bodies develop an offlap system that materializes in the progradation of the coastal systems.

Transgressions also occur in a wide range of possibilities. There are transgressions whose cause is directly linked to erosional processes under stable conditions of relative sea level (TI) or by a rise accompanied by erosion (TII). However, the situation in which coastal sequences can be preserved occurs when transgressions are accompanied by a more or less active depositional regime. Under conditions of sedimentary accretion, the rate of relative sea level rise must

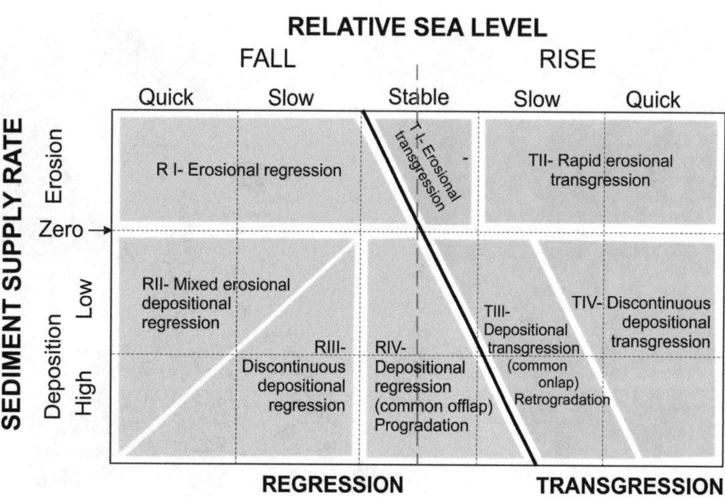

Fig. 25.9 Transgression–regression diagram of Curray [8] following the combinations of the velocity of relative sea level changes and the erosional/depositional rates

exceed the rate of accretion. When the rate of rise is small and is exceeded by the rate of accumulation, depositional transgressions (TIII) occur, and under these conditions the coastal systems move towards the continent, overlapping one another and forming an onlap geometry that materializes the process of retrogradation. On the contrary, when the relative rate of rise exceeds the rate of accumulation, a discontinuous depositional transgression (TIV) occurs.

Coastal systems subjected to depositional transgressions and regressions evolve by developing transgressive and regressive sedimentary sequences (deepening and shallowing sequences). In general terms, it can be said that sedimentary environments are transformed into each other by superimposing their facies in the form of third-order sequences (depositional sequences). There is a transition scheme between the most typical sedimentary environments developed on clastic coasts [9]. In this scheme (Fig. 25.10), the transition between fluviomarine systems, such as deltas and estuaries, through transgressions and regressions can be observed. Transitions between other systems not associated with river mouths and dominated by waves and tides can also be observed.

As for the systems developed at river mouths, this scheme explains why deltas are associated in the scientific literature with regressive systems. They are systems that are easily built during processes of relative sea level fall, since the river must make its way to the new coast, developing the delta from the reworking of submarine materials from the previous stage. However, deltas can also be constructed following depositional filling of estuaries (wave- or tide-dominated) after a period of relative sea level stability. Conversely, estuaries always develop from a situation of relative sea level rise and marine invasion of river valleys.

In systems where fluvial influence is minimal, strandplain environments predominantly originate on well-fed, wave-dominated coasts during regressive processes associated with relative sea level stability or lowering. These systems, given their high sediment availability, evolve easily under transgressive conditions, although part of the sediment in their frontal region can be reworked by the waves to build new systems in their migration landwards.

In non-river-dependent but tidally influenced systems, such as tidal flats and barrier islands, sediment availability is not as high as in the case of strandplains. This means that under transgressive conditions the available material has to be reworked to build new systems (Fig. 25.11a). This process is known as **rollover** [30]. The cohesive character of the tidal sediments, located in the most protected part of the systems, makes them better preserved in the transgressive sequences.

The speed of the transgression may sometimes be greater than the capacity to rework material and the rate of input (transgressions type TIV). Then, the coastal bodies become disconnected from each other and move upwards towards the continent (Fig. 25.11b). Under these conditions, the bodies may be submerged so rapidly that they are subjected to the low-energy regime characteristic of deeper zones and the coastal sequences may be preserved. This process is known as **overstepping** [8].

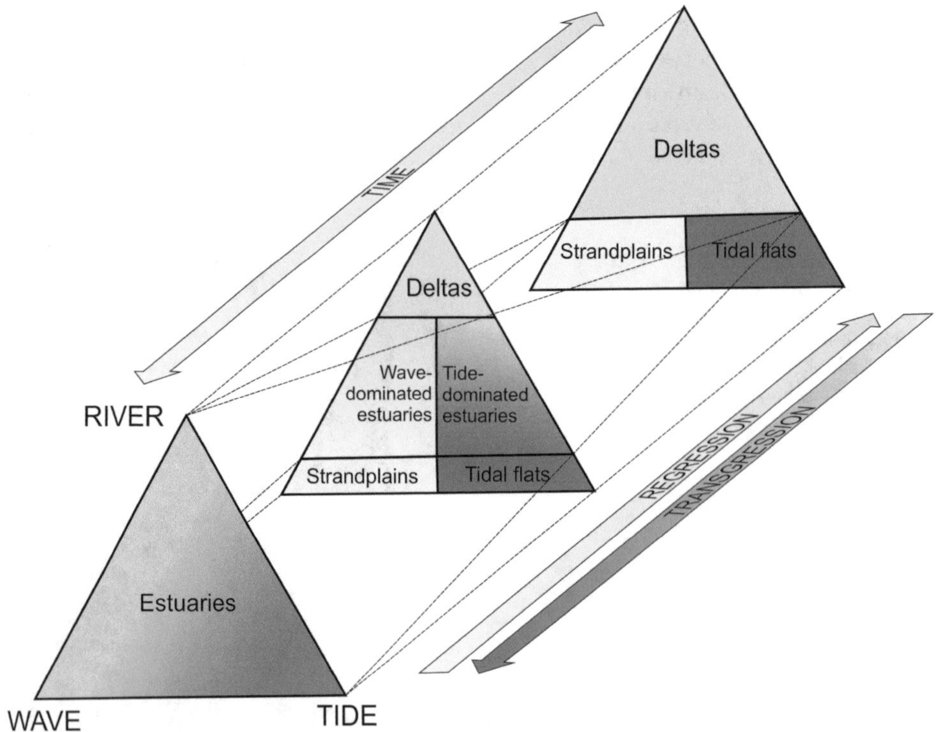

Fig. 25.10 Scheme of evolution of sedimentary environments under transgressions and regressions

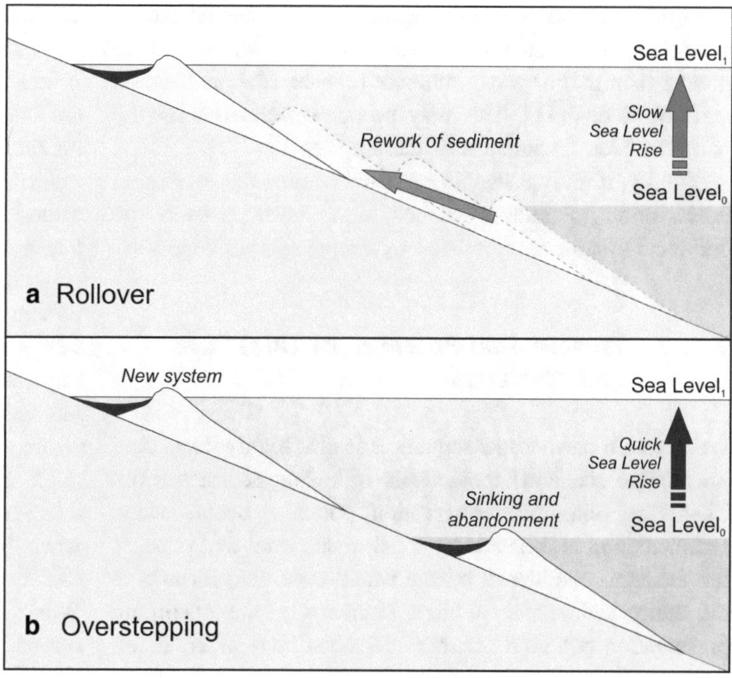

Fig. 25.11 Possible models of shoreline retreat as a response to a transgression (adapted from Mellett and Plater [22])

25.6 Preservation Potential of Coastal Sequences Under Sea Level Movements

The preservation potential of coastal sequences depends entirely on the conditions under which transgressions and regressions occur in relation to the reworking capacity of active coastal processes. The following sections discuss the possibilities for sequences developed in coastal environments to be preserved in the geological record. These possibilities depend largely on the type of environment and its dynamics in relation to the rate of transgressions and regressions.

25.6.1 Preservation Potential of Beaches and Barrier Systems

Considering relative sea level movements, it is necessary to address the problems posed by regressions and transgressions on the preservation potential of barrier island systems. The same processes that act by reworking the sediment during periods of stability also act during periods of relative movement. The preservation of the depositional record in these barrier systems depends largely on the type of sea level movement that follows deposition [14].

A relative drop in sea level tends to preserve the systems in the short term, as the sediments remain out of reach of coastal reworking processes. However, in the long term, coastal deposits are exposed to subaerial erosion processes. Under these conditions, some part of the systems may be spared the more erosional processes. In this context, tidal deltas, as well as estuaries, are particularly susceptible to erosion because they develop in depressions in the terrain. These depressions tend to be occupied by river valleys where runoff from more or less permanent channels will circulate and rework the sediment towards the new shoreline.

Under conditions of sea level rise, sequences will preserve better or worse according to the relationships between the rate of rise and the rate of input. When the relative rise is slow, the shoreline can be prograded if the accretion rate is sufficiently high, according to a depositional regression model (RIV). In this case, the preservation potential of the entire assemblage will be maximal [19].

In the opposite case, when the rate of accumulation does not compensate for the rate of rise, a marine transgression will occur, in which the preservation potential will depend on the transgressive model that is developed (rollover or overstepping).

In the case of a rollover transgression, there is a total or partial reworking of the coastal facies depending on the relationship between the rate of rise and the reworking capacity of the coastal agents. Under conditions of slow rise and high reworking capacity, the entire system is reworked to migrate towards the coast and the preservation potential is null. When the rise is moderate, the inland systems are overtopped by sea level. At first, this implies an increase in the tidal prism and further development of tidal deltas. However, if the wave reworking capacity is greater than the input, the systems subjected to the wave are reworked to move the barrier island towards the mainland. Under these conditions the preservation potential of back-barrier facies, such as lagoons, tidal flats, wetlands and flow deltas, is much greater than that of wave-influenced facies, such as beaches and ebb deltas [11]. This may be one explanation for the paucity of fossil examples of these systems.

Finally, if an overstepping process occurs, the maximum preservation potential is reached, since the deposits of the complete system are fossilized by deeper marine deposits.

25.6.2 Preservation Potential of Tidal Flats and Wetlands

Preservation of intertidal sequences is also highly dependent on relative sea level movements following sedimentation. Thus, the minimum preservation potential occurs when sedimentation is followed by a fall in sea level that exposes the facies to continental agents which eventually dismantle the entire sedimentary edifice. Conversely, the maximum preservation potential occurs under conditions of rapid relative sea level rise, since the sequences are submerged and fossilized by sediments generated in lower energy environments.

During a moderate relative sea level rise, preservation depends on sedimentary input. Thus, under conditions of insufficient input, these environments may be partially reworked to feed the environments that would be placed in increasingly higher topographic positions; under these conditions, the preservation of deepening sequences is rare. In this sense, the subtidal channel-fill facies prove to be the most preservable, as they are protected by the rest of the facies which overlie it and are potentially most easily eroded [10]. Many examples of fossil tidal flats are represented only by these facies. The magnitude of the intertidal facies is usually finer, as it is generally equal to the tidal range of the coast where it was generated.

If, on the other hand, the input is sufficient, changes in the rate of relative sea level rise may be reflected in the preservation of cyclical series of the Milankovitch frequency band (parasequences). In these cases, a repetition of an elementary shallowing sequence consisting of subtidal, intertidal and supratidal facies may develop.

Finally, if the rate of transgression is slow, preservation depends on the location of the barrier island. If the plain is on the backside of a barrier island system, the entire system may be reworked by waves and not be preserved. If, on the other hand, the plain is in a sheltered bay and the rate of input offsets the rate of relative sea level rise, very strong sequences are usually preserved.

In any case, there are numerous examples of preserved tidal flats in the geological record, which suggests that these systems in general have good preservation potential.

25.6.3 Preservation Potential of Estuaries

In transgressive contexts, when river valleys are invaded by the sea, morphology and slope have a significant influence on the establishment of the tidal prism that will regulate sedimentation in estuarine systems. Generally speaking, estuaries become larger with smaller coastal gradients and larger tidal ranges [9]. When incised valleys are transgressed, changes in valley shape can occur which can influence trends in the character of tidal sedimentation. In this sense, the potential for preservation of estuarine sequences will depend on the geometry of the valley as it is inundated, as well as the volume of fluvial and marine sediments during transgression. On the other hand, as estuaries move landward, parts of the sequences may be reworked by the more energetic agents. For example, in wave-dominated estuaries, the barriers that enclose the estuary are subject to strong wave reworking processes. When this occurs, these barriers may or may not be preserved, according to the same criteria discussed in Sect. 25.6.1. Regardless of the type of estuary, in the innermost domain the river-dominated sequence may also be destroyed by erosional incision of tidal and/or fluvial channels [9, 11]. Thus, in wave-dominated estuaries, the central part of the estuary may be the only section preserved in the face of a transgression.

In regressive contexts, the preservation potential may be higher if sea level changes are minimal and depositional regression occurs (RIV type regression). During relative sea level lows, the preservation potential is usually low, since the valley is usually occupied by the fluvial system and fluvial incision processes usually erode and dismantle the estuarine sequences. When this occurs, in many cases only what is deposited on the margins is preserved.

25.6.4 Preservation Potential of Deltas

During transgressive conditions, the preservation of deltas is maximal. This is the reason why deltas are one of the best represented coastal environments in the geological record. However, not all sedimentary sub-environments present in deltas are preserved to the same extent. Of all the deltaic environments, those associated with the delta plain have the highest preservation potential, while the deltaic frontal bars have the lowest because they are more exposed to occasional erosion events. The preservation potential of deltaic systems also depends on the type of delta. In principle, river-dominated deltas have the highest preservation potential, since coastal agents (tides and waves) are capable of reworking the frontal sectors of the delta. Subsidence phenomena related to sediment compaction are responsible for the existence of relative rise and stabilization of the sea level and for the generation of an accommodation space that is compensated for by the sedimentary aggradation of the delta plain with the arrival of new sediments.

During a relative sea level fall, fluvial influence begins to dominate the system. In this context, channel incision can erode deltaic sequences deposited in earlier stages. However, the extent of deltaic systems means that loss through fluvial incision is very localized and the delta can retain a high preservation potential even during periods of relative sea level fall [7].

25.6.5 Preservation Potential of Reefs

In general, reefs have a high degree of preservation and are well represented in the geological record. However, the configuration of reef systems is very sensitive to variations in sea level. In this case, the preservation potential depends on the rates of terrigenous input, turbidity, substrate type and diversity of reef-building fauna. Variations in terrigenous sedimentation rates occur during sea level changes and are related to the relative proximity of continental areas. An influx of siliciclastic sediments normally contributes to killing reefs, but at the same time contributes to their preservation by burial. In this sense, transgressions are usually associated with minimal terrigenous input, while the highest sedimentation rates develop during times of high sea level. In addition to high rates of terrigenous accumulation, there is a greater ability of storms to redistribute locally both terrigenous sediments and fragments stripped from the reef itself. Thus, a rise followed by stabilization usually generates a third-order sequence with a maximum preservation rate [27].

The processes of relative sea level fall lead to the overhang of the riverine bodies in continental areas. Other types of coastal systems are often subject to erosional processes during these periods. However, carbonate systems generally experience periods of cementation when they are affected by inland freshwater. These cementation processes contribute to solidifying the reef and thus make it resistant to erosional processes.

References

1. Allen JRL (1999) Geological impacts on coastal wetland landscapes: some general effects of autocompaction in the Holocene of northwestern Europe. The Holocene 9(1):1–12
2. Andersen OB, Knudsen P (2009) The DNSC08 mean sea surface and mean dynamic topography. J Geophys Res 114:C11
3. Bingham RJ, Haines K, Hughes CW (2008) Calculating the ocean's mean dynamic topography from a mean sea surface and a geoid. J Atmos Oceanic Tech 25:1808–1822
4. Cahoon DR, Hensel PF, Spencer T, Reed DJ, McKee KL, Saintilan N (2006) Coastal wetland vulnerability to relative sea-level rise: wetland elevation trends and process controls. In: Verheven JTA, Beltman B, Bobbink R, Whigham DF (eds) Wetlands and natural resource management ecological studies 190. Springer, New York, pp 271–292
5. Chappell J, Shackleton NJ (1986) Oxygen isotopes and sea level. Nature 324:137–140
6. Cronin TM (2012) Rapid sea-level rise. Quatern Sci Rev 56:11–30
7. Cross TA, Lessenger MA (1998) Sediment volume partitioning: rationale for stratigraphic model evaluation and high-resolution stratigraphic correlation. In: Sandvik KO, Gradstein F, Milton N (eds) Predictive high resolution sequence stratigraphy. Norwegian Petroleum Society Special Publication pp 171–196
8. Curray JR (1964) Transgressions and regressions. In: Miller RL (ed) Papers in Marine Geology. Macmillan, New York, pp 175–203
9. Dalrymple RW, Zaitlin BA, Boyd R (1992) Estuarine facies models: conceptual basis and stratigraphic implications. J Sediment Petrol 62:1130–1146
10. Davis RA Jr (2012) Tidal signatures and their preservation potential in stratigraphic sequences. In: Davis RA, Dalrymple RW (eds) Principles of Tidal Sedimentology. Springer, Heidelberg, pp 35–55
11. Davis RA, Clifton HE (1987) Sea-level change and the preservation potential of wave-dominated and tide-dominated coastal sequences. J Sediment Petrol 41:167–178
12. Dokka RK (2006) Modern-day tectonic subsidence in coastal Louisiana. Geology 34:281–284
13. Emery K, Aubrey D (1985) Glacial rebound and relative sea levels in Europe from tide-gauge records. Tectonophysics 120(3–4):239–255
14. Fisher AG (1961) Stratigraphic record of transgressing seas in light of sedimentation on Atlantic coast of New Jersey. Am Asso Petrol Geol Bull 45:1656–1667
15. FitzGerald DM, Fenster MS, Argow BA, Buynevich IV (2008) Coastal impacts due to sea-level rise. Annu Rev Earth Planet Sci 36:601–647
16. Hanna E, Navarro FJ, Pattyn F, Domingues CM, Fettweis X, Ivins ER, Nicholls RJ, Ritz C, Smith B, Tulaczyk S, Whitehouse PL, Zwally HJ (2013) Ice-sheet mass balance and climate change. Nature 498:51–59
17. Haq BU, Hardenbol J, Vail PR (1987) Chronology of fluctuating sea levels since the Triassic (250 million years ago to present). Science 235:1156–1167
18. Hong B, Shen J (2012) Responses of estuarine salinity and transport processes to potential future sea-level rise in the Chesapeake Bay. Estuar Coast Shelf Sci 104–105:33–45
19. Hubbard DK, Oertel G, Nummedal D (1979) The role of waves and tidal currents in the development of tidal inlets sedimentary structures and sand body geometry: examples from North Carolina, South Carolina and Georgia. J Sediment Petrol 49:1073–1092
20. Konikow LJ (2011) Contribution of global groundwater depletion since 1900 to sea-level rise. Geophys Res Lett 38:L17401
21. Masselink G, Hughes MG, Knight J (2003) Introduction to coastal processes and geomorphology. Routledge, London, 416pp
22. Mellett CL, Plater AJ (2018) Drowned barriers as archives of coastal-response to sea-level rise. In: Moore LJ, Murray AB (eds) Barrier dynamics and response to changing climate. Springer, Heidelberg, pp 57–90
23. Mörner NA (1987) Models of global sea level changes. In: Tooley MJ, Shennan I (eds) Sea level changes. Blackwell, Oxford, pp 332–355

24. Murphy RR, Kemp WM, Ball WP (2011) Long-term trends in Chesapeake Bay seasonal hypoxia, stratification, and nutrient loading. Estuaries Coasts 34:1293–1309
25. Nicholls RJ, Cazenave A (2010) Sea-level rise and its impact on coastal zones. Science 328:1517–1520
26. Pirazzoli PA (1996) Sea-level changes: the last 20,000 years. Wiley, Chichester, 224pp
27. Ray DC, Thomas AT (2007) Carbonate depositional environments, sequence stratigraphy and exceptional skeletal preservation in the Much Wenlock limestone formation (Silurian) of Dudley, England. Palaeontology 50:197–222
28. Rovere A, Stocchi P, Vacchi M (2016) Eustatic and relative sea level changes. Curr Clim Change Rep 2(4):221–231
29. Sinclair HD, Allen PA (1992) Vertical versus horizontal motions in the Alpine orogenic wedge: stratigraphic response in the foreland basin. Basin Res 4:215–232
30. Swift DJP, Phillips S, Thorne JA (1991) Sedimentation on continental margins, V: parasequences. In: Swift DJP, Oertel GF, Tillman RW, Thorne JA (eds) Shelf Sand and sandstone bodies: geometry, facies and sequence stratigraphy, vol 14. IAS Special Publication, pp 153–187
31. Vail PR, Mitchum RM Jr, Todd RG, Widmier JM, Thompson S III, Sangree JB, Bubb JN, Hatlelid WG (1977) Seismic stratigraphy and global changes of sea level. In: Payton CE (ed) Seismic stratigraphy: applications to hydrocarbon exploration, vol 26. American Association of Petroleum Geologists Memoirs, pp 49–212
32. Walters D, Moore LJ, Vinent OD, Fagherazzi S, Mariotti G (2014) Interactions between barrier islands and backbarrier marshes affect island system response to sea level rise: insights from a coupled model. J Geophys Res: Earth Sci 119:2013–2031
33. Willis JK, Chambers DP, Kuo C-Y, Chum CK (2010) Global sea level rise. Oceanography 23:26–35

26 Paleoceanography

26.1 Introduction

Changes in climate and sea level movements throughout Earth's history have been accompanied by major changes in the oceans. On the one hand, there is evidence of numerous modifications in the environmental characteristics (physical, chemical and biological) of ocean waters. On the other hand, plate tectonics have provided a changing framework of the physiography and dimensions of the oceans over geological time. Changes in the position of the continents must necessarily have been accompanied by variations in ocean circulation, which help to explain many of the physical and chemical changes observed in the geological record. These environmental changes have often been accompanied by ecological modifications in the biosphere. The physical, chemical and biological characteristics of ocean water have been recorded as markers in the sediments of the ocean floor at different stages of geological history.

In this regard, it should be noted that the oceanic circulation regime is coupled with that of the atmosphere. For this reason, paleoceanography and paleoclimatology are sciences that go hand in hand in many studies. Thus, many of the climate changes in the past have been studied through geochemical and paleontological markers in the sediments of the ocean floor [2]. The methodology of paleoceanographers to make accurate use of some of these markers is an advance of the last four decades. The development of scientific deep drilling programs and the analytical capability of both geophysical (seismic and magnetic) and geochemical data have been growing in recent decades. As a result, an increasingly long time series of data regarding the present-day oceans has been obtained. In parallel, the application of the knowledge obtained in today's oceans to the geological record of ancient ocean basins, now deformed, has made it possible to obtain data on oceans that have already disappeared.

This chapter will analyze some of the data that paleoceanography can contribute to the study of Earth's history and, above all, how the changes in the coupled ocean–atmosphere system have been reflected in the ancient coasts.

26.2 What is Paleoceanography?

Paleoceanography was conceived as the "*science that studies the past characteristics of the ocean*." However, with the entry on the scene of plate tectonics, it underwent a change in its conception. It is now defined as the "*science that studies the past characteristics of the oceans.*" This change of definition may seem only a nuance, but it hides more important assumptions: that the present oceans have not been the only ones throughout the Earth's history.

A more precise definition would include the study of the following characteristics: physiography, physical and chemical properties of water masses, ocean circulation patterns at different depth levels and biological communities. From this perspective, paleoceanographic studies are heavily weighted with quantitative measurements based on the markers of the sedimentary record mentioned in the previous section. In today's oceans, these quantitative studies can be addressed from the Mesozoic to the present, since the oldest sediments and oceanic crust in the current oceans are of those ages. However, the study of marine and coastal sediments of older ages that outcrop in some mountain ranges can provide data on ocean basins that have disappeared as far back as the Precambrian.

26.3 A Brief History of the Science

Drilling techniques were born in the 1940s and underwent rapid development into the late 1950s. During the beginning of the next decade, ocean drilling began, and numerous undisturbed cores were obtained from deep seafloors. The study of sediments accumulated in the form of long sequences in these cores allowed the characterization of the chemical and ecological properties of the ocean in periods prior to the present time. These could be considered the events that led to the birth of paleoceanography as a discipline of marine geology. However, it was not until some 15 years later that Van Andel et al. [24] used the term for the

J. A. Morales, *Coastal Geology*, Springer Textbooks in Earth Sciences, Geography and Environment,
https://doi.org/10.1007/978-3-030-96121-3_26

first time, and it was not until five more years passed that Schopf [19] wrote the first manual of this discipline.

In the 40 years since that first manual, it can be said that there has been a consolidation of this science, as well as a change of perspective and objectives. Indeed, in the initial approaches to it, both Schopf and his immediate competitor, Kennett [12], overlooked one of the main applications of the knowledge obtained about the oceans of the past: the knowledge of the paleoclimate.

Today, most of the efforts of paleoceanography teams are devoted to the interpretation of climate evolution, taking into account the close link between ocean and atmosphere and with the aim of predicting future climate trends [8].

In this sense, the international collaboration carried out through scientific drilling programs has allowed access to data in realms that would have been unthinkable at the beginning of this science. In short, these data have been a giant step forward in the understanding of the oceanic–climatic system and the history of the Earth [11]. In fact, the first drilling campaigns that provided data on the ocean floor were not carried out with a paleoceanographic objective. The Mohole project, developed in 1961 by the United States, was aimed at understanding the geological nature of the oceanic crust up to the Mohorovicic discontinuity, at the transition with the upper mantle.

It was the Deep Sea Drilling Project (DSDP), which ran from 1966 to 1983, that began to address the study of ocean floor cores, with an objective much closer to what is now considered paleoceanography. This project, developed entirely on the Glomar Challenger (Fig. 26.1), carried out numerous surveys over 17 years in the Atlantic, Pacific and Indian Oceans, as well as in the Mediterranean and Red Seas. The valuable data provided by these results contributed to refining the approach to this science, in addition to achieving an improvement in drilling techniques.

Over the next 20 years, the efforts of the first international drilling program with a purely paleoceanographic predefined aim would materialize: the Ocean Drilling Program (ODP). This program was carried out on the JOIDES Resolution (Fig. 26.2a) until 2003 and conducted 110 expeditions, drilling more than 2000 boreholes in different locations throughout the world's oceans.

From 2003, the ODP was replaced by the Integrated Ocean Drilling Program (IODP). This program was much more ambitious, involving 26 countries and employing more advanced technology. The JOIDES Resolution was revamped to adapt it to new needs, and the Japanese vessel Chikyu (Fig. 26.2b), with much more specialized facilities, joined the campaign. From its entry into operation, the IODP had conducted 52 expeditions by 2013.

At the end of 2013, the IODP partners renewed their collaboration through a new program called the International Ocean Discovery Program (also IODP). This program, which will be active until 2023, has a much more open plan, in which research teams from the 24 participating countries can propose missions with specific objectives. At the time of this book's publication, the program had completed 30 missions with 346 soundings.

But ocean drilling programs cover only a small part of Earth's history in the knowledge of the oceans. Actually, information from the oceans prior to the present has to be studied in the rock record of ancient sediments that were deposited in those paleoceans [23]. Research in this regard has been much less systematic than ocean drilling programs, although in recent decades many papers on sedimentary rocks of ancient formations that had been approached from a purely geological point of view have been revised to obtain information from a paleoceanographic perspective. In this way, it has been possible to obtain information on the composition of now-extinct oceans such as the Panafrican

Fig. 26.1 The first oceanographic explorations: Glomar Challenger (1968–1983) working on a survey for the Deep Sea Drilling Project (DSDP)

Fig. 26.2 Oceanographic vessels employed by the ODP and IODP programs. **a** JOIDES Resolution. **b** Chikyu. (Photographs by JAMSTEC/IODP.)

(Proterozoic), Mirovia (Neoproterozoic), Iapetus (Paleozoic), Rheic (Paleozoic), Panthalassa (Paleozoic, lower Mesozoic), Paleo–Tethys (Carboniferous–Triassic) and Tethys (Permian–Eocene). Many chemical and environmental characteristics of these oceans, including their paleoclimatic context, have already been unraveled. However, other features, such as the oceanic circulation systems of the times when these oceans were active, are yet to be investigated.

26.4 Different Paleoceanographical Foci

Paleoceanography, like oceanography, is nowadays carried out with a multidisciplinary approach. However, in both sciences there are different perspectives that result in different specialized disciplines.

26.4.1 Hydrographic Paleoceanography

The reconstruction of the ancient hydrography of the oceans requires the characterization of the physicochemical properties of the water masses. Thus, the study of paleotemperature and paleosalinity can be approached through a variety of geochemical and paleontological indicators of ancient sediments that provide quantitative data on these properties [7].

The main paleothermal indicator is the isotopic fractionation of oxygen. This technique makes use of the isotopic ratio $\delta^{18}O/\delta^{16}O$ of oxygen that exists in crystalline molecules formed in a given period of the Earth and contained in sediments. It has been shown that the curves of this isotopic ratio obtained in the shells of deep-sea foraminifera are parallel to the thermal curves [5]. In particular, this indicator is used by analyzing the air in bubbles contained in ice or mineral inclusions, as well as in calcite molecules precipitated in the environment or in shells. These studies are based on three premises: (1) the isotope ratio has been changing throughout Earth's history in equilibrium with temperature, (2) carbonates in shells precipitate in equilibrium with water; and (3) the isotopic variation since deposition has been negligible.

Paleosalinity analysis methods are very diverse and could be grouped into three sets: geochemical, mineralogical and paleontological. Between them, up to 11 control parameters can be used to deduce salinity. The following are used as geochemical indicators: boron, bromine, strontium, gallium, sulfur and carbon isotopes. The following can be used as mineralogical indicators: clay minerals, glauconite, phosphates, amino acids and manganese nodules. Paleontological methods are based on the ecological behavior of aquatic organisms and the fact that organisms need to be in osmotic balance with their environment. Thus, the concentration of salts in their cells must be equal to that of the water in which they are immersed. The mere occurrence of certain genera can therefore be used as an indicator of paleosalinity. However, in this regard it should be noted that there are two types of organisms with respect to their tolerance to salinity: **euryhaline** (with high tolerance to salinity changes) and **stenohaline** (with low tolerance). In this case, it is the stenohaline organisms that provide the most information. Sometimes, the application of the different methods gives disparate results, so it is always advisable to make several approaches.

Salinity and temperature variations in today's oceans have been used for climate reconstruction, since there is a close relationship between these properties and the volume of ice at the Poles, and also with the rate of evaporation of oceanic water [21].

26.4.2 Physical Paleoceanography

The discovery of the thermohaline circulation was the result of the first German oceanographic campaigns [25], yet it was not until the last decade of the twentieth century that the knowledge of these currents was combined with the surface ocean circulation to define the "global conveyor belt" and its close relationship with the terrestrial climate [4]. It is now known that, since the opening of the Atlantic Ocean after the breakup of Pangea, this system of currents has undergone variations over time. For example, it is known with certainty that the deep waters of the Atlantic were much warmer during the Miocene [26]. However, there is intense debate about how these changes in temperature of deep water masses affect the overall circulation system [22].

One of the techniques that has been used to approximate the intensity of these currents has been grain size methods. Since current velocity is reflected in the size of grains that can be transported and then deposited, some authors have used grain size variations in deep ocean basins to identify changes in the current system. The first attempts to apply this methodology were made by McCave et al. [14], who interpreted variations in the North Atlantic. Subsequent attempts have been made in other oceans to correlate and detect changes in the overall system.

26.4.3 Ecological Paleoceanography

As discussed in the previous sections, benthic microfossils contained in ocean floor cores have been widely used as indicators of paleosalinity and paleotemperature. In addition to this, planktonic and benthic microfossil associations have provided valuable information on the environmental and ecological conditions of the oceans. Currently, the distribution of planktonic foraminifera on the ocean floor is adapted to the temperature distribution of the water masses, which in turn follows the pattern of ocean currents. This type of pattern has been used to reconstruct these ecological conditions in ancient sediments since their discovery [10] to the present day [18].

There are other geochemical indicators that are excellent markers of ecological characteristics such as biological productivity, nutrient content and dissolved CO_2. In this respect, biogenic barium, some radiogenic isotopes (e.g., carbon, phosphorus and nitrogen), concentrations in some compounds (e.g., nitrates and ammonium) and the ratio between different alkenones are useful in determining ecological properties [7]. Alkenones are long-chain organic compounds produced by planktonic algae that have been preserved in sediments for at least the last 120 million years [16].

26.4.4 Climatic Paleoceanography

Chapters 10 and 24 of this book, as well as the previous sections, have emphasized the relationships between atmospheric and oceanic circulation in climatic cycles of different amplitudes.

Atmospheric circulation, in addition to being the main driver of ocean circulation, is an important source of siliciclastic material input to ocean basins. In this sense, desert dust clouds fall into the ocean and sediments are redistributed by ocean currents. The trace of these particles can be followed in deep ocean sediments and thus be used as an indicator of the amplitude of atmospheric circulation cells. Variations in the concentration and grain size of desert dust particles in sediment cores can be used as a paleoclimate indicator [22]. The distribution of other elements from the continent (e.g., pollen, spores and other organic particles) in bed sediments can be used in the same way [7].

The climate variations and changes described in Chap. 24 have been mostly verified through their reflection in the sedimentary sequences of the ocean floor. In this sense, climate cycles induced by orbital changes (Milankovitch cycles) are clearly reflected in the cyclicity of pelagic sediment laminae. Hays et al. [9] were the first to demonstrate these relationships, which had already been proposed in an elaborate hypothesis by Köppen and Wegener [13]. The identification of climatic cycles of Milankovitch frequency in Mesozoic, Paleozoic and Precambrian ocean floor sediments, now deformed, has extended the knowledge of climate variations to very remote epochs of the planet's geological past.

26.4.5 Oceanic Paleophysiography

A consequence of the change in the position of the continents due to plate tectonics has been the constant change in oceanic physiography. Thus, paleophysiography is closely linked to global paleogeography. In present-day oceans, the information provided by the distribution of magnetic anomalies of ocean floor basalts has made it possible to establish with remarkable precision the physiography of the oceans at different periods of the Earth's history since the beginning of the Mesozoic [15]. From a physiographic point of view, this period represents the transition from a single large ocean basin at the end of the Paleozoic to the current physiography consisting of an ocean mass divided into different smaller oceans.

Paleomagnetic indicators have been shown to be excellent markers of the position of continental masses and, by extension, of the physiography and position of the oceans in pre-Mesozoic times. In this way, the position of the continents with respect to the magnetic poles influenced the entrance inclination of the magnetic beams to the terrestrial surfaces according to latitude. It should be noted that the magnetic minerals of igneous rocks are oriented according to the magnetic beams of the Earth's field, generating a property known as **remanent magnetization**. Thus, the latitude of a rock (and the continent in which it is embedded) at the time of its formation can be known by knowing the angle of the magnetic minerals with respect to the horizontal (Fig. 26.3).

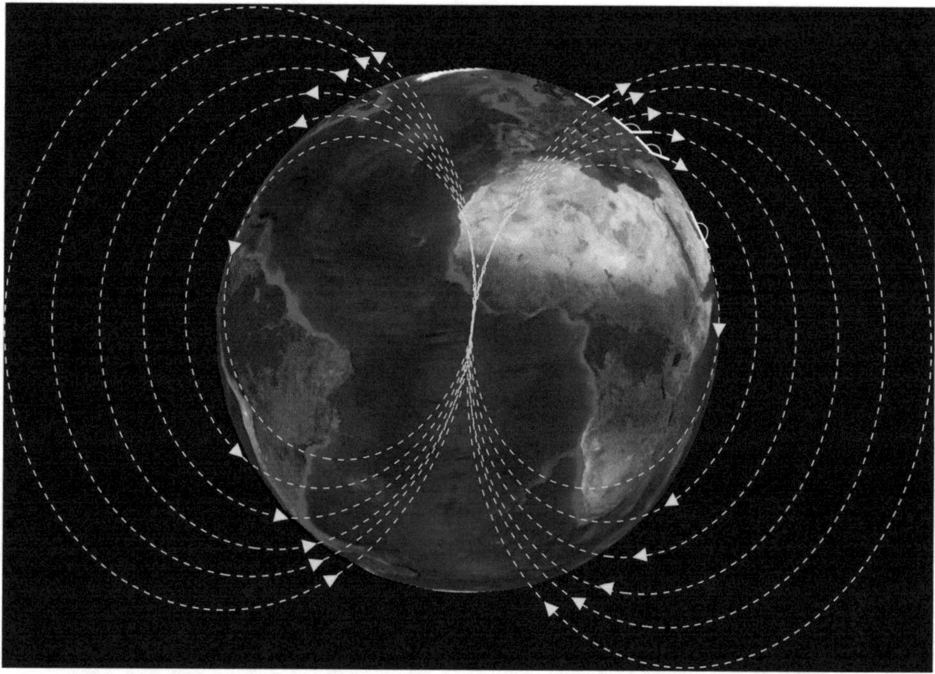

Fig. 26.3 Different angles of interaction of magnetic field beams with the Earth's surface according to latitude

The knowledge of the orientation due to the remaining magnetization in igneous formations of the same age (and in successive epochs) in all continents has made it possible to reconstruct the position of the continents (and the physiography of the oceans) for the times prior to the formation of Pangea. This method has been used to discover the existence of oceans prior to the present ones, as well as their changes in extension throughout geological history.

Not only the dimensions and location of the ancient oceans can be analyzed through gauges, but also their depth. The dissolution of calcite and aragonite crystals depends on temperature and pressure and thus on depth. In this way, there is a certain depth at which aragonite is dissolved, which is called the **lysocline**. Similarly, when calcite dissolves, in this case at greater depths, it reaches the so-called **calcite compensation depth** (CCD). Consequently, calcium carbonate grains from pelagic rain can only reach the bed when the depth is shallower than these dissolution levels. The presence or absence of aragonite and calcite (or fossils of both compositions) in ocean floor sediments can be used as an indicator of paleodepth [17].

One of the phenomena discussed in Chap. 25 has a significant influence on variations in the position of the ocean floor relative to carbonate compensation depths: thermal subsidence. The oceanic crust generated at the ridges is above the compensation levels. In areas close to the ridges, the warm ocean floor is elevated and calcareous muds accumulate there. However, when the emission of new oceanic crust moves these beds away from the ridge, a process of crustal cooling inevitably occurs. This cooling is accompanied by subsidence due to thermal contraction. As the bed depth increases, both the lysocline and the CCD are successively crossed and the calcareous mudstone facies are replaced by abyssal red clays or siliceous mudstone facies (radiolarites and diatomites). The presence in a paleocean of this facies sequence gives an idea of the timing of these depth changes and provides insight into their evolution [24].

The average CCD level is currently at 4500 m. However, being temperature-dependent, compensation levels vary in depth in space and time. On the one hand, latitude has been found to introduce significant changes in CCD depth in all present-day oceans [3]. On the other hand, the depth for the same latitude has changed over geological time in all oceans [1], Fig. 26.4. So, it has changed in time, but also in space—the position of the CCD is estimated to have varied between 3500 and 5500 m.

26.5 Past Oceans

Throughout the Earth's history, plate tectonics have caused continental masses to migrate, collide, be fragmented and then move apart again. In this process, numerous oceans have appeared, evolved and disappeared. In parallel, sediments deposited in the coastal systems surrounding these oceans have consolidated and deformed, becoming part of the sedimentary record among the system tracts deposited on the continental margins. Some of these paleocean systems and their influence on their coasts are characterized below.

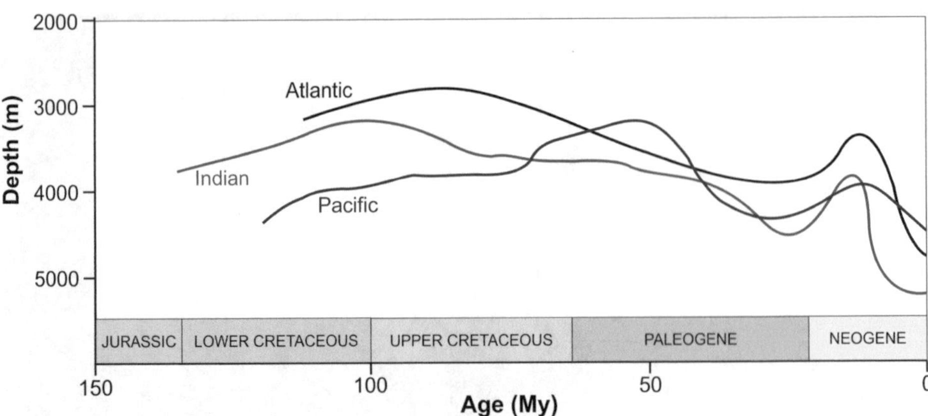

Fig. 26.4 Evolution of the calcite compensation depth (CCD) in the main present-day oceans during the last 150 million years (adapted from Arthur et al. [1])

– *Mirovian Ocean*

Mirovia was probably the first ocean in the history of our planet. Its name comes from the Russian word *mirovoy*, meaning "global." This name was assigned to the only ocean on Earth during the Neoproterozoic, between 1 billion and 750 million years ago. This large ocean surrounded the first megacontinent that grouped all emerged lands, named **Rodinia** (Fig. 26.5a). This must have had extensive coastlines associated with passive margins where transgressive and regressive coastal sequences with significant preservation of tidal signatures were well recorded [6].

– *Panafrican Ocean*

When the breakup of Rodinia began during the late Proterozoic, a rifting process opened a new oceanic basin on the southwestern facade. This basin eventually became the **Panafrican Ocean**. This ocean was located for more than 100 million years between the **Laurentia** and **Gondwana** supercontinents (Fig. 26.5b). The occurrence of a subduction zone on its northern margin caused its closure before the beginning of the Phanerozoic Eon, during the process of collision of the two continents to form a new supercontinent called **Pannotia**. The coasts of the Panafrican Ocean would change from

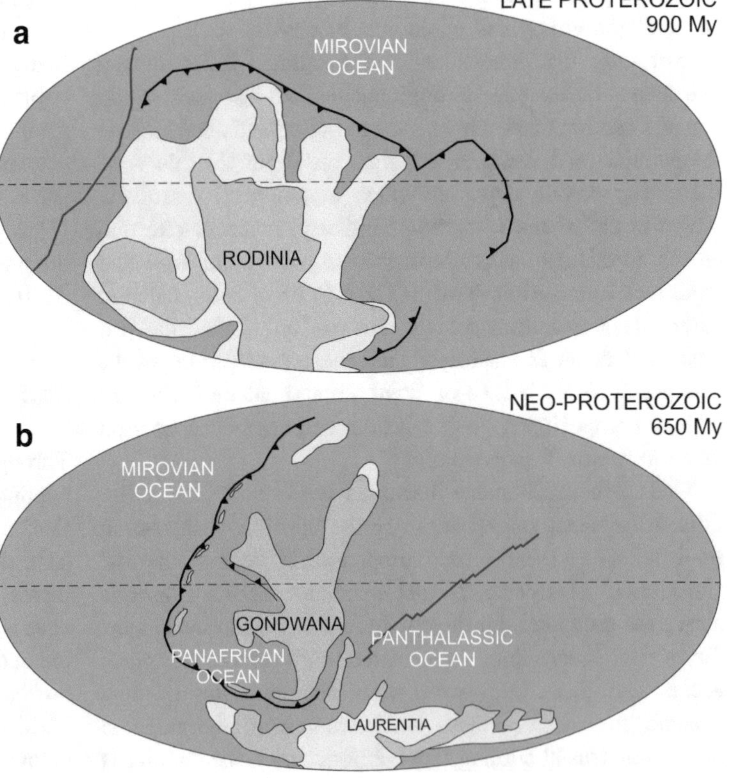

Fig. 26.5 Ancient oceans during the Proterozoic ages. **a** The Mirovian Ocean during the late Proterozoic. **b** The Panafrican Ocean closing during the opening of the Panthalassic Ocean in the Neoproterozoic (Scotese [20])

trailing-edge coasts to collision coasts. Coastal sequences were preserved in the tectonic accretionary prisms during the collision.

– *Panthalassic Ocean*

The Rodinia breakup process continued with the generation of a second rifting in the Neoproterozoic. In this case, it culminated in the emergence of a ridge where the emission of oceanic crust was much more significant than in the case of the Panafrican Ocean. This opening process gave rise to a new ocean called the Panthalassic Ocean or simply **Panthalassa** (Fig. 26.5b). Etymologically, the term comes from a Greek word meaning "all seas." Panthalassa continued its opening process throughout the Paleozoic, until it succeeded in bringing all the continents back together into a megacontinent called Pangea (from the Greek, "all the Earth"). Panthalassa then became the huge global ocean that surrounded **Pangea** until the early Mesozoic (Fig. 26.6a). The coastal systems developed around the Panthalassic Ocean are as diverse as those developed today around the Pacific Ocean. Many of these systems have been preserved in Andean-type subducted margins.

– *Iapetus Ocean*

From the end of the Proterozoic, the megacontinent Rodinia continued to break up into smaller and smaller continents. Between these continental fragments, smaller oceans appeared which remained open until the formation of Pangea at the end of the Paleozoic. The **Iapetus Ocean** was among these oceans. Iapetus was formed by

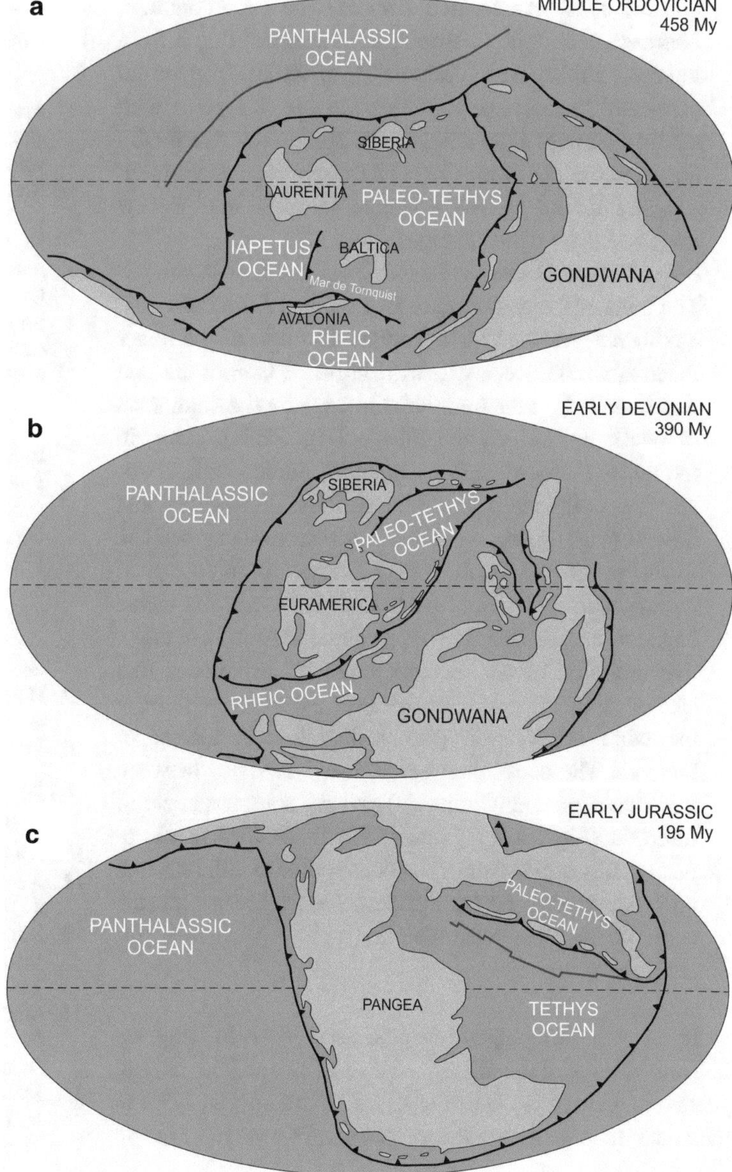

Fig. 26.6 Ancient oceans during the Paleozoic and Mesozoic ages. **a** Panthalassa, Paleo–Tethys, Iapetus and Rheic Oceans during the middle Ordovician. **b** Panthalassa, Paleo–Tethys and Rheic Oceans during the early Devonian, before the collision to build Pangea. **c** Oceans (Panthalassa, Paleo–Tethys and Tethys) during the beginning of the disintegration of Pangea in the early Jurassic

the separation of a supercontinent in the Southern Hemisphere about 600 million years ago. This ocean extended between **Laurentia**, **Baltica** and **Avalonia** (Fig. 26.6a). One of the extensions of this ocean was the Tornquist Sea, located between Avalonia and Baltica and connecting Iapetus with Paleo–Tethys.
This ocean closed as the continents that formed its margins converged during the Caledonian orogeny, forming the continent called **Euramerica** (also Laurussia). The ocean disappeared between the Cambrian and Ordovician, about 400 million years ago. The sediments of its coastal systems were incorporated into the Caledonian, Akkadian and Taconic mountain chains, in the Euramerica continent.

- *Rheic Ocean*
  Another of the oceans that developed in the Paleozoic between the approaching continents was the **Rheic Ocean**. This was located between the supercontinent **Gondwana** and the smaller continents separating it from Iapetus, Baltica and Avalonia (Fig. 26.6a). The ocean continued to exist even after the closure of Iapetus with the formation of Euramerica (Fig. 26.6b), but was finally closed at the end of the Carboniferous when the Variscan orogeny united all the emerged terranes into a new megacontinent called **Pangea**.
- *Paleo–Tethys Ocean*
  The third of the great oceans that separated the Paleozoic continents was the **Paleo–Tethys**. Throughout the lower Paleozoic, this ocean extended north of Gondwana and was bounded on the northwestern edge by the continents of Laurentia, Baltica and Siberia (Fig. 26.6a). Later, in the upper Paleozoic, after Laurentia, Baltica and Avalonia had collided, it was bounded by Euramerica and Siberia (Fig. 26.6b). After the Variscan orogeny and the formation of Pangea, the Paleo–Tethys Ocean continued to exist during the early Mesozoic. However, the ocean began to narrow as a ridge appeared at its southern edge. The emission of new oceanic crust led to stresses that created a subduction zone with an island-arc system at the edge of Laurasia (the northeastern extension of Pangea). The ocean then became an inland sea between this island system (Cimmeria) and the southern coast of Laurasia (Fig. 26.6c). Finally, the Paleo–Tethys disappeared at the end of the Triassic, about 200 million years ago, being replaced by the new Tethys Ocean located south of the Cimmerian islands.
- *Tethys Ocean*

In the late Permian, rifting processes began to break up Pangea. One of these rifting processes, about 250 million years ago, caused the opening of a new mid-ocean ridge at the southern edge of the Paleo–Tethys. This ridge began to emit oceanic crust, starting the opening of the Tethys. Over the next 60 million years, the ridge generated the force that uplifted the Cimmerian islands isolating the old Paleo–Tethys from the new Tethys (Fig. 26.6c). As a result of this process, the new ocean replaced the old one by displacing the Cimmerian islands northward and causing them to collide with Laurasia. At the end of the Tethys Ocean process, it ended up occupying the same position as the Paleo–Tethys.

A similar process occurred between the Tethys, the Indian Ocean and the Mediterranean Sea. After the breakup of Gondwana, the continents of India, Arabia and Africa moved northward, closing Tethys as the Indian Ocean opened and the Mediterranean became isolated. As a consequence of the closure of Tethys, its marine and coastal deposits were deformed to form the Himalayan and Alpine mountain ranges during the Alpine orogeny.

References

1. Arthur MA, Dean WE, Schlanger SO (1985) Variations in the global carbon cycle during the Cretaceous related to climate, volcanism and changes in atmospheric CO_2. In: Sundquist ET, Broecker WS (eds) The carbon cycle and atmospheric CO_2: natural variations Archean to present. Geophysical Monography Series vol 32. Washington, DC, pp 504–529
2. Barnes CR (1999) Paleoceanography and paleoclimatology: an Earth system perspective. Chem Geol 161(1):17–35
3. Berger WH, Be AWH, Sliter WV (1975) Dissolution of deep-sea carbonates: an introduction. Special Publication of Cushman Foundation 13
4. Broecker WS (1991) The great ocean conveyor. Oceanography 4:79–89
5. Emiliani C (1955) Pleistocene temperatures. J Geol 63:538–578
6. Eriksson KA, Simpson E (2012) Precambrian tidal facies. In: Davis RA Jr, Dalrymple RW (eds) Principles of tidal sedimentology. Springer, Heidelberg, pp 397–420
7. Fischer G, Wefer G (eds) (1999) Use of proxies in paleoceanography: examples from the South Atlantic. Springer, Heidelberg, 735pp
8. Gornitz V (2009) Encyclopedia of paleoclimatology and ancient environments. Springer, Heidelberg
9. Hays JD, Imbrie J, Shackleton NJ (1976) Variations in the Earth's orbit: pacemaker of the ice ages. Science 194:1121–1132
10. Imbrie J, Kipp N (1971) A new micropaleontological method for quantitative palaeoclimatology. In: Turekian KK (ed) The late Cenozoic Glacial Ages. Yale University Press, New Haven, pp 71–181
11. IODP (2021). www.iodp.org/about-iodp/history
12. Kennett J (1982) Marine geology. Prentice-Hall, Englewood Cliffs, 813pp
13. Köppen W, Wegener A (1924) Die Klimate der geologischen Vorzeit. Gebrüder Borntraeger Verlagsbuchhandlung, Berlin, 255pp
14. McCave IN, Manighetti B, Beveridge NAS (1995) Circulation in the glacial North Atlantic inferred from grain-size measurements. Nature 374:149–152
15. Müller RD, Seton M (2016) Paleophysiography of ocean basins. In: Harff J, Meschede M, Petersen S, Thiede J (eds) Encyclopedia of marine geosciences. Springer, Dordrecht, pp 638–648

16. Müllet PJ, Kirst G, Ruhland G, Von Storch I, Rosel-Melé A (1998) Calibration of alkenone palaeotemperature index based in core-tops from the eastern South Atlantic and the global ocean. Geochim Cosmochim Acta 62:1757–1772
17. Peterson MNA (1966) Calcite: rates of dissolution in a vertical profile in the central Pacific. Science 154:1542–1544
18. Pflaumann U, Sarnthein M, Chapman MR (2003) Glacial North Atlantic: sea-surface conditions reconstructed by GLAMAP 2000. Paleoceanography 18(3):1065
19. Schopf TJM (1980) Paleoceanography. Harvard University Press, Cambridge, 341pp
20. Scotese CR (2001) Atlas of Earth History. Results of the PALEOMAP project. University of Texas, Arlington, 59pp
21. Shackleton NJ (1987) Oxygen isotopes, ice volume and sea-level. Quatern Sci Rev 6:183–190
22. Thiede J (2016) Palaeoceanography. In: Harff J, Meschede M, Petersen S, Thiede J (eds) Encyclopedia of marine geosciences. Springer, Dordrecht, pp 628–632
23. Torsvik TH, Cocks LRM (2017) Earth history and palaeogeography. Cambridge University Press, Cambridge, 327pp
24. Van Andel TH, Heath GR, Moore TC (1975) Cenozoic history and paleoceanography of the central equatorial Pacific ocean. Geological Society of America memoirs, Boulder vol 143. 134pp
25. Wüst G (1936) Schichtung und Zirkulation des Atlantischen Ozeans.-Wiss. Ergebn. Dt. Atlant. Expedition VFS Meteor 1925–1927: 6 (1). Berlin, pp 1–288
26. Zachos JC, Lohmann KC, Walker JCG, Wise SW (1993) Abrupt climate change and transient climates in the Paleogene: a marine perspective. J Geol 101:193–215

Older Coasts

27

27.1 Introduction

Ancient oceans and paleocontinents have combined to form coastal systems throughout Earth's history. Many of these coastal systems have been preserved in the sedimentary record as stratigraphic formations corresponding to different geological periods. There is an abundance of scientific literature on many of these formations, especially those in Europe, North America and Australia. Compiling the information on all of these ancient coastal systems would be an endless task, well beyond the scope of this book. However, it is possible to synthesize the main characteristics of the most significant coastal systems of each geological period. This chapter analyzes some of the coastal environments that have contributed to a better understanding of the functioning of the Earth system at some key moments in the history of our planet.

27.2 Precambrian Tidalites

Among the most striking coastal geological records are those found in **tidalites**. These are coastal facies with an unequivocal signature that identifies them as sediments deposited by tidal processes. Tidalites are usually characteristic of open tidal flats, but can also occur in environments protected behind barriers and in those associated with other systems such as deltas or estuaries. Although tidalites have been preserved from all periods of Earth's history, perhaps the most significant are the Precambrian tidalites [13].

Known examples of tidalites preserved in the Precambrian record span nearly all of the Archean and Proterozoic eons (Fig. 27.1). The oldest formations are Archean and are found in South Africa [11, 14]. These are formations of the Moodies (−3250 My) and Witwatersrand (−3000–2800 My) Groups. More recent examples are in the Paleo-Proterozoic of Australia [12] represented by the Upper Mount Guide Group (−1800 My, the transition between the Meso- and Neo-Proterozoic of West Africa [9, 37] in the Pelel and Dindefelo Formations of the Segou-Madina Kouta Group (−1200–750 My); and the Neo-Proterozoic of Australia [36] in the Elatina Group (0.6 Ga).

All the tidalites described, whatever their age or location, have a number of common characteristics. Notably, the total absence of organisms with shells in these very early stages of the Earth makes these facies totally devoid of fossils. By the same token, there is also a paucity of trace organics and bioturbation. This fact results in a maximum preservation of physical structures. At the stratotype level, the tidalites are characterized by the presence of parallel tidal bundles (Fig. 27.2) and herringbone cross-stratifications (Fig. 27.3a).

At the outcrop scale, tidalites are organized in repetitive parasequences. Each of these sequences is formed by subtidal bars of sandstones with herringbone-type cross-stratification that may even present tidal bundles in the cross lamellae (Fig. 27.3b). Over these, finer sandstones with decimeter-scale herringbone laminations develop, and over these appear the finely laminated sediments with tidal bundles. The finely laminated layers are usually covered by ripples (Fig. 27.4a and b). The sequence may culminate in layers with bacterial mats (microbially induced sedimentary structures, MISS; Fig. 27.4c) affected by desiccation cracks (Fig. 27.4d). Each of these sequences can range in thickness from 0.5 to 12 m and may represent subtidal, intertidal and supratidal facies succession. The repetition of parasequences is due to the presence of subsiding pulses.

The abundance of tidalites from these early periods of the Earth's history has been attributed to the shorter distance between the Moon and the Earth, meaning greater tidal ranges, which would be characteristic of macrotidal coasts [13]. However, many authors agree on the similarity of these tidal signatures with those that currently appear in tidal flats, ruling out the possibility of generation on megatidal shores.

J. A. Morales, *Coastal Geology*, Springer Textbooks in Earth Sciences, Geography and Environment,
https://doi.org/10.1007/978-3-030-96121-3_27

Eon		Era	Period	Age (My)	Examples
Precambrian	Proterozoic	Neo-proterozoic	Ediacaran	541	Elatina Group (600 Ma)
			Cryogenian	635	
			Tonian	720	
		Meso-proterozoic	Stenian	1000	Segou-Madina Kouta Group (1200-720 Ma)
			Ectasian	1200	
			Calymmian	1400	
		Paleo-proterozoic	Statherian	1600	
			Orosirian	1800	Mount Guide Group (1800 Ma)
			Rhyacian	2050	
			Siderian	2300	
	Archean	Neo-archean		2500	
		Meso-archean		2800	Witwatersrand Group (3000-2800 Ma)
		Palaeo-archean		3200	Moodies Group (3250 Ma)
				3600	

Fig. 27.1 Geological timetable of the Precambrian eon, indicating the age of the best documented examples of Precambrian tidalites

27.3 Late Paleozoic Deltas

The late Paleozoic (Carboniferous and Permian) was a favorable time for the development of extensive siliciclastic coastal systems. During this interval, the final stages of the Variscan orogeny were taking place, and the collision of the great continents to form Pangea was coming to an end. In this context, the dismantling of the new orogenic reliefs provided a good supply of siliciclastic sediment to the coasts surrounding the megacontinent. Several deltaic and estuarine systems developed along the coasts of this new continent, whose supratidal fringes were associated with extensive swamps and marshes. These swamps were the ideal environments for the formation of the coal that gives its name to the main geological period of this time. However, these environments were not only capable of producing coal, but also other natural resources such as petroleum and uranium.

A good example of this type of environment is recorded in the Río Genoa Formation, within the Tepuel Group (Chubut, Argentina). In this example, the detailed sedimentological column correlation study allowed the identification of all the delta plain, delta front and prodelta environments of a river-dominated (lobated) delta developed over the shallow platform facies of the Mojón de Hierro Formation [2]. The Río Genoa Formation consists of 1200 m of mostly sandy facies that include some gravel intercalations, thick shale bodies and abundant coal layers. The sandy facies contains numerous sedimentary structures such as meter-scale cross-stratifications (planar and trough), and current and wave ripples. The facies include a rich fossil record of marine plants and invertebrates that contributed to the biostratigraphic and paleoecological interpretation.

In addition to several distributary channels separated by levees, there are numerous prograding bar-finger sands developed over the fine-sediment prodelta facies. The complete sequence shows the presence of autocycles of progradation and abandonment that are attributed to the existence of switching phenomena of the distributaries and deltaic

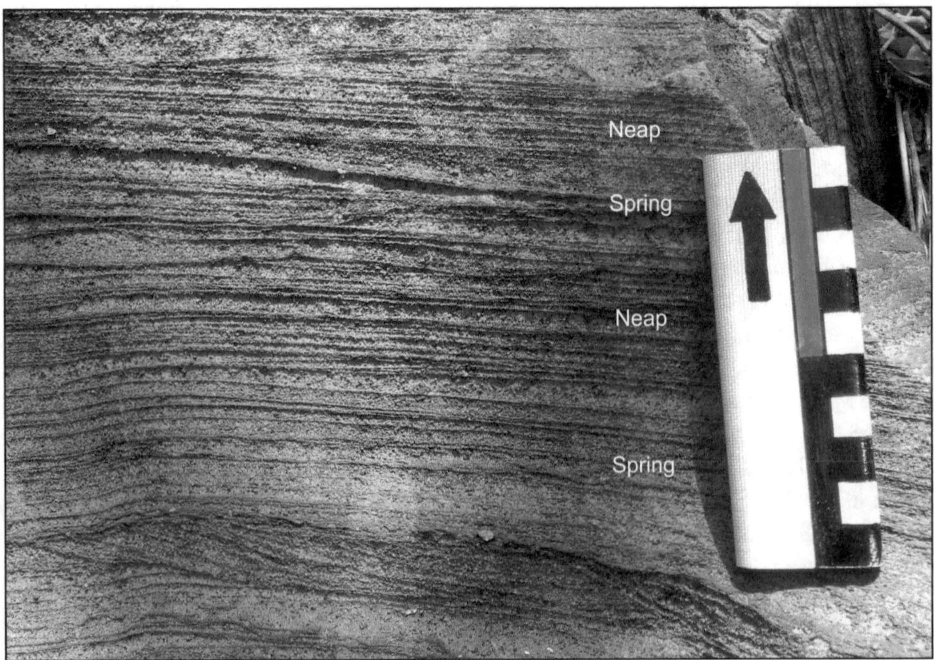

Fig. 27.2 Tidalites of the Stenian Pelel Formation, Segou-Madina Kouta Group (West Africa)

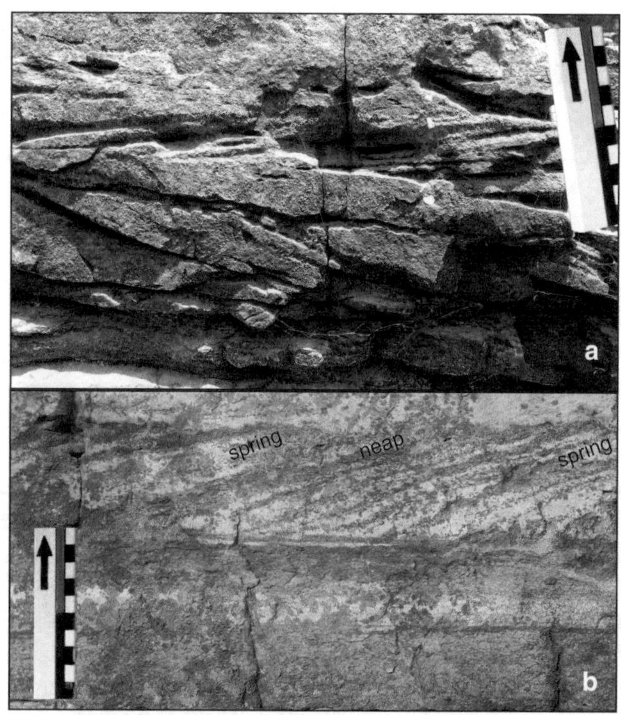

Fig. 27.3 Tidal structures of the Tonian Dindefelo Formation, Segou-Madina Kouta Group (West Africa). **a** Herringbones. **b** Tidal bundles in the cross-bedding

lobes. In any case, it is a highly constructive delta with high rates of progradation and aggradation. The vertical repetition of sequences is controlled by local subsidence phenomena in a stable sea level framework, although two eustatic uplifts have been identified during the Permian.

A very similar system has been described in the Anyang-Hebi coal basin in Henan Province, North China. In this case, 76 boreholes were studied, in addition to the series raised in open-pit workings [30]. The deltaic series is laid out across an unconformity overlying Ordovician carbonate rocks and beneath siliciclastic rocks of Triassic age. The entire sedimentary succession was formed after continental uplift occurred at the end of the early Paleozoic by the collision between the North China plate and the Siberian plate.

At the base of the sequence are the offshore facies, which mainly consist of alternating black shales (with bivalves, gastropods and brachiopods) and carbonates with abundant corals, crinoids, brachiopods, foraminifera and bryozoans (Fig. 27.5a). Overlying the offshore facies are the prodelta facies, which consist of black shales with rhythmic lamination and, above them, the delta front facies consist of a coarsening-upwards sequence of black shales and medium-grained sandstones with abundant flaser structures and small-scale herringbones (Fig. 27.5b). In the delta plain facies, depositional facies corresponding to the different

Fig. 27.4 Sedimentary structures of the upper terms of the tidal sequence in the Dindefelo Formation, Segou-Madina Kouta Group (West Africa). **a** Straight crested ripples. **b** Linguoid ripples. **c** Microbially induced sedimentary structures. **d** Mud cracks

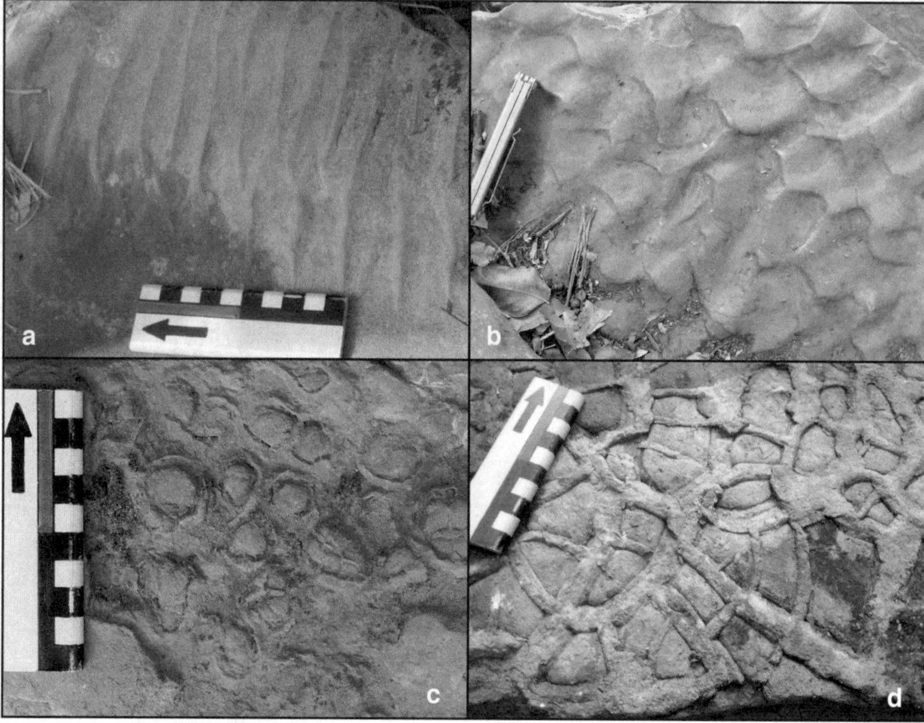

sub-environments can be distinguished. The distributary channel facies consist of lithofacies sequences that range in thickness from 3 to 10 m and are composed of fine-grained sandstones (Fig. 27.5c) with abundant decimeter-scale cross-laminations (ripples). The levee facies consists of the same sandstones that are interbedded with shale laminae with carbonaceous traces (Fig. 27.5d and e). The interdistributary swamp facies consist of carbonaceous shales and coal layers (Fig. 27.5f).

These facies are repeated in regressive sequences that form parasequences, reaching total thicknesses of between 1000 and 10 m. As in the case of the Río Genoa Formation, the repetition of parasequences has been attributed to subsidence pulses separated by periods of stability. It is this subsidence that led to the deposition of a large repetitive succession that allows the exploitation of the coal beds.

Late Paleozoic deltaic successions similar to the two discussed in this chapter are repeated in different basins of the world. Some examples have been described in the Appalachians [26], Scotland [20], Wales [19], Germany [35], Poland [25], Spain [7], Ukraine [23], Southern Africa [5] and India [3], among other places. In many of the outcrops where these sequences have been found, the delta plain parasequences including the coal layers (Fig. 27.6a) and even the clinoforms of prograding units of the delta front (Fig. 27.6b) are observed.

27.4 Late Jurassic–Early Cretaceous Reefs and Carbonate Coasts

During the Jurassic and Cretaceous periods, most of Gondwana (South America and Africa), the southern half of North America and southern Europe were in intertropical latitudes. In these periods, there was a general tendency towards extensional processes, since the fragmentation of Pangea was taking place through rifting processes that gave rise to new marine basins. This process was especially intense in Europe, which was fragmented during the Jurassic in the form of numerous islands separated by shallow seas (Fig. 27.7a). The migration of continents associated with the opening of the Atlantic pushed the continental masses of Europe toward lower latitudes during the Cretaceous (Fig. 27.7b). The coasts of these seas, being at intertropical latitudes, were conducive to the development of reefs and carbonate shores.

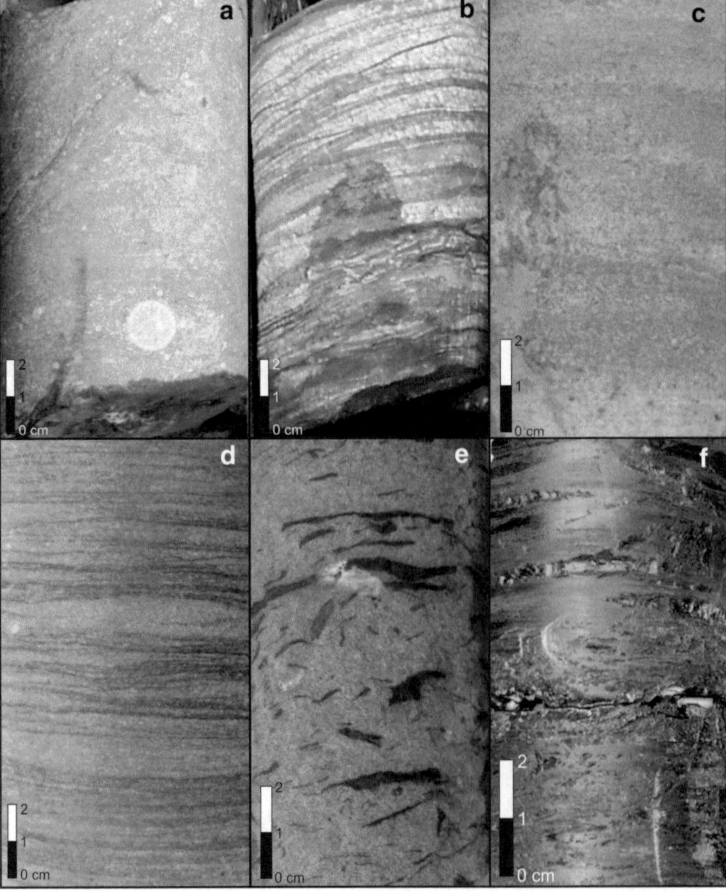

Fig. 27.5 Sedimentary facies observed in the Anyang-Hebi coalfield sequences. **a** Carbonate offshore facies. **b** Small-scale flasers interpreted as prodelta facies. **c** Fine-grained sandstone facies of the distributary channels. **d** Cyclic laminated sandstone and shale facies from levees. **e** Shaley sandstones with coal traces. **f** Coal facies from interdistributary swamps. Photographs by Li et al. [30]

Fig. 27.6 Different aspects of deltaic sequences at an outcrop scale. **a** Coal beds in a delta plain sequence (Namurian Basal Grit Formation, Wales, UK). **b** Prograding sequence in a delta front facies (Westphalian Millstone Grits, Wales, UK)

During the Jurassic period, the distribution of reefs increased significantly. The cause of this increase was in response to several factors, not only those linked to the latitudinal distribution of the coasts of that period. The development of carbonate systems has usually been related to times of highstand [28]. Throughout the Jurassic, the sea level was rising, so that in the late Jurassic it was in a global position that reached 150 m above the present level. In the early Cretaceous, the level dropped rapidly by about 70 m and then rose again by about 20 m above the maximum Jurassic level. This high sea level position caused extensive areas of the continental zones to be flooded, increasing the habitat surface for biohermal organisms and reducing the source areas of terrigenous sediment and, consequently, the volume of siliciclastic sediment. Moreover, the high atmospheric CO_2 levels of the late Jurassic and early Cretaceous exhibited a climatic optimum that was related to these high marine levels. This climatic optimum caused tropical seas to extend above tropical latitudes, reaching 40° latitude. These environmental conditions seem to explain the increase in reef environments during these periods better than the radiation of reef-forming organisms. In any case, all these factors acting together led to the appearance of a large number of different types of reefs linked to different paleogeographic contexts.

In Europe, regional studies of different reefs from this period of Earth's history have identified three types of reef [27]: coral reefs (Fig. 27.8a), siliceous sponge reefs (Fig. 27.7b) and microbial reefs (Fig. 27.7c). In addition, there are mixed-type reefs (Fig. 27.7d). Coral facies dominate the reefs of the Iberian basins. These reefs include a wide variety of coral facies that reflect a diversity of environments associated with the reef bodies.

Siliceous sponge-dominated reefs occur abundantly in the basins of southern Germany (south of the Swabian paleocontinent). These spongiolithic reefs are often associated with microbial mats and algal crusts. Such reefs occur in a coastal belt on the north-European margin of the Tethys from Romania to Portugal and also on the southwestern margin of the Tethys (the Atlas Mountains of Morocco). Sometimes spongiolithic reefs also occur in association with coralline facies. Then, they manage to form large reef bodies that may outcrop at some distance from the coast, forming topographic elevations and reef patches. The distribution of facies in these mixed reefs is closely linked to the bathymetry and the distribution of energies in the environment. In these systems there are also abundant detrital facies, which are linked to bioclastic shallows, washovers and aprons, as well as oolitic tidal deltas [29].

Microbial crusts may be present in the above types of reef, where they combine with siliceous sponges and corals to build large reef structures. But in addition, microbes can also build reefs on their own. In these periods, thrombilitic reefs dominate, which can exceed 30 m in height. Thrombolites are similar to stromatolites in their external form, but their inner structure is not laminated. Pure microbial reefs

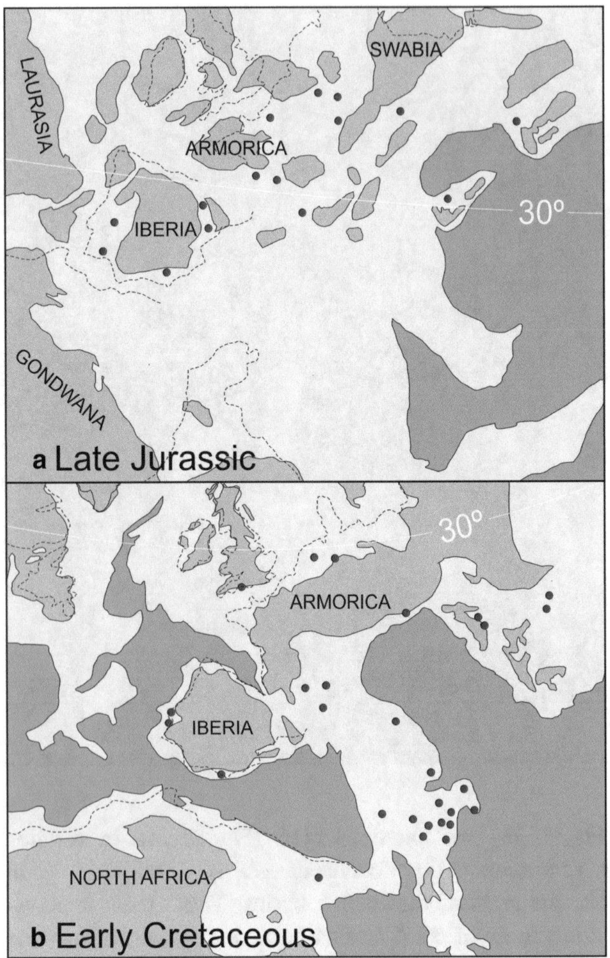

Fig. 27.7 Paleogeographic map of the distribution of reefs (red circles) in the carbonate platforms (light blue) of the late Jurassic (**a**) and early Cretaceous (**b**)

with coralline and spongiolithic influence are greatly abundant in the Mesozoic basins of Iberia. In this context, thrombilitic reefs may even occur in environments with a high influence of siliciclastic input [29].

27.5 Cenozoic Barrier Islands

There are numerous examples of barrier island systems from all periods of the Cenozoic. Eocene barriers are well represented in the Claiborne, Jackson and Wilcox Group Formations of Texas (e.g., Fisher et al. [17]; Davies and Ethridge [8]; Dickinson [10]; Galloway [18]; Johnston and Johnson [24]; Hamilton [21]), as well as in the Misoa Formation in Venezuela [1]. Oligocene barriers are also represented in the Frio Formation of Texas [4]. A good Miocene example has been studied in the Cohansey Formation in New Jersey, USA [6]. More recently, a Pliocene sandy barrier system has been characterized in the Spartizzo–Scandale Formation in Italy [32]. Examples of Pleistocene barrier islands are more abundant. Systems of these ages have been characterized in the Port Dunford Formation in South Africa [22] and on Mockhorn Island in Virginia, USA [16]. In most cases, the unconsolidated character of the facies allows observation of the three-dimensional architecture at outcrop scale, as well as measurement of the migration directions of the bedforms from the dip of the cross-bedding.

Many of the preserved systems have common characteristics. Among them is a similar architecture between barrier/beach/dune bodies and barrier/lagoon/tidal plain bodies. Barrier sequences are characterized by the presence of medium to very coarse sands composed of siliciclastic, bioclastic or mixed grains. Whatever the composition, they are generally constituted of very mature sediments due to the continuous wave reworking. The structures characteristically preserved in these facies are planar, trough (Fig. 27.9a), swaley (Fig. 27.9b) and hummocky cross-stratifications. Regardless of the dominant lithology, the presence of layers with a high concentration of shells is common. These layers have lenticular geometry, suggesting that they are accumulated by storms. Sometimes, the barrier sequences are composed of pebbly sandstone facies, also very mature texturally. In such cases, the characteristic internal structure is seaward-sloping parallel laminations or low-angle landward-sloping cross-stratifications. Barrier sequences may be capped by dune facies. The lithology of these facies corresponds to very well-sorted fine to coarse sands of siliciclastic or bioclastic origin. When bioclasts are present, they usually belong to foraminiferal shells or fragments of mollusks with very light shells. The dune facies characteristically show very complex cross-stratifications (Fig. 27.9c) and root bioturbation is often extensive (Fig. 27.9d).

The back-barrier/lagoon sequences are dominated by the presence of clayey sediments rich in organic matter. These sediments usually have a massive aspect, although they may also appear laminated. Sometimes they are cut by bioclastic debris lenses that correspond to tidal channel bed sediments. Towards the uppermost part of the sequence, flaser/wavy/lenticular structures usually appear, and everything is generally topped by peat layers. All facies show a high degree of bioturbation (Fig. 27.9e), with galleries of the *Ophiomorpha* and *Thalassinoides* types frequently present.

In areas where outcrops allow the contact between these two types of sequences (barrier and lagoon) to be seen, the lagoon sequences may be covered by washover facies (Fig. 27.9f) and dunes, especially in the case of transgressive sequences.

Cenozoic sequences are not usually deformed by compressional tectonic processes, although they may be tilted due to extensional processes. Their sediments may also be affected by early diagenetic processes such as partial dissolution of bioclasts, carbonate cementation and neomorphism.

Fig. 27.8 **a** Mid-Jurassic branching coral facies in Portugal (Photograph: Francisco Félix [15]). **b** Lower Jurassic spongiolite in the Central High Atlas, Morocco (Photograph: Mannani [31]). **c** Lower Jurassic thrombolite in the Central High Atlas, Morocco (Photograph: Mannani [31]). **d** Mixed siliceous sponge–microbialite mound in Portugal (Photograph: Reolid and Duarte[34])

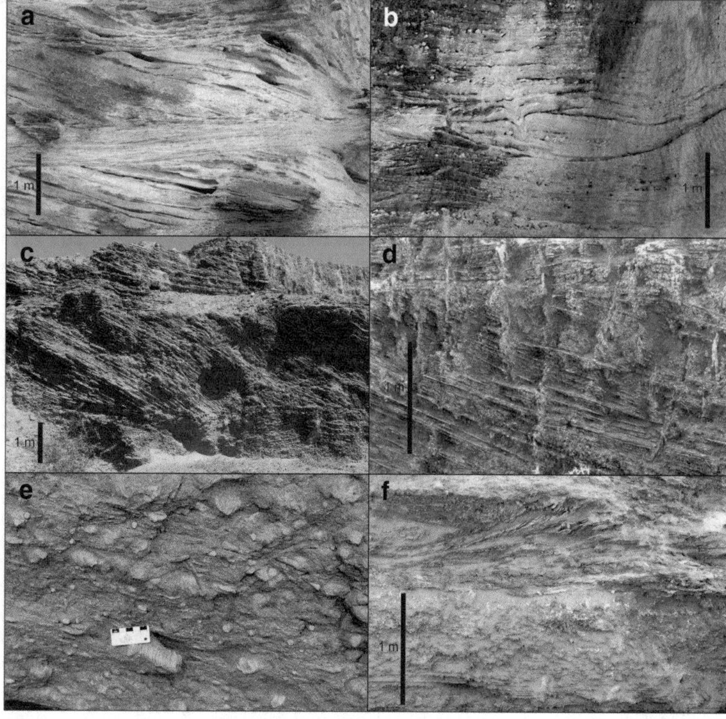

Fig. 27.9 Facies of Cenozoic barrier island systems. **a** Planar cross-bedding and sea-inclined parallel-bedding characteristic of beach facies of a Miocene formation (SW Spain). **b** Swaley cross-bedding in the same formation. **c** Complex cross-bedding characteristic of coastal dune facies of a Pliocene formation (W Morocco). **d** Detail of root bioturbation in the same formation. **e** Burrowed muds characteristics of lagoon facies in a Miocene formation (SW Spain). **f** Bioturbated lagoonal muddy sands covered by cross-bedded washover pebbly sands in a Pleistocene formation (SW Spain)

Despite the obvious similarities, it appears that the Cenozoic sequences preserved in facies are markedly different from the systems functioning today. On the one hand, the depositional facies of many of the environments present in today's systems have not been well preserved in the fossil examples. For example, the facies of the most wave-exposed environments may have been destroyed by processes such as reworking or erosion. On the other hand, the thickness of some sequences may have been increased by sedimentary piling. In general, the dimensions of the sedimentary bodies that constitute the facies architecture preserved in the barrier island systems of the geological record are larger in all three dimensions, because these bodies include migrations of environments over longer time frames than those recorded in present-day systems [33].

References

1. Ambrose WA, Ferrer ER (1997) Seismic stratigraphy and oil recovery potential of tide-dominated depositional sequences in the Lower Misoa Formation (Lower Eocene), LL-652 Area, Lagunillas Field, Lake Maracaibo, Venezuela. Geophysics 62:1483–1495
2. Andreis RR, Cuneo R (1989) Late Paleozoic high constructive deltaic sequences from northwestern Patagonia, Argentine Republic. J S Am Earth Sci 2:19–34
3. Bhattacharya B, Bhattacharjee J, Bandyopadhyay S, Banerjee S, Adhikari K (2018) Early Permian transgressive–regressive cycles: sequence stratigraphic reappraisal of the coal-bearing Barakar Formation, Raniganj Basin India. J Earth Syst Sci 127:29
4. Boyd DR, Dyer BF (1964) Frio barrier bar system of south Texas. Trans Gulf Coast Assoc Geo Soc 14:309–322
5. Cairncross B (2001) An overview of the Permian (Karoo) coal deposits of southern Africa. J Africa Earth Sci 33:529–562
6. Carter CH (1978) A regressive barrier and barrier-protected deposit: depositional environments and geographic setting of the late Tertiary Cohansey sand. J Sediment Res 48:933–949
7. Colmenero JR, Suárez-Ruiz I, Fernández-Suárez J, Barba P, Llorens T (2008) Genesis and rank distribution of Upper Carboniferous coal basin in the Cantabrian Mountains, Northern Spain. Int J Coal Geol 76(3):187–204
8. Davies DK, Ethridge FG (1971) The Claiborne group of central Texas. Trans Gulf Coast Assoc Geol Soc 21:115–124
9. Deynoux M, Duringer P, Khatib R, Villeneuve M (1993) Laterally and vertically accreted tidal deposits in the Upper Proterozoic Madina-Kouta Basin, southeastern Senegal, West Africa. Sed Geol 84:179–188
10. Dickinson KA (1976) Sedimentary depositional environments of uranium and petroleum host rocks of the Jackson Group, south Texas. J Res US Geol Surv 4:615–629
11. Eriksson KA (1977) Tidal deposits from the Archaean Moodies Group, Barberton Mountain Land, South Africa. Sed Geol 18:257–281
12. Eriksson KA, Simpson EL (1990) Recognition of high-frequency sea-level changes in Proterozoic siliciclastic tidal deposits, Mount Isa, Australia. Geology 18:474–477
13. Eriksson KA, Simpson EL (2015) Precambrian tidal facies. In: Davis RA, Dalrymple RW (eds) Principles of tidal sedimentology. Springer, Heidelberg, pp 397–420
14. Eriksson KA, Turner BR, Vos RG (1981) Evidence of tidal processes from the lower part of the Witwatersrand Supergroup. Sed Geol 29:309–325
15. Félix F (2021) Bank of images of science's house in Portugal. https://www.casadasciencias.org/imagens
16. Finkelstein K (1992) Stratigraphy and preservation potential of sediments from adjacent Holocene and Pleistocene barrier-island systems, Cape Charles, Virginia. In: Fletcher III CH and Wehmiller JF (eds) Quaternary coasts of the United States: Marine and Lacustrine Systems, vol 48. SEPM Special Publication, pp 129–140
17. Fisher WL, Proctor CV Jr, Galloway WE, Nagle JS (1970) Depositional systems in the Jackson group of Texas: their relationship to oil, gas, and uranium. Geol Surv Circular 70-74:234–261
18. Galloway WE (1986) Reservoir facies architecture of microtidal barrier systems. Am Asso Petrol Geol Bull 70:787–808
19. George GT (2000) Characterisation and high resolution sequence stratigraphy of storm-dominated braid delta and shoreface sequences from the Basal Grit Group (Namurian) of the South Wales Variscan peripheral foreland basin. Mar Pet Geol 17:445–475
20. Goodlet GA (1959) Mid-Carboniferous sedimentation in the Midland Valley of Scotland. Edinburgh Geol Soc Trans 17:217–240
21. Hamilton DS (1995) Approaches to identifying reservoir heterogeneity in barrier/strandplain reservoirs and the opportunities for increased oil recovery: an example from the prolific oil-producing Jackson-Yegua trend, south Texas. Mar Pet Geol 12:273–290
22. Hobday DK, Jackson MPA (1979) Transgressive shore zone sedimentation and syndepositional deformation in the Pleistocene of Zululand, South Africa. J Sediment Res 49:145–158
23. Izart A, Briand C, Vaslet D, Vachard D, Coquel R, Maslo A (1996) Stratigraphy and sequence stratigraphy of the Moscovian in the Donets Basin. Tectonophysics 268:189–209
24. Johnston DD, Johnson RJ (1987) Depositional and diagenetic controls on reservoir quality in first Wilcox Sandstone, Livingston Field, Louisiana. Am Asso Petrol Geol Bull 71:1152–1161
25. Kedzior A, Grandzinski R, Doktor M, Gmur D (2007) Sedimentary history of a Mississippian to Pennsylvanian coal-bearing succession: an example from the Upper Silesian Coal Basin, Poland. Geol Mag 144(3):487–496
26. Klein GDV, Willard DA (1989) Origin of the Pennsylvanian coal-bearing cyclothems of North America. Geology 17:152–155
27. Leinfelder R (1993) Upper Jurassic reef types and controlling factors. A preliminary report. Profil 5:1–45
28. Leinfelder R (1994) Distribution of Jurassic reef types: a mirror of structural and environmental changes during breakup of Pangea. Can Soc Pet Geol Memoir 17:677–700
29. Leinfelder R (2001) Jurassic reef ecosystems. In: Leinfelder R (ed) The history and sedimentology of ancient Reef systems. Springer, Boston, pp 251–309
30. Li Y, Shaoa L, Fielding CR, Wang D, Mua G (2021) Sequence stratigraphy, paleogeography, and coal accumulation in a lowland alluvial plain, coastal plain, and shallow-marine setting: Upper Carboniferous-Permian of the Anyang-Hebi coalfield, Henan Province, North China. Palaeogeogr Palaeoclimatol Palaeoecol 567:110287
31. Mannani I (2016) The role of micrites in the Sinemurian (Lower Jurassic) sponge–microbialite mounds from Foum Tillicht, central High Atlas, Morocco. Bollettino della Società Paleontologica Italiana 55:157–169
32. Mellere D, Zecchin M, Perale C (2005) Stratigraphy and sedimentology of fault-controlled backstepping shorefaces, middle Pliocene of Crotone Basin, Southern Italy. Sed Geol 176:281–303

33. Mulhern JS, Johnson CL and Martin JM (2019) Modern to Ancient barrier island dimensional comparisons: implications for analog selection and paleomorphodynamics. Front Earth Sci 17
34. Reolid M, Duarte LV (2014) Sponge–microbialite buildups from the Toarcian of the Coimbra region (Northern Lusitanian Basin, Portugal): paleoecological and paleoenvironmental significance. Facies 60:561–580
35. Suess MP, Drozdzewski G, Schaefer A (2007) Sedimentary environment dynamics and the formation of coal in the Pennsylvanian Variscan foreland in the Ruhr Basin (Germany, Western Europe). Int J Coal Geol 69(4):267–287
36. Williams GE (1989) Late Precambrian tidal rhythmites in South Australia and the history of the Earth's rotation. J Geol Soc London 146:97–111
37. Youm CI, Erramji E, Sow EH (2018) Neoproterozoic Dindéfélo waterfall geosite (DCNR, Bassari country, Eastern Senegal): biodiversity and geodiversity between conservation and valorization. J Chem Biol Phys Sci 8(3):197–224

Part V
The Humans in the Coast: Interaction Problems and Coastal Management

I made jetties so they'd catch all the sediment,
removed the rocks and every impediment,
but the tide's rising high to wash away
my island in the night.

"The island"
Bad Religion

28 Human Impacts on Coastal Systems

28.1 Introduction

The majority of the population of most of the world's countries are based in coastal areas, a fact that has been true throughout history [4]. However, since the second half of the twentieth century, human activity has had a greater and faster impact on coastal dynamics. Part of this increased impact on the coast is due to the development of beaches for recreational use, but it should not be forgotten that most economic activity is also located around coastal systems. The protection of human assets requires constant interventions on the natural environment for a variety of reasons—for example, to avoid sediment losses, to replenish sediments or to artificially create new beaches. However, these actions are not the only ones, nor are they the most important ones, that are carried out by humankind on the coasts. If we consider the factors responsible for sediment dynamics on the coast that were studied in Chap. 3 (Fig. 28.1), we see that humans can intervene directly or indirectly in all of them.

- The most direct impact that humans exert on the coasts is the modification of the patterns of action of hydrodynamic agents on the coastal fringe. These interventions include the construction of groins, jetties, breakwaters and seawalls that modify wave refraction patterns. In addition, the restriction of intertidal surfaces by preventing natural drainage in estuaries, deltas and tidal flats modifies their tidal prism. This alters the energy balance at their mouths.
- The least studied of these interventions is the impact that humans can exert on the continental input to the coasts by damming rivers and destroying coastal dunes. This generates a sediment deficit that contributes to the erosion of coastal front systems. Conversely, the creation of artificial beaches through the forced contribution of sand introduces a cumulative modification.
- The human influence on climate has been a matter of debate in recent decades. Today it is assumed that the climate is changing due to anthropogenic emissions into the atmosphere. On the one hand, this means the emission of greenhouse gases from the combustion of hydrocarbons and, on the other hand, the emission of certain aerosols that contribute to global warming and the progressive rise in sea level.

With these considerations in mind, it seems clear that, when humans occupy the coast, they invariably alter the natural environment. Human-induced alterations to coastal dynamics can range from minimal to disastrous and, over decades, the effects have depended more on chance than on planning.

28.2 Direct Human Interventions on the Coast (Coastal Engineering)

In many areas of the planet, coastal areas have been over-occupied without any prior planning or understanding of the dynamics of the coastal segment in which the constructions were carried out. It is true that many of these actions took place before we had the knowledge we have now about the coast. However, even today, when we do have sufficient knowledge, wide stretches of coastline continue to be occupied due to political decisions that do not take this knowledge into account at all. When we speak of human constructions on the coast, the presence of housing developments and tourist facilities quickly comes to mind; however, in addition to these, port facilities and numerous industrial complexes are also found on the coast. In short, the coast has become an important enclave for human economic activities. As a consequence, related infrastructures have for decades themselves been threatened by ordinary coastal processes and even more by high-energy events such as storms [9]. Because of this, coastal engineers have been forced to build structures for the protection of human property. The influence of these activities on the natural dynamics of the coast is analyzed below.

J. A. Morales, *Coastal Geology*, Springer Textbooks in Earth Sciences, Geography and Environment,
https://doi.org/10.1007/978-3-030-96121-3_28

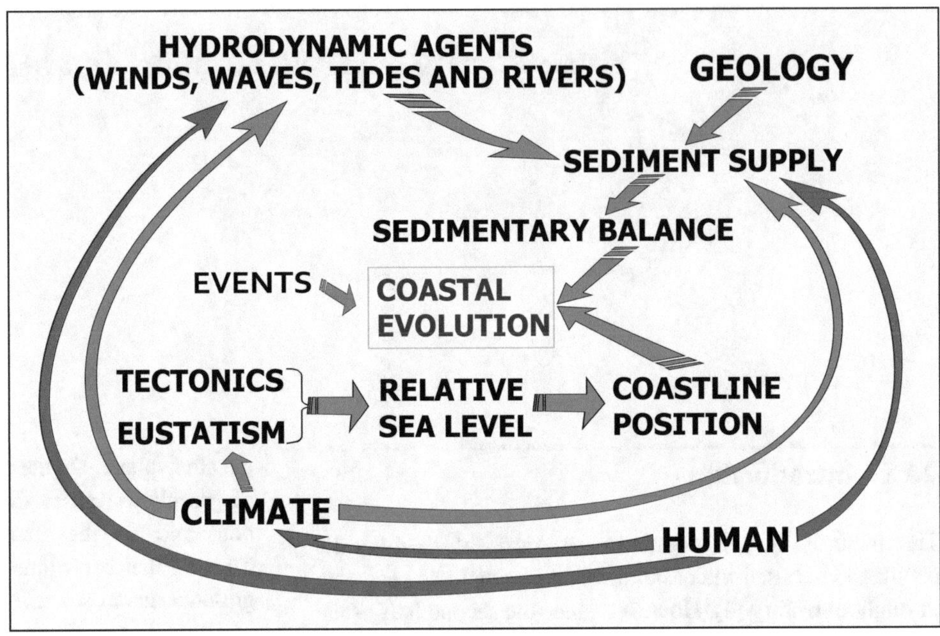

Fig. 28.1 Scheme showing the influence of human activities on different components of coastal dynamics

28.2.1 Building Rigid Structures

The coast is a changing and flexible environment. This flexibility is often incompatible with the presence of human constructions on the coast. Indeed, this is why engineers have created different rigid structures that try to stabilize the position of the coastline. Several types of structures have been created by coastal engineers for this purpose. Some are perpendicular to the coastline, while others are parallel, and some structures even consist of a complete fortification of the coastline.

One of the structures that engineers have classically used to slow down coastal erosion has been **groins** (or groynes). Many beaches, especially those located in front of centers of coastal urbanization, have been converted into a succession of groins in the form of a comb. In other examples, in order to reduce the mobility of the channels that interrupt coastal continuity (tidal inlets, river mouths or even artificial harbor channels), **jetties** have been built on the margins of these channels.

Groins and jetties exert a similar effect. As constructions perpendicular to the coastline, they constitute a barrier to longitudinal sedimentary bypassing (littoral drift). The presence of one of these structures compartmentalizes the coastal fringe by dividing sediment circulation cells. From a dynamic point of view, each structure divides two zones with different behaviors on either side (Fig. 28.2) and causes a change in the wave refraction pattern. On the updrift side of the groin, the desired effect is accumulated sediment, while on the downdrift side a zone of divergence of wave trains is generated [7]. The accumulation of sediment in the updrift segment is achieved because the groin prevents sand outflowing longitudinally into the adjacent sector. Thus, excess sand accumulates on this side of the groin. Conversely, in the downdrift zone, the longitudinal inputs of sediment are cancelled out, generating a sediment deficit that makes this section erosive.

Where groin combs are present, transverse rather than longitudinal dynamics dominate. The groins compartmentalize the beach into multiple sediment circulation cells (Fig. 28.3a). In this way, sand can only transit between different cells when it is eroded during storms and transported transversely to deeper areas. Of particular interest is the situation that occurs when groins are arranged too close together, as each cell between two groins can behave like a reflective beach, generating strong undertow currents in its central sector [3]. These undertow currents can draw significant volumes of sand into the shore face beyond the length of the groins. From this area, the sand can return to the foreshore of the adjoining cell when a fair weather wave acts obliquely to the shore. A situation that is particularly sensitive to coastal erosion occurs when there are steps in the subtidal zone that prevent sand from returning to the foreshore zone.

This behavior shows that the presence of groin combs does not completely prevent beach erosion. In order to prevent the transverse transport of sand into deeper areas, engineers can extend the apex of the breakwater with a longitudinal structure giving it a T-shape (Fig. 28.3b). This type of construction partially prevents the transport of sand into deeper areas, but also prevents the entry of sand into the cell. The construction of such structures is often accompanied by artificial nourishment of the cells between the groins, which function as small pocket beaches.

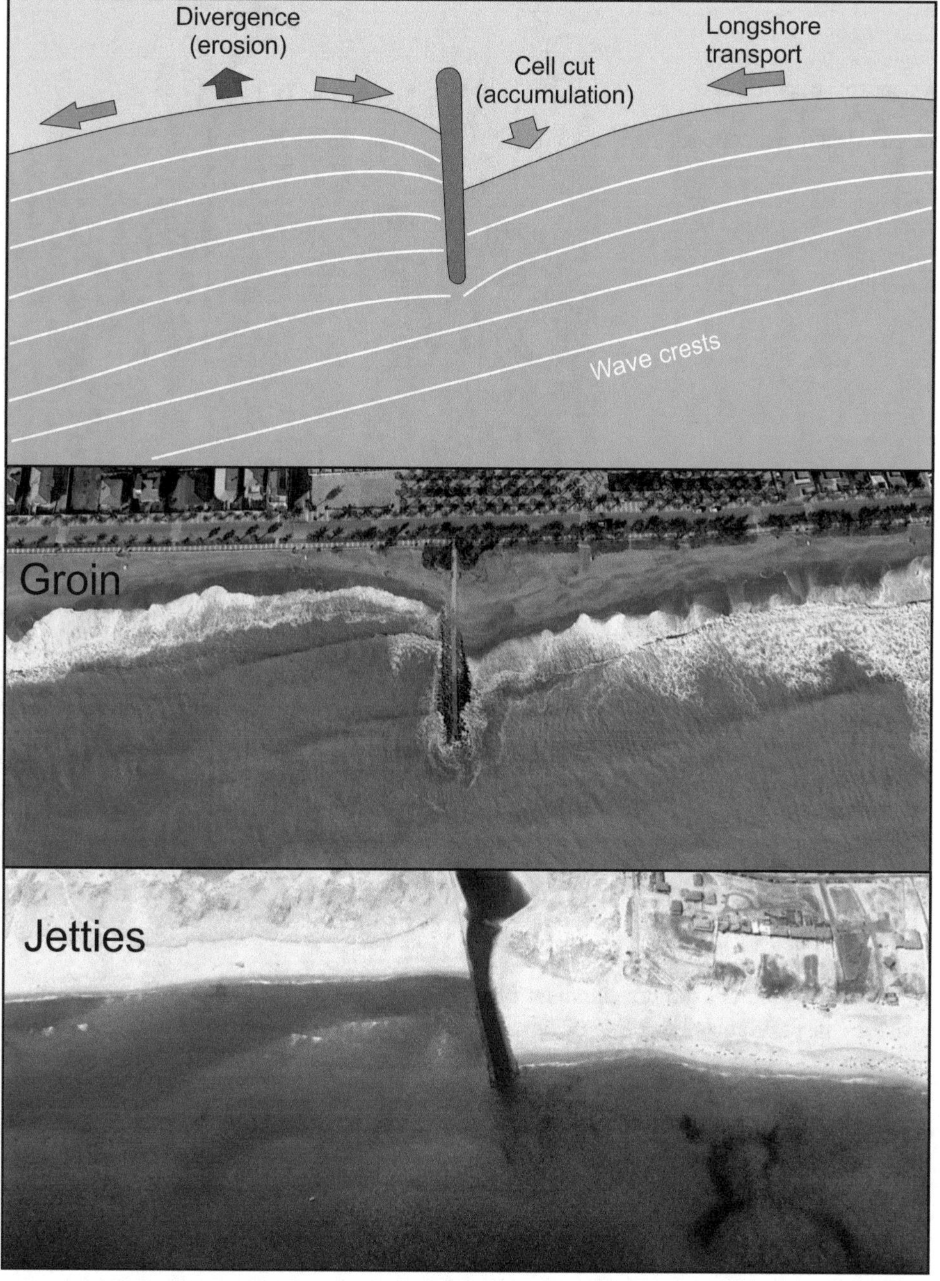

Fig. 28.2 Scheme showing the dynamic modification of the wave trains by a rigid structure built in a perpendicular direction from the coastline and its effect on the sedimentary patterns. Examples shown are of a single groin and two parallel jetties. *Images* Landsat from Google Earth

Breakwaters (or exent dams) are constructions created by linear rock piles, typically less than 100 m in length, which are placed parallel to and separated from the beach. The distance from the breakwater to the shore is calculated using the surf distance. Like the groins, these structures modify the wave action on the coast, damping the waves and affecting the refraction pattern in such a way that wave divergence zones are generated on one side or on both sides of the structure, with a convergence zone behind it (Fig. 28.4). Behind the structure, the sediment volume increases due to drift inputs from the two adjacent cells. However, the generation of these two wave divergence zones outside the protection of the breakwater means an increase in both zones of drift outputs and this, therefore, marks their erosive behavior.

In many cases, these breakwaters are designed in a segmented manner—i.e., several breakwaters are constructed separated by gaps. The design of the length of the gaps in relation to the distance of the breakwater from the coastline is fundamental to understanding the dynamic functioning of the stretch of coastline that will be protected by the breakwater [13]. Normally, the beach area acquires a profile with a succession of concave forms in front of the hollows and convex forms behind the breakwaters. Sometimes the accumulation of sand in the area protected by the breakwaters can even reach the structure, which becomes a kind of tombolo.

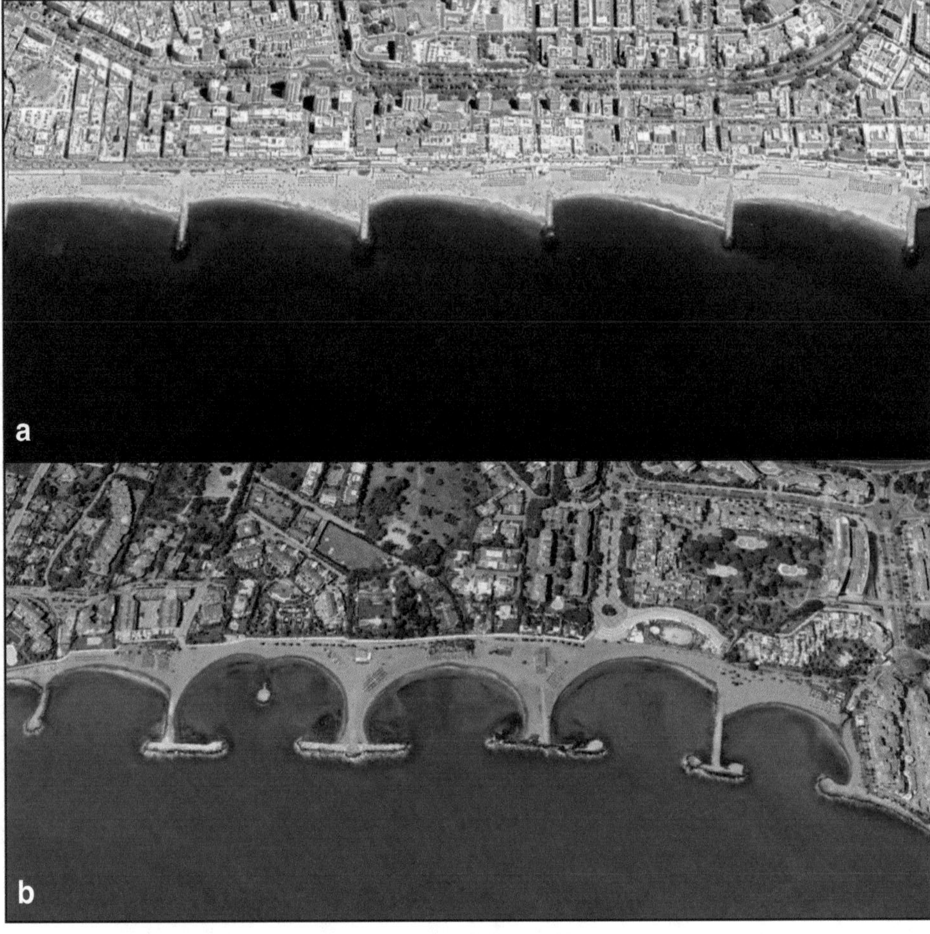

Fig. 28.3 Examples of multiple rigid structures dividing circulation cells in a beach. **a** Comb of straight groins. **b** Multiple T-shaped groins. *Images* Landsat from Google Earth

In other examples, the development of long stretches of coastline has eliminated the line of dunes in order to build promenades. At other times, although the promenades were built after the first line of dunes, they were designed without prior planning and were installed in clearly erosive areas, so that after their construction the line of dunes was eroded. Whatever the causes, in many cases the promenades or the buildings themselves are exposed to wave action, which attacks them especially during high-energy events. In these cases, one of the solutions commonly adopted is the complete protection of the shoreline by building **seawalls**. Another use of seawalls is to stop or slow down the retreat of the cliffs, especially when there are developments located on the cliffs that are threatened by such retreat. In this case, the stabilization of some sections of cliff is associated with a decrease, or even cancellation, of the contribution that the littoral drift supplies from the cliff to other adjacent sectors of the coast.

Seawalls can be constructed in a number of possible ways. Sometimes they consist solely of wooden stakes driven into the waterfront, stacked sandbags, or a combination of both (Fig. 28.5a). These structures are also called **bulkheads**. In this case, it is a temporary structure that is easily and quickly constructed, but can also be easily removed. A more permanent way of constructing a seawall is by cladding. In most cases the revetments are built with rocks extracted from quarries close to the shore (Fig. 28.5b); however, when this is not possible, revetments can be built with precast concrete blocks. The most radical form of seawall is the erection of a vertical concrete wall (Fig. 28.5c).

Normally, these structures protect the buildings behind them, but they do not prevent the erosion of the sediments in front of them. On the contrary, the presence of these structures increases the effect of wave reflection and often causes the beaches in front of them to be completely dismantled [11]. It should be noted that the reflection produced by rock revetments is less than that caused by vertical concrete walls. To minimize the effect of reflection on frontal beach erosion, engineers have devised the creation of revetments with irregularly shaped concrete blocks which create large gaps that disperse wave reflection in many directions, so that the reflected waves counteract each other.

The presence of sea walls has another influence on the sediment balance by preventing the entry of eroded sediment volumes from the supratidal zones into the system, increasing the sediment deficit of these beach segments.

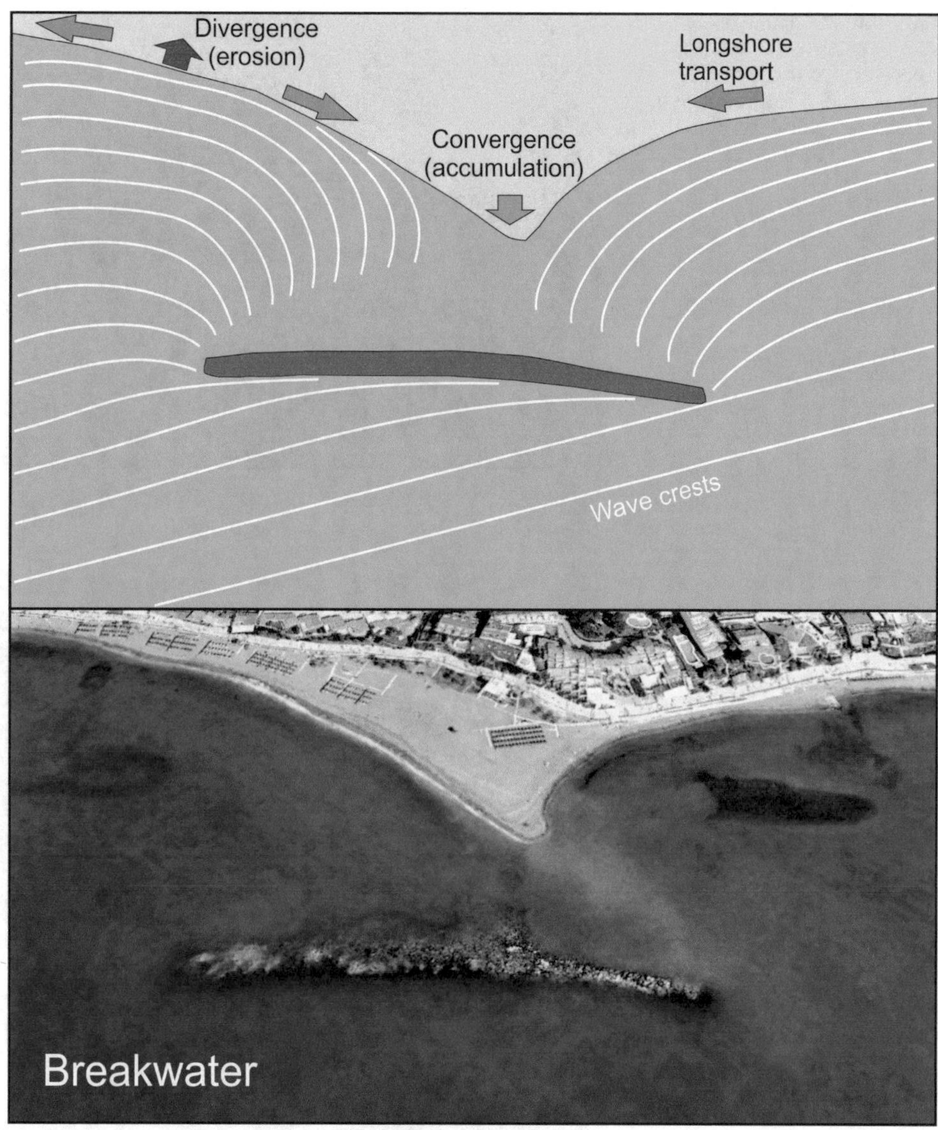

Fig. 28.4 Scheme showing the dynamic modification of the wave trains by a rigid structure built parallel to the coastline and its effect on the sedimentary patterns, with an example of a real case. *Image* Landsat from Google Earth

On the other hand, these structures are frequently attacked and destroyed by waves during storms. In this case, less rigid structures such as wooden stakes, sandbags and rock revetments behave less uniformly against collapse caused by the disappearance of the sand supporting the structure and are more easily reconstructed than vertical concrete walls.

Some of the economic activities that humans carry out on the coast are directly embedded in the tidal systems. For example, activities such as fish farming or the salt industry require the occupation of extensive intertidal areas of coastal systems such as lagoons, deltas or estuaries.

Advanced Box 28.1 Geologists Versus Engineers on Coastal Protection

The anthropic structures that modify coastal systems are designed by coastal engineers according to principles that prioritize functionality and do not always take into account the changes that these structures will have on the dynamics of the natural system. The models applied by the engineers focus on the calculation of the design and the dimensions that the structures must have in order to resist the action of the processes in terms of durability and resilience. The principles governing these approaches can be summarized as follows:

- A large part of human infrastructure and natural resources are located in coastal areas.
- The coasts are highly valued as recreational areas.
- Coastal wetlands are ecologically rich.
- The boundary between land and sea is a permanent danger zone.
- Coasts are easily modifiable.

The conclusion drawn from the sum of all these considerations is that the coast must be protected.

Fig. 28.5 Different types of seawall construction. **a** Using wooden stakes and geotextile bags of sand. **b** Rock revetment. **c** Concrete vertical walls

At the end of the twentieth century, these approaches were strongly contested by coastal geologists, represented by the figure of Orrin Pilkey, one such geologist at Duke University, North Carolina, USA. Geologists understand that most of the time engineers only consider whether a structure *can* be built, but they never consider whether that structure *should* be built or not. The consequence is that the structures, although costly, often have unexpected and undesirable effects on adjacent coastal areas. Pilkey enunciated what are known as "the six truths about the coast" [12]:

(1) Coastal erosion problems do not exist until the presence of human infrastructure forces us to measure them.
(2) The construction of anthropogenic structures on beaches reduces their flexibility and causes erosion.
(3) The interest of the owners should not be confused with the natural interest.
(4) Once you start stabilization, you can't stop.
(5) The cost of saving a property is in the long run much higher than the price of the property itself.
(6) To save the coast, you end up destroying it.

Pilkey's truths are controversial, but perhaps they were the starting point for the development of the integrated coastal zone management (ICZM) concept that governs the principles of coastal occupation today.

28.2.2 Beach Replenishment

Many coastlines in front of major tourist resorts or cities are highly erosive. The demand for beaches for recreational use has led many governments to devote a significant budget to the artificial replenishment of these beaches. The artificial contribution of sand to beaches is known as **replenishment** or **nourishment** (Fig. 28.6).

The choice of grain size usually takes into account the sedimentological characteristics. Normally, the grain size used is intended to be above the entrainment threshold of the dominant wave, in order to as far as possible avoid continued erosion [5]. This, in practice, means the use of an average size coarser than the natural diameter of the replenished beach [14]. However, other grain-size parameters such as sediment sorting are often not taken into account. This, together with the limitations often offered by the lack of a suitable site for dredging the sands for nourishing, means that sometimes suboptimal sediment is used for beach replenishment. Even if the sediment used has an adequate mean grain size, where there is poor sorting the finer sizes are usually washed by waves and wind and are easily extracted from the system. This, in addition to a loss of material volume, leads to a decrease in the quality of the sediment on the beaches, as only the residual fractions remain as a lag.

The erosion of the sediment added by regular replenishments has a clear influence on the feeding of the adjacent non-nourished sectors. On the one hand, the erosion of part of these sediments can mean an increase in sedimentation rates in the shoreface sectors in front of the replenished area. On the other hand, on beaches with a strong longshore component, neighboring beaches may also see an increase in sand input due to littoral drift.

Although these inputs constitute an increase of material into the coastal system, it is an ephemeral input. Since the dynamic equilibrium of the beach is not modified, the input of this material does not produce changes in the sediment balance. This means that if the dynamic agents provoke a

Fig. 28.6 Beach replenishment operations

negative balance on this stretch of coast, it will continue to have erosive behavior even after the artificial contribution of sand. Thus, replenishment is never a definitive solution to an erosion problem, but only a temporary one, and the beach will have to be nourished again in the short or medium term [6].

The cost of beach replenishment in many developed countries runs to hundreds of millions of dollars per year. In many cases, this investment is beneficial because the money returned to the state in the form of taxes from the tourism industry exceeds the initial costs. However, more often than not, the money is not returned and is a sunk cost [8]. Given that replenishments have to be carried out periodically, most of the time, in the long term, the investment in protecting a stretch of beach through nourishment costs far more than the value of the property it is intended to protect [12].

Advanced Box 28.2 Artificial Coasts

For most of the twentieth century, human actions on the coasts were limited to protecting coastal systems with rigid works or sediment replenishments. The development of complex harbor infrastructures and the appearance of construction techniques that allowed building underwater led to large port areas being reclaimed from the sea by the mixed use of rigid structures filled with sedimentary material (Fig. 28.7a). This gave birth to the first artificial shorelines. The increase in port trade at the end of the twentieth century led to an increased need for storage space in many major harbors, which began to expand their facilities by reclaiming land from the sea in this way. Thus, ports such as New York (USA), Rotterdam (the Netherlands) or Barcelona and Algeciras (Spain) were extended with artificial land.

The extreme point of artificial shoreline construction came with the emergence of entire islands with extensions of tens of square kilometers for luxury developments off the coast of the United Arab Emirates. These artificial shorelines have been designed with unique shapes such as palm trees (Fig. 28.7b). In this case, engineers have designed the artificial land revetment with wave-spreading structures to have a minimal effect on wave reflection.

28.2.3 Destruction of the Foredune Systems

The destruction of coastal dune systems for aggregate or beachfront development has been common in the past decades, and is still common in many developing countries. From a dynamic point of view, there is a strong interaction between the foredune and the beach equilibrium. The foredune is the sand reserve of the coast and provides protection for the mainland from extreme erosional effects. Thus, the dune is eroded during storms, but its sand feeds the nearshore and remains there until dynamic conditions favor the wind to pick up the sand to rebuild the dune (Fig. 28.8a). Complete dismantling of the foredune ridge will constitute a decrease in the potential for input to the beach from the mainland, and thus a decrease in the overall sediment volume of the coast.

In addition, the absence of a foredune line leaves coastal buildings exposed to the direct action of storm surges. It should be remembered that storm surges also occur during conditions of meteorological surges. This endangers the coastal buildings themselves. Recent studies have shown that beaches where the dunes have been removed are much more vulnerable (Fig. 28.8b) and have a more pronounced erosive character [9].

28.2.4 Modifications of Tidal Prisms in Restricted Environments

In inland coastal systems such as lagoons, deltas and estuaries, extensive intertidal areas are impounded for uses involving agriculture, aquaculture (Fig. 28.9a) and saline activities (Fig. 28.9b) or industrial waste piling (Fig. 28.9c). The restriction of surfaces that were previously inundated and exposed by water implies a decrease in the tidal prism. This reduction of the tidal prism in turn implies a decrease in the tidal flows drained during each tidal cycle, and thus a decrease in the velocity of currents entering and leaving these systems.

In any of the tidal channels of these systems, the reduction of the mean current velocity will lead to sedimentation and loss of flow. This effect is not a significant change in the case of natural systems, although the loss of draft may affect nautical activities when there is a harbor in these environments. The area where the greatest changes due to the reduction of the tidal prism are observed is at the mouth of the tidal drainage channels. In these areas are located environments such as tidal deltas (in lagoons and estuaries) or frontal bars (in deltas), whose equilibrium depends on the balance of forces between tidal currents and waves. The loss of velocity of the tidal currents will allow greater wave dominance, which is reflected in a landward displacement of the frontal lobes. This phenomenon can lead to the closure of inlets if the decrease of the tidal prism is significant and, in the case of harbors within the systems, constant maintenance dredging is required. This dredging activity also induces modifications of the tidal current velocities, since it implies an increase of the flow section.

28.3 Human Modifications of the Fluvial Sediment Supply to Coastal Systems

Humans can influence the inputs that reach the coast from the mainland by modifying the sedimentary supply from rivers. These changes directly affect the sedimentary balance of a coastal cell.

Fig. 28.7 Examples of artificial coasts. **a** Esplanade docks of the Algeciras harbor (S Spain). **b** Artificial island of Palm Jumeirah (Dubai)

Fig. 28.8 Comparison of two nearby beaches after the same storm. **a** Beach with a partially eroded foredune. **b** Beach without foredune where the storm waves destroyed the sea promenade

Fig. 28.9 Intertidal surfaces restricted by human uses. **a** Fish farms. **b** Industrial salt pans. **c** Industrial waste stockpile

28.3.1 Building of Dams into Rivers

Countless rivers around the world, regardless of their size, have been regulated in order to store water for urban, agricultural or industrial consumption (Fig. 28.10). The presence of a reservoir implies an alteration of the natural flow of water and sediment downstream of the dams. Water flows can be regulated by the partial opening of gates so that rivers maintain an ecological discharge; however, each dam means a cutoff of sedimentary material. The efficiency of some of these dams in retaining sediment is as high as 100% in some cases [17].

On the other hand, each dam represents a significant decrease in the source area of coastal systems. Thus, river

Fig. 28.10 Dam totally cutting the path of water and sediments in a small river

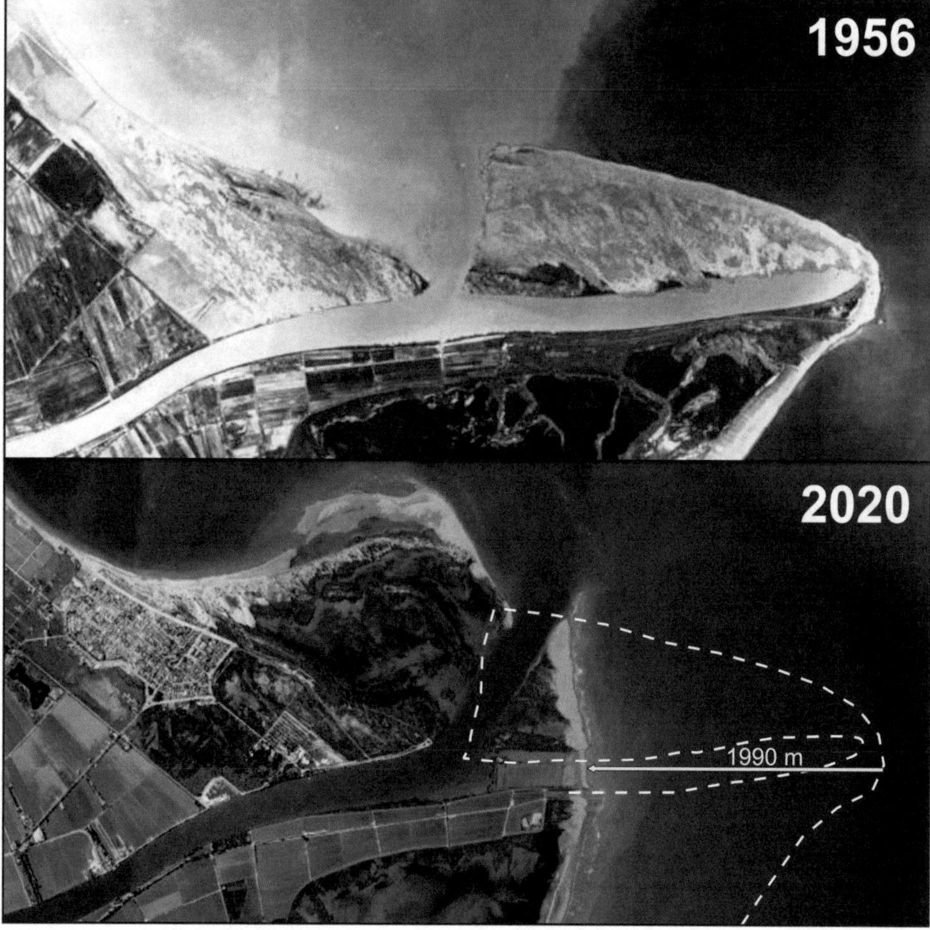

Fig. 28.11 Images showing the strong retreat at the front of the Ebro River Delta (NE Spain) after the building of 70 dams, most of them between 1955 and 1970

basins with a high number of dams are no longer contributing significant volumes of sediment to the coastal systems developed at their mouths. It is estimated that more than a quarter of the sediment delivered by rivers globally has been retained in their reservoirs [15].

This decrease in sediment input implies a change in the morphodynamics and sediment balance of river mouth systems (estuaries and deltas). In estuaries, the water mixing patterns are modified, as less freshwater is received. The mixing zone shifts towards the river and there is usually a tendency for the degree of mixing to increase as the volume of flocculated material increases. In general, the sediment in the central part of estuaries becomes finer and more organic [16].

In deltas, the reduced arrival of sediment in bed load implies a change in the morphology and dimensions of the delta front bars. In general, the reduced supply of fluvial sands to the delta front is accompanied by a retreat due to erosion and subsidence, as fluvial sediments no longer compensate for the wave transport capacity and land subsidence (Fig. 28.11). There are numerous well-documented examples of this effect in many rivers in the Mediterranean (Po and Ebro), Atlantic (Guadiana and Senegal), Indian (Indus and Ganges–Brahmaputra) and Pacific (Mekong, Red River) [2].

28.3.2 Changes in River Catchment

Human-induced changes in the fluvial catchment area are a main cause of modification of the sediment supplied by the rivers to the coastal systems. In this context, deforestation of wide continental areas may cause a dramatic increase in erosional processes and the consequent increment of sediment introduced by rivers in estuarine or deltaic systems. As an example, many deltas developed on the Mediterranean coasts prograded quickly during the last 6,000 years since the vegetation of their basins were eliminated [1]. Another clear example has been described in the Iberian Peninsula, where the massive construction of ships that followed the discovery of America was necessarily accompanied by the cutting down of trees from the vast forests of the interior of the continent. Estuaries tend to respond to this increased sediment input by increasing their fill rate, while deltas normally respond by generating prograding cycles on the delta front.

Conversely, reforestation can reduce the rates of sediment supply from rivers to the coastal systems. This process has also been described by Anthony et al. [1] as occurring in many Mediterranean deltas during the last 50 years. The studied deltas responded to this decrease in contributions with setbacks caused by coastal erosion. Consequently, many coastlines associated with these systems had to be protected by structures.

It is not only the river mouth systems that suffer from sediment deficit, but also the adjacent coasts. Shorelines at the margins of river mouths are often fed with river sediments transported by littoral drift. Sometimes these shores adjacent to river systems continue to receive sediment from the erosion of the delta front. However, erosion and longshore current often fail to supply sediment at a sufficient rate, and beaches also begin to retreat.

28.4 Human Influence on Global Climate

After studying the effect of atmospheric greenhouse gases on global warming in Chap. 24, we can conclude that the increase in these gases has taken place due to the use of fossil fuels since the Industrial Revolution of the nineteenth century. The increase in atmospheric CO_2 levels due to anthropogenic causes has different impacts on the coastal system. These effects have been summarized by Masselink et al. [10], taking into account that many of these effects may have a local character, while others have a global character.

- Increase of dissolved CO_2 in seawater: acidification of coastal water with negative effects on coral reefs.
- Increase in seawater temperature: effect on eutrophication of coastal wetlands; impact on coastal reefs through coral bleaching.
- Modification of the wind regime: effects on the generation of dunes and on the directions in which waves reach the coast.
- Increased number of storms and wider temporal distribution: effects on increased erosion.
- Increased wave size: effect on increased erosion.
- Changes in river flow: effects on sediment input to coastal systems.
- Sea level rise: long-term effects on the evolution of coastal systems.

The consequence of most of the effects is transgression, accompanied by increased erosion and vulnerability of coastal systems, as they are subjected to a change in the maritime climate that tends to increase the energy of hydrodynamic processes.

References

1. Anthony EJ, Marriner N, Morhange C (2014) Human influence and the changing geomorphology of Mediterranean deltas and coasts over the last 6000 years: from progradation to destruction phase? Earth Sci Rev 139:336–361
2. Besset M, Anthony EJ, Bouchette F (2019) Multi-decadal variations in delta shorelines and their relationship to river sediment supply: an assessment and review. Earth Sci Rev 193:199–219

3. Bowen AJ, Inman DL (1969) Rip currents 2: laboratory and field observations. J Geophys Res 74:5479–5490
4. Carter RWG (1988) Coastal environments: an introduction to the physical, ecological, and cultural systems of coastlines. Academic Press, London, 617 pp
5. Dean RG (1974) Compatibility of borrow material for beach fills. In: Proceedings of the 14th coastal engineering conference. American Society of Civil Engineers, pp 1319–1330
6. Dean RG (1983) Principles of beach nourishment. In: Komar PD (ed) Handbook of coastal processes and Erosion. CRC Press, Boca Raton, USA, pp 217–232
7. Dunham JW (1965) Use of long groins as artificial headlands. In: Coastal engineering, Santa Barbara specialty conferences. American Society of Civil Engineers, pp 755–762
8. Houston JR (1996) International tourism and US beaches. Coastal Eng Res Centre (CERC) 96(2):1–3
9. Komar PD (1998) Beach processes and sedimentation. Prentice-Hall Inc., Englewood Cliffs, New Jersey, 429 pp
10. Masselink G, Hughes MG, Knight J (2003) Introduction to coastal processes and geomorphology. Routledge, London, 416 pp
11. McDougal WGM, Krauss NC, Ajiwibowo H (1996) The effects of seawall on the beach. Part II, numerical modeling of supertank seawall tests 12: 702–713
12. Pilkey OH (1981) Geologists, engineers and a rising sea-level. J Coastal Res Northeast Geol 3:150–158
13. Pope J, Dean JL (1986) Development of design criteria for segmented breakwaters. In: Proceedings of the 20th coastal engineering conference. American Society of Civil Engineers, pp 2144–2158
14. Simoen R, Verslype H, Vandenbossche D (1988) The beach rehabilitation project at Ostend, Belgium. In: Proceedings of the 21st coastal engineering conference. American Society of Civil Engineers, pp 2855–2866
15. Syvitski JPM, Kettner A (2011) Sediment flux and the Anthropocene. Philos Trans Royal Soc A: Math Phys Eng Sci 369:957–975
16. Ve ND, Fan D, Vuong BV, Lan TD (2021) Sediment budget and morphological change in the Red River Delta under increasing human interferences. Marine Geology 431:106379
17. Vörösmarty CJ, Meybeck M, Fekete B, Sharma K, Green P, Syvitski JPM (2003) Anthropogenic sediment retention: major global impact from registered river impoundments. Global Planet Change 39:169–190

Coastal Changes and Coastal Hazards

29

29.1 Introduction

Throughout this book it has become clear that the coast is a place of permanent change. "*If there is one thing about the coast that does not change, it is change.*" Many of these changes are cyclical, but others are permanent. Sometimes coastal evolution has specific trends that can be known and assessed, making future changes predictable. All this variability is generically known as **coastal change**.

Humans have to consider that their activities on the coast must allow for the flexibility that the coast requires. However, a significant percentage of the population of civilized countries lives within the influence of coastal change. Houses, hotels, flats, port infrastructure, power plants, industries, refineries, recreational areas and military bases occupy an enormous expanse of coastal segments. And this occupation has increased in the first decades of the twenty-first century. In Spain, for example, the build cost per square meter has multiplied by five in the last 20 years, despite the economic crisis. The point is that all this occupation has been carried out without respecting, and in many cases without even taking into account, coastal changes.

Coastal change is perceived as a hazard by humans, because people and their constructions are vulnerable to these changes. However, there is no real possibility of coastal change affecting humans if there is no exposure to it (Fig. 29.1). In the previous sentence, the concepts of hazard, vulnerability and exposure are conflated. These are different concepts, but they are linked to each other. Two other words that are often associated with these concepts are risk and disaster, which also have different meanings.

- **Hazard** is the probability of a situation occurring that is perceived as a threat to humans. In terms of coastal geology, it is the potential of a coastal process to affect humans.
- **Vulnerability** is the fragility of the natural environment or of human beings to the potential damage that this hazard could produce.
- **Exposure** is the location of a vulnerable person or infrastructure in a place where it can be reached by a hazard.
- **Risk** is the real possibility of a hazardous process affecting people or human infrastructure. Risk is classically defined as hazard × vulnerability × exposure.
- **Disaster** is an event that has caused a lot of damage or destruction—i.e., a destructive reality that has already happened is considered a disaster.

In the above terms, if any of the variables affecting risk (hazard, vulnerability or exposure) is zero, then the risk is zero. For example, if a hazard threatens an area vulnerable to that hazard but that area is not exposed, there is no risk. If a hazard threatens an exposed area, but that area is not vulnerable, the risk is also zero. The problem is that the people living along our coasts are extremely vulnerable to coastal change and its hazards, but they are also exposed to it. It is currently estimated that some ten million people around the world are directly affected by the effects of coastal change associated with events such as storms (where high wave action is coupled with coastal flooding due to storm surges). This number increases for populations at risk from tsunamis.

There are numerous studies in the scientific literature that document, assess and model coastal change, hazards, vulnerability and risk to human exposure to these processes. These studies include research into historical coastal change, geological structure, the dynamic functioning of coastal processes, the influence of sediment input and mobility, the influence of sea level and how extreme storm events affect sedimentation. Social studies are also included, such as the history of occupation of coastal regions and the impacts of coastal change on these populations [22].

Disasters are occurrences when all the adverse conditions of hazard, vulnerability and exposure have come together to produce events that have severely affected certain coastal populations, producing a high economic cost and sometimes costing human lives. Many of these disasters, such as those associated with Hurricane Katrina in 2005 or the tsunami in

J. A. Morales, *Coastal Geology*, Springer Textbooks in Earth Sciences, Geography and Environment,
https://doi.org/10.1007/978-3-030-96121-3_29

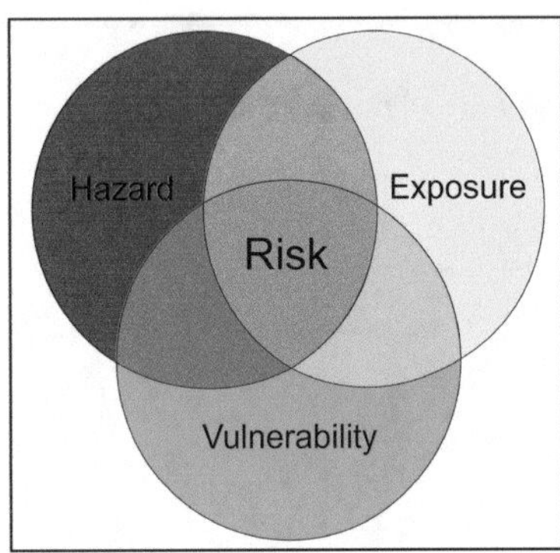

Fig. 29.1 Diagram showing the relationships between hazard, vulnerability and exposure to determine risk

Indonesia in 2004, have been marked in the collective memory by the great damage they caused and the impact of the images that were broadcast around the world by the media.

In some studies, disasters have been classified from a social science point of view. Thus, natural disasters and human disasters have been differentiated [11]. These authors describe "*disasters that are caused by the action of natural phenomena beyond human control*." This category includes disasters related to climatic phenomena (storms and coastal flooding) and tsunamis. Human catastrophes are described as those in which humans are the main cause. These include chemical pollution disasters. A third category could also be included, in which humans can act as a catalyst by accelerating or increasing the effects of a natural disaster. This third category would include catastrophic transformations of coastal dynamics. Today, this approach has been strongly contested to the point that some studies deny the existence of natural catastrophes on the coast [8]. The main argument used is that catastrophes would not occur if humans had not occupied these areas (exposure). By this criterion, a coastal phenomenon acting on an area without humans would never produce a catastrophe. From this point of view, all coastal disasters would belong to the third type described above.

It follows from this argument that human development is one of the main problems in the genesis of coastal risk. We began this chapter by pointing out that building on the coast eliminates the natural flexibility to adapt to coastal change. The construction of houses in flood-prone areas makes them high-risk zones because the buildings are extremely vulnerable, but they are also exposed to coastal processes in general and to more energetic ones in particular. To construct buildings in areas that can be reached by storms is to put not only the buildings at risk, but also the lives of the people who occupy them. It is now known that certain areas of the coast are vulnerable to coastal processes, and yet buildings continue to be constructed there anyway. The result is that, today, more and more developments are being attacked by marine processes, especially in the context of rising sea levels accompanying global change.

Irrespective of the debate between the cause of hazards (natural vs. human), it seems clear for the management and use of coastal zones that there are problems related to two types of phenomena: continuous phenomena and eventual phenomena. These risks are discussed separately below.

29.2 Slow-Rhythm Coastal Changes

Certain coastal processes occur on the coast at a very slow rate. These are known as continuous phenomena. Such processes do not pose a risk to human life, as the slow pace allows humans to get to safety. However, these phenomena do pose a risk to property. Sea level rise and coastal erosion fall into this category. Both phenomena are related to climate change [17].

29.2.1 Sea Level Rise

It is now commonly accepted that sea level is rising. The causes of this rise are well known and are related to climate change, which is evolving towards global warming. The origin of this warming is attributed to the greenhouse effect that has resulted from the extensive use of fossil fuels over the last two centuries. In both hemispheres, the retreat of continental ice masses has been observed, accompanied by a slow increase in oceanic water mass [1]. In addition, global warming implies an increase in ocean water volume due to thermal expansion.

A few decades ago, there was controversy over the magnitude of sea level rise, with differences ranging from 1 mm/year to 2 cm/year, depending on where the measurement took place [2]. Today, there are global measurement networks that estimate mean sea level variations at many coastal locations around the world. Estimating the magnitudes of sea level rise at each of these locations requires at least 30 years of observations. Some stations such as Baltimore (USA) store data from more than 110 years of records, accumulating a 425 mm rise since 1902 [20]. The average estimate of sea level rise must compare the magnitudes observed at stations around the globe. In parallel, satellites recording the dynamic topography of the ocean surface have accumulated global data since 1993 (see Advanced box 25.2 in Chap. 25 of this book). Historical rates of rise of 1.8 mm/year over the last century are now accepted. These rates have accelerated to 3.1–3.4 mm/year when averaged from satellite data [3], but they may have increased even more if the last decade is considered separately, reaching values of 4.6 mm/year (Fig. 29.2).

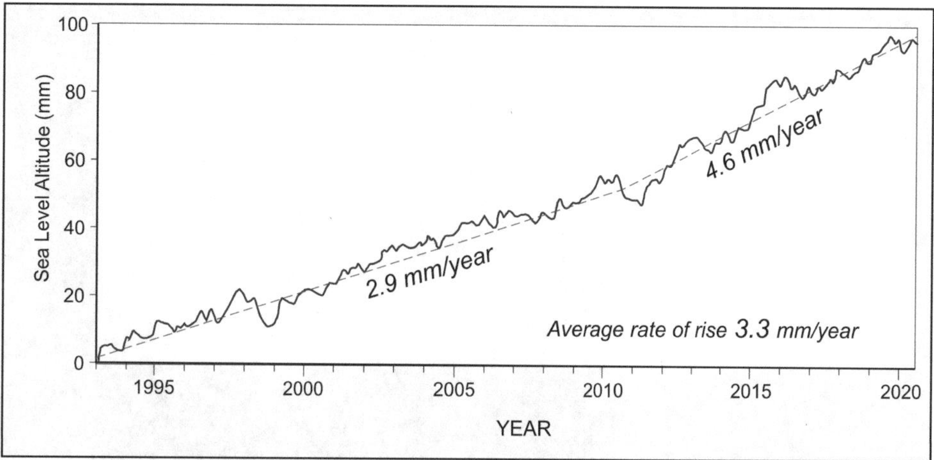

Fig. 29.2 Graph of the averaged sea level evolution from satellite observations since 1993 (NASA Goddard Space Flight Center)

It is not only eustatic rise that is influencing coasts. The effects of subsidence coupled with eustatic rise are accelerating the rate of relative rise in some of the world's most densely populated areas [12]. On the African coasts of the Mediterranean and the coasts of India, China, Korea and Japan, the relative rate of rise is between 10 and 15 mm/year, and in some areas, such as the islands of Indonesia or the Philippines, rates can even exceed 20 mm/year.

The immediate effect of this sea level rise, if not compensated for by sediment inputs, is a migration of the coastline towards the continent. This migration implies inundation of inland coastal areas (Fig. 29.3a) and an increase in the depth of intertidal and subtidal littoral systems.

Under these conditions, coastal systems adapt to the rise through new trade-offs between dynamic processes and sediment input, which depend on the particular coastline and the environments that develop along it [6]. For many cliffs, particularly those made of less resistant materials, sea level rise is accompanied by an increase in the rate of retreat.

In general, barrier island systems move landward through rollover and overstepping processes. In both cases, the rise is accompanied by an increase in overwash processes. Many of these systems are now losing their characteristic geometry as the dune systems are destroyed by wave action, using their sedimentary material to build washovers over the barrier island systems.

At river mouths, the coastal plains are flooded according to the rates of rise. If the rate of sediment aggradation outweighs the rate of rising, there may not be much transformation in these environments. However, if the rate of vertical accretion is less than the rate of rising, the marshes and mangroves associated with estuaries and deltaic plains can be expected to be inundated and drowned. As estuaries are invaded by the sea, tidal exchange at river mouths will increase. In this case, much of the sediment from the barriers will be introduced into the estuarine inlets and preserved within them.

Reef systems are also expected to be inundated. The preservation of reef systems and their position will depend on the relationship between coral growth capacity and the rate of sea level rise. In general terms, reefs are responding to sea level rise with vertical accretion, sometimes accompanied by migration towards the mainland.

All these processes that have been developing naturally in all previous periods of Earth's history are now seriously affecting humans [7, 15]. Urban or industrial areas that mainly develop on coastal plains, such as barrier islands or the interior of fluvio–tidal systems are being affected by marine inundation (Fig. 29.3b).

29.2.2 Coastal Erosion

Coastal erosion is the retreat of the coastline due to a negative sediment balance. We have seen that, in coastal dynamics, there are oscillation margins that include seasonal and inter-annual periods of alternating erosional and depositional conditions. However, this section is not referring to these periodic erosive moments, but to prolonged periods of continuous erosion. Any coastal system can be subject to erosion. Rocky shores are the most erosive of all the systems studied (Fig. 29.4a), but beaches in front of barrier islands (Fig. 29.4b), deltas and estuaries are also susceptible to erosion. Continuous erosion can be measured by comparing the shoreline in the same month between different years. More continuous measurements—e.g., monthly—allow the duration of erosion–accumulation cycles and the return periods of point erosion events to be determined more accurately. There are numerous works in which shoreline displacement and linear erosion rates are determined by remote sensing using geographic information systems (GIS; e.g., [4]). However, these works do not allow for

Fig. 29.3 Results of sea level rise. **a** Effects in a natural environment, at Driftwood Beach, Georgia, USA, **b** effects in a coastal city, at Isla Cristina, SW Spain

calculations of eroded volumes. The calculation of erosion volumes requires the comparison of topographic profiles carried out at regular intervals along the coast (e.g., [24]).

The main cause of erosion is the existence of a negative sediment balance in the coastal section where it occurs. This means that more material flows out of that stretch of coast than comes in. There are two possible causes for this sediment deficit to occur, and they are natural and anthropogenic.

(1) Natural causes
   - Hydrodynamic erosive agents: it is usually the waves that cause erosion. When the carrying capacity of the waves is greater than the volume of material available, the waves remove older material from the shore.
   - Deficit in sedimentary input: most of the time, it is the rivers that present this deficit in their input to the coast, although the deficit can also come from wind transport or from inputs coming from the sea. Among the natural causes of this deficit is the existence of resistant lithologies that do not supply enough sediment to the coast. Another possible cause may be a reduction in inputs in relation to climate change.
   - Increase in the intensity and frequency of storms: this cause involves increased cumulative erosion during times of energy peaks. The overall increase in storm energy is also related to climate change.

(2) *Anthropogenic causes*
   - Modifications of wave refraction and tidal prisms: the existence of human constructions on the coast modifies the balance between hydrodynamic agents and the way in which sediment is distributed. These alterations normally lead to the appearance of segments of coastline where a situation of permanent erosion occurs.
   - River damming: this causes sediment retention in inland areas and a loss of the transport capacity of

Fig. 29.4 Coastal environments under erosion with threat to buildings. **a** Cliff with an urbanized top (Azenhas do Mar, Portugal), **b** Eroded beachfront (Mazagón beach, SW Spain)

rivers to supply sediment to the coast. Both of these are determining factors in generating a sediment deficit in the coastal areas fed by these rivers.

- Artificial sand mining: building in coastal areas requires significant volumes of sand. On many of the world's coasts, this sand has been obtained directly from the dunes. The extraction itself causes a deficit, but the dune also represents a store of sand for coastal dynamics, which is used by the waves during specific moments of the annual cycle. The lack of such a store is often reflected in permanent erosion along that stretch of coastline.

The effects of these erosional phenomena mean a loss of coastal habitats, but above all the loss of recreational areas, buildings and infrastructures and, in short, a loss of security.

When a natural area is subject to erosion, the erosional process does not pose a risk, although a loss of ecosystem value does occur [18]. Where there are no buildings, there is no risk. However, erosion poses a serious risk to areas where humans have built within reach of these processes. It is simply an increase in exposure, with people being responsible for building in areas subject to erosion or even causing erosion with their infrastructure [14]. In these cases, not only buildings located between the seasonal or inter-annual oscillation margins of the coastline would be at risk, but also those located in nearby areas, taking into account the general rate of retreat.

In developed countries, continuous erosion is recognized as a major problem, especially because it is considered to be a phenomenon that has been increasing for decades and is still increasing. As a result, the budget allocated to correcting this impact is often very large, especially in countries that base their economies on tourism. On the Atlantic coast of the United States, the average rate of erosion is 0.9 m per year, and it is even higher on some stretches of the Gulf of Mexico coast, where it exceeds 1.8 m per year.

There are different possible mitigation and management strategies to slow down these erosion processes, which will be discussed in the next chapter of this book.

29.3 High-Energy Coastal Events

Some events concentrate an enormous amount of energy on the coast in a very short space of time. This is the energy generated by the movement of water masses, which is normally used to transport large masses of sediment, but also has great destructive power. These coastal events cause occasional problems on the coast and pose a risk not only to property but also to human life. High-energy events have two distinct origins—meteorological action and seismic activity in the marine environment—and are reflected in the arrival of severe storms and tsunamis on the coast. Both phenomena have been discussed in Chap. 13 from the point of view of the processes and the sediments they leave in the geological record. The following sections describe the effect these processes have on coastal populations.

29.3.1 Extreme Storms

Storms are one of the main threats to coastal areas, especially those that are heavily occupied by humans. Storms represent moments of high wind force and enormous wave energy deployment, which are also accompanied by flooding due to the meteorological surge. The sum of all these causes means that storms represent moments of high erosion.

Chapter 13 considered two types of storms of different intensity: tropical cyclones (hurricanes or typhoons) and mid-latitude storms. In inter-tropical areas, cyclones act on the coast with winds in excess of 200 km/h. These are low-pressure centers surrounded by winds that spiral in a cyclonic pattern. In mid-latitudes, storms have winds of lower intensity than in tropical areas, but the action on the coast is similar due to their longer duration.

The action of storms on the coast is not only related to the direct action of wind on infrastructures. In addition to wind, the energetic discharge represented by large waves and the currents associated with these waves represent a great capacity for sediment movement and an enormous destructive force for the infrastructures located on the coast (Fig. 29.5a). It should be taken into account that the surge that normally accompanies storms can reach up to 6 m, causing the coast to be flooded and the waves to act in areas where they do not normally reach. In addition to the surge, the passage of storms means occasional periods of high rainfall. In coastal systems located at river mouths, high discharge situations join the surge to plug the river inlet, causing fluviomarine flooding in the delta plains and estuarine margins.

The consequences of storm action on natural coastal systems are as follows:

- *Intense erosion of beaches and dunes*: the main effect of storms is erosion along the coast. Larger wave action erodes at a faster rate, with sediment moving to areas below the closure depth, from where it can no longer return to shore. Erosion rates in some coastal segments can exceed 30 meters in a single day.
- *Overwash processes*: on barrier island ridges, the combined action of surges and large waves can exceed the height of the foredunes. Sand from the dunes can be reworked back towards the rear of the system, forming washovers. At the front of tide-dominated systems, these same processes act on the foreshore to displace coarse sediment onto the tidal flats, forming cheniers.
- *Breaching processes*: in barrier island systems, overwash processes can concentrate at certain points, leading to the formation of channels that can evolve into inlets when they become embedded and deepened.

The natural coastal system can partially or fully recover from these storms. On the beaches, thanks to the dynamics of ridges and runnels during periods of fair weather, the sand displaced towards the shoreface during the storm can accumulate again in the foreshore and from there the wind can pick it up to rebuild the foredunes.

However, these processes also cause damage to human infrastructure:

- Direct wind damage;
- Flood damage;
- Damage caused by wave action;
- Damage caused by the accumulation of sediment.

Storms, and especially tropical cyclones, can devastate entire coastal regions, causing extensive damage to property and people. Perhaps the most publicized cases worldwide were Hurricanes Andrew, which struck Florida

Fig. 29.5 Effects of storms on human-occupied coasts. **a** Storm Eleanor acting on the coast of Saint-Malo (NW France), **b** Fluviomarine flood in Chennai (India)

in 1992, and Katrina, which devastated New Orleans in 2005. Andrew was a category five hurricane that cost 23 lives and caused $26.5 billion in property damage. Katrina was the most economically and personally damaging event in the USA, with more than 2,000 deaths and more than $108 billion in property damage. Although the most media-worthy events are those occurring in the USA, there are events as destructive as those described in densely populated areas of Asia. An example of this is the river–sea flooding on the coast of southern India in 2015 (Fig. 29.5b). The passage of a monsoon affected the states of Tamil Nadu and Andhra Pradesh with more than 300 fatalities, 1.8 million people evacuated and damages exceeding $ 3 billion [19].

Records now show that the frequency, duration and average intensity of storms are increasing due to climate change [10, 13]. Meanwhile, the storm season is also expanding, with the first storms arriving earlier and staying later.

29.3.2 Tsunamis

Tsunamis are great waves originated by sudden movements of the ground transferred to the water mass, usually due to tectonic causes (see Chap. 13). These waves are characterized by their kilometric wavelength. The volume of water displaced by a tsunami, and the speed it reaches, are so great that its energy is 70 million times greater than a wind wave of the same height. The direct action of the tsunami wave breaker generates damage in the frontal zone of the coast. On the other hand, the encroachment of the water mass driven towards the inland areas close to the coast causes inundation damage. The extent of tsunami inundation depends on the geographical configuration of the terrain according to variables such as water column in relation to topography, vegetation and the existence of other human obstacles [16]. In certain regions, waves can reach areas more than 10 km inland, especially on coastal plains. At river mouths, the tsunami wave can be channeled and penetrate even further inland. In these flat areas, additional damage can occur related to debris entrainment during the advancing wave, which is densified with a high sediment concentration. In high topographic configurations of funnel-shaped valleys, the water can pile up, amplifying the wave dimensions and reaching heights of more than 100 m.

The effects of a tsunami on the communities it strikes are similar to those of an extreme storm, but much greater in magnitude in all respects:

- *Loss of human life*: the tsunami with the highest death toll was the 2004 Indonesian tsunami. The total number of casualties has not yet been reliably counted, but is estimated to have exceeded 260,000.
- *Loss of infrastructure*: tsunamis cause the destruction of buildings, but also of factories of all kinds, affecting fuel, light and water supplies. In tsunami-hit areas, the effects are total devastation (Fig. 29.6). All this generates very significant economic damage. The event that has generated the most economic damage to date was the Tōhoku 2011 tsunami in Japan. This event, in addition to causing some 27,000 deaths, displaced more than 165,000 people and caused economic damage in excess of $300 billion.
- *Destruction of the economic system*: a less obvious effect is the destruction of the economic system in tsunami-affected areas. In addition to the cost of reconstruction, these areas are affected by the disappearance of businesses that generate economic wealth and capital. Developing countries suffer particularly badly from this effect, as they lack insurance systems and credit systems.

There are societies such as Japan or Chile whose culture and experience make them prepared for tsunamis. However, tsunamis are often unexpected phenomena that reach societies which are totally unprepared. There is a substantial difference in the way a tsunami affects both types of societies. Two events of similar characteristics had very different consequences for the countries of Japan and Indonesia due to their different degrees of development and preparedness. In Japan, society was prepared in terms of defensive infrastructure (coastal walls), warning and evacuation systems, but also in educating its citizens for self-protection. All this was lacking in the case of Indonesia. The result was that the number of casualties in Indonesia was 100 times higher than in Japan, despite the fact that the events affected similar populations.

29.3.3 Other High-Energy Processes Affecting the Coasts

Certain coasts of the world can be affected by singular eventual phenomena that can cause specific imbalances in their dynamics. Coasts located on tectonically active margins

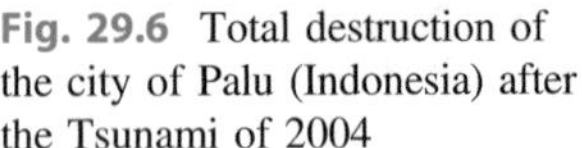

Fig. 29.6 Total destruction of the city of Palu (Indonesia) after the Tsunami of 2004

present a high risk of landslides, both in the emerged area and in the submerged area. In the case of emerged areas, these landslides are usually associated with steep coasts such as cliffs and bluffs. The occurrence of this type of phenomenon causes changes in rhythm in the normal processes of the receding cliff. Landslides in submerged areas can cause tsunamis of moderate size, which do not spread across the oceans but can locally affect the closest coastal areas.

Coasts affected by active volcanism are also subject to changes associated with magmatic activity. The emission of lava flows and pyroclastic waves can advance the coastline towards the sea (creation of new coasts) at the same time that it can modify the volume of sediment input to the coast (epiclastic material). In addition, the action of explosive volcanism in coastal areas can induce a drastic modification of the coast, being associated to phenomena already described as landslides or tsunamis.

29.4 A Multi-hazard Perspective

A combination of some of the hazards that have been discussed separately in this chapter is possible. In fact, the effect of eventual phenomena such as storms and tsunamis can be combined with phenomena that act in the long term. It is clear that the net erosional balance acting continuously on many coasts is greatly accelerated during severe storms. Possible coupled effects between sea level rise and increased frequency of storms, as well as the effects of tsunamis in the scenario of rising sea level, are currently being analyzed.

Rising sea levels and more frequent storms are two phenomena acting together, both of which are associated with climate change. A sea level rise of a few millimeters has a much less visible effect in the short term than the increase in annual storm energy. However, combining the two phenomena in the long term leads to scenarios in which the probability of inundation by surges will progressively increase towards localized areas further and further inland [23]. Moreover, the increase in storm frequency is expected to have an effect on the increase in the overall rate of coastal erosion. This effect, coupled with sea level rise, will mean an increase in coastal retreat in the long term. The problem is that coastal populations continue to grow and are unprepared to adapt to this increase in extreme flooding, especially as storms will increasingly reach higher elevations and may exceed the height of coastal defenses [5].

Unlike storms, sea level rise is not genetically related to tsunami action; however, both phenomena can also act in combination [21]. Theoretically, tsunami action in scenarios of increasing sea level will allow these great waves to reach larger coastal areas. It should be noted that, in the short term, a millimeter rise would have a much smaller effect than the tidal height at the time of the event. However, some studies based on hydrodynamic modeling of possible long-term scenarios estimate that, in certain coastal areas, a moderate rise of tens of centimeters could double the risk of being hit by a tsunami [9]. Such a scenario could occur in about 150 years.

Future strategies for the defense of coastal areas will need to take these combined effects into account.

References

1. Bahr DB, Dyurgerov M, Meier MF (2009) Sea-level rise from glaciers and ice caps: a lower bound. Geophys Res Lett 36 (10):1029
2. Cazenave A, Remy F (2011) Sea level and climate: measurements and causes of changes. Wiley Interdisc Rev: Climate Change 2:647–662
3. Cronin TM (2012) Rapid sea-level rise. Quatern Sci Rev 56:11–30
4. de Sanjosé Blasco J, Gómez-Lende M, Sánchez-Fernández M, Serrano-Cañadas E (2018) Monitoring retreat of coastal sandy systems using geomatics techniques: Somo Beach (Cantabrian coast, Spain, 1875–2017). Remote Sensing 10:1500
5. Dasgupta S, Laplante B, Murray S, Wheeler D (2011) Exposure of developing countries to sea-level rise and storm surges. Clim Change 106:567–579
6. Fitzgerald DM, Fenster MS, Argow BA, Buynevich IV (2008) Coastal impacts due to sea-level rise. Ann Rev Earth Planet Sci 36:601–647
7. Gornitz V (1991) Global coastal hazards from future sea level rise. Palaeogeogr Palaeoclimatol Palaeoecol 89:379–398
8. Hartman C, Squires GD (2006) There's no such thing as a natural disaster. Routledge, UK, p 322
9. Li L, Switzer AD, Wang Y, Chan C-H, Qiu Q, Weiss R (2018) A modest 0.5-m rise in sea-level will double the tsunami hazard in Macau. Sci Adv 4(8):eaat1180
10. Madsen H, Arnbjerg-Nielsen K, Mikkelsen PS (2009) Update of regional intensity-duration-frequency curves in Denmark: tendency towards increased storm intensities. Atmos Res 92:343–349
11. Mileti D, Noji EK (1999) Disasters by design: a reassessment of natural hazards in the United States. Joseph Henry Press, Washington, p 371
12. Nicholls RJ, Lincke D, Hinkel J, Brown S, Vafeidis AT, Meyssignac B, Hanson SE Merkens JL, Fang J (2021) A global analysis of subsidence, relative sea-level rise change and coastal flood exposure. Nat Climate Change 11:338–342
13. Peduzzi P, Chatenoux B, Dao H, De Bono A, Herold C, Kossin J, Mouton F, Nordbeck O (2012) Global trends in tropical cyclone risk. Nat Clim Chang 2:289–294
14. Pilkey OH, Cooper AG (2014) The last beach. Duke University Press, Durham, p 256
15. Pilkey OH, Pilkey O (2019) Sea level rise: a slow Tsunami on America's shores. Duke University Press, Durham, p 208
16. Ramakrishnan D, Ghosh SK, Raja VKM, Vinu Chandran R, Jeyram A (2005) Trails of the killer tsunami: a preliminary assessment using satellite remote sensing technique. Curr Sci 88 (5):709–711
17. Ranasinghe R (2016) Assessing climate change impacts on open sandy coasts: a review. Earth-Sci Rev 160:320–332
18. Roebeling PC, Costa L, Magalhães-Filho L, Tekken V (2013) Ecosystem service value losses from coastal erosion in Europe: historical trends and future projections. J Coast Conserv 17:389–395

19. Sathyanarayana R (2015) Tamil Nadu's deadly rains hit industries hard. Deccan Herald, November 30
20. Sweet WV, Dusek G, Obeysekera J, Marra JJ (2018) Patterns and projections of high tide flooding along the US coastline using a common impact threshold NOAA. Technical Report NOS CO-OPS 086, 56 pp
21. Tursina S, Kato S, Afifuddin M (2021) Coupling sea-level rise with tsunamis: projected adverse impact of future tsunamis on Banda Aceh city, Indonesia. Int J Disaster Risk Reduction 55:102084
22. Viles H, Spence T (1995) Coastal problems: geomorphology, ecology, and society. Edward Arnold, London, p 350
23. Woodruff JD, Irish JL, Camargo SJ (2013) Coastal flooding by tropical cyclones and sea-level rise. Nature 504:44–52
24. Zelaya Wziatek D, Terefenko P, Kurylczyk A (2019) Multi-temporal cliff erosion analysis using airborne laser scanning surveys. Remote Sensing 11:2666

30 Mitigation, Coastal Policies and Integrated Coastal Zone Management

30.1 Introduction

If one thing can be concluded from the previous chapters, it is that human economic and social activities are permanently threatened by coastal processes in both the long and short term. In addition, the natural balance of the coast is in turn threatened by these same human activities. This two-way interaction between humans and natural processes is complex and requires interdisciplinary analysis to obtain an adequate response from coastal management institutions (Fig. 30.1). At present, the possible responses that those responsible for management decisions should make to help people who are menaced by coastal processes, but want to continue to live on the coast, are still under discussion. These responses must respond to strategies that provide security for coastal communities while respecting the balance of natural systems.

This chapter will look at the different measures that governments in some countries are taking to address the challenges posed by human coexistence with the geological hazards induced by coastal change. Some of these measures include the adoption of specific coastal policies and programs. Recently, the comprehensive approach that these policies require is being addressed through an integrated coastal zone management (ICZM) strategy.

30.2 Mitigation

On many coasts around the world, the intersection of geological processes, vulnerability and exposure produces an unacceptable risk to the coastal community. In these areas, coastal managers have typically employed coastal protection measures to reduce risk. These measures together are known as **mitigation**. On many occasions, mitigation has been justified as a form of risk reduction for the environment, but the truth is that most of the time these measures are aimed at protecting human uses of the coast.

Chapter 28 of this book discussed some of the solutions that engineers propose to mitigate the effects of coastal retreat related to sea level rise and erosion. The action that causes the greatest impact is the construction of rigid structures (groins, breakwaters and seawalls). As discussed in Sect. 28.2.1, these structures induce an imbalance in coastal dynamics by modifying wave refraction patterns and the compartmentalization of the coast into different coastal transport cells.

In addition to these, other lower impact defenses have been developed in recent times. These are known as non-traditional defenses [9]:

- *Bulkheads*: these are structures made of metal, wood or sandbags, which extend parallel to the shore to prevent direct wave attack and transverse transport of sand to deeper areas (Fig. 30.2a). Some of these structures are designed either for use at the waterfront on a temporary basis or in inland areas protected from direct ocean wave attack but subjected to smaller waves or wakes.
- *Dewatering systems*: these are drainage systems that lower the water table in the frontal zones of the beach by extracting water from the pores of the sand. In this way, part of the water from the swash is absorbed by the pores, reducing the volume of water drained superficially by the undertow. This decrease implies a lower velocity and, therefore, a lower erosional capacity.
- *Hardening dunes*: this consists of locating a hardened core inside the foredunes. Sometimes the installation is done by artificially constructing the dune over a concrete core, and sometimes the artificial design contains irregular structures designed to increase wind friction by growing the dune over it. During storms, waves can erode the dune front, uncovering the rigid core that acts as a classic defense. After the storm, the wind can rebuild the dune again.
- *Sand-filled geotubes*: these are large bags of strong, porous fabric that can be filled with sand (Fig. 30.2b). Geotubes are designed to behave as rigid structures, although they have the advantage that they can be easily removed by cutting the fabric when no longer needed.

J. A. Morales, *Coastal Geology*, Springer Textbooks in Earth Sciences, Geography and Environment,
https://doi.org/10.1007/978-3-030-96121-3_30

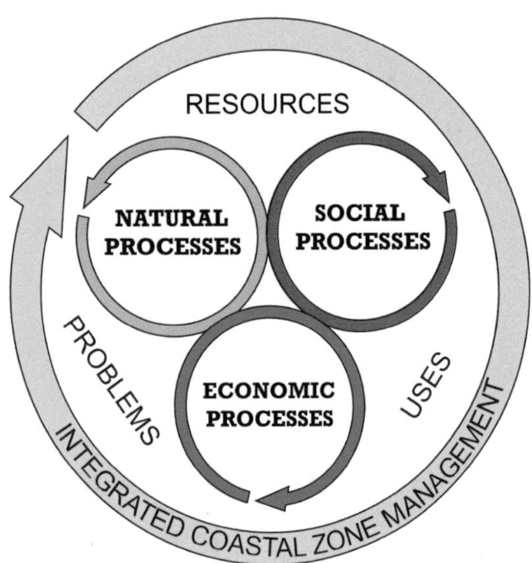

Fig. 30.1 Interaction between natural and human processes (social and economic) that give rise to the need for integrated coastal management for sustainable development

Sometimes, these geotubes are installed as resistant cores in the dunes by mixing this solution with the one described above.

- *Viscous drag mats*: these are covers composed of high-strength plastic sheets woven into a carpet that is fixed to the seabed in deep areas of the waterfront. The plastic fronds create turbulence that interrupts the undertow of the waves and thus reduces erosion. This solution is not suitable for the nearshore, but works very well in deeper waters, and is often used to protect submerged infrastructure in these areas. They have the disadvantage that they degrade over time and contribute to microplastic pollution.
- *Biotextile covers*: these are covers made of biodegradable material that increase the resistance of the soil, favoring the growth of plants on a stable substrate. They are spread over dune systems and backshore areas, and also over the shores of inland systems (tidal channels and lagoons), to cushion the direct impact of waves on the sand. Sometimes, the biotextiles are formed into bags that are filled with sand and distributed over the site to be protected. Over time, the biotextile degrades and disappears, leaving only natural vegetation.

Another mitigation measure adopted from an engineering point of view is beach nourishment. These types of measures have also been extensively discussed in Sect. 28.3.1.

Apart from these mitigation actions that attempt to reduce vulnerability, there are non-engineering ways to combat coastal risk by reducing exposure: pre-construction planning and relocation. Theoretically, the occupation of coastal zones should always follow a proper planning process based on knowledge of the dynamics of geological processes. From this perspective, if it is known that there are processes of continued erosion or threat of encroachment by sea level rise or storm or tsunami attack, the decision should be not to build. Proper **siting** of coastal constructions is considered the best defense (Fig. 30.3a). Sometimes, humans decide to build in a risk zone but adapt their structures to that exposure. One way of doing this is by **elevating structures** (Fig. 30.3b). However, many coasts have already been built inappropriately. Especially in the second half of the twentieth century, many civilized countries massively populated coastal areas without any prior planning or adequate knowledge of the geological processes of the coastline. In such instances, it is often the case that the long-term costs of protecting the built environment are greater than the value of the construction itself. In such cases, relocation should be considered.

When a development is built in an area exposed to coastal hazards, the action of extreme events is unavoidable, despite mitigation and protection measures. In these cases, human preparedness measures are needed to mitigate the effects of the most severe events. If coastal flooding cannot be avoided, it is essential that water drainage systems be kept clear. Generators and electricity transformers should be located on high ground to avoid flooding. Trees susceptible to toppling should be kept away from power lines. Apart from the protection of property, above all, efforts should be made to keep human lives safe. The establishment and marking of safe areas and evacuation routes is vital. It is ultimately of paramount importance to educate the population at risk about self-protection measures in order to minimize loss of life [9].

Such measures should be established on the basis of decisions taken according to coastal strategies and policies. Ideally, these policies should be framed within an integrated coastal management (ICM) program. These aspects will be discussed below.

30.3 Coastal Policies

Decision-making about human development on the coast, as well as the measures taken to protect it, is set within a policy framework. The set of measures taken to make coastal communities sustainable, to reduce human impacts on natural resources and to minimize natural impacts on people and their infrastructure are called **coastal policies.** These measures are informational, economic and legislative in nature (Fig. 30.4) and should, in theory, be aimed at making the coast a desirable and safe place to live.

Coastal policies should be formulated by governmental authorities at different territorial levels. These policies

Fig. 30.2 Some non-traditional mitigation measures. **a** Bulkheads made of sand bags, **b** operations to install a sand-filled geotube in the core of a hardened dune

should provide an orientation and/or a general legislative framework for all bodies and institutions involved in the management and development of the coastal zone, offering security to the citizen user of the coastal zone.

Not many countries have clearly established coastal policies—this is especially true in developing countries, but they are also absent in many developed countries. Three examples where coastal policies are already in place can be highlighted: the USA, the European Union and Australia. Although each of these areas has its own particularities, all of the policies studied have three key underlying principles:

- Appropriate policies can only be based on knowledge. Decisions must be made on the basis of complete and comprehensible information on the dynamics of the geological processes that cause coastal change, the environmental state of the coastal ecosystem and the consequences that different measures may have on the system.
- Coordination between all parties involved is the basis for sustainable development in the coastal zone. The programs should help to identify concordances and contradictions between the actions proposed by the various policies and promote arbitration in the event of conflicts.
- Cooperation between different actors must be organized and maintained. There must be fluid communication between the different sectors of economic activity, as well as between the different levels of territorial administration. Similarly, a continuous exchange of information is recommended, from the scientific community to the administrative level and the community level, and vice versa. In this way, the programs will establish appropriate

Fig. 30.3 Examples of adaptation to the coastal hazards by planning. **a** Siting the first line of promenade and buildings behind the foredunes, **b** Elevating the structure of a building

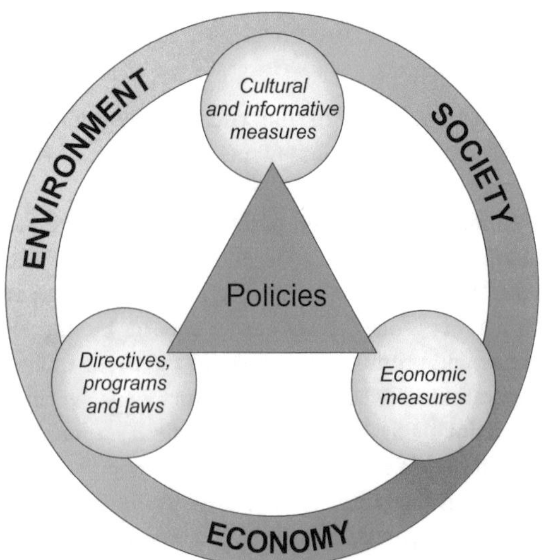

Fig. 30.4 Scheme showing the different interactions of the natural environment, society and economy in influencing decisions on coastal policies

procedures, working methods and forums to ensure dialogue between those involved. This cooperation ultimately leads to the development of a general sense of responsibility.

30.3.1 USA Policies

The National Coastal Zone Management Program (CZMP) was introduced in 1972 and is a world pioneer [10]. This is a federal law that encourages state laws to balance economic development with environmental protection. The objectives of the CZMP are as follows:

- Protection of natural and cultural resources.
- Improvement of coastal water quality.
- Protection of people and property against natural coastal hazards.

- Guarantee of public access to marine and coastal areas.
- Prioritization of the development of uses dependent on the coast.
- Revitalization of the promenades.

It is not a binding law. It is a program that states voluntarily adhere to when enacting their own laws. Although it is not mandatory, 34 of the 35 coastal states currently subscribe to it. The administration and coordination between the coastal management programs of the different states is the responsibility of the National Oceanic and Atmospheric Administration (NOAA)'s Office of Coastal Management. This office is tasked with guiding administrations to bring federal consistency to state programs [12]. Standardization of state programs requires that they include:

- Delimitation and mapping of the coastal zone.
- Definition of permissible land and water uses within the coastal zone.
- Inventory of areas of special interest (economic, cultural, historical and ecological).
- Identification of actors and local authorities.
- Defining the methods by which the state will regulate land and water uses and implement its policies.
- Description of the institutional arrangements and authorities through which the program will be implemented (there are five types of institutional arrangements recognized by the CZMP).
- Definition of planning processes for the siting of installations of community interest.
- Establishment of methods to assess erosion and make decisions on coastal protection and restoration.

30.3.2 European Union Policies

Unlike in the USA, there is no single European model of coastal management or program that firmly regulates the coastal policies of member states [2]. The protection of coasts and marine waters was initially addressed by legislating on some partial aspects affecting marine and coastal areas. Thus, fisheries were regulated through the Common Fisheries Policy (CFP), and coastal water quality control through the Water Framework Directive (WFD). But these laws, although useful for the protection of marine waters, only contribute to the protection of specific issues without having the integrative character that coastal policies require.

With this in mind, the European Union has more recently adopted two instruments, one for the protection of marine areas and the other specifically for coastal areas. For the protection of the marine environment, the Marine Strategy Framework Directive came into force in 2008, while for the protection of Europe's coasts there is the 2002 EU Recommendation on Integrated Coastal Zone Management.

The Marine Strategy Framework Directive (MED) is the first EU law specifically aimed at protecting the marine environment and its natural resources. A framework directive means a law at the European level to which the national laws of the member states must be adapted on a mandatory basis within a certain period of time. In this way, the MED is a European legislative framework for the sustainable use of marine waters whose objectives are set in accordance with a strategy that has a regional focus.

However, for coastal zones there is no framework directive, but only a simple recommendation. The 2002 ICZM Recommendation defines the principles of good coastal planning and management. The objectives of this document are:

- To provide information on the factors favoring or disfavoring sustainable coastal zone management.
- To encourage the exchange of information between actors involved in the management and use of Europe's coastal zones. This debate should generate a consensus on appropriate measures to stimulate ICZM at the European level.

The non-binding nature of this document has made it easy for member countries to continue to fail to adapt their legislation to what is indicated by this recommendation. Thus, despite its existence, coastal planning activities and urban development decisions continue to take place under arbitrary criteria and in a non-unified manner for the entire European territory. This leads to inefficient use of coastal resources, contradictory claims on the use of space, inadequate protection measures and, ultimately, missed opportunities for sustainable coastal development.

30.3.3 Australian Policies

In Australia, there is also no federal legislation or national strategy governing the coastal policies of the different states [7]. Of the seven Australian states, only four have coastal management legislation (New South Wales, Queensland, South Australia and Victoria). Of these four, three of them have put legislation in place within the last decade. Only one of them is devoting any budget to coastal risk zone mapping. The agencies that are supposed to undertake planning and development by implementing coastal policies went into clear decline from 2003 and were virtually non-functional by 2010 [5, 8]. However, all states have developed coastal policies. If there has been coherence between the policies of the different states, it is because the policies were established under the same basic principles. It is only since 2016 that the

Australian government has begun to realize the importance of having a strategy for climate change adaptation and mitigation of coastal vulnerability.

30.3.4 Other Countries

The previous sections make it clear that the laws regulating coastal policies in the most advanced economic systems have more shadows than light. Nevertheless, in the last decade, steps have been taken towards a homogenization of policies not only in these countries, but also in other nations outside the West. For example, the Russian Federation has had a coastal management program for the fishing industry since 1959 and a law for the protection of coastal areas of ecological interest since 1995. In 2002, it regulated the protection of wetlands in a coastal zone management framework, and in 2006 it issued a law on the protection of biodiversity in integrated marine and coastal zone management. These laws are national in scope and enforceable, although they are aimed solely at the protection of natural areas and do not regulate human development in coastal areas or mitigation measures. It should be noted that the Russian Federation has thousands of kilometers of unoccupied coastline.

Another example of Eastern countries with coastal policy legislation is China. China established a marine and coastal management law in 1982 that gave local administrations the ability to manage their coastal concerns. This law basically recognized the inability of the central administration to deal with local problems on a coastline of more than 20,000 km. For decades, local authorities have had neither the technical knowledge, nor the means, nor the money, to meet the challenges of curbing coastal environmental degradation. Moreover, many local authorities have prioritized economic development over environmental sustainability. However, the economic development and international openness of the last two decades have allowed for better funding of environmental policies and international cooperation. This has been reflected in improved coastal management measures [4].

30.4 Integrated Coastal Zone Management

"Integrated Coastal Zone Management (ICZM) is a dynamic, continuous and iterative process designed to promote sustainable coastal zone management. ICZM seeks, over the long term, to balance the benefits of economic development and human uses of the Coastal Zone, the benefits of protecting, preserving and restoring the Coastal Zone, the benefits of minimizing loss of human life and property, and the benefits of public access to and enjoyment of the Coastal Zone, all within the limits set by natural dynamics and carrying capacity." (European [6]).

When the term ICZM is analyzed word for word, the above definition is explicitly reflected [14]. The term **management** refers to the ultimate goal of the process. Although it is management, it actually refers to a process involving research, information gathering, planning and policy-making. In this sense, **planning** is understood as a strategic development that enables the adoption of appropriate policies. Management involves not only decision-making and the establishment of concrete measures, but also the monitoring of the results obtained with the measures taken and the adaptation of new measures to these results. The word **integrated** refers to the need to link multiple sources of information and diverse interests to achieve objectives. In this way, it is necessary to integrate: scientific and technical information on the marine and inland processes found on the coast, the needs of users, the needs of the different economic and social involved sectors, the competences of the different levels of administration (Fig. 30.5). In this process, the participation of all parties involved (scientists, economic agents, social agents, administrative and political agents) is essential.

It should be noted that, from an administrative point of view, there is no precise delimitation of the **coastal zone**. Thus, the term coastal zone is not applied in the same sense as described in Chap. 2 of this book. From a management point of view, the coastal zone is a band of land as well as sea, and can be defined using different criteria based on the interrelationships between humans and the natural environment [1]. These criteria vary depending on geographical characteristics, dynamic behavior, the influence of human activities on the coast or administrative boundaries. In each management unit (countries, states, regions or communities), the seaward limit of the coastal zone may vary from tens of meters to the outer limit of territorial waters. The same is true for the landward boundary, where the coastal zone can be understood as just the fringe affected by coastal processes, or the band up to several kilometers inland where social and economic uses are related to the presence of the coast.

30.4.1 The Evolution of ICZM

The beginnings of coastal occupation are full of cases in which constructions and uses were made without any prior planning. It is only in the last decades that coastal planning activities have started to be developed in some countries. However, even when planning did exist, decisions were taken on a sectoral basis, without an overall strategy and with hardly any links between the plans made by different sectors (urban, port, tourism and so on). This lack of a global

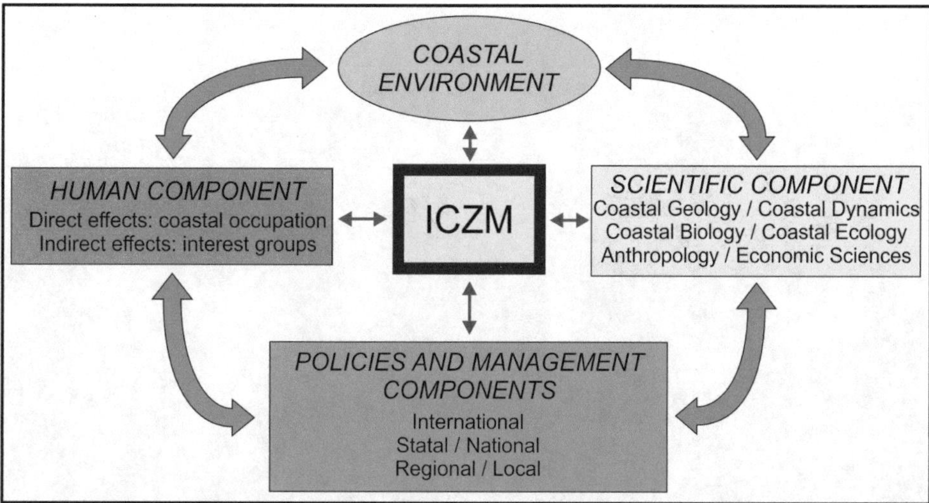

Fig. 30.5 Process of integration of components to develop an adequate coastal zone management program [3]

vision of planning and management led, during the period of greatest human development on the coasts, to an inefficient use of resources. This meant that human activities came into clear conflict with the evolution of the natural environment and the geological processes menacing human constructions. In response to these threats, humans built coastal defenses that further altered the dynamic equilibrium of the coast. Moreover, human uses began to conflict with each other, and there were also inconsistent decisions and claims of some sectors over others. All this created a problematic framework that meant a period of missed opportunities for sustainable coastal development [14].

Within this framework, the entry into force in 1972 of the US Coastal Zone Management Program (CZMP) was a milestone, as it was able for the first time to bring together a comprehensive strategy for human uses of the coast within the framework of knowledge of natural coastal processes. Since the creation of this program, the concept of management has been shifting towards increasingly integrated approaches, as more and more countries have joined in the establishment of ICZM-compliant strategies. After 20 years of policy development within an ICZM framework, this approach was endorsed by the global community at the United Nations Conference on Environment and Development in Rio de Janeiro in 1992. Since then, further steps have been taken, although most national or supranational ICZM strategies, even in the most advanced countries of the world, do not have the status of law and are merely recommendations. Even so, many specific coastal regions have adopted ICZM strategies to develop coastal area development planning. In some areas, such as the north coast of the Dominican Republic, they have thus achieved a coexistence of reefs and beaches with industrial and agricultural zones in which important tourist complexes have been incorporated (Fig. 30.6).

One of the most evolving aspects of the ICZM concept was its focus. While the initial focus was on environmental conservation, later trends have shifted towards economic development and solving human conflicts [15].

30.4.2 Future ICZM

In addition to the conflicts already described in the coastal zone, there are now future challenges such as sea level rise and increased storms related to climate change. Although it is now widely recognized that developing ICZM is the best way for nature and human use to coexist, most ICZM applications have been carried out at local and regional levels. As demonstrated in the section on coastal policies, at the national and supranational levels the legal coverage of these policies has not been sufficiently developed. Even in the case of the United States, the program promotes the policies but does not enforce them, and therefore does not have the status of law. In the European Union, ICZM is contained in a recommendation and has not even acquired the status of a framework directive, which would oblige all member countries to develop binding ICZM laws.

It is true that, at the regional level, laws have been developed to promote ICZM. Also, at the international level, agreements have been signed to develop this management system. For example, in 2011 the Protocol on ICZM in the Mediterranean was put in place, as a legally binding agreement signed by 21 countries located in Europe, Africa and the Middle East. This agreement obliges the European Union to elevate the recommendations on ICZM to the level of a framework directive. The development of these directives obliges the signatory countries to adopt policies under the ICZM prism. This vision will promote the sustainable development of the coast and ensure coherence of coastal strategies and policies.

Fig. 30.6 Panoramic view of Puerto Plata (Dominican Republic). A local plan of ICZM was applied to allow the coexistence of the natural environment with different human uses

In regions where integrated management is in place, there is a tendency to subsequently carry out sustainability measurements to check the success of the program. These measurements are usually complex because they have to take into account many variables; so, indices have been designed to quantify the state of sustainability following the measures undertaken. The indices provide an overview, albeit simplified, and also allow comparison between different areas under integrated management, and are therefore considered a very useful tool [11]. However, there is no single index that is globally accepted, on the contrary, numerous indices have been proposed. These indices often differ when applied at local, regional, international or global level (e.g., [13]). The evaluation of the degree of success of integrated management measures obtained from the use of indicators is used by management authorities for self-adjustment of the policies adopted. All these tools are currently in a state of evolution and their applicability in practice still has much room for improvement.

In the context of the general sea level rise that accompanies current climate change, ICZM is the only option for coexistence between human beings and nature in coastal areas. In response to the invasion of the sea of areas currently occupied by human constructions, in each case the appropriate decision will have to be made, choosing between different options, such as the massive construction of dikes (in the Dutch style), the withdrawal and abandonment of previously occupied zones, or the adaptation of new constructions (floating buildings or ones raised on pillars). All local factors must be taken into account in each case. It will not be easy, but without an adequate ICZM it will be impossible.

References

1. Alhorn F (2018) Integrated coastal zone management. Springer, Heidelberg, p 193
2. Barragán JM (2010) Coastal management and public policy in Spain. Ocean Coast Manag 53:209–217
3. Cantasano N, Pellicone G (2014) Marine and river environments: a pattern of Integrated Coastal Zone Management (ICZM) in Calabria (Southern Italy). Ocean Coast Manag 89:71–78
4. Chen S, Uitto J (2003) Governing marine and coastal environment in China: building local government capacity through international cooperation. China Environ Ser 6:67–80
5. Clarke B (2010) Analysis of coastal policies of Australian state and territory governments. Technical Report, 39 pp
6. Commission E (1999) Towards a European Integrated Coastal Zone Management (ICZM) strategy: general principles and policy options. Office for Official Publications of the European Communities, Luxembourg, p 33
7. Harvey N (2016) Coast to coast conference. Australian Coastal Society.
8. Harvey N, Caton B (2010) Coastal management in Australia. University of Adelaide Press, Adelaide, p 342
9. Herrington TO (2002) Manual for coastal hazard mitigation. New Jersey Sea Grant College Program. Academic report, 108 pp

10. Knecht RW, Archer J (1993) Integration in the US coastal management program. Ocean Coast Manag 21:183–200
11. Meiner A (2010) Integrated maritime policy for the European Union: consolidating coastal and marine information to support maritime spatial planning. J Coast Conserv 14:1–11
12. Olsen SB, Ricci G (2020) US coastal zone management program. http://www.coastalwiki.org/wiki/US_Coastal_Zone_Management_Program
13. Pickaver AH (2009) Further testing of the approved EU indicator to measure the progress in the implementation of integrated coastal zone management in Europe. In: Moksness E, Dahl E, Støttrup J (eds) Integrated coastal zone management. Wiley-Blackwell, Oxford, pp 67–78
14. Schernewski G (2016) Integrated coastal zone management. In: Harff J, Meschede M, Petersen S, Thiede J (eds) Encyclopedia of marine geosciences. Springer, Heidelberg, pp 359–365
15. Turner RK (2000) Integrating natural and socio-economic science in coastal management. J Mar Syst 25(3–4):447–460

Part VI
Final Remarks

The most dedicated research the data,
this info tempo is gatherin momentum.
A thousand rounds of ammo
one of them was spent in Applied science...

"Knock, Knock"
GZA

Future Trends in Coastal Geology

31

31.1 Introduction

Throughout the chapters of this book, the state of current geological knowledge about the coast has been discussed. This knowledge is the result of the last 50 years of research, which in turn built on the concepts and principles described by the pioneers of this science from the end of the nineteenth century. The existence of many scientific journals, as well as publishers dedicated to the dissemination of thematic monographs on coastal geology issues, has contributed to the remarkable increase in detailed knowledge of many coastal systems, both regionally and thematically. But, in addition, the study methodologies that have been established during these years, and the specific techniques that have been developed, have allowed coastal geologists to delve into aspects that could not be observed before. Some of these techniques are among the greatest scientific advances today and also herald trends for the immediate future.

The chapters that make up the last part of this book have shown us that this science becomes especially vital when applied to human problems on the coast. Although the entire book has focused mainly on coastal geology issues, these chapters show us that coastal study in recent times requires an interdisciplinary approach. Even under this approach, knowledge of active geological processes plays a key role. Likewise, the study of sedimentary sequences of Holocene coastal environments allows us to observe their evolution over thousands of years. This knowledge of evolutionary trends is the key to understanding the fragile equilibrium of coastal systems, and the human beings occupying the coasts must be aware of the evolutionary trends of natural systems in order to achieve sustainable development policies in an ICZM framework.

The main contributions of coastal geologists in the last decades fit into three key lines of research: (1) contributions to the general knowledge of the coasts from a dynamic, geomorphological, sedimentological and environmental point of view; (2) contributions about pure methodological development and use of new techniques to improve coastal knowledge; and (3) contributions to environmental diagnosis, showing the effects of some human actions on the coast, mainly focused on planning sustainable coastal management. These same three lines of study highlight the probable trends for the next years, especially in the context of climate change.

31.2 Studies on Coastal Dynamics, Geomorphology and Sedimentology

The general guidelines of knowledge on the dynamic functioning of coastal systems are already firmly established. In addition, the main aspects of coastal geomorphology and sedimentology are well studied from a thematic point of view. However, much remains to be understood in different ways. Here are the main trends that must be among the future research to respond to the immediate challenges.

31.2.1 Underground Records and Architectural Studies

A large part of the published studies has been centered on the surficial distribution of sediment. Studies about sedimentary sequences have been mainly focused on some specific environments such as barrier islands, but others including estuaries have received less attention from this perspective. Future geological research must focus more closely on the stratigraphic record, especially regarding the three-dimensional architectural disposition of the sedimentary facies. These further studies could contribute to determining geometries of sedimentary bodies (Fig. 31.1) and, superimposing a sediment chronology, could help to understand possible slight sea level movements in the last 5000 years. These studies must also be the basis for better interpretation and understanding of ancient stratigraphic coastal sequences by applying the uniformitarian principle.

J. A. Morales, *Coastal Geology*, Springer Textbooks in Earth Sciences, Geography and Environment,
https://doi.org/10.1007/978-3-030-96121-3_31

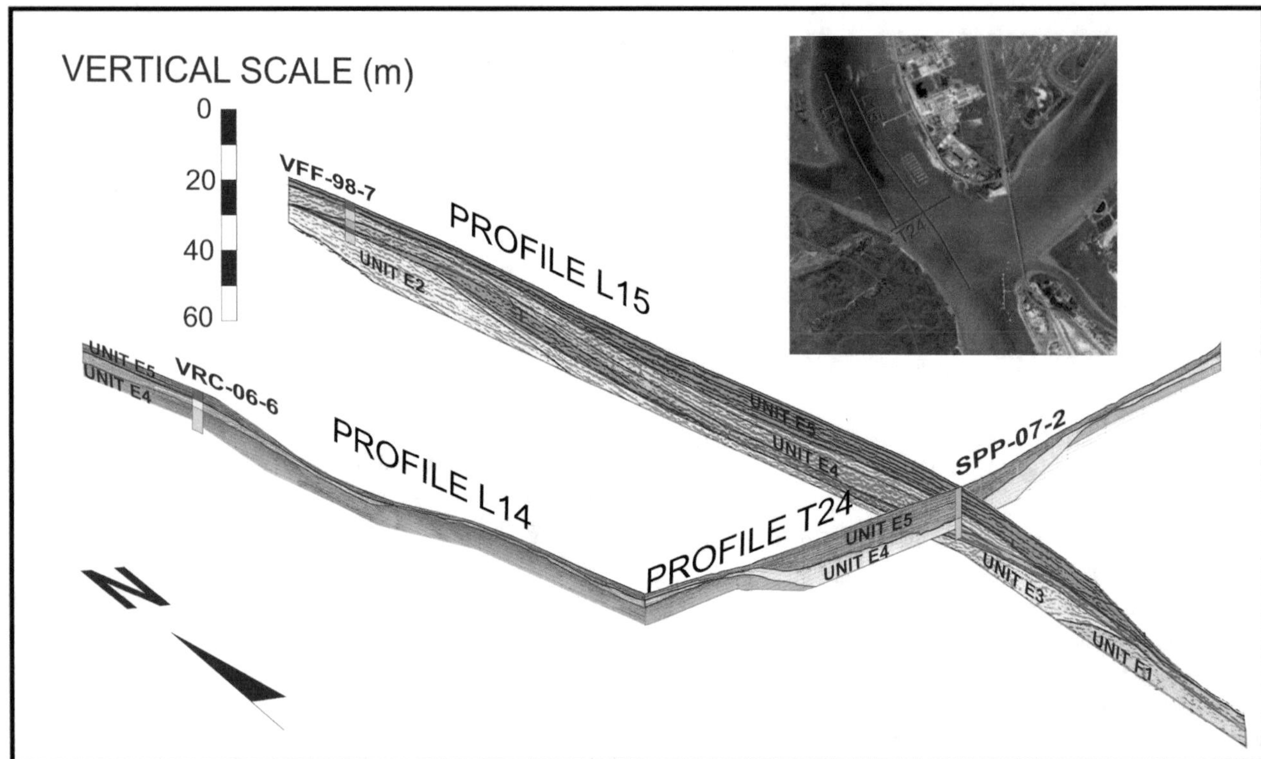

Fig. 31.1 3D architectural scheme based on linked vibracores and seismic profiles of the central part of a mesotidal wave-dominated estuary (Odiel Estuary, SW Spain)

31.2.2 Study of Coastal Events

Coastal events are important energetic and destructive processes that have impacted our coasts in the past, but will also menace us in the future. It will be important for future guidelines to develop detailed studies about the effects of each one of the storms and tsunamis that may reach the coasts in the future. It is also necessary to recognize the records of past events in order to establish event catalogues and determine return periods. This integrated knowledge will be the best tool for developing correct predictions and minimizing future damage.

31.2.3 Extension of Studies to the Submarine Areas of the Coast

Until now, geological study has largely focused on the emerged part of the coastal systems. However, the coast is an integrated system and to get an adequate knowledge about dynamic processes, and also about geomorphology and sedimentology, it is necessary to extend the studies to the sub-littoral areas. For this purpose, some of the new techniques as described in the following section should be adopted.

31.2.4 Upscale and Downscale Studies

Once the mesoscale dynamics of many coasts have been characterized, it is necessary to change the scale in both directions. On the one hand, more detailed and specific studies on a microscale would be useful to understand small nuances in processes and sedimentary distribution. In this sense, detailed sedimentological studies (e.g., facies analysis) will contribute to understanding short-term evolutionary cycles and trends in specific coastal sedimentary environments. On the other hand, studies carried out in adjacent areas must be connected on a macroscale to understand some of the problems in a wider context.

31.2.5 Basic Studies on the Coasts of Developing Countries

The main advances in studies on coastal geology have been developed in only a couple of dozen countries. So, the coasts of the USA, Canada, Western Europe, Russia and Australia are the best known from a coastal geology perspective. In addition, the state of knowledge of other developed countries of South America and Asia, including Argentina, Brazil, Chile, Japan and South Korea, is also high. However, the coasts of the rest of the world's countries remain practically unstudied. These countries have developed human occupation of the coast without planning and without knowledge about their coastal systems, resulting in problems for their buildings and infrastructure, increasing risk of damage and subsequent conflicts. Increasingly urgently, these countries need to develop studies on the geological functioning of their coastal systems.

31.3 Studies on Methodological Development and Use of New Techniques

The development of methodologies emerging from new technologies has been a revolution for coastal geology. Chapter 6 of this book described the main techniques that are beginning to be applied to coastal geology studies. These techniques are advancing at a very rapid pace and are expected to be implemented as a priority in the coming years, and will help to widen knowledge of coastal systems.

31.3.1 Sensors for Process Measurements

During the last years, a new generation of sensors for measuring processes has been incorporated in the equipment used by coastal geologists. The best examples of these include wave gauges, tidal stations, wave radars and Doppler current meters, instruments capable of making continuous measurements and recording them in data loggers. These devices provide geologists with datasets that will contribute to our understanding and ability to quantify a variety of short-term processes which until now were beyond our capacity to see. The application of these kinds of records during subsequent years will allow to recognize the long-term trends of many coastal processes. This will be especially helpful in the changing framework of Earth's climate.

31.3.2 Geophysical Methods

Recent years have seen a growing trend in the use of acoustic techniques such as high-resolution seismic profiles to complete the geomorphological and sedimentological underwater and underground information of coastal systems (Fig. 31.1). These methods have been widely and successfully used by marine and petroleum geologists in deeper areas. The technical development of cheap, smaller and more versatile equipment will contribute to extending the studied areas to the subtidal areas of the coast. In the same way, ground penetration radar is being used to characterize the internal structure of sandy barriers and littoral dunes in the land portion of the coast. The combination of facies descriptions obtained by coring, and the information of the geometry of sedimentary bodies obtained by using these methods, will contribute to establishing 3D architectural facies models in many coastal environments (Sect. 31.2.1).

31.3.3 Remote Sensing

During the last two decades, the digital treatment of data from remote sensing (satellites, airborne sensors and drones) has developed in an exponential way. Presently, most of the coastal evolution that accompanies the dynamic changes of the world's coasts is studied through the use of these techniques. It is foreseeable that the current accelerated technological development that we are experiencing will lead us to the discovery of increasingly sophisticated and precise remote sensing techniques. At the same time, the use of analysis software through online platforms will facilitate more widespread applications and more generalized use.

31.3.4 Bathymetric and Topographic Methods

New acoustic and laser techniques, including side scan sonar or multibeam echosounders such as aerial and drone LIDAR, have allowed high-resolution bathymetric and topographic records to be obtained. These records are contributing to elaborate digital terrain models (DTM) that integrate terrestrial and underwater datasets. The DTM results provide a very useful tool in coastal areas to understand certain geomorphological features. The comparison of high-resolution DTM of different dates allows us to quantify rates of erosion and deposition in coastal environments, thus aiding determination of volumes and surfaces (eroded and filled by sediments).

31.3.5 Numerical Models

Modeling of hydrodynamic and sedimentary processes is a new tool which is now being used by geologists to understand the functioning of coastal systems (Fig. 31.2). Uncalibrated models have proved an excellent method to identify the cause of past effects, but also to get a prediction of future actions on the coast. In the future, the use of these models in coordination with the datasets obtained by using the previously described measuring techniques will allow accurate calibration of the models. These models will then be able to develop reliable predictions which can help coastal managers to take the best decisions and adopt necessary policies.

Fig. 31.2 Vectors and magnitudes of longshore transport obtained from a numerical model (Guadiana Delta front, SW Spain)

31.4 Environmental Studies and Integrated Coastal Zone Management

The beginnings of coastal geology studies were prompted by the work of coastal engineers, since the first aim of these studies was to adapt the coast to human requirements. In fact, the only book entitled Coastal Geology before this one was written by an engineer from this perspective. When geologists began to study the coast, pure science was focused on the correct understanding of the natural coastal system. Now, situated between the scientific aspects studied by coastal geologists are the consequences on coastal dynamics caused by the structures built by the civil engineers. This present state of the subject means a new focus is necessary so that the knowledge can be applied to ICZM. In order to develop the best criteria to take decisions, it will be necessary to study the followings areas.

31.4.1 Mapping Vulnerability

At this moment, a priority objective of main coastal areas is the development of detailed maps of degree of vulnerability to coastal hazards. These kinds of charts are not only needed for natural environments, but especially for the human-occupied coasts. There is not a general agreement about the aspects, variables and parameters that have to be considered to elaborate a map of vulnerability. Nowadays, social, economic, cultural, biological and dynamic factors are all recognized as being equally important. Nevertheless, the main efforts to map coastal vulnerability are currently focusing on geological hazards. Vulnerability maps for individual hazards can be done, but they can also be integrated in a map of general vulnerability. Normally, the mapped element is an index (coastal vulnerability index, or CVI). The most important part of these maps considers the degree of exposure to the main coastal hazards; then, they could be considered maps of risk rather than maps of vulnerability.

31.4.2 Conceptual Modeling

Conceptual models are conceived as a method for visualizing integrated and interpreted data. There are different conceptual models that could be applied in coastal geology, but the most interesting for coastal managers are those that synthesize processes and sedimentary transport as a response to modifications in the parameters (natural or human) that cause the coast to evolve. In this way, conceptual models are a useful way to convey resource information to political bodies, since such systems of representation can be comprehensible to experts and non-experts alike.

31.4.3 Interdisciplinary Studies and Debates

After European Union directives such as Horizon 2020 and US programs that have included the CZMP, scientists have tended to adopt a transversal and interdisciplinary vision of their studies. Taking into account these new rules, coastal geologists are obliged to incorporate the vision of geographers, oceanographers, biologists, ecologists, archaeologists, coastal engineers, mathematicians, physicians, geochemists, economists, social scientists and lawyers in building multi-visionary teams capable of analyzing any aspect of the future ICZM.

31.5 Science for Society and Citizen Science

The last challenge that needs to be addressed is to carry the knowledge to society at large. Since the entire population is and will be affected by coastal processes in the future, coastal scientists are strongly encouraged to communicate their discoveries to the public. Therefore, it is no longer enough to write in scientific publications, but communications that a non-technical audience can understand are needed, as well as reports in webpages and social media.

The concept of **citizen science** goes far beyond the transmission of scientific results to citizens. Citizen science is scientific research involving non-specialists and involving them in the work of scientists and technicians. In the most practical form of citizen science, users are responsible for the collection and/or analysis of data, as well as the process of dissecting the results by researchers. Citizen science is especially useful for coastal geology, as coastal users are most frequently present during the action of coastal processes, which can be measured and photographed by them. In this sense, a widespread use of new technologies, such as simple applications installed on smartphones, will contribute to a broader participation of citizens in scientific data collection. Thus, citizens can actively contribute to the development of science, sharing their knowledge and their own tools and resources to increase experimental data volume. The participants, while adding value to science, acquire local and thematic knowledge framed in an improvement of their vision of the scientific methodology. This way of working is very valuable, especially when it comes to an issue as linked to the development of their lives as coastal geology. In short, citizen science applied to coastal geology is an effective way of promoting scientific education and environmental education, while generating useful knowledge that can be applied by local managers.

Another effective tool for coastal geology is **citizen scientific communication**. This concept refers to a broader and more inclusive way of communicating science, where volunteers engage in citizen scientific journalism. This is a system of dissemination in which the data collected, analyzed by citizens and prepared by scientists are also communicated by the citizen. This process ensures both reliability and transparency.

All these developments of citizen science are especially useful in the debate and exchange of information between professionals, researchers, economic agents, society and policy makers, when it comes to aspects involved in formulating and enacting ICZM.

However, citizen science has so far mainly been a process whereby people join projects designed and conceived by scientists. Currently, there are forms of citizen science in which the citizen contribution is also incorporated into the project approach. In this way, initiatives such as the European Platform for Citizen Science (https://eu-citizen.science/) provide an open forum for the exchange of knowledge, tools, training and resources for citizen scientists. The objective of this space is not so much for citizens to participate in scientific data collection, but, beyond that, it transforms the way citizen science projects are conceived and developed.

31.6 Concluding Remark

Ultimately, coastal geologists have to focus on the future by addressing challenges that will become increasingly difficult, especially in the context of a post-pandemic economic crisis that considerably limits investment in research. However, we will address this task with our best weapons: effort and passion.